Y0-AAX-907

Methods in Enzymology

Volume 216

RECOMBINANT DNA

Part G

METHODS IN ENZYMOLOGY

EDITORS-IN-CHIEF

John N. Abelson Melvin I. Simon

DIVISION OF BIOLOGY
CALIFORNIA INSTITUTE OF TECHNOLOGY
PASADENA, CALIFORNIA

FOUNDING EDITORS

Sidney P. Colowick and Nathan O. Kaplan

Methods in Enzymology

Volume 216

Recombinant DNA

Part G

EDITED BY

Ray Wu

SECTION OF BIOCHEMISTRY
MOLECULAR AND CELL BIOLOGY
CORNELL UNIVERSITY
ITHACA, NEW YORK

ACADEMIC PRESS, INC.
Harcourt Brace Jovanovich, Publishers
San Diego New York Boston
London Sydney Tokyo Toronto

This book is printed on acid-free paper. 

Academic Press, Inc.
1250 Sixth Avenue, San Diego, California 92101-4311

United Kingdom Edition published by
Academic Press Limited
24–28 Oval Road, London NW1 7DX

Library of Congress Catalog Number: 54-9110

International Standard Book Number: 0-12-182117-X

PRINTED IN THE UNITED STATES OF AMERICA

92 93 94 95 96 97 MM 9 8 7 6 5 4 3 2 1

Table of Contents

Section I. Isolation, Synthesis, and Detection of DNA and RNA

Section II. Enzymes and Methods for Cleaving and Manipulating DNA

Section III. Reporter Genes

Section IV. Vectors for Cloning Genes

Contributors to Volume 216

Article numbers are in parentheses following the names of contributors.
Affiliations listed are current.

ROBIN C. ALLSHIRE (51), *MRC Human Genetics Unit, Western General Hospital, Edinburgh EH4 2XU, Scotland*

J. ALTENBUCHNER (40), *Institute of Industrial Genetics, University of Stuttgart, D-7000 Stuttgart 1, Germany*

MICHELLE A. ALTING-MEES (42), *Stratagene Cloning Systems, La Jolla, California 92037*

SHRIKANT ANANT (3), *Department of Genetics, The University of Illinois at Chicago, Chicago, Illinois 60612*

JANET M. BARSOMIAN (23), *New England Biolabs Inc., Beverly, Massachusetts 01915*

ROBERT L. BEBEE (4), *Corporate Research, GIBCO BRL, Life Technologies Inc., Gaithersburg, Maryland 20898*

STEPHAN BECK (15), *Imperial Cancer Research Fund, London WC2A 3PX, England*

ASHOK S. BHAGWAT (21), *Department of Chemistry, Wayne State University, Detroit, Missouri 48202*

ADI D. BHARUCHA (18, 19), *Department of Biochemistry, Faculty of Medicine, Laval University, Ste-Foy, Quebec G1K 7P4, Canada*

WENDY A. BICKMORE (22), *MRC Human Genetics Unit, Western General Hospital, Edinburgh EH4 2XU, Scotland*

ADRIAN P. BIRD (22), *Institute of Cell and Molecular Biology, University of Edinburgh, Edinburgh EH9 3JR, Scotland*

H. C. BIRNBOIM (16), *Ottawa Regional Cancer Centre, and Departments of Biochemistry, Medicine, and Microbiology/Immunology, University of Ottawa, Ottawa, Ontario K1H 8L6, Canada*

KATHRYN J. BOCKHOLD (17), *Département de Biologie Moléculaire, Institut Pasteur, 75724 Paris Cedex 15, France*

JOHAN BOTTERMAN (36), *Plant Genetics Systems, B-9000 Gent, Belgium*

ALLAN R. BRASIER (34), *Division of Endocrinology and Hypertension, University of Texas Medical Branch, Galveston, Texas 77555*

JÜRGEN BROSIUS (41), *Fishberg Research Center for Neurobiology, Mount Sinai School of Medicine, New York, New York 10029*

Z. CAI (10), *Department of Immunology, Mayo Clinic, Rochester, Minnesota 55905*

ALLAN CAPLAN (37), *Department of Bacteriology and Biochemistry, University of Idaho, Moscow, Idaho 83843*

C. THOMAS CASKEY (7), *Howard Hughes Medical Institute, Baylor College of Medicine, Houston, Texas 77030*

FARID F. CHEHAB (14), *Department of Laboratory Medicine, University of California, San Francisco, San Francisco, California 94143*

YAWEN L. CHIANG (8), *Department of Immunology, Genetic Therapy Inc., Gaithersburg, Maryland 20878*

ING-MING CHIU (44), *Departments of Internal Medicine and Molecular Genetics, and Comprehensive Cancer Center, The Ohio State University, Columbus, Ohio 43210*

BRYAN R. CULLEN (31), *Howard Hughes Medical Institute, Section of Genetics, Departments of Microbiology and Medicine, Duke University Medical Center, Durham, North Carolina 27710*

MARC DE BLOCK (36), *Plant Genetics Systems, B-9000 Gent, Belgium*

RUDY DEKEYSER (37), *Instituut ter Aanmoediging van het Wetenschappelijk, Onderzoek in Nijverheid en Landbouw, B-1050 Brussels, Belgium*

SUSANA DE LA LUNA (33), *Centro Nacional de Biotechnología and Centro de Biología Molecular, Universidad Autónoma de Madrid, Campus de Cantoblanco, 28049 Madrid, Spain*

JÜRGEN DENECKE (36), *Department of Molecular Genetics, Swedish University of Agricultural Sciences, S-75007 Uppsala, Sweden*

PETER B. DERVAN (27), *Arnold and Mabel Beckman Laboratory of Chemical Synthesis, Division of Chemistry and Chemical Engineering, Pasadena, California 91125*

JEFFREY R. DE WET (35), *Pfizer Central Research, Pfizer, Inc., Groton, Connecticut 06340*

KATHLEEN D'HALLUIN (36), *Plant Genetics Systems, B-9000 Gent, Belgium*

JAMES EBERWINE (9), *Departments of Pharmacology and Psychiatry, University of Pennsylvania Medical School, Philadelphia, Pennsylvania 19104*

GLEN A. EVANS (46), *Molecular Genetics Laboratory, The Salk Institute for Biological Studies, San Diego, California 92138*

GEORGE R. FEEHERY (23), *New England Biolabs Inc., Beverly, Massachusetts 01915*

RICHARD FINNELL (9), *Department of Veterinary Anatomy/Public Health, Texas A&M University, College Station, Texas 77843*

CARL W. FULLER (29), *Research and Development, United States Biochemical Corporation, Cleveland, Ohio 44122*

GULILAT GEBEYEHU (4), *Molecular Biology Research and Development, GIBCO BRL, Life Technologies Inc., Gaithersburg, Maryland 20898*

CHRISTIANE GOBLET (17), *Département de Biologie Moléculaire, Institut Pasteur, 75724 Paris Cedex 15, France*

JEAN GOULD (30), *Soil and Crop Sciences Department, Texas A&M University, College Station, Texas 77843*

FRANÇOIS GUIDET (1), *GIP Prince de Bretagne Biotechnologie, Penn Ar Prat, 29250 St. Pol De Léon, France*

JOHN D. HARDING (4), *Corporate Research, GIBCO BRL, Life Technologies Inc., Gaithersburg, Maryland 20898*

GARY G. HERMANSON (46), *Molecular Genetics Laboratory, The Salk Institute for Biological Studies, San Diego, California 92138*

PHILIP HIETER (49), *Department of Molecular Biology and Genetics, The Johns Hopkins University School of Medicine, Baltimore, Maryland 21205*

R. M. HORTON (10), *Department of Immunology, Mayo Clinic, Rochester, Minnesota 55905*

DENNIS E. HRUBY (32), *Center for Gene Research and Biotechnology, Department of Microbiology, Oregon State University, Corvallis, Oregon 97331*

JAN JANSSENS (36), *Plant Genetics Systems, B-9000 Gent, Belgium*

Y. W. KAN (14), *Department of Laboratory Medicine, Howard Hughes Medical Institute, University of California, San Francisco, San Francisco, California 94143*

RAJENDRA P. KANDPAL (5), *Department of Genetics, Yale University School of Medicine, New Haven, Connecticut 06510*

DAVID J. KEMP (12), *Menzies School of Health Research, Casuarina, Northern Territory 0811, Australia*

SUN CHANG KIM (26), *Department of Oncology, McArdle Laboratory for Cancer Research, University of Wisconsin, Madison, Wisconsin 53706*

MICHAEL KOOB (2, 28), *McArdle Laboratory for Cancer Research, University of Wisconsin, Madison, Wisconsin 53706*

DAVID LANDRY (23), *New England Biolabs Inc., Beverly, Massachusetts 01915*

PETER LANGRIDGE (1), *Centre for Cereal Biotechnology, The Waite Agricultural Research Institute, University of Adelaide, Glen Osmond, South Australia 5064, Australia*

CHENG CHI LEE (7), *Institute for Molecular Genetics, Baylor College of Medicine, Houston, Texas 77030*

JAN LEEMANS (36), *Plant Genetics Systems, B-9000 Gent, Belgium*

KIRSTEN LEHTOMA (44), *Department of Internal Medicine and Comprehensive Cancer Center, The Ohio State University, Columbus, Ohio 43210*

GEORGES LÉVESQUE (19), *Department of Biochemistry, Faculty of Medicine, Laval University, Ste-Foy, Quebec G1K 7P4, Canada*

ANDREW M. LEW (13), *Walter and Eliza Hall Institute, Melbourne, Victoria 3050, Australia*

KENNETH R. LUEHRSEN (35), *Department of Biological Sciences, Stanford University, Stanford, California 94305*

SCOTT MACKLER (9), *Department of Pharmacology, University of Pennsylvania Medical School, Philadelphia, Pennsylvania 19104*

MICHAEL H. MALIM (31), *Howard Hughes Medical Institute, Departments of Microbiology and Medicine, University of Pennsylvania School of Medicine, Philadelphia, Pennsylvania 19104*

MICHAEL MCCLELLAND (25), *Department of Plant Pathology, University of Nebraska, Lincoln, Nebraska 68583*

KEVIN MIYASHIRO (9), *Department of Pharmacology, University of Pennsylvania Medical School, Philadelphia, Pennsylvania 19104*

DONALD T. MOIR (50), *Department of Human Genetics and Molecular Biology, Collaborative Research, Inc., Waltham, Massachusetts 02154*

MICHAEL NELSON (25), *Department of Plant Pathology, University of Nebraska, Lincoln, Nebraska 68583*

JUAN ORTÍN (33), *Centro Nacional de Biotechnología and Centro de Biología Molecular, Universidad Autónoma de Madrid, Campus de Cantoblanco, 28049 Madrid, Spain*

MICHAEL PANACCIO (13), *Victorian Institute of Animal Science, Attwood, Victoria 3049, Australia*

WILLIAM J. PAVAN (49), *Department of Molecular Biology, Howard Hughes Medical Institute, Princeton University, Princeton, New Jersey 08544*

L. R. PEASE (10), *Department of Immunology, Mayo Clinic, Rochester, Minnesota 55905*

I. PELLETIER (40), *Institute of Industrial Genetics, University of Stuttgart, D-7000 Stuttgart 1, Germany*

SIDNEY PESTKA (20), *Department of Molecular Genetics and Microbiology, University of Medicine and Dentistry of New Jersey, Robert Wood Johnson Medical School, Piscataway, New Jersey 08854*

JAMES C. PIERCE (47), *Cancer Therapeutic Program, The Du Pont Merck Pharmaceutical Company, Wilmington, Delaware 19880*

ANNA J. PODHAJSKA (26), *Department of Microbiology, University of Gdansk, 80-222 Gdansk, Poland*

MATTHEW L. POULIN (44), *Department Molecular Genetics, The Ohio State University, Columbus, Ohio 43210*

EDOUARD PROST (17), *Département de Biologie Moléculaire, Institut Pasteur, 75724 Paris Cedex 15, France*

J. K. PULLEN (10), *Department of Immunology, Mayo Clinic, Rochester, Minnesota 55905*

PETER J. PUNT (39), *Department of Molecular Genetics and Gene Technology, Medical Biological Laboratory, 2280 AA Rijswijk, The Netherlands*

ROGER H. REEVES (49), *Department of Physiology, The Johns Hopkins University School of Medicine, Baltimore, Maryland 21205*

ARLETTE REYNAERTS (36), *Plant Genetics Systems, B-9000 Gent, Belgium*

DAVID RON (34), *Laboratory of Molecular Endrocrinology, Massachusetts General Hospital, Boston, Massachusetts 02114*

J. M. SHORT (42, 43), *Strategene Cloning Systems, La Jolla, California 92037*

DOUGLAS R. SMITH (50), *Department of Human Genetics and Molecular Biology, Collaborative Research, Inc., Waltham, Massachusetts 02154*

ADRIENNE P. SMYTH (50), *Department of Human Genetics and Molecular Biology, Collaborative Research, Inc., Waltham, Massachusetts 02154*

KEN SNIDER (46), *Molecular Genetics Laboratory, The Salk Institute for Biological Studies, San Diego, California 92138*

JAEMOG SOH (20), *Department of Molecular Genetics and Microbiology, University of Medicine and Dentistry of New Jersey, Robert Wood Johnson Medical School, Piscataway, New Jersey 08854*

J. A. SORGE (42, 43), *Strategene Cloning Systems, La Jolla, California 92037*

CORINNE SPENCER (9), *Department of Pharmacology, University of Pennsylvania Medical School, Philadelphia, Pennsylvania 19104*

NAT L. STERNBERG (47), *Cancer Therapeutic Program, The Du Pont Merck Pharmaceutical Company, Wilmington, Delaware 19880*

SCOTT A. STROBEL (27), *Arnold and Mabel Beckman Laboratory of Chemical Synthesis, Division of Chemistry and Chemical Engineering, Pasadena, California 91125*

KIRANUR N. SUBRAMANIAN (3), *Department of Genetics, The University of Illinois at Chicago, Chicago, Illinois 60612*

WACLAW SZYBALSKI (2, 26), *McArdle Laboratory for Cancer Research, University of Wisconsin, Madison, Wisconsin 53706*

KENNETH D. TARTOF (48), *Institute for Cancer Research, Fox Chase Cancer Center, Philadelphia, Pennsylvania 19111*

HIROO TOYODA (11), *Medical Genetics-Birth Defects Center, Department of Medicine and Pediatrics, Cedars-Sinai Medical Center, UCLA School of Medicine, Los Angeles, California 90048*

LEVY ULANOVSKY (6), *Department of Structural Biology, The Weizmann Institute of Science, Rehovot 76100, Israel*

MARC VAN MONTAGU (37), *Laboratorium voor Genetica, Universiteit Gent, B-9000 Gent, Belgium*

CEES A. M. J. J. VAN DEN HONDEL (39), *Department of Molecular Genetics and Gene Technology, Medical Biological Laboratory, 2280 AA Rijswijk, The Netherlands*

M. R. VEN MURTHY (18, 19), *Department of Biochemistry, Faculty of Medicine, Laval University, Ste-Foy, Quebec G1K 7P4, Canada*

P. VIELL (40), *Institute of Industrial Genetics, University of Stuttgart, D-7000 Stuttgart 1, Germany*

VIRGINIA WALBOT (35), *Department of Biological Sciences, Stanford University, Stanford California 94305*

JEFF WALL (14), *Department of Laboratory Medicine, University of California, San Francisco, San Francisco, California 94143*

DAVID C. WARD (5), *Department of Genetics, Yale University School of Medicine, New Haven, Connecticut 06510*

SHERMAN M. WEISSMAN (5), *Department of Genetics, Yale University School of Medicine, New Haven, Connecticut 06510*

ROBERT G. WHALEN (17), *Département de Biologie Moléculaire, Institut Pasteur, 75724 Paris Cedex 15, France*

ELIZABETH M. WILSON (32), *Center for Gene Research and Biotechnology, Department of Microbiology, Oregon State University, Corvallis, Oregon 97331*

GEOFFREY G. WILSON (23, 24), *New England Biolabs Inc., Beverly, Massachusetts 01915*

MICHAEL WITTY (38), *Department of Plant Sciences, University of Cambridge, Cambridge CB2 3EA, England*

ANDREW D. ZELENETZ (45), *Division of Hematologic Oncology/Lymphoma of Memorial Hospital Program in Molecular Biology of the Sloan-Kettering Institute, Memorial Sloan-Kettering Cancer Center, New York, New York 10021*

Preface

Recombinant DNA methods are powerful, revolutionary techniques for at least two reasons. First, they allow the isolation of single genes in large amounts from a pool of thousands or millions of genes. Second, the isolated genes from any source or their regulatory regions can be modified at will and reintroduced into a wide variety of cells by transformation. The cells expressing the introduced gene can be measured at the RNA level or protein level. These advantages allow us to solve complex biological problems, including medical and genetic problems, and to gain deeper understandings at the molecular level. In addition, new recombinant DNA methods are essential tools in the production of novel or better products in the areas of health, agriculture, and industry.

The new Volumes 216, 217, and 218 supplement Volumes 153, 154, and 155 of *Methods in Enzymology*. During the past few years, many new or improved recombinant DNA methods have appeared, and a number of them are included in these new volumes. Volume 216 covers methods related to isolation and detection of DNA and RNA, enzymes for manipulating DNA, reporter genes, and new vectors for cloning genes. Volume 217 includes vectors for expressing cloned genes, mutagenesis, identifying and mapping genes, and methods for transforming animal and plant cells. Volume 218 includes methods for sequencing DNA, PCR for amplifying and manipulating DNA, methods for detecting DNA–protein interactions, and other useful methods.

Areas or specific topics covered extensively in the following recent volumes of *Methods in Enzymology* are not included in these three volumes: "Guide to Protein Purification," Volume 182, edited by M. P. Deutscher; "Gene Expression Technology," Volume 185, edited by D. V. Goeddel; and "Guide to Yeast Genetics and Molecular Biology," Volume 194, edited by C. Guthrie and G. R. Fink.

RAY WU

METHODS IN ENZYMOLOGY

VOLUME I. Preparation and Assay of Enzymes
Edited by SIDNEY P. COLOWICK AND NATHAN O. KAPLAN

VOLUME II. Preparation and Assay of Enzymes
Edited by SIDNEY P. COLOWICK AND NATHAN O. KAPLAN

VOLUME III. Preparation and Assay of Substrates
Edited by SIDNEY P. COLOWICK AND NATHAN O. KAPLAN

VOLUME IV. Special Techniques for the Enzymologist
Edited by SIDNEY P. COLOWICK AND NATHAN O. KAPLAN

VOLUME V. Preparation and Assay of Enzymes
Edited by SIDNEY P. COLOWICK AND NATHAN O. KAPLAN

VOLUME VI. Preparation and Assay of Enzymes (*Continued*)
Preparation and Assay of Substrates
Special Techniques
Edited by SIDNEY P. COLOWICK AND NATHAN O. KAPLAN

VOLUME VII. Cumulative Subject Index
Edited by SIDNEY P. COLOWICK AND NATHAN O. KAPLAN

VOLUME VIII. Complex Carbohydrates
Edited by ELIZABETH F. NEUFELD AND VICTOR GINSBURG

VOLUME IX. Carbohydrate Metabolism
Edited by WILLIS A. WOOD

VOLUME X. Oxidation and Phosphorylation
Edited by RONALD W. ESTABROOK AND MAYNARD E. PULLMAN

VOLUME XI. Enzyme Structure
Edited by C. H. W. HIRS

VOLUME XII. Nucleic Acids (Parts A and B)
Edited by LAWRENCE GROSSMAN AND KIVIE MOLDAVE

VOLUME XIII. Citric Acid Cycle
Edited by J. M. LOWENSTEIN

VOLUME XIV. Lipids
Edited by J. M. LOWENSTEIN

VOLUME XV. Steroids and Terpenoids
Edited by RAYMOND B. CLAYTON

VOLUME XVI. Fast Reactions
Edited by KENNETH KUSTIN

Volume XXXVI. Hormone Action (Part A: Steroid Hormones)
Edited by Bert W. O'Malley and Joel G. Hardman

Volume XXXVII. Hormone Action (Part B: Peptide Hormones)
Edited by Bert W. O'Malley and Joel G. Hardman

Volume XXXVIII. Hormone Action (Part C: Cyclic Nucleotides)
Edited by Joel G. Hardman and Bert W. O'Malley

Volume XXXIX. Hormone Action (Part D: Isolated Cells, Tissues, and Organ Systems)
Edited by Joel G. Hardman and Bert W. O'Malley

Volume XL. Hormone Action (Part E: Nuclear Structure and Function)
Edited by Bert W. O'Malley and Joel G. Hardman

Volume XLI. Carbohydrate Metabolism (Part B)
Edited by W. A. Wood

Volume XLII. Carbohydrate Metabolism (Part C)
Edited by W. A. Wood

Volume XLIII. Antibiotics
Edited by John H. Hash

Volume XLIV. Immobilized Enzymes
Edited by Klaus Mosbach

Volume XLV. Proteolytic Enzymes (Part B)
Edited by Laszlo Lorand

Volume XLVI. Affinity Labeling
Edited by William B. Jakoby and Meir Wilchek

Volume XLVII. Enzyme Structure (Part E)
Edited by C. H. W. Hirs and Serge N. Timasheff

Volume XLVIII. Enzyme Structure (Part F)
Edited by C. H. W. Hirs and Serge N. Timasheff

Volume XLIX. Enzyme Structure (Part G)
Edited by C. H. W. Hirs and Serge N. Timasheff

Volume L. Complex Carbohydrates (Part C)
Edited by Victor Ginsburg

Volume LI. Purine and Pyrimidine Nucleotide Metabolism
Edited by Patricia A. Hoffee and Mary Ellen Jones

Volume LII. Biomembranes (Part C: Biological Oxidations)
Edited by Sidney Fleischer and Lester Packer

Volume LIII. Biomembranes (Part D: Biological Oxidations)
Edited by Sidney Fleischer and Lester Packer

Volume LIV. Biomembranes (Part E: Biological Oxidations)
Edited by Sidney Fleischer and Lester Packer

VOLUME LV. Biomembranes (Part F: Bioenergetics)
Edited by SIDNEY FLEISCHER AND LESTER PACKER

VOLUME LVI. Biomembranes (Part G: Bioenergetics)
Edited by SIDNEY FLEISCHER AND LESTER PACKER

VOLUME LVII. Bioluminescence and Chemiluminescence
Edited by MARLENE A. DELUCA

VOLUME LVIII. Cell Culture
Edited by WILLIAM B. JAKOBY AND IRA PASTAN

VOLUME LIX. Nucleic Acids and Protein Synthesis (Part G)
Edited by KIVIE MOLDAVE AND LAWRENCE GROSSMAN

VOLUME LX. Nucleic Acids and Protein Synthesis (Part H)
Edited by KIVIE MOLDAVE AND LAWRENCE GROSSMAN

VOLUME 61. Enzyme Structure (Part H)
Edited by C. H. W. HIRS AND SERGE N. TIMASHEFF

VOLUME 62. Vitamins and Coenzymes (Part D)
Edited by DONALD B. MCCORMICK AND LEMUEL D. WRIGHT

VOLUME 63. Enzyme Kinetics and Mechanism (Part A: Initial Rate and Inhibitor Methods)
Edited by DANIEL L. PURICH

VOLUME 64. Enzyme Kinetics and Mechanism (Part B: Isotopic Probes and Complex Enzyme Systems)
Edited by DANIEL L. PURICH

VOLUME 65. Nucleic Acids (Part I)
Edited by LAWRENCE GROSSMAN AND KIVIE MOLDAVE

VOLUME 66. Vitamins and Coenzymes (Part E)
Edited by DONALD B. MCCORMICK AND LEMUEL D. WRIGHT

VOLUME 67. Vitamins and Coenzymes (Part F)
Edited by DONALD B. MCCORMICK AND LEMUEL D. WRIGHT

VOLUME 68. Recombinant DNA
Edited by RAY WU

VOLUME 69. Photosynthesis and Nitrogen Fixation (Part C)
Edited by ANTHONY SAN PIETRO

VOLUME 70. Immunochemical Techniques (Part A)
Edited by HELEN VAN VUNAKIS AND JOHN J. LANGONE

VOLUME 71. Lipids (Part C)
Edited by JOHN M. LOWENSTEIN

VOLUME 72. Lipids (Part D)
Edited by JOHN M. LOWENSTEIN

VOLUME 73. Immunochemical Techniques (Part B)
Edited by JOHN J. LANGONE AND HELEN VAN VUNAKIS

VOLUME 74. Immunochemical Techniques (Part C)
Edited by JOHN J. LANGONE AND HELEN VAN VUNAKIS

VOLUME 75. Cumulative Subject Index Volumes XXXI, XXXII, XXXIV–LX
Edited by EDWARD A. DENNIS AND MARTHA G. DENNIS

VOLUME 76. Hemoglobins
Edited by ERALDO ANTONINI, LUIGI ROSSI-BERNARDI, AND EMILIA CHIANCONE

VOLUME 77. Detoxication and Drug Metabolism
Edited by WILLIAM B. JAKOBY

VOLUME 78. Interferons (Part A)
Edited by SIDNEY PESTKA

VOLUME 79. Interferons (Part B)
Edited by SIDNEY PESTKA

VOLUME 80. Proteolytic Enzymes (Part C)
Edited by LASZLO LORAND

VOLUME 81. Biomembranes (Part H: Visual Pigments and Purple Membranes, I)
Edited by LESTER PACKER

VOLUME 82. Structural and Contractile Proteins (Part A: Extracellular Matrix)
Edited by LEON W. CUNNINGHAM AND DIXIE W. FREDERIKSEN

VOLUME 83. Complex Carbohydrates (Part D)
Edited by VICTOR GINSBURG

VOLUME 84. Immunochemical Techniques (Part D: Selected Immunoassays)
Edited by JOHN J. LANGONE AND HELEN VAN VUNAKIS

VOLUME 85. Structural and Contractile Proteins (Part B: The Contractile Apparatus and the Cytoskeleton)
Edited by DIXIE W. FREDERIKSEN AND LEON W. CUNNINGHAM

VOLUME 86. Prostaglandins and Arachidonate Metabolites
Edited by WILLIAM E. M. LANDS AND WILLIAM L. SMITH

VOLUME 87. Enzyme Kinetics and Mechanism (Part C: Intermediates, Stereochemistry, and Rate Studies)
Edited by DANIEL L. PURICH

VOLUME 88. Biomembranes (Part I: Visual Pigments and Purple Membranes, II)
Edited by LESTER PACKER

VOLUME 89. Carbohydrate Metabolism (Part D)
Edited by WILLIS A. WOOD

VOLUME 90. Carbohydrate Metabolism (Part E)
Edited by WILLIS A. WOOD

VOLUME 91. Enzyme Structure (Part I)
Edited by C. H. W. HIRS AND SERGE N. TIMASHEFF

VOLUME 92. Immunochemical Techniques (Part E: Monoclonal Antibodies and General Immunoassay Methods)
Edited by JOHN J. LANGONE AND HELEN VAN VUNAKIS

VOLUME 93. Immunochemical Techniques (Part F: Conventional Antibodies, Fc Receptors, and Cytotoxicity)
Edited by JOHN J. LANGONE AND HELEN VAN VUNAKIS

VOLUME 94. Polyamines
Edited by HERBERT TABOR AND CELIA WHITE TABOR

VOLUME 95. Cumulative Subject Index Volumes 61–74, 76–80
Edited by EDWARD A. DENNIS AND MARTHA G. DENNIS

VOLUME 96. Biomembranes [Part J: Membrane Biogenesis: Assembly and Targeting (General Methods; Eukaryotes)]
Edited by SIDNEY FLEISCHER AND BECCA FLEISCHER

VOLUME 97. Biomembranes [Part K: Membrane Biogenesis: Assembly and Targeting (Prokaryotes, Mitochondria, and Chloroplasts)]
Edited by SIDNEY FLEISCHER AND BECCA FLEISCHER

VOLUME 98. Biomembranes (Part L: Membrane Biogenesis: Processing and Recycling)
Edited by SIDNEY FLEISCHER AND BECCA FLEISCHER

VOLUME 99. Hormone Action (Part F: Protein Kinases)
Edited by JACKIE D. CORBIN AND JOEL G. HARDMAN

VOLUME 100. Recombinant DNA (Part B)
Edited by RAY WU, LAWRENCE GROSSMAN, AND KIVIE MOLDAVE

VOLUME 101. Recombinant DNA (Part C)
Edited by RAY WU, LAWRENCE GROSSMAN, AND KIVIE MOLDAVE

VOLUME 102. Hormone Action (Part G: Calmodulin and Calcium-Binding Proteins)
Edited by ANTHONY R. MEANS AND BERT W. O'MALLEY

VOLUME 103. Hormone Action (Part H: Neuroendocrine Peptides)
Edited by P. MICHAEL CONN

VOLUME 104. Enzyme Purification and Related Techniques (Part C)
Edited by WILLIAM B. JAKOBY

VOLUME 105. Oxygen Radicals in Biological Systems
Edited by LESTER PACKER

VOLUME 106. Posttranslational Modifications (Part A)
Edited by FINN WOLD AND KIVIE MOLDAVE

VOLUME 107. Posttranslational Modifications (Part B)
Edited by FINN WOLD AND KIVIE MOLDAVE

VOLUME 108. Immunochemical Techniques (Part G: Separation and Characterization of Lymphoid Cells)
Edited by GIOVANNI DI SABATO, JOHN J. LANGONE, AND HELEN VAN VUNAKIS

VOLUME 109. Hormone Action (Part I: Peptide Hormones)
Edited by LUTZ BIRNBAUMER AND BERT W. O'MALLEY

VOLUME 110. Steroids and Isoprenoids (Part A)
Edited by JOHN H. LAW AND HANS C. RILLING

VOLUME 111. Steroids and Isoprenoids (Part B)
Edited by JOHN H. LAW AND HANS C. RILLING

VOLUME 112. Drug and Enzyme Targeting (Part A)
Edited by KENNETH J. WIDDER AND RALPH GREEN

VOLUME 113. Glutamate, Glutamine, Glutathione, and Related Compounds
Edited by ALTON MEISTER

VOLUME 114. Diffraction Methods for Biological Macromolecules (Part A)
Edited by HAROLD W. WYCKOFF, C. H. W. HIRS, AND SERGE N. TIMASHEFF

VOLUME 115. Diffraction Methods for Biological Macromolecules (Part B)
Edited by HAROLD W. WYCKOFF, C. H. W. HIRS, AND SERGE N. TIMASHEFF

VOLUME 116. Immunochemical Techniques (Part H: Effectors and Mediators of Lymphoid Cell Functions)
Edited by GIOVANNI DI SABATO, JOHN J. LANGONE, AND HELEN VAN VUNAKIS

VOLUME 117. Enzyme Structure (Part J)
Edited by C. H. W. HIRS AND SERGE N. TIMASHEFF

VOLUME 118. Plant Molecular Biology
Edited by ARTHUR WEISSBACH AND HERBERT WEISSBACH

VOLUME 119. Interferons (Part C)
Edited by SIDNEY PESTKA

VOLUME 120. Cumulative Subject Index Volumes 81–94, 96–101

VOLUME 121. Immunochemical Techniques (Part I: Hybridoma Technology and Monoclonal Antibodies)
Edited by JOHN J. LANGONE AND HELEN VAN VUNAKIS

VOLUME 122. Vitamins and Coenzymes (Part G)
Edited by FRANK CHYTIL AND DONALD B. MCCORMICK

VOLUME 123. Vitamins and Coenzymes (Part H)
Edited by FRANK CHYTIL AND DONALD B. MCCORMICK

VOLUME 124. Hormone Action (Part J: Neuroendocrine Peptides)
Edited by P. MICHAEL CONN

VOLUME 125. Biomembranes (Part M: Transport in Bacteria, Mitochondria, and Chloroplasts: General Approaches and Transport Systems)
Edited by SIDNEY FLEISCHER AND BECCA FLEISCHER

VOLUME 126. Biomembranes (Part N: Transport in Bacteria, Mitochondria, and Chloroplasts: Protonmotive Force)
Edited by SIDNEY FLEISCHER AND BECCA FLEISCHER

VOLUME 127. Biomembranes (Part O: Protons and Water: Structure and Translocation)
Edited by LESTER PACKER

VOLUME 128. Plasma Lipoproteins (Part A: Preparation, Structure, and Molecular Biology)
Edited by JERE P. SEGREST AND JOHN J. ALBERS

VOLUME 129. Plasma Lipoproteins (Part B: Characterization, Cell Biology, and Metabolism)
Edited by JOHN J. ALBERS AND JERE P. SEGREST

VOLUME 130. Enzyme Structure (Part K)
Edited by C. H. W. HIRS AND SERGE N. TIMASHEFF

VOLUME 131. Enzyme Structure (Part L)
Edited by C. H. W. HIRS AND SERGE N. TIMASHEFF

VOLUME 132. Immunochemical Techniques (Part J: Phagocytosis and Cell-Mediated Cytotoxicity)
Edited by GIOVANNI DI SABATO AND JOHANNES EVERSE

VOLUME 133. Bioluminescence and Chemiluminescence (Part B)
Edited by MARLENE DELUCA AND WILLIAM D. MCELROY

VOLUME 134. Structural and Contractile Proteins (Part C: The Contractile Apparatus and the Cytoskeleton)
Edited by RICHARD B. VALLEE

VOLUME 135. Immobilized Enzymes and Cells (Part B)
Edited by KLAUS MOSBACH

VOLUME 136. Immobilized Enzymes and Cells (Part C)
Edited by KLAUS MOSBACH

VOLUME 137. Immobilized Enzymes and Cells (Part D)
Edited by KLAUS MOSBACH

VOLUME 138. Complex Carbohydrates (Part E)
Edited by VICTOR GINSBURG

VOLUME 139. Cellular Regulators (Part A: Calcium- and Calmodulin-Binding Proteins)
Edited by ANTHONY R. MEANS AND P. MICHAEL CONN

VOLUME 140. Cumulative Subject Index Volumes 102–119, 121–134

VOLUME 141. Cellular Regulators (Part B: Calcium and Lipids)
Edited by P. MICHAEL CONN AND ANTHONY R. MEANS

VOLUME 142. Metabolism of Aromatic Amino Acids and Amines
Edited by SEYMOUR KAUFMAN

VOLUME 143. Sulfur and Sulfur Amino Acids
Edited by WILLIAM B. JAKOBY AND OWEN GRIFFITH

VOLUME 144. Structural and Contractile Proteins (Part D: Extracellular Matrix)
Edited by LEON W. CUNNINGHAM

VOLUME 145. Structural and Contractile Proteins (Part E: Extracellular Matrix)
Edited by LEON W. CUNNINGHAM

VOLUME 146. Peptide Growth Factors (Part A)
Edited by DAVID BARNES AND DAVID A. SIRBASKU

VOLUME 147. Peptide Growth Factors (Part B)
Edited by DAVID BARNES AND DAVID A. SIRBASKU

VOLUME 148. Plant Cell Membranes
Edited by LESTER PACKER AND ROLAND DOUCE

VOLUME 149. Drug and Enzyme Targeting (Part B)
Edited by RALPH GREEN AND KENNETH J. WIDDER

VOLUME 150. Immunochemical Techniques (Part K: *In Vitro* Models of B and T Cell Functions and Lymphoid Cell Receptors)
Edited by GIOVANNI DI SABATO

VOLUME 151. Molecular Genetics of Mammalian Cells
Edited by MICHAEL M. GOTTESMAN

VOLUME 152. Guide to Molecular Cloning Techniques
Edited by SHELBY L. BERGER AND ALAN R. KIMMEL

VOLUME 153. Recombinant DNA (Part D)
Edited by RAY WU AND LAWRENCE GROSSMAN

VOLUME 154. Recombinant DNA (Part E)
Edited by RAY WU AND LAWRENCE GROSSMAN

VOLUME 155. Recombinant DNA (Part F)
Edited by RAY WU

VOLUME 156. Biomembranes (Part P: ATP-Driven Pumps and Related Transport: The Na,K-Pump)
Edited by SIDNEY FLEISCHER AND BECCA FLEISCHER

VOLUME 157. Biomembranes (Part Q: ATP-Driven Pumps and Related Transport: Calcium, Proton, and Potassium Pumps)
Edited by SIDNEY FLEISCHER AND BECCA FLEISCHER

VOLUME 158. Metalloproteins (Part A)
Edited by JAMES F. RIORDAN AND BERT L. VALLEE

VOLUME 159. Initiation and Termination of Cyclic Nucleotide Action
Edited by JACKIE D. CORBIN AND ROGER A. JOHNSON

VOLUME 160. Biomass (Part A: Cellulose and Hemicellulose)
Edited by WILLIS A. WOOD AND SCOTT T. KELLOGG

VOLUME 161. Biomass (Part B: Lignin, Pectin, and Chitin)
Edited by WILLIS A. WOOD AND SCOTT T. KELLOGG

VOLUME 162. Immunochemical Techniques (Part L: Chemotaxis and Inflammation)
Edited by GIOVANNI DI SABATO

VOLUME 163. Immunochemical Techniques (Part M: Chemotaxis and Inflammation)
Edited by GIOVANNI DI SABATO

VOLUME 164. Ribosomes
Edited by HARRY F. NOLLER, JR., AND KIVIE MOLDAVE

VOLUME 165. Microbial Toxins: Tools for Enzymology
Edited by SIDNEY HARSHMAN

VOLUME 166. Branched-Chain Amino Acids
Edited by ROBERT HARRIS AND JOHN R. SOKATCH

VOLUME 167. Cyanobacteria
Edited by LESTER PACKER AND ALEXANDER N. GLAZER

VOLUME 168. Hormone Action (Part K: Neuroendocrine Peptides)
Edited by P. MICHAEL CONN

VOLUME 169. Platelets: Receptors, Adhesion, Secretion (Part A)
Edited by JACEK HAWIGER

VOLUME 170. Nucleosomes
Edited by PAUL M. WASSARMAN AND ROGER D. KORNBERG

VOLUME 171. Biomembranes (Part R: Transport Theory: Cells and Model Membranes)
Edited by SIDNEY FLEISCHER AND BECCA FLEISCHER

VOLUME 172. Biomembranes (Part S: Transport: Membrane Isolation and Characterization)
Edited by SIDNEY FLEISCHER AND BECCA FLEISCHER

VOLUME 173. Biomembranes [Part T: Cellular and Subcellular Transport: Eukaryotic (Nonepithelial) Cells]
Edited by SIDNEY FLEISCHER AND BECCA FLEISCHER

VOLUME 174. Biomembranes [Part U: Cellular and Subcellular Transport: Eukaryotic (Nonepithelial) Cells]
Edited by SIDNEY FLEISCHER AND BECCA FLEISCHER

VOLUME 175. Cumulative Subject Index Volumes 135–139, 141–167

VOLUME 176. Nuclear Magnetic Resonance (Part A: Spectral Techniques and Dynamics)
Edited by NORMAN J. OPPENHEIMER AND THOMAS L. JAMES

VOLUME 177. Nuclear Magnetic Resonance (Part B: Structure and Mechanism)
Edited by NORMAN J. OPPENHEIMER AND THOMAS L. JAMES

VOLUME 178. Antibodies, Antigens, and Molecular Mimicry
Edited by JOHN J. LANGONE

VOLUME 179. Complex Carbohydrates (Part F)
Edited by VICTOR GINSBURG

VOLUME 180. RNA Processing (Part A: General Methods)
Edited by JAMES E. DAHLBERG AND JOHN N. ABELSON

VOLUME 181. RNA Processing (Part B: Specific Methods)
Edited by JAMES E. DAHLBERG AND JOHN N. ABELSON

VOLUME 182. Guide to Protein Purification
Edited by MURRAY P. DEUTSCHER

VOLUME 183. Molecular Evolution: Computer Analysis of Protein and Nucleic Acid Sequences
Edited by RUSSELL F. DOOLITTLE

VOLUME 184. Avidin–Biotin Technology
Edited by MEIR WILCHEK AND EDWARD A. BAYER

VOLUME 185. Gene Expression Technology
Edited by DAVID V. GOEDDEL

VOLUME 186. Oxygen Radicals in Biological Systems (Part B: Oxygen Radicals and Antioxidants)
Edited by LESTER PACKER AND ALEXANDER N. GLAZER

VOLUME 187. Arachidonate Related Lipid Mediators
Edited by ROBERT C. MURPHY AND FRANK A. FITZPATRICK

VOLUME 188. Hydrocarbons and Methylotrophy
Edited by MARY E. LIDSTROM

VOLUME 189. Retinoids (Part A: Molecular and Metabolic Aspects)
Edited by LESTER PACKER

VOLUME 190. Retinoids (Part B: Cell Differentiation and Clinical Applications)
Edited by LESTER PACKER

VOLUME 191. Biomembranes (Part V: Cellular and Subcellular Transport: Epithelial Cells)
Edited by SIDNEY FLEISCHER AND BECCA FLEISCHER

VOLUME 192. Biomembranes (Part W: Cellular and Subcellular Transport: Epithelial Cells)
Edited by SIDNEY FLEISCHER AND BECCA FLEISCHER

VOLUME 193. Mass Spectrometry
Edited by JAMES A. MCCLOSKEY

VOLUME 194. Guide to Yeast Genetics and Molecular Biology
Edited by CHRISTINE GUTHRIE AND GERALD R. FINK

VOLUME 195. Adenylyl Cyclase, G Proteins, and Guanylyl Cyclase
Edited by ROGER A. JOHNSON AND JACKIE D. CORBIN

VOLUME 196. Molecular Motors and the Cytoskeleton
Edited by RICHARD B. VALLEE

VOLUME 197. Phospholipases
Edited by EDWARD A. DENNIS

VOLUME 198. Peptide Growth Factors (Part C)
Edited by DAVID BARNES, J. P. MATHER, AND GORDON H. SATO

VOLUME 199. Cumulative Subject Index Volumes 168–174, 176–194 (in preparation)

VOLUME 200. Protein Phosphorylation (Part A: Protein Kinases: Assays, Purification, Antibodies, Functional Analysis, Cloning, and Expression)
Edited by TONY HUNTER AND BARTHOLOMEW M. SEFTON

VOLUME 201. Protein Phosphorylation (Part B: Analysis of Protein Phosphorylation, Protein Kinase Inhibitors, and Protein Phosphatases)
Edited by TONY HUNTER AND BARTHOLOMEW M. SEFTON

VOLUME 202. Molecular Design and Modeling: Concepts and Applications (Part A: Proteins, Peptides, and Enzymes)
Edited by JOHN J. LANGONE

VOLUME 203. Molecular Design and Modeling: Concepts and Applications (Part B: Antibodies and Antigens, Nucleic Acids, Polysaccharides, and Drugs)
Edited by JOHN J. LANGONE

VOLUME 204. Bacterial Genetic Systems
Edited by JEFFREY H. MILLER

VOLUME 205. Metallobiochemistry (Part B: Metallothionein and Related Molecules)
Edited by JAMES F. RIORDAN AND BERT L. VALLEE

VOLUME 206. Cytochrome P450
Edited by MICHAEL R. WATERMAN AND ERIC F. JOHNSON

VOLUME 207. Ion Channels
Edited by BERNARDO RUDY AND LINDA E. IVERSON

VOLUME 208. Protein–DNA Interactions
Edited by ROBERT T. SAUER

VOLUME 209. Phospholipid Biosynthesis
Edited by EDWARD A. DENNIS AND DENNIS E. VANCE

Section I

Isolation, Synthesis, and Detection of DNA and RNA

[1] Megabase DNA Preparation from Plant Tissue

By FRANÇOIS GUIDET and PETER LANGRIDGE

Introduction

Traditional DNA extraction methods yield fragments of about 50 to 100 kilobase pairs (kbp) in length. The largest DNA fragments that can be separated by conventional electrophoresis in an agarose gel are 30 to 40 kbp in size. In contrast, the pulsed-field gel electrophoresis (PFGE) technique allows the separation of DNAs of more than 10,000 kbp (10 Mbp). The different principles involved in PFGE are represented by various acronyms such as FIGE, OFAGE, TAFE, and CHEF. All involve repeated reorientation of the DNA molecules inside the gel matrix due to corresponding changes in electric field parameters (electrode angle, switching time, field inversion, etc.; for a review see Ref. 1). To date the fractionation of DNA molecules has been extended to 12 Mbp,[2] but there does not seem to be any theoretical limit.

To take advantage of these dramatic improvements biologists have designed methods to prepare high molecular weight DNA molecules, so-called megabase DNA (Mbp DNA).[3] For various reasons, plant molecular biologists have been slow to develop specific protocols suitable for preparing Mbp DNA. Without exception, the methods used to prepare plant Mbp DNA involved the preparation of protoplasts as a preliminary step, that is, plant cells are freed of their cell wall by digestion with specific enzymes.[4–8] To circumvent this tedious task, we designed an alternative method that appeared to be both rapid and efficient.[9] The description of an updated version of this method is the subject of the present article.

Principle of Method

The principle of the method is straightforward. It is based on the assumption that grinding leaf tissue in the presence of liquid nitrogen with

[1] R. Anand, *Trends Genet.* **2,** 278 (1986).
[2] M. J. Orbach, D. Vollrath, R. W. Davis, and C. Yanofsky, *Mol. Cell. Biol.* **8,** 1469 (1988).
[3] D. C. Schwartz and C. R. Cantor, *Cell (Cambridge, Mass.)* **37,** 76 (1984).
[4] P. Guzman and J. R. Ecker, *Nucleic Acids Res.* **16,** 11091 (1988).
[5] M. W. Ganal, N. D. Young, and S. D. Tanksley, *Mol. Gen. Genet.* **215,** 395 (1989).
[6] C. Jung, M. Kleine, F. Fischer, and R. G. Herrmann, *Theor. Appl. Genet.* **79,** 663 (1990).
[7] R. A. J. van Daelen, J. J. Jonkers, and P. Zabel, *Plant Mol. Biol.* **12,** 341 (1989).
[8] W. Y. Cheung and M. D. Gale, *Plant Mol. Biol.* **14,** 881 (1990).
[9] F. Guidet, P. Rogowsky, and P. Langridge, *Nucleic Acids Res.* **18,** 4955 (1990).

a mortar and pestle allows the preparation of plant cells, either isolated or in small aggregates. These plant cells, surrounded by a more or less damaged cell wall, contain intact organelles and their membranes are amenable to digestion by the combined action of a detergent (sarkosyl) and a proteolytic enzyme (proteinase K). The integrity of the DNA molecules is maintained by the addition of a chelating agent (ethylenediaminetetraacetic acid; EDTA), which helps protect the DNA from nucleases. Most important, the plant material (powder) is embedded in agarose prior to digestion, thus avoiding any mechanical shearing during subsequent treatments. Once the various cell membranes have been dissolved, the DNA molecules are liberated from their associated proteins by the proteinase K. The entire treatment is done at 53–55°, which is still within the optimal temperature range of action for the proteinase K but well out of the active range for most plant nucleases. The DNA remains in the cavities created inside the agarose plugs by the original plant cells while solutes and small products of cell wall degradation diffuse out of the plugs. The DNA is still accessible to DNA-modifying enzymes such as restriction endonucleases and can be subjected to molecular biological manipulation.

Materials and Reagents

The plant materials used are either green leaves of 10-day-old seedlings, seeds, or commercial flour. Wheat–rye recombinant plants have been described in Rogowsky *et al.*[10] and are obtained from Ken W. Shepherd (Waite Institute, South Australia). Seeds from alfalfa, lentils, and soybeans are from a local shop. Rye (cv. 'South Australian') flour is from W. Thomas Company (Port Adelaide, South Australia).

Low melting temperature (LMP) and LE agarose are both from FMC BioProducts (Rockland, ME), proteinase K and restriction enzymes are from Boehringer (Mannheim, Germany), and radiolabeled dCTP and the transfer membrane HyBond N^+ are from Amersham (Arlington Heights, IL). The PFGE system used was a CHEF DR II from Bio-Rad Laboratories (Richmond, CA).

Solutions used to treat or store the plugs include a lysis solution (10 m*M* Tris-HCl, pH 8.0, 500 m*M* EDTA, 1% (v/v) sarkosyl, 1 mg/ml proteinase K), 1× ET (1 m*M* Tris-HCl, pH 8.0, 50 m*M* EDTA), and TE (10 m*M* Tris-HCl, pH 8.0, 1 m*M* EDTA).

After the electrophoretic runs the gels are stained with ethidium bromide (1 μg/ml) for 45 min and destained extensively to optimize the signal-

[10] P. Rogowsky, F. Guidet, P. Langridge, K. W. Shepherd, and R. M. D. Koebner, *Theor. Appl. Genet.* **82,** 537 (1991).

to-background ratio. Destaining of up to 20 hr does not visibly affect the sharpness of the DNA bands. The gels are then photographed and irradiated for 1 min with 254-nm UV light, depurinated in 0.25 *M* HCl for 15 min, treated with alkali (1.5 *M* NaCl, 0.5 *M* NaOH) twice for 15 min, and equilibrated in the alkali transfer solution (1.5 *M* NaCl, 0.25 *M* NaOH) for 15 min prior to setting up the capillary transfer system. The transfer lasts for 24 hr. The hybridization conditions have been reported in Rogowsky *et al.*[10] The probe pAW173 detects a moderately repeated rye-specific element.[11]

Method

About 0.4 g of young green leaves are ground to a fine powder in liquid nitrogen using a pestle and mortar. The powder is transferred to a crucible preheated to 50°, mixed with 2 ml of 0.7% (w/v) LMP agarose in 1× ET, and gently stirred with a sterile spatula to obtain a homogeneous mixture (alternatively the powder can be mixed with 1 ml of 1× ET and then added to 1 ml of 1.4% agarose solution). The mixture is then poured directly into the mold (Bio-Rad CHEF DR II mold), shaking it gently while pouring to maintain the homogeneity of the mixture. It is then allowed to set at 4° for 20 min in a lying position to avoid a deposit of debris at the bottom of each agarose plug.

The plugs are transferred into petri dishes and incubated in the lysis solution; we use 10 ml of lysis solution per 10 plugs (each plug is about 250 μl agarose mixture). The incubation is done at 53–55° on a rocking platform in an oven or by floating the petri dishes in a water bath (to be on the safe side it is wise to float the petri dishes inside a plastic box with a minimum of water, the box itself floating in the water bath). At the end of the treatment the plugs are stored at 4° in 1× ET.

We have extended the method and used flour or crushed seeds instead of the leaf material.[12] The seeds are crushed in a mortar and pestle without liquid nitrogen. Most of the results presented here have been obtained by using crushed seeds or flour.

Results

Source of Material

The present method of direct Mbp DNA isolation was developed because high yields were obtained rapidly and without the elaborate tech-

[11] F. Guidet, P. Rogowsky, C. Taylor, W. Song, and P. Langridge, *Genome* **34,** 81 (1991).
[12] F. Guidet and P. Langridge, *C.R. Acad. Sci. Paris, Ser. 3* **314,** 7 (1992).

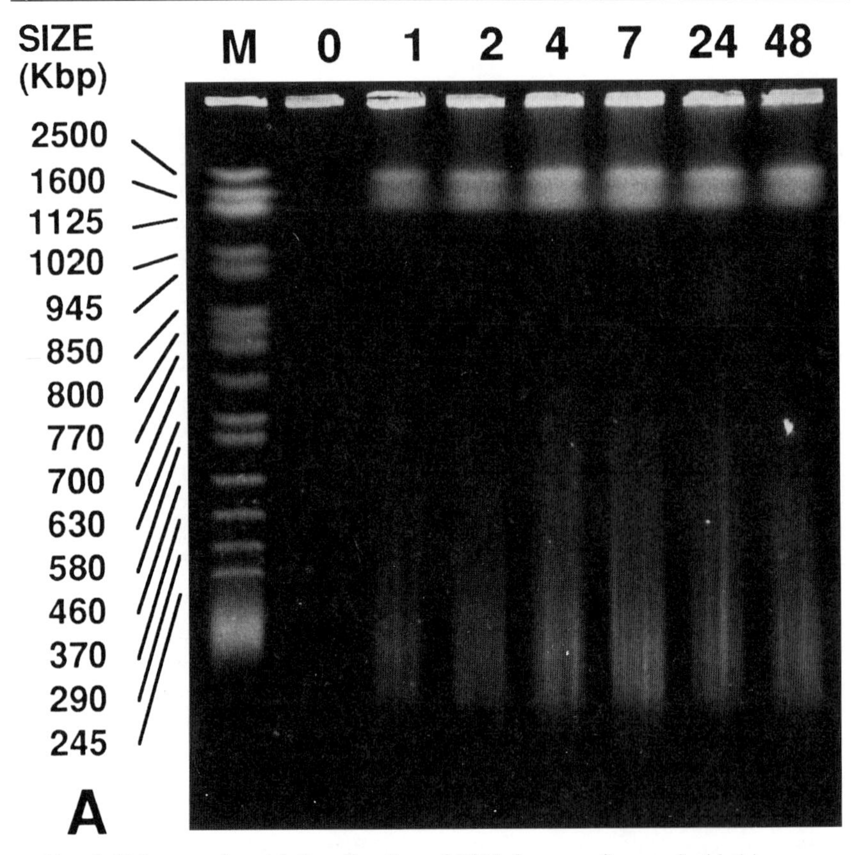

FIG. 1. Release and restriction digestion of DNA from rye flour embedded in agarose. Two series of plugs were incubated with lysis solution at 55° for the periods indicated at the top of each lane (time is in hours). One series (A) was electrophoresed directly (run conditions: time ramp 50 to 90 sec at 200 V, in a 1% agarose gel in 0.5× TBE buffer for 24 hr). The other series was incubated with *Hin*dIII and subsequently electrophoresed (B) (run conditions: time ramp 1 to 6 sec at 125 V, in a 1.5% agarose gel in 0.5× TBE buffer for 18 hr). The arrow in (B) indicates relic DNA that hybridizes with an rDNA probe. The gel from (B) was transferred and probed with pAW173 (C). (A) and (B) are ethidium bromide-stained gels; (C) is an autoradiogram.

niques involved in the preparation of nuclei or protoplasts. Like other authors[6,7] we were not able to isolate Mbp DNA from nuclei, although we obtained good-quality DNA from protoplasts prepared from young leaves or suspension cultures. However, the yield of leaf protoplasts, especially in the case of cereals, is very low; that is, only a small fraction of the leaf cells can be turned into protoplasts and are amenable to lysis. Cell

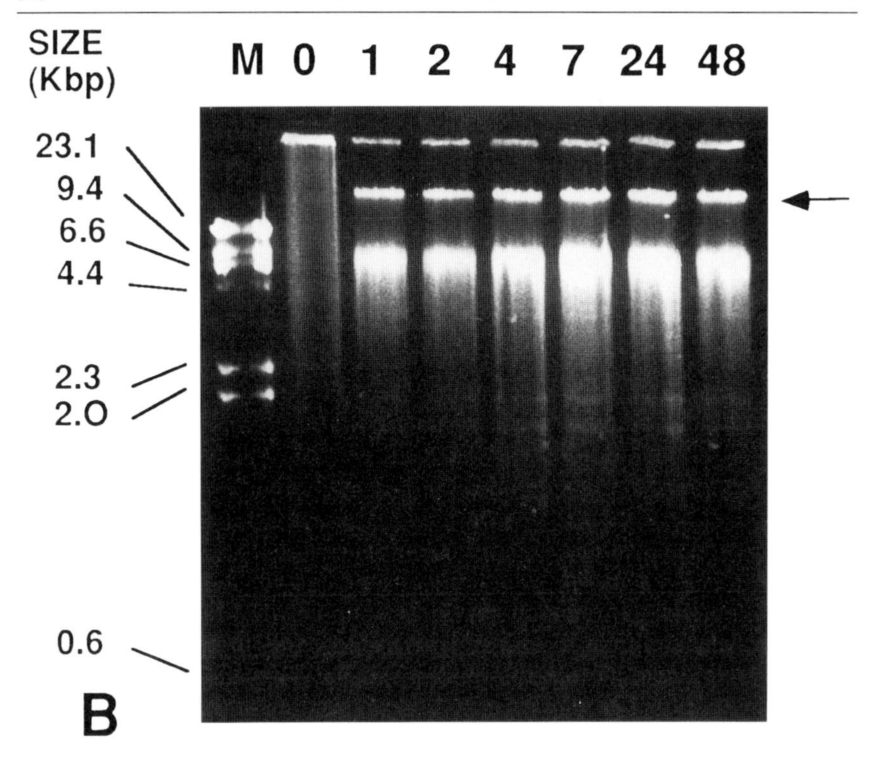

FIG. 1. (*continued*)

suspension cultures give excellent yields of protoplasts and Mbp DNA but are not useful for genetic analysis due to a high chance of chromosomal aberration. As mentioned above, our standard method was successfully extended to the flour or crushed seeds of a number of plant species.

Lysis Treatment Time Course

The efficiency of the lysis treatment and the accessibility of the Mbp DNA to restriction enzymes can be tested by a time course of treatment (Fig. 1). Rye flour plugs are taken out of the lysis solution (proteinase K, sarkosyl, EDTA) after 0, 1, 2, 4, 7, 24, and 48 hr of incubation at 53°. The plugs are then either electrophoresed directly (Fig. 1A) or incubated with the restriction endonuclease *Hin*dIII prior to electrophoresis (Fig. 1B). The latter gel is transferred and hybridized with the dispersed repetitive probe pAW173 (Fig. 1C).

Lysis appears to occur very rapidly. After only 1 hr, the characteristic doublet band of high molecular weight DNA is visible. As shown by the

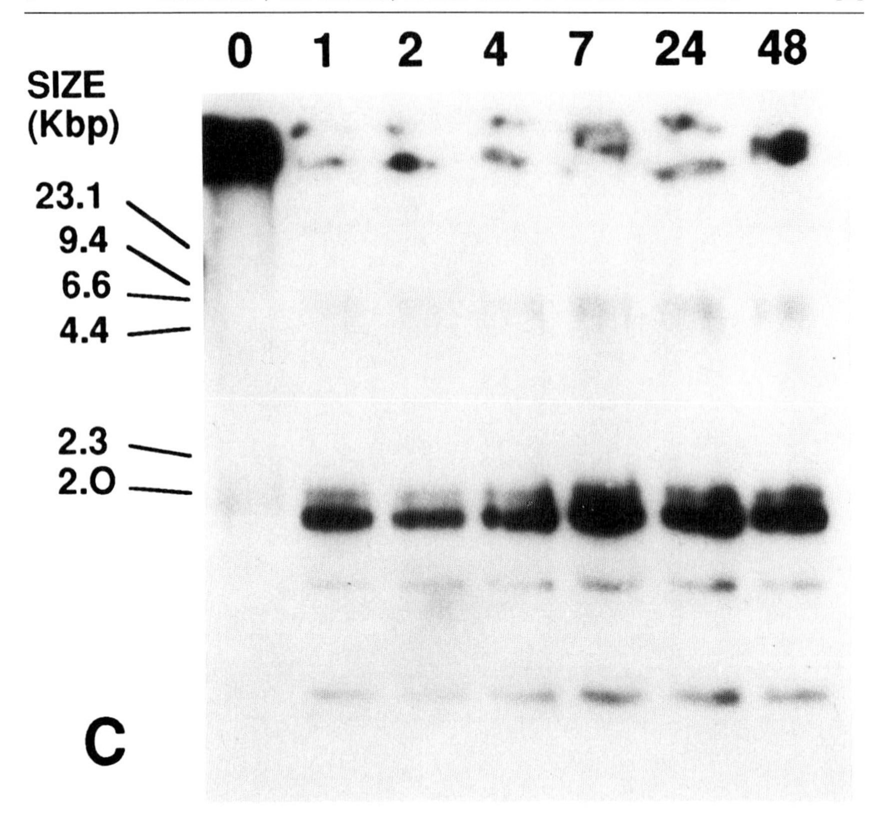

FIG. 1. (*continued*)

ethidium bromide-stained gel picture (Fig. 1B), the appearance of this double band coincides with the accessibility of the DNA preparation to specific endonuclease digestion. The hybridization with the radiolabeled probe pAW173 of the DNA digested with *Hin*dIII confirms that the specificity of the digest is already established after 1 hr of lysis treatment (Fig. 1C). The only effect of the length of the treatment is on the amount of DNA amenable to restriction enzyme. The hybridization pattern obtained is identical to the one observed with conventional DNA preparation and electrophoresis[11] and indicated that *in situ* digestion of the DNA in agarose plugs proceeds to completion and with the desired specificity. The upper band (marked with an arrow on Fig. 1B) corresponds to relic DNA and hybridizes strongly to an rDNA probe (data not shown). A similar rate of lysis was observed with isolated protoplasts by Van Daelen *et al.*[7] and Cheung and Gale.[8]

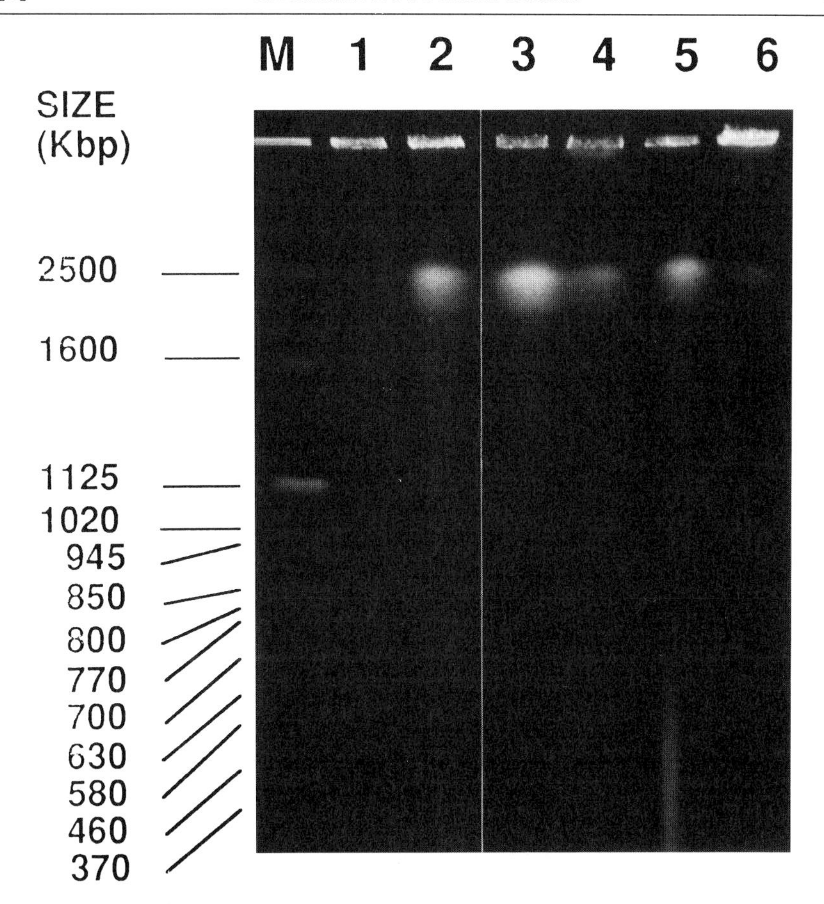

FIG. 2. Unrestricted DNA from various plant species and prepared from different materials. Lane 1, Wheat/rye translocation 1DL/1RS; lane 2, wheat/rye addition line (+5R); lane 3, rye; lane 4, alfalfa; lane 5, lentil; lane 6, soybean. DNAs 1 and 2 were prepared from crushed green leaves; DNA 3 from flour; DNAs 4–6 from crushed seeds. The electrophoresis run conditions were as follow: time ramp 50 to 200 sec at 150 V, in a 0.6% agarose gel in 0.5× TBE buffer for 40 hr. The gel has been stained with ethidium bromide.

Trial on Different Species

We have successfully applied the method to crushed seeds from a variety of plant species from the Gramineae and Leguminosae (Fig. 2). However, the amount of plant material embedded in a given volume of agarose must be adapted for each species. For instance, 5 to 10 times more starting material (i.e., flour) is needed in the case of rice, barley, and oats as compared to rye or wheat (between 0.1 and 0.3 g for rye and wheat,

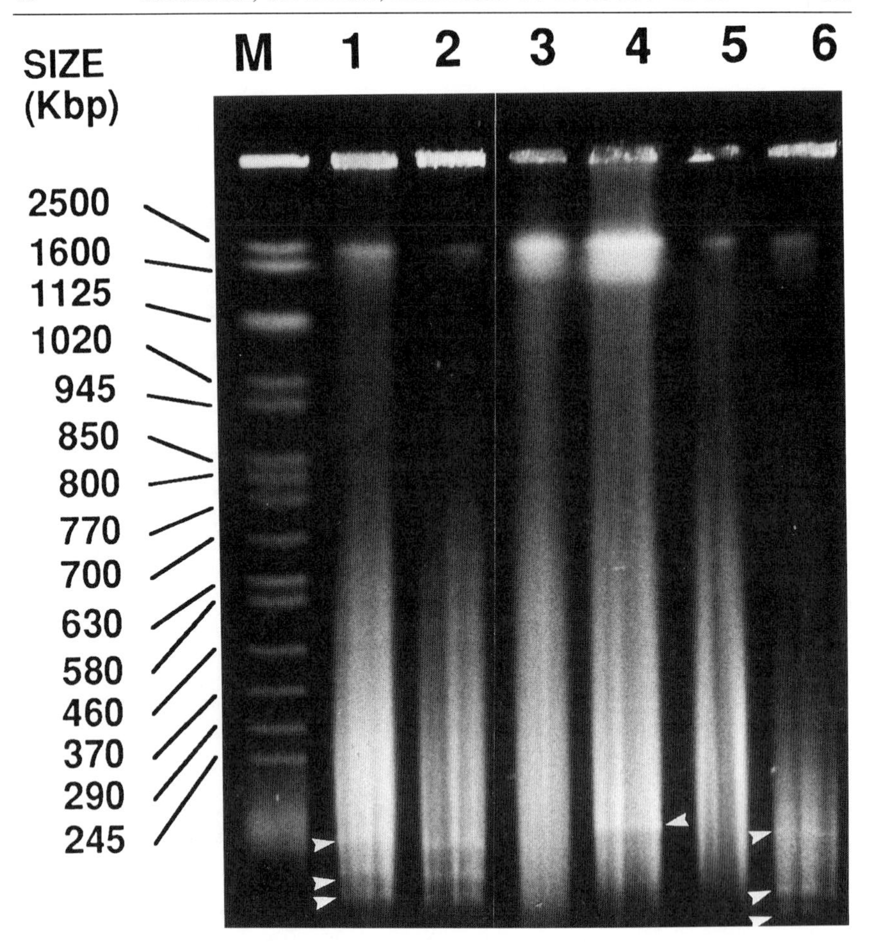

FIG. 3. Restriction digestion of DNA from various plant species and prepared from different materials. Lanes 1 to 6 are as in Fig. 2. The DNA was incubated overnight at 37° with 50 units of *Mlu*I and electrophoresed (run conditions: time ramp 50 to 90 sec at 200 V, in a 1% agarose gel in 0.5× TBE buffer for 24 hr). The gel has been stained with ethidium bromide. White arrowheads indicate specific banding patterns.

depending on the type of flour). The undigested DNA from all the samples appears at the separation limit of the gel, indicating a length of greater than 2.5 Mbp, but a major portion of the high molecular weight DNA remains in the wells due to its very large size.

Restriction Digestion

The restriction digest protocol is as follows. After the lysis treatment the plugs are stored at 4° in 1× ET; the DNA seems to be stable for months

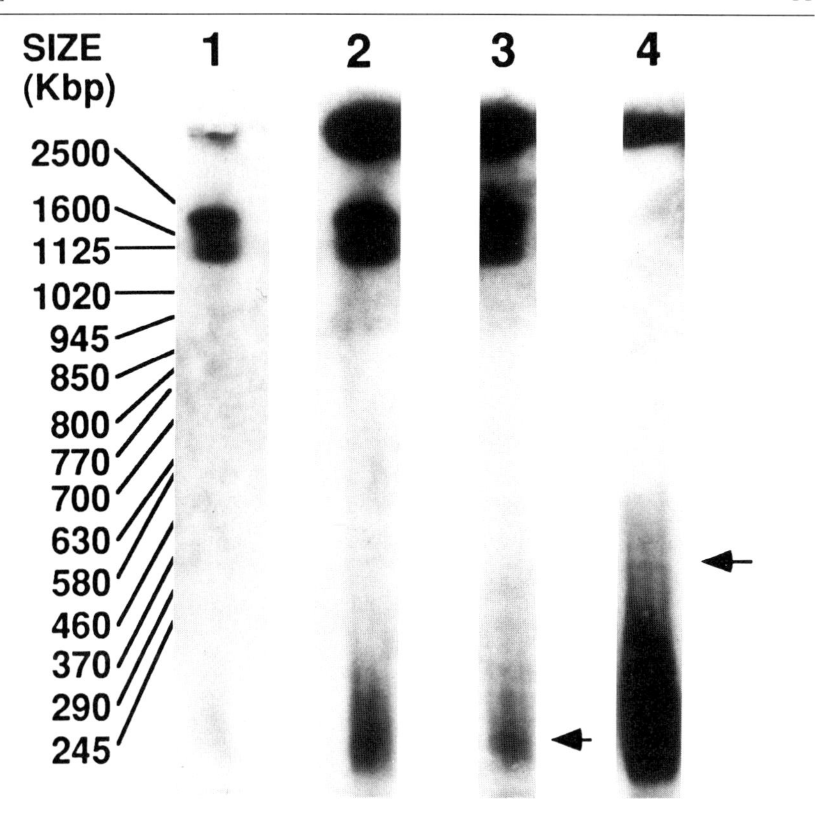

FIG. 4. Autoradiograph of DNA from rye flour embedded in agarose probed with pAW173. Agarose-embedded rye flour was lysed for 48 hr and the released DNA was digested with various restriction enzymes. Lane 1, unrestricted DNA; lane 2, *Not*I; lane 3, *Xho*I; lane 4, *Sal*I. The electrophoresis run conditions were as follow: time ramp 50 to 90 sec at 200 V, in a 0.8% agarose gel in 0.5× TBE buffer for 24 hr. The gel was transferred to membrane and hybridized with the radiolabeled probe pAW173. The arrows indicate discrete bands.

and possibly years under these conditions. The day before the digestion, the plugs are sliced to a suitable size, soaked in 10 ml of TE in a petri dish, and stored overnight at 4°. The next day the plugs are extensively washed for 3–4 hr at room temperature on a rocking platform with about three changes of TE and then individually transferred to Eppendorf tubes with 500 μl of restriction buffer for 2 hr at room temperature. They are finally transferred to tubes containing 100 μl of restriction buffer including 1 μl of 10 mg/ml acetylated nuclease-free bovine serum albumin (BSA) and 50 to 60 units of restriction enzyme. They are kept on ice for 30 min in order for the enzyme to penetrate the agarose and are then incubated overnight at the appropriate temperature. The reaction is stopped by the addition of

500 μl of 1× ET to the tubes. The plugs can be stored at 4° (for a few weeks) or electrophoresed immediately. It is worth noting that, unlike many published procedures, we have not found it necessary to use the antiprotease phenylmethylsulfonyl fluoride (PMSF) prior to digestion.

Under these conditions, good results are obtained for the Mbp DNA prepared from leaves, seeds, or flour (Fig. 3). Although the enzyme action is sometimes difficult to assess, the presence of a discrete, specific banding pattern within the smeary background of the digested DNA (marked with arrowheads in Fig. 3) is a good indication of precise restriction digestion.

Southern Blot Hybridization

Southern blot hybridization on pulsed-field gels is more difficult to perform than on normal gels. The reasons are not clear, but factors such as DNA quality, electrophoretic separation, and gel treatment may have a big influence. It seems of great importance to depurinate the large DNA fragments for successful transfer. An example of a Southern blot is shown in Fig. 4. As pAW173 hybridizes to a repetitive DNA sequence in rye, individual bands are difficult to visualize in the lane smears (some discrete bands are indicated by arrows in Fig. 4).

Concluding Remarks

The procedure described for producing very large DNA fragments from plant tissue has many advantages, in particular its simplicity, rapidity, and efficiency. The extension of this method to the use of seeds and flour makes it even more "user friendly." These materials can be stored for years and are always ready to use. Once a certain type of flour has been assessed, it always keeps its characteristics, in terms of quality of DNA and quantity to use for an optimum result.

Acknowledgments

The authors would like to thank Dr. Peter Rogowsky for critical reading of the manuscript and friendly suggestions. This work has been supported by the Australian Research Council.

[2] Preparing and Using Agarose Microbeads

By MICHAEL KOOB and WACLAW SZYBALSKI

Introduction

Protocols for preparing and manipulating large DNA molecules without breakage are required for the routine analysis of genomes by pulsed-field gel electrophoresis (PFGE). To protect genomic DNA from shear, cells are typically embedded in agarose blocks before lysis and deproteinization.[1] A less commonly used alternative means of preparing stable genomic DNA was devised by Cook to study the chromatin structure of human chromosomes.[2,3] This process starts with whole cells being embedded in agarose microbeads,[4,5] which are essentially minute agarose blocks, and is followed by cell lysis and release of DNA, which remains trapped in the microbeads. Several adaptations of Cook's original microbead procedure to pulsed-field gel applications have been described.[6–8]

Because of their small size, agarose microbeads potentially offer several advantages over agarose blocks, both in ease of handling and in the speed with which the DNA can be prepared and enzymatically manipulated. Despite this, however, microbeads have not been widely considered to be satisfactory replacements for agarose plugs. This is due for the most part to difficulties many researchers have reported when working with microbead-embedded DNA. Once washed, agarose microbeads tend to stick to plasticware, pipette tips, and to each other, thus complicating manipulations. When loaded in wells and electrophoresed, they often produce bands that are diffuse or streaked. Finally, when cells are embedded at too high a density, trapping and background smearing tend to occur more frequently as compared with DNA embedded in agarose plugs.

In the course of our work to develop new genomic cleavage techniques, we have developed a comprehensive, simple set of procedures for prepar-

[1] D. C. Schwartz and C. R. Cantor, *Cell (Cambridge, Mass.)* **37,** 67 (1984).

[2] P. R. Cook, *EMBO J.* **3,** 1837 (1984).

[3] D. A. Jackson and P. R. Cook, *EMBO J.* **4,** 913 (1985).

[4] K. Nilsson, W. Scheirer, O. W. Merten, L. Ostberg, E. Liehl, H. W. D. Katinger, and K. Mosbach, *Nature (London)* **302,** 629 (1983).

[5] K. Nilsson, W. Scheirer, H. W. D. Katinger, and K. Mosbach, this series, Vol. 121, p. 352.

[6] G. F. Carle and M. V. Olson, this series, Vol. 155, p. 468.

[7] M. McClelland, this series, Vol. 155, p. 22.

[8] P. J. Piggot and C. A. M. Curtis, *J. Bacteriol.* **169,** 1260 (1987).

ing and enzymatically manipulating genomic DNA in agarose microbeads that reproducibly overcome the major problems typically associated with these techniques.[9] Incubation steps for lysis, deproteinization, and enzymatic modification are short, the microbeads are easy to handle, and the quality of the resulting pulsed-field gels is usually superior to those typically obtained with agarose plugs. Although the details in the protocol given are for the preparation of *Escherichia coli* and *Saccharomyces cerevisiae* genomic DNA, these procedures can be readily applied to most types of cells. Adaptation of the appropriate steps is discussed. Similarly, the protocol for the enzymatic digestion of the microbead-embedded DNA can be easily adapted to almost any other enzymatic modification (e.g., methylation[9]).

Preparing and Using Agarose Microbeads

Solutions

Escherichia coli lysis buffer (ELB): Prepare a buffer containing 1 *M* NaCl, 0.1 *M* ethylenediaminetetraacetic acid (EDTA; pH 8.0/25°), 10 m*M* Tris-HCl (pH 8.0/25°), and 1% (w/v) sodium *N*-lauroylsarcosine with sterile stock solutions and store at room temperature. Add 1 mg lysozyme (Sigma, St. Louis, MO) and 2 μl RNase (10 mg/ml DNase-free stock, stored at 4°) per milliliter of lysis buffer just before use

Yeast spheroplast buffer (YSB): Prepare a solution of three parts sterile SCE (shown below) and two parts sterile 0.5 *M* EDTA (pH 8.0/25°) and store at room temperature. Resuspend microbead-embedded yeast in 4 ml of this buffer and just before incubation at 37° add 1 mg lyticase (about 1000 units; Sigma) and 200 μl 2-mercaptoethanol (about 14.4 *M*; Sigma) to this suspension. (The Corex tube should be covered with Parafilm during incubation because of the 2-mercaptoethanol)

ES: Add 1% (w/v) sodium *N*-lauroylsarcosine to sterile 0.5 *M* EDTA (pH 8/25°) and store at room temperature

ESP: Add 1 mg proteinase K (Sigma) per milliliter ES and incubate at 37° for 0.5 hr to eliminate DNases. Prepare this solution just before use

SCE: Prepare a buffer consisting of 1.0 *M* sorbitol, 0.1 *M* sodium citrate, and 60 m*M* EDTA, adjust to pH 7.0/25° with HCl, sterilize, and store at room temperature

TE: 10 m*M* Tris-HCl, 1 m*M* EDTA, pH 8.0/25°

[9] M. Koob and W. Szybalski, *Science* **250,** 271 (1990).

STE: Dilute sterile 5 *M* NaCl stock to 1 *M* TE

PMSF/TE: Prepare a 100 m*M* phenylmethylsulfonyl fluoride (PMSF) stock by dissolving 26 mg PMSF in 1.5 ml ethanol and store at $-20°$. Dilute to 1 m*M* with TE just before use. (PMSF is toxic and should be handled with care)

TEX, 0.1EX, and 0.5EX: Add 0.01% (v/v) Triton X-100 (Sigma) to sterile TE, 0.1 *M* EDTA (pH 8.0/25°), and 0.5 *M* EDTA (pH 8.0/25°), respectively, and store at room temperature

Cell Preparation

Cells are grown and prepared for embedding in microbeads in essentially the same manner as for agarose plugs. In every case they are washed once in a buffer in which they are stable and then concentrated to twice the final desired concentration in the same buffer.

Escherichia coli. Inoculate 50 ml LB (5 g yeast extract, 10 g tryptone, 10 g NaCl, distilled water to 1 liter) with 1 ml overnight culture and shake at 37° until the OD_{550} is 0.2–0.3. Add chloramphenicol to a final concentration of 180 μg/ml to stop chromosomal replication and aid in cell lysis. Again incubate at 37° until growth has stopped (0.5–1 hr). Chill on ice, determine the final OD_{550}, and calculate the number of cells per milliliter by assuming that there are 1×10^8 cells/ml at an OD_{550} of 0.24. Concentrate $1–2 \times 10^9$ cells by centrifugation, wash once with 5 ml STE, and resuspend in STE to a final volume of 2 ml. (*Not*I-digested DNA from microbeads made with 1×10^9 cells will give fine, light DNA bands on PFGE, and that from microbeads made with 2×10^9 cells will give intense, heavy bands.)

Saccharomyces cerevisiae. Grow cells by inoculating 10 ml YPD (10 g yeast extract, 20 g peptone, 20 g glucose, distilled water to 1 liter) with a fresh culture and shake overnight at 30°. Wash the cells once with 5 ml SCE and resuspend in SCE to a final volume of 2 ml.

Embedding Cells in Microbeads

1. Add 5 ml of paraffin oil (or light mineral oil) to a sterile 25-ml Pyrex flask and warm to 42°. Prepare a 1% (w/v) solution of low-melting-point (LMP) agarose (e.g., InCert agarose; FMC, Rockland, ME) in glass-distilled sterile H_2O and cool to 50° [see (a) in the following section].

2. Warm the 2 ml of washed cells to room temperature and add them to 2 ml 50° agarose. Swirl briefly to mix, and pour into the warm paraffin oil.

3. Vortex the oil/agarose vigorously [see (b) in the following section] until a fine, milky emulsion has formed (30–60 sec). Immediately swirl the

flask in an ice/salt water bath (1 min) to cool the oil quickly and solidify the agarose beads.

4. To remove the oil, add 5 ml 0.1EX to the oil/microbeads emulsion, vigorously swirl the flask, and pour the suspension into a 15-ml Corex centrifuge tube. Pellet the microbeads by centrifugation (5 min at 5000 rpm in a Sorvall (Norwalk, CT) SS-34 rotor, 4°). Carefully pour the oil and buffer off the pellet.

Notes on Embedding Cells

a. When enzymatic reactions will be performed that require incubation above 55°, a highly purified normal-melting-point agarose (e.g., GTG agarose; FMC) may be used in place of LMP agarose in the above procedure. However, when this is done, the temperature of the cells, oil, and agarose should be increased to 37, 50, and 50°, respectively. This keeps the agarose from setting before the microbeads have formed. In addition, deproteinization (below) can be carried out at 65°.

b. Vigorous vortexing and rapid cooling (step 3) are the most critical steps in making uniform preparations of small microbeads. The best results are obtained when the flask is held nearly horizontal with its base pressed firmly to the vibrating nub, causing the liquid to froth violently. To entrap larger cells, such as those from human tissue culture, less vigorous mixing should be used to produce slightly larger microbeads.

c. Attempting to embed cells at unusually high concentrations will result in large numbers of cells trapped on the microbead surface and free in solution. This in turn will lead to severe clumping and, following cell lysis, an unusually viscous solution. Discard preparations that show these symptoms and repeat the embedding process at a lower cell concentration.

Cell Lysis and Deproteinization

1. To remove the cell walls from the embedded cells, resuspend the microbead pellet in 4 ml of the appropriate buffer (ELB or YSB—note that the lyticase and 2-mercaptoethanol are added only at this point) and incubate at 37° for 0.5 hr (*E. coli*) or 1 hr (yeast). Pellet the microbeads and pour off the buffer [see (a)–(c) in the following section].

2. Resuspend the pellet in 4 ml ESP, incubate at 52° for 1 hr, and then remove the ESP [see (c) in the following section].

3. Wash the deproteinized pellet once with 8 ml TE, twice with 4 ml PMSF/TE, once or twice with 8–10 ml TE, and then once with 2 ml 0.5EX. Take care to remove all of the EX from above the pellet after the last wash.

4. To check the quality of the preparation, draw 200 μl of the micro-

beads from the pellet into an uncut Gilson P-200 disposable pipette tip (Rainin, Emeryville, CA) and then expell them back to the pellet. They should pipette smoothly and have a consistency similar to that of a heavy oil. If they do not, they either (1) have clumps of microbeads, or (2) are too dry. To eliminate small clumps of microbeads, repeatedly draw them into and forcefully expell them from the pipette tip until they flow smoothly [see (d) in the following section]. In the case of an overly dry pellet, add enough 0.5EX back to the pellet to allow the microbeads to be pipetted.

5. Cover the Corex tube with Parafilm and store the microbeads at 4° [see (e) in the following section].

Notes on Preparing Genomic DNA

a. The enzymes and buffers used in step 1 should be appropriately modified for removing the cell wall from other types of cells. Microbead-embedded cells without cell walls are treated directly with ESP (i.e., a separate lysis step is not necessary).

b. For incubation during the lysis and deproteinization steps, the Corex tube containing the resuspended microbeads is placed in the appropriate water bath. Shaking is not necessary.

c. To change a buffer, pellet the microbeads by centrifugation (5 min at 5000 rpm in a Sorvall SS-34 rotor, 4°) and pour the buffer carefully off the pellet. Add the new buffer and completely resuspend the pelleted microbeads by vigorous vortexing. Clumps of microbeads that do not separate during vortexing should be broken by pipetting up and down through a disposable transfer pipette.

d. In addition to making the microbeads easy to pipette, disrupting microbead clumps allows for more uniform and efficient enzymatic treatment of the DNA in the microbeads and minimizes trapping during pulsed-field gel electrophoresis. The Triton X-100, which does not inhibit most enzymatic reactions at low concentrations, prevents the microbeads from sticking to themselves and to the plastic and glassware with which they came in contact. Incorporation of small amounts of this nonionic detergent in many of the solutions throughout this procedure (as indicated) is critical.

e. Although genomic DNA prepared in this way can be stored at 4° and used successfully for over a year, the best results are obtained from microbeads less than a few months old.

Enzymatic Digestion of Microbead-Embedded DNA

1. Pipette 20–30 μl of microbeads into a clear or light-colored Eppendorf microfuge tube [see (a) in the following section].

2. Wash the beads once with 200 μl TEX and then twice with 200 μl

of the supplier's recommended digestion buffer without bovine serum albumin (BSA) [see (b) in the following section].

3. Add 0.1 vol BSA (2–3 μl of sterile 1 mg/ml stock solution) and 0.5–1 μl restriction enzyme to the buffer-equilibrated pellet and mix thoroughly with the end of the pipette tip. Place the microfuge tube in the appropriate water bath and incubate for 1 hr (shaking is not necessary).

4. If the microbeads will be loaded on a gel immediately after the digestion, wash the pellet with 200 μl TEX. For samples that are to be stored for longer than a few hours before loading, add an equal volume of ES to the digest, incubate at 52° for 15 min to inactivate all nucleases completely, and store the sample at 4°.

Notes on Enzymatic Manipulation of DNA in Microbeads

a. Using Gilson P-200 disposable pipette tips, from each of which approximately 5 mm of the end has been removed (use a razor blade), transfer small amounts of microbead suspensions. The larger bore of such pipette tips allows the slow-flowing microbead suspension to be more easily and accurately pipetted.

b. To wash the microbead pellet, forcefully pipette TEX or buffer into the tube (the force of the buffer injected into the tube is sufficient to resuspend the microbeads completely), centrifuge for 1 min in a microfuge at maximum speed, and remove the buffer down to the microbead pellet by pipetting with an uncut tip. Although microbeads made from yeast have a slightly milky appearance, those made from *E. coli* are clear and difficult to see. To distinguish the clear microbead pellet from the buffer, hold the Eppendorf tube up to a light to make the pellet/buffer boundary visible by the difference in diffraction. The buffer remaining in the microbead pellet after the washing process is sufficient for the enzymatic reaction.

c. The above protocol can be used for sequential digestions and for enzymatic manipulations other than digesting. In each case, wash the pellet once with TEX (to remove the previous buffer and to coat the microbeads with Triton X-100) and then twice with the appropriate buffer. Add the BSA, enzyme, and other necessary reagents directly to the buffer-equilibrated pellet.

Pulsed-Field Gel Electrophoresis (PFGE)

1. Prepare (a) an agarose solution suitable for PFGE in a 250-ml Pyrex Erlenmeyer flask and (b) a small amount of a 0.5% solution of the same agarose and place them in a 50° water bath until needed [see (a) in the following section].

2. Add 10 μl TE to the TEX-washed microbead pellet and mix with

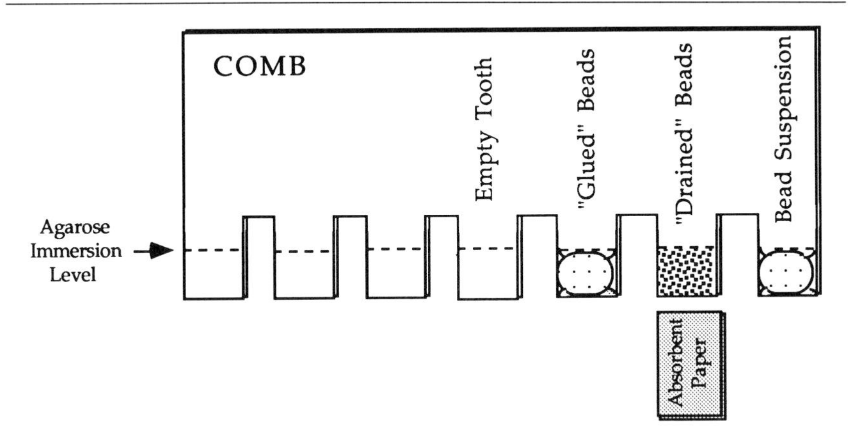

FIG. 1. Loading microbeads on the teeth of a comb (top view).

the pipette tip. Load the TE/microbead suspension directly on a tooth of the (horizontal) comb using a cut pipette tip [see (a) in the preceding section]. Form a uniform, flat droplet completely covering the portion of the tooth face that will be immersed in agarose (see Fig. 1). Repeat for each sample.

3. Once all samples have been loaded on the teeth, remove the TE by touching the lower edge of the microbead/TE droplet with the edge of a small piece of absorbent paper towel cut to the width of the tooth. Hold the towel against the edge of the tooth until all of the TE has been absorbed from the sample into the towel (about 15 sec) and the texture of the microbeads is visible. If necessary, stray microbeads may be pushed into place with the paper towel at this time. Add 5–10 μl of the 0.5% agarose solution [50°; see step 1 in this section and (a) in the following section] to the microbead layer [see (b) in the following section]. Repeat this process for each sample.

4. When the agarose "glue" has set (1 min), place the comb in the gel mold with the sample facing in the direction the DNA is to be electrophoresed. Pour the agarose gently into the mold and allowed it to thoroughly cool (0.5 hr). To avoid disturbing the microbead layer when the comb is removed, push the comb gently but firmly away from the embedded microbeads (until a small space is visible between the side of each well and tooth) and then pull straight up.

5. Run the gel as usual for PFGE.

Notes on Pulsed-Field Gel Electrophoresis

a. The 0.5% agarose solution, which will serve as a "glue" to hold the microbeads on the comb, should not be above 55° when it is used or it will melt microbeads made from LMP agarose.

b. In step 3, the agarose replaces the TE that has been blotted off and the microbead layer will again appear smooth. The agarose should be added in droplets from the top and bottom edges of the microbead mass to avoid displacing the microbeads. Do not allow the microbeads to dry on the comb before the agarose "glue" is added, as this will adversely affect the quality of the pulsed-field gel.

c. We have found that loading microbeads on the comb gives the best results for PFGE. As an alternative to this method, the microbeads can be loaded directly into the wells after the gel has set and then sealed in place with agarose. However, we typically have a higher background, more diffuse bands, and more problems with "streaking" when we use this latter approach.

Concluding Remarks

High-quality, intact genomic DNA can be rapidly prepared and digested in agarose microbeads with the protocols described here. Complete lysis and deproteinization are achieved with a combined incubation time of 2 hr or less. Furthermore, digestion with all restriction enzymes tested has been complete within 1 hr, the time typically allowed for the digestion of DNA in solution.

Microbeads not only protect large DNA molecules from shear, but also act as giant DNA-carrying "cells," thus converting DNA into a "solid state" and allowing its easy and rapid transfer to various solutions by sedimenting, washing, and resuspending the microbeads.

In addition to *E. coli* and *S. cerevisiae,* we have successfully applied these microbead protocols to *Pseudomonas aeruginosa, Trypanosoma brucei rhodesiense, Trypanosoma cruzi, Plasmodium* species, and *Magnaporthe grisea* and we are confident that they can be adapted to most, if not all, types of cells.

[3] Isolation of Low Molecular Weight DNA from Bacteria and Animal Cells

By Shrikant Anant and Kiranur N. Subramanian

Introduction

Procedures are described for the rapid isolation of low molecular weight DNAs, such as plasmids from bacteria and animal cells. Plasmids are covalently closed, circular double-stranded DNA molecules that can replicate as extrachromosomal elements in bacteria. The major use of plasmids is to serve as vectors for the cloning of foreign DNA molecules or segments thereof, with the cloning leading to a wide variety of applications in molecular biology. The simplest vector constructions contain a plasmid origin of replication, one or two antibiotic resistance genes, which enable selection in antibiotic-containing medium, and a multiple cloning site (MCS) containing a number of unique restriction cleavage sequences. Additional features might be included depending on the need. An example is a viral origin of replication (like that of the small animal DNA viruses simian virus 40, polyomavirus, or bovine papillomavirus type 1), which is included if replication is desired in animal cells providing the viral and cellular factors required for replication.

It will be desirable to make a successful verification of cloning as quickly as possible so that one can go on to perform the subsequent studies for which the cloning is intended. A preliminary screening for bacteria harboring recombinant clones can be done by the performance of colony hybridization using a labeled probe specific for the cloned DNA.[1] Colony hybridization would be especially useful if the cloning efficiency is expected to be low. Cloning efficiencies have increased considerably over the years due to improvements such as the employment of an appropriate ratio of vector to DNA fragment to be cloned, the use of linearized vectors dephosphorylated at the ends, or by cloning between a pair of dissimilar termini produced by cleavages at two different restriction sites. We have usually circumvented colony hybridization (saving valuable time and effort), and proceeded straightaway to the inevitable task of small-scale isolation of plasmids from bacterial colonies (or overnight cultures) and their analysis, performed generally within a few hours.

Certain applications may require extraction of low molecular weight DNA from animal cells in culture. Examples are experiments performed with shuttle vectors capable of replication in both bacteria and animal cells, replication assays performed with plasmids containing viral origins of replication, and rescue of integrated DNA containing a viral origin of replication by fusion with cells permissive for its replication.

We describe here procedures for the rapid isolation of plasmids (or other low molecular weight DNAs) on a small scale from bacteria and animal cells.

[1] M. Grunstein and J. Wallis, this series, Vol. 68, p. 379.

Principle

Two of the most popular procedures for the small-scale isolation of plasmid DNA from bacteria are the alkaline lysis method[2] and the boiling method.[3] The alkaline lysis method is based on the principle that in the pH range of 12.0 to 12.6, small, covalently closed circular DNA (such as plasmid DNA) does not denature, but high molecular weight chromosomal DNA does and forms an insoluble network on renaturation, enabling one to selectively extract the soluble plasmid DNA.[2] The alkaline lysis method has been described previously in this series,[4] and will not be discussed here. The boiling method, described originally by Holmes and Quigley,[3] has been modified subsequently by Riggs and McLachlan.[5] We describe here a modified version of the boiling method successfully employed in our laboratory. We also describe another method developed in our laboratory based on one-step lysis and extraction of bacteria with a phenol–chloroform mixture.[6] We have found both these methods to be fast, reliable, and consistent.

The boiling method is based on the observation that when bacteria, the cell walls of which are removed by lysozyme treatment, are lysed with a nonionic detergent, boiled for a brief period, and cooled, the chromosomal DNA forms an insoluble clot along with denatured protein and other cellular debris. The clot is pelleted by centrifugation, leaving the soluble plasmid DNA in the supernatant.[3] The boiling might also inactivate the bacterial deoxyribonucleases and facilitate the isolation of plasmid DNA in undegraded form. The phenol–chloroform lysis–extraction procedure is based on the observation that the lysis, insolubilization of chromosomal DNA, and removal of protein and cell debris can be achieved in one step by vigorously shaking a suspension of bacteria with a phenol–chloroform mixture in the presence of 1.25 *M* NaCl.[6] On centrifugation to separate the aqueous and organic layers the plasmid DNA is found to reside in the aqueous layer, while the interphase between the layers contains the chromosomal DNA, denatured proteins, and cell debris.[6]

The method described for the isolation of plasmids or other low molecular weight DNA from animal cells is based on a procedure developed originally by Hirt for the selective extraction of the small, covalently closed circular, double-stranded DNA genomes of papovaviruses.[7] This

[2] H. C. Birnboim and J. Doly, *Nucleic Acids Res.* **7,** 1513 (1979).

[3] D. S. Holmes and M. Quigley, *Anal. Biochem.* **114,** 193 (1981).

[4] H. C. Birnboim, this series, Vol. 100, p. 243.

[5] M. G. Riggs and A. McLachlan, *BioTechniques* **4,** 310 (1986).

[6] D. C. Alter and K. N. Subramanian, *BioTechniques* **7,** 456 (1989).

[7] B. Hirt, *J. Mol. Biol.* **26,** 365 (1967).

method is based on the observation that when cells are lysed with an ionic detergent and incubated in the cold with 1.0 *M* NaCl, the low molecular weight viral (or plasmid) DNA remains soluble while the high molecular weight chromosomal DNA forms an insoluble precipitate along with denatured proteins and cell debris, enabling the isolation of the soluble viral (or plasmid) DNA by a simple centrifugation step.[7]

Materials and Reagents

Bacterial Strains and Plasmids. Escherichia coli strains DH5α[8] and MC1061[9] are used as hosts for the propagation of the plasmids. An *E. coli* strain such as DH5α is especially desirable as a host because it lacks endodeoxyribonuclease I, the major nonspecific deoxyribonuclease of *E. coli.*[10] This property is beneficial for the efficient isolation of undegraded plasmid after the bacteria are lysed. However, other strains such as *E. coli* MC1061, containing endodeoxyribonuclease I, can also be used because the boiling step during plasmid isolation helps in the inactivation of the host deoxyribonucleases.

Plasmids pSP65 or pGEM-1 (Promega Corp., Madison, WI) or recombinant clones derived from them are used in the methods for plasmid extraction from bacteria. These plasmids contain the ColE1 origin of replication, the β-lactamase gene conferring resistance to ampicillin, and an MCS region for cloning. Plasmid pSV2-*cat* is used in the method for extraction of low molecular weight DNA from animal cells. This plasmid, in addition to the ColE1 origin and the β-lactamase gene, contains the bacterial chloramphenicol acetyltransferase gene under the control of the complete simian virus 40 (SV40) early promoter including the viral origin of replication.[11]

Transformation of *E. coli* host strains with the plasmids is carried out in the presence of $CaCl_2$ as described.[12] Bacteria are grown in Luria-Bertani (LB) broth medium.[13]

Animal Cells. The SV40-transformed monkey kidney cell line COS-1, expressing the SV40 replication initiator protein large tumor (T) antigen,[14] is used as the host for the replication in animal cells of SV40 origin-containing plasmid pSV2-*cat*. The introduction of the pSV2-*cat* into

[8] D. Hanahan, *J. Mol. Biol.* **166,** 557 (1983).
[9] M. Casadaban and S. N. Cohen, *J. Mol. Biol.* **138,** 179 (1980).
[10] H. Durwald and H. Hoffmann-Berling, *J. Mol. Biol.* **34,** 331 (1968).
[11] C. M. Gorman, L. Moffat, and B. H. Howard, *Mol. Cell. Biol.* **2,** 1044 (1982).
[12] M. Mandel and A. Higa, *J. Mol. Biol.* **53,** 159 (1970).
[13] S. Luria and J. W. Burrous, *J. Bacteriol.* **74,** 461 (1957).
[14] Y. Gluzman, *Cell (Cambridge, Mass.)* **23,** 175 (1981).

COS-1 cells is carried out by the DEAE-dextran transfection procedure.[15] The cells are maintained as monolayers in 100-mm plastic tissue culture dishes in Dulbecco's modified Eagle's minimal essential medium supplemented with 10% (v/v) fetal bovine serum, with the dishes incubated in the presence of 5% CO_2 in a humidified atmosphere in a CO_2 incubator (such as model 3158; Forma Scientific, Marietta, OH).

Equipment Needed

Two benchtop microcentrifuges (termed microfuge hereafter) (e.g., HERMLE Z230M; National Labnet Co., Woodbridge, NJ): Each with a 24-slot standard rotor for 1.5-ml capped polypropylene tubes. One of these is kept in the laboratory and used for room-temperature applications, such as spinning down bacteria and separation of aqueous and organic layers. The other is kept in the cold room, and is used primarily for spinning down alcohol-precipitated DNA

One bench-top clinical centrifuge (e.g., IEC model HNS-II; International Equipment Co., Needham Heights, MA): Equipped with a standard horizontal rotor and swinging buckets for eighteen 15-ml capped plastic tubes. This centrifuge is used for the low-speed centrifugation of animal cells

One refrigerated superspeed centrifuge (e.g., Sorvall RC-5B; Du Pont Co., Wilmington, DE): Equipped with an SA-600 rotor that holds twelve 15-ml Corex glass tubes with adaptors

Three variable volume pipettors (e.g., Micropipette; VWR Corp., Philadelphia, PA): For pipetting liquids in the ranges of 1 to 50 μl, 50 to 200 μl, and 200 to 1000 μl

Power supply for gel electrophoresis (e.g., EC model 458; E-C Apparatus Corp., St. Petersburg, FL): Supplying power up to 400 V

Horizontal agarose slab gel electrophoresis unit (e.g., model H4; Bethesda Research Laboratories, Gaithersburg, MD): Including buffer tank and gel platform

Test tube racks: To hold the various centrifuge tubes

Pasteur pipettes (drawn to a fine tip in a flame): For aspiration of liquids following centrifugation

Pipette tips: For use with variable volume pipettors

Suction flask: Connect to the laboratory vacuum line for aspiration of supernatants following centrifugation

Reagents. Agarose such as SEAKEM LE (supplied by FMC Bioproducts, Rockland, ME) is used for gel electrophoresis of DNA. Good-quality restriction endonucleases can be bought from a number of sources.

[15] J. H. McCutchan and J. S. Pagano, *Natl. Cancer Inst. Monogr.* **41,** 351 (1968).

We buy ours mainly from New England BioLabs, Inc. (Beverly, MA). Pancreatic ribonuclease A, proteinase K, and lysozyme are from Sigma Chemicals (St. Louis, MO). The ribonuclease solution [10 mg/ml in 10 m*M* Tris-HCl (pH 7.5), 15 m*M* NaCl] is made deoxyribonuclease free by heating for 10 min in a boiling water bath.

Methods

Method I: Boiling Method for Isolation of Plasmid DNA from Bacteria

1. Inoculate a plasmid-harboring bacterial colony into 5 ml of LB broth medium supplemented with the appropriate antibiotic in a 15-ml culture tube. Grow overnight at 37° with constant shaking. Transfer 1.5 ml of the fully grown bacterial suspension to a 1.5-ml microfuge tube. Centrifuge at room temperature for 30 sec at full speed (12,000 *g*) in a microfuge. Remove the supernatant carefully and completely by vacuum suction using a drawn Pasteur pipette.
2. Resuspend the bacterial pellet in 100 μl of solution I [16% (w/v) sucrose, 20 m*M* Tris-HCl (pH 8.0)] by vortexing, until no clumps remain.
3. Continue vortexing and add 100 μl of solution II to the tube. This ensures total mixing of the two solutions with the suspended bacteria. [Solution II: Combine one part of a freshly made 10 mg/ml solution of lysozyme in 20 m*M* Tris-HCl (pH 8.0) with nine parts of a solution of Triton X-100 (1%, v/v, in water).]
4. Place in a boiling water bath for 40 sec. Spin immediately at room temperature for 5 min at full speed in a microfuge. Using a toothpick, carefully remove the white, gelatinous, clotlike pellet that contains cellular debris along with chromosomal DNA.
5. Add 23 μl of 3.0 *M* sodium acetate (pH 5.2 or 7.0), and precipitate the plasmid DNA by adding either 500 μl of ethanol or 250 μl of 2-propanol and mixing well. Store at −70° for 30 min.
6. Pellet down the precipitate by centrifugation at full speed in a microfuge for 15 min at 4°. The pellet will be yellowish at the bottom with a thin, white coating at the top. Carefully aspirate the supernatant. Add 50 μl of sterile 10 m*M* Tris-HCl (pH 7.8) and tap the tube to disperse the yellowish pellet. Do not vortex.[16] This preparation contains about 3 to 5 μg of plasmid

[16] Gentle tapping of the tube dissolves the yellowish pellet containing plasmid DNA, leaving the white coating at the top (which consists mainly of protein) floating intact in the solution. Vortexing, on the other hand, disperses this white material into the DNA solution; the dispersed particles hinder the migration of the DNA into an agarose gel in the preliminary identification of recombinant clones by gel electrophoresis performed in the next step.

DNA and about 75 to 100 μg of RNA, and is suitable for transformation of bacteria.[12]

7. If the screening is intended to verify cloning, the presence of the cloned insert in the plasmid can be tested in a preliminary manner by subjecting 5 μl of the sample to electrophoresis on a 0.6% (w/v) agarose gel (0.3 × 20 × 20 cm) (alongside vector DNA as a control) at 150 V for 3 hr using 0.5× TEA buffer [1× TEA buffer: 40 m*M* Tris base, 20 m*M* sodium acetate, 2 m*M* ethylenediaminetetraacetic acid (EDTA), adjusted to pH 7.8 with acetic acid]. Stain the gel with a 0.5 μg/ml solution of ethidium bromide in water and visualize DNA bands under ultraviolet light. Clones carrying inserts will be retarded in their electrophoretic migration compared to the vector.[17]

8. If the samples need to be characterized further by enzymatic analysis (such as diagnostic restriction digestions to verify cloning or to check plasmid structure), the samples (especially those identified as positives in the preliminary test in step 7) need to be purified further as follows. Add 40 μl of 3.0 *M* sodium acetate (pH 7.0) and sterile distilled water to bring the volume up to 400 μl. Add 400 μl of a 1 : 1 mixture of phenol and chloroform.[18] Emulsify by vigorous vortexing. Spin at full speed for 5 min in a microfuge at room temperature to separate the aqueous and organic layers. Transfer the upper aqueous layer to another 1.5-ml microfuge tube, and extract with 400 μl of chloroform. Spin at full speed for a few seconds in a microfuge at room temperature. Transfer the upper aqueous layer to another 1.5-ml microfuge tube. Precipitate the nucleic acids by adding 800 μl ethanol and mixing well. Store for 30 min at −70°. Spin down for 15 min at full speed in a microfuge at 4°.

9. Dissolve the DNA pellet in 20 μl of TE [TE: 10 m*M* Tris-HCl (pH

[17] Please note that the uncut plasmids tested here will exhibit multiple bands corresponding to different topological forms. The two major bands are the faster moving, covalently closed circular form (termed form I) and the slower moving, relaxed circular form (termed form II). In addition, there will be minor bands (which move more slowly than form II), corresponding to multimeric forms (especially dimers). The compact, fast-migrating form I DNA exhibits the electrophoretic mobility expected of a linear DNA molecule that is smaller in size by about 40%. For example, the form I DNA of the 2865-base pairs (bp) plasmid pGEM-1 exhibits the mobility of a linear DNA of approximately 1720 bp in size. By comparing the mobilities of form I DNA bands of pGEM-1 and recombinant clones derived thereof, we can easily detect retardation of migration caused by inserts as low as 200 bp in size. For inserts smaller than 200 bp this preliminary screening method may not be useful, and one may have to go directly to diagnostic restriction mapping or sequencing as mentioned in step 9.

[18] The phenol contains 0.1% (w/v) 8-hydroxyquinoline to prevent oxidation, and is equilibrated with 1.0 *M* Tris-HCl (pH 7.8) followed by 0.1 *M* Tris-HCl (pH 7.8), prior to formation of a 1 : 1 mixture with chloroform.

7.8), 1 mM EDTA]. The yield of plasmid DNA is about 3 μg/1.5 ml of bacterial culture. This DNA preparation is suitable for diagnostic restriction digestions. Add pancreatic ribonuclease A at a concentration of 20 μg/ml to the digestion mixture to remove the RNA.[19] Alternatively, if a sequence analysis of the cloned insert is to be done, skip the restriction digestion but remove RNA as described above.[20] Purify plasmid DNA by phenol–chloroform extraction followed by ethanol precipitation as described in step 8. The resulting DNA preparation is suitable for double-stranded DNA sequencing by the Chen and Seeburg procedure,[21] using oligodeoxynucleotide primers flanking the MCS region (available commercially).

Method II: The Phenol–Chloroform Lysis–Extraction Method for Isolation of Plasmids from Bacteria

1. Grow bacteria overnight and pellet down in a 1.5-ml microfuge tube as described in method I, step 1. Resuspend the bacterial pellet thoroughly by vortexing in 200 μl of TE buffer. Place in a boiling water bath for 1 min.[22]

2. Add 200 μl of TNE [TNE: 10 mM Tris-HCl (pH 7.8), 1 mM EDTA, 2.5 M NaCl]. Vortex for 15 sec. Add 400 μl of a 1 : 1 phenol–chloroform mixture.[23] Vortex vigorously for 15 sec. Spin at full speed in a microfuge for 15 min at room temperature.

3. Transfer the upper aqueous layer to another 1.5-ml microfuge tube. Add an equal volume of chloroform, vortex for 10 sec, and spin at room temperature for 30 sec to separate the two layers. Transfer the upper aqueous layer to another 1.5-ml microfuge tube. Precipitate the plasmid DNA by mixing with an equal volume of ice-cold 2-propanol. Spin for 15 min at 4° in a microfuge to pellet down the precipitate.

4. Dissolve the pellet in 50 μl of TE. If the intention of the screening is to verify cloning, subject 5 μl of the DNA preparation to electrophoresis

[19] Removal of RNA may be required prior to analysis of the restriction digests by electrophoresis on a 2% agarose gel and visualization of the DNA bands by staining with ethidium bromide to prevent masking of DNA bands smaller than 500 bp in length by the diffuse RNA band.

[20] RNA should be removed prior to sequencing because fragments of RNA could interfere with the sequencing reaction by annealing with the template and causing incorrect priming.

[21] E. Y. Chen and P. H. Seeburg, *DNA* **4,** 165 (1985).

[22] The boiling step, which was not a part of the original protocol,[6] has been included here because we find that it facilitates the lysis of bacteria with the phenol–chloroform mixture in step 2, presumably by making the bacteria more susceptible to lysis.

[23] The phenol used in this mixture is previously equilibrated with 1.0 M Tris-HCl (pH 7.8) and 1.0 M NaCl, and contains 0.1% (w/v) 8-hydroxyquinoline to prevent oxidation.

on a 0.6% (w/v) agarose gel (as described in method I, step 7), to carry out a preliminary identification of recombinant clones containing inserts. To the remaining DNA solution of each positive clone identified in this manner, add 40 μl of 3.0 *M* sodium acetate and water to bring the volume up to 400 μl. Add 800 μl of ethanol. Mix well. Store at $-70°$ for 30 min. Pellet down the precipitate by centrifugation in a microfuge at 4° for 15 min.

5. Dissolve the pellet in 20 μl of TE. The yield of plasmid DNA is about 2 μg/1.5 ml of bacterial culture. This DNA preparation is suitable for diagnostic restriction digestions for purposes such as the verification of cloning, taking care to include ribonuclease A in the digestion mixture to remove the RNA present in the DNA preparation.[19] If the DNA preparation is intended for sequence analysis, remove RNA by ribonuclease treatment[20] and purify plasmid DNA by phenol–chloroform extraction followed by ethanol precipitation as described in method I, step 9.

Method III: Isolation of Low Molecular Weight DNA from Animal Cells

1. Animal cells harboring low molecular weight DNA (such as COS-1 monkey kidney cells transfected with the SV40 origin-containing plasmid pSV2-*cat*, as mentioned under Materials and Reagents) are used as host cells in this method. Aspirate the culture medium. Wash the cell monolayer twice with 5 ml of phosphate-buffered saline (PBS) [PBS: 8.1 m*M* Na_2HPO_4, 1.5 m*M* KH_2PO_4, 140 m*M* NaCl, 2.7 m*M* KCl (pH 7.4)]. Add 5 ml of PBS. Scrape off the cells with a rubber policeman and transfer the cell suspension to a 15-ml centrifuge tube. Wash the plate with another 5 ml of PBS to pick up remaining cells and add to the tube. Spin down cells at 1000 *g* in a clinical centrifuge at room temperature for 10 min.

2. Pour off the supernatant. Resuspend the cells in 0.5 ml of a solution of 10 m*M* Tris-HCl (pH 7.8) and 10 m*M* EDTA. Vortex vigorously to unclump the cells. Transfer the cell suspension to a 15-ml Corex tube. Add 50 μl of 10% sodium lauryl sulfate (SDS), resulting in 0.9% final concentration of SDS. Mix gently but thoroughly to lyse the cells. Do not vortex. Keep at room temperature for 10 min, but not exceeding 15 min.

3. Add 140 μl of 5.0 *M* NaCl, resulting in a final NaCl concentration of 1.0 *M*. Mix gently but thoroughly, taking care not to shear the chromosomal DNA. Cover the tube with Parafilm. Store at 4° overnight or for at least 8 hr. Do not freeze the sample.

4. Spin at 15,000 *g* in a Sorvall refrigerated centrifuge in an SA-600 rotor at 4° for 30 min. Transfer the supernatant containing the low molecular weight DNA to a 1.5-ml microfuge tube.

5. Add 30 μg yeast tRNA per milliliter of DNA solution.[24] Add 100 μg of proteinase K and incubate at 50° for 1 to 2 hr. Extract twice with equal volumes of a 1 : 1 mixture of phenol–chloroform,[18] and once with an equal volume of chloroform alone. Separate the layers by centrifugation at full speed in a microfuge at room temperature, saving the aqueous layer at each stage for the subsequent step.

6. Following these organic extractions, transfer the aqueous layer to a 1.5-ml microfuge tube. Add two volumes of ethanol and mix well. Store at −70° for 30 min. Spin in a microfuge at 4° for 15 min to pellet down the DNA precipitate.

7. Resuspend the pellet in 400 μl of TE containing 0.3 *M* sodium acetate (pH 7.0), and precipitate the DNA again, by mixing with 2 vol of ethanol. Store at −70° for 30 min. Spin in a microfuge at 4° for 15 min to pellet down the DNA precipitate.

8. Dissolve the pellet in 40 μl TE. The solution contains low molecular weight DNA along with some RNA. The DNA preparation is suitable for diagnostic restriction digestions followed by analysis involving Southern blot hybridization using a labeled probe specific for the DNA.[25] If the isolated low molecular weight DNA is a bacterial plasmid, this DNA preparation is also suitable for reintroduction into bacteria by transformation.[12]

Acknowledgment

We thank Julie Yamaguchi and Daniel C. Alter for help with standardization of the boiling method and development of the phenol–chloroform lysis–extraction method, respectively, and Dr. Cho-Yau Yeung for helpful discussions. We thank Linda Cardenas for clerical assistance. This work was supported by grants from the National Institutes of Health, the American Cancer Society, and the University of Illinois at Chicago.

[24] The tRNA is added as a carrier at this stage to minimize loss of DNA during the organic extractions and to help in the precipitation of the DNA at later steps.

[25] E. M. Southern, *J. Mol. Biol.* **98,** 503 (1975).

[4] DNA Isolation Using Methidium–Spermine–Sepharose

By JOHN D. HARDING, ROBERT L. BEBEE, and GULILAT GEBEYEHU

Introduction

Rapid, quantitative isolation of DNA from complex biological samples is required for many protocols in molecular biology and molecular diagnostics. Standard methods using organic solvents such as phenol are reliable, but are time consuming and involve the use of toxic chemicals. Therefore many alternative methods for DNA isolation have been examined over the past several years.

In designing novel reagents for rapid isolation of DNA from complex samples, we reasoned that a very efficient "capture reagent" would consist of an intercalating dye attached to a solid support by a molecular linker.[1,2] The intercalator binds the DNA with high affinity and the solid support allows rapid separation of the bound material from unwanted contaminants.

In this chapter we describe protocols for isolating DNA from a variety of complex samples using a DNA capture reagent consisting of the intercalator, methidium, attached to a Sepharose bead by a spermine linker. DNA is released from the reagent in 0.1 to 0.5 *N* KOH or NaOH and is characterized by procedures such as dot-blot hybridization, sequencing, or polymerase chain reaction analysis.

Materials

DNA capture reagent (methidium–spermine–Sepharose) and DNA extraction buffer [Cat. No. 80885A; Bethesda Research Laboratories (Life Technologies, Inc., Gaithersburg, MD)]: The DNA capture reagent is synthesized as described in Harding *et al.*[1]

Proteinase K solution: 50 mg/ml proteinase K (Bethesda Research Laboratories) in 0.2 *M* Tris-HCl, 0.1 *M* Na_2EDTA, 3% (v/v) Brij 35; pH 7.5

TE buffer: 10 m*M* Tris-HCl, 1 m*M* Na_2EDTA; pH 7.5

Methods

Protocol 1: Isolation of DNA from Serum or Urine and Characterization by Dot-Blot Analysis

1. To 50 μl of serum or urine in a 1.5-ml microcentrifuge tube add 40 μl of DNA extraction buffer and 10 μl of proteinase K solution. Incubate for 60 min at 65°.

2. Vortex the stock tube of DNA capture reagent vigorously for a few seconds to suspend the capture reagent uniformly and pipette 50 μl of the slurry into the sample tube. Vortex the sample tube vigorously for a few

[1] J. Harding, G. Gebeyehu, R. Bebee, D. Simms, and L. Klevan, *Nucleic Acids Res.* **17,** 6947 (1989).

[2] G. Gebeyehu, L. Klevan, and J. Harding, U.S. Patent 4,921,805 (1990).

seconds and immediately place it on a rotator (such as a Labquake rotator, Labindustries Inc., Berkeley, CA) for 30 min at room temperature.

3. Spin the tube for 30 sec in a microcentrifuge (12,000 *g*) to pellet the capture reagent–DNA complex. Carefully remove the supernatant with a micropipette and discard it appropriately (serum and urine may be biohazards).

4. Add 100 μl of TE buffer, vortex vigorously, and spin in a microcentrifuge for 30 sec. Carefully remove the supernatant with a micropipette and discard appropriately.

5. To the capture reagent pellet add 100 μl of 0.5 *N* NaOH. Vortex vigorously for a few seconds and place on a rotator for 10 min at room temperature.

6. Spin out the capture reagent in a microcentrifuge for 30 sec and carefully remove the supernatant, which contains the DNA, with a micropipette.

7. For alkaline dot blotting, presoak the membrane (nitrocellulose or nylon) in 0.5 *N* NaOH for 5 min with gentle agitation. Place the membrane on a dot-blot apparatus. Pass the sample, eluted directly from the capture reagent, through the membrane. Wash each dot-blotted sample with 500 μl of 0.5 *N* NaOH. For nitrocellulose, bake the membrane for 1 hr at 80° in a vacuum oven. For nylon, treat the membrane according to the instructions of the manufacturer. Prehybridize and hybridize the membrane by standard protocols or using conditions previously optimized for a particular probe.

8. An alternative dot-blotting technique that has worked equally well is as follows. To the sample eluted from the capture reagent, add an equal volume of 2 *M* ammonium acetate. Ammonium acetate is prepared by dissolving the solid salt to a 2 *M* final concentration; the pH is not adjusted. Before applying the sample, soak the membrane briefly in distilled water and then for 5 min in 1 *M* ammonium acetate with gentle agitation. Place the membrane on a dot-blot apparatus. Pass the DNA sample through the membrane; wash each dot blotted sample with 500 μl of 1 *M* ammonium acetate. If the membrane is to be baked prior to hybridization (e.g., nitrocellulose and some nylons), incubate it in 20× SSC (3 *M* NaCl, 0.3 sodium citrate, pH 7.0) with gentle agitation for 5 min at room temperature prior to the baking step.

Protocol 2: Isolation of DNA from Whole Blood or Cultured Cells and Analysis by Polymerase Chain Reaction

1. Dilute 1 to 10 μl of whole blood, collected in ethylenediaminetetraacetic acid (EDTA) as an anticoagulant, to a total volume of 50 μl with TE buffer in a 1.5-ml microcentrifuge tube. If cultured cells are to be

analyzed, pellet the cells from the culture medium and resuspend the pellet in 50 μl of TE buffer. To either sample, add 40 μl of DNA extraction buffer and 10 μl of proteinase K solution and incubate at 60° overnight.

2. Add 50 μl of well-suspended DNA capture reagent slurry to the sample, vortex vigorously, and place on a rotator for 30 min at room temperature.

3. Spin the tube for 30 sec in a microcentrifuge (12,000 *g*) to pellet the capture reagent–DNA complex. Carefully remove the supernatant with a micropipette and discard it appropriately (blood may be a biohazard).

4. Add 100 μl of TE buffer, vortex vigorously, to wash the reagent and spin in a microcentrifuge for 30 sec. Carefully remove the supernatant with a micropipette and discard appropriately.

5. Repeat step 4 two more times.

6. To the pelleted capture reagent–DNA complex, add 50 μl of 0.1 *M* KOH. Vortex vigorously and place on a rotator for 10 min at room temperature.

7. Spin the reagent in a microcentrifuge for 30 sec. Carefully pipette the supernatant, containing the DNA, into a clean microcentrifuge tube.

8. To the supernatant, add 25 μl of 7.5 *M* ammonium acetate and 200 μl of absolute ethanol. Incubate the sample on ice for 10 min; pellet the precipitated DNA in a microcentrifuge, dry the pellet in a vacuum centrifuge, and suspend it in 64 μl of distilled water.

9. Set up a 100-μl polymerase chain reaction (PCR) (using all 64 μl of sample) as described in the instructions to the Cetus-Perkin Elmer (Norwalk, CT) Gene-Amp kit.[3] Following the reaction, the PCR products are ethanol precipitated, suspended in 15 μl of TE buffer, electrophoresed in a 4% (w/v) horizontal agarose gel (5.7 × 8.3 cm in a Bethesda Research Laboratories Horizon 58 apparatus), and visualized by ethidium bromide staining.

Protocol 3: Isolation of DNA from M13 Phage Lysates and Analysis by DNA Sequencing

1. As described in Sambrook *et al.*,[4] precipitate phage particles from 1.2 ml of a cleared supernatant from an M13-infected culture by adding 0.3 ml of 2.5 *M* NaCl containing 20% (w/v) polyethylene glycol (PEG 8000).

2. Incubate for 15 min at room temperature and spin the tube in a microcentrifuge for 10 min. Remove and discard the supernatant.

[3] R. Saiki, D. Gelfand, S. Stoffel, S. Scharf, R. Higuchi, G. Horn, K. Mullis, and H. Erlich, *Science* **239,** 487 (1988).

[4] J. Sambrook, E. Fritsch, and T. Maniatis, "Molecular Cloning, A Laboratory Manual." Cold Spring Harbor Laboratory, Cold Spring Harbor, New York, 1982.

3. Resuspend the pelleted phage particles in 50 μl of TE buffer and add 40 μl of DNA extraction buffer and 10 μl of proteinase K solution. Incubate for 60 min at 60°.

4. Add 100 μl of capture reagent slurry, vortex vigorously, and place on a rotator for 30 min at room temperature.

5. Proceed as in steps 3 through 6 of protocol 1, above.

6. To the supernatant, containing the eluted DNA (in 50 μl of 0.5 *N* NaOH), add 25 μl of 7.5 *M* ammonium acetate and 200 μl of absolute ethanol. Incubate the sample on ice for 10 min, pellet the precipitated DNA in a microcentrifuge, dry the pellet in a vacuum centrifuge, and suspend it in 10 μl of TE buffer. Use 1 to 5 μl for sequencing according to the particular procedure that is being used.

Comments on Protocols

The protocols described above have worked consistently in our hands and can be used as starting points for other applications. The basic protocols can often be shortened even further by reducing the initial capture reagent binding and final alkali elution steps to 5 min each and by eliminating the TE buffer wash of the reagent after the initial binding step. Likewise, the proteinase K digestion of the sample can sometimes be reduced. The success of these modifications depends on the particular type of sample being assayed.

For best results, two specific points should be kept in mind. First, the DNA capture reagent must be uniformly dispersed in the sample during the binding step. The investigator should use a rotator that turns the tubes completely end over end, rather than a shaking platform or other mixing device. Second, proteins in the sample must be digested thoroughly by proteinase K before the capture reagent is added to the sample. Use of the DNA extraction buffer helps assure efficient degradation of proteins. The capture reagent will bind undigested proteins, but not proteinase K digestion products.[1] Undigested proteins eluted from the reagent with the DNA can deleteriously affect dot-blot and sequencing assays.

One potential limitation of the reagent should also be noted. Treatment with dilute alkali is the only effective means of releasing nucleic acid from the DNA capture reagent. Thus, we have analyzed the released DNA using procedures that do not require native DNA.

Results

Basic Features of Capture and Release of DNA

We initially performed experiments to examine basic features of the capture and release protocols using radioactive DNA added to buffer or to human serum treated with proteinase K, as described in detail in Harding *et al.*[1] In summary these experiments indicated that (1) capture of DNA is independent of salt concentration up to at least 3 *M* NaCl or KCl; (2) capture is independent of EDTA concentration up to at least 0.5 *M*; (3) capture occurs in the presence of detergents such as 0.1 to 1% sodium dodecyl sulfate or 1% (v/v) Triton X-100; (4) relatively small amounts of DNA are captured from large sample volumes, e.g., 10 ng of DNA from 0.5 ml of sample; (5) relatively large amounts of DNA are captured from small volumes, e.g., 5 μg of DNA from 30 μl of sample; (6) DNA is undegraded by the capture and release protocols; and (7) RNA can also be captured, although the utility of this feature is obviated by the requirement for alkaline release of nucleic acid from the capture reagent.

Capture and Characterization of DNA from Complex Biological Samples

Serum and Urine. Results of an experiment demonstrating capture and dot-blot quantitation of viral DNAs in serum or urine, as performed by protocol 1 above, are shown in Fig. 1.

In the experiment (columns 1–4, Fig. 1), various amounts of a plasmid containing a cloned hepatitis B viral genome (HBV) were added to normal (uninfected) serum, captured, released, dot-blotted, and probed with an HBV RNA probe. As seen in Fig. 1 (columns 2 and 4, row c), 0.5 pg of the HBV target DNA was detected on nylon or nitrocellulose membranes, respectively. As controls, alkali-denatured plasmid was dot-blotted directly onto the filters in Fig. 1 (columns 1 and 3). Comparison of the intensities of the spots (measured by laser densitometry) in columns 1–4 in Fig. 1 indicate that the signal is about 30% as intense for the samples captured from serum as for the control samples spotted directly onto the filter.

In row f of columns 2 and 4 (Fig. 1), the serum that was treated with capture reagent contained no added HBV DNA. The absence of signals in these rows indicates that proteins or other potential contaminants in the serum do not cause spurious signals on the dot-blot.

The results shown in columns 7–9 of Fig. 1 indicate that HBV DNA present in virus particles can be captured and quantitated. The source of HBV-infected serum was a positive control from a commercially available HBV test kit (HepProbe; Life Technologies, Inc.). A hybridization signal was obtained from 50 μl of infected serum (Fig. 1, columns 8 and 9, row a) and from 50 μl of a 1 : 10 mixture of infected serum and normal serum (Fig. 1, columns 8 and 9, row b). Normal serum alone gave no signal (Fig. 1, columns 8 and 9, row e). Comparison of the intensity of the spots from

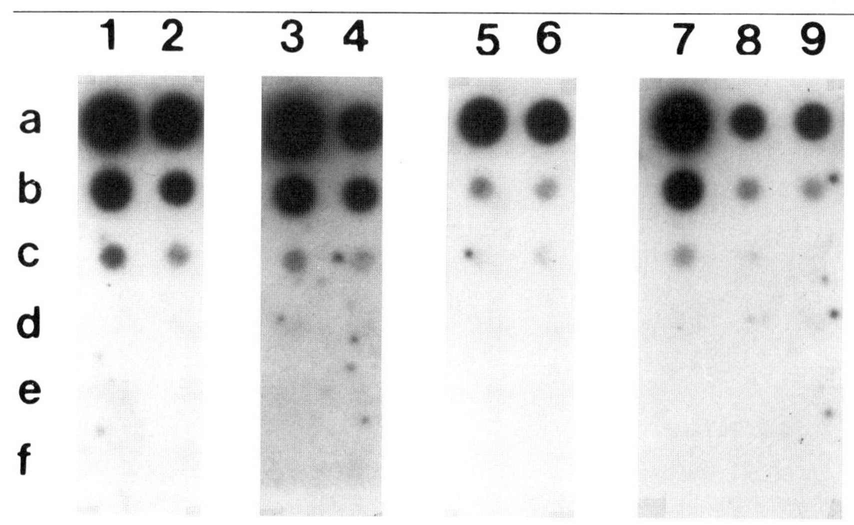

FIG. 1. Dot-blot analysis of viral DNAs captured from human serum or urine. Column 1: Standard amounts of plasmid pHBVT702 DNAs (containing a cloned hepatitis B virus sequence) were added to 200 μl of 0.5 *N* NaOH and applied directly to a nylon membrane (Biotrans, Pall Corp., Glen Cove, NY). Column 2: Standard amounts of plasmid DNAs were added to normal serum, captured, and released as in protocol 1, applied to the filter, and hybridized with an HBV RNA probe. A 7-day exposure of the autoradiograph is shown. Columns 3 and 4: The same as columns 1 and 2, respectively, except that a nitrocellulose filter was used. A 4-day exposure of the autoradiograph is shown. Row a, 50-pg HBV target; row b, 5-pg target; row c, 0.5-pg target; row d, 0.25-pg target; row e, 0.05-pg target; row f, no target. Column 5: As for column 1 except that plasmid pT7T3-19CMV DNAs (containing a cloned cytomegalovirus sequence) were applied to a Biotrans membrane. Column 6: Standard amounts of CMV plasmid DNAs captured from human urine. Row a, 67-pg target; row b, 6.7-pg target; row c, 0.67-pg target; row d, 0.33-pg target; row e, 0.067-pg target; row f, no target. The filter was hybridized with a CMV RNA probe; a 5-day exposure of the autoradiograph is shown. Column 7: The same as column 1 (plasmid pHBVT702). Columns 8 and 9: Serum containing HBV virus (see text) was diluted with normal serum (where appropriate) and incubated with capture reagent as described in protocol 1. DNA released from the reagent was applied to a Biotrans membrane and hybridized with an HBV RNA probe. Row a, undiluted, infected serum; row b, infected serum diluted 1 : 10 with normal serum; row c, 1 : 100 dilution of infected serum; row d, 1 : 200 dilution of infected serum; row e, normal serum. A 7-day exposure of the autoradiograph is shown. Hybridization of each filter was performed as follows: The filter was prehybridized for 30 min at 65° in 10% (v/v) formamide, 10% (w/v) dextran sulfate, 5× SSPE (1× SSPE: 0.18 *M* NaCl, 10 m*M* sodium phosphate, 1 m*M* Na_2EDTA; pH 7.4), 5% (w/v) sodium dodecyl sulfate, and 100 μg/ml sheared herring sperm DNA. (Herring sperm DNA was not included when nylon membranes were used.) The filter was hybridized with 1 ml of hybridization solution (prehybridization solution containing 10^7 cpm of RNA probe) for 3 hr at 65°. Following hybridization, the filter was washed briefly at room temperature with 2× SSPE, incubated for 5 min at room temperature with 5 μg/ml RNase A in 2× SSPE, and washed three times for 5 min in 0.1× SSPE, 0.1% sodium dodecyl sulfate at 65°. The filter was dried and autoradiographed at −70° with two intensifying screens. (Data are reproduced from Harding *et al.*[1] by permission of Oxford University Press.)

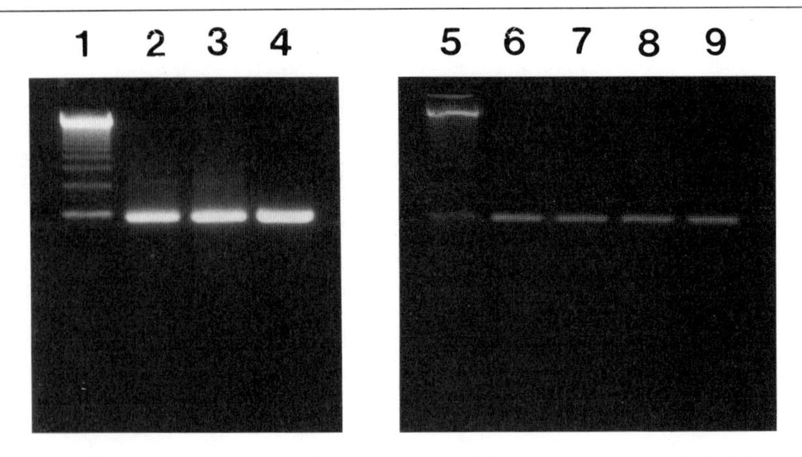

FIG. 2. Polymerase chain reaction analysis of DNA captured from whole blood and cultured cells. Human genomic DNA was isolated using the capture reagent either from whole blood or from a HeLa cell culture as in protocol 2. The polymerase chain reaction was performed using the GH18 and GH19 primers of Scharf *et al.*,[5] complementary to specific human β-globin gene sequences, for 30 cycles. The reaction products were electrophoresed on a 4% agarose gel as described in protocol 2. Ethidium bromide-stained gels are shown. Lane 1: The 123-bp ladder (Bethesda Research Laboratories) size markers. The fragment of greatest mobility is 123 bp in size. PCR reaction products are from DNA isolated from 10 μl of whole blood (lane 2), 1 μl of whole blood (lane 3), and from a control plasmid containing the β-globin sequence (lane 4). Lane 5: 123-bp ladder size markers. PCR reaction products are from DNA isolated from 10,000 HeLa cells (lane 6), 1000 HeLa cells (lane 7), 100 HeLa cells (lane 8), and control plasmid (lane 9). Lanes 1–4 and 5–9, respectively, are from different gels. (Data are reproduced from Harding *et al.*[1] by permission of Oxford University Press.)

the infected serum with the control spots of column 7 in Fig. 1, together with the result obtained above indicating that the hybridization assay is about 30% efficient, indicates that 50 μl of the infected serum contains about 15 pg of HBV DNA.

The experiment in columns 5 and 6 (Fig. 1) indicates that cytomegalovirus DNA (CMV) added to urine can also be detected using protocol 1, above. Approximately 0.7 pg of DNA was detected in the CMV-positive sample in column 6, row c (Fig. 1), and signal was not obtained from urine lacking CMV DNA (Fig. 1, column 6, row f).

Whole Blood and Eukaryotic Cell Cultures. The results of the experiment in Fig. 2 demonstrate that DNA can be captured from cultured human cells or whole blood, according to protocol 2 (above), and characterized by the polymerase chain reaction (PCR).[5]

Lanes 2 and 3 of Fig. 2 show that human β-globin PCR reaction prod-

[5] S. Scharf, G. Horn, and H. Erlich, *Science* **233,** 1076 (1986).

ucts are readily detected from genomic DNA captured from 10 or 1 μl of whole blood, respectively. Lanes 6–8 of Fig. 2 show PCR reaction products obtained from 10,000, 1000, or 100 cultured human cells, respectively.

For both experiments of Fig. 2, control reactions were performed in which template DNAs were not included in the PCR reactions. No PCR reaction products (including β-globin products and "primer–dimers") were synthesized in these reactions, indicating that, as expected, the PCR was template dependent (data not shown).

The PCR reactions of both experiments in Fig. 2 show relatively little dose response. This probably reflects two factors. First, the PCR reaction itself is not linear under the particular conditions used in these experiments. Second, to assure a gel band that reproduced well for publication, we applied the entire PCR reaction to the gel. If smaller amounts of reaction products are run on the gel, a nonlinear dose response is observed.

M13 Phage Lysates. The results of Fig. 3 demonstrate that high-quality single-stranded phage M13 sequencing templates are prepared using protocol 3 (above). The first four lanes on the left-hand side of Fig. 3 are sequencing reactions obtained from single-stranded M13 DNA isolated by standard phenol extraction protocols[4] and the four lanes on the right-hand side of Fig. 3 are obtained from M13 DNA isolated using the capture reagent. Inspection of Fig. 3 indicates that the same sequence can be read from both sets of lanes, demonstrating the utility of capture reagent for preparing sequencing templates.

Conclusions

The DNA capture reagent has two notable features that make it a useful alternative to traditional methods for DNA isolation. First is its ease of use. DNA is captured from serum or whole blood in a 45-min procedure that does not involve organic solvents. In addition, the reagent is amenable to batch processes. For example, 24 samples in microcentrifuge tubes (enough to fill a microcentrifuge rotor) can readily be analyzed simultaneously by one operator.

The second important feature of the capture reagent is its versatility. It can be used to isolate DNA from many different biological samples. Capture is independent of the salt concentration of the sample and occurs in the presence of detergents, proteinase K, and EDTA, which are often used in protocols for the isolation of viral and cellular DNAs from complex samples. Very small amounts of DNA can be isolated from large volumes of sample and analyzed using a sensitive assay such as hybridization or the polymerase chain reaction.

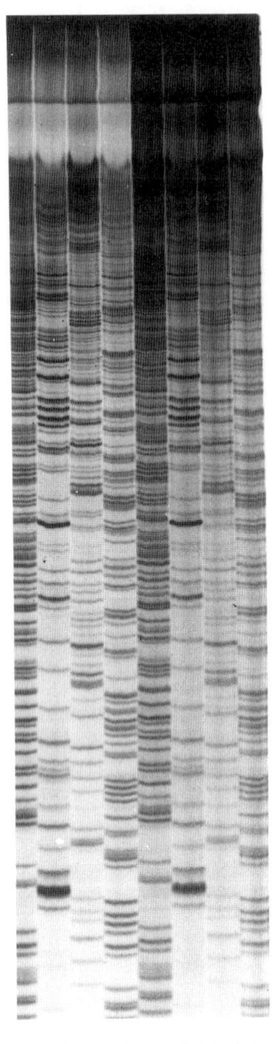

FIG. 3. Sequencing reactions performed on M13 DNA captured from cell lysates. M13 DNA was isolated as in protocol 3. Sequencing reactions were performed using the Kilo-Base Sequencing System (Bethesda Research Laboratories) according to the manufacturer's instructions. Reaction products were resolved on a 6% polyacrylamide–8 *M* urea gel at 60 W in Tris–borate EDTA buffer. The first four lanes (left-hand side) are reactions using DNA isolated by standard phenol extraction procedures (Sambrook *et al.*).[4] The other four lanes (right-hand side) are reactions using DNA captured and released from the capture reagent. (Data are reproduced from Harding *et al.*[1] by permission of Oxford University Press.)

These features should facilitate analysis of DNA from pathogens or other DNAs of interest in complex biological samples.

Acknowledgments

We thank our colleagues, Leonard Klevan and Dietmar Rabussay, for support and encouragement and Jim D'Alessio, Sherry Challberg, and Jim Hartley for providing plasmids and HBV-infected serum.

[5] Chromosome Fishing: An Affinity Capture Method for Selective Enrichment of Large Genomic DNA Fragments

By RAJENDRA P. KANDPAL, DAVID C. WARD, and SHERMAN M. WEISSMAN

Several technological advances have made it feasible to construct large-scale restriction maps of complex genomes. A combination of various techniques, i.e., pulsed-field gel electrophoresis (PFGE),[1-3] the construction of linking and jumping libraries,[4-7] and cosmid[8] and yeast artificial chromosome (YAC)[9] cloning and selection methods, is available to expedite genome mapping. The isolation of specific sequences from a plasmid library also has been achieved by avidin–biotin affinity chromatography after a RecA-mediated search for homologous sequences[10] by an appropriate probe.[11] Despite the advantages these methods provide, there have been technical difficulties and limitations in their application. Conventional methods of mapping involve the isolation of end probes from plasmid or phage clones and successive screening of libraries to obtain overlapping

[1] D. C. Schwartz and C. R. Cantor, *Cell (Cambridge, Mass.)* **37,** 67 (1984).
[2] G. F. Carle, M. Frank, and M. V. Olson, *Science* **232,** 65 (1986).
[3] G. Chu, D. Vollrath, and R. W. Davis, *Science* **234,** 1982 (1986).
[4] C. L. Smith, S. K. Lawrance, G. A. Gillespie, C. R. Cantor, S. M. Weissman, and F. S. Collins, this series, Vol. 151, p. 461.
[5] M. R. Wallace, J. W. Fountain, A. M. Bereton, and F. S. Collins, *Nucleic Acids Res.* **17,** 1655 (1989).
[6] F. S. Collins and S. M. Weissman, *Proc. Natl. Acad. Sci. U.S.A.* **81,** 6812 (1984).
[7] A. M. Poustaka and H. Lehrach, *Trends Genet.* **2,** 174 (1986).
[8] J. Collins and B. Hohn, *Proc. Natl. Acad. Sci. U.S.A.* **75,** 4242 (1978).
[9] D. T. Burke, G. F. Carle, and M. V. Olson, *Science* **235,** 806 (1987).
[10] B. Rigas, A. A. Welcher, D. C. Ward, and S. M. Weissman, *Proc. Natl. Acad. Sci. U.S.A.* **83,** 9591 (1986).
[11] C. M. Radding, *Annu. Rev. Genet.* **16,** 405 (1986).

sets of clones. The extensive time requirement for such "walking" was partly overcome by the advent of cosmid cloning,[8] which allows an average walk of 20 kilobases (kb) in one step. Given the large size of a chromosome [$\sim 1.3 \times 10^8$ base pairs (bp)], walking at 20 kb per step is still a slow and tedious process for any sizable mapping project. The presence of either repetitive sequences in the probe or unclonable sequences would terminate the walks in many instances. Some of the gaps between contiguous DNA fragments (contigs) can be filled by analysis of YAC clones, which can contain DNA fragments greater than 100 kb in size. Although YAC cloning has the capacity for obtaining large insert clones, the occurrence of chimeric YACs and the possibility of sequence rearrangements in YACs can be problematic. In addition, the low yield of DNA per YAC colony makes the preparation of DNA and screening of libraries technically demanding.

In order to supplement and complement the existing methods, we developed an affinity capture protocol for selective enrichment of large-size genomic DNA fragments.[12] The method requires a probe originating from the end of the target fragment and involves limited resection of the DNA by using a strand-specific exonuclease, hybridization of a biotinylated RNA probe to the resected end, selective retention of the hybrid molecules on an avidin matrix, and specific release of the captured DNA by treatment with ribonuclease. The eluted fragment, which is expected to be highly enriched, can then be used for preparing a fragment-specific sublibrary, and as a probe for screening phage, cosmid, or YAC libraries as well as for selecting cDNAs encoded by the fragment. Thus the enriched fragment will permit isolation as well as mapping of the expressed sequences. As mentioned before, this protocol requires an end probe to isolate the cognate fragment. Such probes are available from linking libraries. To facilitate the availability of regionally localized linking clones further, we also developed a polymerase chain reaction (PCR)-assisted method for constructing linking and jumping libraries.[13] In this chapter, we present details of methods for affinity capture of large DNA fragments and PCR-based construction of jumping and linking libraries.

Principle of Affinity Capture

The general steps of the method are described in Fig. 1. Genomic DNA prepared in agarose plugs is digested by incubating the plug in an appropriate restriction enzyme reaction mixture for 6–8 hr. The ends of

[12] R. P. Kandpal, D. C. Ward, and S. M. Weissman, *Nucleic Acids Res.* **18,** 1789 (1990).

[13] R. P. Kandpal, H. Shukla, D. C. Ward, and S. M. Weissman, *Nucleic Acids Res.* **18,** 3081 (1990).

the DNA fragments are then resected by treatment with λ-exonuclease. At first we hybridized a biotinylated RNA transcribed from an end-specific DNA probe to the resected DNA, and then retained the RNA–DNA hybrid on a Vectrex–avidin (Vector Laboratories, Burlingame, CA) matrix. However, the biotinylated RNA probe hybridizing to the resected DNA interacted with such a high affinity with the avidin matrix that suitable conditions for releasing the hybrid molecule were not found. Biotinylated nucleotide analogs containing a spacer arm with a cleavable thio linkage are suitable substrates for RNA transcription.[14] Unfortunately, an analog with a sufficiently long spacer arm to achieve efficient cleavage of the -S–S- linkage with dithiothreitol after avidin–biotin interaction was not readily available commercially. The strategy we devised, in which ribonuclease is used for the release of the RNA–DNA hybrid with minimum perturbation of the matrix, is as follows.

The probe fragment is cloned downstream of a simian virus 40 (SV40) fragment (AB) in a pBluescript (Stratagene, La Jolla, CA) vector (construct 1), in which T3 and T7 RNA promoters can be used for transcribing either strand of the probe. A second clone containing SV40 DNA fragment B is constructed in a pBluescript vector (construct 2). When a transcript of construct 1 obtained with T3 RNA polymerase and a biotinylated transcript of construct 2 obtained with T7 RNA polymerase are annealed together, an RNA-biotinylated RNA hybrid is formed (Fig. 1). This hybrid RNA molecule can then be hybridized to a resected genomic DNA fragment, thus forming a tripartite hybrid molecule. We term the transcript from construct 1 the "bridge RNA." The bridge RNA serves to connect the large genomic DNA fragment on one end to the biotinylated RNA on the other end, separated by a stretch of single-stranded RNA in the middle. This single-stranded stretch of RNA can be readily digested by ribonuclease and this causes specific release of the captured fragment from the avidin matrix. This enzymatic method of release is efficient and specific. Other alternative strategies, i.e., use of commercially available cleavable biotin analogs and copper chelate chromatography,[15,16] were not as successful.

Preparation of Genomic DNA in Agarose Plugs

Cells are harvested from a healthy growing culture, washed with phosphate-buffered saline (PBS), and resuspended in PBS at a concentration

[14] P. R. Langer, A. A. Waldrop, and D. C. Ward, *Proc. Natl. Acad. Sci. U.S.A.* **78,** 6633 (1981).

[15] T. Herman, E. La Fever, and M. Shimkus, *Anal. Biochem.* **156,** 48 (1986).

[16] A. A. Welcher, A. R. Torres, and D. C. Ward, *Nucleic Acids. Res.* **14,** 10027 (1986).

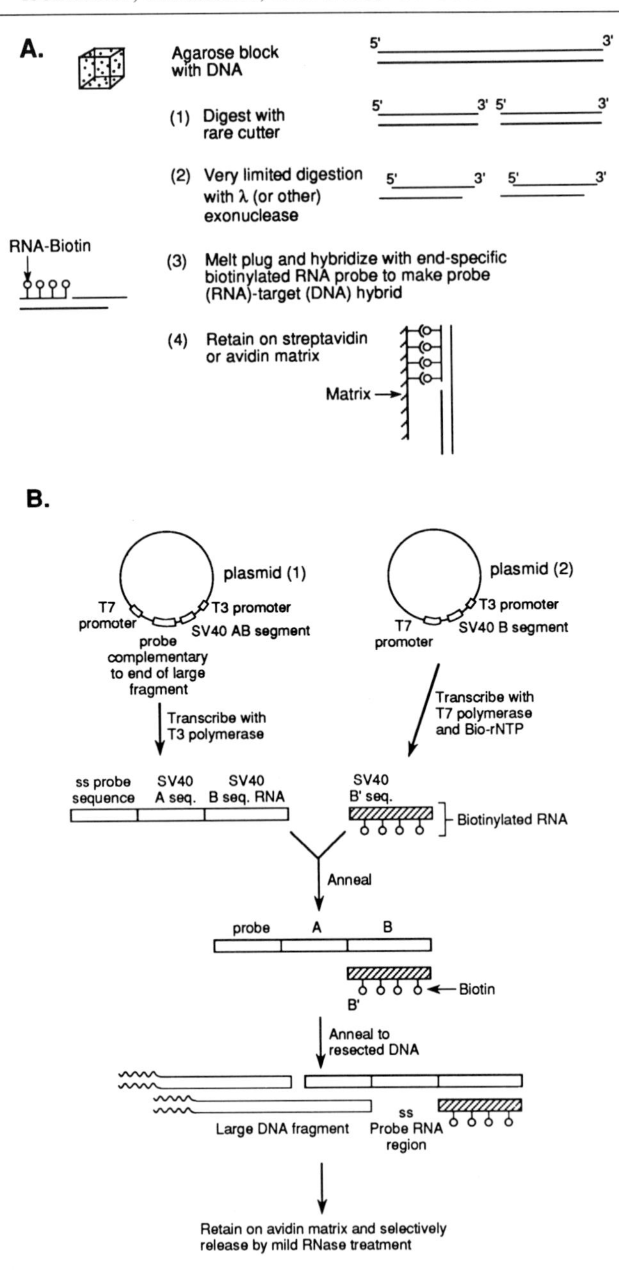

FIG. 1. Principle of the affinity purification procedure. (A) Fishing protocol. A general description showing the digestion of the genomic DNA with a rare cutter, limited digestion with λ-exonuclease, hybridization with a bridge RNA and biotinylated probe RNA, and retention of hybrid molecule onto a Vectrex–avidin matrix. (B) Structure of hybridizing

of 2×10^7 cells/ml. The cell suspension is mixed with an equal volume of 1% (w/v) low melting temperature agarose made up in 125 m*M* ethylenediaminetetraacetic acid (EDTA) and the solution dispensed into a mold that yields 120-μl plugs. The mold is left on ice for 30 min to let the agarose solidify. The plugs are then pushed out of the mold into a solution containing 0.5 *M* EDTA (pH 9.0), 1% (w/v) sodium lauroylsarcosine, and 1 mg/ml proteinase K (ESP), and incubated at 50° for 48 hr to lyse the cells. The plugs appear clear after complete lysis has been achieved. We reduced the agarose concentration from 1 to 0.7%; the resulting final agarose concentration of 0.35% in the plugs allowed them to melt readily at 70°.

The agarose plugs are treated with 10 m*M* Tris, pH 7.5, 0.1 m*M* EDTA (TE) containing 1 m*M* phenylmethylsulfonyl fluoride (PMSF) at room temperature three times for 3 hr each followed by incubation in TE three times for 3 hr each. Restriction enzyme digestion is carried out in a 250-μl volume in buffers recommended by the manufacturers, supplemented with 100 μg/ml bovine serum albumin. The incubation is carried out at the desired temperature for 6–8 hr. After the digestion the plug is incubated in ESP at 50° for 2 hr, followed by incubations in TE containing PMSF and then in TE.[17]

Resection with λ-Exonuclease

λ-Exonuclease is our preferred resection enzyme. However, any strand-specific exonuclease can be used effectively as long as the enzyme preparation is free of nonspecific nuclease contamination. The purity of commercially available λ-exonuclease (Bethesda Research Laboratories, Gaithersburg, MD) was checked by incubating it with genomic DNA plugs and subsequent PFGE analysis of the DNA. We observed no detectable

[17] C. L. Smith, P. E. Warburton, A. Gaal, and C. R. Cantor, *in* "Genetic Engineering" (J. K. Setlow and A. Hollander, eds.), Vol. 8, p. 45. Plenum, New York, 1986.

species. A pBSM13(+) plasmid containing a probe DNA subcloned from the end of a 150-kb *Sfi*I fragment and two SV40 DNA fragments A and B (plasmid 1). Plasmid 2 contains probe SV40 DNA fragment B in a pBSM13(+) vector. The RNA transcript of plasmid 1 with T3 RNA polymerase and biotinylated RNA transcript of plasmid 2 with T7 RNA polymerase, when hybridized together, yield an RNA-biotinylated RNA hybrid. This hybrid can be hybridized to λ-exonuclease-resected genomic DNA and retained onto an avidin matrix. The stretch of single-stranded RNA in the tripartite hybrid molecule facilitated the release of the captured large fragment by its susceptibility to ribonuclease action. [Reproduced from R. P. Kandpal, D. C. Ward, and S. M. Weissman, *Nucleic Acids Res.* **18,** 1789 (1990).]

internal cleavage of the large-size genomic DNA. A major advantage of using λ-exonuclease is its high processivity in the stepwise release of mononucleotides from the 5′ end of DNA.[18] Thus a large excess of the enzyme can be used to saturate all the available ends of the DNA prior to initiating enzyme digestion, resulting in a relatively homogeneous species of the resected DNA. While λ-exonuclease is reported to utilize DNA with a blunt end as the preferred substrate,[19] we have used DNA containing both 5′-recessed or 3′-recessed ends as substrates.

The conditions for the λ-exonuclease reaction were standardized by using a known plasmid DNA along with genomic DNA in an agarose plug. The extent of resection was checked by hybridizing a radiolabeled probe complementary to the resected end of the plasmid. The first time point of a resection reaction that leads to strong hybridization was considered to be the optimum resection time. The agarose plug containing restriction enzyme-digested genomic DNA was incubated with λ-exonuclease buffer without $MgCl_2$ (70 m*M* glycine–KOH, pH 9.3) and 25 units of λ-exonuclease in a volume of 250 μl at room temperature for 20 min, to permit the diffusion of λ-exonuclease into the plugs. Following this incubation, the temperature was reduced to 15° and the reaction was started by adding $MgCl_2$ to a final concentration of 3 m*M*; incubation at 15° proceeded for 10 min. Although the optimal temperature for λ-exonuclease is 37°, a reaction at 15° better limits and controls the resection. The reaction can be carried out at 37°; however, the time of incubation must be reduced considerably. After the resection at 15°, the reaction is stopped by adding EDTA to a final concentration of 20 m*M* and the λ-exonuclease is digested with 100 μg proteinase K at 37° for 30 min. The proteinase K is inactivated by incubating the plug in TE containing 1 m*M* PMSF, followed by incubation in TE with three buffer changes at 3-hr intervals. Experiments reveal that λ-exonuclease resection can be better synchronized by performing the reaction at 4° with a larger amount of enzyme for more than 60 min (W. D. Bonds, unpublished observations, 1992).

Preparation of Transcripts

The RNA probe transcripts are prepared in a reaction mixture containing 40 m*M* Tris-HCl (pH 8.0), 25 m*M* NaCl, 0.5 m*M* each of ATP, CTP, GTP, and UTP, 8 m*M* $MgCl_2$, 2 m*M* spermidine, 50 units ribonuclease inhibitor (Promega, Madison, WI), 10 m*M* dithiothreitol, 200 units of T3 or T7 polymerase, and 10 to 20 μg of appropriately digested construct 1

[18] C. M. Radding and D. M. Carter, *J. Biol. Chem.* **246,** 2513 (1971).

[19] J. W. Little, *in* "Gene Amplification and Analysis" (J. G. Chirikjian and T. S. Papaps, eds.), Vol. 2, p. 136. Elsevier, New York, 1981.

DNA in a total volume of 200 μl. Constructs 1 and 2 are linearized with *Eco*RI and *Hin*dIII, respectively. The reaction mixture is incubated at 37° for 60 min. The DNA template is then digested with pancreatic DNase (1 unit/μg DNA) at 37° for 15 min and the mixture is extracted with phenol and precipitated with ethanol. The RNA is stored as ethanol suspension at −20° until required. Biotinylated RNA corresponding to SV40 fragment B is synthesized under the same conditions by replacing UTP with 1 m*M* bio-11-UTP (Enzo Biochemicals, New York, NY) in the reaction mixture. Extraction with phenol is avoided, because biotinylated RNA is extracted into the phenol–water interphase. Therefore, inactivation of DNase is carried out by heating the reaction mixture at 70° for 15 min. Following DNase inactivation, the reaction mixture is passed through a Sephadex G-50 spin column, and the eluate is precipitated with ethanol.

Polymerase Chain Reaction

A standard PCR[20] is carried out in a mixture containing 50 m*M* KCl, 10 m*M* Tris-HCl, pH 8.6, 1 m*M* $MgCl_2$, 0.2 m*M* each of dATP, dCTP, dGTP, and dTTP, 50–100 ng of primers, 0.5 unit *Taq* polymerase, and DNA template (10–50 ng) in a reaction volume of 50 μl. These reactions are run in a Cetus Perkin-Elmer (Norwalk, CT) Thermocycler using the following parameters: 1 min at 94°; 2 min at 55°; 1 min at 72°; repeated for a total of 30 to 60 cycles. These parameters may have to be modified, depending on the template and primers used. It also is useful to standardize the Mg^{2+} concentration optimum for every pair of primers tested. We routinely add dNTPs after heating the reaction mixture to 80°, to reduce nonspecific amplification to a large extent. Another modification that we include is to heat the template DNA at 94° for 3 min before incubating at 80°. The reaction mixture containing buffer, primers, dNTPs, and *Taq* polymerase, prewarmed to 80°, is then added to the template solution. We observed that this protocol yields greater amplification than standard procedure, especially in cases where the template DNA contains a high percentage of G and C content.

Protocol for Affinity Capture

Annealing of Bridge RNA and Biotinylated Capture Probe

The bridge RNA (20 μg) is annealed to an eight-fold molar excess of biotinylated capture probe RNA in a hybridization buffer containing

[20] R. K. Saiki, D. H. Gelfend, S. Stoffel, S. J. Clark, R. Higuchi, G. T. Horn, K. B. Mullis, and H. A. Erlich, *Science* **239,** 487 (1988).

0.5 *M* NaP_i (pH 7.2), 4% (w/v) sodium dodecyl sulfate (SDS), and 5 m*M* EDTA at 65° for 45 min. The extent of annealing can be checked by analyzing the hybridization mixture on a 1% agarose gel. The biotinylation of the capture RNA is checked by preparing the transcript in the presence of bio-11-UTP and [^{32}P]CTP and mixing the biotinylated [^{32}P]RNA with an appropriate amount of Vectrex–avidin. We observed more than 95% retention of the biotinylated RNA on the affinity matrix. A similar test is performed on the retention of the RNA-biotinylated RNA hybrid. In these experiments the bridge RNA is formed in the presence of [^{32}P]UTP, and subsequently hybridized to a molar excess of biotinylated capture RNA. The hybridization mixture is then added to a suspension of Vectrex–avidin matrix. Such studies indicate that more than 90% of the ^{32}P-labeled RNA is retained on the matrix, suggesting nearly complete annealing of the bridge RNA to the capture RNA. It is important that nearly all the bridge RNA used for hybridizing to the resected target DNA should be present as a bridge RNA-biotinylated RNA hybrid. Any free bridge RNA remaining in the hybridization mixture will form an RNA–DNA hybrid, which cannot be retained on the Vectrex matrix. It is absolutely necessary that all the reagents and solutions used in these manipulations be free of ribonuclease. We have included 4% (w/v) SDS in the hybridization mixture to inactivate any traces of ribonuclease.

Hybridization of Annealed RNA-Biotinylated RNA to Resected DNA: Its Capture on and Release from Affinity Matrix

The agarose plug containing 30–40 μg of restriction enzyme-digested and λ-exonuclease-resected DNA is melted at 70° and mixed with the RNA-biotinylated RNA hybrid in a hybridization buffer containing 0.5 *M* NaP_i (pH 7.2), 4% (w/v) SDS, and 5 m*M* EDTA. The hybridization reaction is carried out in 1.5-ml volume in a 15-ml tube at 65° for 1 hr. The contents of the hybridization solution are mixed by gently rolling the tube along its side. It is important that the melting of the agarose plug, the mixing of DNA with probe RNA, and the hybridization manipulations be performed as gently as possible to minimize breakage of large DNA molecules. After hybridization, the mixture is diluted to 10 ml by adding 8.5 ml of 10 m*M* Tris-HCl (pH 7.5) and 0.3 m*M* EDTA, to a final concentration of 75 m*M* NaP_i, 10 m*M* Tris (pH 7.5), 0.6% (w/v) SDS, and 1.0 m*M* EDTA. The diluted hybridization mixture is then mixed gently with a 1.0 ml suspension of 350 mg pretreated Vectrex–avidin. Nonspecific binding of DNA to Vectrex–avidin is reduced by incubating the matrix in 150 m*M* NaCl, 100 m*M* Tris (pH 7.5), and 25 μg/ml of sonicated and denatured salmon sperm DNA for 30 min at room temperature. Excess DNA is removed by washing the matrix three or four times with the same buffer devoid of salmon sperm

DNA. The diluted hybridization mixture is incubated with the treated Vectrex–avidin at room temperature (20–23°) for 60 min with intermittent mixing achieved by rolling the tube. The matrix is allowed to settle under gravity and the supernatant collected. The matrix is washed successively with 10 ml of buffer A [150 m*M* NaCl, 100 m*M* Tris, pH 7.5, 0.5% (w/v) SDS, 1 m*M* EDTA] at room temperature, 10 ml of buffer B [15 m*M* NaCl, 100 m*M* Tris, pH 7.5, 0.5% (w/v) SDS, 1 m*M* EDTA] at room temperature, 10 ml of buffer B at 65° for 20 min, 10 ml of buffer C (80 m*M* NaCl, 10 m*M* Tris, pH 7.5) at room temperature, and finally the captured DNA is specifically eluted by incubating the matrix with 2 ml of buffer C containing 100 μg/ml ribonuclease at 37° for 60 min.

Assessment of Affinity-Captured DNA

The ribonuclease eluate is then tested for its purity, integrity, enrichment, and recovery. We found it convenient to employ PCR for these tests. As shown in Fig. 2, in a model experiment, in which a 150-kb *Sfi*I fragment containing the β-globin fragment was captured by using a 452-bp-long *Sfi*I to *Sac*I piece as probe, we made use of two sets of primers, one set originating from the middle of the fragment and the other set specific for the distal region. Those primers helped us assess the integrity and the recovery of the captured DNA. The purity of the fragment is assessed by amplifying the captured DNA with primer sets unrelated to the isolated fragment. The presence or absence of such nontargeted sequences in the ribonuclease eluate will indicate the status of enrichment. Thus, PCR is a rapid method for assessing the general quality of an affinity capture experiment. We routinely run parallel experiments with resected plasmid DNA target and the bridge RNA-biotinylated capture RNA, and process it through all the steps. The plasmid recovery can then be assessed using PCR. These experiments ensure that all the intermediate manipulations, which can be checked easily, are proceeding in the expected manner.

Prior to performing PCR on the ribonuclease-eluted DNA, the eluate is extracted with phenol and phenol–chloroform and precipitated with ethanol in the presence of 10 μg glycogen as carrier. We observed that glycogen may inhibit the activity of *Taq* polymerase. It is, therefore, necessary to remove glycogen from the reaction mixture. Purification of the DNA by using Geneclean glass milk (obtained from Bio 101, La Jolla, CA) removes the glycogen, and also concentrates the DNA.

Parameters Affecting Hybridization Efficiency

The following parameters affect the extent of hybridization in a major way: (1) amount of starting genomic DNA, (2) amount of bridge RNA, (3) amount of biotinylated capture RNA, and (4) time of hybridization.

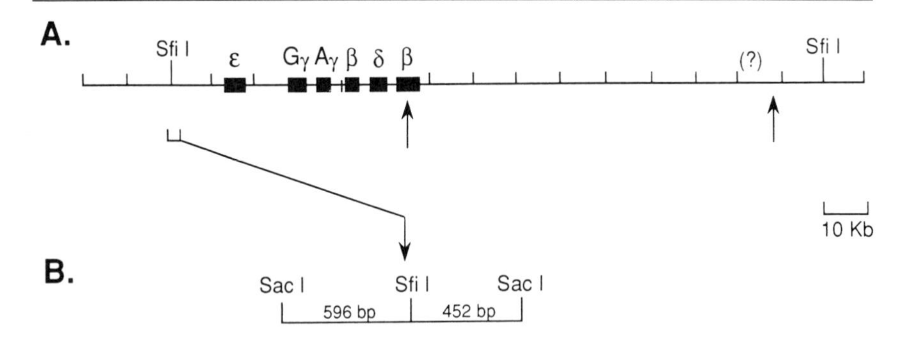

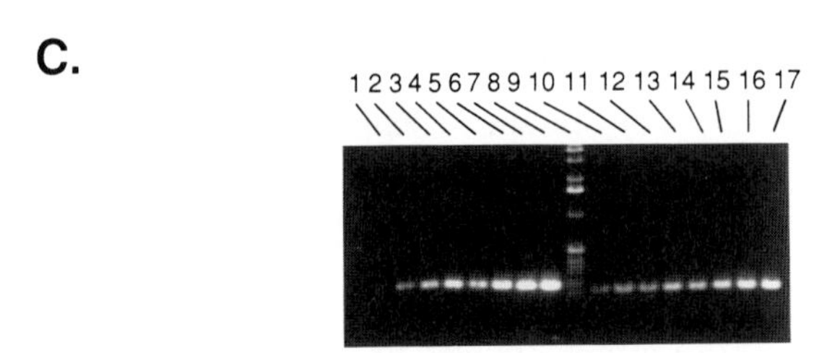

FIG. 2. Structure of the 150-kb *Sfi*I genomic fragment and time course and RNA concentration dependence of hybridization with genomic DNA. (A) The relative position of β-like globin genes in the 150-kb *Sfi*I fragment. Arrows indicate the primers that are specific for the β-globin gene, and a 500-bp *Hin*dIII fragment located toward the extreme 3′ end of the large fragment. Both sets of primers amplify 200-bp-long DNA segments. (?) indicates a stretch of DNA that had been difficult to clone by conventional techniques. (B) The 1-kb DNA piece from which a 452-bp *Sfi*I to *Sac*I fragment was isolated and subcloned into a pBSM13(+) plasmid vector containing a 215-bp-long *Hin*dIII DNA fragment from SV40. (C) PCR amplification of ribonuclease eluates from experiments in which the amount of RNA and the time of hybridization were changed. Forty to 60 μg of *Sfi*I-digested 3.1.0 DNA was resected with λ-exonuclease and hybridized to varying amounts of RNA transcripts for either 1 or 15 hr at 65°. In the following, the numbers in parentheses represent the amount of bridge RNA and time of hybridization. Lane 1, 10 (5 μg, 1 hr); lane 2, 11 (5 μg, 15 hr); lane 3, 12 (40 μg, 1 hr); lane 4, 13 (40 μg, 15 hr); lane 5, 14 (60 μg, 1 hr); lane 6, 15 (60 μg, 15 hr); lane 7, 16 (150 μg, 1 hr); lane 8, 17 initial DNAs amplified and one-tenth fraction applied to gel. Lanes 1–8, primers specific for β-globin; lanes 10–17, primers specific for 3′ downstream *Hin*dIII fragment. Lane 9 contains size markers. Amplification was carried out for 30 cycles. [Reproduced from R. P. Kandpal, D. C. Ward, and S. M. Weissman, *Nucleic Acids Res*. **18,** 1789 (1990).]

We have observed that 20–30 μg of restriction-digested and resected genomic DNA yields no detectable nonspecific DNA in the captured fraction, as judged by PCR. When the amount of genomic DNA used is significantly higher (100 μg), the presence of nonspecific DNA in the ribonuclease eluate can be detected by PCR. We recommend performing several hybridization experiments with 20–30 μg of genomic DNA to obtain a larger amount of specific DNA. Typically, the yield of specific DNA in an experiment is between 2 and 5%, which translates to a yield of 20–50 pg of a 150-kb fragment from 20 μg of genomic DNA. This amount does not include nonspecific DNA present in the captured fraction. Other indirect calculations, e.g., number of plaques obtained after ligating the *Eco*RI-digested captured DNA to λ arms and packaging, suggest that the total DNA present in the ribonuclease eluate is approximately 1 ng, indicating enrichment factors of 400- to 1000-fold.

Other parameters that can improve the recovery are (1) higher amount of RNA-biotinylated RNA and (2) longer time of hybridization. In a series of experiments we increased the RNA concentration from 5 to 150 μg and carried out the hybridization for either 1 or 15 hr at 65° (Fig. 2C). The results indicated that the recovered fragment increased from 2% at 5 μg RNA (15-hr hybridization) to 15% at 150 μg RNA (15-hr hybridization). The recovery began to plateau at around 40–60 μg of RNA. Although prolonged hybridization leads to higher recovery, it may also cause networking of DNA due to exposure of repetitive sequences at the resected ends. Thus, longer duration of hybridization may have deleterious effects on the quality of the recovered DNA, in terms of both overall size and purity. We have routinely used 20 μg of RNA and performed a 1-hr hybridization to obtain 1000-fold enrichment.

Another factor that can reduce nonspecific hybridization is the stringency of washing. We have included a wash at 65° for 20 min in the presence of 15 m*M* NaCl. Although the extent and number of 65° washes can be increased, this decreased the yield of the specific fragment.

Cloning Affinity-Enriched DNA

One convenient way to preserve the captured DNA is to clone it in a phage vector. As mentioned above, from a 20-μg genomic DNA hybridization 1–5 ng of the DNA can be captured onto the affinity matrix and specifically eluted. This amount of DNA can be digested with a suitable restriction enzyme, subsequently ligated to a λ vector, and packaged with high efficiency.

We have cloned a fraction of the ribonuclease eluate efficiently into a λgt10 vector. The ribonuclease eluate is extracted with phenol and phenol–chloroform, and digested with *Eco*RI. The digested DNA is precipitated with ethanol in the presence of glycogen. It is then ligated to 100 ng of dephosphorylated arms of λgt10 vector in a total volume of 4 μl at 16° for 10 hr. The ligated mixture is then packaged into phage heads using Gigapack gold packaging extract (Stratagene, La Jolla, CA). A total of 330,000 plaques was obtained in 1 such cloning experiment. Thus, the amount obtained from a single experiment is sufficient to make a representative library enriched for sequences of the selected fragment.

Alternative Applications of Affinity-Enriched DNA

The ribonuclease-eluted DNA, after extraction with phenol and phenol–chloroform and precipitation with ethanol, can be amplified by using *Alu* or Line primers.[21,22] These primers will amplify sequences bounded by appropriately aligned *Alu* or Line sequences and provide signature PCR products of the enriched DNA. The enriched DNA will contain a relatively small amount of randomly trapped contaminating DNA. Thus PCR with *Alu* or Line primers will primarily amplify DNA corresponding to the enriched fragment.

To enrich for telomere-specific sequences from *Not*I-digested mouse genomic DNA, we have used telomeric repeat sequences as a "fishing" probe. The affinity captured DNA was then amplified by *Alu* and Line primers and used as probe to screen a cosmid library. Such a screening has resulted in the isolation of cosmid clones specific to telomeric regions of several mouse chromosomes (D. Rounds and D. C. Ward, unpublished observations, 1992). These results suggest that this approach can be used for isolating phage, cosmid, or YAC clones corresponding to any large-size DNA fragment of interest.

Polymerase Chain Reaction-Based Construction of Linking and Jumping Libraries

A prerequisite for the fishing protocol is to have a "hook" or a probe originating from the end of the large fragments. Such fragments are generally isolated as linking fragments. The conventional method for constructing a linking library involves at least two cloning steps and it is relatively time consuming. We have made use of PCR in a rapid method for constructing jumping and linking libraries. Every individual clone from such

[21] D. L. Nelson, S. A. Ledbetter, L. Corbo, M. F. Victoria, R. Ramirez-Solis, T. D. Webster, D. H. Ledbetter, and C. T. Caskey, *Proc. Natl. Acad. Sci. U.S.A.* **86,** 6686 (1989).

[22] S. A. Ledbetter, D. L. Nelson, S. T. Warren, and D. H. Ledbetter, *Genomics* **6,** 475 (1990).

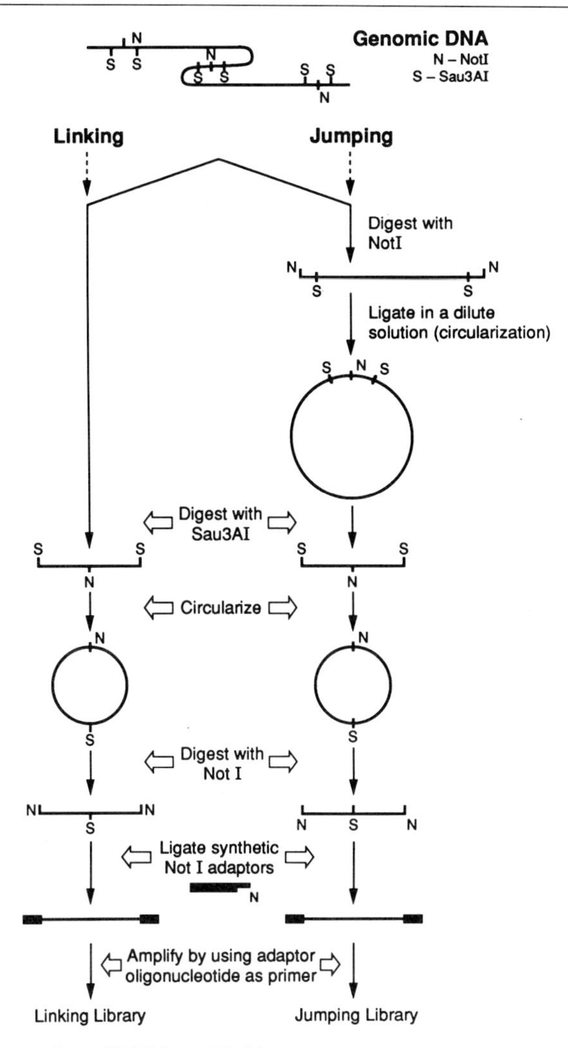

FIG. 3. Construction of PCR-based linking and jumping libraries. A schematic flow chart for various steps involved in the construction of linking and jumping libraries. [Adapted from R. P. Kandpal, H. Shukla, D. C. Ward, and S. M. Weissman, *Nucleic Acids Res.* **18,** 3081 (1990).]

libraries can then be used to fish out its corresponding large fragment. Although we have briefly described this methodology,[13] the following is a more detailed protocol for constructing such libraries.

The principle of this method is schematically depicted in Fig. 3. Genomic DNA in an agarose plug is digested with *Not*I at 37° for 6–8 hr.

The enzyme is inactivated by proteinase K and proteinase K in turn is inactivated by incubating the plug in TE containing 1 m*M* PMSF followed by incubation in TE alone. A slice of the plug containing ~5 μg of digested DNA is melted in the presence of 1 m*M* spermidine and diluted to a concentration of 0.2 μg/ml in a mixture containing 50 m*M* Tris-HCl (pH 7.8), 10 m*M* $MgCl_2$, 20 m*M* dithiothreitol (DTT), and 1 m*M* ATP. Care should be taken not to shear the DNA at the melting and dilution step. The DNA solution should not be pipetted at this point under any circumstances. The solution is mixed by gently rocking it back and forth. After preincubating the ligation mixture at 4° for 4 hr, T4 DNA ligase (New England BioLabs, Beverly, MA) is added to a final concentration of 4 units/μl, and incubation is continued for 12 hr. An additional aliquot of ligase is then added and further incubated for 12 hr. The ligase is inactivated by heating the solution at 65° for 30 min. The circularized mixture is then digested with *Sau*3AI or *Bst*YI briefly at 37 or 60°, to cut the DNA into smaller pieces. The digested DNA is concentrated and desalted by using Centriprep-30 (Amicon, Beverly, MA). Complete digestion of the DNA with *Sau*3A or *Bst*YI is ensured by further digesting it at this stage. The digested DNA is then recircularized at a concentration of 1 μg/ml with 4 U (units)/μl of T4 DNA ligase. The ligase is again digested with proteinase K at 50° for 30 min and extracted with phenol and phenol–chloroform. The circularized DNA mixture is concentrated in a Centriprep-30. Circularized DNA (200 ng) is digested with 10 U *Not*I in a 100-μl volume at 37° for 1 hr, followed by inactivation of the enzyme at 68° for 30 min. The digested DNA is precipitated with ethanol and ligated to 50 ng of *Not*I adaptors (28 nucleotides) at 16° for 12 hr.

The fraction of the *Bst*YI or *Sau*3AI circles thus formed that have internal *Not*I sites will consist of *Not*I to *Bst*Y/*Sau*3AI fragments originating from the two ends of the same *Not*I fragment. The collection of all such circles/clones is termed a "*Not*I jumping library." In addition to using these circles for constructing a jumping library, they may be used for performing a directional jump. If the sequence around one end of a *Not*I fragment is known, then by synthesizing appropriate primers and using them for inverse PCR amplification, the other end of the jumping clone can be isolated.

The protocol for constructing a linking library is the same as for the jumping library, except that the first step of digesting the DNA plug with *Not*I is omitted. Instead, the process is started by digesting the genomic DNA with *Sau*3AI or *Bst*YI and carrying through the protocol to form *Bst*YI or *Sau*3AI circles. The circles containing internal *Not*I sites will represent a *Not*I linking DNA and the collection of all such linking DNAs is termed a "*Not*I linking library."

After forming the small circles and ligating on the *Not*I adaptors, the ligase is removed from the reaction mixture by treatment with proteinase K (50 μg) at 50° for 30 min. The mixture is then extracted with phenol and phenol–chloroform and run on a 1.2% (w/v) Seaplaq (FMC Bioproducts, Rockland, ME) low melting agarose gel to remove the dimers of *Not*I adaptors. The DNA is isolated from agarose by melting the gel, extraction with phenol and phenol–chloroform, and concentration on a Centricon 30 column (Amicon).

*Not*I adaptor-ligated DNA (50 ng) is amplified with an appropriate oligonucleotide primer of the *Not*I adaptor under the following conditions. The template DNA is first heated at 94° for 4 min and then allowed to incubate at 80°. A reaction mixture containing *Taq* buffer, dNTPs, primer, and *Taq* polymerase maintained at 80° is added to the DNA template to a final concentration of 10 m*M* Tris-HCl (pH 8.3), 1.5 m*M* $MgCl_2$, 0.2 m*M* dNTPs, 50 ng primer, and 1 unit *Taq* polymerase in a total volume of 50 μl. Our experiments suggest that the following modifications in the PCR conditions can improve the yield, especially when the template DNA contains a relatively higher proportion of G and C bases: (1) 2% (v/v) dimethyl sulfoxide, (2) 0.04 m*M* 7-deaza-dGTP, 0.16 m*M* dGTP, and 0.2 m*M* each of dATP, dCTP, and dTTP, and (3) 0.2 m*M* each of dATP and dTTP, and 0.4 m*M* each of dCTP and dGTP. The amplification is carried out for 40 cycles with incubations at 94° (1 min, 15 sec), 55° (1 min, 30 sec), and 72° (2 min). An aliquot of the amplified mixture is applied to a 1.2% (w/v) agarose gel to check the extent of amplification. The amplified DNA from these experiments represents a collection of linking or jumping DNAs.

After amplification, the jumping or linking DNA can be used in a variety of ways. It can be cloned in a λ vector after digesting with *Not*I. The resulting clones from one type of library can then be used to hybridize against the clones from the other type of library. Such hybridization screening permits the isolation of linking clones corresponding to a particular jumping clone and vice versa. Repeated screenings will allow ordering of various *Not*I sites along the chromosome.

Conclusions

Conventional methods, such as cloning, linkage mapping, and *in situ* hybridization, have permitted assignment of genes responsible for many genetic disorders to specific chromosomes. Technical advances leading to new methodologies, namely PFGE, YAC cloning, and high-resolution *in situ* mapping, have resulted in the cloning of additional genes responsible for genetic diseases. The methods described here supplement existing techniques and provide alternative approaches to identifying other yet unknown genes.

By obtaining a *Not*I restriction map utilizing clones from linking and jumping libraries, it may be possible to isolate contiguous genomic DNA fragments by the "fishing" protocol. This would facilitate microanalysis of any macrochromosomal region of interest. Although the affinity capture method needs further refinement to realize its potential for isolating megabase size fragments, modifications of the basic technique suitable for such purposes are being explored. These methods provide yet another tool to investigate the molecular structure of mammalian chromosomes.

Acknowledgments

We wish to thank Mary Wallace for excellent technical preparation of the manuscript. This work was supported by an Outstanding Investigator Grant CA 42556 of the National Cancer Institute to Sherman M. Weissman, and a National Institutes of Health Grant GM 40633 to David C. Ward.

[6] Trapping Electrophoresis of End-Modified DNA in Polyacrylamide Gels

By LEVY ULANOVSKY

Introduction and Principle of Method

Field inversion gel electrophoresis (FIGE) dramatically improves the separation of large double-stranded (ds) DNA fragments (e.g., in the size range of yeast chromosomes) on agarose gels, as compared to constant field electrophoresis.[1] However, relatively little improvement by FIGE has been found for short single-stranded (ss) DNA (hundreds of bases) on polyacrylamide gels.[2–4] This chapter discusses the mechanism of a manyfold improvement in short ssDNA separation achieved by attaching a small protein, streptavidin, to one end of the DNA.

Why does FIGE work well for long DNA fragments, but not for short

[1] G. F. Carle, M. Frank, and M. V. Olson, *Science* **232,** 65 (1986).

[2] E. Lai, N. Davi, and L. Hood, *Electrophoresis* **10,** 65 (1989).

[3] B. W. Birren, M. I. Simon, and E. Lai, *Nucleic Acids Res.* **18,** 1481 (1990).

[4] D. L. Daniels, L. Marr, R. L. Brumley, and F. R. Blattner, *in* "Structure and Methods" (R. H. Sarma and M. H. Sarma, eds.), Vol. 1, p. 29. Adenine Press, New York, 1990.

ones? Although the nature of the FIGE phenomenon is still largely a mystery, there is little disagreement that the FIGE mechanism should require some kind of orientation of the DNA by the electric field. Optical techniques have made it possible to measure how much the electric field orients DNA in gels.[5] It was observed that the degree of the DNA orientation in constant-field gel electrophoresis decreases with the fragment size. The degree of orientation is a compromise between the influences of the electric field and of the thermal motion, kT. While the electric field tends to orient (to align) the DNA chain along the field, the thermal motion tends to disorient it into a random coil. The actual shape of the DNA chain is in a dynamic equilibrium controlled by these two factors. The transition from a completely oriented configuration to a completely random one requires that a certain "orientation energy" be overcome. The longer the DNA, the larger is the orientation energy. Therefore, for large molecules the equilibrium is shifted toward the elongated configuration, while for small ones the shape is closer to the random coil.

The DNA orientation (elongation) during constant-field gel electrophoresis is probably what gives rise to the well-known phenomenon of "limiting mobility." The essence of this phenomenon is that above a certain size of DNA, the mobility of the fragment no longer depends on its length and the electrophoresis loses its resolution. These large fragments of different sizes comigrate and are usually seen on gels as one thick band above the bands of shorter fragments resolved below it (e.g., see Fig. 1, lane C). Existing theories of DNA electrophoresis[6,7] invoke a transition from "small" to "large" DNA to explain the limiting mobility phenomenon. For small DNA molecules the thermal motion energy, kT, is higher than the orientation energy and thus randomizes the configuration of the DNA. The electrostatic force component that pulls a DNA molecule through the pores of the gel is proportional to the end-to-end distance of the fragment projected onto the field direction. For randomly coiled small fragments, the end-to-end distance is proportional to the square root of the DNA length L. This predicts a $1/L$ DNA size dependence for the mobility of small molecules, as described by the "biased reptation" theory (see Refs. 8 and 9; for a simple explanation see the appendix at the end of this article). At a large enough size, however, naked DNA moves through the gel in an elongated shape, kT being too low to disrupt the configuration. For these

[5] G. Holtzwarth, C. B. McKlee, S. Steiger, and G. Crater, *Nucleic Acids Res.* **15,** 10031 (1987).

[6] O. J. Lumpkin, P. Dejardin, and B. H. Zimm, *Biopolymers* **24,** 1573 (1985).

[7] G. Slater and J. Noolandi, *Biopolymers* **25,** 431 (1986).

[8] L. Lerman and H. Frisch, *Biopolymers* **21,** 995 (1982).

[9] O. J. Lumpkin and B. H. Zimm, *Biopolymers* **21,** 2315 (1982).

large fragments, extended along the field, the end-to-end distance is proportional to their contour length. Therefore, both the electrostatic force and the opposing friction force applied to the large fragments are proportional to the fragment length, and thus the (limiting) mobility is size independent but voltage dependent.

An interesting and less understood fact regarding the DNA orientation gave rise to the idea behind the present work. It is that, even within the size range of the limiting mobility (where the fragments are all presumably oriented along the electric field to some extent) the degree of orientation falls rapidly with decreasing DNA size.[5] It indicates that while the phenomenon of the limiting mobility may occur regardless of the degree of orientation of DNA (as long as it is oriented at all, rather than randomized), the FIGE effect is probably dependent on this degree.

If the DNA orientation is what facilitates the separation by FIGE, we may wish to consider why the electric field has such a hard time orienting small DNA chains. One of the physical reasons is that along its length DNA has a rather uniform distribution of charge, mass, and other physical characteristics determining the electrophoretic behavior. Then a question arises: can one make FIGE work by introducing an artificial nonuniformity, an asymmetry into the molecule? The asymmetry can be achieved by making one end substantially different from the other, for example by binding a protein or another entity to one end of the DNA, but not to the other. The difference between the two ends would determine which end leads the way, and which end trails behind. Then the orientation is no longer a matter of chance.

Indeed, it has been found that such an asymmetric modification, achieved by binding streptavidin to one end, results in a dramatic change in the pattern of DNA mobility.[10] Surprisingly, the change was observed not only in FIGE, but even in constant field. This phenomenon is discussed below and interpreted as DNA trapping in the gel (see Results and Discussion, below).

It should be noted that the whole phenomenon of the electrophoretic trapping of end-modified ssDNA has not yet been investigated thoroughly enough for practical purposes, in particular for DNA sequencing. Therefore, the present chapter should perhaps be regarded as a guide for further experiments, rather than as an established standard for practical applications.

Technical Notes, Materials, and Methods

Streptavidin can be attached to one end of ssDNA by means of a noncovalent but very strong bond between streptavidin and biotin. Biotin

[10] L. Ulanovsky, G. Drouin, and W. Gilbert, *Nature (London)* **343,** 190 (1990).

is a small molecule (M_r 244) that can be covalently linked to a nucleotide in a DNA fragment. There are at least three ways to make end-biotinylated ssDNA molecules: (1) filling in the recessed 3′ end with the incorporation of a biotinylated nucleotide in one position and then denaturing; (2) enzymatic extension of 5′ end-biotinylated primer and then denaturing; and (3) 3′ end extension by terminal transferase. We will discuss the first reaction in detail and then the other two briefly.

Filling in Recessed 3′ End. An example of such a reaction would be filling in by Sequenase (USB, Cleveland, OH) of the recessed 3′ end left by *Hin*dIII restriction cleavage:

```
5′—· · · · · NNN—3′
   | | | | | | | |
3′—· · · · · NNNCTAG—5′
```

For a final reaction volume of 24 μl, take 10 μl of 1.0 μ*M* DNA ends in TE buffer. Add 3 μl 5× reaction buffer for Sequenase (200 m*M* Tris-HCl, pH 7.5, 100 m*M* $MgCl_2$, 250 m*M* NaCl) and 1.5 μl 0.1 *M* dithiothreitol (DTT). The first base to be filled in, guanine, is currently not available as a biotinylated triphosphate (biotinylated dGTP), but the other three are. So, unmodified dGTP (0.5 μl, 1.0 m*M*) is added for filling in this position. Next comes dATP, which in contrast to dGTP is available in a biotinylated form; so we add 1.0 μl of 0.4 m*M* biotinylated dATP. Radioactive dTTP (6.0 μl, 3.3 μ*M* being the sum of both hot and cold dTTP) is incorporated next. Its concentration is that of the commercially available stock. If no radioactive labeling is required, cold dTTP (0.5 μl, 1.0 m*M*) is used instead. The last nucleotide should be an unmodified dNTP, in this case dCTP (0.5 μl, 1.0 m*M*). This is recommended in view of the exonuclease activity of polymerases (even a low level), because unmodified dNTPs are available in a high concentration.

The radioactive nucleotide must follow the biotinylated one and not vice versa. Otherwise, there is a chance that the radioactive label may be incorporated without biotinylation of the DNA. This could happen due to a relatively low incorporation rate of biotinylated dNTP by polymerases, as compared to the unmodified dNTP.

If [^{32}P]dNTP is used as a label and incorporated in the 3′ direction of the biotinylated nucleotide, as above, the incorporation of the biotinylated dNTP can be determined by the following test. First, as a control for the biotinylation reaction, it is advisable to carry out a reaction, parallel to the above, except that the biotinylated dNTP is replaced by the unmodified dNTP. Then a drop of 1 to 3 μl of each of the two reaction products is put on a small Whatman (Clifton, NJ) DE-81 paper circle. The radioactivity can be measured with any Geiger counter sensitive to [^{32}P]. If the counter goes off scale, the measurement is performed with the DE-81 circle kept at a fixed distance from the detector tube, e.g., 5 cm. The DNA immediately binds to the paper. After washing three or four times with 0.5 *M* Na-phosphate buffer, pH 7.0 (3 min for each wash), the unincorporated dNTPs are removed while the DNA remains bound to the paper. Then, after these washes are completed, another reading is taken with the same Geiger counter at the same distance as before. The reading of the biotinylated DNA should be compared to that in the control experiment, in which the unmodified dNTP is used instead of the biotinylated one. A comparison of the two readings shows the incorporation efficiency of the biotinylated dNTP. Another control is also recommended, in which no polymerase is added and thus no incorporation occurs. This control shows the efficiency of washing out the unincorporated dNTPs. The unwashed dNTP should give less than 1% of the counts before washing.

The acceptance of a biotinylated dNTP by DNA polymerases is much lower than that of an unmodified dNTP. Even the most efficient polymerases, like Sequenase, require much higher concentrations of biotinylated dNTP than that of unmodified dNTP to match their incorporation efficiencies. Therefore, one should keep the biotinylated dNTP concentration as high as possible. In practice, commercially available biotinylated dNTPs are supplied in 0.3 to 0.4 m*M* concentration, the final concentration in the reaction being even lower. In contrast to the biotinylated dNTPs (in the above example biotinylated dATP), unmodified dNTPs (in the above example dGTP and dCTP) are available in concentrations that are higher by orders of magnitude. However, the concentration of the unmodified dNTP (dGTP and dCTP above) in the reaction should not exceed that of biotinylated dNTP (dATP above) by more than two orders of magnitude. Otherwise, the incorrect incorporation of unbiotinylated dGTP or dCTP in place of biotinylated dATP by the DNA polymerase may be too frequent. If one follows these rules, the biotinylation should be on the order of 100%. Unbiotinylated DNA molecules should not be detected on the autoradiogram in any event (if the base order is kept as recommended above), because they lack the radioactive label.

Enzymatic Extension of Biotinylated Primer. Most techniques for biotinylated primer synthesis produce a mixture of biotinylated and nonbiotinylated primers of the same sequence. These can be separated on polyacrylamide gel electrophoresis (PAGE), the mobility difference being equivalent to about one nucleotide difference in length (if only one nucleotide per primer is biotinylated). The biotinylated primer is then purified and used for a regular primer extension reaction and, in particular, an enzymatic sequencing reaction with the dideoxy terminators. Apart from a dideoxy sequencing ladder, one can produce a size marker with one biotinylated end by enzymatically extending the same biotinylated primer with no dideoxy terminators, and then cutting the DNA with a restriction enzyme of a given fragment size pattern. This technique is an alternative to the method of filling in the sticky ends described above.

3' End Extension by Terminal Transferase. This reaction can result in an incorporation of more than one biotin molecule into a single end. To avoid this "overbiotinylation" one can use a very low ratio of biotinylated to nonbiotinylated dNTP. However, in this way a relatively low percentage of the DNA molecules may become biotinylated. To overcome this problem, one possibility would be to use radioactive streptavidin, thus leaving the unbiotinylated molecules undetectable.

Storage and Streptavidin Binding. After the DNA is end biotinylated by one of these three methods, or in some other way, it can be stored at −20°. After thawing, the binding of streptavidin is performed by incubation of the biotinylated DNA with streptavidin at room temperature for 20 min.

Excess of Streptavidin. How can the binding of more than one DNA fragment to a streptavidin molecule be avoided? Streptavidin consists of four identical subunits. It has a tetrahedral symmetry, and thus four equivalent binding sites for biotin. To avoid multiple binding of biotinylated DNA fragments to streptavidin one can use a large molar excess (30–100 times) of streptavidin over biotin, both incorporated and unincorporated. This leaves the three extra binding sites unoccupied in each streptavidin molecule. These sites can probably be used for binding various biotinylated groups and biotinylated small molecules to streptavidin. It may be done to vary the parameters of the trapping process and thus change the electrophoretic behavior of the complex, as well as to attach fluorescent, radioactive, or similar label.

Denaturing of DNA. How can the DNA be denatured without disrupting the biotin–streptavidin bond? In the presence of urea or formamide, boiling for only a few minutes may disrupt the biotin–streptavidin bond. Therefore, to denature the DNA in 70–90% formamide, the samples should be heated at 65° for about 5 min. This way the DNA is well denatured and the biotin–streptavidin link is preserved. However, with no denaturing agent, such as urea, formamide, or sodium dodecyl sulfate (SDS), boiling for a few minutes does not damage the biotin–streptavidin link.

Diffusion of Streptavidin in Gel. In the polyacrylamide gel, streptavidin can migrate across the lanes and bind to biotinylated DNA in adjacent lanes. In view of this phenomenon, the problem arises of how to run control lanes that are supposedly carrying nonstreptavidinated DNA. If this control DNA carries biotin, then one way to circumvent the problem is to keep the control lane at a distance from the lanes carrying free streptavidin. If, on the other hand, the control lane should be next to a streptavidinated one, the control DNA should carry no biotin. The latter

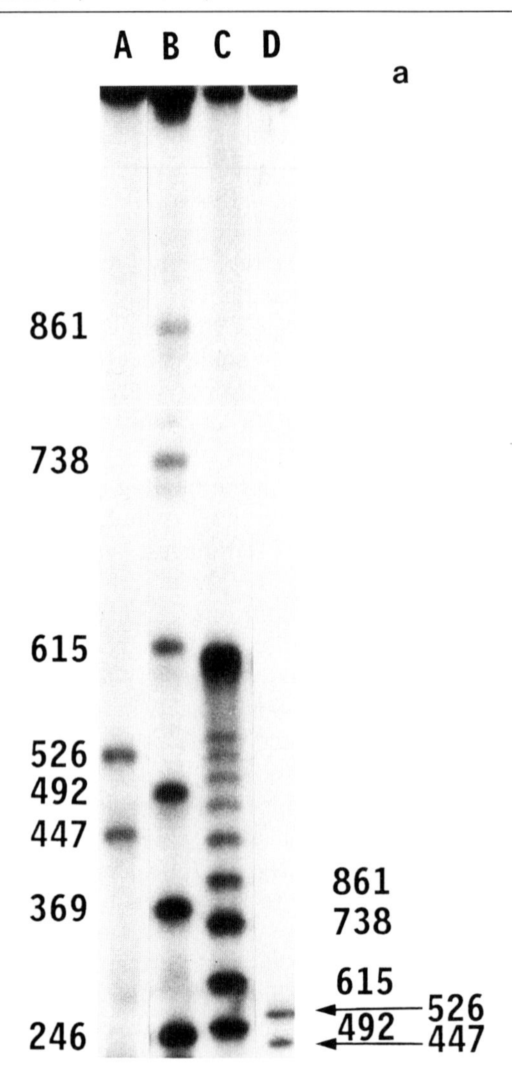

FIG. 1. Autoradiogram of constant-field electrophoresis of end-streptavidinated and unmodified ssDNA on 6% (a) and 8% (b) polyacrylamide gels.[10] The electric field strengths are 60 and 100 V/cm, respectively. In (a) and (b) the samples in the lanes are as follows. Lane A, end-streptavidinated SV40/*Hin*dIII DNA digest; lane B, end-streptavidinated 123-base ladder; lane C, unmodified 123-base ladder; lane D, unmodified SV40/*Hin*dIII DNA digest. The numbers on the left- and right-hand sides show the lengths (in bases) of the streptavidinated and unmodified DNA fragments, respectively. The arrows indicate the size markers in the two outer lanes. The gels were 13.5 cm long (slot to bottom) and 0.5 mm thick. Cooling by circulating water resulted in a temperature of 15° during the 8-hr run (a), and in a rise from 14 to 20° during the 60-min run (b).

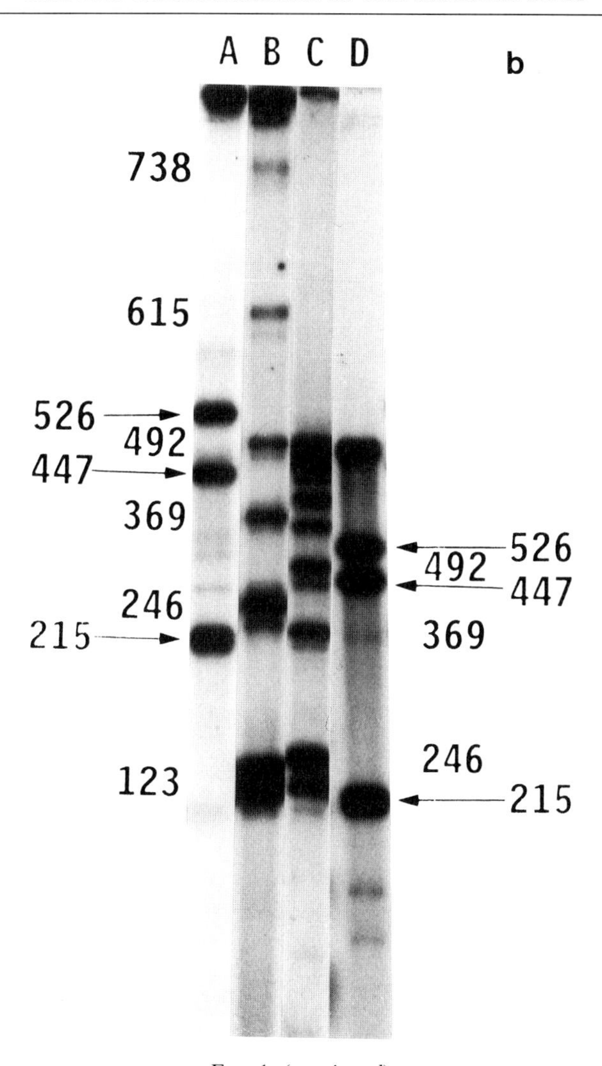

FIG. 1. (*continued*)

is preferable and implies that an unbiotinylated sample should be prepared in parallel to the biotinylated one.

Results and Discussion

Constant-Field Electrophoresis of End-Streptavidinated ssDNA. The autoradiograms in Fig. 1 show how streptavidin attached to one end of

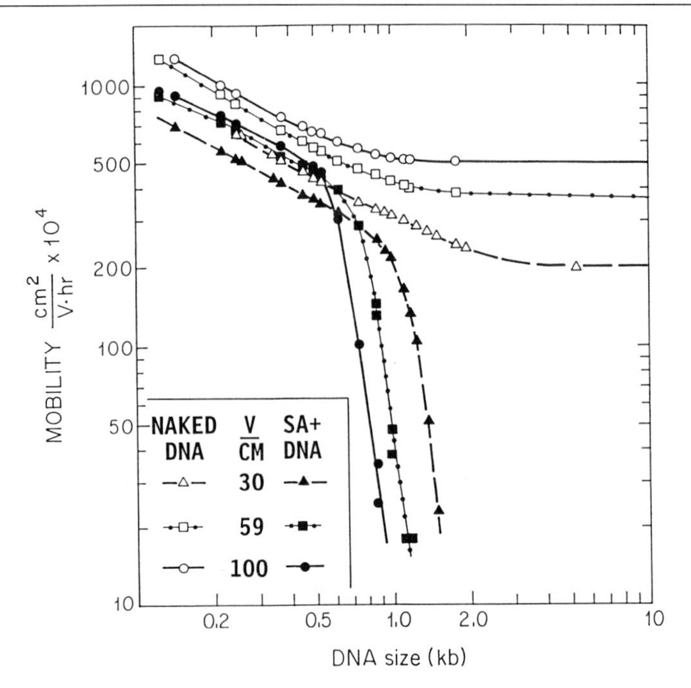

FIG. 2. Plots of electrophoretic velocity (cm/hr) per unit electric field (V/cm) in constant-field electrophoresis, versus ssDNA fragment length for end-streptavidinated and unmodified ssDNA.[10] The polyacrylamide gel concentration is 6%. The three types of curves relate to three different electric field strengths (see inset). SA, streptavidin.

ssDNA retards this complex in constant electric field PAGE in comparison with the same ssDNA lacking streptavidin. Figure 2 shows plots of DNA size dependence of the mobility of both streptavidinated and unmodified DNA. At small DNA sizes, the streptavidin–DNA complex is only moderately retarded as compared to the unmodified DNA of the same size. The mobility curves in the size range between 120 and 500 bases look like almost parallel straight lines. This parallel appearance, however, arises from the log–log nature of the plot. But in fact, in this size range, the absolute mobility gap narrows with increasing size, as it should, because the relative contribution of streptavidin to the total friction of the streptavidin–DNA complex decreases with DNA size. One would expect this mobility gap between the streptavidinated and unmodified DNA to keep narrowing. This is what, indeed, happens to end-streptavidinated dsDNA in agarose gels, where the gap eventually vanishes (for large DNA) at the sizes approaching the limiting mobility plateau (L. Ulanovsky, 1989, unpublished data).

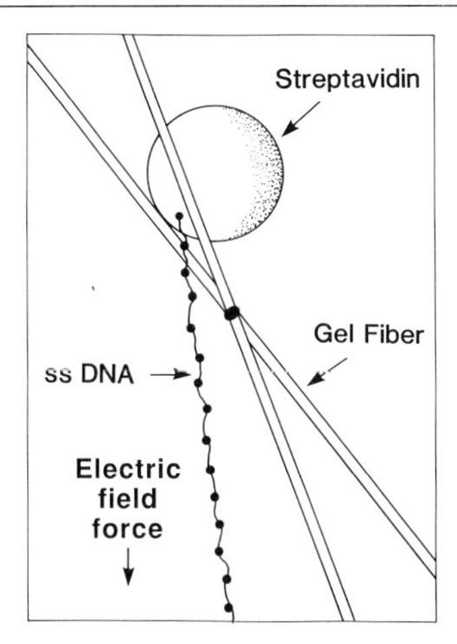

FIG. 3. One possible type of DNA trapping model (schematic, for illustration only). Because the DNA charge is negative, the electric force direction shown here is opposite to the actual electric field direction.

Contrary to these expectations, however, no merging of the two curves occurs for ssDNA in PAGE (Fig. 2). Instead, the streptavidinated DNA mobility curve drops down steeply at about the size where the unmodified DNA approaches the limiting mobility. Above this size, the separation between bands of streptavidinated DNA is several times wider than between bands of unmodified DNA (compare lanes B and C in Fig. 1a or b). Moreover, above a certain DNA size the streptavidin–DNA complex does not seem to migrate in the gel at all (e.g., above roughly 1000 bases in Fig. 1a).

An interesting aspect of this effect is related to the voltage dependence of the mobility pattern. The mobility curves for three different voltage gradients [30, 59, and 100 V/cm (constant in time)] are shown in Fig. 2. As the voltage grows, the size of the DNA fragments at which the steep drop of the mobility curve occurs decreases.

Trapping Model. A physical model that seems to agree with all these data, and which also finds support in FIGE experiments (see below), is the DNA trapping model. In this model, the electric field pulls the end-streptavidinated DNA into a trap formed by the fibers of the polyacrylamide gel. Figure 3 shows schematically one possible type of trap. The total charge of the hanging DNA chain and thus the energy of trapping depend on the DNA length, each nucleotide carrying one electron charge. In other words, the longer the DNA (and/or the higher the electric field strength), the larger is the energy required to lift the streptavidin–DNA

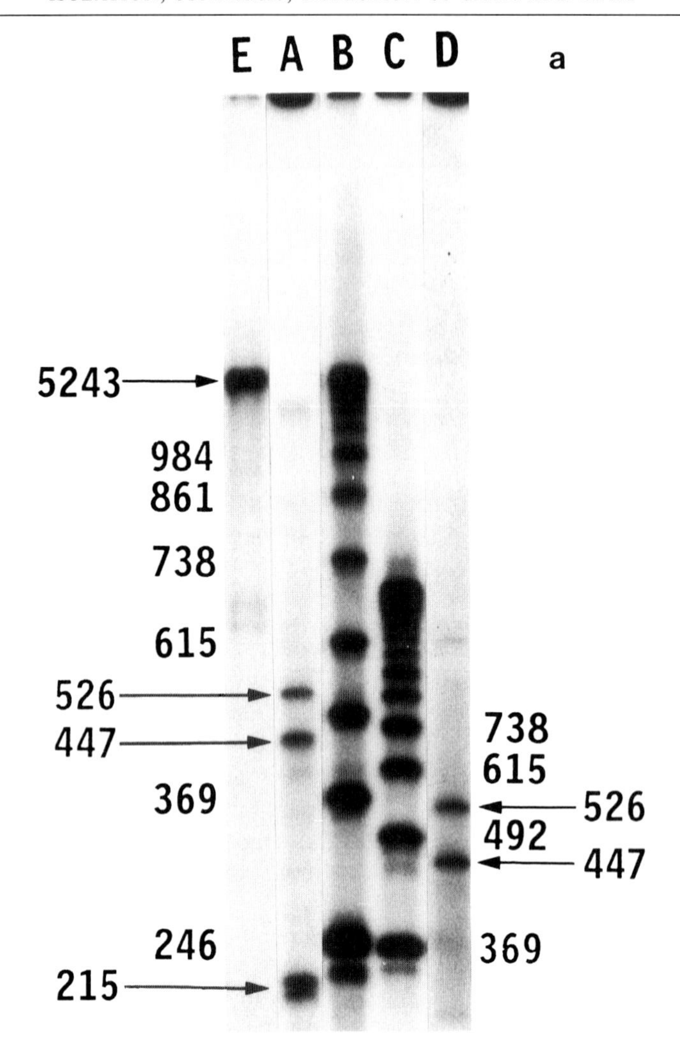

FIG. 4. Autoradiogram of FIGE of end-streptavidinated and unmodified ssDNA.[10] The electric field strength is 60 V/cm. The forward/reverse pulse durations are 800/200 msec in (a), and 800/20 msec in (b). The samples in lanes A–D are as in Fig. 1. The samples in lanes E and F are SV40/*Bam*HI (5243 bases) and pUC/*Hin*dIII (2686 bases), respectively. The polyacrylamide gel concentration is 6%.

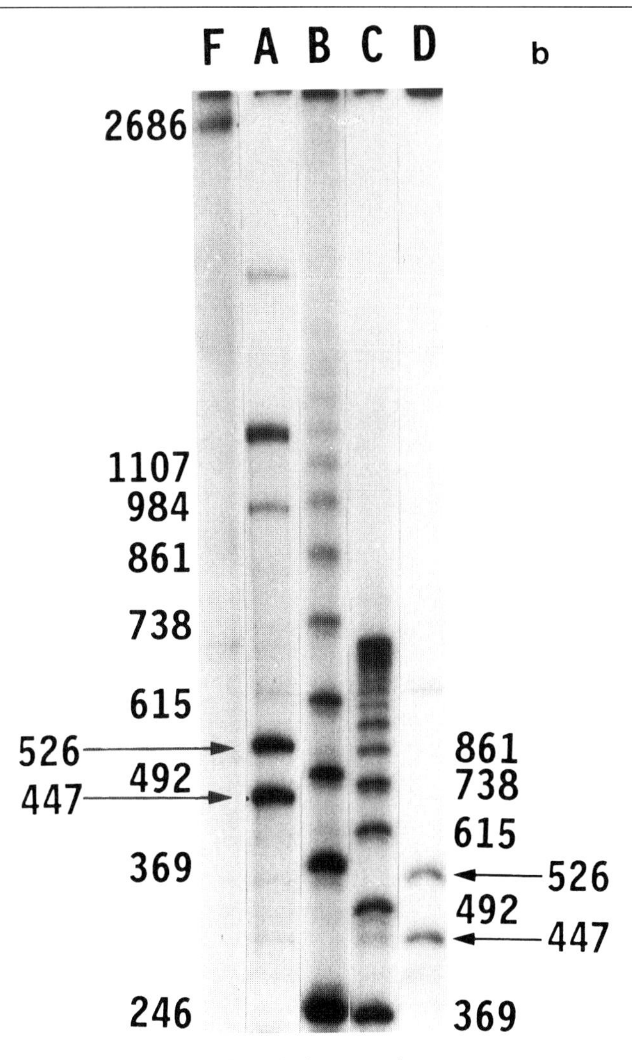

FIG. 4. (*continued*)

complex and release it from the trap. If the thermal energy can do that to a smaller streptavidinated DNA, it may have a harder time with a longer one. Therefore, the probability of the release of the complex in a unit of time depends on the DNA size. A long streptavidinated DNA may spend most of its time in the traps. Depending on the DNA size, the complex may or may not be able to jump from one trap to another. A small DNA would jump much more often, and thus spend more time moving down

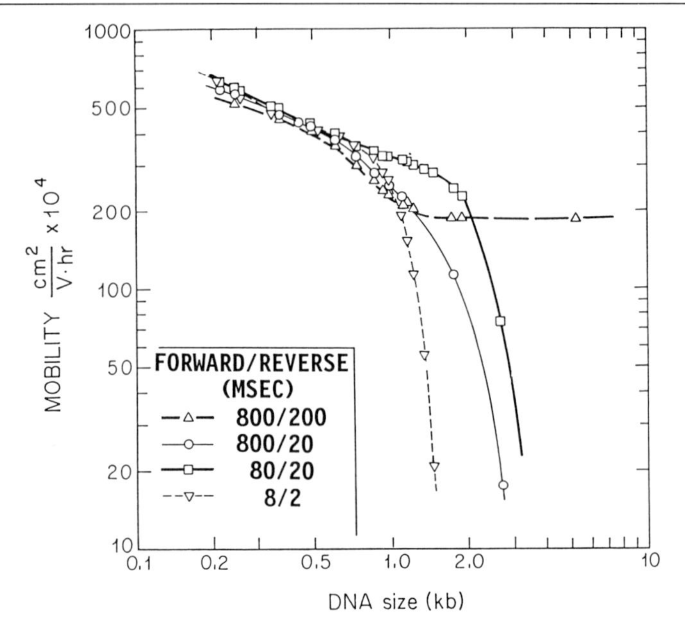

FIG. 5. Plots of electrophoretic velocity (cm/hr) per unit electric field (V/cm) versus ssDNA fragment length in different field inversion regimes (see inset) for end-streptavidinated and unmodified ssDNA.[10] The 6% (w/v) polyacrylamide gels were run at 60 V/cm at a constant temperature of 15°. The four types of curves relate to different pulse durations (see inset). For the purpose of velocity calculations, the net running time was the sum of the duration of all forward pulses minus that of all reverse pulses in the run.

the field from trap to trap. This mechanism can explain the sharp DNA size dependence of protein–DNA complex mobility on the gel plotted in Fig. 2.

The effect of voltage on the DNA size dependence of the mobility (Fig. 2) also fits this trapping model. The higher the voltage, the smaller the DNA size above which the trapping of the streptavidin–DNA complex begins. This effect is somewhat similar to the trapping of open circular dsDNA in agarose gels.[11]

In a more schematic manner one can imagine a coat hanging on a hook. The force of gravity would represent the electric field. A gusty wind would represent the thermal motion. If the coat is very light, a gust may have enough energy to blow the coat off the hook. If, however, the coat is heavy, the probability of its release by the wind is much lower. This setup can be viewed as a method of "coat separation by weight."

[11] S. D. Levene and B. H. Zimm, *Proc. Natl. Acad. Sci. U.S.A.* **84,** 4054 (1987).

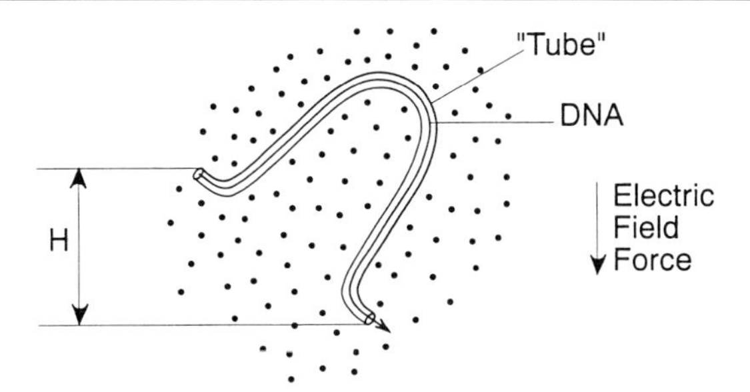

FIG. A.1. Reptation model of DNA migration in gel electrophoresis. According to the model, the DNA is moving through the gel as a snake through grass, in the sense that the obstacles (gel fibers) confine the motion of the molecule to its own path, and the sideways motion is not allowed by the obstacles (gel fibers). The obstacles, therefore, form an imaginary "tube" along the DNA path. Regarding the field direction, see Fig. 3 caption.

Field Inversion Gel Electrophoresis of End-Streptavidinated DNA. The trapping model is supported by streptavidinated DNA behavior in FIGE as well. Indeed, three different observations indicate this. First, the reverse pulse enhances the mobility of the streptavidinated DNA as compared to the constant field regime at the size range of the trapping (where the steep drop of the mobility curve occurs in the constant field; compare Fig. 2 with Fig. 5). The interpretation is that the reverse pulse helps to release the streptavidin–DNA complex from the trap by lifting the complex for a while. Second, the longer the reverse pulse, the larger the size of the streptavidinated DNA that is allowed to migrate in the gel (Figs. 4 and 5). This behavior can be expected of an *elastic* chain hanging on a hook in terms of the probability of release. Third, a comparison of the 800/20- and 80/20-msec curves on Fig. 5 shows that streptavidinated DNA of the same size migrates faster when the 20-msec reverse pulse is applied every 0.01 sec (the release occurs more often), than when it is every 0.1 sec.

Appendix: Biased Reptation Theory

Simplified Derivation of 1/L Dependence of DNA Mobility on Fragment Size

Until 1982 it was a mystery why DNA was separated by size in gel electrophoresis. Simple logic suggested that the DNA migration velocity should depend on the ratio of the charge to the friction coefficient of the fragment. Because both are proportional to the DNA length, there seemed to be no reason why the velocity should be size dependent. Indeed, DNA electrophoresis in solution, in contrast to that in gel, exhibits a size-independent velocity. The breakthrough in our understanding came in 1982 with the emergence of the biased reptation theory.[8,9]

The basic assumption of the biased reptation theory is that the DNA molecule in a gel "reptates" through the pores, i.e., moves in a snakelike fashion. The DNA chain moves along its curving path, its "tube," but not sideways (Fig. A.1). For short enough DNA fragments, both their shape in the gel and their movement are dominated by the thermal motion, to which the electric field provides only a small bias. Therefore, the shape of such a molecule is essentially a random coil. The molecule is pushed back and forth along its random path in a Brownian manner with a small net velocity in the direction of the electric field force. This velocity will be calculated here.

The total electric force acting on the DNA chain is QE, where Q is the charge of the DNA fragment, and E is the electric field strength. The longitudinal component of this force that pulls the fragment along its reptation path is QEH/L, where H is the projection of the end-to-end distance on the field direction, and L is the total length of the fragment. The corresponding longitudinal velocity u (along the tube axis) is

$$u = QEH/Lx$$

where x is the friction coefficient of the fragment. When $H = 0$, there is no movement, except Brownian. When H is not zero, then the net movement due to the field is toward the end that is downstream relative to the field force. This process may be schematically visualized as a flow of liquid in a capillary in Fig. A.1.

As the DNA chain moves ahead by one nucleotide, it is equivalent to a transfer of one nucleotide from the tail to the head by length L along the chain, or by distance H in projection on the field direction. Therefore, to convert the longitudinal velocity u into the vertical velocity V, we should add another H/L factor:

$$V = uH/L = QEH^2/L^2x$$

While QE/x does not depend on the chain length L, because both Q and x are proportional to L, the other factor, H^2/L^2, does. Averaged over time, H^2 is proportional to L, due to the random walk properties of the DNA coil. Therefore, the averaged migration velocity V is proportional to $1/L$, as it is indeed observed.

Acknowledgments

I thank Dr. Walter Gilbert for support, guidance, and stimulating discussions.

[7] Direct Complementary DNA Cloning Using Polymerase Chain Reaction

By CHENG CHI LEE and C. THOMAS CASKEY

Introduction

A major approach in complementary DNA (cDNA) cloning involves screening a cDNA library with oligonucleotide probes encoding a known peptide sequence. However, the degeneracy of the genetic code for all amino acids except methionine and tryptophan makes a perfect codon selection for the synthetic oligonucleotide probe difficult.[1] Oligonucleotide probes lacking complementarity to their target sequences frequently generate spurious hybridization signals and often fail to identify authentic clones, rendering cDNA library screening a complicated process. Several strategies designed to overcome these problems involve the selection of multiple oligonucleotide probes, inclusion of the most frequently used codons,[2-4] and replacement of ambiguous nucleotides with deoxyinosine and 5-fluorodeoxyuridine.[5,6] These approaches are frequently unsatisfactory because actual codon usage in the target sequence is heterogeneous. To overcome these limitations, we have developed a procedure based on the polymerase chain reaction (PCR) for cDNA cloning using the amino acid sequence as the primary information for the gene of interest. The technology for mixed oligonucleotide-primed amplification of cDNA (MOPAC) uses a mixture of degenerate primers corresponding to the amino acid sequence to generate the authentic cDNA by PCR.[7]

Principle of Method

This procedure is based on the idea that if one could "reverse translate" an amino acid sequence to its authentic cDNA codon usage, then

[1] F. H. C. Crick, *J. Mol. Biol.* **38,** 367 (1968).
[2] R. Grantham, C. Gautier, M. Gouy, M. Jacobson, and R. Mercier, *Nucleic Acids Res.* **9,** 43 (1981).
[3] R. Lathe, *J. Mol. Biol.* **183,** 1 (1985).
[4] J. H. Yang, J. H. Ye, and D. C. Wallace, *Nucleic Acids Res.* **12,** 837 (1984).
[5] J. F. Habener, C. D. Vo, D. B. Le, G. P. Gryan, L. Ercolani, and A. H. Wang, *Proc. Natl. Acad. Sci. U.S.A.* **85,** 1735 (1988).
[6] E. Ohtsuka, S. Matsuki, M. Ikehara, Y. Takahashi, and K. Matsubara, *J. Biol. Chem.* **260,** 2605 (1985).
[7] C. C. Lee, X. Wu, R. A. Gibbs, R. G. Cook, D. M. Muzny, and C. T. Caskey, *Science* **239,** 1288 (1988).

stringent hybridization conditions can be used when screening a cDNA library with this cDNA as a probe. A technology that was highly suitable for such a purpose was the polymerase chain reaction.[8] To overcome the major problem of selecting a perfect complementary PCR primer set for the reaction, the synthesis of all the possible primer combinations coding for the selected peptide sequence was chosen instead. The fundamental assumption behind this approach was that the authentic sequence primer would selectively anneal to its target complementary sequence, outcompeting the less complementary primers during the annealing process. In practice this strategy works well, and there are interesting features to it.

Interesting Features of Procedure

One such feature was the finding that a "perfect match" primer need not be present in the degenerate oligonucleotide population for the procedure to be successful. There is a tolerance for base-pair mismatch between the primer and the cDNA template during PCR.[9] Usually the amplified product possesses different primer sets, but the newly synthesized DNA is unique and will correspond to the amino acid sequence of interest. Studies show that this technology can tolerate a high level of degeneracy in the primer mixture. A success with a degeneracy of 262,144 in combination with another primer degeneracy of 8192 has been described.[10] Other studies using primers of less degeneracy are more frequently described.[11–13] However, the increase in primer degeneracy frequently led to an increase in nonspecific priming, resulting in nonauthentic DNA being amplified.[13] With *Taq* polymerase, this problem can be minimized by careful selection of the annealing condition in the PCR. Another efficient way to eliminate nonspecific products calls for the use of a "guessmer" oligonucleotide probe corresponding to the DNA to be synthesized.[7,9] This probe, when used in low-stringency hybridization, will identify the authentic cDNA on a Southern blot.

[8] K. B. Mullis and F. A. Faloona, this series, Vol. 155, p. 335.

[9] C. C. Lee and C. T. Caskey, *in* "Genetic Engineering" (J. K. Setlow, ed.), Vol 11, p. 159. Plenum, New York, 1989.

[10] S. J. Gould, S. Subramani, and I. E. Scheffler, *Proc. Natl. Acad. Sci. U.S.A.* **86,** 1934 (1989).

[11] Z. Zhao and R. H. Joho, *Biochem. Biophys. Res. Commun.* **167,** 174 (1990).

[12] L. D. Griffin, G. R. MacGregor, D. M. Muzny, J. Harter, R. G. Cook, and E. R. B. McCabe, *Am. J. Hum. Genet.* **43,** A185 (1988).

[13] S. I. Girgis, M. Alevizaki, P. Denny, G. J. M. Ferrier, and S. Legon, *Nucleic Acids Res.* **16,** 10371 (1988).

Other Considerations

One important consideration is the length of the primer for PCR, because the length of the amino acid sequence available is probably limited. Our studies show that the degenerate primer PCR can be affected by the length of the priming oligonucleotides. Shorter primers with greater degeneracy do not compensate well for a longer primer in the efficiency of authentic product generation.[9] Oligonucleotide primer length corresponding to six amino acids is ideal, with a minimum of five amino acids. Another consideration involves the addition of a restriction enzyme site on the primers. For efficient cloning of the PCR products into a cloning plasmid this addition is advantageous. An addition of two extra nucleotides beyond the 5′ end of the restriction site will ensure efficient digestion by the restriction enzyme.

It is recommended that the RNA for cDNA synthesis be from the same animal species as the peptide sequence. This would eliminate the possible differences in codon usage between animals of different species for the same gene. There is no need to purify the total RNA into poly(A)$^+$ RNA for these reactions. Our studies show that the procedure will work as efficiently with total RNA as with poly(A)$^+$ RNA.[9] Priming the mRNA for cDNA synthesis with either the antisense primer or random primer is preferred over the oligo(dT) primer, especially for large genes. The efficiency for 5′ DNA synthesis by reverse transcriptase decreases as the length of the cDNA increases. Thus with peptide sequence from the amino terminal end of the protein, the cDNA synthesized with oligo(dT) priming would be a disadvantage. A size selection should also be carried out on the PCR product before cloning into a plasmid vector. This would remove most of the nonspecific products that have been amplified.

Protocols

First-Strand cDNA Synthesis

For a 50-μl reaction use 50 m*M* Tris-HCl, pH 8.3, 75 m*M* KCl, 1 m*M* dithiothreitol (DTT), 15 m*M* $MgCl_2$, 1 m*M* dNTP, 400 ng of antisense primer or random primer, and 20 μg of total RNA. Heat the RNA at 65° for 3 min and quickly cool on ice before use. The cDNA synthesis is initiated by adding 400 units of M-MLV (Moloney murine leukemia virus) reverse transcriptase (GIBCO-Bethesda Research Laboratories, Gaithersburg, MD) and incubated at 37° for 1 hr. *Optional:* Single-strand cDNA (sscDNA) can be generated by treating the cDNA reaction with 75 μl of 0.7 *M* NaOH, 40 m*M* ethylenediaminetetraacetic acid (EDTA), and incubating the reaction at 65° for 10 min. The cDNA is precipitated in 200 m*M* ammonium acetate, pH 4.5, and 2.5 vol absolute ethanol. Recover the cDNA by centrifugation in a microcentrifuge for 10 min at maximum speed. Wash the nucleic acid pellet in 70% ethanol, lyophilize, and resuspend in 50 μl of sterilized distilled H_2O.

Mixed Oligonucleotide Amplification of cDNA

For a 50-μl reaction use 10 m*M* Tris-HCl, pH 8.0, 50 m*M* KCl, 1.5 m*M* $MgCl_2$, 200 μ*M* dNTP, 1 μ*M* concentrations of each degenerate primer mixture, 2 μl of synthesized cDNA, and 2.5 units of *Taq* polymerase. Add a drop of light paraffin oil to prevent evaporation during PCR. The PCR can be run on a thermal cycler under the following conditions: Denature at 94° for 30 sec, anneal at 48° for 30 sec, and elongate at 72° for 3 min. After 35 cycles, analyze 10 μl of the reaction on a TBE [89 m*M* Tris–borate, 89 m*M* boric acid, and 0.2 m*M* EDTA (pH 8.0)]-agarose gel. If the expected product is less than 300 base pairs, use a 4% (w/v) NuSieve (FMC Corp., Marine Colloids Div., Rockland, ME) agarose gel for the analysis. This should be followed by a Southern blot using the radiolabeled "guessmer" oligonucleotide as a probe. Hybridize and wash the blot at low stringency and the probe should identify a DNA fragment corresponding to the length of amino acid sequence. Note that the annealing temperature suggested here will not necessarily work for all degenerate PCR primers. It is important to increase or decrease the annealing temperature if the suggested conditions did not work.

Fractionation of Polymerase Chain Reaction Product

Fractionate the PCR products in an 8% (w/v) TBE polyacrylamide gel. Gel electrophoresis can be carried out at 80 V until the dye front reaches the bottom. Soak the gel in TBE buffer containing 0.02% (v/v) ethidium bromide for 20 min, and isolate the expected DNA size fragment under UV light. Elute the DNA from the gel by soaking the gel in sterilized, distilled H_2O overnight at 4°. The DNA eluate can be further purified by ethanol precipitation. Clone the DNA into a standard plasmid vector or radiolabel the DNA and use it as a probe.

Acknowledgments

C.T.C. is an Investigator of the Howard Hughes Medical Institute.

[8] Direct Complementary DNA Cloning and Screening of Mutants Using Polymerase Chain Reaction

By YAWEN L. CHIANG

Introduction

Conventional cDNA cloning and screening for mutations can be tedious and time consuming. In the early 1970s a few papers described DNA amplification using polymerase[1-3] and Khorana *et al.*[2] suggested the fundamental theory behind the polymerase chain reaction (PCR). In 1984, K. Mullis developed the PCR and filed a patent in 1985 on the "process for amplifying nucleic acid sequences"; the patent was issued in 1987. In 1986, K. Mullis, J. Larrick, and I discussed the possibility of using the technique for direct cDNA cloning and for screening mutants for the variable region of immunoglobin heavy and light chains. We successfully developed the method, establishing a new approach for cloning antibody cDNA.

The PCR amplifies defined regions of the genome by using oligomer-directed primer extension. The technique depends on the DNA replication enzyme (DNA polymerase) and the chain reaction made during the exponential amplification procedure. The reaction is efficient; only small numbers of cells and minute amounts of DNA are needed. Thus it is feasible to clone an antibody gene directly from small numbers of hybridoma cells. Because the PCR is very sensitive, there are numerous contamination considerations (such as the stability and control of reagents, primer design, and reaction sensitivity) that must be addressed.

Many new methods and instruments for the PCR have been developed during the past few years. We will continue to find novel applications of the PCR—it is a powerful technique for modern molecular biology.

Principle of Polymerase Chain Reaction Used for Direct cDNA Cloning

The PCR is based on repeated cycles of high-temperature template denaturation, oligonucleotide primer annealing, and polymerase-depen-

[1] K. Kleppe, E. Ohtsuka, R. Kleppe, I. Molineux, and H. G. Khorana, *J. Mol. Biol.* **56,** 341 (1971).

[2] H. G. Khorana, K. L. Agarwal, H. Buchi, M. H. Caruthers, N. K. Gupta, K. Kleppe, A. Kumar, E. Ohtsuka, U. L. RajBhandary, J. H. van de Sande, V. Sgaramella, T. Terao, H. Weber, and T. Yamada, *J. Mol. Biol.* **72,** 209 (1972).

[3] A. Panet and H. G. Khorana, *J. Biol. Chem.* **249,** 5213 (1974).

dent extension. One can synthesize large quantities of a particular target DNA sequence *in vitro*.[4–6] By using reverse transcriptase, RNA can be transcribed into cDNA, which is then amplified using the *Taq* DNA polymerase. In our experiments, we also added the appropriate cloning site sequences with PCR primers to facilitate easy cDNA cloning into expression vectors.

Because of the amplification capabilities and extreme sensitivity of the PCR, minute changes of nucleotides in clones and cell lines can be detected by using direct sequencing[7] immediately following cDNA amplification. Chamberlain *et al.*[8] detected the deletion of the Duchenne muscular dystrophy locus via multiplex DNA amplification, and Wrischnick *et al.*[9] also used direct sequencing of enzymatically amplified DNA to determine the length mutations in human mitochondrial DNA.

The revolutionary changes in molecular biology, resulting from the development of sequencing and PCR techniques, have allowed us to efficiently clone the immunoglobulin heavy and light chain variable regions from hybridoma cells. The development of our reproducible experiments, combined with the knowledge of the immunoglobulin constant region sequences provided by Kabat *et al.*,[10] has made it possible to produce manmade antibodies without using conventional hybridoma techniques. (The immunoglobulin constant region is named as such because of its characteristic of conserved sequences.)

Experimental Section

Materials and Methods

A summary of the method is outlined in Fig. 1. A murine hybridoma cell line, 269-10F7 (IgG_1, κ), that produces anti-human C5a monoclonal antibodies[11] was used. From N-terminal amino acid microsequencing data,

[4] R. K. Saiki, S. Scharf, F. Faloona, K. B. Mullis, G. T. Horn, H. A. Erlich, and N. Arnheim, *Science* **230,** 1350 (1985).

[5] S. J. Scharf, G. T. Horn, and H. A. Erlich, *Science* **233,** 1076 (1986).

[6] K. B. Mullis and F. A. Faloona, this series, Vol. 155, p. 335.

[7] U. B. Gyllensten and H. A. Erlich, *Proc. Natl. Acad. Sci. U.S.A.* **85,** 7652 (1988).

[8] J. S. Chamberlain, R. A. Gibbs, J. E. Rainer, P. N. Nguyen, and C. T. Caskey, *Nucleic Acids Res.* **16,** 11141 (1988).

[9] L. A. Wrischnick, R. G. Higuchi, M. Stoneking, H. A. Erlich, N. Arnheim, and A. C. Wilson, *Nucleic Acids Res.* **15**(2), 529 (1987).

[10] E. A. Kabat, T. T. Wu, M. Reid-Miller, H. M. Perry, and K. S. Gottesman (eds), "Sequences of Protein of Immunological Interest," 4th Ed., U.S. Department of Health and Human Services, Washington, D.C., 1987.

[11] Y. L. Chiang, R. Sheng-Dong, M. A. Brow, and J. W. Larrick, *BioTechniques* **7,** 360 (1989).

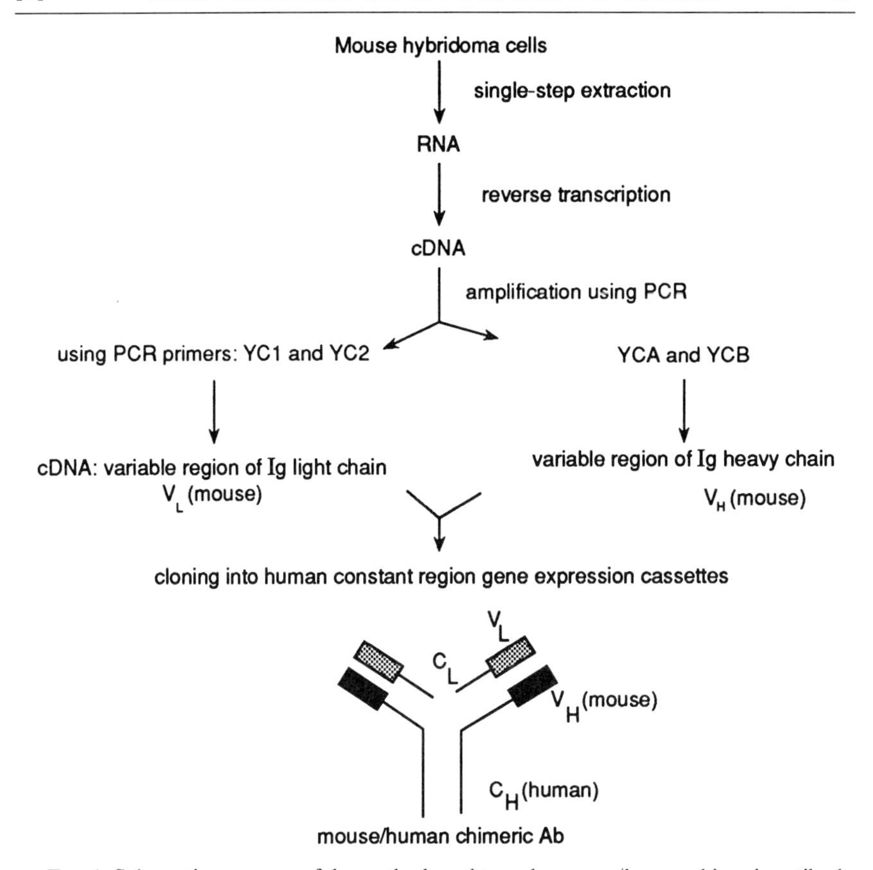

FIG. 1. Schematic summary of the method used to make mouse/human chimeric antibody (Ab).

we determined that 269-10F7 belongs to mouse κ light chain subgroup III and mouse IgG_1 heavy chain subgroup IA. From the database of Kabat *et al.*,[10] stretches of invariant N-terminal amino acid residues for both heavy and light chains were recognized. Oligonucleotide primers for the DNA sequence coding for this region were designed and prepared for PCR. The primer sequences were determined according to the usual codon of the specific immunoglobulin subgroups. Depending on the experimental design, additional nucleotide sequences for convenient cloning sites can be used if necessary.

We applied the rules and considerations for synthesizing oligomers and created two 5′ end primers (sense primer), YC1 and YCA. YC1 was used for the immunoglobulin light chain gene. It is a 39-bp oligomer, 5′-

CCCGAATTCGACATTGTGCTGACCCAATCTCCAGCTTCT-3′. YCA was used for the immunoglobulin heavy chain gene. It is a 36-bp oligomer, 5′-CCCGAATTCGATGTGCATCTTCAGGAGTCGGGACCT-3′.

We then designed the 3′ end primers (antisense primer) for the PCR. YC2 and YCB were synthesized. YC2 was used for the light chain gene. It is a 38-bp oligomer, 5′-CCGAATTCGATGGATACAGTTGGTGCAG-CATCAGCCCG-3′. YCB was used for the heavy chain gene. It is a 39-bp oligomer, 5′-CCCGAATTCTGGATAGACAGATGGGGGTGTCGTTT-TGGC-3′.

By employing YC1 and YC2 in the DNA amplification reaction, the entire variable-region sequences of the immunoglobulin light chain gene for hybridoma 269-10F7 were made. Likewise, by using YCA and YCB, the entire variable-region sequences of the immunoglobulin heavy chain gene were made. These cDNA fragments will be ligated with DNA from the human light and heavy chain construct regions, and the DNA construct will be placed into an expression vector to produce mouse/human chimeric antibody. This approach to create cDNA fragments and to use it to produce antibody is much quicker than the conventional method of constructing a cDNA library containing the mouse light and heavy chain variable region. The method involves the following steps.

Single-Step Preparation of RNA. Total cellular RNA is prepared by an acid guanidinium thiocyanate–phenol–chloroform single-step extraction method.[12] The RNA_{zol} method established by C1NNA/BIOTECX Laboratories International, Inc. (Friendswood, TX) provides a good yield of clean RNA prepared from small amounts of tissue or cells within 3 hr.

Reverse Transcription of RNA to Synthesize cDNA. First-strand cDNA synthesis is accomplished by priming the RNA with 1 μM oligo(dT) (12- to 18-mer). A 10-μl reaction volume contains a dried pellet of RNA, 200 units of Moloney murine leukemia virus (M-MLV) reverse transcriptase (Bethesda Research Laboratories, Gaithersburg, MD), 0.1 mg/ml of bovine serum albumin (BSA), 0.5 mM concentrations of each dNTP (dGTP, dATP, dTTP, and dCTP), 0.25 units of RNasin, 50 mM Tris-HCl, pH 7.5, 75 mM KCl, 10 mM dithiothreitol (DTT), 3 mM $MgCl_2$, and 50 μg/ml of actinomycin D. The reaction is carried out in a 37° water bath for 1 hr.

cDNA Amplification Using the Polymerase Chain Reaction. In the same tube, cDNA amplification using the PCR is carried out in 25 to 35 cycles. The sense and antisense primers mentioned above (1 μM each), 0.5 mM dNTP, and 1 unit of *Taq* polymerase in reaction buffer are added and the volume brought to 50 μl. The final concentration of the reaction

[12] P. Chomczynski and N. Sacchi, *Anal. Biochem.* **162,** 156 (1987).

buffer is 50 m*M* KCl, 1.5 m*M* $MgCl_2$, 10 m*M* Tris-HCl (pH 8.4), and 0.02% (w/v) gelatin. The repeated PCR cycles involve denaturing at 95° for 1 min, primer annealing at 55° for 1 min, and *Taq* polymerase reaction at 72° for 3 min.

Polymerase Chain Reaction Product Analysis. One-tenth of the final volume is loaded onto an ethidium bromide-stained, 3% (w/v) Nusieve agarose, and 1% (w/v) Seakem agarose (FMC BioProducts, Rockland, ME) minigel and analyzed (Figs. 2 and 3).

Specific Experimental Considerations

Some important practical procedures should be considered when using PCR techniques for direct cDNA cloning. It is important to design appropriate primers for the PCR reaction. Lathe[13] reviewed some theoretical and practical considerations to be used when choosing synthetic oligonucleotide probes deduced from amino acid sequence data. Girgis *et al.*[14] generated DNA probes for peptides with highly degenerate codons using mixed primers in the PCR reaction.

To minimize contamination problems, use a positive displacement pipetter and follow good laboratory procedures (including frequent changing of globes, careful aliquoting of reagents, and physically separating the pre- and post-PCR reactions). One needs to assess the results critically, repeat experiments to obtain consistent results, and select good positive controls.

The sensitivity, stability, and control of reagents, and the selection of equipment, are important. Choose credible sources when purchasing commercially available standard reagents and equipment. All experimental parameters including annealing, extension, temperature, numbers of cycles, and enzyme-to-primer ratios must be optimized on an individual basis. One should avoid using excess DNA in the reaction. DNA from other, unrelated genes extracted from other cells should be included as a negative control. Other controls such as distilled H_2O or Tris–ethylenediaminetetraacetic acid (EDTA) buffer should be used in reactions where there are no target sequences. A reaction that contains just the primer sequences should also be run. To verify the data, use different pairs of primers. To determine if there is other DNA contamination, primers and probes outside the 5′ or 3′ region of the PCR product should be made and checked using the Southern blot analysis data.

cDNA, derived from all different types of RNA (mRNA, tRNA, rRNA, total RNA, and viruses) and from very low copy number RNA target

[13] R. Lathe, *J. Mol. Biol.* **183,** 1 (1985).

[14] S. I. Girgis, M. Alevizaki, P. Denny, G. J. Ferrier, and S. Legon, *Nucleic Acids Res.* **16,** 10371 (1988).

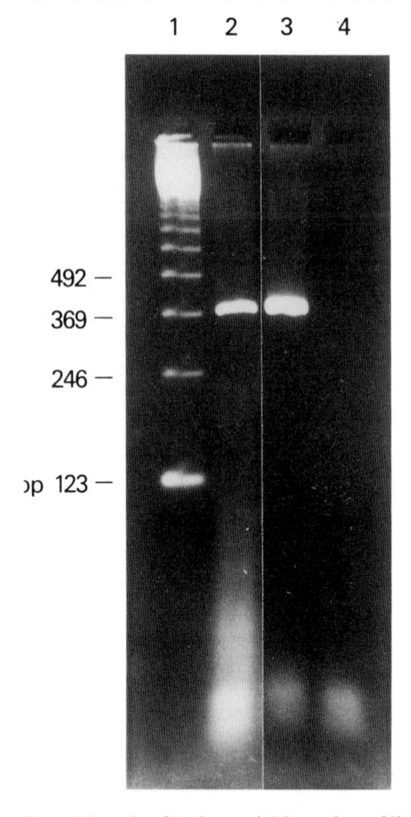

FIG. 2. PCR products from 25 cycles for the variable region of light chain. Lane 1, 123-bp ladder marker; lane 2, PCR product from 10^6 cells; lane 3, PCR product from mouse Ig light chain cDNA clone obtained from conventional cDNA library; lane 4, no cells, only primers contained.

sequences can be amplified. Wang *et al.*[15,16] used the reverse transcription reaction procedure published in 1987 and developed a GeneAmp thermostable γTth reverse transcriptase RNA PCR kit (Perkin Elmer Cetus, Emeryville, CA). By using this kit a cDNA product up to 1.3 kilobases (kb) in length can be generated.

Other choices of methods and materials are available. A special technology section has been published[17] that describes other options for the

[15] A. M. Wang, M. V. Doyle, and D. F. Mark, *Proc. Natl. Acad. Sci. U.S.A.* **86,** 9717 (1989).

[16] E. S. Kawasaki and A. M. Wang, "PCR Technology" (H. A. Erlich, ed.), p. 89. Stockton Press, New York, 1989.

[17] Special Technology Section: Methods and Materials: Amplification of Nucleic Acid Sequences: The Choices Multiply, *J. NIH Res.* **3,** 81 (1991).

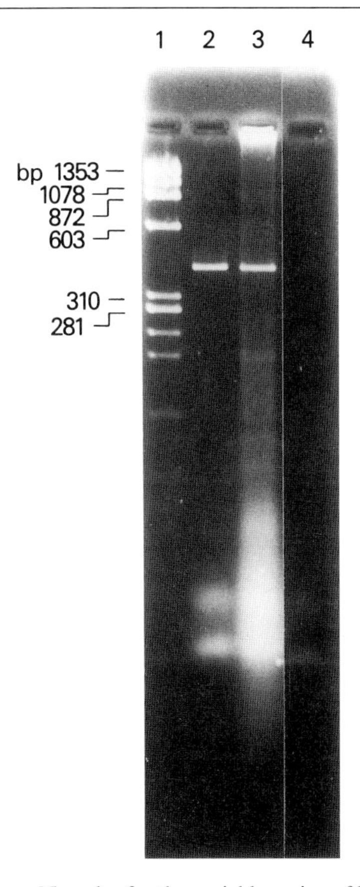

FIG. 3. PCR products from 35 cycles for the variable region of heavy chains. Lane 1, 300 ng ϕX174 *Hae*III marker; lane 2, PCR product from mouse Ig heavy chain cDNA clone obtained from conventional cDNA library; lane 3, PCR product from 10^6 cells; lane 4, no cells, only primers contained.

amplification of nucleic acid sequences as well as alternative thermal cyclers, primers and probes, thermostable DNA polymerase, reagents and reagent kits, miscellaneous products, and manufacturers. It is not necessary to use conventional M13 subcloning for sequencing. The method of direct sequencing of the PCR product is established.[9,18] Other advances in PCR have been reported elsewhere.[19]

[18] C. Wong, C. E. Dowling, R. K. Saiki, R. G. Higuchi, H. A. Erlich, and H. H. Kazazian, Jr., *Nature* (*London*) **330,** 384 (1987).

[19] H. A. Erlich, D. Gelfand, and J. J. Sninsky, *Science* **252,** 1643 (1991).

Results

A clear 369 bp expected band for Ig light chain cDNA fragment with the cloning site (Fig. 2) and a 390 bp expected band for Ig heavy chain cDNA fragment (Fig. 3) were obtained. The PCR products were subcloned into M13 followed by subsequent sequencing.[20] The cDNA cloning of the variable regions of heavy and light chains, were obtained by constructing a conventional cDNA library, has been published elsewhere.[11] These cDNA clones were also sequenced.

No mutations were found in the sequences obtained by either the cDNA clone from the conventional cDNA library or from the cDNA amplified fragments produced directly by the PCR of the hybridoma cell RNA.

We are presently developing a method to synthesize chimeric mouse/human monoclonal antibodies by using synthetic long oligomers ligated together containing just the mouse CDRs (complementarity-determining regions), and to synthesize mouse/human MAb containing fewer mouse sequences.

Acknowledgments

I would like to thank R. Dong and M. A. Brow for providing technical assistance, P. J. Lee for providing the amino acid sequencing, the Cetus DNA Synthesis group for supplying the oligonucleotides, and J. Larrick for helpful advice during the time we worked together at Cetus. I would also like to thank D. McLaughlin for word processing assistance and G. K. Lee for editing of the manuscript.

[20] J. Vieira and J. Messing, *Gene* **19,** 259 (1982).

[9] Complementary DNA Synthesis *in Situ:* Methods and Applications

By JAMES EBERWINE, CORINNE SPENCER, KEVIN MIYASHIRO, SCOTT MACKLER, and RICHARD FINNELL

Synthesis of complementary DNA (cDNA) has traditionally been performed on RNA isolated from large amounts of cells or tissue. cDNA probes and cDNA libraries have been successfully constructed from purified RNA; however, the use of isolated RNA from a large tissue source results in the loss of cellular resolution. This is an important consideration given that some RNAs are confined to distinct cell types within a specific tissue. In such situations, the RNA from these few cells represents only a small percentage of the whole-tissue total RNA. Therefore, by using a tissue as a source for RNA isolation, specific low-abundance RNAs from a few cells will be diluted by the RNA from the surrounding cells. The key to isolation of these specific mRNAs is to enrich for the cells that express the mRNAs. Enrichment schemes encompass many technologies, including fluorescence-activated cell sorting and panning procedures. Alternatively it is possible to enrich for these cells capitalizing on knowledge of their anatomical localization. This can be accomplished by performing cDNA synthesis *in situ*, incorporating radioactivity into the cDNA, and examining the autoradiographic distribution of the resultant radiolabeled cDNA. Such a procedure involving cDNA synthesis *in situ* has been developed and is referred to as *in situ* transcription (IST).[1]

In situ transcription was originally used as an mRNA localization technique. Briefly, IST is performed by annealing a specific oligonucleotide primer to the mRNA in a tissue section, washing away the unhybridized primer, and initiating cDNA synthesis by addition of reverse transcriptase and deoxynucleotide triphosphates. This is followed by washing away of the unincorporated triphosphates and autoradiographic exposure of the tissue section. The ability to synthesize cDNA *in situ* allows radiolabeled deoxynucleotides to be incorporated into the growing cDNA molecule in an anatomically defined region. This technology has several applications, including cellular localization of specific mRNAs, assessing transcriptional regulation by pharmacological agents, expression profiling of mRNA populations from anatomically defined areas, cloning of mRNA populations, determining the degree of translational control of specific mRNAs, and analysis of gene expression in live cells. Of the forementioned IST applications, the only topic that will not be discussed extensively in this chapter is the use of IST in cellular localization studies, which has been covered in detail elsewhere.[2] Each of the other applications will be presented in the context of a specific experimental system. We also present the coupling of IST to a nucleic acid amplification procedure, antisense RNA amplification (aRNA), which facilitates the cloning of mRNA populations from tissue sections as well as the mRNA from a single cell. Many of the procedures for these different topics share procedural steps, therefore repeated reference to these common elements will be apparent in this chapter.

[1] L. H. Tecott, J. D. Barchas, and J. H. Eberwine, *Science* **240,** 1661 (1988).
[2] I. Zangger, L. Tecott, and J. Eberwine, *Technique* **1,** 108 (1989).

In Situ Transcription in Study of Gene Expression in Developing Mouse Embryos

In Situ Transcription

Developmental biology constitutes one of the most exciting and prolific areas of life science research currently conducted worldwide. As in many research disciplines, advances in the understanding and appreciation of gene expression in developing organisms has been limited by the availability of sufficient quantities of desirable embryonic material from which to isolate mRNA and to develop cDNA libraries to screen. Such limitations have been overcome by the application of *in situ* transcription coupled with aRNA amplification technology (discussed below) to the study of gene expression in mouse embryos during the period of neural tube closure. Specifically, we were interested in determining the relative ratio of specific mRNAs following embryonic exposure to maternal treatments that have previously been shown to disrupt neural tube morphogenesis and result in the production of lethal congenital defects of the nervous system.[3,4] The following section details the necessary procedural considerations for this type of application.

At the desired gestational stage, the dam is killed and the uterus is removed into a petri dish containing ice-cold Krebs buffer. Starting at the ovarian end, the decidual capsule is torn open with watchmaker's forceps under a dissecting microscope. The embryos are then carefully dissected free from their extraembryonic membranes, with extreme care taken not to prematurely "pop" the embryo out of the chorionic sac. Those embryos that remain intact following removal of the amnion are prepared for histological sectioning. We have successfully utilized two different fixation protocols. The first procedure involved placing the embryos for at least 48 hr in Bodian's fixative (80% ethanol, 37% formaldehyde, and 5% glacial acetic acid). Similarly, we have also used 4% paraformaldehyde as a fixative (for 5 min) with excellent results, noting however that paraformaldehyde-fixed tissue is more difficult to work with and the integrity of the tissue can be compromised at any washes above room temperature. This is in contrast to adult tissue, in which paraformaldehyde fixation is often the method of choice. Once fixed, the embryos are lightly stained with Fast Green[5] to facilitate orientation of the embryos for sectioning. This staining procedure does not interfere with the cDNA synthesis, and is

[3] R. H.Finnell, S. P. Moon, L. C. Abbott, J. A. Golden, and G. F. Chernoff, *Teratology* **33,** 247 (1986).

[4] R. H. Finnell, G. D. Bennett, S. B. Karras, and V. K. Mohl, *Teratology* **38,** 313 (1988).

[5] D. R. Hilbelink and S. Kaplan, *Stain Technol.* **53,** 261 (1978).

quite useful for discriminating regions of the embryo for localization experiments. After standard paraffin embedding, 8-μm-thick sagittal sections of the embryos are cut, mounting three sections on a slide. The slides are then stored in slide boxes at $-80°$ to prevent nucleic acid degradation.

Prior to performing *in situ* transcription on the embryonic slides, it is necessary to remove the paraffin from the sections. This procedure involves several washes to remove the paraffin, followed by a postfixation step. The slides are placed in Coplin jars and initially washed with xylene three times, replacing the solvent every 5 min. The sections are then slowly rehydrated in ethanol, with two 5-min washes of 100, 90, and finally 70% ethanol. The remaining ethanol is washed off with sterile water. The slides should next be washed twice in a phosphate-buffered saline (PBS), changing the PBS after 5 min. After two 5-min fixation steps in 4% paraformaldehyde, the PBS washes are repeated and the slides are then allowed to dry completely, at which point they may be restored at $-80°$, or directly used for *in situ* transcription studies.

The first step in performing *in situ* transcription from fixed tissue (embryo) sections is the prehybridization step. The slides used for the study must be at room temperature and completely dry. The individual embryo sections are encircled with a ring of rubber cement to create a well. The size of the well is important, for the larger it is the greater the volume of hybridization buffer needed to fill the well, thus increasing the cost. The rubber cement should therefore be as close to the section as possible, without covering any portion of it. This process can easily be accomplished using a 5-ml syringe and a large gauge (16-gauge) needle. The application of the rubber cement is repeated two to three times to create a well of the appropriate depth. Once the rubber cement has dried, the prehybridization buffer can be applied. This buffer is composed of 25% formamide in 5 $\times$ standard sodium citrate (SSC). The slides are placed on prehybridization buffer-moistened 3-mm paper in a petri dish or other humidified chamber. The appropriate volume of hybridization buffer is added to cover the section. In our embryonic studies, anywhere from 18 to 25 μl has been sufficient. With a typical slide containing three embryo sections, two will receive 100 ng of the primer. In these studies a modified oligonucleotide primer that will hybridize to all of the poly(A)$^+$ RNA within the cell was chosen. This oligonucleotide was modified so that it can direct the amplification of the cDNA. The oligonucleotide primer is an oligo(dT_{24})-T7 amplification oligonucleotide, which contains 24 thymidine residues that will hybridize to polyadenylated mRNA molecules. It is important in this procedure to perform the same *in situ* transcription reaction save the addition of the primer on at least one tissue section. This serves as a control for determining the effectiveness of the hybridization and reverse transcription step, as well as the level of endogenous background activity. The sections are allowed to prehybridize at either room temperature or 37° overnight. For embryonic tissue fixed directly in 4% paraformaldehyde, the prehybridization should be at room temperature.

At the completion of the prehybridization step the buffer is removed from the wells, being careful not to damage the section with the pipette tip. The sections are then placed in Coplin jars and washed in 2× SSC at room temperature, replacing the fluid after 15 min. The second wash in 0.5× SSC is for 1 hr and is performed at either 40° or room temperature, depending on the fixative used for the tissue section. At the completion of the washes, the slides must again be allowed to dry completely and the rubber cement should be reapplied to provide wells for the hybridization step. The IST reaction mixture, based on a 50-μl total volume, is as follows: 5 μl of 10× transcription buffer (1× = 50 m*M* Tris-HCl, pH 8.3, 6 m*M* $MgCl_2$, and 0.12 *M* KCl), 3.75 μl of 100 μM dithiothreitol (DTT), 1.25 μl of dATP (10 m*M*), 1.25 μl of dGTP (10 m*M*), 1.25 μl of TTP (10 m*M*), 1 μl of [α-^{32}P]dCTP (approximately 800 μCi/25 μl), 4 μl of reverse transcriptase (at least 20 units/section) and the remaining volume is sterile water (32.5 μl in this 50-μl example). No nonlabeled dCTP is added in this reaction in order to enhance the autoradiographic signal. Indeed, one in four nucleotides of the synthesized cDNA will be radiolabeled. The disadvantage of not including nonlabeled dCTP in the reaction is that the cDNA transcripts tend to be shorter (because of low substrate concentration); however, the autoradiographic signal is still specific. Longer cDNA transcripts can be obtained by the addition of nonlabeled dCTP to a final concentration of 25 μM. For each section, approximately 18–25 μl of reaction mixture is applied and the humidified petri dish is incubated at 37° for 1 hr. This is best accomplished by floating the dish in a water bath set to the desired temperature. At the completion of the incubation period, the radioactive hybridization buffer is removed from the sections and the slides are rinsed to remove unincorporated deoxynucleotides. The slides can be placed in staining trays that are suspended in a diethyl pyrocarbonate (DEPC)-treated beaker (2000 ml) containing 2× SSC for the first hour and then 0.5× SSC for the next 6 to 18 hr. All of the rinses should be at room temperature.

Once the slides have been rinsed and air dried, they are ready to be exposed to X-ray film. Prior to this step, the rubber wells should be removed to permit proper apposition of the sections to the film. The slides can be taped to a cardboard backing and the films can be exposed to the slides either at −80° with an intensifying screen from 5 to 120 min, or at room temperature without a screen for varying lengths of time. Figure 1 represents a typical autoradiographic result of gestational day 10 SWV mouse embryos that were processed for *in situ* transcription. On the left,

(A) GD10 Embryo
33 Somites

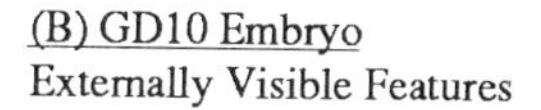

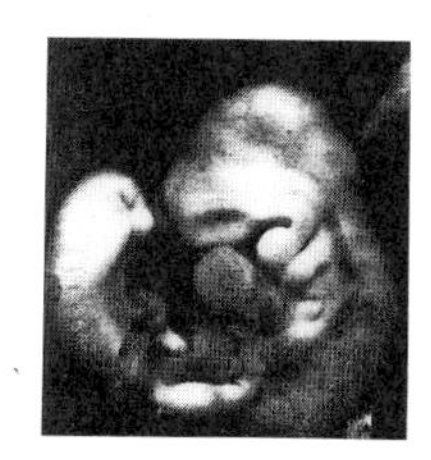

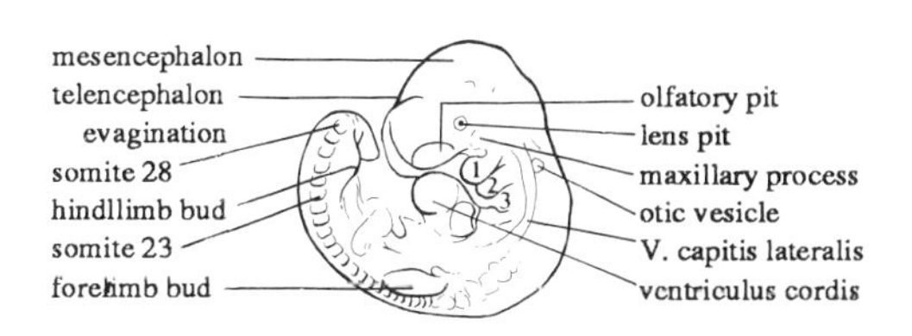

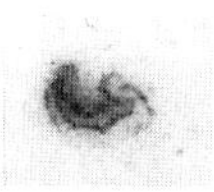

(C) GD SWV Embryo
VPA(600) Treated
No Primer

(D) GD10 SWV Embryo
VPA(600) Treated
Oligo-(dT)-T7 Primed

FIG. 1. *In situ* transcription in gestational day (GD) 10 SWV embryos. (a) Photograph and (b) schematic of a GD-10 embryo with approximately 28–35 somite pairs (K. Theiler, "The House Mouse: Atlas of Embryonic Development" Springer-Verlag, New York, 1989). Sections from embryos at this stage of development were taken through the IST procedure. In (c) the section did not receive the oligo(dT)-T7 primer, demonstrating the endogenous level of background binding. In (d) the embryo was hybridized in the presence of the oligo(dT)-T7 primer. Note the intense staining when compared to (c) in areas of rapid cellular proliferation.

where no primer was added to the section, the signal is very faint, indicative of the endogenous background level of hybridization. The embryo on the right was allowed to hybridize in the presence of the oligo(dT)-T7 amplification oligonucleotide primer, which hybridizes to all of the poly$(A)^+$ mRNA transcripts. It should be noted that the intense staining, in comparison to the unprimed embryo, occurs most prominently in areas of rapid cellular proliferation, such as the heart and branchial arches.

Amplified Antisense RNA Synthesis

The cDNA that has been synthesized directly on the embryonic section may be removed for further studies by alkaline denaturation of the mRNA–cDNA hybrids that were formed. This is accomplished by carefully adding 20 μl of 0.5 *N* NaOH and 0.1% (w/v) sodium dodecyl sulfate (SDS) to each section and repeatedly pipetting up and down until the entire section is removed into a 1.5-ml Eppendorf tube. To this is added 20 μl of

1 *M* Tris (pH 7.0) and 1 μl of tRNA (10 mg/ml) for use as a carrier. After a phenol–chloroform extraction and ethanol precipitation step, the single-stranded cDNA is allowed to self-prime (hairpinning) to initiate its second strand synthesis. This is accomplished by resuspending the dry pellet of cDNA in 10 μl of water, to which the following are added: 2 μl of 10× React 2 buffer (1× = 50 m*M* Tris, pH 7.0, 50 m*M* NaCl, 5 m*M* $MgCl_2$), 1 μl each of deoxynucleotides (dATP, dGTP, TTP; 10 m*M*), 2 μl of [α-^{32}P]dCTP, and 1 μl each of T4 DNA polymerase and Klenow polymerase. This is allowed to incubate for 2 hr at 37°. The next step involves a brief heating at 65° for 3 min to inactivate the DNA polymerase, followed by the addition of 81 μl of 1× S1 nuclease buffer (30 m*M* sodium acetate, pH 4.5, 50 m*M* NaCl, 1 m*M* zinc acetate). One unit of S1 nuclease enzyme is added and the reaction is incubated at 37° for 3 min. Following a phenol–chloroform extraction and ethanol precipitation, the now double-stranded cDNA is ready for blunt ending. To accomplish this, the dry pellet is resuspended in 4.5 μl of sterile water to which 1 μl of the following are added: React 2 buffer (10×), a mix of all four deoxynucleotides (each at 2.5 m*M*), and 0.5 μl of T4 DNA polymerase. This mixture is incubated at room temperature for 20 min. At the end of this "filling-in" reaction the sample is phenol–chloroform extracted and ethanol precipitated. The double-stranded cDNA is dissolved in 12 μl of sterile water followed by drop dialysis against sterile water on a Millipore (Bedford, MA) HAWP filter for no less than 3 hr to remove any unincorporated deoxynucleotides. At this point the double-stranded cDNA is ready to be cloned or amplified for use as a probe.

The amplification of the embryonic cDNA into antisense aRNA can be exceptionally useful in the analysis of gene expression. We have successfully amplified cDNA obtained from *in situ* transcribed cDNA from embryonic tissue sections, essentially creating riboprobes that were used to screen Southern blots containing cDNA clones of interest. To do this, we utilized a Promega (Madison, WI) Riboprobe Gemini System II buffer kit, which provided all of the necessary components for the amplification, except the radiolabeled CTP. In a 20-μl total volume reaction, the following reagents are utilized: 4 μl transcription buffer (5×), 4 μl of cDNA, 1 μl of CTP (100 μ*M*), 1 μl each of ATP, GTP, and UTP (10 m*M*), 2 μl of 0.1 *M* DTT, 0.5 μl of RNasin, 1.5 μl of water, 3 μl of [α-^{32}P]CTP, and 1 μl of concentrated T7 RNA polymerase. Prior to adding the enzyme, 0.5 μl of the mixture is spotted on a small piece of 1-mm Whatman (Clifton, NJ) paper for scintillation counting. The sample is incubated for 3 hr at 37°, after which a second 0.5-μl sample is spotted on the Whatman paper, the nucleic acids precipitated in 10% (w/v) trichloroacetic acid (TCA), and the sample counted. At this point the aRNA, which has been amplified several thousandfold (yet maintains a high degree of fidelity relative to the original mRNA found in the embryo section), can either be used to probe Southern blots or taken through a second round of amplification. The latter involves several procedural steps described elsewhere.[6,7] After the amplification process is complete, the volume of the aRNA probe is brought up to 60 μl with sterile water, heat denatured for 5 min, placed on ice for a few minutes, and then applied directly to the blot of interest.

Figure 2 represents the hybridization pattern of an aRNA probe made from a single section of a gestational day 10 SWV mouse embryo. The Southern blot contained 500 ng of the following cDNA clones: Hox 7.0, heat shock protein (HSP) 70, HSP 68, retinoic acid receptor subunits (α, β, and γ), calcium channel, and potassium channel. By quantitating the autoradiographic intensity of the aRNA hybridization to these cDNA clones (taking into account cDNA size and aRNA size), it is possible to measure the relative abundance of the mRNAs of interest. Furthermore, in keeping with our interest in developmental defects, by utilizing embryos that have been exposed *in utero* to various teratogens it is possible to observe fluctuations in gene expression in response to the teratogenic treatment.

In Situ Transcription in Study of Low-Abundance mRNAs from Adult Tissues

Experiments investigating the presence and regulation of mRNAs are critical in understanding behaviors that result from interactions between several cell types. However, these studies when performed *in situ* may be complicated by at least three factors. The relative abundance of different mRNAs varies severalfold and studies of mRNAs of low abundance usually require the use of large amounts of tissue to ensure the isolation of sufficient quantities of these mRNAs. The amount of the mRNAs of interest may be further decreased by dilution from mRNAs present in surrounding cells (as exemplified by the heterogeneity of neurons and glia within the nervous system). Finally, if *in situ* hybridization is the technique of choice for assessment of mRNA regulation it is essential to have prior knowledge of at least a portion of the nucleic acid sequence for the design of nucleotide probes. As previously mentioned, the combination of *in situ*

[6] J. Eberwine, H. Yeh, K. Miyashiro, Y. Cao, S. Nair, R. Finnell, M. Zettel, and P. Coleman, *Proc. Natl. Acad. Sci. U.S.A.* **89,** 3010 (1992).

[7] J. Eberwine and S. Mackler, *in* "Molecular Analysis of Cellular Response to Opiate Challenge, Fidia Symposia in Neurosciences" (E. Costa and T. Joh, eds.), (in press), Academic Press, New York, 1992.

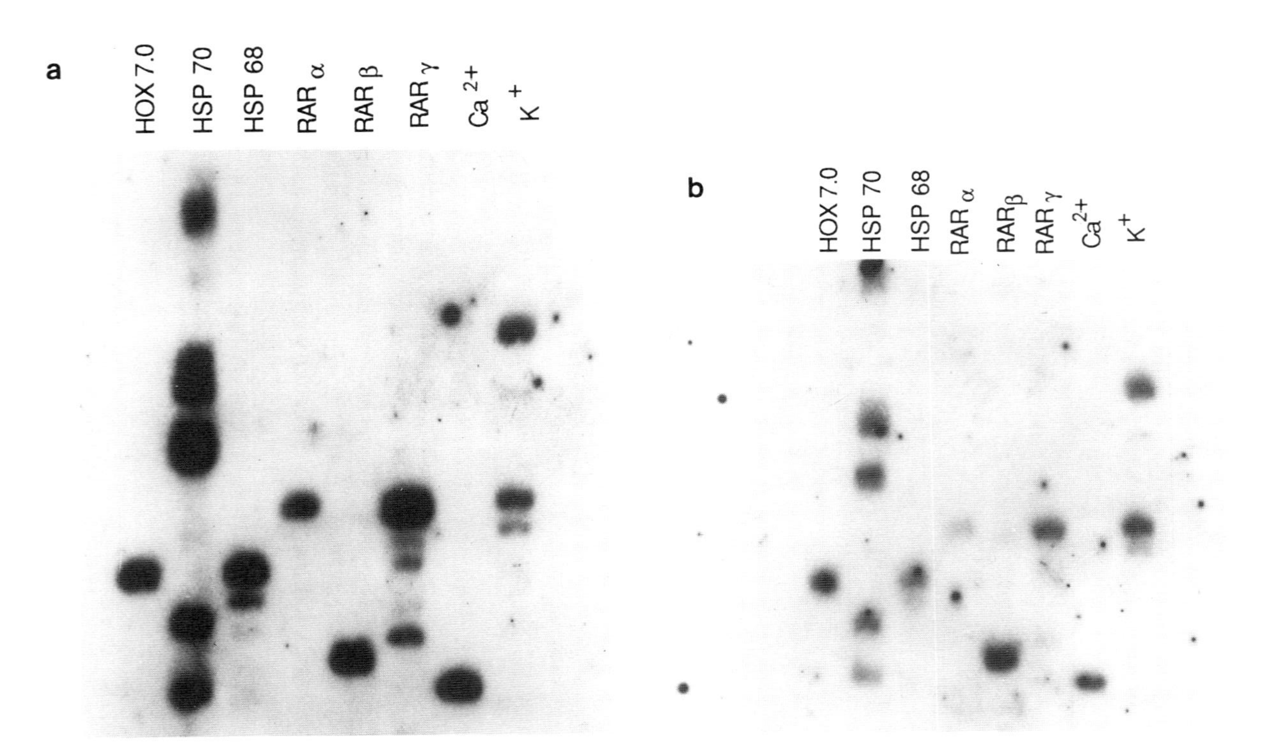

FIG. 2. Genetic expression profiles of GD 10 : 0 SWV embryos exposed at a critical period of morphogenesis to either a hyperthermic (43°) insult (a) or a 38° control treatment (b). The IST-aRNA-derived probes were hybridized to cDNA clones for Hox 7.0 (homeobox), HSP 70 (heat-shock protein), HSP 68, RARα (retinoic acid receptor, α subunit), RARβ (β subunit), RARγ (γ subunit), Ca (calcium channel), and K (potassium channel).

transcription (IST) and antisense RNA amplification (aRNA) has been developed to study both known and novel mRNA molecules in an anatomically defined region of the adult central nervous system (CNS) as well as in tissue culture cell lines. This procedure offers distinct advantages, compared to *in situ* hybridization, and can minimize the aforementioned complicating factors.

In these studies the only significant difference in performing IST on adult tissues, as compared to that described in the previous section for embryonic tissue, is the actual fixation procedure. *In situ* transcription is performed by cutting 10- to 15-μm-thick sections from fresh frozen tissue (coronal sections including the rat striatum are an example). The sections should be fixed for 5 min in 3% neutral buffered paraformaldehyde followed by several washes in 2× SSC and dehydration through graded ethanols (70, 85, 95, and 100%). These fresh-frozen postfixed sections can be stored at $-80°$ for months to years without significant degradation of the nucleic acids. All solutions should be prepared so that they are RNase free. When ready for use, the sections are slowly brought to room temperature and the tissue is encircled with a thin layer of rubber cement. Surrounding regions of the section that are not important for further study may first be removed with a sterile razor blade. As an example, in studies of the striatum the cerebral cortex can be carefully cut away, leaving the striatum on the slide for IST.

The remainder of the steps in performing IST for this tissue are the same as described in the section *In Situ* Transcription (above). The IST-cDNA can also be removed from the slide and amplified as previously described, or used directly as a probe. We have utilized this procedure to screen for the presence of selected gene families in defined tissue types. Using the rat pituitary as a tissue source, we have annealed a 154-fold degenerate oligonucleotide primer directed against membrane-spanning region 2 of the β-adrenergic receptor to 11-μm-thick fresh frozen postfixed pituitary sections from rats that had been treated with haloperidol (dopamine receptor antagonist) or bromocriptine (dopamine receptor agonist) (Fig. 3). The cDNA transcripts have been removed and used as a probe to screen Southern blots of genomic DNA. It was assumed that specificity of hybridization would be conferred by the primer and that the IST-cDNAs would be derived from just a few mRNAs that would hybridize to this primer. Indeed, the results presented in this autoradiogram (Fig. 4) show that there are only a few discrete genomic DNA bands hybridizing to the IST-cDNA. While many of the bands were of equal intensity, nonetheless there were other bands whose intensity varied significantly. This suggests that there were differing initial amounts of mRNA corresponding to these genomic DNA bands. It is reasonable to speculate that these cDNAs were

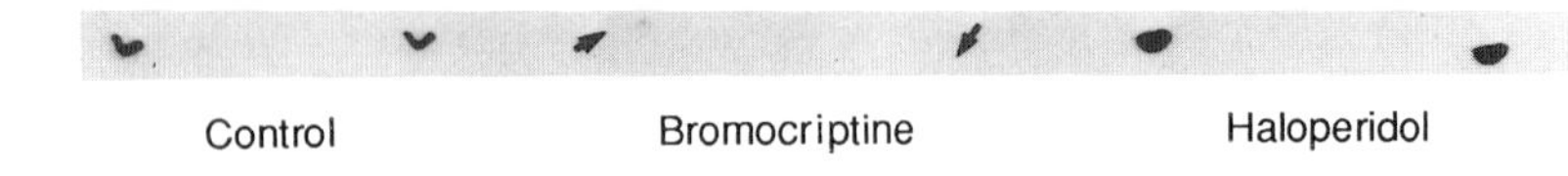

FIG. 3. *In situ* transcription in dopamine-treated rat pituitaries. The IST was performed in rat pituitary sections from animals that were injected with saline (control), bromocriptine (2 mg/kg), or haloperidol (2 mg/kg). Animals were injected once daily for 4 days. The primer in these reactions was directed to the proopiomelanocortin (POMC) mRNA sequence. *In situ* transcription was performed in the presence of radiolabeled dCTP. Exposure of the film autoradiogram was for 10 min. The arrows point to the pituitaries after treatment with bromocriptine, which reduces the amount of POMC mRNA. The increase in signal in the haloperidol-treated animals results from an increase in POMC mRNA.

derived from mRNAs whose abundance was altered by the administered dopaminergic agents. These IST-cDNAs or genomic DNA bands can now be cloned for further characterization.

Electrophoretic Analysis of *in Situ* Transcription-Derived cDNAs and Assessment of Translational Control

For IST to take place, cDNA synthesis must occur. While cDNA synthesis can be detected by autoradiography of the tissue section after incorporation of radiolabeled nucleotides, this anlaysis does not provide any characterization of the synthesized cDNAs. An analysis of the IST-derived cDNAs was initiated to optimize the reaction conditions, with the rationale that the longer the cDNAs, the greater the signal-to-noise ratio. With this rationale in mind, a strategy was developed for characterization of the cDNA transcripts. The IST-derived cDNAs that are removed from the tissue section (as described in *In Situ* Transcription in Study of Low-Abundance mRNAs from Adult Tissues, above) can be electrophoresed in a denaturing polyacrylamide gel (Fig. 5). The autoradiographic signal obtained from this analysis shows a distinct banding pattern (see Fig. 6). This banding pattern is not generated by alkaline hydrolysis of cDNA because several different denaturing agents, including KOH, NaOH, guanidinium hydrochloride, and guanidinium isothiocyanate, have been used to remove cDNA from the tissue sections. The banding pattern produced from cDNA transcripts is the same for the different denaturants. When proopiomelanocortin (POMC) mRNA from rat and mouse pituitaries is *in situ* transcribed and the transcripts examined by DNA sequencing gel analysis, the banding pattern can be correlated with the sequence of the mRNA. The sequence differences between rat and mouse POMC mRNAs are readily discernible from the gel.[8] Comparison of the sequence with the bands indicates that the bands correspond primarily with G residues of

[8] J. Eberwine, D. Newell, C. Spencer, and A. Hoffman, submitted for publication.

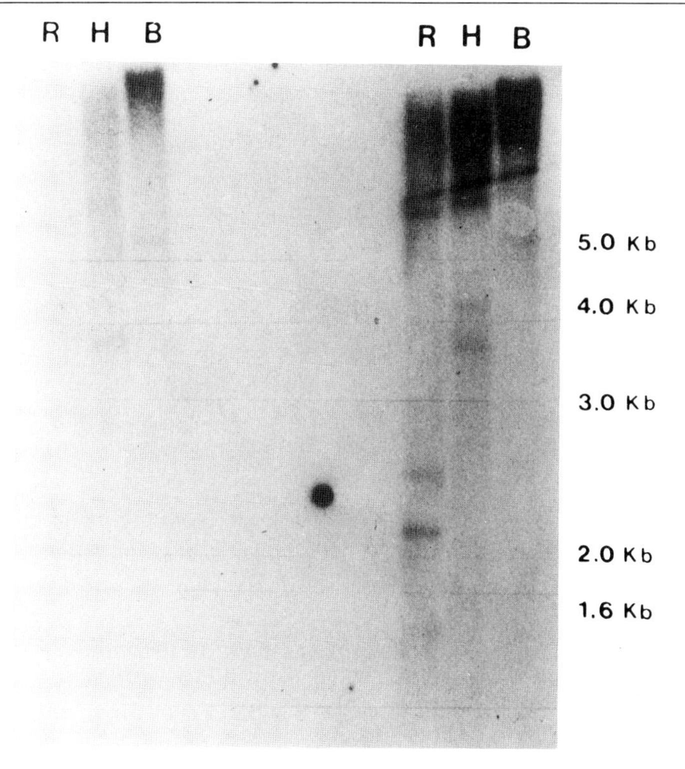

FIG. 4. Rat genomic DNA Southern blot probed with IST-cDNA. Ten micrograms of rat genomic DNA was cut with *Eco*RI (R), *Hin*dIII (H), or *Bam*HI (B). *In situ* transcription was performed on pituitary sections using a degenerate primer directed to the second membrane-spanning region of the G protein-coupled receptors. *Left:* Autoradiogram resulting from probing with IST-cDNA taken from bromocriptine (increases dopamine action)-treated animals. *Right:* Autoradiogram of the haloperidol (HALDOL)-treated animals. The differences in band intensity are reflective of the amount of IST-cDNA in the probe which in turn is reflective of the amount of individual hybridizing mRNAs in the tissue section. (Data were first presented in Eberwine and Mackler,[8] and is reproduced here with the permission of Academic Press.)

the mRNA and to a lesser extent with A, C, and U. We were initially concerned that these bands resulted from a fixation artifact due to cross-linking of the mRNA to other molecules by the paraformaldehyde treatment of the section.[9] This is not the case because IST performed on unfixed tissue shows the same banding pattern as that found in fixed tissue (Fig. 6). Additionally, the banding pattern is not a random occurrence due solely to interaction of the primer with the mRNA. We know this because in previous experiments, when primer extension is performed on mRNA in solution, a smear of termination sites is observed; some bands form that

[9] M. Feldman, *Proc. Nucleic Acid Res. Mol. Biol.* **13,** 1 (1973).

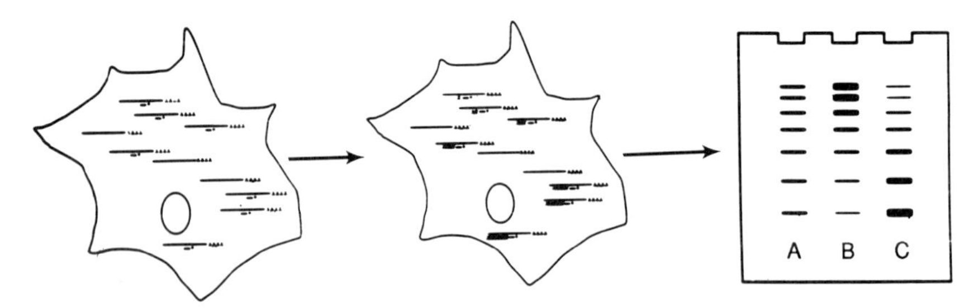

FIG. 5. Use of IST in the analysis of translational control. The experimental methodology of IST includes the *in situ* hybridization of a radiolabeled oligonucleotide primer to fixed cells (left), the production of cDNA *in situ* using IST (middle), and the removal of the transcripts from the section and analysis on a sequencing gel (right). The gel schematic shows the type of autoradiographic intensity shifts that are seen when the translation of an mRNA is decreased (B) or increased (C) relative to control (A).

do not correspond to the same bands isolated from the mRNA in a tissue section (Fig. 6). It is apparent that the specific banding pattern observed for POMC results from an interaction of POMC mRNA *in situ* with the tissue section.

We investigated this banding phenomenon in the mouse AtT-20 pituitary tumor-derived cell line. AtT-20 cells produce POMC mRNA and protein in a regulated manner similar to the corticotroph of the anterior pituitary.[10] This cell line was selected because of the ability to grow large numbers of cells, which provides a constant and consistent source of POMC-producing tissue. For this procedure, AtT-20 cells are plated at an initial density of 1×10^5 cells/well onto sterile poly(L-lysine)-coated glass circular coverslips in 24-well plates. The glass coverslips are first cleaned for 3 min in 0.2 *N* HCl, followed by 3 min in distilled water and 3 min in acetone, and air dried overnight. The coverslips are then coated in poly(L-lysine) (10 min in 100 mg/500 ml), rinsed in distilled water for 20 sec, and then dried at room temperature for 2 days. The coverslips are rinsed in chloroform in the hood to sterilize them. After 48 hr in culture, the cells are briefly washed twice in phosphate-buffered saline (PBS) at 4°, and then fixed directly in their wells for 20 min in 4% neutral-buffered paraformaldehyde. The cells are then washed briefly in 2× SSC and stored in their wells for future use in 70% ethanol at 4°. When ready for use, the coverslips should be removed from the wells and allowed to dry. They can then be affixed to microscope slides with rubber cement, which not only facilitates handling but also is useful to form a well around the cells. Approximately 100 μl of hybridization solution (4× SSC, 40% formamide,

[10] V. Hook, S. Helsler, S. Sabol, and J. Axelrod, *BBRC* **106,** 1364 (1982).

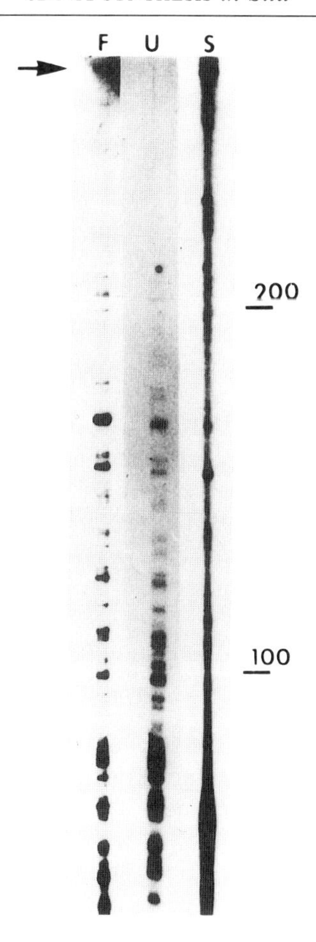

FIG. 6. *In situ* transcription-derived POMC cDNA from rat pituitaries. The POMC IST-cDNA banding patterns generated from fixed (F) and unfixed (U) rat pituitary sections are shown. These banding patterns are identical. Additionally, the banding pattern generated from rat pituitary RNA in solution (S) is shown. While there are bands that appear in lane S they are distinct from those that appear in lanes F and U. Arrow indicates the top of the gel. Numbers on the right-hand side indicate the range of size, in nucleotides, of the cDNA bands.

30 ng of primer/100 μl of hybridization solution) is needed to cover the coverslips. If the primer is labeled with a single ^{32}P moiety using the kinase reaction and no radioactivity is incorporated into the IST cDNA, then the observed intensity of the bands is reflective of the abundance of the band after the IST reaction. The kinase reaction is performed in a 10-μl final volume with 100 ng of primer, 300 μCi of [γ-^{32}P]ATP, 1 μl of 10 $\times$ T4 kinase buffer, and 0.5 μl of T4 polynucleotide kinase. The reaction is

incubated at 37° for 60 min followed by phenol–chloroform extraction and three ethanol precipitations. These reaction conditions usually result in specific activities of between 8×10^8 and 2×10^9. After hybridization, the sections are washed for 1 hr in 2× SSC followed by 5 hr in 0.5× SSC at room temperature. The sections are then washed for 10 min in IST reaction buffer to exchange the SSC salts with IST buffer salts. The IST reaction is performed with dNTPs at a concentration of 250 μM for 90 min at 37°. The sections are washed briefly (15 min in 0.5× SSC) and then air dried. X-Ray film autoradiography can be performed to confirm that hybridization and IST occurred. An additional advantage of doing film autoradiography is that the anatomical localization of the IST reaction can be confirmed and may help in directing the removal of the IST derived cDNAs. The IST-derived cDNA transcripts can be removed as described previously. The cDNA is phenol–chloroform extracted and ethanol precipitated. For electrophoretic analysis the cDNA from three coverslips is resuspended in 4 μl of 3% (w/v) bromphenol blue, 3% (w/v) xylene cyanol, 5 m*M* ethylenediaminetetraacetic acid (EDTA), and 95% formamide. The cDNA is then electrophoresed on a 6% (w/v) acrylamide gel containing 7 *M* urea (data not shown).

Interestingly, the autoradiographic intensities of the bands in the banding pattern can be altered by pharmacological manipulation. The intensity of the bands corresponding to larger DNA fragments is increased relative to that of bands corresponding to smaller fragments when the cells were treated with NaF. The relative intensities of the bands are distinct from those generated from untreated cells. Because the specific activity of each band is the same (each band is labeled once at the 5′ end of the oligonucleotide primer), the presence of differing intensities for the cDNA bands indicates that there are different molar amounts of cDNAs at each of these bands. NaF, among other things, is reported to inhibit ribosome binding to mRNA (initiation). We tested the ability of NaF to alter the distribution of mRNA in polysomes versus monosomes using sucrose gradient purification of polysomes and found that NaF decreases the number of POMC mRNAs in the polysome fractions of the gradient. This result confirms the hypothesis that a redistribution of mRNA from the highly translated polysome fraction to the low translational monosome fraction is induced by NaF. The parallel to the IST banding pattern is significant, in that if translation of POMC mRNA is decreased, fewer ribosomes would be expected on the mRNA. One interpretation of this result is that the shift in intensity of the IST banding pattern may result from the inability of the reverse transcriptase, which is moving along the mRNA molecule, to progress further down the mRNA before encountering any steric hindrance resulting from binding of multiple ribosomes to the mRNA molecule. The converse of this argument would be that with more ribosomes on a mRNA, i.e., more translation, the ratio of the intensities of the high molecular weight bands to the intensities of the lower molecular weight bands would be decreased.

This hypothesis was tested by adding known modulators of POMC protein production to AtT-20 cells. When 8-bromo-cAMP was given to the cells for 1 hr to stimulate protein synthesis, the ratio of high molecular weight to low molecular weight IST bands was low, while 1 hr of dexamethasone treatment, which decreases POMC protein levels, resulted in reversal of the ratio of the banding pattern. The time course of these treatments was such that no measurable changes in POMC mRNA levels could be detected. These results represent acute responses of the expression of POMC to modulators of POMC production.

Additionally, the differing intensities of the cDNA bands is probably reflecting characteristics of the structure of the mRNA within the cell. With further analysis of the banding pattern and changes that are produced by various modulators, we hope to be able to rigorously correlate IST banding patterns with the cell biology of mRNA interaction with cellular constituents as well as with its translational state.

In Situ Transcription in Single Live Cells

Dissociated Neuronal Cells

The diversity and complexity of mRNAs found in either tissue sections or cells in culture have provided many initial insights into the regulation of gene expression at the transcriptional level. However, one difficult task in molecular biology has been elucidating the diversity and complexity of gene expression within a single cell. Analysis of the total of gene expression in single cells requires that the low level of endogenous mRNA be "amplified" to easily manipulated amounts. Current technology such as the polymerase chain reaction (PCR), while providing a means for amplifying the endogenous signal, does have significant drawbacks when applied to populations of RNAs in single cells. These drawbacks consist of base pair copying errors occurring approximately once every 500 bases and the inability of the enzyme *Taq* DNA polymerase to amplify different lengths of cDNA transcripts linearly with the same efficiency. As a result of this latter observation, the amplified cDNA population is often biased to a population of shorter cDNA transcripts. With these technical obstacles in PCR technology, the use of *in situ* transcription and amplified antisense RNA technologies is more suited to amplifying the mRNA population from tissue sections (as previously described) or single cells.

In situ transcription can be performed in single live neurons.[11] The biggest obstacle to performing this reaction initially was the delivery of the IST reagents into the single cell. This hurdle was overcome by adopting the tools of electrophysiology. Using a whole-cell patch electrode, IST of the cellular RNA within a live cell can be accomplished by perfusing the oligo(dT)-T7 primer, dNTPs, and reverse transcriptase from the patch electrode into the cell. The concentrations of these varied components are as follow: 1 μl oligo(dT)-T7 primer (100 ng/μl) [24-nucleotide (nt) poly(T) sequence and 33-nt sequence containing the T7 RNA promoter), 1 μl each of dNTPs (10 mM) (dATP, dCTP, dGTP, and TTP), 2 μl 10× amplification buffer (1× = 154 mM KCl, 6 mM NaCl, 5 mM $MgCl_2$, 0.1 μM $CaCl_2$, 10 mM HEPES, pH 8.3), 12 μl distilled H_2O, and 1 μl reverse transcriptase (20 units/μl).

The perfusion takes approximately 10 min as determined by monitoring the diffusion of fluorescently tagged molecules from the patch electrode throughout the cell. The viability of the cell is also monitored by measuring the resting membrane potential of the cell using the whole-cell patch recording configuration with the IST buffer in place of a standard recording solution. The most variable aspect of this procedure is in the cellular response to the osmolality of the amplification solution. If cell viability is compromised, the osmolality of the solution as determined by the concentration of primer, dNTPs, and amplification buffer can be reduced at least fivefold below that listed by dilution of the amplification mix. Following perfusion, the contents of the single cell are aspirated into the pipette and incubated at 37° for 60 min. Following incubation, the pipette contents are ejected from the pipette and diluted into 50 μl of TE (10 mM Tris, pH 8.0, 1 mM EDTA). One microgram of tRNA is added as carrier, and the sample is phenol–chloroform extracted followed by ethanol precipitation of the aqueous phase. The cDNA is pelleted, air dried, and resuspended in 20 μl of distilled H_2O.

Second-strand synthesis is accomplished using DNA hairpinning to produce a primer for DNA synthesis as described in *In Situ* Transcription (above). Specifically, 10 μl of the first-strand DNA–RNA hybrid, 25 μl of 2× second-strand buffer [800 mM Tris, pH 7.6, 1.0 M KCl, 50 mM $MgCl_2$, 100 mM $(NH_4)_2SO_4$, 50.0 μg/ml bovine serum albumin (BSA)], 2 μl each of 10 mMdNTPs (dATP, dCTP, dGTP, and TTP), 5 μl distilled H_2O, and 1 μl each of Klenow and T4 DNA polymerase are combined. This reaction is performed at 37° for 1 hr. As with the first-strand reaction, the newly synthesized double-stranded cDNA is phenol–chloroform extracted, etha-

[11] R. N. VanGelder, M. E. vonZastrow, A. Yool, W. C. Dement, J. D. Barchas, and J. H. Eberwine, *Proc. Natl. Acad. Sci. U.S.A.* **87,** 1663 (1990).

nol precipitated twice, and again resuspended in 20 μl of distilled water after being air dried.

The DNA is next treated with S1 nuclease for 5 min at 37° to eliminate the hairpin formed in the second-strand synthesis reaction. The S1 nuclease reaction is performed by mixing 10 μl second-strand DNA, 2 μl 10× S1 nuclease buffer (1× = 30 m*M* sodium acetate, pH 4.6, 50 m*M* NaCl, 1 m*M* zinc acetate, 7 μl distilled H_2O, and 1 μl S1 nuclease (1 unit/μl). After S1 nuclease treatment the DNA is phenol–chloroform extracted, ethanol precipitated, and resuspended in 10 μl of distilled H_2O after air drying.

The next step in the procedure is for the cDNA to be drop dialyzed on a Millipore membrane (0.025-μm pore size) against distilled H_2O for 4 hr. This step removes the free deoxynucleotide triphosphates remaining from the prior first- and second-strand reactions. If time is of the essence, this step can be accomplished by instead performing two ethanol precipitations of the cDNA. With each ethanol precipitation, approximately 90% of the free triphosphates will be washed away. Of course, with each ethanol precipitation a small amount of cDNA will also be lost. With this in mind, because the quantity of starting material is only between 0.1 and 1.0 pg, any loss of the dsDNA may amount to losing a significant proportion of the population and may result in a less accurate representation of the original population. For this reason, the preferred method of free triphosphate elimination is the use of drop dialysis.

With the completion of this step, the dsDNA has been constructed to contain the T7 RNA polymerase promoter such that antisense RNA will be made when T7 RNA polymerase is added to initiate RNA synthesis. While it is convenient to use the Promega Riboprobe Gemini System II buffer kit, all the solutions and buffers can be made up independently. It is crucial that all solutions and buffers be RNase free. RNA amplification is performed by adding the following reagents: 4 μl 5× buffer (200 m*M* Tris-HCl, pH 7.9, 30 m*M* $MgCl_2$, 10 m*M* spermidine, 50 m*M* NaCl), 4 μl of double-stranded cDNA to be amplified, 1 μl of 100 m*M* DTT, 1 μl each of 10 m*M* stocks of NTPs, 1.5 μl distilled H_2O, 2 μl [^{32}P]CTP (10 mCi/μl), 0.5 μl RNasin (50 units/μl) and 1 μl T7 RNA polymerase (2000 units/μl). The aRNA amplification is carried out at 37° for 2–4 hr. To determine molar amplification, 0.5 μl of the reaction mix is spotted on 1-mm Whatman paper immediately after the addition of the T7 RNA polymerase and a second time after the reaction is complete. The TCA (CCl_3COOH)-precipitable radioactivity will not only give indications of how well the reaction worked but also the molar amplification. After the reaction is finished, the product is phenol–chloroform extracted and ethanol precipitated. The pellet is resuspended in 20 μl of distilled H_2O and is ready for use.

While the molar amplification of the cDNA often varies due to the reaction conditions, up to 2000-fold amplification can be routinely achieved. The aRNA population synthesized using this procedure is proportional in abundance to the initial mRNA population. This has been determined by comparing the relative abundance of several specific mRNAs by single-cell expression profiling with the abundance of multiple cells using the standard Northern blotting procedure.

First, the amplified RNA population is now antisense to the original mRNA population.[11] In such form, it can be used directly as a probe either for screening blots or libraries. Second, the population of mRNAs is present in the same proportions as in the original population. There appears to be little skewing of the amplification process. Additionally, the aRNA population size range is the same as the cDNA and not biased toward shorter sequences as is true for the PCR.

In spite of molar amplification up to 2000-fold, the amplification of the mRNA from the single cell still represents between only 0.2 and 2 ng of RNA. While this limited amount of material may be adequate for screening blots and libraries, a second round of amplification is often necessary to construct libraries from single cells and to examine low-abundance messages that may not exhibit significant hybridization with the 2000-fold amplified RNA. For this purpose a second round of cDNA synthesis is performed (see Amplified Antisense RNA Amplification, above). The aRNA resulting from two rounds of amplification of the original material from a single cell results in sufficient material to perform expression profiling, screening, and construction of cDNA libraries.

Single Cells within Thick Tissue Slices

The study of gene expression in individual cells may also be performed in slice preparations. The major advantage of this approach is that the cell under investigation remains in an immediate environment that has not been experimentally perturbed. This means, in studies of the nervous system, that neurons maintain many of their normal synaptic connections while the experiment is being performed. The application of patch clamp technology to tissue sections of varying thickness[12] makes this approach feasible. Preliminary experiments in our laboratory have included studies of the hippocampus isolated from newborn rats.

After decapitation and gross dissection of the brain on ice (rat pups ranging in age from 5 to 28 days), sections of 100 to 200 μm in thickness were cut on a Vibratome set at maximum vibration. Several slices are

[12] F. A. Edwards, A. Konnerth, B. Sakmann, and T. Takahashi, *Pfluegers Arch.* **414,** 600 (1989).

preincubated at 37° in physiological saline solution (125 mM NaCl, 2.5 mM KCl, 2 mM $CaCl_2$, 1 mM $MgCl_2$, 25 mM glucose, 26 mM $NaHCO_3$, and 1.25 mM NaH_2PO_4) for 1–4 hr. An individual slice is mounted onto a modified microscope stage that includes a water immersion lens and differential phase-contrast optics. A broken micropipette is positioned directly above pyramidal cells (any cells that are recognized by morphological criteria may be chosen). The superficial surface above the cell is partially removed via gentle suction and blowing. This cleaning step ensures the development of a tight seal with the patch clamp electrode. An RNase-free microelectrode is next used in the formation of a whole-cell clamp. The interior of the microelectrode contains the necessary reagents for first-hand cDNA synthesis. The contents of this microelectrode are processed in a manner identical to that described previously for dissociated cells in culture (see Dissociated Neuronal Cells, above). The average size of amplified aRNA obtained from a hippocampal pyramidal neuron in the CA1 region in this experiment was 800 bases, with the highest size visible by autoradiography being approximately 3 kb. In expression profiling this aRNA hybridized to cDNAs for the muscarinic and γ-aminobutyric acid (GABA)-A receptors, which were immobilized on nitrocellulose paper.

Summary

In situ transcription is the synthesis of cDNA within cells. This chapter has illustrated some of the applications of IST to the study of gene expression in complex cell environments. While the importance of transcription in modulating cellular activity has been long appreciated, the role of translational control mechanisms in regulating central nervous system functioning is just beginning to be recognized. Previous limitations in the availability of tissue have made it difficult to construct cDNA libraries from defined cell populations, to examine translational control, and to quantitate differences in the amount of mRNA for many distinct mRNAs in the same sample. *In situ* transcription facilitates all of these procedures, making it possible to characterize aspects of gene regulation that were previously difficult. Indeed, taken to its furthest extreme it is now possible to characterize gene expression in single live cells. This level of analysis allows basic questions, such as How different morphologically identical cells are at the level of gene expression, and How synaptic connectivity and glial interactions influence gene expression in single cells, to be experimentally approached. The ability to characterize gene expression in small amounts of tissue and single cells is critical to gaining an understanding of the contribution of specific cell types to the physiology of the central nervous system.

Acknowledgments

The critical reading of this manuscript by Hae Young Kong and Jennifer Phillips was greatly appreciated. The retinoic acid receptor cDNA clones were provided by P. Chambon, HOX7 by B. Robert, HSP clones by L. Moran, calcium channel by T. Snutch, and potassium channel by L. Kaczmarek. Larry Tecott helped in the preparation of Fig. 6. These studies were supported by Grant ES04326 to R.F. and BRSG05415, Mahoney Fellowship and AG09900 to J.H.E.

[10] Specific Amplification of Complementary DNA from Targeted Members of Multigene Families

By Z. CAI, J. K. PULLEN, R. M. HORTON, and L. R. PEASE

Introduction

The polymerase chain reaction (PCR) provides convenient strategies using cDNA to study the structural and functional properties of proteins. However, the coexistence of transcripts in many cell types from structurally related members of multigene families complicates the identification and analysis of specific mRNA species. Polymerase chain reaction amplification artifacts have been reported when more than one homologous transcript serves as the template for the PCR primers.[1-3] Typical among these artifacts are recombinants in which sequences from transcripts derived from different loci have been scrambled. The frequency of such recombination has been reported as 8 and 5%.[1,2] Excision repair of heteroduplexes during cloning could lead to higher frequency of recombination (25%) in cloned PCR fragments.[3]

Two strategies have been employed to overcome this problem.[1,4] The first is to sequence many clones from independent amplification reactions to establish the consensus sequence of each cDNA. Typically, nonrecombinant clones represent the majority of sequences identified and specific recombinant motifs tend to be unique, although secondary structures in the amplified templates can lead to the independent generation of identical mutants. The second approach has been to identify sequences in the mRNA, preferably in the 5′ and 3′ untranslated regions of the transcript, that provide locus specificity. This approach can yield highly specific

[1] P. D. Ennis, J. Zemmour, R. D. Salter, and P. Parham, *Proc. Natl. Acad. Sci. U.S.A.* **87,** 2833 (1990).

[2] A. Meyerhans, J.-P. Vartanian, and S. Wain-Hobson, *Nucleic Acids Res.* **18,** 1687 (1990).

[3] R. Jansen and F. D. Ledley, *Nucleic Acids Res.* **18,** 5153 (1990).

[4] Z. Cai and L. R. Pease, *Immunogenetics* **32,** 456 (1990)

amplification products with no detectible recombination artifacts.[4,5] Only a few clones (usually three) need be characterized to determine the structure of the mRNA. Here we describe how a locus-specific strategy was employed in the analysis of class I alleles of the mouse *H-2* complex.

Thirty to 40 members of the class I multigene family are located in five regions (*K, D/L, Qa, Tla,* and *Hmt*) within the *H-2* complex.[6] *H-2* haplotypes differ with respect to the number of genes in each of the regions. All the known haplotypes encode a single *K* gene, while there is variability found in the number of *D/L* genes.[6] Several haplotypes encode a single *D/L* gene while others encode two or three.

Our analysis of *H-2* cDNA illustrates situations where primers were successfully defined to amplify unique members of the multigene family, and also illustrates situations where more than one transcript was amplified by the same primers within a single PCR. In cases where only one gene was targeted, a single sequence was identified in a straightforward manner.[3-5] In cases where more than one homologous transcript was targeted, consensus sequences for each of the transcripts could be discerned along with a number of unique recombinant clones.[7] These observations have significant implications for developing PCR strategies for probing the eukaryotic genome for sequences contained within multigene families.

Approach

The identification of locus-specific sequences within multigene families is a critical step in establishing an effective PCR strategy for the analysis of alleles from targeted loci. We have taken advantage of deletions, insertions, and localized regions of sequence diversity for this purpose. Oligonucleotides were designed containing priming sequences that span the junctions of insertions or deletions or that have their 3′ ends nested in a region of nonhomology. If it is not possible to design primers that have absolute specificity for a particular locus within a highly homologous family, partially restricting the specificity of primers at each end of the fragments targeted for amplification can lead to a specific product.

In the following example, our goal was to develop a specific cloning strategy to characterize the cDNA derived from the *K*, *D*, and *L* loci of the major histocompatibility complex of a variety of inbred mouse lines. Complicating the problem further is the fact that these 3 loci are a subset

[5] R. M. Horton, W. H. Hildebrand, J. M. Martinko, and L. R. Pease, *J. Immunol.* **145,** 1782 (1990).
[6] L. R. Pease, R. M. Horton, J. K. Pullen, and Z. Cai, *Crit. Rev. Immunol.* **11,** 1 (1991).
[7] Z. Cai and L. R. Pease, *J. Exp. Med.* **175,** 583 (1992).

of a larger, highly homologous family of approximately 30 genes. Many of the family members are transcribed and are represented in the RNA isolated from spleen cells. We have shown previously that the 5′ coding sequences of the mRNA from *K*, *D*, and *L* alleles are not distinguishable from each other by the presence of locus-specific sequence motifs.[6] The 3′ coding sequences of these genes are highly conserved and did not present good opportunities for designing locus-specific primers.

Locus specificity was achieved by taking advantage of localized sequence divergence in the 5′ untranslated region of the mRNA and the presence of a specific deletion within a B2 SINE repetitive element that is found specifically inserted into the 3′ untranslated region of messages encoded by the *D* and *L* loci.[8]

K sequences differ from *D* and *L* sequences by 10–12 substitutions in the 21 bases of the 5′ untranslated region, immediately 5′ of the ATG translational initiation site (Table I). Oligonucleotides representing both the *K* and *D*/*L* motifs were synthesized for use as the 5′ PCR primers (Table II).

The 3′ PCR primers were synthesized taking advantage of an insertion of a B2 SINE repetitive element at the 3′ untranslated region of *D*/*L* genes (Fig. 1). The *D*/*L* gene-associated B2 SINE sequences are absent from the *K* genes and differ from other B2 SINE sequences in the mouse genome by an 11-base pair (bp) deletion (Fig. 1). Therefore, an oligonucleotide sequence bridging the 11-bp deletion at the B2 SINE sequence provides strong specificity for *D*/*L* genes (Table II). Although *D*/*L* and *K* genes share 3′ noncoding sequences (NC2) that are downstream from the B2 SINE insertion, the presence of a polyadenylation signal (AATAAA) within the *D*/*L*-associated B2 SINE sequence leads to the posttranscriptional cleavage of downstream sequences of *D*/*L* gene transcripts on polyadenylation. Therefore a primer specific for sequences in the 3′ end of the untranslated region of the class I genes was used as a "*K*-specific" 3′ PCR primer for cDNA amplification (Table II). One or more restriction enzyme sites were included on the 5′ ends of the primers, allowing the amplified PCR products to be cloned routinely into pUC 18.

We have used these primers to amplify single-stranded cDNA from several different major histocompatibility (MHC) genotypes. Seven *K* and seven *D*/*L* genes were isolated using this approach. In the cases where only one targeted gene was expressed, a single sequence was identified in a straightforward manner and observed PCR mutations were rare (1/2000 bp). In one case where multiple *D*/*L* sequences were simultaneously ampli-

[8] M. Kress, Y. Barra, J. G. Seidman, G. Khoury, and G. Jay, *Science* **226,** 974 (1984).

TABLE I
SEQUENCES OF 5′ UNTRANSLATED REGION OF CLASS I GENES[a]

	* *
D consensus	5'-ACTCAGACACCCGGGATCCCAG**ATG**GGGGCG**ATG**-3'
D^b	----------------------------------
D^d	---------------------
D^k	---TT------------------------------
D^p	-----------------------------------
D^q	------------
L^d	-----------------------------------
L^q	------------
K^b	--------AGT-GC-A---G-C--CA--T------
K^d	--------AGT-GCTA---G-C--CCA-T------
K^k	--------AGT-TC-A---G-C--CC--T------
K^q	--------AGT-TC-A---G-C--C---T------
KI^k	--------TG---C-----------A--TTG----
$D2^d$	-------------T---------------------
Q4	-G-----TG---A-A---------G---A--A---
27.1	-G----T-GA----A------CC-G---A--A---
Q10	---------G---C---------------------
TLa 17.3A	-AGACTTT-TA-AACT--GG-TC-CCCTAAC---
$PH2^d$ 37	TTCAG-TTC-T-ACAGA-----GGA-T-AG----
Thy 19.4	-ACAGAGACAGACTCGCT---A--CCA-AGA---

[a] Specific primers were devised from sequences between the positions marked by asterisks.

fied by our primers, three independent sequences and two unique recombinants between the three sequences were identified. The observed recombination rate was about 15% (2 recombinant clones in 13 clones sequenced). Successful targeting of specific cDNA for amplification by selecting locus-specific primers significantly reduced the complexity of the analysis of the *H-2* gene sequences.

TABLE II
POLYMERASE CHAIN REACTION PRIMERS USED TO AMPLIFY AND CLONE *K*, *D*, AND *L* ALLELES[a]

Primer	Sequence
K	
5′	5′-GGATCCTGCAGAATTCAGAAGTCGCGAATCG-3′
3′	5′-CTGCAGAATTCTAGAAGCTTAACAAGAAATCAGCCC-3′
D/*L*	
5′	5′-AGACACCCGGGATCCCAGATG-3′
3′	5′-CGCGGATCCGAAGAGCAGTCAGCGCTGAG-3′

[a] Restriction enzyme recognition sites useful for cloning amplified fragments are underlined.

Procedure

RNA Isolation

Total cellular RNA was prepared from spleen cells using a guanidine isothiocyanate extraction method followed by cesium chloride purification.[9] We have found that the quality of RNA is critical in determining the success of cDNA synthesis and PCR amplification. Many isolation procedures can be used to obtain suitable RNA. The procedure we have used is as follows.

1. One or two mouse spleens are directly placed into a polypropylene tube (17 × 100 mm) with 2 ml phosphate-buffered saline (PBS), and then dispersed with a tissue grinder (cultured cells can be used directly). The cell pellet is lysed with 8 ml GIT buffer (4 *M* guanidine isothiocyanate, 0.025 *M* sodium acetate, 0.1 *M* 2-mercaptoethanol) and the mixture is passed through a 25-gauge needle to shear the DNA until there is no visible viscosity. Alternatively, spleen or other tissues can be snap frozen in liquid nitrogen and ground into a powder with mortar and pestle. The powder is then lysed with GIT buffer and passed through the 25-gauge needle.

2. Spin the lysate for 20 min at 10,000 *g* at 20° in a midspeed centrifuge. Carefully layer the supernatant from the lysate onto 4 ml of 5.7 *M* CsCl, 0.025 *M* sodium acetate in a 14 × 89 mm polyallomer tube or 16 × 76 mm polyallomer Quick-Seal tube (Beckman, Palo Alto, CA). Ultracentrifuge 22 hr at 80,000 *g* at 20° in a swinging bucket rotor (SW41 Beckman rotor

[9] J. M. Chirgwin, A. E. Przybyla, R. J. MacDonald, and W. J. Rutter, *Biochemistry* **18,** 5294 (1979).

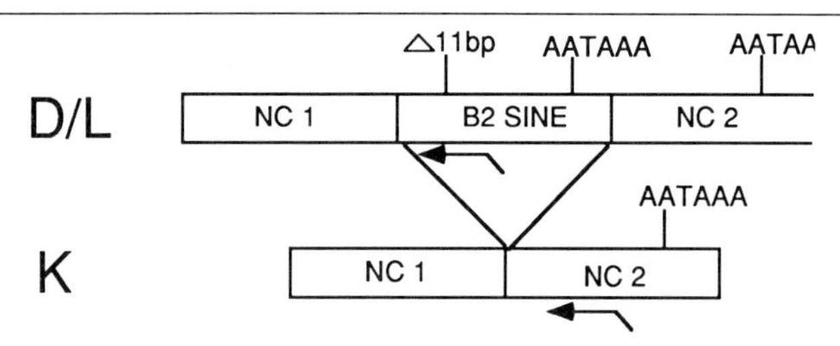

FIG. 1. Location of 3′ locus-specific primers. The 3′ untranslated regions of the *K* and *D/L* genes of the mouse *H-2* complex are shown. The *D/L* genes have a repetitive element (B2 SINE) inserted into this region; other class I genes do not. An 11-bp deletion distinguishes this B2 SINE sequence from other members of the family in the mouse genome. The positions of the locus-specific oligonucleotides are indicated. The nonhomologous tails of the oligonucleotides are composed of restriction endonuclease recognition sites used for cloning the amplified fragments.

at 25,000 rpm) or 16 hr at 142,000 *g* in a fixed-angle rotor (75TI Beckman rotor at 40,000 rpm), pelleting the RNA.

3. Aspirate most of the supernatant from the bottom of the tube, leaving about 1 ml. Pour off the remaining supernatant by rapidly inverting the tube. Blot the end of the tube and dry with a sterile cotton-tipped applicator. With the tube inverted, cut off the top with a razor blade, approximately 10 mm from the bottom. This is done to avoid RNase contamination from the side of the tube.

4. Dissolve the RNA pellet in 200 μl of 0.3 *M* sodium acetate. Transfer it to a 1.5-ml microcentrifuge tube. Rinse the cup once with an additional 100 μl 0.3 *M* sodium acetate. Precipitate the RNA with 700 μl of 95% ethanol overnight at $-70°$ or for 2 hr in a dry ice/ethanol bath.

5. Spin the precipitated RNA sample in a microcentrifuge for 15 min at 4° at maximum speed ($\sim$12,000 *g*). Aspirate the supernatant and wash the pellet once with 70%, ice-cold ethanol. Dry the pellet in a Speed-Vac for 5–10 min. Do not overdry. Resuspend the pellet in 100 to 200 μl RNase-free deionized H_2O, and quantify at OD_{260}. Store 20-μg aliquots of the RNA under ethanol at $-70°$.

RNA degradation and contamination of genomic DNA have been the two common problems associated with failure at later steps. If the RNA has been prepared correctly, the major rRNA species will be clearly visible on electrophoresis in a 1.2% (v/v) formaldehyde agarose gel and staining with ethidium bromide. If the integrity of these bands is not obvious, our recommendation is that RNA be reisolated before proceeding. Each time we have ignored this criterion, subsequent steps have failed. Conversely, when the RNA was intact at this step, we have routinely isolated the targeted cDNA successfully.

cDNA Synthesis

The cDNAs were generated by extension of a poly(T) primer using reverse transcriptase and 20 μg of total cellular RNA as a template. All the reagents used in the cDNA synthesis step were provided in a cDNA-synthesizing system (Amersham, Arlington Heights, IL).

1. Total cellular RNA (20 μg) is ethanol precipitated. The pellet is washed twice with 70% ethanol. Dry the pellet in a Speed-Vac. Do not overdry.
2. Dissolve the RNA pellet with 14 μl of deionized, RNAse-free H_2O (the volume of water can be adjusted according to the concentration of reagents in the kit). Add 5 μl of 5× first-strand synthesis buffer, 1 μl of sodium pyrophosphate solution, 1 μl of human placental ribonuclease inhibitor (HPRI), 2 μl of dNTP mix, 1 μl of oligo(dT) primer (random primer can also be used), and 1 μl (20 units) of reverse transcriptase to the RNA solution. The final volume of reaction solution is 25 μl.
3. Mix the reagents and incubate the solution at 42° for 90 min to synthesize the first-strand cDNA. Dilute the cDNA reaction mix with 75 μl deionized H_2O and store it at −20° until needed.

Polymerase Chain Reaction Amplification

Polymerase Chain Reaction Conditions

Standard PCR conditions are employed in 0.6-ml polypropylene tubes (Robbins Scientific, Sunnyvale, CA) and cycling reactions are carried out in a DNA thermal cycler (Perkin-Elmer Cetus, Norwalk, CT) for 25 or 30 cycles.

Denaturation: 1 min at 94°

Annealing: 2 min at 50° (60° has also been used as an annealing temperature to amplify the cDNA of *D/L* genes)

Elongation: 1 to 3 min (determined by an enzyme rate of 1000 base/min) at 72°. For the class I cDNA, 1.5 min is sufficient

Reaction Solution

The PCR reaction is usually carried out in a 100-μl volume, which includes 3 μl of diluted cDNA reaction mix and 2.5 units of *Taq* DNA polymerase (Perkin-Elmer Cetus). The final concentration of each reagent is as follows:

Locus-specific primers, 1 μ*M* each

dNTP mix (ultrapure; Pharmacia, Piscataway, NJ), 200 μ*M*

KCl, 50 mM
Tris-HCl (10 mM), pH 8.3
MgCl, 0.5–2.5 mM (empirically determined)
Deionized H_2O to a final volume of 100 μl

After mixing the reaction solution, 75 μl of light mineral oil (Sigma, St. Louis, MO) is added in each reaction tube to prevent the evaporation during the PCR reaction.

While there are many variables that influence the efficiency and specificity of PCR, we have found Mg^{2+} concentration to be among the most discriminating. Consequently we routinely titrate Mg^{2+} concentration over a range of 0.5 to 2.5 mM in 0.5-mM increments. This can be conveniently accomplished by preparing 10× PCR buffers containing 500 mM KCl, 100 mM Tris-HCl, pH 8.3 with five different concentrations of $MgCl_2$ (5.0 to 25 mM $MgCl_2$). We have found 0.1% (w/v) gelatin, often added to stabilize enzymes in the reaction, to be optional. The optimal Mg^{2+} concentration can vary dramatically among primer–template combinations. The 10 mM $MgCl_2$ buffer system works best in the example discussed here.

Cloning Steps

Following PCR, the reaction mixtures are extracted twice with phenol–chloroform (1 : 1) to remove the mineral oil and proteins. Sodium acetate (pH 5) is added to the aqueous phase, bringing the final concentration to 0.3 M. tRNA (10 μg) is added as carrier, and the mixture is precipitated with 2.5 vol of 95% ethanol. The pelleted nucleic acid is dried, resuspended in deionized H_2O, and digested with appropriate restriction enzymes. The digested products are separated by electrophoresis in 1.2% (w/v) agarose (SeaKem GTG; FMC BioProducts, Rockland, ME). The full-length cDNA amplification products are isolated using glass beads (GeneClean, Bio101, La Jolla, CA). At this point, the isolated fragments are ready to be cloned into the vector of choice (in our case, pUC 18) using standard procedures.

Conclusion

Specific amplification of cDNA from targeted members of a multigene family was achieved by designing oligonucleotide PCR primers that are complementary to localized segments of diversity within the 5′ and 3′ untranslated regions of the encoded mRNAs. In cases where the primers differentiated among all the different species of cDNA, only a single amplification product was recovered. However, in cases where more than one homologous mRNA species was amplified in the same PCR, multiple products were recovered. Among these products were the targeted cDNAs, as well as related recombinant DNA species. The presence of the chimeric amplification products highlights a major limitation of using regions of homology to design PCR primers for the analysis of multigene families. Even under conditions where there are only two or three homologous species of cDNA amplified by a set of primers, the presence of recombinants among the amplification products significantly complicated the analysis.

[11] Long Synthetic Oligonucleotide Probes for Gene Analysis

By HIROO TOYODA

Introduction

Synthetic oligonucleotide probes have been utilized widely in cloning many biologically important genes.[1] Two general approaches have been described: (1) a mixed-probe method[1,2] and (2) a unique long-probe method.[3–5] The mixed-probe method employs a mixture of oligonucleotides [15–20 nucleotides (nt) long] representing all possible coding sequences for a short peptide sequence of a given protein. The radiolabeled probe mixture is used to screen colonies under hybridization conditions stringent enough so that only the perfectly matched oligonucleotide sequence forms a stable duplex with the target. This method has been applied successfully to identify cDNA clones for class I and II major histocompatibility complex (MHC) genes, as well as other biologically important genes.[1,6,7] On the other hand, the unique long-probe method uses a longer (over 30 nt), single oligonucleotide sequence derived either

[1] K. Itakura, J. J. Rossi, and R. B. Wallace, *Annu. Rev. Biochem.* **53,** 323 (1984).

[2] R. B. Wallace, M. J. Johnson, T. Hirose, T. Miyake, E. H. Kawashima, and K. Itakura, *Nucleic Acids Res.* **7,** 879 (1981).

[3] S. Anderson and I. B. Kingston, *Proc. Natl. Acad. Sci. U.S.A.* **80,** 6838 (1983).

[4] M. Jaye, H. de la Salle, F. Schamber, A. Balland, V. Kohli, A. Findeli, P. Tolstoshev, and J.-P. Lecocq, *Nucleic Acids Res.* **11,** 2325 (1983).

[5] R. Lathe, *J. Mol. Biol.* **183,** 1 (1985).

[6] S. V. Suggs, R. B. Wallace, T. Hirose, E. H. Kawashima, and K. Itakura, *Proc. Natl. Acad. Sci. U.S.A.* **78,** 6613 (1981).

[7] Y. Kajimura, H. Toyoda, M. Sato, S. Miyakoshi, S. A. Kaplan, Y. Ike, S. M. Goyert, J. Silver, D. Hawke, J. E. Shively, S. V. Suggs, R. B. Wallace, and K. Itakura, *DNA* **2,** 175 (1983).

METHODS IN ENZYMOLOGY, VOL. 216

from a cloned gene segment or designed from a peptide sequence by following favored codon usage. This unique long-probe approach has been employed to clone various genes, including the insulin receptor and insulin-like growth factor I,[4,8,9] for which the available peptide sequence contains amino acids with degenerate codons. Some general considerations for the use of long synthetic oligonucleotides for gene cloning have been reported in a previous volume of this series.[5,10] We examined several factors influencing hybridization specificity of long synthetic oligonucleotide probes (over 37 nt long) used for gene analysis.[11] We studied HLA class II β genes of the human MHC as a model system, because class II genes comprise a family of closely related genes having a high degree (75%) of nucleotide sequence homology.[12] This chapter describes procedures for using long synthetic oligonucleotide probes to identify individual genes within a family of closely related sequences.

Methods

Genomic DNA Isolation

High molecular weight DNA was isolated from peripheral blood leukocytes (PBLs), Epstein-Barr virus (EBV)-transformed B cell lines, and tissues according to published methods, with minor modifications.[13] Briefly, PBLs isolated from 5 ml of blood collected into a tube containing ethylenediaminetetraacetic acid (EDTA) (purple top), or cells (10 to 15 × 10^6 cells), were resuspended in 400 μl of TNE buffer (10 m*M* Tris-HCl, pH 8.0, 150 m*M* NaCl, 10 m*M* EDTA, pH 8.0) containing 1% (w/v) sodium dodecyl sulfate (SDS) and proteinase K (500 μg/ml; Boehringer Mannheim, Indianapolis, IN), and the reaction solution was incubated at 60° overnight. An equal volume of phenol (buffered with Tris-HCl at pH 8.0) was added and the solution was vortexed, then centrifuged. Then 400 μl of CIA (chloroform–isoamyl alcohol, 24 : 1) was added, and the solution was vortexed, then centrifuged. The aqueous phase was transferred to a new tube and was again extracted with an equal volume of CIA. DNA was

[8] A. Ullrich, C. H. Berman, T. J. Dull, A. Gray, and J. M. Lee, *EMBO J.* **3,** 361 (1984).

[9] A. Ullrich, J. R. Bell, E. Y. Chen, R. Herrera, L. M. Petruzzeli, T. J. Dull, A. Gray, L. Coussens, Y.-C. Liao, M. Tsubokawa, A. Mason, P. H. Seeberg, C. Grunfeld, O. M. Rosen, and J. Ramachandran, *Nature (London)* **313,** 756 (1985).

[10] W. I. Wood, this series, Vol. 152, p. 443.

[11] Y. Kajimura, J. Krull, S. Miyakoshi, K. Itakura, and H. Toyoda, *Gene Anal. Tech. Appl.* **7,** 71 (1990).

[12] R. C. Giles and J. D. Capra, *Adv. Immunol.* **37,** 1 (1985).

[13] H. Toyoda, M. Onohara-Toyoda, J. Krull, C. M. Vadheim, J. C. Bickal, W. J. Riley, N. K. Maclaren, K. Itakura, and J. I. Rotter, *Dis. Markers* **7,** 215 (1989).

precipitated from the aqueous phase with 2 vol of cold 100% ethanol and resuspended in 100 to 500 μl of TE buffer (1 m*M* Tris-HCl, pH 8.0, 0.1 m*M* EDTA, pH 8.0).

For isolation of DNA from tissue, 0.1 to 0.5 g of tissue was homogenized gently in 5 ml of TNE buffer containing 1% (w/v) SDS and proteinase K (500 μg/ml) in a Dounce tissue homogenizer using the B pestle (Wheaton, Millville, NJ). Subsequent treatment was exactly as for cells (described above), except the final resuspension volume was 0.5 to 1.0 ml TE.

Oligonucleotide Synthesis and Purification

Synthetic oligonucleotide probes were synthesized on a Coder 300 (Du Pont, Wilmington, DE) by the phosphoramidite method using the "trityl off" program.[14] Probes were cleaved from the columns and dried under vacuum according to manufacturer's directions.

The dried crude oligonucleotides were purified by polyacrylamide gel electrophoresis[15] with minor modifications. The dried oligonucleotide was resuspended in water (100 to 200 μl) and an aliquot (5 to 10 total OD_{260} units, 1 OD_{260} = 30 μg/ml) was electrophoresed on a 20% (w/v) polyacrylamide gel containing 7 *M* urea and dye markers [Bromophenol blue (BPB) and Xylene cyanol (XC)] in 0.5× TBE (1× TBE = 89 m*M* Tris-HCl, 89 m*M* boric acid, 2.5 m*M* EDTA, pH 8.3) at 300 to 400 V for 2 to 3 hr. Under these conditions, BPB migrates with 10-nt and XC migrates with 27-nt oligomers. Bands visualized by shadowing on a fluorescent silica plate with an ultraviolet (UV) light (254 nm) were excised, placed in 1.5-ml Eppendorf tubes containing 0.5 to 1.0 ml water, and incubated at 50° overnight. Tubes were spun briefly, the supernatant was removed, and the pellet dried and resuspended in TE buffer.

Southern Blot Analysis

Two methods were used for Southern blot analysis: standard Southern blotting, and in-gel Southern blotting.[7,16] It is important to note that occa-

[14] M. H. Caruthers, *in* "Methods of DNA and RNA Sequencing" (S. M. Weisman, ed.), p. 1. Praeger, New York, 1983.

[15] Oligonucleotide can be purified through commercially available columns, such as OPC (Applied Biosystems, Foster City, CA) and NENSORB PREP (New England Nuclear–Du Pont). If these columns are used for purification, oligonucleotide synthesis should be carried out using a "trityl-on" program, so that oligonucleotide will be retained in the hydrophobic column resin.

[16] B. J. Conner, A. A. Reyes, C. Morin, K. Itakura, R. L. Teplitz, and R. B. Wallace, *Proc. Natl. Acad. Sci. U.S.A.* **80,** 278 (1983).

sionally more than 50% of the DNA fragment is lost from the gel during drying with the in-gel method. Standard Southern blotting was carried out as previously described using nylon membrane (MSI magna nylon 66, 0.45 μm; Fisher, Pittsburgh, PA).[13]

Hybridization

Hybridization with synthetic oligonucleotide was carried out as previously described in "Seal-a-meal" bags (Dazey Corp., Industrial Airport, KS)[2,6,7] using a shaker bath. Prehybridization was carried out in 5× SSPE (1× SSPE = 0.18 *M* NaCl, 10 m*M* NaH_2PO_4, 1 m*M* EDTA, pH 7.7), 5× Denhardt's solution [1× Denhardt's = 0.02% (w/v) bovine serum albumin (BSA), 0.02% (w/v) Ficoll, 0.02% (w/v) polyvinylpyrrolidone], 0.1% (w/v) SDS, and 100 μg/ml yeast tRNA (Clontech, Palo Alto, CA) from 4 hr to overnight at 55°. Hybridization with a ^{32}P-labeled synthetic probe (specific activity: 1–5 × 10^8 cpm/μg probe DNA) was in 20 ml hybridization solution containing 100 to 500 ng of labeled probe. Up to three membranes were placed in the same bag.

Probes were end labeled with T4 polynucleotide kinase (Bethesda Research Laboratories, Gaithersburg, MD).[2,6,7] Briefly, probe DNA (100 to 500 ng) in 20 μl reaction buffer [66 m*M* Tris-HCl, pH 7.5, 10 m*M* magnesium acetate, 10 m*M* dithiothreitol (DTT)] was incubated with [γ-^{32}P]ATP (>7000 Ci/mmol; New England Nuclear, Boston MA) plus 10 units of T4 polynucleotide kinase (Bethesda Research Laboratories) at 37° for 30 min. Labeled probe DNA was purified on an NACS-Prepac column (Bethesda Research Laboratories) according to the protocol of the manufacturer. The final volume of the purified labeled probe DNA solution was approximately 200 μl. Hybridization was done at 55° for 12 hr, and then the membrane was washed with 2× SSC (1 × SSC = 150 m*M* NaCl, 15 m*M* Na_3 citrate) for 2 hr at room temperature, followed by washes at various higher stringency conditions as detailed below. Finally, membranes were exposed to Kodak (Rochester, NY) XAR5 film for 5–10 days at −70° with two intensifying screens (Du Pont).

Results

Long synthetic oligonucleotide probes are most useful when genes to be examined are members of a family of closely related genes, and when each gene may exist in several different allelic forms. The best example of this situation may be the MHC gene family, in which genes (alleles) in different subregions possess a high degree of sequence homology; for example, the β chain genes of the *DR* and *DQ* subregions. To design probes specific for each subregion gene, the following factors should be considered: (1) the probe should detect only genes belonging to one subregion, but not genes belonging to other subregions, and (2) the probe should detect all allelic forms of the same subregion gene. Probe length is an important factor; previous studies have shown that an oligonucleotide probe less than 25 nt long will discriminate between a single base-mismatched sequence and a perfectly matched sequence.[1,2,5]

We examined optimum probe length for detecting subregion-specific gene fragments using *DR*-β genes in the human MHC as a model system. The *DR*-β genes show a high degree of sequence homology to the *DQ*-β genes in the *D* region of the human MHC.[12] As shown in Fig. 1, we designed two oligonucleotides, one 37 nt long (37-mer) and one 48 nt long (48-mer), representing the *DR*-β genes. Overall sequence homology in this probe region between the *DR*- and the *DQ*-β genes is around 62% for the 37-mer and 65% for the 47-mer. For the 37-mer probe, the first 7 nucleotides are identical to the *DQ*-β gene sequence, whereas in the 48-mer, the first 15 consecutive nucleotides are identical. Southern blots using genomic DNA isolated from cell lines and digested with *Bam*HI showed that the 48-mer probe cross-hybridized with some of the *DQ*-β gene fragments[11] even under high stringency conditions (2× SSC at 60° for 5 min, twice). This result suggests that if a probe happens to contain a long enough continuous run of nucleotides perfectly matching the sequence of another member of the family, cross-hybridization will occur. Conditions suitable for hybridization using oligonucleotide probes between 30 to 40 nt in length are as follows: (1) the hybridization temperature should be 55° and (2) the membrane should be washed in 6× SSC at 65° for 5 min, twice. For probes longer than 40 nt, the hybridization temperature should be increased to

			Matched Portion								Mismatched Portion					
Residue	78	79	80	81	82	83	84	85	86	87	88	89	90	91	92	93
AA Sequence	Y	C	R	H	N	Y	G	V	V	E	S	F	T	V	Q	R
DNA Sequence (DR-beta)	TAC	TGC	AGA	CAC	AAC	TAC	GGG	GTG	GTG	GAG	AGC	TTC	AGA	GTG	CAG	GGG
		***	***	***	***	***	*	**	* *		**		*	**	***	***
AA Sequence	V	C	R	H	N	Y	Q	L	E	L	R	T	T	L	Q	R
DNA Sequence (DQ-beta)	GTG	TGC	AGA	CAC	AAC	TAC	CAG	TTG	GAG	CTC	CGC	ACG	ACC	TTG	CAG	GGG
DR Probe: 37-mer					---	---	---	---	---	---	---	---	---	---	---	---
48-mer	---	---	---	---	---	---	---	---	---	---	---	---	---	---	---	---

FIG. 1. Comparison of sequence homology between two target genes. Sequence comparison between *DR*- and *DQ*-β genes in the human MHC is shown. Two probes of different length were designed to examine the effect of length on the hybridization specificity. Asterisk (*) indicates matched nucleotides between two target genes.

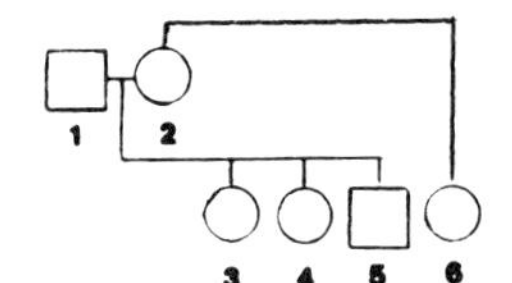

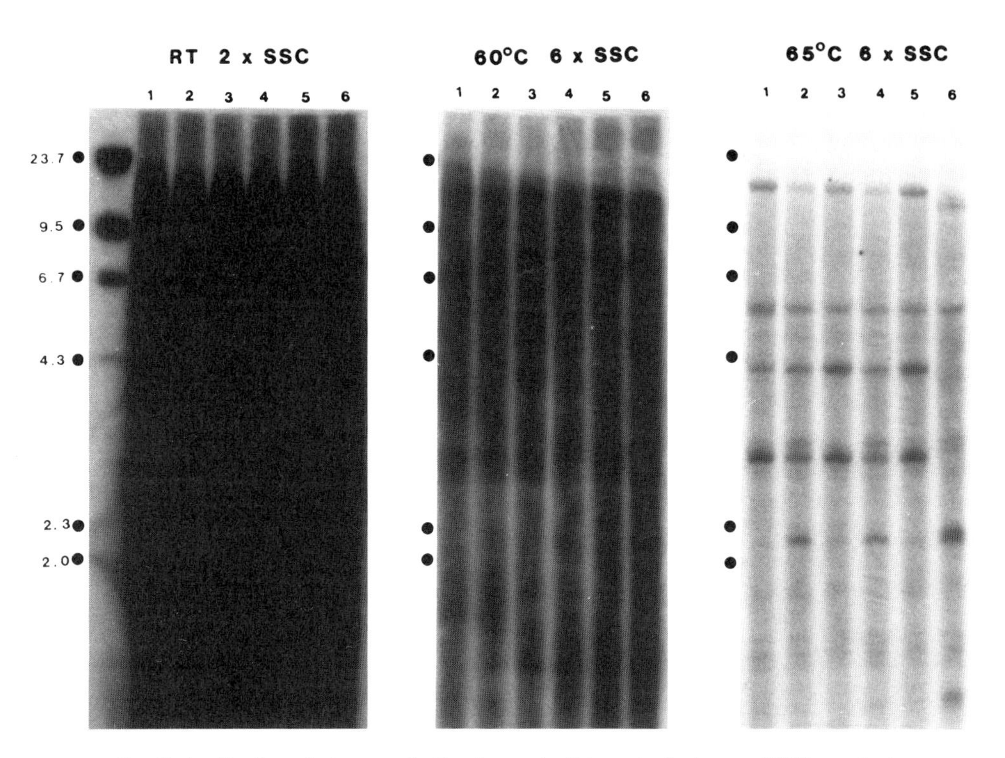

FIG. 2. Application of a long synthetic oligonucleotide probe for human MHC genotyping. The probe specific for the *DR-β* gene (37-mer) was used to genotype the human class II genes of the MHC. It should be noted that the alleles represented by bands present in both parents were transmitted to their offspring. RT, room temperature.

65° and the membrane should be washed in 6× SSC at 65° for 5 min, then in 2× SSC at 65° for 5 min, twice. However, because hybridization and washing conditions depend strongly on probe sequence, it may be necessary to adjust these conditions to obtain the best results.

We employed a long oligonucleotide probe for MHC class II genotyping. DNA isolated from members of a family was blotted and probed to examine the segregation to offspring of each parental *DR-β* genotype. We used the 37-mer probe specific for the *DR-β* genes as described in Fig. 1. As shown in Fig. 2, washing conditions are critical. We observed many fragments after the membrane was washed in 6× SSC at 60° for 5 min, twice. When the washing temperature was increased to 65°, only fragments specific for the *DR*-β genes were detected.

Comments

Long unique synthetic oligonucleotide probes are useful for gene cloning and analysis, but several factors are important in designing the probes and in the process of hybridization. On the basis of the studies described here and from previous studies, the following guidelines are recommended: (1) minimize the secondary structure of the probe, which might influence the stability of the duplex; (2) avoid designing a probe with high G-C content (e.g., over 75%); and (3) use washing conditions that are of high enough stringency to prevent detection of related, but unwanted, sequences. In general, washing in 6× SSC at 65° for 10 min, twice, provides conditions stringent enough when the probe length is around 40 nt. If the probe length is greater than 45 nt, it is preferable to wash in 2× SSC at 65° for 10 min, twice. These conditions have been used successfully for gene analysis in our laboratory. Ullrich *et al.* describe conditions for hybridization in formamide when using long synthetic oligonucleotides for gene cloning.[8] They add formamide (10 to 40%) to the hybridization solution and incubate at 42° in 5× SSC, 5× Denhardt's solution, 50 m*M* sodium phosphate (pH 6.8), 1 m*M* sodium pyrophosphate, 100 μM ATP, and 50 μg/ml sonicated salmon sperm DNA, followed by washes at 37° in 0.2× SSC with 0.1% sodium dodecyl sulfate (SDS).

The thermal stability of DNA duplexes in SSC solution is highly dependent on nucleotide sequence[17] and detailed physicochemical studies regarding duplex stability have been published.[5] If oligonucleotide probes of widely varying dissociation temperature (T_d)[18] are used to screen for related genes, specific stringency conditions must be empirically determined for each probe, a time-consuming and tedious task. An alternative is to use tetramethylammonium chloride [$(CH_3)_4NCl$] in the hybridization solution; this has been reported to eliminate the dependence of T_d on probe sequence. Hybridization conditions then depend only on probe length.[17,19]

In addition to appropriate washing conditions, it is necessary to have uninterrupted runs of nucleotides matching the target sequence in the

[17] W. I. Wood, J. Gitschier, L. Lasky, and R. M. Lawn, *Proc. Natl. Acad. Sci. U.S.A.* **82,** 1585 (1985).

[18] S. V. Suggs, T. Hirose, T. Miyake, E. H. Kawashima, M. J. Johnson, K. Itakura, and R. B. Wallace, *ICN-UCLA Symp. Dev. Biol.* **23,** 683 (1981).

[19] K. A. Jacobs, R. Rudersdorf, S. D. Neill, J. P. Dougherty, E. L. Brown, and E. F. Fritsch, *Nucleic Acids Res.* **16,** 4637 (1988).

middle portion of the probe, although this factor depends on the overall probe length. Probes containing clusters of nucleotides matching the target gene sequence may form more stable duplexes than probes having the same overall percentage of homology but in which the matching nucleotides are dispersed along the probe.[5] The optimum length for an uninterrupted perfect match is around 14 nt for probes less than 45 nt in length. To design probes with greater homology, it is advisable to employ favored codon usage for amino acid sequence, and to allow G-T mismatch rather than A-C or C-T mismatch, because these latter mismatches significantly destabilize the duplex.[20] Some reports of gene cloning illustrate the success of employing longer (over 50 nt) synthetic probes.[21-23] Although long synthetic oligonucleotides can be useful for gene analyses, they are expensive to synthesize, and it takes time to find suitable conditions for their use.

In summary, short oligonucleotide probes have been used in DNA diagnostic studies in various genetic diseases, where it is important to distinguish a 1-bp mismatch between normal and affected individuals.[24] In contrast, long unique synthetic oligonucleotide probes can be useful, both for restriction length polymorphism analysis and for gene cloning, especially where multiple closely related genes must be distinguished.

Acknowledgments

I gratefully acknowledge Dr. J. Prehn for helpful discussions and review of the manuscript. This work was supported by the American Diabetes Association Feasibility Grant, Juvenile Diabetes Foundation International, the Stuart Foundations, and National Institutes of Health Grant GM25658.

[20] S. Ikuta, K. Takagi, R. B. Wallace, and K. Itakura, *Nucleic Acids Res.* **15,** 797 (1987).

[21] N. K. Robakis, N. Ramakrishna, G. Wolfe, and H. M. Wisniewski, *Proc. Natl. Acad. Sci. U.S.A.* **84,** 4190 (1987).

[22] D. Zeng, J. K. Harrison, D. D. D'Angelo, C. M. Barber, A. L. Tucker, Z. Lu, and K. R. Lynch, *Proc. Natl. Acad. Sci. U.S.A.* **87,** 3102 (1990).

[23] P. W. Gray, K. Barrett, D. Chantry, M. Turner, and M. Feldman, *Proc. Natl. Acad. Sci. U.S.A.* **87,** 7380 (1990).

[24] S. H. Orkin, *in* "The Metabolic Basis of Inherited Disease" (C. R. Scriver, A. L. Beaudet, W. S. Sly, and D. Valle, eds.), Vol. 1, p. 165. McGraw-Hill, New York, 1989.

[12] Colorimetric Detection of Polymerase Chain Reaction-Amplified DNA Segments Using DNA-Binding Proteins

By DAVID J. KEMP

Introduction

The polymerase chain reaction (PCR)[1,2] system should provide a useful alternative to conventional procedures in many areas of mass screening such as the detection of pathogens, epidemiology, and human genetic applications, e.g., HLA typing and screening for genetic diseases. However, the commonly used laboratory procedures for detection of the products of PCR reactions are not well suited to mass screening, as they generally require gel electrophoresis. Furthermore, artifactual DNA molecules resulting from such events as dimerization of the primers[3] or misincorporation of primers into irrelevant sequences can readily arise. We have therefore developed an assay system for detecting DNA amplified by PCR that is highly specific, rapid, readily applicable to mass screening, and suitable for any known sequence.[4,5] The amplified DNA assay (ADA) described here was designed for colorimetric detection with equipment already undergoing general use in diagnostic laboratories, but could readily be adapted to more sensitive fluorescence technology. The ADA readily detected human immunodeficiency virus (HIV) sequences in DNA purified from blood samples of acquired immunodeficiency syndrome (AIDS) patients.[5]

Principle of Amplified DNA Assay

The ADA is conceptually and technically based on a standard enzyme-linked immunosorbent assay (ELISA) reaction as performed in many diagnostic laboratories for the detection of antigens by monoclonal anti-

[1] R. K. Saiki, S. Scharf, F. Faloona, K. B. Mullis, G. T. Horn, H. A. Erlich, and N. Arnheim, *Science* **230,** 1350 (1985).
[2] R. K. Saiki, D. H. Gelfand, S. Stoffel, S. J. Scharf, R. Higuchi, G. T. Horn, K. B. Mullis, and H. A. Erlich, *Science* **239,** 487 (1988).
[3] H. A. Erlich (ed.), "PCR Technology." Stockton Press, New York, 1989.
[4] D. J. Kemp, D. B. Smith, S. J. Foote, N. Samaras, and M. G. Peterson, *Proc. Natl. Acad. Sci. U.S.A.* **86,** 2423 (1989).
[5] D. J. Kemp, M. J. Churchill, D. B. Smith, B. A. Biggs, S. J. Foote, M. G. Peterson, N. Samaras, N. J. Deacon, and R. Doherty, *Gene* **94,** 223 (1990).

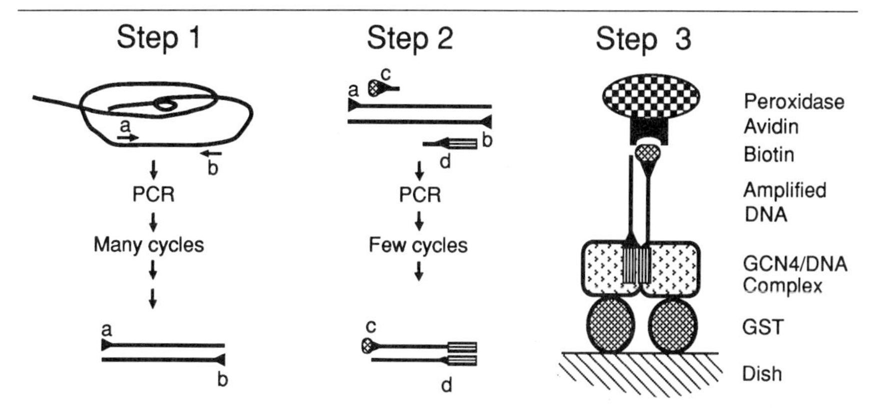

FIG. 1. Schematic diagram of the ADA reaction. Oligonucleotides *a* and *b* are primers for PCR with the target sequence of interest. Oligonucleotides *c* and *d* are nested between *a* and *b*; *c* bears a 5′-biotin, while *d* bears the sequence 5′-GGATGACTCA that can bind (when double stranded) to GCN4. After the PCRs, the amplified DNA bearing these ligands is bound to the wells of a microtiter dish coated with the fusion polypeptide from clone GST-GCN4 3.12.[4] The amplified DNA is detected by binding an avidin peroxidase conjugate to the biotin, followed by reaction with a chromogenic substrate.

bodies. It relies on binding the amplified DNA via a sequence in one oligodeoxynucleotide to a DNA-binding protein such as GCN4 (a regulatory molecule from yeast[6]) coated on the wells of a microtiter dish (Fig. 1). Avidin peroxidase is then bound to biotin at the 5′ end of the other oligonucleotide and detected colorimetrically using the same reagents as for an ELISA.[4,5] The PCRs may be performed in the wells of a microtiter dish and the amplified DNA can then be captured via GCN4 immobilized on an array of 96 prongs or beads attached to the lid of a microtiter dish.[5] Hence in theory it is only necessary to pipette each DNA sample once, and 96 samples can then be handled simultaneously through amplification, capture, and detection.

The DNA-binding protein TyrR from *Escherichia coli* has also been shown to work satisfactorily in this assay.[7] The existence of many other DNA-binding proteins that recognize different sequences could facilitate detection of multiple different sequences after a single multiplex PCR. A similar approach using the *lac* repressor as a DNA-binding protein has been developed independently.[8] An alternative approach has used biotin

[6] A. G. Hinnebusch, *Proc. Natl. Acad. Sci. U.S.A.* **81,** 6442 (1984).

[7] T. Triglia, V. P. Argyropoulos, B. E. Davidson, and D. J. Kemp, *Nucleic Acids Res.* **18,** 1080 (1990).

[8] J. Lundeberg, J. Wahlberg, M. Holmberg, U. Pettersson, and M. Uhlen, *Cell Biol.* **9,** 287 (1990).

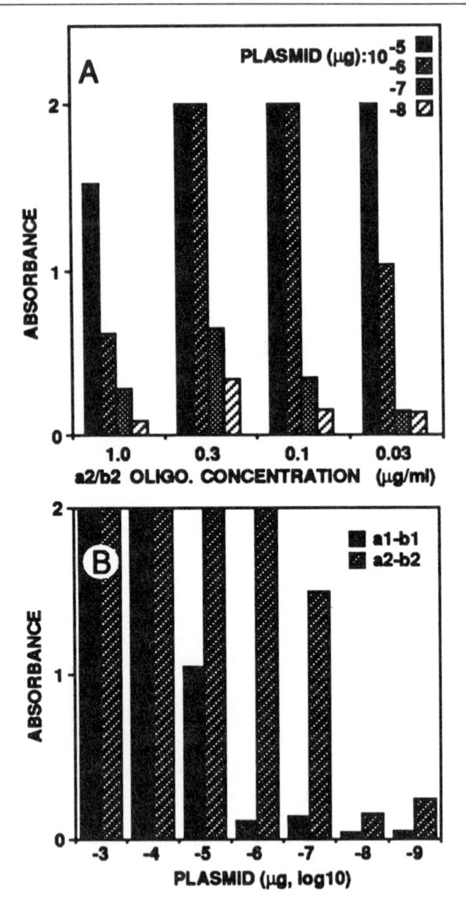

FIG. 2. Sensitivity of ADA reactions with different oligonucleotides. (A) Effect of oligonucleotide concentration. Two-step PCRs on the *gag* gene of HIV were carried out with oligonucleotides *a2* and *b2* (identical concentrations, as indicated at the bottom) and *c* and *d* (5 μg/ml) with 10^{-8} to 10^{-5} μg of HIV plasmid pHXBL2 [J. Sodroski, R. Partarca, C. Rosen, F. Wong-Staal, and W. Haseltine, *Science* **229,** 74 (1985)] as indicated in a 20-ml reaction, cycled 30 times (95°/30 sec, 65°/60 sec) and then 12 times (95°/30 sec, 40°/60 sec, 65°/30 sec). Five microliters of the product was then analyzed in an ADA with a one-step binding reaction. The absorbances of the colorimetric reactions are shown for each of the four different input plasmid DNA concentrations at each *a2*/*b2* concentration. (B) Sensitivity of ADA reactions using *a* and *b* oligonucleotides located different distances from *c* and *d*. The oligonucleotides *a1, a2, b1,* and *b2* were at 0.3 μg/ml, while *c* and *d* were at 5 μg/ml. Five microliters of the product was analyzed in an ADA with a one-step binding reaction. Consensus oligonucleotides corresponding to sequences from the *gag* gene of HIV were selected after aligning seven available sequences from the 1988 HIV database. The oligonucleotides are defined below: The numbers in parentheses refer to the position in the HIV genome and the direction of the arrows indicates whether the sequence has been reversed and complemented.

and digoxygenin as the ligands.[9] A feature of double-stranded DNA-binding proteins as anchors for the DNA is that unincorporated oligonucleotide primers are not recognized because they are single stranded. Unlike other ligands that have been used, the unincorporated primers therefore cannot compete for binding with the PCR product.

Overcoming Problems of Primer Dimers

In many PCRs carried out for a sufficient number of cycles, especially in the absence of target DNA, primer dimers are an inevitable by-product.[3] In assays such as the ADA it is necessary to discriminate between primer dimers and genuine PCR products, as they are a serious potential source of false positives. To achieve this, we have used two successive PCRs.[4,5] The target DNA was first amplified by PCR with oligonucleotide primers designated *a* and *b* (Fig. 2). A second nested set of oligonucleotides bearing the ligands (biotin and the GCN4-binding site; oligonucleotides *c* and *d* in Fig. 2) were then incorporated by a smaller number (3–12) of PCR cycles, insufficient to generate primer dimers. The two successive PCRs can be performed in the one reaction mixture with all four oligonucleotides simultaneously present using the thermal stabilities of oligonucleotides of different lengths to separate the two reactions.[5] The external oligonucleotides are 20 or more nucleotides long while for the internal ones the regions complementary to the template are 15 nucleotides long. The PCR is cycled first at high stringency (annealing at 65°) to incorporate the longer outside oligonucleotides and then at low stringency (annealing at 40°) to incorpo-

[9] D. A. Nickerson, R. Kaiser, S. Lappin, J. Stewart, L. Hood, and U. Landegren, *Proc. Natl. Acad. Sci. U.S.A.* **87,** 8923 (1990).

---

a1 ATGAGAGAACCAAGGGGAAG
(1470 → 1489)

a2 GGGGGACATCAAGCAGCCATGCAAATG
(1362 → 1388)

b1 TTGGTCCTTGTCTTATGTCCAGAATGC
(1656 ← 1630)

b2 ACTCCCTGACATGCTGTCATCATTTCTTC
(1846 ← 1818)

c 5′-Biotin-CAGGAACTACTAGTA
(1498 → 1512)

d GGATGACTCATAGGGCTATACATTC
(1625 ← 1611)

rate the shorter inside oligonucleotides bearing the ligands. This system works effectively, but the yield of the *c–d* product bearing the ligands is severely depressed if there is an excess of *a* and *b* oligonucleotides. This yield effect does not depend on the lengths of the oligonucleotides. Preliminary results suggest that it is caused by the 5′→3′-exonuclease nick translation activity of *Taq* polymerase—elongation of a strand primed by oligonucleotide *a* could destroy a biotin-bearing *c*-primed strand already synthesized in the same cycle. Hence it is vital to titrate the concentration of the oligonucleotides *a* and *b*. A polymerase that does not have 5′→3′-exonuclease activity may provide an alternative solution.

Materials and Reagents

Fused polypeptide GST-GCN4: To provide an abundant source of GCN4, a GCN4 coding sequence was fused to the glutathione *S*-transferase gene in vector pGEX-2T (clone GST-GCN4 3.12) and expressed in *E. coli*.[4] The glutathione *S*-transferase moiety retains the capacity to bind to glutathione covalently attached to agarose so that it can be affinity purified conveniently. Purified GST-GCN4 is available from AMRAD Corporation (Kew Victoria, Australia)

Oligonucleotides with GCN4-binding sites: Oligonucleotides for PCR were synthesized by standard procedures in an Applied Biosystems (Foster City, CA) synthesizer. The GCN4-binding site is added onto the 5′ end of the PCR oligonucleotide, so that the general structure is 5′-GCN4 site–PCR oligonucleotide-3′. GCN4-binding sequences that have been used successfully so far are as follows:

1. 5′-AAGTGACTCAAGTGACTCAA–PCR oligonucleotide-3′
2. 5′-AGCGGATGACTCATTTTTTT–PCR oligonucleotide-3′
3. 5′-GGATGACTCA–PCR oligonucleotide-3′

The sequence specificity of GCN4 binding has been studied in more detail[10]

5′-Biotinylated oligonucleotides: These were synthesized as described.[11] All other reagents are readily available either as PCR or ELISA agents and are listed below

Dishes: Flexible polystyrene microtiter plates (Dynatech Laboratories, Inc., Chantilly, VA) have been used most commonly but other plates tested have proved to be satisfactory

[10] A. R. Oliphant, C. J. Brandl, and K. Struhl, *Mol. Cell. Biol.* **9,** 2944 (1989).

[11] L. Wachter, J.-A. Jablonski, and K. L. Ramachandran, *Nucleic Acids Res.* **14,** 7985 (1986).

Methods

General Protocol for Amplified DNA Assay with One-Step Binding Reaction

We have now developed the following simplified ADA procedure.[5] A stock of PCR mix that contains all four oligonucleotides in PCR buffer is made (see below). This mix can be stored frozen as the *Taq* polymerase does not lose significant activity on thawing. We make the stock in 90% of the final volume but it could be made at any suitable concentration.

The protocol for the PCR is then to mix 18 μl of PCR mix plus 2 μl of DNA sample. Cover with paraffin oil and cycle, e.g., 35 times at high stringency and then 12 times at low stringency.

The protocol for the ADA is then to add 50 μl of the avidin peroxidase "binding mix" (see below) to the wells of a GST-GCN4-coated microtiter dish. Add 5 μl of the amplified DNA sample and incubate at room temperature for 20 min. During this time, the DNA binds to the dish and also binds avidin/peroxidase. Wash and add substrate to allow color development. For strong positives the color can be seen in a few seconds. If required, a standard curve for the colorimetric reactions can be generated using known amounts of PCR-amplified DNA.

Nested Polymerase Chain Reaction with Two Steps Thermally Separated

Composition of Polymerase Chain Reaction Mix. The final composition of the PCR mix used was 50 m*M* KCl, 10 m*M* Tris-HCl, pH 8.4, 2.5 m*M* $MgCl_2$, 0.25 m*M* dNTP, and *Taq* polymerase (0.5 units/100 μl) but this was not investigated in detail. The concentration of the oligonucleotides is important, however.[5]

The optimal concentration of the external oligonucleotides (*a* and *b*) in the cases that we have studied is about 0.3 μg/ml (Fig. 2A). It is important to titrate this for each system as it can have a drastic effect on the sensitivity. This appears to be the most important variable, probably because of the nick translation effect that can subsequently remove the biotin from the final PCR product as described above.

The internal oligonucleotides bearing the biotin and the GCN4 recognition sequence have been used at 5 μg/ml. A vital point is that if the subsequent ADA is to be carried out by simultaneously binding the PCR product to GCN4 and avidin peroxidase, then avidin peroxidase must be added in excess.

Location of Oligonucleotides. We have found in the HIV system that the sensitivity is considerably greater if the *c* and *d* oligonucleotides are located approximately 100 rather than just a few bases internally from the *a* and *b* oligonucleotides (Fig. 2B). It is not known whether this is a general effect.

Length of Oligonucleotides and Thermal Cycling Conditions. Oligonucleotides *a* and *b* that are 20–27 bases long have been used successfully. Oligonucleotides *c* and *d* are chosen to have considerably lower thermal stabilities than *a* and *b*. In the examples studied so far, *c* and the region of *d* complementary to the template were reduced to 15 bases.

With a thermal cycling regimen of only two steps per cycle (95° for 30 sec/65° for 60 sec, i.e., both annealing and extension at 65°) the larger oligonucleotides were efficiently incorporated while the shorter ones were not. After a suitable number of cycles (generally 30–35) an annealing step of 40° is introduced and the shorter *c* and *d* oligonucleotides are then incorporated. Six to 15 such cycles (95° for 30 sec, 40° for 60 sec, 65° for 30 sec) have been used.

Procedure for Amplified DNA Assay with One-Step Binding Reaction

Coating Plates with GST-GCN4

1. Dilute the GST-GCN4 to 5 μg/ml in phosphate-buffered saline (PBS).
2. Add 100 μl/well.
3. Incubate at 37° for 1 hr.
4. Tip out the GST-GCN4.
5. Add 200 μl "blocking mix"/well [blocking mix is 10% (w/v) nonfat powdered milk in PBS].
6. Incubate 10 min at room temperature or leave until ready.

Binding Polymerase Chain Reaction-Amplified DNA

1. Tip out the blocking solution.
2. Add 50 μl/well of "binding mix" [binding mix is 10% (w/v) nonfat powdered milk, 4 μg/ml sonicated double-stranded salmon DNA, 0.05% (v/v) Tween 20, and 50 μg/ml horseradish peroxidase–avidin D conjugate (Vector Laboratories, Inc., Burlingame, CA) in PBS].
3. Add samples of the PCR (5 ml).
4. Incubate at room temperature for 20 min, preferably on a rocking platform.
5. Tip out the reaction mixture.
6. Wash with PBS–Tween 20 two to four times, with PBS two to four times, with H_2O once.
7. Thoroughly blot the inverted plate by tapping firmly on absorbent paper towels.

Color Development

1. Add 100 μl of substrate [substrate is freshly prepared 1 m*M* 2,2-azinobis(3-ethylbenzthiazoline-6-sulfonic acid) (ABTS), 0.1% (v/v) H_2O_2 in fresh 0.1 *M* trisodium citrate, pH 4.2]. The ABTS is stored as a 100× concentrate in dimethyl sulfoxide (DMSO) at 4°. The diluted solution should be prepared during the DNA-binding step.
2. Incubate at room temperature for 20 min.
3. Read absorbance. We have used a Titertek Multiskan MCC/340 scanner (Lab Systems, Helsinki, Finland) on mode 2 using filters of 414 and 492 nm.

The procedure as described above was used to detect HIV sequences in DNA purified from HIV antibody-positive patients.[5]

Procedure for Amplified DNA Assay with a Two-Step Binding Reaction for Assaying Polymerase Chain Reactions Performed in Microtiter Dishes

The ADA system can be further simplified by performing the PCRs in a microtiter dish and then capturing the amplified DNA from each of the 96 wells simultaneously on an array of 96 GST-GCN4-coated beads, such as those attached to the lid of a Falcon (Becton Dickinson, Rutherford, NJ) FAST ELISA dish. The beads are coated by immersion in GST-GCN4, blocked with powdered milk–DNA mix, and then immersed in the PCR samples. Subsequently the beads are washed, exposed to avidin peroxidase in another dish, washed, and placed in another dish containing substrate for color development. The color intensity was about threefold lower than with GST-GCN4-coated wells with the same samples, presumably because of the lower surface area of the beads. Hence the more sensitive substrate, tetramethylbenzidine (TMB), was used. Passage of the GST-GCN4-coated beads through the layer of paraffin oil did not appear to interfere with capture of the PCR products.

Coating Beads with GST-GCN4

1. Dilute the GST-GCN4 to 5 μg/ml in PBS.
2. Add 50 μl GST-GCN4/well in a flexible polystyrene dish (Dynatek Industries, McLean, VA).
3. Place the array of beads in the dish of GST-GCN4.
4. Incubate at 37° for 1 hr.
5. Block with blocking mix containing 4 μg/ml salmon DNA and 0.05% (v/v) Tween 20 in another dish at room temperature for 10 min.
6. Drain off excess solution on a paper towel for a few seconds.

Capturing Amplified DNA

1. Place the array of beads in the microtiter dish used for the PCR.
2. Incubate at room temperature for 20 min; the DNA binds to the GST-GCN4-coated beads during this step.
3. Wash with PBS–0.05% (v/v) Tween 20 at least four times. This is conveniently done in the lid of a conventional microtiter dish.
4. Drain off excess solution on a paper towel for a few seconds.

Color Development

1. Add 100 μl/well of TMB substrate to a flexible polystyrene dish (0.4 mM TMB plus 1.4 mM H_2O_2 in 0.1 M sodium acetate, pH 5.5; as for the ABTS, the TMB is stored as a 100$\times$ concentrate in DMSO and diluted just before use).
2. Place the array of beads in the dish of substrate.
3. Incubate at room temperature for 20 min and then remove the beads. Read absorbance with the Titertek Multiscan MCC/340 set on mode 1 using filter number 7.

Procedure for Amplified DNA Assay with DNA-Binding Protein TyrR

The only other DNA-binding protein that we have tested in the ADA so far is the *E. coli* regulatory protein TyrR.[7] The ADA procedure used was identical to that described with GCN4, except that 0.2 mM ATP was present in the binding mix. The oligonucleotide with the TyrR-binding site was 5′-GGTGTGTAAATATATATTTACACA–PCR oligonucleotide-3′. Both GST-GCN4 and TyrR bound specifically only to the PCR product containing the correct recognition sequence.[7]

Applications and Problems

The ADA has been applied to the detection of DNA from a number of pathogens, including HIV (see Fig. 3), *Plasmodium falciparum* (D. J. Kemp, unpublished observations, 1991), infectious laryngotracheitis virus,[12] and *Mycoplasma pneumoniae* (R. Harris, personal communication, 1991). With purified DNA, the limit of detection for HIV and *P. falciparum* has been estimated to be less than 100 molecules of the sequence under examination. The sensitivity of course relies on the amplification by the PCR. The actual sensitivity of colorimetric detection of the amplified DNA

[12] M. W. Shirley, D. J. Kemp, M. Sheppard, and K. J. Fahey, *J. Virol. Methods* **30,** 251 (1990).

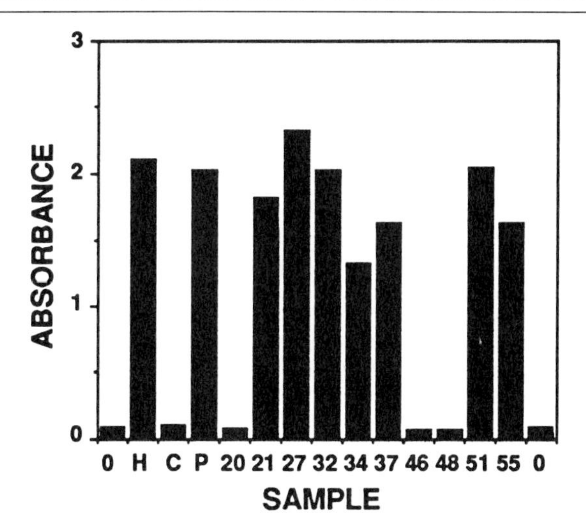

FIG. 3. Detection of HIV sequences in DNA from infected human peripheral white blood cells. Blood samples were taken from patients as well as from normal volunteers with a low risk of HIV infection. Conditions for the PCRs and ADAs were as in Fig. 2B except that 35 and 12 cycles were used for the 2 PCR steps, respectively. The DNA samples in the PCRs were as follow: 0, none; H, HIV-infected CEM cells; C, uninfected CEM cells; P, HIV plasmid; 20 and 46, control DNA from HIV-negative volunteers; 21–37, 51, and 55, DNA from HIV-infected patients; 48, DNA from a seronegative bisexual man.

product using ABTS is about the same as detection of the DNA on an agarose gel by ethidium bromide fluorescence and photography with short-wave (302 nm) UV light. The sensitivity of colorimetric detection can be increased severalfold by using TMB as substrate. Presumably, the use of oligonucleotides bearing multiple biotin molecules would increase the sensitivity further. A method for quantitation based on coamplfication of target DNA with a cloned DNA into which a DNA-binding sequence has been introduced is directly applicable to the ADA.[13]

The sensitivity could be increased further by using sensitive chemiluminescent or fluorescence-based detection systems but these systems are not compatible with a standard microtiter plate scanner. Hence it seems most useful to optimize the sensitivity of the PCR. One problem described above is that in the nested oligonucleotide system for avoiding primer dimers, an excess of the outside oligonucleotides inhibits accumulation of biotinylated *c–d* PCR product. Probably the most important problem, however, is in the preparation of DNA from blood or other biological samples. While the sensitivity that can be achieved with purified DNA is

[13] J. Lundeberg, J. Wahlberg, and M. Uhlen, *BioTechniques* **10,** 68 (1991).

much greater than with whole blood, the preparation of DNA becomes rate limiting. Automation or simplification is clearly still required in this area but this is a general problem in PCR-based diagnostics.

It is important to realize that DNA-binding proteins such as GST-GCN4 have both a general DNA-binding activity and a sequence-specific DNA-binding activity. We have found that the general DNA-binding activity can be outcompeted with sonicated double-stranded salmon DNA, and so the assay as described for the *gag* gene of HIV is specific for the GCN4-binding site supplied by the oligonucleotide.[4] However, it should be noted that the generalized DNA-binding activity is potentially very useful in its own right for anchoring any DNA molecule to a solid phase, particularly as the binding is reversed and the DNA released at 65°. One other application that utilizes the sequence-specific binding of immobilized GCN4 is a method for screening cloned libraries by PCR, using a vector oligonucleotide and an oligonucleotide specific for the sequence of interest and bearing a GCN4-binding site.[14,15] However, only nanogram amounts of DNA are actually bound to GST-GCN4 in a microtiter well or minicentrifuge tube. It would be useful to increase this to microgram levels to carry out other manipulations of PCR-amplified DNA.

Acknowledgments

This work was supported by the National Health and Medical Research Council of Australia, AMRAD Corporation, Ltd., the John D. and Catherine T. MacArthur Foundation, and the Commonwealth AIDS Research Grants Committee. I thank H. Saunders for typing the manuscript.

[14] A. M. Lew and D. J. Kemp, *Nucleic Acids Res.* **17,** 5859 (1989).

[15] A. M. Lew, V. M. Marshall, and D. J. Kemp, this series, Vol. 218, (in press).

[13] Preparation of DNA from Blood for Polymerase Chain Reaction in Microtiter Dish

By ANDREW M. LEW and MICHAEL PANACCIO

Preparation of Samples: Limiting Step for Polymerase Chain Reaction

The polymerase chain reaction (PCR)[1,2] has the advantage of not requiring highly purified DNA as the substrate source. Nevertheless, the reaction is inhibited by the presence of organic material such as blood. For small numbers of cells (e.g., $<10^3$) and small amounts of organic matter, simply boiling the sample in distilled water often suffices. These conditions are seldom met, however, for most biological sources, e.g., nasal mucus, blood, and synovial fluid. This is particularly so when the amount of target DNA is only a small fraction of total DNA, e.g., diagnosis of a low level of parasitemia. Hence, some sample preparation for partial purification (to eliminate inhibiting substances or to release the DNA) is often required, e.g., treatment with a nonionic detergent to obtain a nuclear pellet followed by proteinase K digestion.[3] This normally entails a centrifugation step and hence is not expedient for a microtiter dish procedure. Proteinase K digestions and centrifugation steps also mean more handling and more time consumed for sample preparation. The preparation of DNA is often a rate-limiting step in obtaining a PCR product from a biological source. A strategy was devised to minimize the number of manipulations and to obviate the need for centrifugation. Anti-histone antibodies are adsorbed to a tube or well of a microtiter dish. This technique has been designated immuno-PCR. Although we believe this technique to have general applications, the example we describe here is the preparation of *Plasmodium falciparum* DNA from malaria-infected human blood.

Principle of Immuno-Polymerase Chain Reaction

Eukaryotic cells have their DNA intimately but noncovalently bound to histones to form chromatin. Antibodies to histones can be adsorbed to plastic (in the form of a microcentrifuge tube or a microtiter well) and used

[1] K. B. Mullis and F. A. Faloona, this series, Vol. 155, p. 335.

[2] R. K. Saiki, D. H. Gelfand, S. Stoffel, S. J. Scharf, R. Higuchi, G. T. Horn, K. B. Mullis, and H. A. Erlich, *Science* **239,** 487 (1988).

[3] R. Higuchi, *in* "PCR Technology" (H. A. Erlich, ed.), p. 31. Stockton Press, New York, 1975.

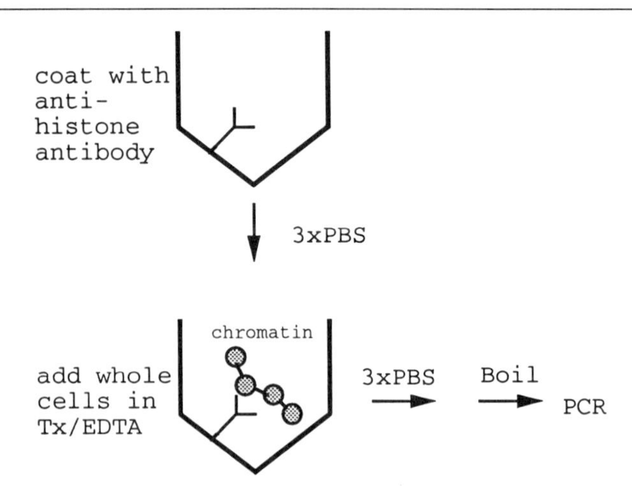

FIG. 1. Outline of strategy of immuno-PCR. Tx, Triton X-100.

to capture chromatin from cells (Fig. 1). Unbound material can be "flick-washed" away without the need to centrifuge. DNA is then released from the surface of the well or tube in the first cycle of PCR. Therefore antibodies to histones are coated on wells of a microtiter dish and used to capture DNA as chromatin and PCR done on the same wells.

To increase the specificity further, it is also theoretically possible to use an antibody (e.g., a monoclonal antibody) that is specific for *Plasmodium* histones to select for *Plasmodium* chromatin in a mixture of human and *Plasmodium* chromatin.

Several conditions must be satisfied before chromatin is readily accessible to antibody. First, the cell membrane needs to be lysed by 0.5% (v/v) Triton X-100. The nuclear membrane needs to be disrupted with excess ethylenediaminetetraacetic acid (EDTA) and/or hypotonicity. EDTA also chelates the divalent cations that would otherwise keep the chromatin insoluble. Whole blood contains 2.4 m*M* Ca^{2+} and 1.8 m*M* Mg^{2+}. Enough salt is needed to keep the chromatin soluble yet too much salt would release the histones from the DNA.

Material and Reagents

Anti-Histone Antibodies

Two sources of anti-histone antibodies were obtained.

Mouse Anti-Plasmodium falciparum Histones. Mouse anti-*P. falci-*

parum histones were isolated using a modification of the procedure of Caplan.[4]

Four milliliters of packed, malaria-infected human red blood cells with a parasitemia of 10% were washed in 0.01 *M* phosphate-buffered saline (PBS), pH 7.2, and lysed in 0.5% (v/v) Triton X-100 (Sigma, St. Louis, MO) for 1 hr at 4° and nuclei were sedimented at 4000 *g* for 15 min. After washing in 0.5% (v/v) Triton X-100 (4000 *g* for 15 min) the nuclear pellet was suspended in 3 ml of 0.14 *M* NaCl/0.05 *M* $NaHSO_3$ and stirred gently for 15 min. Three washes in the same solution at 4000 *g* for 10 min resulted in a chromatin pellet. Histones were extracted with 1.5 ml of 0.4 *M* HCl for 20 min with gentle stirring. Two more extractions with 0.75 ml of 0.4 *M* HCl followed. The addition of 10 vol of cold acetone to the pooled supernatant precipitated the histones after overnight incubation at $-20°$. The histone precipitate was resuspended in 1 ml of 0.3 *M* NaCl.

CBA × BALB/c mice were given two intraperitoneal injections (a month apart) of 50 μg of histones emulsified in Freund's incomplete adjuvant. Serum was obtained 3 weeks after the last injection. Immunoglobulin was purified on protein A-agarose beads.[5]

Human Anti-Histone Antibodies. Serum was obtained from a patient who had been treated with hydralazine and had developed autoantibodies to histones. This was a gift from Dr. S. Whittingham (Walter and Eliza Hall Institute, Melbourne; Australia). An immunoglobulin fraction was purified on protein A-agarose beads.

Robot Arm Polymerase Chain Reaction Machine

This was purchased from Innovonics (Melbourne, Australia) and contains three water baths that can be adjusted to the desired temperatures. An automatic robot arm that has been adapted to carry microtiter dishes lowers the dish sequentially, for predetermined times, into each of the water baths. These microtiter dishes were specially made with high walls to prevent water from the water baths entering the wells and air was not allowed to be trapped between the wells (Fig. 2). Temperature changes are effected rapidly. Thus, for 30 cycles with 1 min in each water bath, PCR is completed in less than 2 hr.

Methods

Optimum Concentration of Immunoglobulin for Coating

Fifty microliters of various concentrations of Ig (0, 5, 10, 30, and 90 μg/ml PBS) is used to coat microtiter dishes for 2 hr at room temperature.

[4] E. B. Caplan, *Biochim. Biophys. Acta* **407,** 109 (1975).

[5] P. L. Ey, S. J. Prowse, and C. R. Jenkin, *Immunochemistry* **15,** 429 (1978).

FIG. 2. Polymerase chain reaction machine robot arm carrying the modified microtiter dish.

After three PBS washes (solutions are dislodged from the dish by inverting the dish in a flicking action), an enzyme-linked immunosorbent assay (ELISA) is performed using peroxidase-conjugated sheep anti-human Ig to determine the lowest concentration of Ig in the coating solution for optimum binding.[6] This has been found to be 10 μg/ml (Fig. 3).

Immuno-Polymerase Chain Reaction

Fifty microliters of the immunoglobulin fraction of anti-histone serum is coated onto wells of a microtiter dish at 10 μg/ml in PBS for 2 hr at room temperature or 4° overnight. After three washings in PBS, 50 μl of various buffers is tried to optimize the fracture of nuclei without interfering

[6] A. Lew, *J. Immunol. Methods* **72,** 171 (1984).

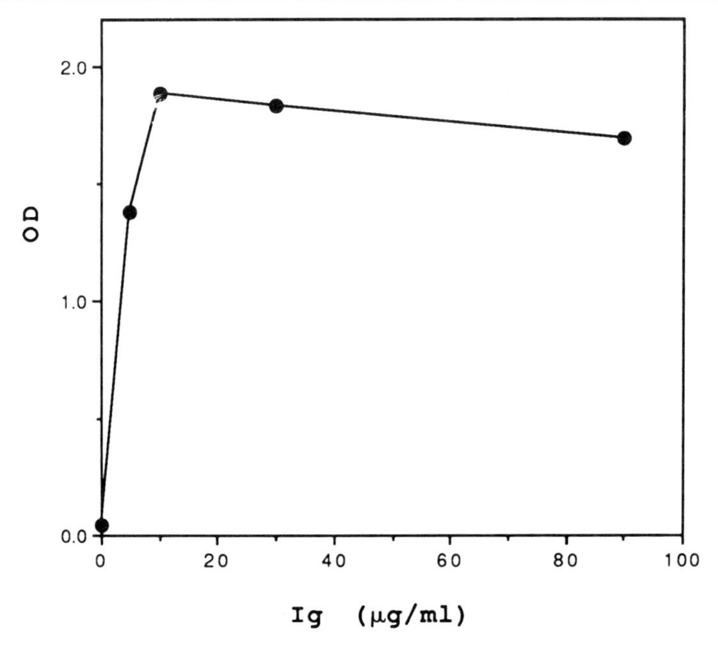

FIG. 3. The optimum concentration of immunoglobulin for coating polypropylene tubes. Protein A-purified fractions of human anti-histone serum at 0, 5, 10, 30, and 90 μg/ml in PBS were added to polypropylene tubes, incubated for 2 hr at room temperature, and washed three times in PBS. An ELISA was performed using peroxidase-conjugated sheep anti-human Ig.

with the histone–DNA interaction or the antigen–antibody reaction. Five microliters of whole blood (heparinized, 10 IU/ml) containing *P. falciparum* parasites is added and mixed. After a 2-hr incubation, wells are washed three times with PBS and PCR is carried out (as described in Ref. 2) except that the total volume is 50 μl and 30 cycles of 95° for 1 min and 70° for 1 min are performed in a robot arm machine. Specific *Plasmodium* oligonucleotides[7] are synthesized and used for priming. The PCR product is electrophoresed in a 1% (w/v) agarose gel containing 0.1 μg/ml ethidium bromide for visualization.

Effect of Buffer Constituents of Immuno-Polymerase Chain Reaction

Because of the considerations outlined above, the following are added to the human anti-histone antibody-coated wells or wells coated with

[7] J. A. Smythe, M. G. Peterson, R. L. Coppel, A. Saul, D. J. Kemp, and R. F. Anders, *Mol. Biochem. Parasitol.* **39,** 227 (1990).

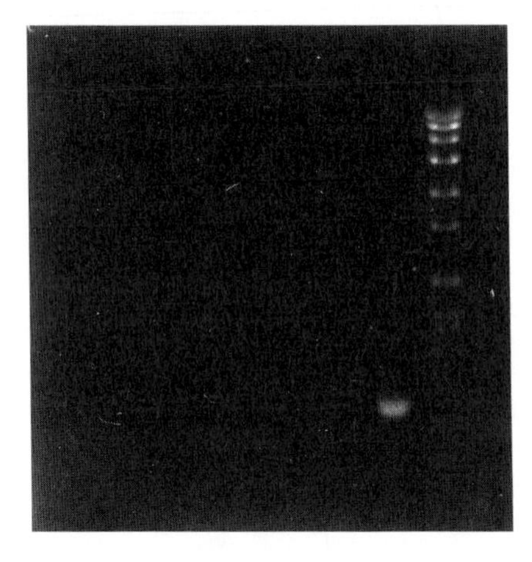

FIG. 4. The specificity of immuno-PCR. Polypropylene tubes were coated with human anti-histone antibody (Ab^+) or normal Ig (Ab^-). Blood infected with *P. falciparum* malaria (Ag^+) or uninfected blood (Ag^-) was added to the tube containing 0.5% (v/v) Triton X-100 in TE. Additional salt was (10× PBS^+) or was not (10× PBS^-) added. The PCR was performed and the products electrophoresed in a 1% (w/v) agarose gel containing 0.1 μg/ml ethidium bromide for staining DNA. The far right-hand lane shows the DNA size markers (*Eco*RI-cut *spp*-1).

normal human Ig: 45 μl of 0.5% (v/v) Triton X-100 in 10 m*M* Tris/1 m*M* EDTA, pH 8.0, and 5 μl of human blood at approximately 50% hematocrit and with a 20% parasitemia of *P. falciparum* cultured *in vitro*. After mixing and standing for 10 min, 5 μl of a 10× PBS solution is added to some wells. After a 2-hr incubation at room temperature followed by three PBS washes, PCR is done directly in the wells. As shown in Fig. 4, only the well that is coated with anti-histone antibody, contains infected blood, and also receives the extra salt produces a DNA fragment of the expected 700 bp. If normal Ig, normal blood, or no salt is used, no visible band is obtained.

If the 10× PBS is added initially with the blood and Triton X-100 buffer, there is no visible DNA fragment (data not shown), presumably because the nuclear envelope is not disrupted and hence the chromatin is not accessible. If 0.5% (v/v) deoxycholate is used instead of 0.5% (v/v)

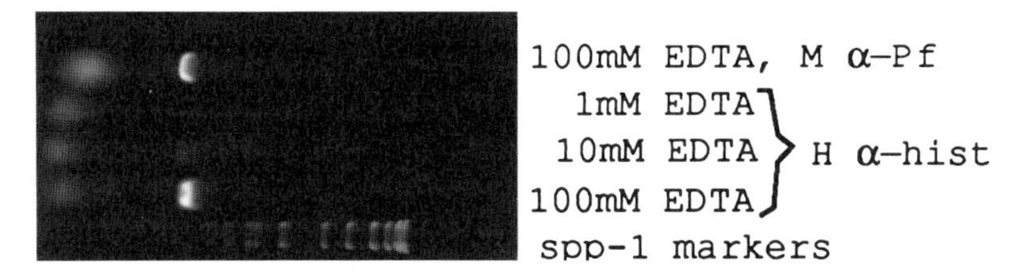

FIG. 5. Varying the EDTA concentration in immuno-PCR. Conditions as in Fig. 3 except that 1, 10, and 100 m*M* EDTA were used instead of TE, and in one tube mouse anti-*P. falciparum* histone (M *α-Pf*) was used instead of human anti-histone autoantibody.

Triton X-100, the DNA is released from the histones[8] and immuno-PCR is unsuccessful (data not shown).

For a field study, especially for hazardous material, a minimum of handling is important and fresh samples of blood from needle pricks are often collected. Thus, a buffer was sought that had high enough salt to waive the 10× PBS step and yet was able to destabilize the nuclear membrane. As shown in Fig. 5, 100 m*M* EDTA is the optimal solution to use, although with 10 m*M* EDTA a PCR product is visible on ethidium bromide staining. This is somewhat surprising, as 100 m*M* EDTA, pH 8.0, would contain about 200 m*M* Na^+ and hence be isotonic. Perhaps this excess of chelating agent destabilizes nuclei enough (by making divalent cations completely unavailable) to allow chromatin to escape.[9] Also shown in Fig. 5 is that either the human autoantibody, or the mouse anti-*Plasmodium* histone, is effective in immuno-PCR. Subsequently, 0.5% (v/v) Triton X-100 in 100 m*M* EDTA is used in experiments.

Sensitivity of Immuno-Polymerase Chain Reaction

The results described above used cultured *P. falciparum* with a parasitemia ranging from 5 to 20%. To determine how sensitive the immuno-PCR assay was, a *P. falciparum* culture with a parasitemia of 20% was diluted in fresh human whole blood to a parasitemia of 1, 0.2, and 0.04%. The parasites were mainly in the ring stage of development to mimic the clinical situation. Five microliters of this was then added to the antibody-coated tube containing 0.5% (v/v) Triton X-100 and 100 m*M* EDTA and

[8] R. J. Delange and E. L. Smith, *in* "The Proteins" (H. Neurath and R. L. Hill, eds.), Vol. 4, 3rd Ed., p. 119. Academic Press, New York, 1979.

[9] D. B. Roodyn, *in* "Subcellular Components, Preparation and Fractionation" (G. D. Birnie, ed.), 2nd Ed., p. 15, Butterworth, London, 1972.

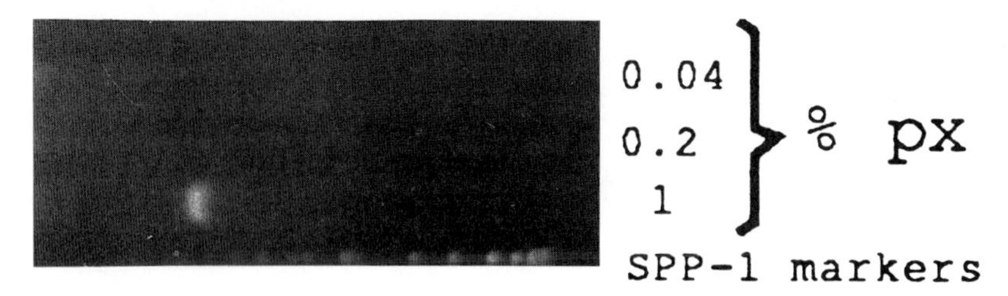

FIG. 6. Sensitivity of immuno-PCR. Using tubes coated with mouse anti-*P. falciparum* histones, blood containing 0.04, 0.2, and 1% parasitemia was used as the source of DNA for immuno-PCR and the products treated as in Fig. 3.

immuno-PCR was performed as above. When the tubes were coated with mouse anti-*Plasmodium* histone antibody, a visible DNA fragment is obtained at 1% parasitemia (Fig. 6) but not at 0.2 or 0.04% parasitemia.

Discussion

Possibilities for Field Use

Immuno-polymerase chain reaction would seem to be easily adaptable for field use:

1. Large numbers of samples can be processed rapidly in a microtiter dish.
2. Few manipulations are needed.
3. No centrifugation or freezing is needed.
4. The volume of sample required is small (5 μl). Fingerprick bleeds will suffice. For frozen samples of primary isolates, immuno-PCR has been successful by scratching the surface contents of a vial without thawing the entire precious sample.
5. Antibody-coated dishes can be stored for at least 1 week (longer times were not tested) in the cold in PBS/1 mg/ml azide.
6. Once the chromatin has been captured on the dish, it can be stored dried for at least a week before PCR.

Sensitivity and Possible Improvements

Immuno-polymerase chain reaction using anti-*Plasmodium* histone antibody was effective when the starting material was at 1% parasitemia. This level of parasitemia corresponds to the equivalent amount of DNA expected in the leukocyte fraction of blood. (*Note:* The amount of DNA per *Plasmodium* cell is one-hundredth of that per mammalian cell.) Therefore this level of sensitivity is probably appropriate for genetic studies but not when the amount of target DNA in blood is very low [e.g., detection of human immunodeficiency virus (HIV) proviral DNA in blood].

There are several ways to increase the sensitivity of immuno-PCR:

1. Increase the avidity of antibody; hyperimmunized sera could be used.
2. Increase the efficiency of coating, e.g., using covalent coupling.
3. Increase the amount of antibody coated. Affinity-purified antibodies could be used instead of just an immunoglobulin fraction. Alternatively, a monoclonal antibody would increase the proportion of antibody to total immunoglobulin.
4. Orient the antibody molecules. By coating the tubes first with proteins that bind the Fc (e.g., protein A, protein G, antibodies to Fc), then all the anti-histone antibodies would have then antigen-binding regions free to capture the chromatin.

Future Directions

We have used anti-histone antibodies to capture the chromatin but other agents could be used to capture DNA. These include antibodies to control elements that bind DNA for prokaryotes, antibodies to DNA itself, and antibodies to the target microorganisms itself. Ligands other than antibodies could be used, e.g., nonspecific DNA-binding proteins. Given our experience with immuno-PCR a critical factor to investigate in these systems would be the buffer composition.

[14] Amplification and Detection of Specific DNA Sequences with Fluorescent PCR Primers: Application to ΔF508 Mutation in Cystic Fibrosis

By Farid F. Chehab, Jeff Wall, and Y. W. Kan

Introduction

The polymerase chain reaction (PCR) is now a well-established technology in various disciplines of biology. Following PCR, a detection method is needed to analyze the amplification products. Denaturing gradient gel electrophoresis,[1] DNA sequencing[2] and single-strand conformation polymorphism[3] are examples of detection methods that have been applied to characterize mutations in human DNA,[4–6] often involving a single base substitution. The choice of detection system usually depends on specific applications and whether the procedure is performed in a clinical or research environment. The use of radioactivity in the research laboratory may not be as problematic as in the clinical laboratory, where radioactivity, because of its potential hazard and strict regulations concerning shielding and disposal, is not desirable. For example, radioactive DNA sequencing of the amplified product, although highly informative, is labor intensive, hazardous, and expensive; especially when the labor of the clinical assay must be translated into a bill to the patient. Our role in the implementation of molecular biological techniques within a clinical molecular diagnostics laboratory has led us to search for a simple, quick, and inexpensive assay to detect specific DNA sequence variation following amplification by the polymerase chain reaction. We have described an allele-specific color PCR assay based on the conjugation of fluorescent dyes to the PCR primers.[7] This system has proven to meet the requirements for a DNA diagnostic assay in terms of simplicity, accuracy, cost, labor, and reliability. We describe in this chapter the technical aspects of this assay and its application to the ΔF508 mutation in cystic fibrosis.[8]

Background and Principle of Assay

In any PCR amplification, two primers, each directed to a specific site on opposite DNA strands, are used to amplify exponentially a target DNA sequence flanked by the two primers. In an "allele-specific PCR" aimed at detecting subtle nucleotide variations, one of those two primers is positioned at the site of the sequence alteration such that extension of this allele-specific primer results only when the primer and template are

[1] V. C. Sheffield, D. R. Cox, L. S. Lerman, and R. M. Myers, *Proc. Natl. Acad. Sci. U.S.A.* **1,** 232 (1989).
[2] U. B. Gyllensten and H. A. Erlich, *Proc. Natl. Acad. Sci. U.S.A.* **20,** 7652 (1988).
[3] M. Orita, Y. Suzuki, T. Sekiya, and K. Hayashi, *Genomics* **5,** 874 (1989).
[4] S. P. Cai and Y. W. Kan, *J. Clin. Invest.* **85,** 550 (1990).
[5] R. A. Gibbs, P. N. Nguyen, L. J. McBride, S. M. Koepf, and C. T. Caskey, *Proc. Natl. Acad. Sci. U.S.A.* **86,** 1919 (1989).
[6] M. Dean, M. B. White, J. Amos, B. Gerrard, C. Stewart, K. T. Khaw, and M. Leppert, *Cell (Cambridge, Mass.)* **61,** 863 (1990).
[7] F. F. Chehab, and Y. W. Kan, *Proc. Natl. Acad. Sci. U.S.A.* **86,** 9178 (1989).
[8] B. Kerem, J. M. Rommens, J. A. Buchanan, D. Markiewicz, T. K. Cox, A. Chakravarti, M. Buchwald, and L. C. Tsui, *Science* **245,** 1073 (1989).

perfectly matched, thus producing a discrete PCR product. For each mutation to be tested two identical amplification reactions are set up, one containing a primer identical to the wild-type DNA sequence and the other containing a primer identical to the mutant sequence. Following PCR, the diagnostic presence of a PCR product is identified by gel electrophoresis and ethidium bromide staining. Thus, a homozygous normal individual would amplify only with the normal primer, a heterozygous with both primers, and a homozygous mutant only with the mutant primer. Because the absence of amplification product is also of diagnostic value, two primers that amplify an internal control target DNA sequence are also included in the reaction. We have applied this scheme to the detection of a 4-bp deletion at codons 41–42 of the β-globin gene.[9] A similar yet different assay has been described[10] wherein the two allele-specific primers directed to the same DNA strand are included in a single reaction along with a third primer that anneals to the opposite DNA strand. Thus a competitive oligonucleotide priming reaction results, favoring the hybridization and extension of only the perfectly matched primer onto the template. However, to identify which allele-specific primer has been extended in the amplification, two identical reactions are set up. In one reaction only the wild-type allele-specific primer is labeled with ^{32}P and in the other reaction the mutant allele-specific primer is radiolabeled. The PCR products extended from each primer in each reaction are differentiated from each other by autoradiography and ethidium bromide staining.

We have adapted this competitive oligonucleotide priming approach to a fluorescent dye system in which no radioactivity is involved, no internal control is needed, and only one reaction per sample tested is required. In addition, the scoring of the assay is a simple visual inspection of the colors following PCR. In this system, one allele-specific primer is labeled with fluorescein, another with rhodamine, and the third primer is unlabeled. After amplification, the PCR products are separated by gel electrophoresis from the unextended primers, and the color of the PCR products is visualized by ultraviolet irradiation of the gel. A green or red color reflects an amplification derived from either allele-specific primer whereas a yellow color, resulting from the complementation of red and green under ultraviolet light, indicates that both fluorescent primers have been utilized in the reaction. Another alternative to detect the diagnostic color, and which has relevance in automating this procedure, is by real-time laser scanning of gel electrophoresis.

[9] F. F. Chehab, S. P. Cai, and Y. W. Kan, *Blood* **70,** 73a (1987).

[10] R. A. Gibbs, P. N. Nguyen, and C. T. Caskey, *Nucleic Acids Res.* **17,** 2437 (1989).

Materials and Reagents

Two hundred and fifty milligrams of either Aminolink 2 (Applied Biosystems, Foster City, CA) or Aminomodifier II (Clontech Laboratories, Palo Alto, CA) is dissolved in 100 μl acetonitrile and attached to the extra port of an Applied Biosystems DNA synthesizer. The N-hydroxysuccinimide esters of fluorescein (5 mg) or rhodamine (5 mg) from Applied Biosystems and termed FAM and ROX, respectively, are dissolved in 60 μl of dimethyl sulfoxide (Fisher Scientific, Pittsburgh, PA). Sephadex G-25 gel filtration prepacked columns (NAP-25) and 100 m*M* solutions of deoxynucleotide triphosphates are obtained from Pharmacia (Uppsala, Sweden). *Taq* DNA polymerase is from Perkin-Elmer Cetus (Norwalk, CT). TEAA is 0.1 *M* triethylamine acetate buffer, sodium bicarbonate buffer consists of 0.22 *M* $NaHCO_3/Na_2CO_3$, pH 9.0, and TE buffer is 10 m*M* Tris, pH 8.0, 1 m*M* ethylenediaminetetraacetic acid (EDTA).

Synthesis of the Oligonucleotide Primers

The synthesis of oligonucleotides is performed by the phosphoramidite chemistry either on an Applied Biosystems or a Biosearch DNA synthesizer. An amino group is attached to the last base at the 5′ end of the oligonucleotide. This step is performed by programming the instrument to add the amino linker (Aminolink II or Aminomodifier II) as the last "base" in the synthesis. The oligonucleotide is then recovered in ammonium hydroxide, deprotected overnight at 55°, dried down, and resuspended at a concentration of 0.2 A_{260} units/μl in sterile distilled water.

Conjugation and Purification of Dye to Amino-Oligonucleotide

A 1-μmol synthesis yields approximately 100 A_{260} units of a 20-mer oligonucleotide. Fifteen microliters (3 A_{260} units) of the above crude amino-oligonucleotide is mixed with 6 μl of a sodium bicarbonate buffer, pH 9.0, and 6 μl of the dye [5 mg/60 μl dimethyl sulfoxide (DMSO)]. The reaction is incubated 3 hr to overnight at room temperature in the dark (conveniently placed in a drawer). The aim of the next step is to separate the free dye from the mixture of dye-conjugated and nonconjugated amino-oligonucleotides. The reaction is diluted to 1 ml with 0.1 *M* TEAA, and loaded on a prepacked NAP-25 column equilibrated with 0.1 *M* TEAA as the elution buffer. The sample is allowed to penetrate into the column, then two 2-ml volumes of elution buffer are added onto the column. After the second application, 2 ml of colored eluate is *immediately* collected. This eluate consists of a mixture of dye-conjugated and unconjugated amino-oligonucleotides. The final purification consists of an high-performance liquid chromatography (HPLC) step to separate the dye-conjugated amino-oligonucleotide from the unconjugated amino-oligonucleotide based on the hydrophobic property that the dye confers to the oligonucleotide. We have used either an Applied Biosystems or a Perkin-Elmer HPLC equipped with a UV detector, an Aquapore 300 C_8 reversed-phase column (Applied Biosystems), and a 2-ml sample loop. After equilibration of the column with 8% acetonitrile in 0.1 *M* TEAA, the eluate recovered from the previous purification step is injected onto the reversed-phase column and a first gradient of 8–20% acetonitrile in 0.1 *M* TEAA over 25 min is started. This separates the unlabeled primer from the fluorescein-labeled primer. The rhodamine primer that is more hydrophobic than the fluorescein primer elutes in a second consecutive gradient of 20–40% acetonitrile in 0.1 *M* TEAA over 10 min. We recover the fluorescent primers manually at the start of each peak. The colors should be clearly discernible to the naked eye. Faint colors indicate poor conjugation. Typical conjugations of the dye to the oligonucleotide vary from 20 to 40%. The purified fluorescent oligonucleotides are dried down and resuspended in sterile distilled water at a concentration of 35 pmol/μl.

Amplification of the ΔF508 Mutation in Cystic Fibrosis

The DNA sequence of the wild-type fluorescein allele-specific ΔF508 primer is 5′-GAAACACCAAAGATGATA-3′ and that of the rhodamine mutant cystic fibrosis primer is 5′-GGAAACACCAATGATATT-3′. The unconjugated PCR primer is 5′-GCAGAGTACCTGAAACAGGAA-3′. The amplification protocol[11] consists of 30 pmol of each primer in a 50-μl reaction consisting of 10 m*M* Tris, pH 8.3, 50 m*M* KCl, 2.5 m*M* $MgCl_2$, 100 μ*M* TTP, dATP, dCTP, and dGTP, 0.5 μg genomic DNA, and 5 units of *Taq* DNA polymerase. The reaction is heated for 2 min at 95°, then taken through 35 cycles on a Perkin-Elmer Cetus thermal cycler, each cycle consisting of 30-sec incubation steps at 95, 55, and 72°. The amplification yields PCR products of 304 bp (wild type) and 301 bp (mutant).

Gel Electrophoresis and Photography

After amplification, a 20-μl aliquot of each PCR is mixed with 10 μl of a Ficoll-bromphenol blue loading dye and loaded into a 5 × 5 × 2 mm slot of a 10% (w/v) polyacrylamide gel (10.5 × 10 × 0.2 cm) in 45 m*M*

[11] R. K. Saiki, D. H. Gelfand, S. Stoffel, S. J. Scharf, R. Higuchi, G. T. Horn, K. B. Mullis, and H. A. Erlich, *Science* **239,** 487 (1988).

Tris–borate, 1 m*M* EDTA. Following a short electrophoresis at 150 V, 35 mA for 20–30 min just to separate the unextended color primers from the colored PCR products, the gel is illuminated directly without any prior staining by a short-wavelength ultraviolet light transilluminator purchased from Fotodyne (New Berlin, WI). No attempt is made to separate the two PCR products because a green or a red color indicates, respectively, that either the wild type or the mutant product has been amplified (Fig. 1). A yellow band reflects amplification from both alleles and results from the superimposition of the red and green fluorophores, which complement to yellow under ultraviolet light. We found that the excitation wavelength from the transilluminator varies from different vendors, so either a short- or long-wavelength transilluminator may be adequate. Although the dyes are excited at a low efficiency by UV lighting, the accumulation of enough PCR products makes the color visible to the naked eye. Photography of the gels was performed with an MP-4 Polaroid camera and a Polaroid color film type 668 (80 ASA). Because this is a very slow film, long exposure times have turned out to give the best color photographs. The gel is illuminated with UV light and two exposures (5 min each at *f*4.5) are taken for each picture. One exposure is performed with a gelatin filter No. 8, then the shutter is closed, the filter is changed to a 23A, and the shutter is opened for another 5-min exposure. The gel should not be moved in between exposures. The film is allowed to develop for 2 min. Satisfactory photographs have also been obtained with only the No. 8 filter and one exposure only. For fainter fluorescent bands, 10-min exposures for each filter have also given good results.

Laser Scanning

Another detection method that we have evaluated is one that replaces polyacrylamide gel electrophoresis and color photography and allows convenient data entry and peak integration. Although gel electrophoresis is still performed, the fluorescence is detected at a much higher efficiency by a laser that continuously scans the gel close to the bottom. Such an instrument was made available to us by Applied Biosystems and is essentially similar to their fluorescent DNA sequencer (model 373A), except that the gel is horizontal, shorter, and made of agarose. Also, because of the optimal excitation and emission of the fluorophores, it is possible to detect amplified products after 16 PCR cycles, thus reducing the amplification of PCR contaminants that usually show up in later cycles. When an aliquot of the color PCR is fractionated on the gel, the fluorescent PCR products migrate to the position scanned by the laser and the fluorescent emission of the excited dyes is collected onto a photomultiplier and trans-

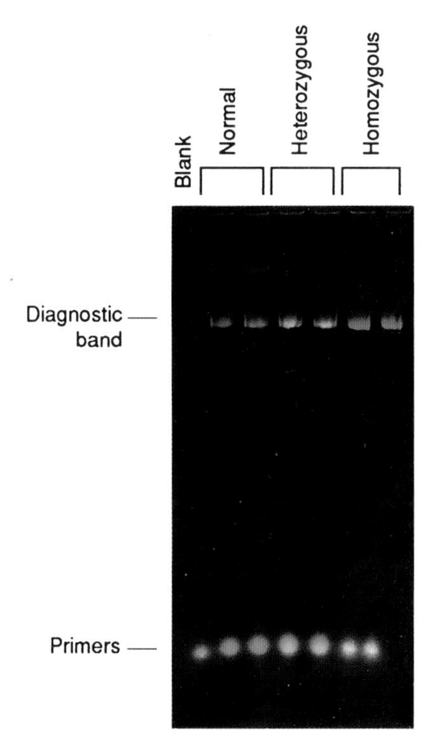

FIG. 1. Electrophoretic separation of an aliquot from the ΔF508 color PCR assay. The 10% (w/v) polyacrylamide gel was irradiated by a transilluminator and the diagnostic color PCR band visualized. The green color reflects amplification from the normal primer, the red from the ΔF508 primer, and the yellow from both. The unextended primers are shown at the bottom of the gel. Color photography was as described in the text.

mitted to a computer that records and stores the results. Because each fluorescent dye is different in its excitation–emission spectrum, multiple dyes that cannot be differentiated by the human eye can be simultaneously used in the amplification and still be discriminated from each other by the laser. At the end of the electrophoretic run, the results are plotted and analyzed by the computer. The advantage of such a system is that it automates the reading of the colors in the gel and allows the use of different fluorescent dyes in a multiplex PCR format for the purpose of DNA markers or internal controls within the same lane on the gel. In a typical electrophoretic run, a 1- to 5-μl aliquot from the amplification reaction is mixed with 1–2 μl of a 30% (w/v) Ficoll (M_r 8000) without bromphenol blue and loaded onto a 2.5% (w/v) agarose gel. The electrophoresis is allowed to proceed for 2.5 hr at 100 V. The results are then plotted by the computer. Fluorographs generated by the ΔF508 assay are shown in Fig. 2. The unextended primers migrate faster than the PCR products and are therefore first detected by the laser while the slower migrating diagnostic PCR products are subsequently identified.

Discussion

We have used the ΔF508 color assay for the detection of the most common cystic fibrosis mutation from clinical blood specimens, amniotic fluid, and chorionic villi tissue samples. To test the reliability of the assay, we have backed it up with both a heteroduplex method[12] and a reverse dot-blot assay.[13] We found full concordance with the three methods and thus adopted the color assay, as it is simple, quick, reliable, and economical. This technique may also be adapted for visualization of the diagnostic color without the need for electrophoresis. Thus, the amplified products can be separated from the unextended primers by centrifugation through a microfiltration device (Centricon 100 from Amicon, Danvers, MA) and the color of the PCR products can be visualized under either ultraviolet light or by fluorometry. Because in some instances the PCR yields spurious bands, this approach may be limiting, although in our hands, once a PCR assay is optimized to yield unique bands, it is highly reproducible. On the other hand, the adaptation of fluorophores with distinct excitation–emission spectra in a multiplex PCR would allow the discrimination among PCR products with identical mobilities through the gel by virtue of their

[12] J. Rommens, B. S. Kerem, W. Gree, P. Chang, L. C. Tsui, and P. Ray, *Am. J. Hum. Gen.* **46,** 395 (1990).

[13] R. K. Saiki, P. S. Walsh, C. H. Levenson, and H. A. Erlich, *Proc. Natl. Acad. Sci. U.S.A.* **86,** 6230 (1989).

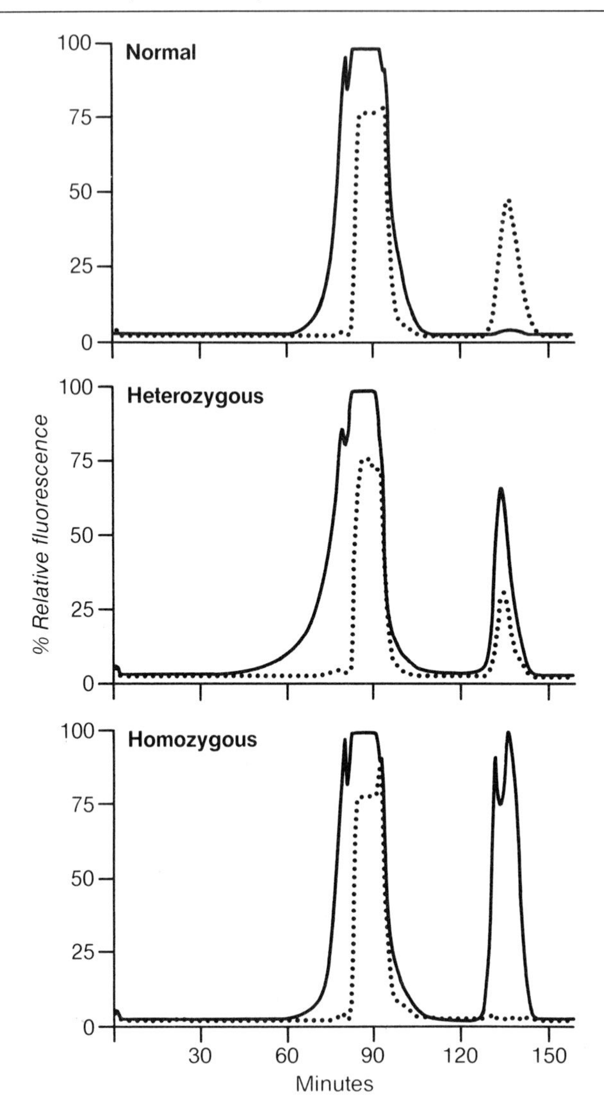

FIG. 2. Plots resulting from the electrophoretic separation and laser scanning of the color ΔF508 PCR products in normal, heterozygous, and homozygous individuals. The 2.5% agarose gel is continuously scanned near the bottom with a dual-wavelength laser. The x and y coordinates of the plots show, respectively, the running time of electrophoresis and the relative fluorescence of the PCR products. The dotted line represents fluorescein detection (green, normal primer) while the continuous line represents rhodamine (red, mutant primer) fluorescence. The fluorescent unextended primers are first detected by the laser and are represented by the two superimposed peaks in each amplification. The diagnostic peaks are then detected. A normal individual shows a diagnostic peak for fluorescein, the heterozygous fluorescein and rhodamine, the homozygous mutant only rhodamine.

different colors. This could be valuable for various situations, such as the DNA diagnostics of Duchenne/Becker muscular dystrophy, in which the multiplex PCR yields different PCR products that must be resolved by gel electrophoresis.[14] Also, the PCR detection of point mutations does not yield amplified products that can be resolved by conventional gel electrophoresis; but with the inclusion of fluorescent primers, the product from each allele can be discriminated. Thus, the sickle cell, IVS1 position 110, and the β^E single-base substitutions in the β-globin gene have been detected by this assay.[15-17] The automation of the fluorescent ΔF508 assay for cystic fibrosis screening would ease its implementation in future DNA diagnostics clinical laboratories if no gel electrophoresis is required. Thus it is possible that a small instrument could be devised that would couple both the amplification and the color read-out of the amplified products after removal of the unextended fluorescent primers.

[14] J. S. Chamberlain, R. A. Gibbs, J. E. Ranier, P. N. Nguyen, and C. T. Caskey, *Nucleic Acids Res.* **16,** 11141 (1988).
[15] F. F. Chehab and Y. W. Kan, *Lancet* **335,** 15 (1990).
[16] F. F. Chehab and Y. W. Kan, *Proc. Natl. Acad. Sci. U.S.A.* **86,** 9178 (1989).
[17] S. H. Embury, G. L. Kropp, T. S. Stanton, T. C. Warren, P. A. Cornett, and F. F. Chehab, *Blood* **76,** 619 (1990).

[15] Nonradioactive Detection of DNA Using Dioxetane Chemiluminescence

By STEPHAN BECK

Introduction

The use of nonradioactive DNA detection systems is becoming increasingly popular in molecular biology research. Currently, the most commonly used systems are based on either fluorescence, colorimetric staining, or chemiluminescence. This chapter will describe three closely related strategies for the detection of membrane-immobilized DNA using enzymatically catalyzed chemiluminescence. Several chemiluminescent systems, including the dioxetane system, have been known for some time and have been further developed (for a review, see Ref. 1). However, none of these systems has been able to find general acceptance for routine use. The reason for this is likely to be found in the failure to adequately meet

[1] S. Beck and H. Köster, *Anal. Chem.* **62,** 2258 (1990).

most of the crucial criteria that are needed to develop sensitive, but robust, protocols. These criteria are (1) high quantum yield (Φ_{CL}) of the chemiluminescent reaction, (2) high stability of the involved compounds, (3) long lifetime of the reaction, and (4) low complexity of the reaction. Stabilized, enzyme-triggerable 1,2-dioxetanes[2,3] seem able to overcome these problems.[4] The chemical and physical properties of such 1,2-dioxetanes have been described in detail in Ref. 1, and therefore only the principle of the phosphatase/dioxetane reaction will be discussed here. The protocols described here have been successfully tested on dot blots, Southern blots, and DNA sequence blots.

Principle of Method

The protocol for chemiluminescent detection begins after the target DNA has been transferred and immobilized (e.g., by UV irradiation) to a membrane. In a multistep reaction the catalyzing enzyme (e.g., alkaline phosphatase) is specifically attached to the target DNA. This can be achieved by various strategies, three of which are described here and are shown schematically in Fig. 1 and in Table I. In strategies A and B the biotin–(strept)avidin affinity system is employed, but other affinity systems (e.g., the digoxigenin system) are likely to work as well. In strategy A the target DNA itself is biotinylated at one or more sites. Alkaline phosphatase is attached to the biotin via a streptavidin bridge. In strategy B the target DNA is detected by hybridization with a biotinylated probe to which alkaline phosphatase is attached via a streptavidin bridge. Finally, in strategy C the target DNA is detected by hybridization with a probe to which alkaline phosphatase is directly attached. After alkaline phosphatase has been attached to the target DNA by one of the strategies, the light reaction is initiated by adding the chemiluminescent substrate. Figure 2 shows the principle of the phosphatase/dioxetane reaction. Cleavage of the protecting phosphate group (Fig. 2a) by alkaline phosphatase produces the destabilized dioxetane anion (Fig. 2b), which rapidly decomposes into the chemiluminescence-generating compound methyl *m*-oxybenzoate (Fig. 2c). During the transition of methyl *m*-oxybenzoate to the electronic ground state, light is emitted with a maximum at 470 nm. The rate-limiting step of the reaction is the decomposition of the dioxetane anion (Fig. 2b), which has a half-life ($t_{1/2}$) of 2–30 min depending on the chemical environment. The emitted photons can be detected and captured in several

[2] A. P. Schaap, M. D. Sandison, and R. S. Handley, *Tetrahedron Lett.* **28,** 1159 (1987).
[3] I. Bronstein, B. Edwards, and J. C. Voyta, *J. Biolum. Chemilumin.* **4,** 99 (1989).
[4] S. Beck, T. O'Keeffe, J. M. Coull, and H. Köster, *Nucleic Acids Res.* **17,** 5115 (1989).

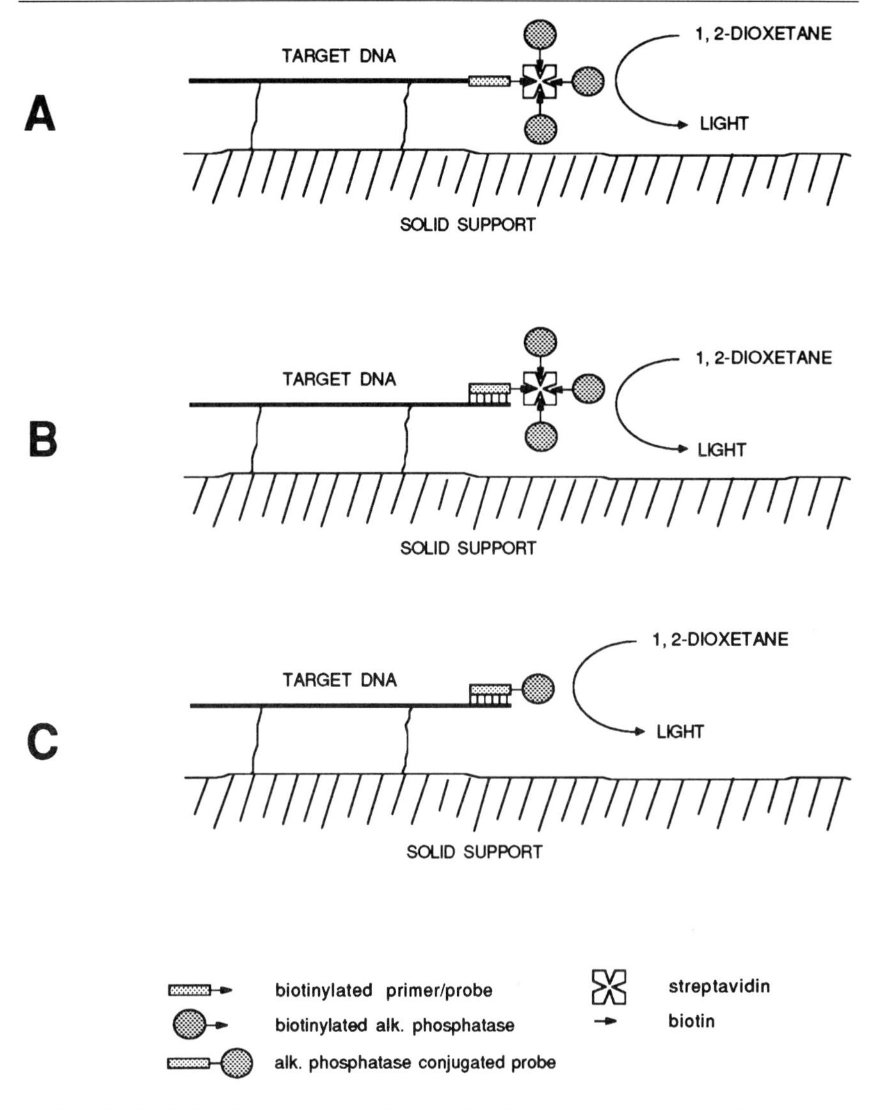

FIG. 1. Strategies for enzyme-catalyzed, chemiluminescent detection of membrane-immobilized DNA. [Adapted with permission from Ref. 1. Copyright (1990) American Chemical Society.] (A) Detection of immobilized, biotinylated target DNA. (B) Detection of immobilized target DNA by hybridization with a biotinylated probe. (C) Detection of immobilized target DNA by hybridization with an enzyme-conjugated probe.

TABLE I
FLOW CHARTS FOR CHEMILUMINESCENT DETECTION STRATEGIES A, B, AND C[a]

Step	Time (min)	Temperature[b] (°C)	Strategy		
			A	B	C
Prehybridization (1× buffer I)	30	*		√	√
Hybridization (1× buffer I)	60	*		√	√
Wash (1× buffer II)	10	*		√	√
Wash (1× buffer II)	10	*		√	
Wash (1× buffer II)	10	*		√	
Wash (1× buffer II)	10	*	√	√	
Streptavidin (10× buffer II)	10	21	√	√	
Wash (1× buffer II)	10	21	√	√	
Wash (1× buffer II)	10	21	√	√	
Wash (1× buffer II)	10	21	√	√	
Wash (1× buffer II)	10	21	√	√	
Phosphatase (10× buffer II)	10	21	√	√	
Wash (1× buffer III)	10	21	√	√	√
Wash (1× buffer III)	10	21	√	√	√
Wash (1× buffer III)	10	21	√	√	√
Wash (1× buffer III)	10	21	√	√	√
Wash (1× buffer III)	10	21			√
Lumiphos	10	21	√	√	√
Exposure	*	21	√	√	√
Strip (1× buffer II)	15	80		√	√
Wash (1× buffer II)	10	21		√	√
Wash (1× buffer II)	10	21		√	√
Storage (1× buffer II)	*	21		√	√

[a] Using a copy of Table I was found to be helpful for keeping track of the experiment by checking off the individual steps.

[b] *, Variable, depending on experiment. See text for examples.

ways: (1) by eye, after several minutes of dark adaptation (the simplest way), (2) by exposure to Polaroid or X-ray film (currently the most applicable way), and (3) by scanning with a light-sensitive CCD (charge-coupled device) camera (a more future-oriented way).

Materials

Heat sealer (e.g., HM-1300; Hulme-Martin, London, England): 50-cm sealing capacity

Heat-sealable plastic bags (e.g., Spec-Fab Co., Riverton, NJ): 38 × 45 cm

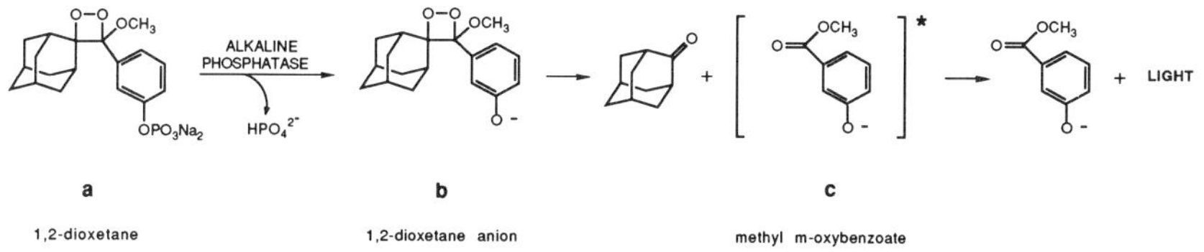

FIG. 2. Scheme of the phosphatase/dioxetane reaction. (a) Protected 1,2-dioxetane {4-methoxy-4-(3-phosphate phenyl) spiro[1,2-dioxetane-3,2-adamantane], disodium salt}; (b) 1,2-dioxetane anion; (c) electronically excited methyl *m*-oxybenzoate. [Adapted with permission from Ref. 1. Copyright (1990) American Chemical Society.]

Thermostatted (ambient to 80°) orbital shaker (e.g., G25; New Brunswick Scientific, Ltd., Hatfield, England)

Orbital shaker (e.g., PR70; Hoefer Scientific Instruments, Newcastle, England)

Nylon membrane (e.g., BNRG3R; Pall Biosupport, Portsmouth, England): 0.2 μm

UV cross-linker (e.g., UVC 1000; Hoefer Scientific Instruments)

X-Ray film (e.g., Fuji Medical), film cassette, and film development facility

Reagents

Alkaline phosphatase (in 30 m*M* Triethanolamine, pH 7.5, 3 *M* NaCl, 1 m*M* $MgCl_2$, 0.1 m*M* $ZnCl_2$), biotin conjugate (Cat. No. S-927; Molecular Probes, Eugene, OR)

Streptavidin [in 10 m*M* PO_4 buffer, pH 7.2, 15 m*M* NaCl, 0.05% (w/v) NaN_3] (Cat. No. 5532LB; Life Technologies, Inc., Uxbridge, England)

Lumiphos 480 (without fluorophore) (Lumigen, Inc., Detroit, MI)

Buffer I (1×, pH 7.2):

- Polyethylene glycol 8000, 100.00 g/liter
- NaCl, 7.30 g/liter
- Na_2HPO_4, 2.36 g/liter
- $NaH_2PO_4 \cdot 2H_2O$, 1.31 g/liter
- Sodium dodecyl sulfate [can be of low grade, e.g., Sigma (St. Louis, MO) Cat. No. L-5750], 50.00 g/liter

Buffer II (10×, pH 7.2):

- NaCl, 7.30 g/liter
- Na_2HPO_4, 2.36 g/liter
- $NaH_2PO_4 \cdot 2H_2O$, 1.31 g/liter
- Sodium dodecyl sulfate (as above), 50.00 g/liter

Buffer III (10×, pH 9.5):

Tris-HCl (pH 9.5), 100 mM
NaCl, 100 mM
$MgCl_2$, 10 mM

Strategy A: Detection of Biotinylated Target DNA

1. Wash membrane in a heat-sealed plastic bag with ~250 μl 1× buffer II/cm^2 of membrane for 10 min at room temperature with moderate shaking. Drain the bag.

2. Add ~50 μl (per cm^2 of membrane) 10× buffer II containing 0.5 μg/ml streptavidin. Incubate for 10 min at room temperature with moderate shaking. Drain the bag.

3. Wash membrane with ~250 μl 1× buffer II/cm^2 of membrane for 10 min at room temperature with moderate shaking. Drain the bag. Repeat the wash three times.

4. Add ~50 μl (per cm^2 of membrane) 10× buffer II containing 0.5 μg/ml biotinylated alkaline phosphatase. Incubate for 10 min at room temperature with moderate shaking. Drain the bag.

5. Wash membrane with ~250 μl 1× buffer III/cm^2 of membrane for 10 min at room temperature with moderate shaking. Drain the bag. Repeat the wash three times. Drain the bag/membrane to completion or blot membrane moist-dry between two sheets of Whatman (Maidstone, England) 3MM filter paper.

6. Add ~50 μl Lumiphos/cm^2 membrane. Incubate for 10 min at room temperature with moderate shaking. Drain the bag/membrane to completion or blot membrane moist-dry between two sheets of Whatman 3MM filter paper. Alternatively, under a hood spray Lumiphos uniformly over the surface of the membrane, using a vaporizing spray bottle.

7. Seal the membrane in a new bag or cover with plastic wrap and expose to X-ray film. The length of the exposure time varies between experiments and is under Concluding Remarks (below). Start with a 10-min exposure time.

Strategy B: Detection by Hybridization with Biotinylated Probe

1. Prehybridize the membrane with ~250 μl 1× buffer I/cm^2 of membrane in heat-sealed plastic bag for 30 min at ambient or hybridization temperature with moderate shaking. Drain the bag.

2. Add ~50 μl (per cm^2 of membrane) preheated 1× buffer I containing 1 nM biotinylated probe (final concentration). Hybridize the membrane at an appropriate temperature (e.g., 42° for 20-mers) for 1 hr with moderate shaking. Drain the bag.

3. Wash the membrane with ~250 μl 1× buffer II/cm^2 of membrane

for 10 min with moderate shaking. The temperature of the wash can vary depending on the probe and required stringency (e.g., 21° for 20-mers). Drain the bag. Repeat the wash three times.

4. Add ~50 μl (per cm^2 of membrane) 10× buffer II containing 0.5 μg/ml streptavidin. Incubate for 10 min at room temperature with moderate shaking. Drain the bag.

5. Wash the membrane with ~250 μl 1× buffer II/cm^2 of membrane for 10 min at room temperature with moderate shaking. Drain the bag. Repeat the wash three times.

6. Add ~50 μl (per cm^2 of membrane) 10× buffer II containing 0.5 μg/ml biotinylated alkaline phosphatase. Incubate for 10 min at room temperature with moderate shaking. Drain the bag.

7. Wash the membrane with ~250 μl 1× buffer III/cm^2 of membrane for 10 min at room temperature with moderate shaking. Drain the bag. Repeat the wash three times. Drain the bag/membrane to completion or blot membrane moist-dry between two sheets of Whatman 3MM filter paper.

8. Add ~50 μl Lumiphos per cm^2 of membrane. Incubate for 10 min at room temperature with moderate shaking. Drain the bag/membrane to completion or blot membrane moist-dry between two sheets of Whatman 3MM filter paper. Alternatively, under a hood spray Lumiphos uniformly over the surface of the membrane, using a vaporizing spray bottle.

9. Seal the membrane in a new bag or cover with plastic wrap and expose to X-ray film. The length of the exposure time varies between experiments and is discussed under Concluding Remarks (below). Start with a 10-min exposure time.

10. Remove the probe by adding ~250 μl preheated 1× buffer II/cm^2 of membrane. Incubate in a heat-sealed plastic bag for 15 min at 80° with moderate shaking. Drain the bag quickly and completely before the solution cools.

11. Wash the membrane with ~250 μl 1× buffer II/cm^2 of membrane for 10 min at room temperature with moderate shaking. Drain the bag. Repeat the wash once.

12. Store the membrane in ~100 μl 1× buffer II/cm^2 of membrane in a heat-sealed plastic bag for subsequent probings.

Strategy C: Detection by Hybridization with Phosphatase-Conjugated Probe

1. Prehybridize membrane with ~250 μl 1× buffer I/cm^2 of membrane in a heat-sealed plastic bag for 30 min at ambient or hybridization temperature with moderate shaking. Drain the bag.

2. Add ~50 μl (per cm^2 of membrane) 1× buffer I solution containing

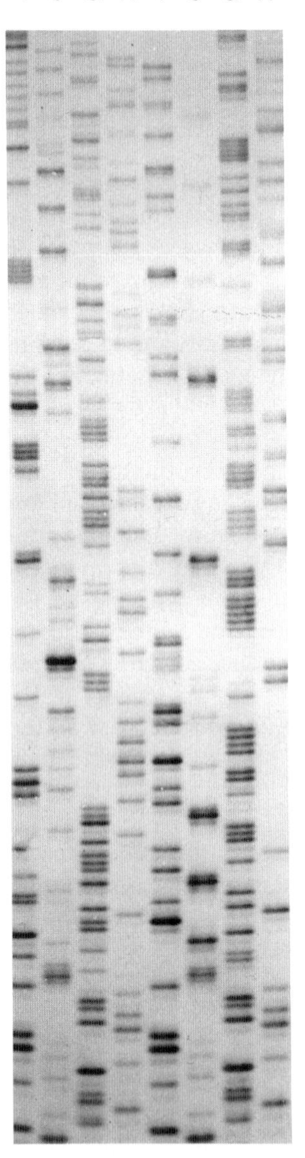
a
T C G A T C G A

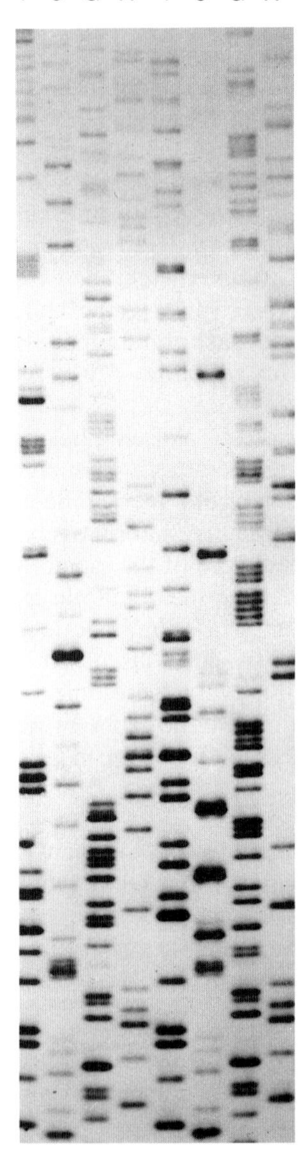
b
T C G A T C G A

1 n*M* phosphatase-conjugated probe (final concentration). Hybridize the membrane at an appropriate temperature (e.g., 42° for 20-mers) for 1 hr with moderate shaking. Drain the bag.

3. Wash the membrane with ~250 μl 1× buffer II/cm^2 of membrane for 10 min with moderate shaking. The temperature of the wash can vary depending on probe and required stringency (e.g., 21° for 20-mers). Drain the bag.

4. Wash the membrane with ~250 μl 1× buffer III/cm^2 of membrane for 10 min with moderate shaking. The temperature of the wash can vary depending on the probe and required stringency (e.g., 21° for 20-mers). Repeat the wash four times. Drain the bag/membrane to completion or blot the membrane moist-dry between two sheets of Whatman 3MM filter paper.

5. Add ~50 μl Lumiphos per cm^2 of membrane. Incubate for 10 min at room temperature with moderate shaking. Drain the bag/membrane to completion or blot the membrane moist-dry between two sheets of Whatman 3MM filter paper. Alternatively, under a hood spray Lumiphos uniformly over the surface of the membrane, using a vaporizing spray bottle.

6. Seal the membrane in a new bag or cover with plastic wrap and expose to X-ray film. The length of the exposure time varies between experiments and is discussed in the next section. Start with a 10-min exposure time.

7. Remove the probe by adding ~250 μl preheated 1× buffer II/cm^2 of membrane. Incubate in a heat-sealed plastic bag for 15 min at 80° with moderate shaking. Drain the bag quickly and completely before the solution cools down.

8. Wash the membrane with ~250 μl 1× buffer II/cm^2 of membrane for 10 min at room temperature with moderate shaking. Drain the bag. Repeat the wash once.

9. Store the membrane in ~100 μl 1× buffer II/cm^2 of membrane in a heat-sealed plastic bag for subsequent probings.

Concluding Remarks

The described protocols for nonradioactive detection of DNA using enzyme-catalyzed chemiluminescence provide equal, or superior, sensi-

---

FIG. 3. Comparison of a radioactive and chemiluminescent detected DNA sequence blot. Human genomic DNA was subcloned into M13, colabeled with ^{35}S and biotin, and sequenced as described in ref. 5 and (S. Beck, this series, Vol. 184, p. 612) (except that dATP and dGTP were replaced by 7-deaza dATP and 7-deaza dGTP). The lane order is, from left to right: T, C, G, A. (a) Radioactive detection with [^{35}S]dATP after an 8-hr exposure to X-ray film (Fuji). (b) Dioxetane chemiluminescent detection (according to strategy A, using the described spray technique) of the same blot after a 4-min exposure to X-ray film (Fuji).

tivity when compared to radioactive systems (see Fig. 3). The following tips and observations should prove helpful to achieve results with the same consistency and reliability is possible with radioactive systems.

The right choice of membrane is crucial for the dioxetane chemiluminescent detection. The use of nitrocellulose and polyvinylidene difluoride (PVDF) membrane is not recommended in this context because of their property to quench light and therefore also any chemiluminescent signal. Nylon membranes, in general, are excellent supports for use with dioxetane chemiluminescent detection. They are available in a variety of pore sizes and modifications, such as unsupported, supported, neutral, and charge modified. In most cases, supported nylon membranes have higher mechanical strength than unsupported ones. Neutral nylon membranes tend to have less unspecific background problems than charge-modified ones due to their overall lower binding capacity. The recently introduced polysulfone membranes have not yet been tested. The membrane used in Fig. 3 was Biodyne-A (Pall Biosupport, Portsmouth, England), a supported, neutral nylon-66 membrane.

Another crucial step is the immobilization of the target DNA to the nylon membrane, which is usually achieved by UV irradiation. The kinetics and efficiency of UV cross-linking are greatly influenced by the amount of moisture in the membrane (T. O'Keeffe and S. Beck, unpublished data, 1989). Under- or over-cross-linking can lead to loss of signal. It is therefore recommended that the membrane be completely dried (e.g., by baking for 15 min at 80°) prior to cross-linking. The short baking of the membrane at 80° may also help to enhance the signal-to-noise ratio.[5] The optimum cross-linking time, or energy,[6] is best determined experimentally for each setup by titration (e.g., of dot blots) followed by hybridization. The example shown in Fig. 3 was baked for 15 min at 80° and cross-linked for 30 sec with ~1500 $\mu W/cm^2$.

The buffer system used for membrane blocking and hybridization is essentially the same as described previously by Church and Gilbert.[6] Replacement of the phosphate buffer by a citrate buffer (1× SSC) works equally well. Sodium dodecyl sulfate (SDS) is used to block unspecific binding sites of the membrane. However, it is important to be aware that SDS needs to be washed off completely prior to the application of the dioxetane substrate for two reasons: (1) SDS inhibits the enzymatic triggering of the chemiluminescent reaction and (2) some dioxetane formulations (e.g., Lumiphos) are stabilized by the addition of the anionic detergent cetyltriammonium bromide (CTAB), which forms a precipitate in conjunc-

[5] S. Beck, *Anal. Biochem.* **164,** 514 (1987).

[6] G. M. Church and W. Gilbert, *Proc. Natl. Acad. Sci. U.S.A.* **81,** 1991 (1984).

tion with SDS. Replacement of SDS with CTAB as blocking agent did not work under the conditions used. In some experiments using phosphatase-conjugated DNA probes according to strategy C it was observed that the presence of polyethylene glycol (PEG) in the hybridization buffer caused increased background. In such cases PEG is simply omitted. The described buffer system limits strategies B and C to the use of DNA probes ≥12 nucleotides in length because of the insolubility of SDS at temperatures below 15°. This limitation in combination with dioxetane chemiluminescent detection may be overcome by a sarkosyl-containing buffer system[7] and/or chemically modified (e.g., methylated or brominated) DNA probes.[8,9] However, under the conditions described in strategy A, replacement of SDS by sarkosyl resulted in considerably higher background. Occasionally unspecific spots are observed after the chemiluminescent detection. In most cases these will disappear by making fresh solutions and/or filtering them through a 0.2- to 0.45-μm filter. The example shown in Fig. 3b was done with 2-month-old, unfiltered solutions.

The given protocols call for heat-sealable plastic bags for all membrane-processing steps. Instead of the bags, roller bottles can, of course, also be used; they are becoming increasingly popular because of their potential for automation.[5,10]

The average exposure time is less than 1 hr and for some applications several minutes are quite sufficient. Exposure time depends greatly on the detection medium: a comparison of X-ray (Fuji Medical) versus Polaroid 57 film revealed a 50 : 1 difference in exposure time. In contrast to Polaroid films, the manufacturers of X-ray films provide no information on sensitivity (such as ASA or dynamic range) and therefore more sensitive X-ray films for this particular application may exist. Once triggered, the light emission coming off the membrane reaches a plateau after about 15 hr and remains constant for up to several days.[1,11] Due to interaction with the nylon membrane, the emission wavelength is shifted by 10 nm from λ_{max} = 470 to 460 nm.[1,11] Despite the "limitations" mentioned above, images such as the DNA sequence blot shown in Fig. 3b can still be captured on X-ray film within minutes due to the high turnover rate (4.1×10^3 molecules/sec) of the phosphatase/dioxetane system.

[7] R. Drmanac, Z. Strezoska, I. Labat, S. Drmanac, and R. Crkvenjakov, *DNA Cell Biol.* **9,** 527 (1990).

[8] J. D. Hoheisel and H. Lehrach, *FEBS Lett.* **274,** 103 (1990).

[9] J. D. Hoheisel, A. G. Craig, and H. Lehrach, *J. Biol. Chem.* **265,** 16656 (1990).

[10] P. Richterich, C. Heller, H. Wurst, and F. M. Pohl, *BioTechniques* **7,** 52 (1989).

[11] R. Tizard, R. L. Cate, K. L. Ramachandran, M. Wysk, J. C. Voyta, O. J. Murphy, and I. Bronstein, *Proc. Natl. Acad. Sci. U.S.A.* **87,** 4514 (1990).

[16] Extraction of High Molecular Weight RNA and DNA from Cultured Mammalian Cells

By H. C. BIRNBOIM

Introduction

Although a large number of techniques have been described for the isolation of pure high molecular weight RNA and DNA from mammalian cells, no method is universally applicable because of different problems encountered with different cell and tissue types and different uses of the purified nucleic acids. In general, it is difficult to prepare both RNA and high molecular weight DNA from the same extract,[1,2] because each presents separate difficulties. The most common problem encountered in the isolation of intact RNA is the presence of stable ribonucleases inside and outside of cells. Because RNA is single stranded, one cut by a ribonuclease may destroy its usefulness. Thus, effective ribonuclease inhibitors are needed from the moment of initial cell lysis. RNA is not shear sensitive but it is very alkali labile, so the pH must be controlled at the pH of maximal stability (6.5–7.0), particularly if it is to be heated. Other factors need to be considered to isolate high molecular weight DNA. Deoxyribonucleases are less of a problem because, compared to ribonucleases, they are generally more labile and inhibitable by chelators. Furthermore, DNA is a duplex molecule, so that cleavage of one of the two strands usually does not destroy its experimental usefulness. Its high viscosity makes it difficult to redissolve after alcohol precipitation and it is sensitive to degradation by shearing. Preparation of intact, chromosome-size pieces of DNA requires special procedures such as embedding cells in agarose,[3] which will not be considered here. The present chapter describes methods for the isolation of highly purified RNA and DNA in near-quantitative yield from small numbers of cultured mammalian cells and white blood cells. These nucleic acids are suitable for Southern and Northern transfers and for other applications. The RNA extraction procedure is slightly modified from one described earlier.[4]

[1] F. D. Kury, C. Schneeberger, G. Sliutz, E. Kubista, H. Salzer, M. Medl, S. Leodolter, H. Swoboda, R. Zeillinger, and J. Spona, *Oncogene* **5,** 1403 (1990).
[2] J. Karlinsey, G. Stamatoyannopoulos, and T. Enver, *Anal. Biochem.* **180,** 303 (1989).
[3] D. C. Schwartz and C. R. Cantor, *Cell (Cambridge, Mass.)* **37,** 67 (1984).
[4] H. C. Birnboim, *Nucleic Acids Res.* **16,** 1487 (1988).

Extraction of RNA

The moment of initial disruption of a cell is the time that ribonucleases are released and can cause degradation of cellular RNA. Cell extraction should therefore be carried out in a solution containing strong denaturing agents and/or other inhibitors of ribonucleases. The solution should be at pH 6.5–7.0 (the pH of optimal stability for RNA).[5] Although guanidinium thiocyanate is useful as a denaturing agent and ribonuclease inhibitor,[6–10] a mixture of sodium dodecyl sulfate (SDS) and a low concentration of urea[11] is an effective inhibitor of potent leukocyte ribonucleases, which are more difficult to inhibit than pancreatic ribonuclease.[4] At the same time, this mixture effectively activates proteinase K.[12] Once most of the protein is digested, nucleic acids can be concentrated by precipitation with alcohol. This strategy minimizes the volume of phenol required for subsequent extractions and improves recovery. DNA can be eliminated almost completely by precipitation of high molecular weight RNA with LiCl, obviating the need for ribonuclease-free deoxyribonuclease or buoyant density sedimentation.

Reagents For RNA Extraction and Denaturation

RES-1: 0.5 *M* LiCl, 1 *M* urea, 1.0% SDS, 0.02 *M* sodium citrate, and 2.5 m*M* cyclohexanediaminetetraacetate (CDTA) (final pH 6.8). Filter before addition of SDS and store at room temperature

Proteinase K: 5 mg/ml in water; store at −20°

Phenol/chloroform: 10 g phenol crystals (analytical grade) dissolved in 10 ml of chloroform. Store at room temperature, protected from light. The solution should be colorless. The use of recrystallized phenol is unnecessary

Sodium acetate (pH 5.2) (2 *M* in sodium ion): For 100 ml of solution, 0.2 mol of sodium acetate is adjusted to pH 5.2 (as measured at 0.05 *M*) with acetic acid and made up to volume. Filter, autoclave, and store at room temperature over chloroform

Acetic acid: 2.0 *M*

[5] H. C. Birnboim, unpublished observation (1974).

[6] J. H. Chirgwin, A. E. Przybyla, R. J. MacDonald, and W. J. Rutter, *Biochemistry* **18,** 5294 (1979).

[7] G. Cathala, J. F. Savouret, B. Mendez, B. L. West, M. Karin, J. A. Martail, and J. D. Baxter, *DNA* **2,** 329 (1983).

[8] S. A. Krawetz, J. C. States, and G. H. Dixon, *J. Biochem. Biophys. Methods* **12,** 29 (1986).

[9] P. Chomczynski and N. Sacchi, *Anal. Biochem.* **162,** 156 (1987).

[10] S. L. Berger and J. M. Chirgwin, this series, Vol. 180, p. 3.

[11] R. Soeiro, M. H. Vaughan, and J. E. Darnell, *J. Cell Biol.* **36,** 91 (1968).

[12] H. Hilz, U. Wiegers, and P. Adamietz, *Eur. J. Biochem.* **56,** 103 (1975).

LiCl/ethanol: Combine 3 vol of 5 *M* LiCl (which has been filtered and autoclaved) with 2 vol of 95% ethanol. Store at room temperature

CCS: 1 m*M* sodium citrate, 1 m*M* CDTA, 0.1% SDS, adjusted to pH 6.8. Autoclave before addition of SDS. Filter and store at room temperature

FFP (formaldehyde/formamide/phosphate): FFP [100 μl 37% formaldehyde, 94 μl formamide, 6.6 μl 1 *M* sodium phosphate (pH 6.8)] is freshly prepared before use. RNA will be partially degraded during heating prior to Northern analysis if the pH is <6.0. Formamide tends to become acidic on standing and the highest grade of formamide available (e.g., Omnisolv grade from BDH Chemicals, Poole, England) should be used. It can be stored at 4° over mixed-bed resin. Formaldehyde may also be acidic. The pH of formamide, formaldehyde, or mixtures is conveniently monitored using 2 vol bromthymol blue (0.25 mg/ml in 0.5 m*M* NaOH) as a color indicator

Where indicated, solutions are filtered through a 0.45-μm membrane filter. Cyclohexanediamine tetraacetate is a chelator that, compared to ethylenediaminetetraacetate (EDTA), binds metal ions 100–1000 more strongly, is much more soluble in acid and alcohol, and is easier to prepare in concentrated solutions. It is available from Sigma Chemical Co. (St. Louis, MO).

RNA Extraction Procedure for Cultured Cells

Volumes given are for $3–5 \times 10^6$ cells grown as a monolayer in a 10-cm petri dish and can be readily scaled up or down according to the cell number and surface area. The dish is chilled on an aluminum plate in contact with ice. The culture medium is quickly removed by aspiration and the cell layer washed twice with cold balanced salt solution. Two milliliters of RES-1 is added briskly over the entire surface. The dish may be set on a rocking platform if several samples are being processed simultaneously. The thick lysate is scraped to one side using the edge of a disposable plastic weigh boat and transferred to a 15-ml centrifuge tube. The dish is rinsed with 1 ml of RES-1 and this is combined with the first lysate. For cells grown in suspension culture, the washed cell pellet is first suspended in a small volume of balanced salt solution and then RES-1 is added quickly at the same RES-1-to-cell ratio.

The lysate is sonicated at low power for 5 sec to shear the DNA and reduce its viscosity. Proteinase K (30 μl) is added and the mixture incubated at 50° for 30 min. After cooling to room temperature, 0.2 ml sodium acetate and 7 ml cold 95% ethanol are added. The mixture is allowed to stand for 20 min at −20° for a precipitate of nucleic acids to form. After centrifugation at 12,000 *g* for 10 min at 0°, the supernatant is removed and the precipitate is dissolved in 0.9 ml RES-1. The dissolved pellet of nucleic acids is transferred to a 1.5-ml polypropylene microcentrifuge tube.

Proteinase K (10 μl) is added and the sample is again incubated at 50° for 30 min. When cool, 0.1 ml of phenol/chloroform is added and the sample vortexed repeatedly over a 3-min period. The aqueous phase is recovered after centrifugation in a microcentrifuge for 5 min at room temperature. The organic layer is back extracted with 0.1 ml of RES-1 by vortexing for about 10 sec and centrifuging for 2 min. The combined aqueous fractions are then extracted once with 0.1 ml of chloroform to remove phenol by vortexing for 20 sec and centrifuging for 2 min. After this stage, to avoid contamination with exogenous ribonucleases it is necessary to employ precautions, such as wearing disposable gloves and using autoclaved reagents, tubes, and pipette tips.

The aqueous phase is transferred to a 2-ml polypropylene microcentrifuge tube (Sarstedt, St. Laurent, Quebec, Canada). High molecular weight RNA is selectively precipitated with 7.5 μl 2.0 *M* acetic acid and 1 ml LiCl/ethanol. The sample is mixed thoroughly and held on ice for at least 4 hr (preferably 16 hr). Precipitated RNA is collected by centrifugation for 2 min; the supernatant (containing DNA and low molecular weight RNA) is carefully and completely removed using a fine pipette tip, followed by brief centrifugation to remove any liquid left on the walls of the tube. The walls can be washed with 0.5 ml of a 1 : 1 mixture of LiCl/ethanol and water; the solution should be added carefully, so as not to dislodge the pellet of RNA, and then removed by aspiration.

The final pellet containing purified RNA is dissolved in CCS (0.4 ml). Note that LiCl-precipitated RNA is slow to dissolve. To lower the salt concentration further in preparation for electrophoresis, RNA is reprecipitated by the addition of 35 μl of sodium acetate and 1.0 ml of cold ethanol. The volume of CCS, sodium acetate, and ethanol should be reduced if the original cell number was low ($<3 \times 10^6$ cells/10-cm dish) to reduce losses at the ethanol precipitation stage. After allowing the precipitate to form at −20° for 20–30 min, the RNA is collected by centrifugation for 2 min. The RNA is reprecipitated once more from 0.2 ml of CCS with sodium acetate and ethanol. Depending on its intended use, the RNA may be dissolved in CCS or in sterile citrate/CDTA without SDS. The final preparation is stored frozen.

Typical recovery of high molecular weight RNA is about 25 $\mu g/10^6$ cultured cells. The content of RNA in leukocytes is much lower and the expected recovery is about 0.5 $\mu g/10^6$ cells. Note that CDTA absorbs significantly below 260 nm and so $A_{263\,nm}$ should be used to estimate the quantity of RNA recovered.

Denaturation of RNA prior to Northern Analysis

Gel electrophoresis of "denatured" RNA samples can be carried out using 1.2% (w/v) agarose in 20 m*M* sodium phosphate, 2 m*M* CDTA, 1 *M* formaldehyde. RNA samples in CCS are denatured by mixing 10 μl RNA (3–10 μg), 5 μl formamide, and 5 μl FFP and heating at 55° for 30 min. Note that RNA will be degraded by this heating step if the pH is too low (<pH 6.0) due to acidic formamide or formaldehyde.

Extraction of DNA

Initial steps in the procedure are similar to those used for RNA extraction. A similar extraction solution (different in that the pH is 8, the pH for maximal stability of DNA) is used and proteinase K is included to remove the bulk of protein. DNA is alcohol precipitated while still high in molecular weight to improve recovery. Subsequently, the precipitate is dissolved in a small volume to keep the solution viscous, which affords protection against degradation by shearing. Note that concentrated solutions of high molecular weight DNA may take a day or two to dissolve completely.

Additional Reagents For DNA Extraction

DES (DNA extraction solution): 1 *M* LiCl, 1 *M* urea, 0.2% SDS, 50 m*M* Tris-HCl, 5 m*M* CDTA (final pH 8.0). Filter before addition of SDS and store at room temperature

Ribonuclease: 1 mg/ml pancreatic ribonuclease in H_2O, boiled for 5 min to inactivate deoxyribonuclease. Store frozen

CT: 1 m*M* CDTA, 10 m*M* Tris-HCl, pH 7.5. Autoclave, then filter; store at room temperature

DNA Extraction Procedure for Cultured Cells

As for RNA, the volumes given are for 3–5 $\times$ 10^6 cells grown as a monolayer in a 10-cm petri dish and can be scaled up or down accordingly. After washing with cold balanced salt solution, add 1.5 ml DES to cover the entire surface. The viscous lysate is scraped to one edge of the plate and then transferred to a disposable 15-ml graduated centrifuge tube. The surface of the plate is rinsed with 0.5 ml DES, which is combined with the first lysate. Proteinase K (40 μl) is added and the mixture is shaken gently by hand until fairly homogeneous.

The lysate is incubated at 50° for 60 min. After cooling to room temperature, 0.55 vol 2-propanol is added and the sample mixed by inversion until a clot of DNA forms. The clot is collected on a fine glass rod or by low-speed centrifugation. The supernatant is removed as completely as possible. The DNA is dissolved in 0.1 ml of CT containing 50 μg/ml ribonuclease. It is shaken gently at first and then more vigorously as the solution becomes more viscous. When most is dissolved, the sample is incubated for 20 min at 37° to allow the ribonuclease to partially digest the RNA. The viscous solution is transferred to a 1.5-ml polypropylene tube and the original tube rinsed with 0.1 ml of DES. An equal volume of phenol/chloroform is added and the sample extracted by vortexing repeatedly over a 5-min period or by shaking on a mechanical shaker for 20–30 min.

After centrifugation for 2 min in a microcentrifuge, the aqueous layer and any interface material is transferred to a fresh tube. Fresh phenol/chloroform is added and the extraction is repeated. The aqueous layer is recovered and clarified by a 5-min centrifugation, leaving behind any insoluble material.

DNA is precipitated with 1 vol 95% ethanol (at room temperature) and the precipitate collected by centrifugation. The pellet is washed twice by suspending it in 1 ml ethanol with vortexing over several minutes to extract traces of phenol and SDS. Following brief centrifugation, the supernatant is discarded. Residual ethanol is removed under vacuum.

The final pellet is dissolved in CT to give an estimated final concentration of about 500 μg DNA/ml. The solution should be very viscous (viscoelastic) and may take 24 hr or longer to dissolve completely. Store at 4°. Note that removal of precise volumes requires care because of the high viscosity of the solution. As for RNA, $A_{263\,\text{nm}}$ is used instead of $A_{260\,\text{nm}}$ to estimate the quantity of DNA recovered. RNA contamination in DNA is conveniently estimated using a spectrofluorometer. About 1 μg/ml DNA is dissolved in 0.5 μg/ml ethidium bromide in CT. A fluorescence reading [excitation $(\text{Ex})_{520}$, emission $(\text{Em})_{590}$] is taken and then 10 μg/ml ribonuclease is added. Fluorescence readings are taken again at 5-min intervals until readings are stable. A decrease in fluorescence indicates RNA is present; the percentage contamination can be estimated knowing that the specific fluorescence of ribosomal RNA is about one-half that of duplex DNA.

Problems that May Be Encountered

Purification of DNA from cells that secrete large amounts of extracellular matrix material such as collagen, hyaluronic acid, or proteoglycans has presented problems, giving rise to material that coprecipitates with nucleic acids on alcohol precipitation. The same difficulty is not encountered during purification of RNA from the same cells using the procedure as described. In the case of DNA, extensive treatment of the first lysate with equal volumes of phenol/chloroform before alcohol precipitation and proteinase K treatment may minimize the problem, but at a cost of a poorer yield of DNA.

Concluding Remarks

Extraction of RNA and of DNA from mammalian cells each present different problems. For RNA, the most serious potential difficulty is degradation by endogenous ribonucleases. For DNA, poor yield and degradation by shearing can result if the concentration is too low. In the described procedure, potentially hazardous substances such as guanidinium thiocyanate, phenol, and chloroform are avoided or minimized. The general strategy for both RNA and DNA purification is to treat with proteinase K at an early step so that alcohol precipitation can be used to concentrate the samples and allow further processing in microcentrifuge tubes. This decreases the volume of phenol needed and also ensures near-quantitative recovery of material. Both procedures give highly purified DNA or RNA in good yield from cultured cells and from blood cells. The applicability of the procedures to the extraction of nucleic acids from tissues and organs has not been established.

Acknowledgments

This work was supported by a grant from the Medical Research Council of Canada.

[17] One-Tube versus Two-Step Amplification of RNA Transcripts Using Polymerase Chain Reaction

By CHRISTIANE GOBLET, EDOUARD PROST, KATHRYN J. BOCKHOLD, and ROBERT G. WHALEN

Introduction

Since the first report of specific DNA amplification using the polymerase chain reaction (PCR) by Saiki *et al.*,[1] the number of different applications has grown steadily, as have the modifications of the basic method. The capacity to amplify specific segments of DNA, made possible by PCR, has transformed the way we approach both fundamental and applied biological problems. The polymerase chain reaction is an *in vitro* method

[1] R. Saiki, S. Scharf, F. Faloona, K. Mullis, G. Horn, H. A. Erlich, and N. Arnheim, *Science* **230,** 1350 (1985).

for the enzymatic synthesis of specific DNA sequences, using two oligonucleotide primers that hybridize to opposite strands and flank the region of interest in the target DNA. A repetitive series of cycles involving template denaturation, primer annealing, and the extension of the annealed primers by DNA polymerase results in the exponential accumulation of a specific fragment, the termini of which are defined by the 5′ ends of the primers. Because the primer-extension products synthesized in one cycle can serve as a template in the next, the number of target DNA copies approximately doubles at every cycle. Moreover, the substitution of a thermostable DNA polymerase isolated from *Thermus aquaticus* (*Taq*)[2] for the previously used Klenow fragment of *Escherichia coli* DNA polymerase I[1,3] has allowed repeated cycling at the high temperatures (>90°) required for strand separation, without having to add Klenow DNA polymerase after each cycle.

For the purpose of detecting transcripts in total RNA, two general ways of performing the experiment are possible. In the first, a complementary DNA strand (cDNA) is prepared from the RNA using reverse transcriptase. This preparation is carried out independent of the PCR experiment, and the cDNA is synthesized and purified according to standard protocols. Once the cDNA is obtained, DNA amplification is performed with specific reaction conditions particular to the PCR step with the *Taq* polymerase enzyme. This protocol is called the two-step method. A second, alternative way of amplifying transcripts from RNA samples consists of performing the reverse transcription and the PCR reactions under the same reaction conditions. This method, which we described in 1989, is called the one-tube method.[4] In this chapter, we describe the advantages and limits of the one-tube amplification protocol versus the two-step method.

Methods and Results

RNA Preparations and Oligonucleotides Used

Total RNA is prepared from either 4- to 6-week-old Wistar rats (Iffa Credo, Lyon, France). Different tissues are used: hindlimb skeletal muscles (essentially the gastrocnemius muscle), cardiac ventricles and atrial from rats and hindlimb skeletal muscles (essentially gastrocnemius) from

[2] R.Saiki, D. Gelfand, S. Stoffel, S. Scharf, R. Higuchi, G. Horn, and H. A. Erlich, *Science* **239,** 487 (1988).

[3] K. B. Mullis and F. Faloona, this series, Vol. 155, p. 335.

[4] C. Goblet, E. Prost, and R. G. Whalen, *Nucleic Acids Res.* **17,** 2144 (1989).

mice. The RNAs used here are extracted according to the method of Chirgwin *et al.*[5] In other experiments not presented the RNA is extracted using the acidic phenol method[6]; no differences are apparent between the two methods of preparation as regards the effectiveness of the one-tube PCR procedure.

Specific oligonucleotide primers are selected for detection of the mRNAs encoding the fast IIB isoform of myosin heavy chain (MHCIIB) from rat and mouse, or the β_1 isoform of the thyroid hormone receptor (β_1-T_3R). The sequences and location of the primers are given in Fig. 1. The primers are synthesized on a Gene Assembler Plus (Pharmacia–LKB, Rockville, MD), using the published sequence for the rat MHCIIB,[7] the unpublished sequence from Dr. Shin'ichi Takeda (personal communication) for mouse MHCIIB, and the published sequence for the β_1-T_3R.[8]

Two-Step Amplification Method

The first-strand cDNA synthesis is carried out as follows. While keeping all solutions on ice, mix 10 μg of total cellular RNA, 15 units of RNasin RNase inhibitor (Cat. No. 27-0815-01; Pharmacia, Piscataway, NJ), 2.0 μl of pd$(N)_6$ primers (random hexamers, Cat. No. 27-2166-01; Pharmacia) dissolved at 5 mg/ml, 4 μl of a 5× buffer [composition: 250 m*M* Tris-HCl, pH 8.3 at 42°, 40 m*M* $MgCl_2$, 150 m*M* KCl, 50 m*M* dithiothreitol (DTT)], and finally sterile diethyl pyrocarbonate (DEPC)-treated H_2O to give a final volume of 20 μl. This mix is heated at 95° for 1 min on a Hybaid Intelligent heating block apparatus, chilled on ice, mixed, and spun for 5 sec in a microfuge. Then add, at room temperature, 2.0 μl of dNTPs (25 m*M*), 15 units of RNasin, and 24 units of reverse transcriptase [avian myeloblastosis virus (AMV), Cat. No. 120 111; Appligène]. This mix is incubated at 42° for 1 hr to allow cDNA synthesis to proceed. At the end of this period add 30 μl of 0.7 *M* NaOH, 40 m*M* ethylenediaminetetraacetic acid (EDTA), mix gently, and incubate at 65° for 10 min to destroy the remaining RNA. To precipitate the synthesized cDNA, add 5 μl of 2 *M* ammonium acetate, pH 4.5 and 130 μl of chilled absolute ethanol. This mix is spun at 14,000 rpm for 30 min at 4° in a microfuge. The supernatant

[5] J. M. Chirgwin, A. E. Przybyla, R. J. Mac Donald, and W. J. Rutter, *Biochemistry* **18,** 5294 (1979).

[6] P. Chomczynski and N. Sacchi, *Anal. Biochem.* **162,** 156 (1987).

[7] B. Nadal-Ginard, R. M. Medford, H. T. Nguyen, M. Periasamy, R. M. Wydro, D. A. Hornig, R. Gubits, L. I. Garfinkel, D. Weiczorek, E. Bekesi, and V. Mahdavi, *in* "Muscle Development, Molecular and Cellular Control" (M. L. Pearson and H. F. Epstein, eds.), p. 143. Cold Spring Harbor Laboratory, Cold Spring Harbor, New York, 1982.

[8] R. J. Koenig, R. L. Warne, G. A. Brent, J. W. Harney, P. R. Larsen, and D. D. Moore, *Proc. Natl. Acad. Sci. U.S.A.* **85,** 5031 (1988).

POSITION OF THE PRIMERS ON THE RAT MHCIIB mRNA (3' coding and non-coding sequence):

GAC CTA GTG GAT AAA TTA CAG ACT AAA GTG AAA GCC TAC AAG AGA
CAA GCA GAG GAG GCT GAG GAA CAA TCC AAC GTC AAC CTG GCC AAG
TTC CGC AAG ATC CAG CAC GAG CTG GAG GAA GCC GAG GAG CGG GCC
GAC ATC GCC GAG TCC CAG GTC AAC AAG CTG CGG GTG AAG AGC CGA
GAG GTT CAC ACC AAA GTC ATA AGC GAA GAA TAG CTCAATTCCTTCTGT
TGAAAGGTGACAGAAGAAATCACACAATGTGACGTTCTTTGTCACTGTCCTGTATATCA
AGGAAATAAAGCTGCAGATAATTTTGCAAAAAAAAAAAAAAA

POSITION OF THE PRIMERS ON THE MOUSE MHCIIB mRNA (3' coding and non-coding sequence):

TTG GTG GAC AAA CTA CAG ACT AAA GTG AAA GCC TAC AAG AGA CAG
GCT GAG GAG GCT GAG GAA CAA TCC AAT GTC AAC CTG GCC AAG TTC
CGT AAG ATC CAG CAC GCG CTG GAG GAA GCC GAG GAG CGG GCT GAC
ATC GCG GAG TCC CAG GTC AAC AAG CTG CGG GTG AAG AGC CGA GAG
GTT CAC ACT AAA GTC ATA AGC GAA GAA TAA TCCATCTTTCTGTTGAGAG
GTGACAGGAGAAATCACAAAATGTGACGTTCTTTGTCACTGTCCTGTATATCACGGAAA
TAAATTCTGCAGATAATTTTGCAATCTAAAAAAAAAA

POSITION OF THE PRIMERS ON THE RAT β1-T3R mRNA (5' non-coding sequence):

ATGACTCCTAACAGTATGACAGAAAAATGCCTTCCAAGCCTGGGACAAACAGAAGCCCC
ATCCAGACCGAGGCCAGGACTGGAAGCTGGTAGGAATGTCCGAAGCCTGCCTGCACAGG
AAGAGCCATGTGGAGAGGCGTGGTGCATTGAAGAATGAGCAAACATCCTCACACCTCAT
CCAGGCCACTTGGGCGAGCTCTATATTCCATCTGGATCCAGACGATGTGAACGATCAGA
GCGTCTCAAGCGCCCAGACTTTCCAGACCGAAGAGAAGAAATG

FIG. 1. The sequences of the various primers used in the experiments reported here for the detection of mRNA transcripts were derived from the cDNA sequences of rat MHCIIB,[7] mouse MHCIIB (S. Takeda, unpublished communication, 1991), and the rat β_1-T_3R.[8] The positions of the primer sequences are underlined.

is removed, and the precipitate is resuspended in approximately 1 ml of 70% ethanol and spun again at 14,000 rpm for 30 min in the cold. The supernatant is discarded and the precipitate dried for approximately 10 min in a vacuum system (Speed-Vac concentrator; Savant, Hicksville, NY). The resulting dry pellet is dissolved in 50 μl sterile H_2O, and subsequently 5 μl of the cDNA product, equivalent to the 1 μg of RNA used in the one-tube method (see below), is used as a template for PCR.

The PCR experiment is then performed as follows on a Hybaid Intelligent Heating Block apparatus. A mixture of the cDNA, 100 ng of the chosen primers, and water to give a total volume of 30 μl is heated at 94°

for 5 min and then cooled on ice. A mix of 20 μl of a 5× PCR buffer [335 mM Tris-HCl, pH 8.8 at 25°, 83 mM $(NH_4)_2SO_4$, 33.5 mM $MgCl_2$, 50 mM 2-mercaptoethanol, 0.85 mg/ml bovine serum albumin (BSA)], 0.75 unit *Taq* DNA polymerase (AmpliTaq DNA polymerase, N801-0060; Cetus), and 500 μM dNTPs (Boehringer Mannheim Biochemicals, Meylan, France) is then added to each tube and overlayed with 100 μl paraffin oil to avoid evaporation. Forty cycles of PCR are performed. A single cycle consists of 1 sec at 94° for denaturation, 30 sec at 55° for annealing, and 2 min at 72° for elongation; the fortieth cycle has a 3-min elongation to extend incomplete DNA fragments fully. Twenty microliters are electrophoresed in an agarose gel [1.4% (w/v) Nusieve (FMC, Rockland, ME) and 0.6% (w/v) type II agarose (Sigma, St. Louis, MO)].

One-Tube Amplification Method

In this method, the AMV reverse transcriptase and the *Taq* DNA polymerase are both present in the same tube, and the reaction parameters are altered accordingly. One microgram of total RNA, 100 ng of the appropriate primers, and water in a total volume of 30 μl are mixed, the solution is heated at 65° for 15 min to denaturate the RNA, and is then cooled on ice. A 5× PCR buffer with the same composition as above, 0.75 unit *Taq* DNA polymerase, 1 unit of AMV reverse transcriptase, and 500 μM dNTPs are added. To allow cDNA synthesis, we introduce a new step in the heating block program: the reaction is first incubated at 42° for 15 min to allow the reverse transcriptase to function, and then 40 cycles are performed as described above.

Results

Figure 2 illustrates the amplified DNA fragments obtained from the MHCIIB (lane a) and the β_1-T_3R (lane b) mRNA transcripts, using total RNA prepared from rat gastrocnemius muscle as a substrate. Lanes c and d of Fig. 2 show the products for the MHCIIB and β_1-T_3R transcripts, respectively, from a PCR experiment using gastrocnemius muscle tissue from mouse. The expected size of the amplified fragments of the MHCIIB transcripts is 220 bp in rat and 240 bp in mouse; the expected size for the β_1-T_3R transcripts is 193 bp in rat (the expected size of the mouse transcript is unknown). The results show that the amplified fragments from a relatively abundant transcript in muscle (MHCIIB) are at the expected size and are easily visualized on a background of few extraneous bands, in both the rat (lane a, Fig. 2) or the mouse (lane c, Fig. 2) samples. On the contrary, the amplified products from a considerably less abundant

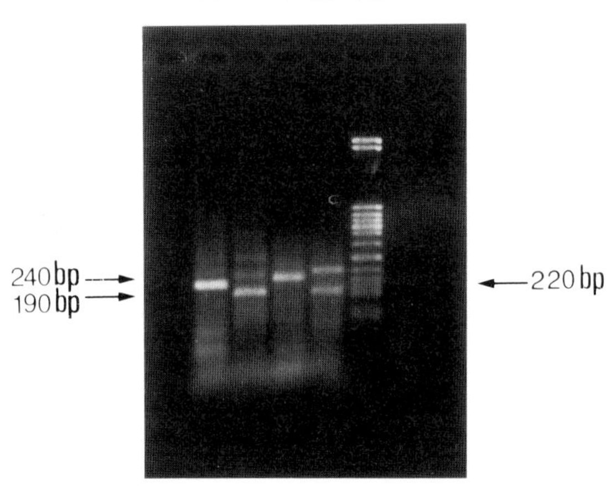

FIG. 2. Electrophoretic analysis of PCR-amplified fragments. The RNA samples in lanes (a) and (b) are from rat hindlimb muscle, and those in lanes (c) and (d) are from mouse hindlimb muscle. PCR amplification was carried out using primers for the rat MHCIIB mRNA in lanes (a) and (c), or using the primers for the rat β_1-T_3R mRNA in lanes (b) and (d). Lane (m) is the molecular weight marker, pBR322/*Hae*III + pBR322/*Taq*I.

transcript such as β_1-T_3R exhibit several bands; this type of result could be related to the source and/or possibly the quality of the RNA preparation. However, even in these circumstances bands of the expected size can nonetheless be readily observed. In this experiment, in which the PCR reaction was carried out for 40 cycles, the MHCIIB bands in the products are only marginally more visually intense than the β_1-T_3R bands, indicating that in these conditions, no quantitative assessment can be made in comparing different samples.

The abundance of the particular mRNA species in a given tissue doubtless contributes to the quality of the results obtained using the one-tube method. As an example (Fig. 3), the β_1-T_3R transcript was amplified from total RNA samples prepared from skeletal muscle (lane a), cardiac ventricle (lane b), and cardiac atrium (lane c) of the rat; the expected size of the amplified fragments is 193 bp. Only in the case of certain tissues (the atrial sample in this experiment) is a profile obtained that is relatively free of extra bands.

Figure 4 shows a comparison between the one-tube protocol and the two-step method. In lanes a to d (Fig. 4), we attempted to amplify a rare transcript (β_1-T_3R) in rat skeletal (lanes a and b, Fig. 4) and cardiac ventricular muscle tissue (lanes c and d, Fig. 4). The products amplified

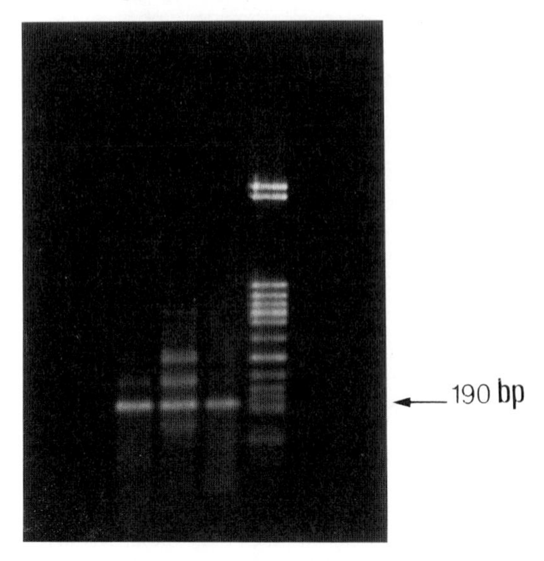

FIG. 3. Electrophoretic analysis of PCR-amplified fragment from the rat β_1-T_3R transcript. The RNA samples are from rat skeletal muscle (lane a), rat ventricular muscle (lane b), or rat atrial muscle (lane c). Lane (m) is the molecular weight marker, pBR322/*Hae*III + pBR322/*Taq*I.

by the one-tube method show, as previously seen in Fig. 3, artifactual bands. No amplified product can be detected by the two-step method. However, after hybridization with a specific probe, in the one-tube method only the band at the expected size was detected. No other bands were seen, implying that the other bands seen in ethidium bromide staining were indeed artifactual. The absence of PCR products in the two-step protocol in this particular experiment remains unexplained (variability with these relatively rare transcripts is frequently a problem), but the results presented serve to illustrate that the one-tube method is at least as reliable as the two-step procedure.

Lanes e and f in Fig. 4 represent the amplified products from the abundant MHCIIB transcript in rat skeletal muscle, obtained by the one-tube and the two-step method, respectively. The quality of the results is similar in both instances, indicating that for this major transcript no obvious artifacts are observed using the simpler one-tube method, even under these conditions of 40 PCR cycles.

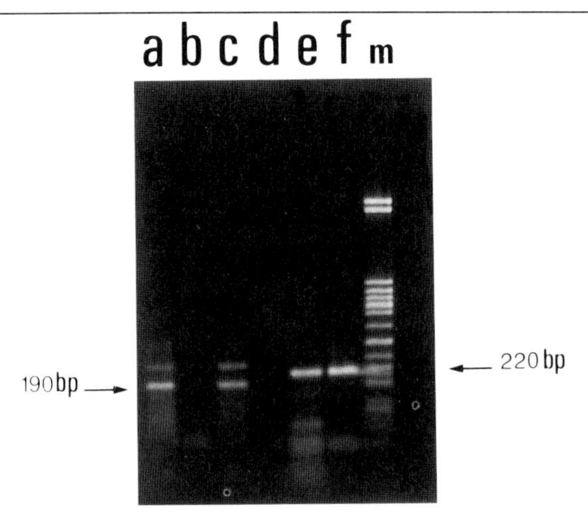

FIG. 4. Electrophoretic analysis of PCR-amplified fragments. Transcripts from the β_1-T_3R mRNA were amplified from RNA prepared from rat skeletal muscle (lanes a and b) and from cardiac ventricular muscle (lanes c and d). The amplification was carried out using the one-tube method (lanes a and c) or the two-step protocol (lanes b and d). Transcripts from the rat MHCIIB mRNA were amplified from RNA prepared from rat skeletal muscle using the one-tube method (lane e) or the two-step protocol (lane f). Lane (m) is the molecular weight marker, pBR322/*Hae*III + pBR322/*Taq*I.

Discussion and Conclusions

From our experience in using the two different approaches to the PCR detection of mRNA transcripts in total cellular RNA, illustrated by the results presented here, we conclude that in some experimental conditions the one-tube amplification method is as efficient as the two-step method but much less time consuming. It is a method particularly suitable for routine amplification of a large number of RNA samples. The conditions required for amplification of specific bands in the absence of satellite bands involve either the abundance of the RNA in the tissue tested or some feature of the RNA itself as obtained from different tissues. Indeed, as we demonstrate here, a rare transcript, the β_1-T_3R mRNA, is easily amplified by this method in RNA from some tissues (e.g., cardiac atrial, *vide infra*) but not in some other muscle tissues (cf. Fig. 3).

In the case of an abundant transcript such as the MHCIIB mRNA in skeletal muscle, the one-tube method reproducibly yields a single major band after one-tube PCR amplification. In the results presented here, the high number of PCR cycles almost certainly causes transcript production to reach a plateau (cf. Fig. 2 and above), and thus our conclusions concern only the qualitative aspects of the results; this will be sufficient for certain applications of mRNA detection by PCR. However, from the results obtained in certain favorable conditions (i.e., the presence of an abundant transcript) we suggest that the difficulties in carrying out a quantitative analysis will not be greater with the one-tube method compared to the two-step procedure, which requires a prior and independent cDNA preparation Because quantitative studies will normally involve the comparison of large numbers of samples, the one-tube procedure may have distinct advantages in this experimental situation.

Acknowledgments

This work was supported by the Association Française contre les Myopathies, the Institut National de la Santé et de la Recherche Médicale, the National Institutes of Health, and the Muscular Dystrophy Association of America. This laboratory is financed by the Pasteur Institute and the Centre National de Recherche Scientifique. E. Prost and K. J. Bockhold were recipients of fellowships from the Association Française contre les Myopathies, and C. Goblet and R. G. Whalen are Chargé de Recherche and Directeur de Recherche, respectively, of the Centre National de Recherche Scientifique.

[18] Characterization of Polysomes and Polysomal mRNAs by Sucrose Density Gradient Centrifugation Followed by Immobilization in Polyacrylamide Gel Matrix

By ADI D. BHARUCHA and M. R. VEN MURTHY

Fractionation of mRNAs is generally carried out by gel electrophoresis using conditions based on differences in charge, nucleotide composition, size of the molecule, or conformational properties.[1] mRNAs are composed of functionally different elements, a sequence that codes for a specific protein, and other sequences in the 5′ and 3′ regions that are implicated in the regulation of translation of the coding sequence. Different mRNAs of comparable lengths may have different proportions of coding and non-coding nucleotides. Electrophoresis is not useful in separating mRNAs based on the lengths of the coding sequence alone, because it cannot distinguish between the different functional regions of mRNA, which are all made up of the same four ribonucleotides. On the other hand, inside the cell, mRNAs are associated with ribosomal subunits to give rise to

[1] J. Sambrook, E. F. Fritsch, and T. Maniatis, "Molecular Cloning: A Laboratory Manual," 2nd Ed., Vol. 1. Cold Spring Harbor Laboratory, Cold Spring Harbor, New York, 1989.

polysomal aggregates in which the size of the polysome is proportional to the length of the coding region of the mRNA molecule. Sucrose density gradient (SDG) centrifugation, which is extensively employed for the size fractionation of polysomes, therefore offers a very useful method for the separation of mRNAs based on the length of the coding region. Generally, following centrifugation, the SDG is scanned in an ultraviolet (UV) spectrophotometer using a flow cell and the polysome peaks are recovered separately using a fraction collector. The polysomal aggregates are sedimented from each of these pooled fractions by ultracentrifugation, when further characterization of these particles and their mRNA is desired. The numerous manipulations required in this approach not only involve long delays, but often lead to low yields of polysomes and mRNA due to increased chance of contamination with RNase. Diffusion of the bands following centrifugation and flow turbulences during the passage of the gradient fluid through the flow cell of the spectrophotometer also result in poor resolution.

We have developed a novel method for the fractionation of polysomes and mRNA that combines the resolving capacity of SDG centrifugation with the analytical versatility of a gel medium.[2] In principle, the gradient in which polysomes are fractionated by centrifugation contains, in addition to the usual ingredients, other components that, by themselves, have no effect on the sedimentation of polysomes, but are able to solidify the gradient when an appropriate signal is given at the desired time. This is achieved by including in the medium a mixture of acrylamide, bisacrylamide (cross-linker), *N,N,N′,N′*-tetramethylethylenediamine (TEMED; polymerization accelerator), and riboflavin (polymerization catalyst). In contrast to the TEMED–ammonium sulfate system commonly employed for gelling acrylamide, the TEMED–riboflavin system initiates polymerization only on exposure to direct light and in the presence of traces of oxygen. The construction of the gradient, sample deposition, and centrifugation are, therefore, all carried out in diffuse light. Immediately after centrifugation, the gradients are gelled by exposing them to fluorescent light, thus entrapping the separated polysome bands in the gel matrix. Gelation occurs rapidly (20–30 min), with no disturbances in the boundaries of the bands due to diffusion or convection currents. This method not only serves to stabilize the polysomes for as long as needed, but permits easy scanning and quantitation of the separated polysome aggregates. The gel cylinder containing the entrapped polysomes is sliced longitudinally and the polysome bands in the slices are revealed using protein

[2] M. R. V. Murthy, A. D. Bharucha, and R. Charbonneau, *Nucleic Acids Res.* **14,** 6337 (1986).

or nucleic acid stains. The stained polysome bands are then quantitated using a gel scanner set to an appropriate wavelength.

The separation and entrapment of polysomes of different size classes in a gel matrix in the above fashion opens the way for the potential fine analysis of polysomes and their constituents (mRNA, rRNA, proteins) by the highly versatile techniques of polyacrylamide gel systems. For example: (1) polysome bands can be cut out from the slices and the proteins and mRNAs isolated by phenol extraction or by electroelution; (2) proteins and mRNAs in the bands can be further analyzed by one-dimensional or two-dimensional gel electrophoresis by inserting the bands directly into the sample slots of the gel; (3) the gel slices may be subjected to Western blot and the nascent proteins in the polysome bands may be probed with specific labeled antibodies in order to localize the polysome aggregates involved in the synthesis of the corresponding antigens; (4) following Northern blot, the association of a given mRNA with any of the polysome bands may be verified using appropriate cDNA or synthetic oligonucleotide probes.

Reagents

Acrylamide, bisacrylamide, TEMED, and riboflavin are from Eastman Organic Chemicals, (Rochester, NY). Coomassie Brilliant blue and methylene blue are from BDH, Ltd. (Poole, England) and from Fisher Scientific Company (Pittsburgh, PA), respectively. Sucrose (ultrapure) is from Schwarz/Mann (Orangeburg, NY) and glycerol (USP) from Anachemia Chemicals, Ltd. (Montreal, Canada).

Solutions

HKM buffer (5×): 0.25 *M* HEPES (pH 7.4), 0.5 *M* potassium acetate, and 25 m*M* magnesium acetate

Acrylamide (20%, w/v): 19 g of acrylamide and 1 g of bisacrylamide dissolved in 100 ml of water and the solution filtered successively through 40- and 22-μm Millipore (Bedford, MA) filter

TEMED (1%, v/v): 1 ml TEMED in 100 ml water

Riboflavin (0.005%, w/v): 5 mg riboflavin dissolved in 100 ml of water and stored in a dark bottle

Homogenization medium: 0.25 *M* sucrose, 5 m*M* mercaptoethanol, and 500 μg/ml heparin, all dissolved in 1× HKM buffer. This is prepared just before use

Sedimentation medium: 2.0 *M* sucrose and 5 *M* mercaptoethanol dissolved in 1× HKM buffer. This is prepared just before use

Polymerizing gradient medium: 20 ml HKM buffer, 25 ml of 20% acrylamide, 2.5 ml of 1% TEMED, 10 ml of 0.005% riboflavin, and water to 100 ml. The solution is prepared in a dark bottle just before use without exposure to direct light

Coomassie blue staining solution: 1.0 g Coomassie blue, 500 ml methanol, 100 ml glacial acetic acid, and water to 1 liter

Coomassie blue destaining solution: 45 ml methanol, 100 ml ethanol, 25 ml glycerol, 100 ml glacial acetic acid, and water to 1 liter

Methylene blue staining solution: 2.0 g methylene blue, 32.8 g anhydrous sodium acetate, 10 ml glacial acetic acid, and water to 1 liter (pH 4.7)

Ethidium bromide staining solution: 5 μg/ml of ethidium bromide in 0.5 *M* ammonium acetate

Preparation of Polysomes

Tissues are homogenized in 5 vol of homogenization medium.The homogenate is centrifuged at 12,000 *g* for 10 min (all operations are performed at 4°). The supernatant is layered over one-third the volume of sedimentation medium and spun at 38,000 rpm for 4.5 hr in the Beckman (Palo Alto, CA) SW-40 rotor. The free polysome pellets are rinsed with HKM buffer and either stored at −80° until use or immediately dissolved in the same buffer to a concentration of 150 A_{260} units/ml.

Density Gradient Centrifugation of Polysomes

We have used both linear and isokinetic sucrose or glycerol gradients for the fractionation of polysomes. Appropriate amounts of sucrose or glycerol are dissolved in the HKM buffer or the polymerizing gradient medium as required. Linear density gradients (20–45% sucrose) are constructed with the help of a Buchler (Nuclear Chicago Corp., Fort Lee, NJ) gradient former. The isokinetic gradients (15–33% sucrose or 25–55% glycerol) are prepared as described by Noll.[3] Polymerizing gradients are made in subdued light and 16.5 ml is superposed over a 0.6-ml cushion of 60% sucrose. Polysomes are suspended in 10% sucrose in HKM buffer and 30–50 μl of the suspension containing 4.5–7.5 A_{260} units is layered on each gradient. The tubes are centrifuged in a Beckman SW 27 rotor at 27,000 rpm for 4.5 hr in the case of linear gradients and for 3 hr in the case of isokinetic gradients.

Gelling of Gradient

Following centrifugation, the tubes containing the polymerizing gradients are placed upright in a support and are illuminated by a circular

[3] H.Noll, *in* "Techniques in Protein Biosynthesis" (P. N. Campbell and J. R. Sargent, eds.), Vol. 2, p. 123. Academic Press, New York, 1969.

fluorescent tube light in such a manner that all parts of the gradient are uniformly exposed to light. The gradient gels in about 30 min except for the 60% sucrose cushion at the bottom and the sample solution at the top of the tube, which do not contain acrylamide. The cylindrical gel can be released from the tube easily by cutting the bottom of the tube with a sharp razor and sliding out the gel by means of a syringe filled with sterile distilled water.

Longitudinal Slicing of the Gelled Gradient

We have used two types of apparatus for cutting the gradient into longitudinal slices, depending on the quantities of polysomes required for further biochemical manipulations. Figure 1 shows a gel slicer whose construction and method of employment have been described previously.[4] The cutting part of the apparatus consists of a thin, stainless steel wire mesh through which the gel is forced lengthwise by application of light pressure from a water-filled syringe. The wire mesh is sandwiched between an entry tube and an exit tube that protect the gel during its passage through the mesh. The gel slices leaving the exit tube are collected in a beaker containing an appropriate solution compatible with the next desired treatment. The slices are easily separated by teasing with a stainless steel spatula. Most of the slices, except those originating from the periphery of the gel, have uniform 2 × 4 mm rectangular cross-sections and, with proper care, 16–20 such slices can be recovered by this procedure.

An alternative sectioning devise that gives rise to fewer, but wider, gel slices is shown in Fig. 2. The apparatus consists of a small rectangular Plexiglas tray with 0.5-mm-thick walls and the following internal dimensions: 1.5 cm wide, 1.8 cm deep, and 8.5 cm long. The narrow walls of the tray are cut in the form of combs with a number of parallel and vertical teeth reaching to the floor of the tray. The spaces between the teeth are just wide enough to accommodate a thin nylon thread. The positions of teeth on the opposite walls are aligned so that a nylon thread passing through two corresponding opposite spaces is parallel to the length of the tray.

Molten agarose solution (1.6%, w/v) is poured into the Plexiglas tray to a height of 2 mm and allowed to set at room temperature. The leakage of agarose through the combs is prevented by blocking with rectangular pieces cut from a thin plastic sheet (for example, photographic film). The polyacrylamide cylinder containing the polysome bands is laid over the agarose layer and more hot agarose (1.6%) is added to cover the gel 2–3

[4] M. R. V. Murthy and P. De Grandpré, *Anal. Biochem.* **152,** 35 (1986).

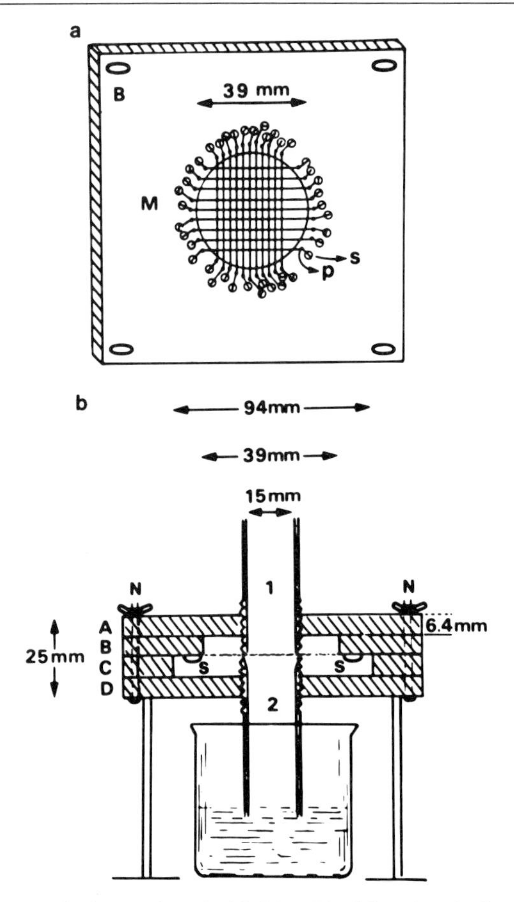

FIG. 1. Stainless steel wire mesh embedded in a Plexiglas plate for longitudinal slicing of cylindrical polyacrylamide gels containing immobilized polysome bands. (a) top view; (b) cross-sectional view. M, Wire mesh; S, screws holding the cross-wires of the mesh; P, tension pins; N, wing nuts and bolts holding the Plexiglas plates A–D; 1 and 2, entry and exit tubes for the gel. For details of construction and operation of the apparatus, see Ref. 4.

mm beyond the upper surface of the gel. The polyacrylamide gel cylinder is thus embedded inside the agarose, which protects it against inadvertant damage during the slicing step. The plastic blockers are removed and a length of nylon thread (Trilene XL 0.9-kg nylon fishing line; Berkley Ltd., Manitoba, Canada) is inserted between opposite interteeth spaces. The thread is held firmly under tension and the gel is sliced by making to-and-fro motions like a dental floss, until the thread cuts all the way through the gel. This is repeated with the other spaces until the gel is cut into the

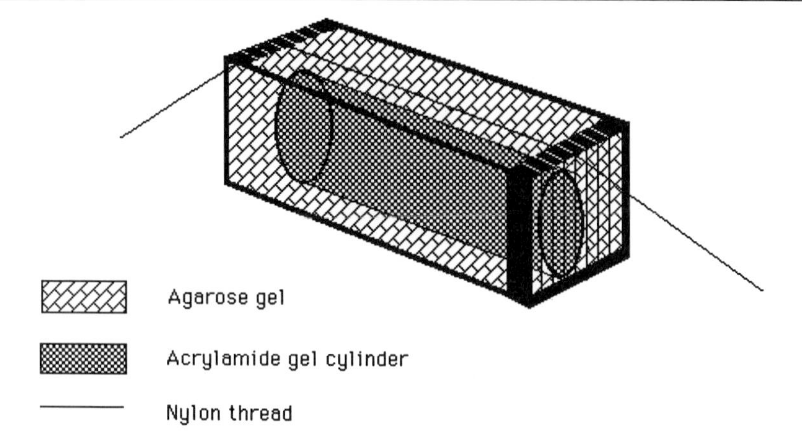

FIG. 2. Apparatus for longitudinal slicing of cylindrical gels using nylon thread.

maximum number of slices. Because the cutting pressure is mainly applied at the extremities, one must be careful that the nylon thread is maintained horizontal and taut throughout the procedure so that the slices have uniform thickness. It is possible to color the nylon thread by staining it in 5% (w/v) Coomassie blue for 30 min and rinsing off the excess dye. This permits visualization of the thread during its passage through the gel, so that the progress of the slicing may be followed. The sliced gel is then transferred to a petri dish containing an appropriate buffer, and the agarose borders are removed with the help of a spatula.

Revelation of Polysome Bands in Gel Slices

Polysome bands in the gel slices can be visualized by staining with one of the following reagents.

Coomassie blue: The entire gelled gradient or gel slices are fixed in 12.5% trichloroacetic acid (TCA) for 45 min. The gels are rinsed in distilled water, soaked in Coomassie blue staining solution for 6–8 hr, and are then destained in Coomassie blue destaining solution until all polysome bands are clearly revealed.

Methylene blue: The gels are fixed in 1 *M* acetic acid for 20 min and then transferred to the methylene blue staining solution for 1.5 hr. Excess stain is washed off with distilled water.

Silver stain: The staining is carried out essentially according to the method of Oakley *et al.*[5]

[5] B. R. Oakley, D. R. Kirsch, and N. R. Morris, *Anal. Biochem.* **105,** 361 (1980).

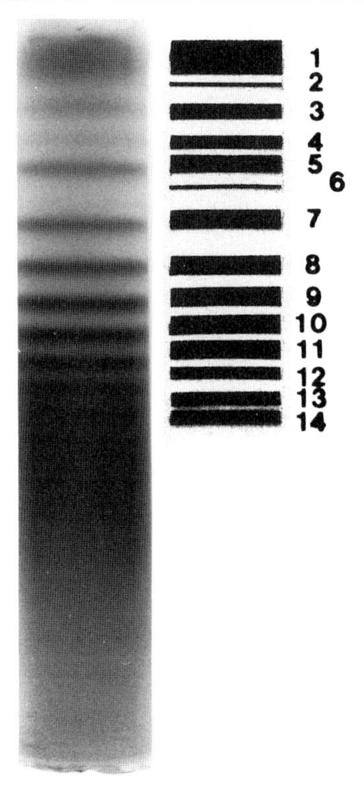

FIG. 3. Polysomes separated by sucrose density gradient, gelled, and stained with Coomassie blue. Free polysomes were prepared from 10-day-old rat brains and were fractionated on a linear sucrose gradient (20–45%) in a polymerizing medium as described in the text. Following centrifugation, the gradient was gelled by exposure to light and stained with Coomassie blue. Numbers 1–14 indicate the bands of polysomal aggregates from monomers to tetradecamers.

Ethidium bromide: The gels are immersed in the ethidium bromide staining solution for 15–45 min, rinsed briefly in water, and visualized in UV light.

Examples of gels stained with each of the above reagents are shown in Figs. 3 and 4. Staining with Coomassie blue requires a much longer time for staining and destaining as compared to the other procedures, particularly if the gel is thick, but it gives clear and stable dark blue bands with high contrast, which can be scanned and quantitated densitometrically even after prolonged storage (Figs. 3 and 4). Methylene blue staining takes much less time because it does not require extensive destaining, but it is less sensitive and the color fades more quickly. Silver staining is highly

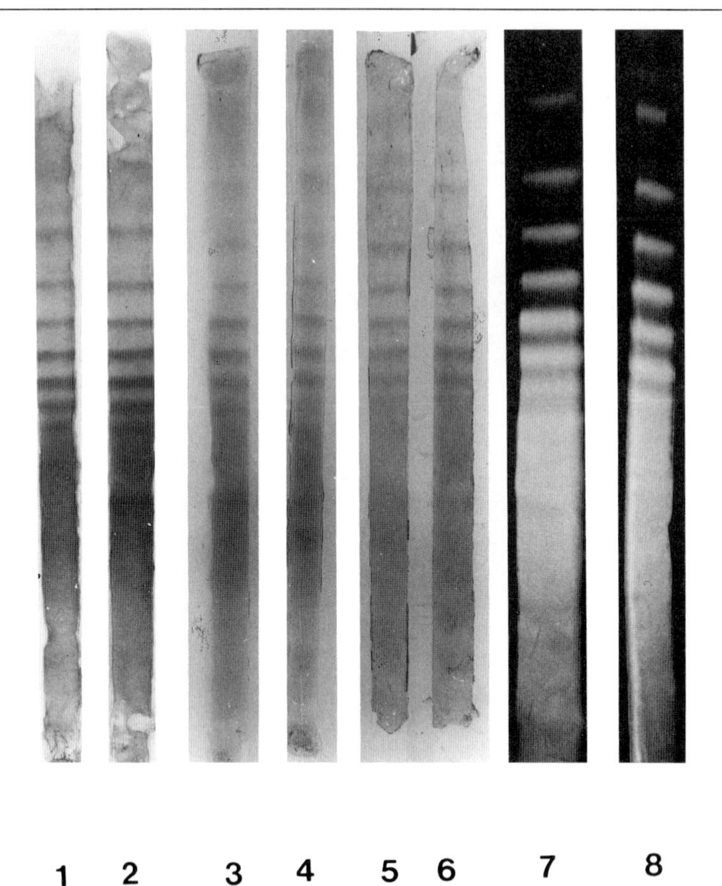

FIG. 4. Immobilized polysome bands stained with Coomassie blue (lanes 1 and 2), Methylene blue (lanes 3 and 4), silver stain (lanes 5 and 6), and ethidium bromide (lanes 7 and 8). Free polysomes were prepared from 10-day-old rat brains, fractionated, and stained as described in the text.

sensitive, but the color of the bands is not always proportional to the quantity of polysomes present in the band. Ethidium bromide reveals sharp, distinct bands under UV light after short treatment, but the bands are difficult to quantitate. However, this last method is very useful for a quick verification, using one or two slices, that the polysomes have been satisfactorily resolved on the gradient, before experimenting further with the remaining slices.

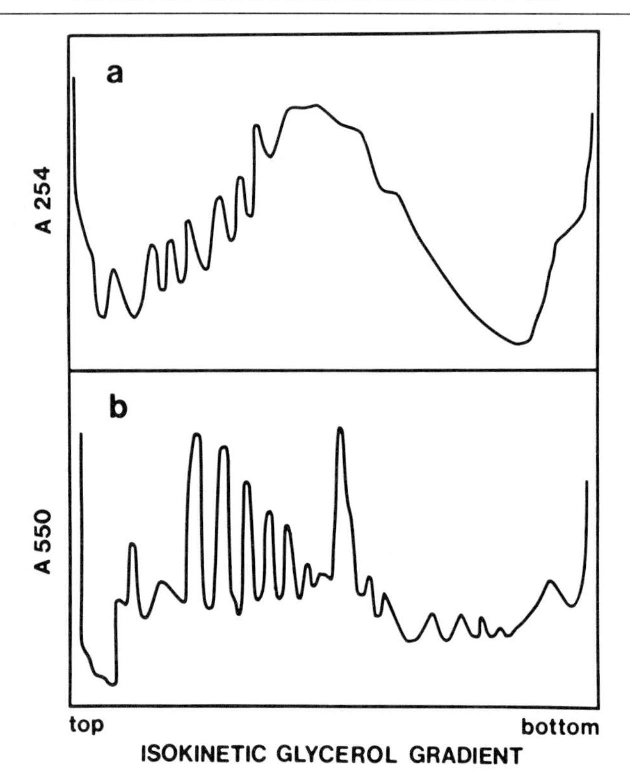

FIG. 5. Scanning of polysomes fractionated on (a) nonpolymerizing and (b) polymerizing isokinetic glycerol gradients (25–55%). Free polysomes were prepared from 10-day-old rat brains. Polysomes fractionated on the nonpolymerizing gradient were scanned using an ISCO spectrophotometer equipped with a flow cell and set to 254 nm. Polysomes fractionated on the polymerizing gradient were gelled, sliced, stained with Coomassie blue, and then scanned at 550 nm as described in the text.

Scanning and Quantitation of Polysome Bands

Liquid sucrose gradients that do not contain acrylamide are monitored by piercing the bottom of the tube with a needle, which is connected to the flow cell of an ISCO (Lincoln, NE) recording spectrophotometer set at 254 nm. The gel slices are scanned in a Gilford (Oberlin, OH) spectrophotometer set at 550 nm and equipped with a linear transport gel scanner attachment. Quantitation is made by integration of the area of each peak.

Polysome profiles obtained by the densitometric scanning of gel slices compare favorably with the standard flow cell UV absorbance analysis of liquid gradients. Figure 5 shows separation of polysomal aggregates on

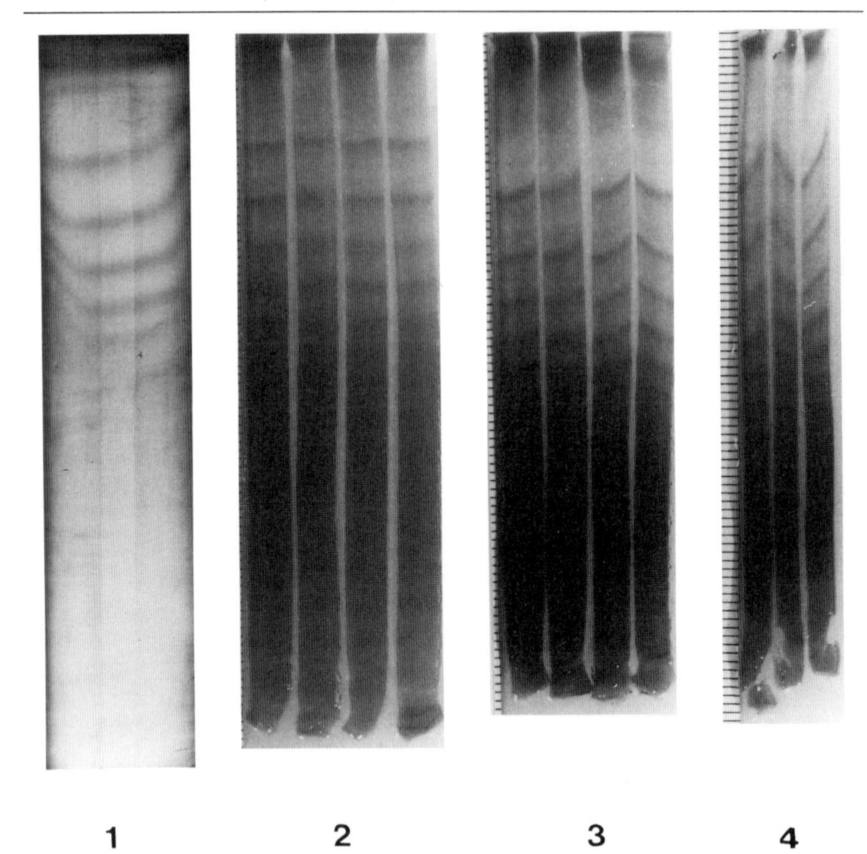

FIG. 6. An example of uneven sedimentation of polysomes in sucrose density gradient. 1, The gel cylinder showing distorted polysome bands; 2, slices from the center of the gel cylinder; 3, slices originating from the neighborhood of the gel center; 4, slices from the extreme periphery of the gel cylinder. Free polysomes from 10-day-old rat brains were fractionated in linear polymerizing sucrose gradient (20–45%), sliced, and stained with Coomassie blue as described in the text.

nonpolymerizing or polymerizing isokinetic glycerol gradients. More discrete bands are obtained by the gel-scanning method as compared to flow cell UV scanning of liquid gradient. This is presumably due to the fact that the polysome bands are immobilized immediately after centrifugation while they are free to diffuse in a liquid medium, particularly if there is a long delay before scanning. In addition, the flow cell scanning, by itself, is a serious source of mixing because of turbulence in narrow tubes.

Sedimentation of particles through a liquid medium is influenced by a number of factors, including the nature of the liquid–liquid and liquid–solid interfaces involved, the dimensions of the liquid medium through which the particles sediment, the relative densities, viscosities, and frictional forces, composition of the buffer, sample load, etc. Under appropriate conditions, centrifugation of particles of different sizes such as the polysomes may give rise to discrete horizontal bands parallel to each other, as in Fig. 3. However, occasionally, skewed bands may be formed, because of differential velocities of sedimenting particles in different regions of the gradient. When such a gradient is scanned by using a flow cell, the resolution of absorbance peaks is generally very unsatisfactory and useless for the recovery and treatment of specific polysomal aggregates. However, when a gradient of this type can be gelled, sliced, and stained, it is possible to obtain slices in which the bands are well separated, even if they are not horizontal (Fig. 6). This would permit scanning and quantitation as well as recovery of the bands by extraction or electroelution.

[19] Biochemical Manipulations of Minute Quantities of mRNAs and cDNAs Immobilized on Cellulose Paper Discs

By GEORGES LÉVESQUE, ADI D. BHARUCHA, and M. R. VEN MURTHY

Preparation and biochemical manipulation of mRNAs and cDNAs constitute a major portion of the experimental work in molecular biology. Very often, a single operation requires completion of a large number of successive steps in which the intermediate molecules must be separated from the other reactants and recovered by extraction, precipitation, centrifugation, etc., before the subsequent step can be undertaken. This invariably results in unpredictable losses of the desired product, which can be particularly serious and frustrating when one is working with scarce biological materials. Immobilization of the critical reactants on insoluble matrices is an excellent means by which secondary products may be removed from a given reaction. This may, in some cases, even enhance the rates and yields of reaction. A number of methods have been developed to bind RNAs and DNAs to solid supports, such as Sephadex, agarose, cellulose, fiberglass, acrylic polymers, nitrocellulose, and nylon. Gilham[1] and Erhan *et al.*[2] have described both aqueous and nonaqueous reactions for the covalent linking of nucleic acids and oligonucleotides to either cellulose powder or to cellulose sheets. Oligo(dT) linked to cellulose pow-

[1] P. T. Gilham, this series, Vol. 21 [10] p. 191.
[2] S. Erhan, L. G. Northrup, and F. R. Leach, *Proc. Natl. Acad. Sci. U.S.A.* **53,** 646 (1965).

der is extensively used for the separation and purification of messenger RNAs from other nucleic acids and proteins.[3] Poly(U) covalently bound to diazothiophenyl paper has also been used for isolation of mRNA from total cytoplasmic RNA.[4] An excellent early review of the matrices used for the attachment of nucleic acids is provided by Weissbach and Poonian.[5] Current applications of membranes and solid supports in blotting and affinity chromatography of nucleic acids are described in several practical manuals on molecular biology.[6,7]

It is known that mRNAs and DNAs immobilized on many of the solid matrices described above conserve their ability to hybridize with complementary nucleic acid sequences, although in many cases the nature of the interaction between the nucleic acids and the solid supports is not clear. It is also not clear whether the nucleic acids bound in this manner can function efficiently as substrates in enzymatic reactions normally performed in solution. Vitek *et al.*[8] were able to reverse transcribe mRNA bound to oligo(dT) cellulose powder, indicating that cellulose did not interfere in this reaction. Our experiments using mRNA and cDNA bound to filter paper disks indicate that binding to cellulose does not obstruct the nucleic acids from acting as efficient substrates in such diverse reactions as translation, reverse transcription, polymerization, or enzymatic amplification by polymerase chain reaction (PCR).

In principle, cDNA is prepared directly from small amounts of mRNA trapped on oligo(dT) cellulose paper disks, using the oligo(dT) of the paper as primer for reverse transcription. The newly formed cDNA, covalently linked to the disk, is then employed as such in further reactions.[9] This procedure (1) permits manipulations of very small amounts of mRNA and cDNA, (2) reduces the processing time of serial reactions considerably, because the bound nucleic acid may be purified by rinsing in an appropriate buffer rather than by precipitation and centrifugation, and (3) reduces losses of nucleic acid products that might otherwise occur during each of the recovery steps.

[3] H. Aviv and P. Leder, *Proc. Natl. Acad. Sci. U.S.A.* **69,** 1408 (1972).

[4] D. H. Wreschner and M. Herzberg, *Nucleic Acids Res.* **12,** 1349 (1984).

[5] A. Weissbach and M. Poonian, this series, Vol. 34 [54] p. 463.

[6] J. Sambrook, E. F. Fritsch, and T. Maniatis, "Molecular Cloning: A Laboratory Manual," 2nd Ed. Cold Spring Harbor Laboratory, Cold Spring Harbor, New York, 1989.

[7] P. D. G. Dean, W. S. Johnson, and F. A. Middle, "Affinity Chromatography: A Practical Approach." IRL Press, Oxford, 1985.

[8] M. P. Vitek, S. G. Kreissman, and R. H. Gross, *Nucleic Acids Res.* **8,** 1191 (1981).

[9] M. R. V. Murthy, A. D. Bharucha, and R. Charbonneau, *Nucleic Acids Res.* **14,** 7134 (1986).

Preparation of Pyridine Salt of Thymidine 5′-Monophosphate

Thymidine 5′-monophosphate (TMP) is polymerized and covalently linked to filter paper disks essentially according to the nonhydrous reaction described by Gilham[1] for use on cellulose powder. A Bio-Rad (Richmond, CA) Econo-column of 3-ml capacity is packed with Dowex-X4 ion-exchange resin (100–200 mesh; total ion-exchange capacity, 1.4 mEq/ml; J. T. Baker Chemical, Phillipsburg, NJ) and washed extensively with distilled water (~60–70 ml) until the pH of the washing, which is initially close to 1, rises to 6. The column is loaded with 1 mmol of 5′-TMP–$Na_2 \cdot H_2O$ (386.2 mg) dissolved in 10 ml of water and the column is washed with 3 ml of water. The nucleotide is then eluted down with water and the eluate (approximately 12 ml) is neutralized to pH 6.6 by dropwise addition of pyridine (approximately 6 ml). The mixture is lyophilized overnight in a round-bottomed quick-fit flask.

Polymerization of Thymidine 5′-Monophosphate to Oligo(dT)

The lyphilized TMP–pyridine salt is mixed with 2 ml of freshly distilled pyridine and 2 mmol of *N,N′*-dicyclohexylcarbodiimide (412 mg; Sigma Chemical Co., St. Louis, MO) and shaken for 5 days at room temperature in the presence of a few glass beads. Because the mixture assumes a viscous consistency during this procedure, the agitation should be vigorous enough to ensure proper mixing of the ingredients. Anhydrous conditions should be strictly observed in regard to the solvents throughout this and subsequent reactions.

Immobilization of Oligo(dT) on Cellulose Disks

Small round disks of 4-mm diameter (average weight of each disk, 3.4 mg) are cut out from Whatman (Clifton, NJ) 3MM filter paper and the disks are washed repeatedly with water and methanol, and then dried *in vacuo* at 100° overnight.

To the flask containing oligothymidylate are added 25 ml of dry pyridine, 1 g of *N,N′*-dicyclohexylcarbodiimide, and 2 g of cellulose disks dried as above. The whole mixture is again shaken at room temperature for 5 days under adequately sealed anhydrous conditions. The cellulose disks are then collected by filtration on a Büchner funnel, washed twice with dry pyridine, and then kept suspended in 50% aqueous pyridine overnight. The disks are washed extensively with warm ethanol and distilled water to remove excess unreacted reagents and by-products. A final washing is carried out using 1 *M* NaCl and 10 m*M* sodium phosphate (pH 7) until the A_{260} of the eluate is less than 0.03. The disks are placed in 95% ethanol and stored at $-20°$. Just before use, the disk is dried in a vacuum desiccator for removing ethanol and then equilibrated with the binding buffer [0.5 *M* NaCl, 10 m*M* Tris-HCl, pH 7.5, 0.2% (v/v) mercaptoethanol].

Determination of Binding Capacity of Oligo(dT) Cellulose Disks

Oligo(dT) cellulose disks are placed individually in polyethylene tubes containing 50 μl of binding buffer and 4 μg of [^{3}H]poly(A) [~25,000 disintegrations per minute (dpm)]. The mixture is shaken at room temperature for 5 min, the unreacted solution is aspirated out, and the disks are washed three times with the binding buffer. The bound poly(A) is then released with the eluting buffer (10 m*M* Tris-HCl, pH 7.5). The washings and eluates are counted from which the proportion of bound poly(A) is calculated. Generally we obtain a poly(A) binding capacity of 1.5–2.0 μg/disk, which is comparable to the values reported for the commercial preparations of oligo(dT) cellulose powders [20–40 A_{260} units of poly(A)/*g*].

Preparation of Oligo(dT) Cellulose mRNA Disks

Aliquots of total RNA equivalent to 5–10 mg of tissue are dissolved in 10 μl of binding buffer and deposited on oligo(dT) cellulose disks under slow suction. The disks are rinsed twice with the binding buffer and twice with 75% ethanol to remove the salts in the binding buffer and then dried in a vacuum desiccator.

Translation

Each disk is wetted with a reaction mixture made up of 6.5 μl of reticulocyte lysate (code N90; New England Nuclear, Boston, MA) and 1.5 μl (10 μCi) of [^{35}S]methionine. They are incubated at 30° for 0–40 min in sealed tubes, rinsed in 10% (w/v) trichloroacetic acid containing 0.5% (w/v) cold methionine, and then counted. The kinetics of incorporation is linear for 30–40 min and the efficiency of translation of mRNA is about 40–60% of that in solution. If desired, the disk containing the newly synthesized polypeptides may be directly inserted into the sample slots of polyacrylamide gels for electrophoretic analysis.

Reverse Transcription

Single-Stranded cDNA. Five micrograms of total cellular RNA is dissolved in 19 μl of reverse transcription mixture with the following final composition: 50 mM Tris-HCl (pH 8.3), 75 mM KCl, 3 mM $MgCl_2$, and 1 mM each of dATP, dTTP, dCTP, and dGTP. The tube is heated at 90° for 2 min to denature the mRNA and cooled immediately in ice. One microliter (200 U) of Moloney murine leukemia virus (M-MLV) reverse transcriptase (Bethesda Research Laboratories, Gaithersburg, MD) is added and the mixture is absorbed on a disk of oligo(dT) cellulose in a small Eppendorf tube. Disks prepared in this manner are incubated at 42° for up to 90 min. The disks are rinsed in 50% (v/v) formamide, 0.1% (w/v) sodium dodecyl sulfate (SDS) and 10 mM Tris-HCl (pH 7.6) at 60°, rinsed again in 95% ethanol, dried, and counted. The reaction is linear for 20–45 min, depending on the concentration of mRNA. Approximately 80–90% of radioactivity is released from the disk by treatment with S1 nuclease, indicating that the polymerized product is indeed a single-stranded cDNA extending from the oligo(dT) primer attached to the cellulose disk.

Double-Stranded cDNA. The second strand synthesis can be carried out on single-stranded cDNA bound to the cellulose disk either by self-priming[10] of the RNA-free cDNA in the presence of the Klenow fragment of *Escherichia coli* DNA polymerase I or by nick translation of the mRNA–cNA hybrid.[11] However, the intact ds-cDNA is not quantitatively released from the disk for further manipulation (e.g., for cloning into a plasmid vector). Treatment with S1 nuclease releases only 25–30% of the molecules with an average size of approximately 500 nucleotides, while the others are degraded or remain bound to the disk. A possible solution to this difficulty might be to insert an oligonucleotide recognized by a restriction enzyme, between the cellulose disk and the oligo(dT) so that the ds-cDNA may be cleaved from the disk, endowed with the proper sticky ends for subsequent ligation to a vector.

Amplification of Selected mRNAs by Polymerase Chain Reaction

Reverse transcription is carried out on oligo(dT) cellulose disks as described above. Following incubation at 42° for 40 min, the disks are rinsed twice with the PCR buffer (40 mM Tris-HCl, pH 8.3, 47.5 mM KCl, 2.2 mM $MgCl_2$, 50 μg/ml bovine serum albumin) to eliminate all reaction ingredients. Small amounts of proteins, DNA, and ribosomal RNAs, which are often present as contaminants in mRNA preparations, are also removed during this step, leaving behind only the cDNA that is covalently linked to the cellulose disk. The washed disks are then suspended in 50 μl

[10] H. Land, H. Grez, H. Hauser, W. Lindenmaier, and G. Schutz, *Nucleic Acids Res.* **9,** 2251 (1981).

[11] U. Gubler and B. J. Hoffman, *Gene* **25,** 263 (1983).

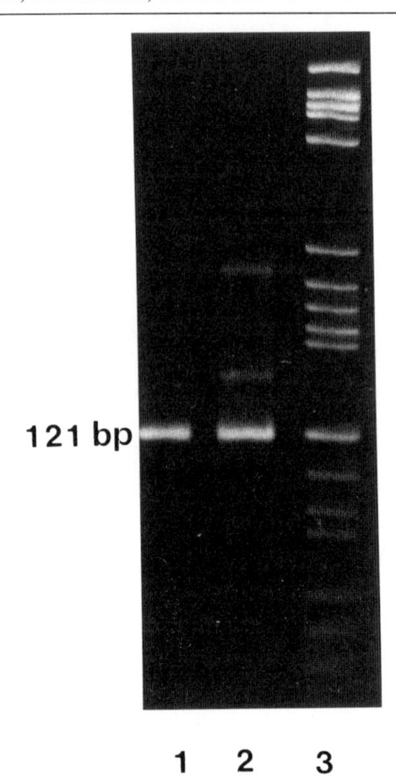

FIG. 1. The PCR amplification of a segment of human β-globin cDNA. The amplified segment (nucleotides 222–342, numbered from the first nucleotide of the initiator codon of β-globin cDNA[12]) spans a region of cDNA overlapping exons 2 and 3. Reverse transcription and amplification were carried out as described in the text, in the presence of the following PCR primers: P1 = 5′-CACAGACCAGCACGTTGCCCAG-3′; P2 = 5′-GATGGCCTGGCTCACCTGGACA-3′. The reaction products were analyzed by polyacrylamide gel electrophoresis amplification of cDNA bound to oligo(dT) cellulose (lane 1) or free cDNA in solution (lane 2). Lane 3: Oligonucleotide size markers (*Hae*III digest of pBR322 DNA).

of PCR buffer containing 5 units of *Taq* DNA polymerase (Amersham, Arlington Heights, IL) and 25 ng each of two oligonucleotide primers corresponding to the region of the cDNA to be amplified. Amplification is carried out in 20–40 cycles of denaturation at 94° for 90 sec, annealing at 55° for 90 sec, and polymerization at 72° for 2 min. The tubes are cooled to 4°, centrifuged, and the supernatants stored at 4° for analysis.

Figure 1 shows PCR amplification of a 121-bp segment spanning exons

[12] R. M. Lawn, A. Efstratiadis, C. O'Connell, and T. Maniatis, *Cell* (*Cambridge, Mass.*) **21,** 647 (1980).

2 and 3 of β-globin cDNA. cDNA was prepared by reverse transcription of human globin mRNA using cellulose-bound oligo(dT) (lane 1) or free oligo(dT) (lane 2) as primer. Lane 1 (Fig. 1) shows a discrete single band with the expected size of 121 bp, while several faint bands of higher size are also seen in lane 2, in addition to the 121-bp band. The larger amplified products in lane 2 are presumably due to nonspecific binding of PCR primers to DNA molecules present as impurities in the RNA preparation. Immobilization of globin cDNA on cellulose disks permits removal of DNA and other contaminants during the washing step prior to amplification. Experiments in our laboratory have also demonstrated that the oligo(dT)-cellulose binding method can be used for the reverse transcription and PCR amplification of extremely small quantities of specific mRNAs, even in the presence of a large number of other mRNAs and rRNAs.[13] This may find application in the detection of low-level expression of cell-specific genes in heterologous cells, a phenomenon that has been termed "illegitimate transcription."[14]

Screening for Specific Transcripts

A target mRNA of interest in a given population of mRNA can be detected or even quantitated by the dot-blot procedure[6] using an appropriate probe. But the extreme susceptibility of mRNA to degradation by chance contamination with RNases is a practical drawback in the storage and manipulations of RNA samples. However, once reverse transcribed and immobilized on cellulose disk, the cDNAs act as stable and durable records of their corresponding mRNAs and can be repeatedly screened by the same or different probes, as needed.

Enrichment of Target mRNAs by Hybridization Subtraction

The mRNA profiles of a cell may undergo significant changes in response to a variety of genetic, physiological, pathological, chemical, or environmental signals. This may involve the transcription of a large number of genes or may result in the induction or suppression of a limited number of unique genes. The method of hybridization subtraction[6] is often used to enrich selectively the transcripts of the affected genes for eventual cloning and characterization. Vitek *et al.*[8] have described a procedure for the enrichment of ecdysterone-induced *Drosophila* mRNAs by chromatog-

[13] G. Levesque, A. Bharucha, and M. R. V. Murthy, unpublished results.

[14] J. Chelly, J. P. Concordet, J. C. Kaplan, and A. Kahn, *Proc. Natl. Acad. Sci. U.S.A.* **86,** 2617 (1989).

raphy on uninduced cDNA-cellulose columns. The cDNA cellulose disks are easier to manipulate in such reactions and may be useful for the processing of small amounts of mRNA by hybridization subtraction. However, the optimum conditions of hybridization and elution would have to be worked out depending on the qualitative and quantitative differences between the two populations of mRNA under comparison.

[20] Hybrid Selection of mRNA with Biotinylated DNA

By JAEMOG SOH and SIDNEY PESTKA

Introduction

Hybrid selection of mRNA has been used successfully to purify specific mRNA and to screen large numbers of recombinant DNA molecules.[1-6] This method utilizes hybridization of mRNA to immobilized single-stranded DNA or hybridization in solution followed by selection of mRNA · DNA hybrids. In this chapter, a procedure to make the biotinylated single-stranded DNA with the use of an asymmetric polymerase chain reaction (PCR) is described for use in a high-efficiency hybrid selection procedure. After hybridization of the biotinylated DNA with mRNA in solution, streptavidin agarose is used to trap the hybrid duplex of mRNA · DNA onto the solid matrix. The selected mRNA is then eluted from the streptavidin agarose.

Procedures

Preparation of Primers

5′-Aminoalkyl deoxyoligonucleotides were synthesized by the phosphoramidate method on a DNA synthesizer (model 380B; Applied Biosys-

[1] B. M. Paterson, B. E. Roberts, and E. L. Kuff, *Proc. Natl. Acad. Sci. U.S.A.* **74,** 4370 (1977).
[2] N. D. Hastie and W. A. Held, *Proc. Natl. Acad. Sci. U.S.A.* **75,** 1217 (1978).
[3] R. P. Ricciardi, J. S. Miller, and B. E. Roberts, *Proc. Natl. Acad. Sci. U.S.A.* **76,** 4927 (1979).
[4] R. Jagus, this series, Vol. 152, p. 567.
[5] S. Maeda, R. McCandliss, M. Gross, A. Sloma, P. C. Familletti, J. M. Tabor, M. Evinger, W. P. Levy, and S. Pestka, Proc. Natl. Acad. Sci. U.S.A. **77,** 7010 (1980).
[6] S. Nagata, H. Taira, A. Hall, L. Johnsrud, M. Streuli, J. Ecsödi, W. Boll, K. Cantell, and C. Weissman, *Nature (London)* **284,** 316 (1980).

tems, Foster City, CA) with *O*-methoxyphosphoramidates[7] to yield the 5′-aminoalkyl oligonucleotide containing the amino linker $NH_2(CH_2)_2PO_4$-CGACTCACTATAGGGAGACC (PA1) that is complementary to the plasmid vector pGEM4 (Promega, Madison, WI). To attach biotin to this aminoalkyl oligonucleotide, 1.0 A_{260} unit (35 μg, 5 nmol) of gel-purified oligomer was placed into a 1.5-ml conical polypropylene tube and then dried under vacuum. For biotinylation, 0.3 mg of solid sulfo-NHS-LC-biotin (NHS, *N*-hydroxysuccinimide; LC, long chain, $(CH_2)_9$; No. 21335, Pierce, Rockford, IL) was added to the tube followed by addition of 20 μl of fresh carbonate buffer (50 m*M* $NaHCO_3$/100 m*M* Na_2CO_3, pH 9.0) to dissolve the aminoalkyl oligonucleotide and sulfo-NHS-LC-biotin. The reaction mixture was incubated at room temperature for 4 hr, then dried under vacuum, and the tube contents redissolved in 20 μl of deionized formamide. Electrophoresis on a 20% (w/v) polyacrylamide/8 *M* urea denaturing gel was then used to purify the biotinylated oligonucleotide, PB1.[7] The upper band was cut from the gel under short-wavelength UV light, the gel crushed, and the PB1 oligonucleotide eluted into 1.0 ml of TE buffer [10 m*M* Tris-HCl, 1 m*M* ethylenediaminetetraacetic acid (EDTA), pH 8] by shaking overnight at 37°. Finally, PB1 primers were purified with an NENSORB 20 cartridge (Du Pont, Wilmington, DE) according to the manufacturer's protocol. The second primer consisted of a conventional oligodeoxynucleotide complementary to the second strand of the vector on the opposite side of the inserted sequence (GGTGACACTATAGAATACAC, P1). It was prepared with a DNA synthesizer as above.

Asymmetric Polymerase Chain Reaction

The plasmid DNA used was pLF2-19 derived from pGEM4 (Promega). It contains the Hu-IFN-αA cDNA sequence[5,8] between *Eco*RI and *Xba*I sites under control of the SP6 promoter. The asymmetric PCR was performed as described.[9] The reaction mixture contained 100 pg of target DNA, 50 m*M* KCl, 10 m*M* Tris-HCl (pH 8.4), 5.0 m*M* $MgCl_2$, 50 pmol PB1 primer, 0.5 pmol P1 primer, 200 μ*M* of each of the four deoxynucleotide triphosphates, 200 μg/ml gelatin (EIA purity grade; Bio-Rad, Richmond, CA), and 2.5 units of *Taq* polymerase (Perkin-Elmer Cetus, Norwalk, CT). The PCR was carried out for 40 cycles: 94° for 1.5 min, 40° for 2 min, and 72° for 2 min. The amplified DNA was precipitated by adding 2.5 vol of ethanol and 0.1 vol of 3 *M* sodium acetate (pH 5.2) and permitting

[7] A. Chollet and E. H. Kawashima, *Nucleic Acids Res.* **13,** 152 (1985).
[8] S. Pestka, *Arch. Biochem. Biophys.* **221,** 1 (1983).
[9] U. B. Gyllensten and H. A. Erlish, *Proc. Natl. Acad. Sci. U.S.A.* **85,** 7652 (1988).

the mixture to stand for 20 min at −70°. The DNA was pelleted by centrifugation (Eppendorf microcentrifuge, 15,000 rpm, 15 min), washed with 75% ethanol, dried under vacuum, and then dissolved in distilled water. The biotinylated single-stranded DNA (ssDNA, the asymmetric PCR product) is designated DNA-B for brevity.

In Vitro Transcription

The mRNA was prepared by *in vitro* transcription as described.[10] The reaction mixture contained 1.0 μg *Hin*dIII-linearized pLF2-19 template DNA, 1 unit/μl RNasin (Promega), 10 m*M* dithiothreitol (DTT), 40 m*M* Tris-HCl (pH 7.5), 6 m*M* $MgCl_2$, 2 m*M* spermidine, 10 m*M* NaCl, 0.5 m*M* each of ATP, CTP, and GTP, 50 μ*M* UTP, 50 μCi [α-^{32}P]UTP (>800 Ci/mmol; Amersham, Arlington Heights, IL), and 20 units of SP6 polymerase (Promega). The reaction mixture was incubated at 37° for 1 hr followed by DNase I treatment with 16 units of pancreatic DNase I (code DPRT; Worthington Diagnostics, Freehold, NJ) at 37° for 30 min. After extraction with equal volumes of phenol–chloroform and chloroform, the transcripts were precipitated with ethanol at −70° for 20 min after addition of 3 vol of 100% and 0.1 vol of 3 *M* sodium acetate (pH 5.2). The RNA pellets were dried and dissolved in 100 μl of water treated with diethyl pyrocarbonate (DEPC). The above was passed through Sephadex G-50 spin-column (QuickSpin columns; Boehringer Mannheim, Indianapolis, IN) to remove the free nucleotides. The capped mRNA was made as above except that the reaction mixtures contained 0.5 m*M* 7-methyl-GpppG, 0.5 m*M* ATP, 0.5 m*M* CTP, 0.5 m*M* UTP, 50 μ*M* GTP, and 200 μg bovine serum albumin (BSA)/μl.

Hybridization

Hybridization was performed in 50 μl that contained 0.1 μg of DNA-B for [^{32}P]mRNA or 0.4 μg of DNA-B for capped mRNA, about 80 ng of [^{32}P]mRNA or 0.25 μg of capped mRNA transcript encoding Hu-IFN-αA, 1 *M* NaCl, 50 m*M* piperazine-*N*,*N*′-bis(ethanesulfonic acid) (PIPES; pH 7.0), and 2 m*M* EDTA. The mixture was incubated at 85° for 2 min and then at 55° for 1 hr (to a C_0t of 2–8 × 10^{-2} mol of nucleotides × sec/liter).

Binding of mRNA · DNA-B Hybrid to Streptavidin Agarose

Sixty microliters of streptavidin agarose suspension (No. S-1638; Sigma, St. Louis, MO) was washed three times with 0.5 ml of 250 m*M*

[10] C. S. Kumar, G. Muthukumaran, L. J. Frost, M. Noe, Y.-H. Ahn, T. M. Mariano, and S. Pestka, *J. Biol. Chem.* **264,** 17939 (1989).

NaCl/10 mM Tris-HCl (pH 8.0) buffer, then 100 μl of the same buffer containing 5 μg calf liver tRNA (Boehringer Mannheim) was placed into the washed streptavidin agarose and the suspension shaken on a rotary shaker for 15 min at 300 rpm to block nonspecific binding. All steps involving binding and elution were performed at room temperature unless otherwise noted. The above 50 μl of hybridization mixture, which contained the mRNA · DNA-B hybrid, and 50 μl of distilled water were then added to the streptavidin suspension and the reaction mixture of 200 μl shaken for 90 min at room temperature at 300 rpm. After the incubation, the agarose was separated from the suspension in a Spin-X tube (Costar, Cambridge, MA) by centrifugation for 3 min in a microfuge. The isolated agarose was washed two times with 150 μl of distilled water. Finally, the mRNA · DNA hybrid containing the biotinylated DNA bound to the streptavidin agarose was recovered on the Spin-X filter. Alternatively, it is possible to centrifuge and to wash the agarose in a standard conical polypropylene centrifuge tube.

Elution of mRNA from Streptavidin Agarose

To the streptavidin agarose 200 μl of elution buffer (10 mM Tris-HCl, 30% deionized formamide, pH 7.8) was added. The agarose suspension was incubated for 10 min at 60°, immediately frozen in a dry ice–ethanol bath, and thawed in an ice bath. The agarose suspension in the Spin-X tube was centrifuged for 3 min in a microfuge at room temperature to yield the eluted mRNA in the flow-through fraction. If a standard centrifuge tube is used instead of a Spin-X tube, after eluting the mRNA the agarose is pelleted and the supernatant containing the eluted mRNA is aspirated. One-tenth volume of 3 M sodium acetate (pH 5.2), 2.5 vol of ethanol, and 5 μg of calf liver tRNA as carrier were added to precipitate the mRNA at −70° for 20 min. The ethanol precipitates were pelleted as above, washed two times with 75% ethanol, dried under vacuum, and dissolved in 1 to 10 μl of distilled water for further experiments. The overall scheme is outlined in Fig. 1.

S1 Nuclease Digestion

To check the resistance of [^{32}P]mRNA to nuclease after hybridization and elution from the hybrids, S1 nuclease digestion was carried out at 37° for 15 min in a 30-μl reaction mixture that contained 50 units of S1 nuclease (Amersham), 30 mM sodium acetate (pH 4.5), 100 mM NaCl, 1 mM $ZnSO_4$, 0.25 mg/ml denatured DNA, and [^{32}P]mRNA (about 15,000–30,000 cpm).

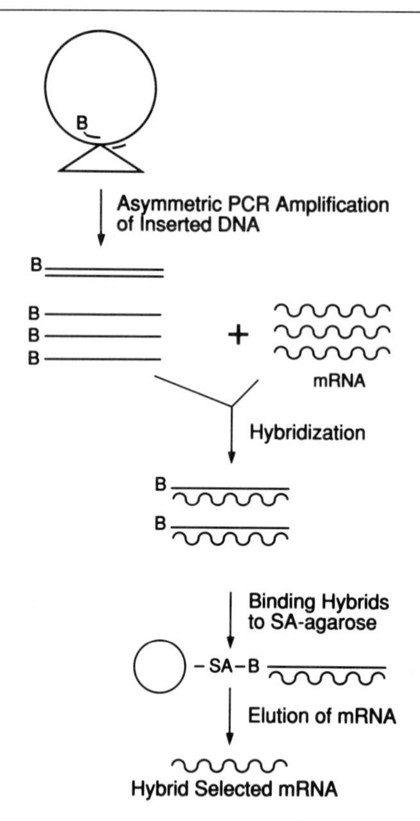

FIG. 1. Schematic digram for hybrid selection of mRNA with biotinylated DNA (DNA-B). Biotinylated ssDNA and dsDNA were made by asymmetric PCR. The single-stranded biotinylated DNA (present in excess of the dsDNA) was used to trap the specific mRNA after hybridization followed by binding to streptavidin agarose. Subsequent melting of mRNA from the hybrid releases the selected mRNA for assay. B, Biotin; SA, streptavidin.

The radioactivity associated with undigested RNA was determined by precipitation with trichloroacetic acid as described.[11]

Microinjection of mRNA into Xenopus laevis Oocytes

Oocytes were prepared and injected with the capped transcript as described.[12,13] After injection, the oocytes were incubated in 0.5× Leibowitz L-15 medium (10 μl/oocyte) containing theophylline (0.5 m*M*) and

[11] S. Pestka, *J. Biol. Chem.* **243,** 2810 (1968).

[12] R. L. Cavalieri, E. A. Havell, J. Vilek, and S. Pestka, *Proc. Natl. Acad. Sci. U.S.A.* **74,** 3287 (1977).

[13] C. S. Kumar, T. M. Mariano, M. Noe, A. K. Deshpande, P. M. Rose, and S. Pestka, *J. Biol. Chem.* **263,** 13493 (1988).

protease inhibitors (1 μg/ml aprotinin, 1 μg/ml leupeptin, and 1 μg/μl pepstatin A) at 18° for about 20 hr. Ten frog oocytes were used for each sample for antiviral assay.

Antiviral Assay

At the end of incubation the medium was removed and fresh L-15 Leibowitz medium (10 μl/oocyte) with protease inhibitors was added. The oocytes were homogenized with a micropipette, and pelleted in a microfuge at 14,000 rpm for 3 min to remove debris and fat. The clear solution was collected and diluted for antiviral assay. The antiviral activity was measured on MDBK cells by a cytopathic effect inhibition assay[14] with vesicular stomatitis virus (VSV) as the challenge virus.

Hybrid Selection with Filter-Binding Method

Diazobenzyloxymethyl (DBM) paper was prepared and diazotized exactly as described.[4] For comparison with our procedure, hybridization of mRNA to filter-bound DNA and elution was carried out as described.[5,15] Briefly, two 0.25-cm^2 pieces of DBM paper with 100 or 400 ng of Hu-IFN-αA cDNA were hybridized as a sandwich[16] with 10 μl of hybridization medium, which contains 0.02 *M* PIPES (pH 6.8), 0.2% (w/v) sodium dodecyl sulfate (SDS), 0.9 *M* NaCl, 1 m*M* EDTA, 1 μg/μl yeast tRNA, 50% (v/v) formamide, and 10^4 cpm [^{32}P]mRNA or 250 ng capped mRNA, respectively. After hybridization at 48° for 15 hr, the filters were washed two times at 37° for 20 min with 1× SSC (0.15 *M* sodium citrate, 0.15 *M* NaCl), 0.2% (w/v) SDS, 1 m*M* EDTA, and 50% (v/v) formamide. Then the mRNA was eluted at 70° with 150 μl of elution buffer containing 0.02 *M* PIPES (pH 6.8), 2 m*M* EDTA, and 95% (v/v) formamide, and precipitated with ethanol as described above.

Hybridization of Biotinylated Single-Stranded DNA (DNA-B) to mRNA

Almost all [^{32}P]mRNA (94%) was protected from S1 nuclease digestion after hybridization with DNA-B (Table I). Only 2.6% of the [^{32}P]mRNA was nonspecifically bound, whereas 94% of the [^{32}P]mRNA was specifically hybridized to DNA-B.

Summary of Binding of mRNA · DNA-B Hybrid to Streptavidin Agarose

About 90% binding of mRNA · DNA-B to the streptavidin agarose was achieved in 90 min at room temperature (Fig. 2) The hybrid molecules,

[14] P. C. Familletti, S. Rubinstein, and S. Pestka, this series, Vol. 78, p. 387.
[15] R. McCandliss, A. Sloma, and S. Pestka, this series, Vol. 79, p. 618.
[16] J. Casey and N. Davidson, *Nucleic Acids Res.* **4,** 1539 (1977).

TABLE I
S1 NUCLEASE RESISTANCE BEFORE AND AFTER HYBRIDIZATION OF [^{32}P]mRNA ENCODING Hu-IFN-αA WITH DNA-B[a]

mRNA	S1 nuclease resistance (%)
[^{32}P]mRNA only	2.6
[^{32}P]mRNA–DNA hybrid	94.0

[a] DNA and mRNA were hybridized as described in text, precipitated with ethanol, washed with 75% ethanol, digested with S1 nuclease, and precipitated with trichloroacetic acid (TCA) to determine nuclease resistance.

which have the 5′-aminoalkyl linkage without biotin, bind to streptavidin-agarose negligibly (about 1.5%) irrespective of incubation time. Vectrex–avidin D (Vector Lab, Burlingame, CA) and biotin-cellulose (Pierce) were tested as potential matrices. Maximum binding (about 90%) of the hybrid to Vectrex–avidin D took place in a short time (5–10 min) and nonspecific binding was as low as with streptavidin agarose. However, it was difficult to elute mRNA after binding to the Vectrex–avidin D. The binding of the hybrid to biotin-cellulose incubated with avi-

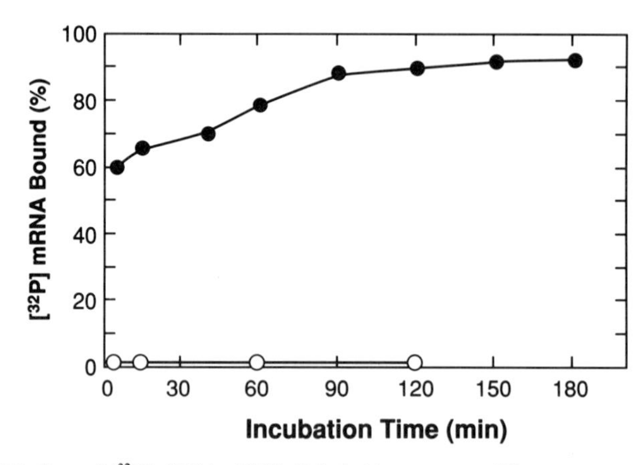

FIG. 2. Binding of [^{32}P]mRNA · DNA-B hybrid to streptavidin agarose as a function of incubation time. The conditions are described in text. Solid circles (●) represent binding of the mRNA · DNA-B complex to streptavidin agarose; open circles (○) represent nonspecific binding of [^{32}P]mRNA-DNA (amino alkyl instead of biotin) to the matrix.

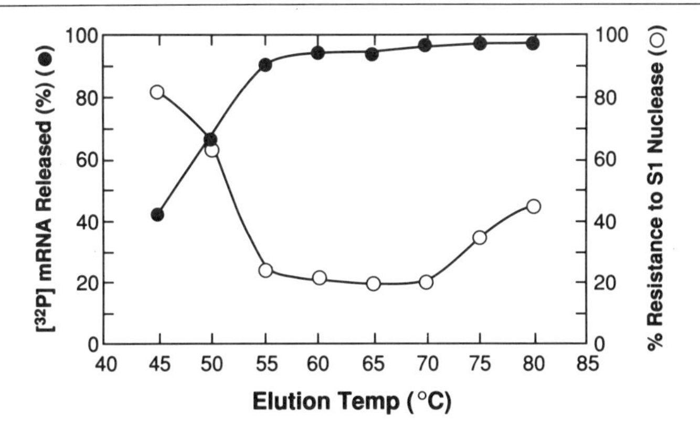

FIG. 3. Elution of [^{32}P]mRNA encoding Hu-IFN-αA from the hybrids bound to streptavidin agarose as a function of temperature. Elution of [^{32}P]mRNA was carried out at various temperatures with 200 μl of elution buffer [30% (v/v) formamide, 10 mM Tris-HCl, pH 7.8]. Eluted [^{32}P]mRNA was precipitated with 2.5 vol ethanol, 0.1 vol of 3 M sodium acetate (pH 5.2), and 5 μg of calf liver tRNA, then digested with 50 units of S1 nuclease (Amersham). Details are described in text. Solid circles (●) represent elution of [^{32}P]mRNA; open circles (○) represent S1 nuclease resistance of eluted [^{32}P]mRNA.

din[17] showed a high nonspecific binding of about 17% [8% in the presence of 0.1% (w/v) SDS].

Elution of mRNA from mRNA · DNA-B · Streptavidin Agarose

About 95% of [^{32}P]mRNA and [^{32}P]mRNA · DNA-B were released above 60° with 30% (v/v) formamide elution buffer (Fig. 3). The increase of S1 nuclease resistance of the eluted mRNA above 75° indicates that intact mRNA · DNA hybrids were removed from the column at high temperatures as a result of the low thermal stability of the biotin–streptavidin complex under these conditions (Fig. 3). Therefore, the optimum elution of [^{32}P]mRNA in the presence of 30% formamide in 10 mM Tris-HCl (pH 7.8) was between 60 to 70°. Under these conditions, about 77% of intact mRNA was eluted from the matrix Table II). This recovery is at least two times higher than filter-binding methods.[5,6,15] Figure 4 demonstrates that the eluted mRNA retained its physical integrity throughout the procedure. Moreover, the mRNA retained its biological activity, which was checked by injection of the mRNA into oocytes followed by antiviral assay of expressed Hu-IFN-αA (Table II). The overall yield of biological activiy was consistently about 30%.

[17] M. L. Shimkus, P. Guaglianone, and T. M. Herman, *DNA* **5,** 247 (1986).

TABLE II
RECOVERY OF INTACT [^{32}P]mRNA AND BIOLOGICAL ACTIVITY[a]

	Percentage recovered			
	DNA-B		DBM paper	
Step	[^{32}P]mRNA	Biological activity	[^{32}P]mRNA	Biological activity
Original	100 (100)	100	100 (100)	100
Hybridization	94 (94)			
Binding to streptavidin agarose or DBM paper	91 (86)		32 (32)	
Hybrid-selected mRNA	77 (66)	33	88 (28)	<1

[a] Recovery of [^{32}P]mRNA is based on the comparison of the yield of disintegrations per minute after each step. Biological activity was checked by antiviral assay of Hu-IFN-αA after translation of capped mRNA in frog oocytes injected with capped mRNA. The initial amount of capped mRNA was 250 ng in 1 μl. After the procedure, the resultant hybrid-selected capped mRNA was dissolved in 1 μl of distilled water. Fifty nanoliters (12.5 ng of untreated mRNA/50 nl) was injected into each oocyte (15 oocytes for each sample). The numbers outside the parentheses represent the recovery from step to step; the numbers within parentheses represent the overall recovery of [^{32}P]mRNA. DNA-B and DBM paper represent hybrid selection with biotinylated DNA and DBM paper-bound DNA method, respectively.

Concluding Comments

The hybrid selection method with ssDNA-B has several advantages over the filter-binding techniques. The asymmetric PCR with excess of biotinylated primer makes it possible to generate the ssDNA-B, which has a biotin molecule with an extended arm, which in turn facilitates the binding of the hybrids to the streptavidin agarose without steric hindrance. In addition, ssDNA-B can be used in solution hybridization, which is more rapid and efficient than filter hybridization. With the biotinylated specific primers complementary to vector sequences, any inserted DNA can be easily amplified and biotinylated. The biotinylated primer (PB1) was designed to be complementary to the coding strand of the inserted DNA so that it can hybridize with its corresponding mRNA. Elution of mRNA from the hybrid could be accomplished because the biotin–streptavidin linkage was retained under conditions of elution with 30% formamide/10 m*M* Tris-HCl between 60 and 70° (Fig. 3).

The ssDNA-B can be used for *in situ* hybridization after labeling with

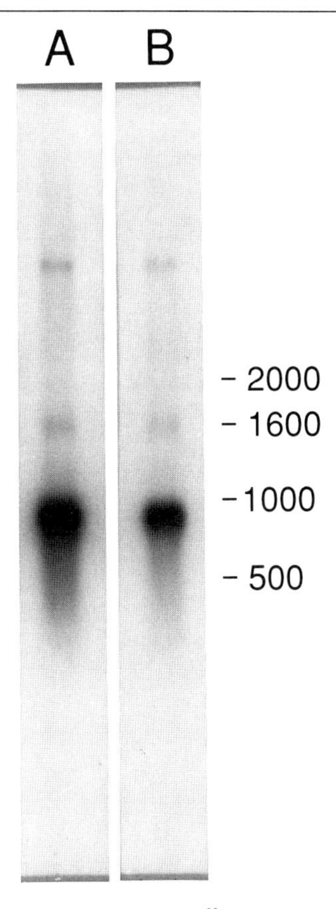

FIG. 4. Autoradiography of hybrid-selected [^{32}P]mRNA. The mRNA was eluted from mRNA · DNA-B · streptavidin-agarose with 30% formamide/10 m*M* Tris-HCl (pH 7.8) at 60° for 10 min, precipitated with ethanol, dissolved in 5 μl of distilled water, and analyzed on a formaldehyde denaturing agarose gel [1.0% (w/v), 11 × 11 cm, 10 V/cm). Lane A, control [^{32}P]mRNA; lane B, hybrid selected [^{32}P]mRNA.

a fluorochrome instead of biotin[18,19] for direct solid-phase sequencing[20,21] as well as for hybrid selection of mRNA as described here. This procedure can be utilized in various additional applications such as selecting a set of mRNAs with a fragment of DNA with complementary sequence, enrichment of homologous mRNA to the probe DNA-B, and for hybrid depletion

[18] R. Zawatzky, J. De Maeyer-Guignard, and E. De Maeyer, this series, Vol. 119, p. 474.

[19] J. H. J. Hoeijmakers, P. Borst, J. Van Den Burg, C. Weissmann, and G. A. M. Cross, *Gene* **8,** 391 (1980).

[20] T. Hultman, S. Stahl, E. Hornes, and M. Uhlén, *Nucleic Acids Res.* **17,** 4937 (1989).

[21] M. Uhlén, *Nature (London)* **340,** 733 (1989).

assays. The procedure can be improved further with avidin-coated magnetic beads such as Dynabeads M-280 streptavidin (Dynal, Great Neck, NY), although the beads were not evaluted for this purpose in this study.[22]

Acknowledgment

This study was supported in part by U.S. Public Health Service Grants CA46465 and AI-25914 to Sidney Pestka. Jaemog Soh was a Becton-Dickinson predoctoral fellow during this period.

[22] E. Hornes and L. Korsnes, *Gene Anal. Tech. Appl.* **7,** 145 (1990).

Section II

Enzymes and Methods for Cleaving and Manipulating DNA

[21] Restriction Enzymes: Properties and Use

By ASHOK S. BHAGWAT

Introduction

Most experiments in molecular biology involve the use of restriction enzymes[1] (REs) at some stage of the experiment. The ability of these enzymes to cut DNA at a specific sequence of bases has greatly stimulated the growth of recombinant DNA technology. The popularity of these enzymes is, in part, due to their wide commercial availability and the simplicity of their use. Over 1900 REs are known and of these 275 are available from companies based around the world. The purpose of this chapter is to list all commercially available enzymes, to describe their general properties, the basic procedures for their use, and ways of troubleshooting problems encountered in their use. Although the number of enzymes being discussed is large, a few simple rules apply to virtually all the enzymes. One of the aims of this chapter is to show that most enzymes can be used successfully without the help of premade kits, colorful charts, complex buffers, or specialized equipment. Some of this ground has been covered in previous reviews by Fuchs and Blakesley,[2] Brooks,[3] and Roberts and Macelis.[4]

Availability, Selection, and Storage

Table I lists all commercially available REs, the sequences they recognize, and commercial sources for the enzymes. They are available in concentrations ranging from 1 to 100 units/μl. One unit of the enzyme is defined as the amount that can digest 1 μg of bacteriophage λ DNA to completion in 1 hr in a 50-μl reaction using the specified buffer and at the specified temperature. For enzymes that do not cleave λ DNA, the unit definition uses adenovirus-2 or some other DNA. Most vendors of REs specify buffers and temperatures for their use and often supply the buffers (but see below). The catalogs of many vendors [New England BioLabs (Beverly, MA) and United States Biochemicals, (Cleveland, OH) among

[1] In this chapter only the typeII restriction endonucleases are described. Technological applications have not been found for other types of restriction enzymes.

[2] R. Fuchs and R. Blakesley, this series, Vol. 100, p. 3.

[3] J. E. Brooks, this series, Vol. 152, p. 113.

[4] R. J. Roberts and D. Macelis, *Nucleic Acids Res.* **19**(Suppl.), 2077 (1991).

TABLE I
COMMERCIALLY AVAILABLE RESTRICTION ENZYMES

Number	Name	Sequence[a]	Source[b]	Isoschizomers available	Buffer[c]	Methylation interference[d]
1.	*Aat*I	AGG^CCT	OU	139, 259		Partially blocked by Dcm
2.	*Aat*II	GACGT^C	EGLMNOPRSUVX	—	K	
3.	*Acc*I	GT^MKAC	ABEGIKLMNOPRSUVX	—	K	
4.	*Acc*II	CG^CG	AEGKVX	69, 89, 197, 263		
5.	*Acc*III	T^CCGGA	AEGKR	63, 77, 175, 190		Partially blocked by Dam
6.	*Aci*I	CCGC (−2/−2)	N	—		
7.	*Acy*I	GR^CGYC	EMRVX	12, 44, 163		
8.	*Afa*I	GT^AC	K	107, 231		
9.	*Afl*II	C^TTAAG	ABGKNU	52		
10.	*Afl*III	A^CRYGT	BGMNUX	—		
11.	*Age*I	A^CCGGT	N	—		
12.	*Aha*II	GR^CGYC	GN	7, 44, 163		
13.	*Aha*III	TTT^AAA	E	113		
14.	*Alu*I	AG^CT	ABEFGIKLMNOPRSUVX	—	M	
15.	*Alw*I	GGATC (4/5)	NU	—		Completely blocked by Dam
16.	*Alw*21I	GWGCW^C	F	30, 161		
17.	*Alw*26I	GTCTC (1/5)	F	68		
18.	*Alw*44I	G^TGCAC	FORU	23, 250		
19.	*Alw*NI	CAGN3^CTG	NU	—		
20.	*Aoc*I	CC^TNAGG	E	37, 94, 109, 134, 194, 237		
21.	*Aos*I	TGC^GCA	GX	35, 152, 155, 194		
22.	*Apa*I	GGGCC^C	BEGIKLMNOPRUVX	73	N	Partially blocked by Dcm
23.	*Apa*LI	G^TGCAC	AEGKNUX	18, 250	N	
24.	*Apy*I	CC^WGG	M	87, 144, 196	N	Not blocked by Dcm
25.	*Ase*I	AT^TAAT	NU	26, 267	H	

26.	*Asn*I	AT^TAAT	M	25, 267		
27.	*Asp*I	GACN^NNGTC	M	264		
28.	*Asp*700I	GAANN^NNTTC	M	274		
29.	*Asp*718I	G^GTACC	M	174		Partially blocked by Dcm
30.	*Asp*HI	GWGCW^C	M	16, 161		
31.	*Asu*I	G^GNCC	R	102, 238		
32.	*Ava*I	C^YCGRG	ABEGIKLMNOPRSUVX	135, 211	M	
33.	*Ava*II	G^GWCC	ABEGIKMNPRSVX	129, 247	K	Partially blocked by Dcm
34.	*Ava*III	ATGCAT	G	147, 209		
35.	*Avi*II	TGC^GCA	M	21, 152, 155, 194		
36.	*Avr*II	C^CTAGG	N	55		
37.	*Axy*I	CC^TNAGG	GV	20, 94, 109, 134, 194, 237		
38.	*Bal*I	TGG^CCA	ABGIKRSVX	191	N	Partially blocked by Dcm
39.	*Bam*HI	G^GATCC	ABEFGIKLMNOPRSUVX	84	H	Not blocked by Dam; not blocked by Dcm
40.	*Ban*I	G^GYRCC	EGIMNOPUVX	132		
41.	*Ban*II	GRGCY^C	EGIKLMNOPRSUVX	126	K	Partially blocked by Dcm
42.	*Ban*III	ATCGAT	OU	62, 71, 93, 109		
43.	*Bbe*I	GGCGC^C	AK	148, 173, 199, 215		
44.	*Bbi*II	GR^CGYC	AK	7, 12, 163		
45.	*Bbr*PI	CAC^GTG	M	133, 222, 223		
46.	*Bbs*I	GAAGAC	N	—		
47.	*Bbu*I	GCATG^C	R	217, 252		
48.	*Bbv*I	GCAGC (8/12)	EGINX	—		
49.	*Bcg*I	(10/12) GCAN6TCG (12/10)[e]	N	—		
50.	*Bcl*I	T^GATCA	BEGILMNOPRSUVX	151		Completely blocked by Dam
51.	*Bcn*I	CC^SGG	AFK	200		

(*continued*)

TABLE I (*continued*)

Number	Name	Sequence[a]	Source[b]	Isoschizomers available	Buffer[c]	Methylation interference[d]
52.	*Bfr*I	C^TTAAG	M	9		
53.	*Bgl*I	GCCN4^NGGC	BEFGILMNOPRSUVX	—	K	Not blocked by Dcm
54.	*Bgl*II	A^GATCT	ABEFGIKLMNOPRSUVX	—	M	Not blocked by Dam
55.	*Bln*I	C^CTAGG	K	36		
56.	*Bmy*I	GDGCH^C	M	74, 242		
57.	*Bpu*1102I	GC^TNAGC	F	97, 149		
58.	*Bsa*I	GGTCTC (1/5)	N	127		
59.	*Bsa*AI	YAC^GTR	N	—		
60.	*Bsa*BI	GATNN^NNATC	N	182		Partially blocked by Dam
61.	*Bsa*JI	C^CNNGG	N	—		Not blocked by Dcm
62.	*Bsc*I	AT^CGAT	L	42, 71, 93, 104		
63.	*Bse*AI	T^CCGGA	X	5, 77, 175, 190		
64.	*Bsi*CI	TT^CGAA	U	72, 85, 108, 178, 212, 245		
65.	*Bsi*YI	CCN5^NNGG	U	—		
66.	*Bsi*WI	C^GTACG	N	253		
67.	*Bsm*I	GAATGC (1/−1)	EGLMNOUVX	—		
68.	*Bsm*AI	GTCTC (1/5)	NU	17		
69.	*Bsp*50I	CG^CG	F	4, 89, 197, 263		
70.	*Bsp*68I	TCG^CGA	F	208, 254		
71.	*Bsp*106I	AT^CGAT	E	42, 62, 93, 104		Partially blocked by Dam
72.	*Bsp*119I	TT^CGAA	F	64, 85, 108, 178, 212, 245		
73.	*Bsp*120I	G^GGCCC	F	22		
74.	*Bsp*1286I	GDGCH^C	EGKNRUX	56, 242		
75.	*Bsp*AI	^GATC	L	112, 183, 203, 239		
76.	*Bsp*CI	CGAT^CG	E	228, 275		
77.	*Bsp*EI	T^CCGGA	N	5, 63, 175, 190		Partially blocked by Dam

78.	*Bsp*HI	T^CATGA	NU	232	K	Partially blocked by Dam
79.	*Bsp*MI	ACCTGC (4/8)	NU	—		
80.	*Bsr*I	ACTGG (1/−1)	N	—		
81.	*Bss*I	GGNNCC	X	206		
82.	*Bss*HII	G^CGCGC	BEGLMNOUVX	—		
83.	*Bss*XI	GCNGC	G	153		
84.	*Bst*I	G^GATCC	GPV	39		Not blocked by Dam
85.	*Bst*BI	TT^CGAA	N	64, 72, 108, 178, 212, 245	K	
86.	*Bst*EII	G^GTNACC	BEGLMNOPRSUVX	88, 36, 141	H	Not blocked by Dcm
87.	*Bst*NI	CC^WGG	ENX	24, 144, 196	H	Not blocked by Dcm
88.	*Bst*PI	G^GTNACC	K	86, 136, 141		
89.	*Bst*UI	CG^CG	NU	4, 69, 197, 263		
90.	*Bst*VI	C^TCGAG	G	96, 218, 270		
91.	*Bst*XI	CCAN5^NTGG	BEGKLMNORUVX	—	H	
92.	*Bst*YI	R^GATCY	NU	185, 271	N	Not blocked by Dam
93.	*Bsu*15I	AT^CGAT	F	42, 62, 71, 104		
94.	*Bsu*36I	CC^TNAGG	N	20, 37, 109, 134, 194, 237		
95.	*Bsu*RI	GG^CC	FG	158, 219		
96.	*Ccr*I	C^TCGAG	X	90, 218, 270		
97.	*Cel*II	GC^TNAGC	LM	57, 149		
98.	*Cfo*I	GCG^C	BILMRV	162, 164, 165		
99.	*Cfr*I	Y^GGCCR	F	119		
100.	*Cfr*9I	C^CCGGG	FOU	225, 248, 272		
101.	*Cfr*10I	R^CCGGY	AFKMNOU	—		
102.	*Cfr*13I	G^GNCC	AFKOU	31, 238		
103.	*Cfr*42I	CCGC^GG	F	176, 189, 235, 258		
104.	*Cla*I	AT^CGAT	ABEGKMNPRSVX	42, 62, 71, 93	M	Partially blocked by Dam
105.	*Cpo*I	CGGWCCG	K	106, 233		
106.	*Csp*I	CG^GWCCG	R	105, 233		
107.	*Csp*6I	G^TAC	F	8, 231		

(*continued*)

TABLE I (*continued*)

Number	Name	Sequence[a]	Source[b]	Isoschizomers available	Buffer[c]	Methylation interference[d]
108.	*Csp*45I	TT^CGAA	R	64, 72, 85, 178, 212, 245		
109.	*Cvn*I	CC^TNAGG	B	20, 37, 94, 134, 194, 237		
110.	*Dde*I	C^TNAG	BEGILMNOPRUVX	—		
111.	*Dpn*I	GA^TC	ABEGILMNRSUVX	—	H	Cleaves only if adenine is methylated
112.	*Dpn*II	GATC	UN	75, 112, 203, 239		Completely blocked by Dam
113.	*Dra*I	TTT^AAA	ABEFGIKLMNOPRSUVX	13	M	
114.	*Dra*II	RG^GNCCY	EGM	142, 226		
115.	*Dra*III	CACN3^GTG	EMNUX	—	M	
116.	*Drd*I	GACN4^NNGTC	N	—		
117.	*Dsa*I	C^CRYGG	M	—	H	Partially blocked by Dam
118.	*Dsa*V	^CCNGG	M	241	K	
119.	*Eae*I	Y^GGCCR	EGKLMNUVX	99		Partially blocked by Dcm
120.	*Eag*I	C^GGCCG	N	125, 131, 273	H	
121.	*Eam*1104I	CTCTTC (1/4)	F	123, 177		
122.	*Eam*1105I	GACN3^NNGTC	F	—		
123.	*Ear*I	CTCTTC (1/4)	N	121, 177		
124.	*Ecl*136II	GAG^CTC	F	234, 257		
125.	*Ecl*XI	C^GGCCG	M	120, 131, 273		
126.	*Eco*24I	GRGCY^C	F	41		
127.	*Eco*31I	GGTCTC (1/5)	F	58		
128.	*Eco*32I	GAT^ATC	F	145		
129.	*Eco*47I	G^GWCC	FOU	33, 247		
130.	*Eco*47III	AGC^GCT	AFKLMORU	—		
131.	*Eco*52I	C^GGCCG	AEFKOU	120, 125, 273		

132.	*Eco*64I	G^GYRCC	F	40		
133.	*Eco*72I	CAC^GTG	F	45, 222, 223		
134.	*Eco*81I	CC^TNAGG	AFKOU	20, 37, 94, 109, 194, 237		
135.	*Eco*88I	C^YCGRG	F	32, 211		
136.	*Eco*91I	G^GTNACC	F	86, 88, 141		
137.	*Eco*105I	TAC^GTA	FOU	249		
138.	*Eco*130I	C^CWWGG	FU	146, 260		
139.	*Eco*147I	AGG^CCT	F	1, 259		
140.	*Eco*NI	CCTNN^N3AGG	N	—		
141.	*Eco*O65I	G^GTNACC	GK	86, 88, 136		
142.	*Eco*O109I	RG^GNCCY	AFGKLNOUVX	114, 226		Partially blocked by Dcm
143.	*Eco*RI	G^AATTC	ABEFGIKLMNOPRSUVX	—	M	
144.	*Eco*RII	^CCWGG	BEGOUV	24, 87, 196	M	Completely blocked by Dcm
145.	*Eco*RV	GAT^ATC	ABEGIKLMNOPRSUVX	128	H	
146.	*Eco*T14I	C^CWWGG	AK	138, 260		
147.	*Eco*T22I	ATGCA^T	KOU	34, 209		
148.	*Ehe*I	GGC^GCC	FOU	43, 173, 199, 215		
149.	*Esp*I	GC^TNAGC	EGU	57, 97		
150.	*Esp*3I	CGTCTC (1/5)	F	—		
151.	*Fba*I	TGATCA	K	50		
152.	*Fdi*II	TGC^GCA	U	21, 35, 155, 194		
153.	*Fnu*4HI	GC^NGC	N	83		
154.	*Fok*I	GGATG (9/13)	AEGIKMNRUVX	—		Partially blocked by Dcm
155.	*Fsp*I	TGC^GCA	EGNS	21, 35, 152, 194	M	
156.	*Gsu*I	CTGGAG (16/14)	F	—		
157.	*Hae*II	RGCGC^Y	ABEGIKLMNOPRSUX	—		
158.	*Hae*III	GG^CC	ABGIKLMNOPRSUVX	95, 219	M	Partially blocked by Dcm
159.	*Hap*II	C^CGG	AGIK	171, 193		
160.	*Hga*I	GACGC (5/10)	NUX	—		

(*continued*)

TABLE I (*continued*)

Number	Name	Sequence[a]	Source[b]	Isoschizomers available	Buffer[c]	Methylation interference[d]
161.	*Hgi*AI	GWGCW^C	NX	16, 30	H	
162.	*Hha*I	GCG^C	ABEGKNOPRSUX	98, 164, 165		
163.	*Hin*1I	GR^CGYC	FOU	7, 12, 44		
164.	*Hin*6I	G^CGC	F	98, 162, 165		
165.	*Hin*P1I	G^CGC	NX	98, 162, 164		
166.	*Hinc*II	GTY^RAC	ABEFGIKLNOPRSUVX	167	M	
167.	*Hind*II	GTY^RAC	M	166	M	
168.	*Hind*III	A^AGCTT	ABEFGIKLMNOPRSUVX	—	M	
169.	*Hinf*I	G^ANTC	ABEGIKLMNOPRSUVX	—	M	
170.	*Hpa*I	GTT^AAC	ABEGIKLMNOPRSUVX	—	K	
171.	*Hpa*II	C^CGG	BEFGLMNOPRSUVX	159, 193	M	
172.	*Hph*I	GGTGA (8/7)	NUVX	—		Partially blocked by Dam; not blocked by Dcm
173.	*Kas*I	G^GCGCC	N	43, 148, 215		
174.	*Kpn*I	GGTAC^C	ABEFGIKLMNOPRSUVX	29	N	Not blocked by Dcm
175.	*Kpn*2I	T^CCGGA	F	5, 63, 77, 190		Not blocked by Dam
176.	*Ksp*I	CCGC^GG	EM	103, 189, 235, 258		
177.	*Ksp*632I	CTCTTC (1/4)	M	121, 123		
178.	*Lsp*I	TT^CGAA	LV	64, 72, 85, 108, 212, 245		
179.	*Mae*I	C^TAG	M	230		
180.	*Mae*II	A^CGT	M	—		
181.	*Mae*III	^GTNAC	M	—		
182.	*Mam*I	GATNN^NNATC	M	60		Partially blocked by Dam
183.	*Mbo*I	^GATC	BEGIKNOPRSVX	75, 112, 203, 239	H	Completely blocked by Dam
184.	*Mbo*II	GAAGA (8/7)	BGIKNPRSUVX	—		Partially blocked by Dam
184a.	*Mcr*I	CGRYCG	M	—		

185.	*Mfl*I	R^GATCY	AK	92, 271		
186.	*Mlu*I	A^CGCGT	ABEFGIKLMNOPRSUVX	—		
187.	*Mly*I	GASTC	L	—		
188.	*Mnl*I	CCTC (7/7)	EGNUX	—		
189.	*Mra*I	CCGCGG	G	103, 176, 235, 258		
190.	*Mro*I	T^CCGGA	MOU	5, 63, 77, 175		
191.	*Msc*I	TGG^CCA	NU	38	M	Partially blocked by Dcm
192.	*Mse*I	T^TAA	NU	—		
193.	*Msp*I	C^CGG	ABEFGIKLMNOPRSUVX	159, 171	M	
194.	*Mst*I	TGC^GCA	X	21, 35, 152, 155		
195.	*Mst*II	CC^TNAGG	EX	20, 37, 94, 109, 134, 237		
196.	*Mva*I	CC^WGG	AFKMOU	24, 87, 144		Not blocked by Dcm
197.	*Mvn*I	CG^CG	M	4, 69, 89, 263		
198.	*Nae*I	GCC^GGC	EGKLMNOUVX	—		
199.	*Nar*I	GG^CGCC	BEGMNOPUVX	43, 148, 173, 215		Not blocked by Dcm
200.	*Nci*I	CC^SGG	BEGLMNOUVX	51	K	
201.	*Nco*I	C^CATGG	ABEFGIKLMNOPRUVX	—	H	
202.	*Nde*I	CA^TATG	BEFGKLMNPSUVX	—	H	
203.	*Nde*II	^GATC	BGM	75, 112, 183, 239		Completely blocked by Dam
204.	*Nhe*I	G^CTAGC	BEGKLMNOPRUVX	—	M	
205.	*Nla*III	CATG^	NU	—		
206.	*Nla*IV	GGN^NCC	N	81		
207.	*Not*I	GC^GGCCGC	ABEFGIKLMNOPRSUVX	—		
208.	*Nru*I	TCG^CGA	ABEGIKLMNOPUVX	70, 254	K	Partially blocked by Dam
209.	*Nsi*I	ATGCA^T	BELMNRVX	34, 147	H	
210.	*Nsp*I	RCATG^Y	AKMU	214		
211.	*Nsp*III	C^YCGRG	PV	32, 135		
212.	*Nsp*V	TTCGAA	ABGKOPU	64, 72, 85, 108, 178, 245		
213.	*Nsp*BII	CMG^CKG	GU	—		

(*continued*)

TABLE I (*continued*)

Number	Name	Sequence[a]	Source[b]	Isoschizomers available	Buffer[c]	Methylation interference[d]
214.	*Nsp*HI	RCATG^Y	G	210		
215.	*Nun*II	GG^CGCC	G	43, 148, 173, 199		
216.	*Pac*I	TTAAT^TAA	N	—		
217.	*Pae*I	GCATG^C	F	47, 252		
218.	*Pae*R7I	C^TCGAG	NX	90, 96, 270		
219.	*Pal*I	GG^CC	EPV	95, 158		
220.	*Pfl*MI	CCAN4^NTGG	NU	266	H	Partially blocked by Dcm
221.	*Ple*I	GAGTC (4/5)	NU	—		
222.	*Pma*CI	CAC^GTG	K	45, 133, 223		
223.	*Pml*I	CAC^GTG	NU	45, 133, 222	H	
224.	*Ppu*MI	RG^GWCCY	NU	—		Partially blocked by Dcm
225.	*Psp*AI	C^CCGGG	E	100, 248, 272		
226.	*Pss*I	RGGNC^CY	I	114, 142		
227.	*Pst*I	CTGCA^G	ABEFGIKLMNOPRSUVX	—	H	
228.	*Pvu*I	CGAT^CG	ABEFGKLMNOPRSUVX	76, 275	H	Not blocked by Dam
229.	*Pvu*II	CAG^CTG	ABEFGIKLMNOPRSUVX	—	M	
230.	*Rma*I	C^TAG	N	179		
231.	*Rsa*I	GT^AC	ABEGILMNOPRSUVX	8, 107	N	
232.	*Rsp*XI	T^CATGA	G	78		
233.	*Rsr*II	CG^GWCCG	BEGNUX	105, 106		
234.	*Sac*I	GAGCT^C	AEGIKLMNOPRSUVX	124, 257	N	
235.	*Sac*II	CCGC^GG	EILNOPRUVX	103, 176, 189, 258	N	
236.	*Sal*I	G^TCGAC	ABEFGIKLMNOPRSUVX	—	H	
237.	*Sau*I	CC^TNAGG	M	20, 37, 94, 109, 134, 195		
238.	*Sau*96I	G^GNCC	BEGLMNRVX	31, 102		Partially blocked by Dcm
239.	*Sau*3AI	^GATC	ABEGIKLMNOPRSUVX	75, 112, 183, 203	H	Not blocked by Dam
240.	*Sca*I	AGT^ACT	ABEFGIKLMNOPRSUVX	—	H	

241.	*Scr*FI	CC^NGG	EGMNOSUVX	118	K	Partially blocked by Dcm
242.	*Sdu*I	GDGCH^C	F	56, 74		
243.	*Sfa*NI	GCATC (5/9)	NUX	—		
244.	*Sfi*I	GGCCN4^NGGCC	BEGILMNOPRSUVX	—		Not blocked by Dcm
245.	*Sfu*I	TT^CGAA	M	64, 72, 85, 108, 178, 212		
246.	*Sgr*AI	CR^CCGGYG	M	—		
247.	*Sin*I	G^GWCC	LRSV	33, 129		
248.	*Sma*I	CCC^GGG	ABEFGIKLMNOPRSUVX	100, 225, 272	K	
249.	*Sna*BI	TAC^GTA	EGKLMNVX	137		
250.	*Sno*I	G^TGCAC	LMV	18, 23		
251.	*Spe*I	A^CTAGT	BEGKLMNORSUX	—		
252.	*Sph*I	GCATG^C	ABEGIKLMNOPRSUVX	47, 217	H	
253.	*Spl*I	C^GTACG	AK	66		
254.	*Spo*I	TCG^CGA	R	70, 208		Not blocked by Dam
255.	*Sse*8387I	CCTGCA^GG	AK	—		
256.	*Ssp*I	AAT^ATT	BEGKLMNORUVX	—	M	
257.	*Sst*I	GAGCT^C	B	124, 234		
258.	*Sst*II	CCGC^GG	B	103, 176, 189, 235		
259.	*Stu*I	AGG^CCT	ABEGIKLMNPRVX	1, 139	M	Partially blocked by Dcm
260.	*Sty*I	C^CWWGG	BEGMNRUVX	138, 146		
260a.	*Swa*I	ATTT^AAAT	M			
261.	*Taq*I	T^CGA	BEFGILMNOPRSUVX	265	H	Partially blocked by Dam
262.	*Tfi*I	GAWTC	N	—		
263.	*Tha*I	CG^CG	BI	4, 69, 89, 197		
264.	*Tth*111I	GACN^NNGTC	AEGIKNPUVX	27	H	
265.	*Tth*HB8I	T^CGA	AK	261		Partially blocked by Dam
266.	*Van*91I	CCAN4^NTGG	F	220		
267.	*Vsp*I	AT^TAAT	FK	25, 26		

(continued)

TABLE I (*continued*)

Number	Name	Sequence[a]	Source[b]	Isoschizomers available	Buffer[c]	Methylation interference[d]
268.	*Xba*I	T^CTAGA	ABEFGIKLMNOPRSUVX	—	M	Partially blocked by Dam
269.	*Xcm*I	CCAN5^N4TGG	NU	—		
270.	*Xho*I	C^TCGAG	ABEFGIKLMNOPRSUVX	90, 96, 218	H	
271.	*Xho*II	R^GATCY	EGMVX	92, 185	N	Not blocked by Dam
272.	*Xma*I	C^CCGGG	EINRUVX	100, 225, 248	M	
273.	*Xma*III	C^GGCCG	B	120, 125, 131		
274.	*Xmn*I	GAANN^NNTTC	EGNUX	28	N	
275.	*Xor*II	CGAT^CG	B	76, 228		Not blocked by Dam

[a] Site of enzyme cleavage is indicated by a caret (^). In cases where the site of cleavage lies outside the recognition sequence, numbers in parentheses indicate the distance from the 3′ end of the recognition sequence. A positive number indicates cleavage to 3′ side of the end of the recognition sequence; a minus preceding the number indicates cleavage to 5′ side of the end of the recognition sequence (i.e., within the recognition sequence). Some letters represent ambiguous bases: R, G or A; Y, C or T; M, A or C; K, G or T; S, G or C; W, A or T; H, A or C or T; D, G or A or T; and N, A or C or G or T.

[b] The list of vendors is *not* complete. It is also by no means fixed; new enzymes are constantly being added and some deleted from most vendors catalogs. It lists most major sellers of the enzymes. For names of some other companies see Ref. 4 and "Guide to Biotechnology

Products and Instruments'' (*Science 151* suppl., pp. G78–G80). Names of the companies have been abbreviated. This list is derived from information provided by most of the vendors as of February 1991 and from Ref. 4. Key to the names and the company addresses is given below. A telephone number follows each address. A, Amersham Corporation, 2636 South Clearbrook Drive, Arlington Heights, IL 60005, Tel. 800-323-9750; B, GIBCO-BRL, Life Technologies, Inc., 3175 Staley Road, Grand Island, NY 14072-0068, Tel. 800-828-6686; E, Stratagene, 11099 North Torrey Pines Road, La Jolla CA 92037, Tel. 800-424-5444; F, Fermentas, Fermentu 8, Vilnius 232028, Lithuania, Tel. 641279; G, Bioexcellence, Ltd., Whitehall House, Whitehall Road, Colchester, Essex, England C028HA, Tel. 206866007; I, International Biotechnologies, Inc., P.O. Box 9558, 25 Science Park, New Haven, CT 06535, Tel. 203-786-5600; K, Takara Biochemical, Inc., 719 Allston Way, Berkeley, CA 94710, Tel. 800-544-9899; L, Northumbria Biologicals, Ltd., Nelson Industrial Estate, Cramlington, Northumberland NE23 9B1, United Kingdom, Tel. (0670) 732992; M, Boehringer Mannheim Corporation, 9115 Hauge Road, P.O. Box 50414, Indianapolis, IN 46250-0414, Tel. 800-428-5433; N, New England Biolabs, 32 Tozer Road, Beverly, MA 01915-5599, Tel. 800-632-5227; O, Toyobo, 1185 Avenue of the Americas, New York, NY 10036, Tel. 212-869-8200; P, Pharmacia LKB Biotechnology, Inc., 800 Centennial Ave., P.O. Box 1327, Piscataway, NJ 08855-1327, Tel. 800-526-3593; R, Promega, 2800 Woods Hollow Road, Madison, WI 53711-5399, Tel. 800-356-9526; S, Sigma Chemical Company, P.O. Box 14508, St. Louis, MO 63178, Tel. 800-325-8070; U, United States Biochemical, P.O. Box 22400, Cleveland, OH 44122, Tel. 800-321-9322; V, Serva, 50 A&S Drive, Paramus, NJ 07652, Tel. 800-645-3412; X, New York Biolaboratories, 30 Austin Blvd., P.O. 167, Commack, NY 11725, Tel. 516-543-3800.

[c] Refers to buffers used by the members of the author's laboratory. Recipies for these buffers are listed in Table V.

[d] Inhibition of cleavage by methylation in *E. coli*. Dam methylates adenine in GATC, while Dcm methylates the second cytosine in CCWGG. Where only a subset of the sequences recognized by the restriction enzyme is blocked by the methylation, this is described as a partial block.

[e] *Bcg*I cleaves the two strands on both sides of the recognition sequence. The cleavage is such that a two nucleotide 3′ overhang is created on either side of the sequence.

others] also contain excellent, although often undocumented, compilations of information regarding thermostabilities of the enzymes, compatibility of ends generated for cloning, suitability of the enzymes for genomic mapping, and methylation interference.

While the sequence of the DNA at hand or the design of the experiment may influence the choice of the enzyme, the following items should also be considered. For most base sequences, more than one enzyme is available that recognizes that sequence. These enzymes are identified as isoschizomers in Table I. Different experimental goals may necessitate the use of different isoschizomers. An enzyme that generates a four-nucleotide overhang may be preferable over its isoschizomer that creates a two-nucleotide overhang (*Kas*I over *Nar*I) if the ends are to be ligated to similarly generated cohesive ends. In contrast, if the experiment requires ligation of ends generated by enzymes that recognize different sequences, enzymes that generate blunt ends would be preferable over those that create an overhang (*Sma*I over *Xma*I). An enzyme that generates a 5′ overhang is preferable over its isoschizomers that generate a 3′ overhang (*Asp*718I over *Kpn*I) if the DNA is to be radiolabeled at the 5′ end. Some enzymes are available in "mapping grade" only and are unsuitable for cloning experiments.

Table II lists all palindromic four- and six-nucleotide sequences. Out of the 16 possible palindromic tetranucleotide sequences, 12 are recognized by at least 1 RE. For each of these 12 sequences 1 example of an RE is listed in Table II, except where different isoschizomers have different cleavage specificity. In the latter case one example of each cleavage specificity is listed (examples are *Mbo*I and *Dpn*I). Out of the 64 possible hexameric palindromic sequences, 48 are recognized by at least 1 RE. Once again, one enzyme per hexameric sequence is listed except in those cases where different enzymes cleave the same sequence differently. There are several uses for such a tabulation. It can be used to identify enzymes with compatible ends for cloning. For example, if *Mbo*I was used to generate DNA fragments, the fragments may be ligated to ends generated by *Bgl*II, *Bam*HI, or *Bcl*I, but not by *Pvu*I or other enzymes. Clearly, "true" isoschizomers (i.e., enzymes with the same sequence and cleavage specificity) can replace *Bgl*II, *Bam*HI, or *Bcl*I in the cloning experiment. Such isoschizomers can be found in Table I.

In some experiments it is useful to introduce a restriction site at or near a particular DNA sequence. For example, the RE recognition site may mark the site of a mutation. Sophisticated computer programs exist that identify sequences that may be changed to create a new restriction site without breaking constraints such as the amino acid sequence or a consensus protein-binding site. A simple alternative to the computer

TABLE II
Enzymes with Palindromic Recognition Sequences[a]

Sequence	AATT	ACGT	AGCT	ATAT	CATG	CCGG	CGCG	CTAG	GATC	GCGC	GGCC	GTAC	TATA	TCGA	TGCA	TTAA
N↓◇◇◇◇N									*Mbo*I							
N◇↓◇◇◇N		*Mae*II				*Hpa*II		*Mae*I		*Hin*6I		*Csp*6I		*Taq*I		*Mse*I
N◇◇↓◇◇N			*Alu*I				*Acc*II		*Dpn*I		*Hae*III	*Rsa*I				
N◇◇◇↓◇N										*Hha*I						
N◇◇◇◇↓N					*Nla*III											
A↓◇◇◇◇T			*Hind*III				*Mlu*I	*Spe*I	*Bgl*II							
A◇↓◇◇◇T														*Cla*I		*Vsp*I
A◇◇↓◇◇T				*Ssp*I						*Eco*47III	*Stu*I	*Sca*I				
A◇◇◇↓◇T																
A◇◇◇◇↓T					*Nsp*7524I										*Nsi*I	
C↓◇◇◇◇G					*Nco*I	*Xma*I		*Avr*II			*Xma*III	*Spl*I		*Xho*I		*Afl*II
C◇↓◇◇◇G				*Nde*I												
C◇◇↓◇◇G		*Pma*CI	*Pvu*II			*Sma*I										
C◇◇◇↓◇G							*Sac*II		*Pvu*I							
C◇◇◇◇↓G												*Pst*I				
G↓◇◇◇◇C	*Eco*RI						*Bss*HII	*Nhe*I	*Bam*HI		*Bsp*120I	*Asp*718		*Sal*I	*Apa*LI	
G◇↓◇◇◇C								*Nar*I								
G◇◇↓◇◇C			*Ecl*136I	*Eco*RV		*Nae*I										*Hpa*I
G◇◇◇↓◇C																
G◇◇◇◇↓C		*Aat*II	*Sac*I		*Sph*I					*Bbe*I	*Apa*I	*Kpn*I				
T↓◇◇◇◇A					*Bsp*HI	*Acc*III		*Xba*I	*Bcl*I							
T◇↓◇◇◇A																
T◇◇↓◇◇A		*Sna*BI					*Nru*I			*Mst*I	*Bal*I					*Aha*III
T◇◇◇↓◇A																
T◇◇◇◇↓A																

[a] Commercially available enzymes with 4- or 6-bp recognition are listed. The top line shows the central four bases in the sequences, while the first column shows the two outside bases and the cleavage specificity. The site of cleavage is shown by an arrow. N in the first column represents any base, while diamonds represent the appropriate bases in each column.

TABLE III
ENZYMES WITH LARGER THAN 6-bp RECOGNITION SEQUENCE[a]

Name	Recognition sequence[b]	Available isoschizomers
*Cpo*I	CGGWCCG	*Csp*I and *Rsr*II
*Csp*I	CG^GWCCG	*Cpo*I and *Rsr*II
*Not*I	GC^GGCCGC	—
*Pac*I	TTAAT^TAA	—
*Rsr*II	CG^GWCCG	*Cpo*I and *Csp*I
*Sfi*I	GGCCN4^NGGCC	—
*Sgr*AI	CR^CCGGYG	—
*Sse*8387I	CCTGCA^GG	—
*Swa*I	ATTT^AAAT	—

[a] Only those enzymes with larger than 6-bp "effective" recognition sequences are listed. For example, *Ppu*MI, which recognizes RGG-WCCY, effectively recognizes 5.5 bp due to its three ambiguous positions.

[b] Abbreviations regarding the recognition sequences are the same as in Table I.

programs is to generate manually palindromic 6-bp sequences at the target site within the constraints. With 75% of all possible hexameric palindromic sequences being recognized by at least one enzyme, chances of success for this strategy are high.

Enzymes that recognize sequences longer than 6 bp are potentially useful in genome mapping because they generate large fragments. These enzymes are listed in Table III. *Not*I and *Sfi*I have been particularly useful in this respect. McClelland *et al.*[5] have noted that if the %G + C of the DNA of the organism under study is known, then REs can be chosen based on the %G + C of their recognition sites to generate larger (or smaller) restriction fragments. For example, *Staphyloccocus aureus* DNA has 34% G + C content and a 2.2×10^6-bp genome and has only 16 *Sma*I (CCCGGG) sites and approximately 20 *Sac*II (CCGCGG) sites. It may also be possible to select REs for genome mapping based on the rarity of certain trinucleotides or tetranucleotides in the genome.[5] In bacterial genomes the tetranucleotide CTAG is rare. As a result, REs such as *Xba*I (TCTAGA) and *Avr*II (CCTAGG) have few sites in bacterial genomes. Other REs may also be useful for genome mapping in some cases (see below).

Two enzymes that recognize sequences that are significantly longer than 8 bp have become available commercially. Neither is a restriction

[5] M. McClelland, R. Jones, Y. Patel, and M. Nelson, *Nucleic Acids Res.* **15,** 5985 (1987).

enzyme. A protein required for the packaging of bacteriophage λ called "terminase" recognizes a *cos* site in the bacteriophage genome and creates a double-strand break. This enzyme has been used in a kit to map inserts in a cosmid vector (Takara Biochemicals, Berkeley, CA). Some yeast mitochondrial introns code for enzymes that allow transposition of the intron. One such enzyme, the ω-nuclease from yeast, recognizes a 18-bp sequence (TAGGGATAA↓CAGGGTAAT) and is commercially available (Boehringer Mannheim, Indianapolis, IN). As yet, wide use of the last two enzymes in recombinant DNA work has not been reported.

Another class of REs that are of special interest includes enzymes that recognize asymmetric base sequences. These often cleave DNA outside the site of recognition and hence making the ends generated by these enzymes flush using a DNA polymerase and deoxynucleotide triphosphates or ligating DNA fragments generated by these enzymes to other DNA fragments does not destroy the recognition sites. This property has been put to use in several applications, including the generation of deletions of increasing length[6] and mapping the sequence specificity of DNA modification.[7] Enzymes of this kind are listed in Table IV.

Restriction enzymes should be stored at −20° or below, preferably in non-"frost-free" freezers. Most are stable for several months, some for many years. We store them in Styrofoam-lined boxes designed to fit several tubes. Often the concentration of the supplied enzyme is very high and diluting the supplied preparation is desirable. It is possible to dilute enzymes in a glycerol-based buffer [10 m*M* Tris-HCl, pH 7.4, 50 m*M* KCl, 1 m*M* dithiothreitol, 100 mg/ml bovine serum albumin, 50% (v/v) glycerol] for storage at −20°.

Reaction Buffers and Conditions

Restriction enzymes require Mg^{2+} for catalysis, but no other cofactors or coenzymes. In addition, monovalent cations (Na^+ or K^+) and reducing agents (2-mercaptoethanol or dithiothreitol) are frequently necessary for efficient DNA cleavage. Optimal pH ranges from 7.2 to 7.8. The optimal temperature for the reaction varies from 30 to 70°, but most enzymes function at 37°. Typical durations of incubation are 1 to 2 hr, but with some enzymes it may be economical to incubate with a small amount of enzyme overnight. Like most enzymes, addition of serum albumin or some other inert protein to the reaction mixture stabilizes REs and is helpful during longer incubations.

[6] N. Hasan, S. C. Kim, A. J. Podhajska, and W. Szybalski, *Gene* **50,** 55 (1986).
[7] G. Posfai and W. Szybalski, *Gene* **69,** 147 (1988).

TABLE IV
ENZYMES WITH ASYMMETRIC RECOGNITION SEQUENCES

Name	Sequence[a]	Available isoschizomers
*Aci*I	CCGC (−2/−2)	—
*Alw*I	GGATC (4/5)	—
*Alw*26I	GTCTC (1/5)	*Bsm*AI
*Bbs*I	GAAGAC	—
*Bbv*I	GCAGC (8/12)	—
*Bcg*I	(10/12) GCAN6TCG (12/10)	—
*Bsa*I	GGTCTC (1/5)	*Eco*31I
*Bsm*I	GAATGC (1/−1)	—
*Bsm*AI	GTCTC	*Alw*26I
*Bsp*MI	ACCTGC (4/8)	—
*Bsr*I	ACTGG (1/−1)	—
*Eam*1104I	CTCTTC (1/4)	*Ear*I, *Ksp*632I
*Ear*I	CTCTTC (1/4)	*Eam*1104I, *Ksp*632I
*Eco*31I	GGTCTC (1/5)	*Bsa*I
*Esp*3I	CGTCTC (1/5)	—
*Fok*I	GGATG (9/13)	—
*Gsu*I	CTGGAG (16/14)	—
*Hph*I	GGTGA (8/7)	—
*Ksp*632I	CTCTTC (1/4)	*Ear*I, *Eam*1104I
*Mbo*II	GAAGA (8/7)	—
*Mnl*I	CCTC (7/7)	—
*Ple*I	GAGTC (4/5)	—
*Sfa*NI	GCATC (5/9)	—

[a] Abbreviations regarding the recognition sequence are the same as in Table I.

Reaction buffers are prepared at 10× concentration and stored frozen at −20° in 1-ml aliquots. Four different buffers are prepared and these may be used with most commercially available enzymes (Table V). We have tested these buffers with over 25% of the available enzymes and have been successful in obtaining complete digestion of DNA. These are noted in Table I. Although the remaining enzymes in Table I have not been tested with these buffers, it should be possible to use one of these buffers or a mixture thereof for those enzymes as well. Following pointers may be helpful in the choice of the buffer: some enzymes work only in the presence of K^+ and do not work well with Na^+ (e.g., *Sma*I). Hence the monovalent cation in the buffer recommended by the manufacturer should guide the choice. While some enzymes do not require the presence of 2-mercaptoethanol in the buffer for their action (example, *Sal*I), most are not inhibited by it (an exception is *Sau*3AI). If the recommended pH for the reaction differs from 7.6 by more than 0.5 units, it is preferable to

TABLE V
REACTION BUFFERS (10×)

Item[a]	Composition			
	H (high Na^+)	M (medium Na^+) (mM)	N (no salt) (mM)	K (medium K^+) (mM)
NaCl or KCl	1.5 M	500	—	500
Tris-HCl (pH 7.8)[b]	100 mM	100	100	100
$MgCl_2$	100 mM	100	100	100
2-Mercaptoethanol[c]	100 mM	100	100	100

[a] If long incubations are desired, then adding nuclease-free bovine serum albumin (BSA) to 100 μg/ml is recommended. Preparing the 10× buffers with BSA is *not* recommended, because of the tendency of BSA to come out of solution on repeated freeze-thawing.
[b] At 25°.
[c] The first three items are mixed in 10- to 100-ml volumes, divided in 1-ml aliquots, and the aliquots frozen at −20°. Before first use, 7 μl of 2-mercaptoethanol is mixed into a tube to make the complete buffer. This can be refrozen after use and can be used again over several days. After 2-mercaptoethanol evaporates (about 1 month) the tube is discarded and a new tube thawed.

make a separate buffer for the enzyme. Using these simple rules it may be possible to avoid filling the freezer with confusing color-coded or number-coded buffers from different manufacturers. Another simple alternate scheme for buffers has been suggested by McClelland *et al.*[8]

Reactions are carried out in 0.5- or 1.5-ml microfuge tubes. For long incubations at 37° and for all incubations above 37°, layering a small amount of mineral oil over the reaction mixture is recommended. Enzymes isolated from thermophilic bacteria show optimum activity at 50° or higher, but most also show significant levels of activity at 37° (>30%). Therefore, it is often possible to avoid the inconvenience of setting up several water baths at different temperatures by using a moderate excess of the enzymes.

The choice of bacteriophage λ DNA as the substrate in the definition of unit activity has the unfortunate consequence that 1 μg of DNA from other sources will often not be cut to completion by 1 unit of an enzyme under the recommended conditions. This is partly because of the nonrandom distribution of bases in any piece of DNA. In a 50-kbp piece of DNA (roughly the size of bacteriophage λ) with a random base composition a 6 bp-recognizing enzyme would be expected to have 12 target sites. In λ DNA this number varies from 1 (for *Xba*I and *Xho*I, among others) to 46 (for *Bsm*I) for different 6 bp-recognizing enzymes. Hence the number of

[8] M. McClelland, J. Hanish, M. Nelson, and Y. Patel, *Nucleic Acids Res.* **16,** 364 (1988).

units of an enzyme required for cutting at all the sites in 1 μg of experimental DNA depends on the frequency of occurrence of its sites in both λ and the experimental DNA. There is no straightforward solution to this problem. Consequently, when dealing with an unfamiliar piece of DNA we routinely use enzymes at a concentration 5- to 10-fold in excess of what would be required by the unit definition. There are additional reasons why a piece of DNA may not be digested by an RE at the expected rate. These are discussed in the Troubleshooting Section, below.

Enzyme Inactivation

A simple, reversible way to stop the restriction digest is to add ethylenediaminetetraacetic acid (EDTA) to the reaction. Each EDTA molecule will chelate two Mg^{2+}, preventing catalysis. An irreversible method for inactivating REs is phenol extraction. The residual phenol must be removed before this DNA can be used for other enzymatic manipulations. Ether extraction and ethanol precipitation are often used for this purpose. We use a rapid procedure that removes phenol, residual RE, and small molecules from the reaction. The procedure uses a "spin column" and is quite versatile. In addition to inactivating enzymes we use it to change reaction buffers, to purify miniprep DNAs, and for the removal of unincorporated linkers and primers.

Spin Column Preparation. For small volumes of DNA (<50 μl) the column is prepared using three 1.5-ml microfuge tubes (Brinkmann, Westbury, NY); for larger volumes, two 16-ml polypropylene tubes (Falcon 2059 tubes; Becton-Dickinson Labware, Oxnard, CA), and one Poly-prep column (Bio-Rad, Richmond, CA) are used. For the smaller column one microfuge tube is punctured with a hot 23-gauge needle (Becton-Dickinson) and the resulting hole is blocked with a 25-μl suspension of glass beads. The glass beads (425- to 600-μm diameter; Sigma, St. Louis, MO) are washed in concentrated HCl and in TE (10 m*M* Tris-HCl, pH 7.6, 1 m*M* EDTA) and resuspended in TE before use. Sepharose CL-6B (Pharmacia-LKB, Piscataway, NJ) is washed with TE and resuspended in TE. Up to 0.5 ml of the slurry is added to the tube and the tube inserted inside another microfuge tube (Fig. 1). In the case of the Poly-prep column, the Sepharose slurry (up to 2 ml) is directly added to the column and the column inserted inside a polypropylene tube (Fig. 1). In either case, the tubes are spun at room temperature at 1700 *g* for 4 min in a swinging bucket rotor (e.g., Omni spin centrifuge with H4211 rotor; Sorvall, Norwalk, CT). Following centrifugation, the lower tube is discarded. At this point the column tubes can be capped and stored at 4° for several weeks.

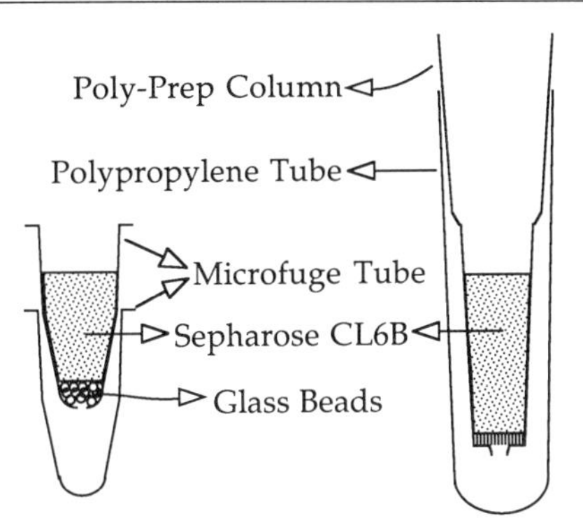

FIG. 1. Setup for the spin columns.

DNA Purification Using Spin Columns. It is advisable to add a small amount of dye (e.g., 100 μg/ml bromphenol blue) to the solution containing DNA to monitor the success of the procedure. If removal of proteins is also desirable, sodium dodecyl sulfate (SDS) is added to 0.1% (w/v) to prevent aggregation of the proteins. The sample is layered onto the column bed and the tube (or column) inserted inside a sterile microfuge tube (or a larger, 16-ml polypropylene tube). These columns are spun at 1700 *g* for 4 min as in the preparation of the columns. DNA appears in the lower tube and the dye, phenol, SDS and other small molecules are retained by the column. Although the exclusion limit of these columns has not been studied carefully, tRNA molecules and many REs are removed by these columns. We have also used them to remove small restriction fragments such as the 51-bp linker fragment from M13mp series of vectors from solutions. The maximum volumes of DNA solutions that can be purified in this way are 50 and 200 μl for the smaller and larger columns, respectively.

Comments. The yield of this procedure is 50–80%. The yield of DNA in the small-scale procedure may be further increased by the use of siliconized glass beads. The main precautions to take are as follows: (1) In preparing the column, make sure that the column is not wet. A wet column will lead to an increase in the volume of the DNA solution during the spin and will not remove small molecules efficiently. (2) Load the sample on the column bed and not along the sides of the tube. (3) Spin the tubes immediately (within 5 min) after loading the sample. A longer wait will lead to significant diffusion of the sample in the column and hence an ineffective removal of the small molecules.

Analysis of Digested DNA.

Agarose gel electrophoresis is the most commonly used technique for separating DNA fragments. DNA is stained with ethidium bromide and visualized with near-UV light. These procedures have been adequately explained in many sources.[9,10]

Troubleshooting

Incomplete or No Digestion of DNA

This is the most common problem encountered in the use of REs. Sometimes the sequence of the DNA being studied predicts the presence of a restriction site, but restriction digestion with the appropriate enzyme does not reveal its presence. There are a number of different possible reasons that could cause this problem. These are discussed below, starting with the most simple ones.

Poor Quality of DNA. "Mini-prep" DNAs or other quickly prepared DNAs may have salts, EDTA, proteins, and other contaminants, some of which can inhibit REs. One of the simplest ways to remove contaminants is to perform the spin column procedure outlined above. Ethanol precipitation can also remove small molecules that inhibit the enzymes. Phenol extraction removes proteins from the solution and is *not* very useful to alleviate the inhibition. For larger quantities of DNA (>100 μg) CsCl gradients or chromatographic methods may be useful for the purification.

Inactive Enzyme. The enzyme may have been inactivated during shipping or storage. We use bacteriophage λ DNA prepared from methylation-deficient bacteria (see below) to test the activity of the enzyme.

Methylation Interference. DNA from most organisms is naturally modified at many sites. This can sometimes lead to inhibition of restriction. In the extreme case of bacteriophage T4, where DNA contains glucosylated hydroxymethylcytosine in place of normal cytosine, the DNA is resistant to most REs. Cases in which the DNA is resistant to only some enzymes are also known. Cytosines within many CG dinucleotides in mammalian DNA are methylated. CCGG sites within mammalian DNA are sensitive to *Hpa*II only if the CG dinucleotides within them are unmethylated. For this reason, it is useful to know if the genomic DNA being studied is methylated and what this methylation may be. It is sometimes possible to avoid methylation interference by using an isoschizomer. *Msp*I is an isoschizomer of *Hpa*II that is insensitive to methylation within the CG

[9] F. Ausubel, R. Brent, R. E. Kingston, D. D. Moore, J. G. Seidman, J. A. Smith, and K. Struhl (ed.), "Current Protocols in Molecular Biology." Greene Publ., and Wiley (Interscience), New York, 1987.

[10] J. Sambrook, E. F. Fritsch, and T. Maniatis, "Molecular Cloning: A Laboratory Manual." Cold Spring Harbor Laboratory, Cold Spring Harbor, New York, 1989.

dinucleotide. Nelson and McClelland[11] have compiled a list of all known examples of methylation interference. This list may be used for the search of methylation-insensitive isoschizomers.

Inhibition of REs by methylation can be useful. Isoschizomers with differential sensitivity to methylation can be used to map sites of DNA methylation in a genome. A popular pair for mapping sites of methylation in mammalian genomes is *Hpa*II and *Msp*I. A comparison of fragments generated by the two enzymes is used to identify the methylated CCGG sequences. Endogenous methylation can also make a frequently cutting enzyme into an infrequently cutting enzyme. Several 6 bp-recognizing enzymes contain one or two CG dinucleotides. Most are inhibited by the methylation of cytosine and hence cleave only those sites in the genome where such methylation is absent. Examples of such enzymes include *Mlu*I (ACGCGT), *Bss*HII (GCGCGC), *Sna*BI (TACGTA), and *Xho*I (CTCGAG).[11] This effectively increases the specificity of these enzymes and results in fragments that are much larger than would be predicted by the base composition. These fragments can be useful in long-range genome mapping and cloning.

Purified methylases that are part of restriction-modification (R-M) systems have been used to increase the specificity of restriction enzymes. McClelland and colleagues (reviewed in [25] this volume) have used a number of methylases to increase the effective recognition sequences of a number of restriction enzymes to up to 12 bp. Two groups have used a novel strategy to cleave at a single *Hae*II or a single *Eco*RI site within the yeast genome.[12,13] First, all the sites for the RE except the target site were methylated with the appropriate methylase. Methylation of the target site was prevented by physically blocking it with a sequence-specific DNA-binding protein (*lac* repressor)[12] or by the formation of a triple helix with a synthetic oligomer.[13] Following methylation, the methylase and the blocking agent were removed and the DNA digested with the appropriate RE to cleave DNA at the target site.

It should also be remembered that the K/12 strain of *E. coli* contains three types of DNA methylations, EcoK, Dam, and Dcm. Hence DNA propagated in *E. coli* may contain one or more of the methylation patterns. EcoK methylation occurs at A^{Me}ACN6GTGC and GC^{Me}ACN6GTT sequences, where ^{Me}A is the methylated adenine. Because of the infrequency of this methylation, it rarely interferes with restriction digests. Some strains of the bacterium used for molecular cloning lack EcoK methylation,

[11] M. Nelson and M. McClelland, *Nucleic Acids Res.* **19**(Suppl.), 2045 (1991).

[12] M. Koob and W. Szybalski, *Science* **250,** 271 (1990).

[13] S. A. Strobel and P. B. Dervan, *Nature (London)* **350,** 172 (1991).

but contain the other two methylations. Sometimes this is unavoidable. Strains used for gene cloning are made *recA*⁻ to assure the stability of the cloned DNA. Introducing a *dam*⁻ mutation in this strain is impossible because of a poorly understood lethal phenotype of the double mutant. Dam protein methylates adenine within the sequence GATC. Methylation of this adenine inhibits many enzymes whose recognition sequences contain GATC or overlap with it (e.g., *Bcl*I; Table I). It is necessary to propagate the DNA in a *dam*⁻ strain to obtain cleavage by the enzyme being interfered with. A list of such strains has been published by Raleigh[14] and is also available from many vendors. However, it is useful to remember that not all enzymes are inhibited by internal or overlapping Dam methylation (e.g., *Bam*HI) and use of a *dam*⁻ strain is not always necessary.

Dcm protein methylates the second cytosine in the sequences CCAGG and CCTGG. This methylation inhibits several REs (e.g., *Eco*RII; Table I). As is true for Dam methylation, not all enzymes are inhibited by internal or overlapping Dcm methylation (e.g., *Bst*NI). Strains *dcm*⁻ and *dcm recA* are available[14] and the resistant clones can be propagated through these strains to make the DNA sensitive to REs. Reported incidents of inhibition of cleavage due to Dam of Dcm methylations, and incidents where the two methylation are not inhibitory, are noted in Table I.

Finally, if base modifications in a DNA molecule of known sequence are found to inhibit many enzymes, the polymerase chain reaction (PCR) may be used to prepare DNA that is sensitive to the enzymes. Primers from sequences flanking the target sequence can be used to amplify the target sequence using normal deoxynucleotide triphosphate precursors. The resulting DNA should be sensitive to REs. One advantage of this method is that it can be performed on very small amounts of DNA and without extensive purification. Hence in situations where the presence or absence of a small number of restriction sites in a small piece (less than a few kilobase pairs) of DNA is to be tested, performing PCR on the DNA prior to RE digest is attractive. This strategy is used to test individuals for genetic diseases.

Site Selectivity by Restriction Enzymes. This is a broad and poorly understood category that includes examples where factors such as flanking sequence, the number of sites in the substrate, and the physical nature of the substrate (supercoiled vs. linear DNA) appear to influence the cleavage efficiency. For example, it has been shown that two *Eco*RI sites in bacteriophage λ DNA are cleaved at rates differing by an order of magnitude. Similar examples have been documented for *Pae*R7I, *Nae*I, *Nar*I, *Eco*RII, and other enzymes. The reasons underlying this phenomenon are not

[14] E. A. Raleigh, this series, Vol. 152, p. 130.

clearly understood, but some solutions to alleviate this problem have emerged. (1) Once again, isoschizomers can be useful. While *Pae*R7I cleaves only one of its two sites in the adenovirus-2 DNA efficiently, its isoschizomer *Xho*I cleaves both the sites well[15]; (2) for some of the enzymes for some of the sites, including a "sensitive" DNA in the reaction can stimulate cleavage of the resistant substrate. Bacteriophage T7 DNA contains two sites for *Eco*RII that are poorly cleaved by the enzyme. pBR322, on the other hand, contains six sites for the enzyme, most of which are susceptible to cleavage. Mixing T7 and pBR322 DNA in the reaction leads to cleavage of both the DNAs.[16] Similar activation of resistant or slow cutting sites has also been seen for *Nae*I, *Bsp*MI, *Nar*I, *Hpa*II, and *Sac*II[17]; (3) in the presence of spermidine *Nae*I can cleave otherwise resistant sites. Similar activation of other enzymes by spermidine has not been reported.[18]

Relaxed Specificity ("Star Activity")

Under certain reaction conditions, some REs cleave noncanonical sequences. These altered conditions include replacement of Mg^{2+} by Mn^{2+}, low ionic strength, high pH, and the presence of organic solvents such as glycerol or ethylene glycol. The altered activity of the enzyme is indicated by an asterisk and hence is referred to as star activity. While *Eco*RI endonuclease recognizes the sequence GAATTC, *Eco*RI* recognizes a variety of sequences of the type NAATTC, GNATTC, etc., where N is any base.[19] Similarly, while *Bam*HI recognizes GGATCC, *Bam*HI* (also called *Bam*HI.1) recognizes sequences of the type GGNTCC, GGANCC, and GNATCC.[20]

Although it may be possible to relax the specificities of many REs by altering reaction conditions, this happens infrequently under normal experimental conditions. Even under the conditions where star activity is found, cleavage at the canonical site is most rapid and cleavage at other sites is a small fraction of the products. The most common reasons for obtaining star activity are using incorrect buffer and using too high a volume of stock RE in the reaction. Mistakes may be made during the

[15] T. R. Gingeras and J. A. Brooks, *Proc. Natl. Acad. Sci. U.S.A.* **80,** 402 (1983).

[16] D. H. Krüger, G. J. Barcak, M. Reuter, and H. O. Smith, *Nucleic Acids Res.* **16,** 3997 (1988).

[17] A. Oller, W. V. Broek, M. Conrad, and M. D. Topal, *Biochemistry* **30,** 2543 (1991).

[18] M. Conrad and M. D. Topal, *Proc. Natl. Acad. Sci. U.S.A.* **86,** 9707 (1989).

[19] B. Polisky, P. Greene, D. E. Garfin, B. J. McCarthy, H. M. Goodman, and H. W. Boyer, *Proc. Natl. Acad. Sci. U.S.A.* **72,** 3310 (1975).

[20] G. Nardone and J. G. Chirikjian, *in* "Gene Amplification and Analysis" (J. G. Chirikjian, ed.), p. 147. Elsevier Science, New York, 1987.

preparation of the reaction buffer, resulting in relaxed specificity. Most pH electrodes do not measure Tris buffers accurately, and the pH of Tris buffers varies considerably with temperature. All REs are shipped in 50% (v/v) glycerol. Hence using a large volume of the stock enzyme in the reaction leads to a relaxation of the enzyme specificity due to glycerol. Keeping the enzyme volume one-tenth or less of the total reaction volume (glycerol concentration $< 5\%$, v/v) avoids this effect of glycerol.

DNA Degradation

Occasionally, a restriction digest of DNA will result in degradation. In nearly every case, contaminants in the DNA are the culprits. Cells including *E. coli* contain a number of nonspecific nucleases. If these are not eliminated during the preparation of DNA, degradation of DNA will result during the digest. This can be tested by incubating the DNA in the buffer at the appropriate temperature in the absence of any added RE. Phenol extraction followed by removal of the phenol will alleviate the nuclease contamination.

Acknowledgments

The author would like to thank R. J. Roberts for providing data on restriction enzymes. The information provided by various companies about their products is also appreciated. The format of Table III was developed by New England Biolabs and the author is grateful to I. Schildkraut for permission to use it. Members of the author's research group, M. Dar, S. Gabbara, J. Gopal, and M. Wyszynski, made several useful contributions to the manuscript. The author is most grateful to R. M. Blumenthal and J. A. Brooks for their comments on the manuscript. The author is the recipient of Research Career Development Award from the National Institutes of Health (NIH) and the research in his laboratory is supported by NIH Grant GM40576 and by a grant from Boehringer Mannheim, GmbH, Penzberg, Germany.

[22] Use of Restriction Endonucleases to Detect and Isolate Genes from Mammalian Cells

By WENDY A. BICKMORE and ADRIAN P. BIRD

Introduction

The genomes of mice and humans are being studied intensively to identify and characterize genes. In many laboratories the approach is to focus on genetic loci of particular medical and biological interest to locate and isolate relevant genes. Often this must be achieved without the benefit of any information about the RNA or protein products of the gene. The main consideration in a quest of this kind is the sheer size of the field of search. The haploid mammalian genome comprises 3×10^9 bp of DNA and only a few percent of this codes for protein. The vast majority of DNA is therefore nongenic. A further complication is the fragmented (exonic) nature of nearly all genes. An mRNA of 1 kilobase (kb) can be derived from a primary transcript many times this length. Thus even when the approximate position of the gene is already known, it is not a simple matter to locate it definitively. Fortunately, in mammals at least, evolution has lessened the difficulty by leaving unusual sequences at the 5′ ends of most genes. These sequences can be located using certain types of restriction endonucleases, thereby facilitating the mapping and cloning of the gene of interest.

The 5′ sequences mentioned above are unusual in three respects[1,2]: (1) they are nonmethylated at CpG sites, whereas the rest of the genome is heavily methylated at CpG; (2) they contain the sequence CpG at the frequency expected from their base composition, whereas in the rest of the genome CpG is present at less than one-quarter of its expected frequency; (3) their base composition is generally more than 60% G + C, compared to an average of 40% G + C for the rest of the genome. Taken together, the G + C richness and lack of CpG deficiency means that CpG is usually 10 times more frequent in these regions than in the rest of the genome; hence their most common name is "CpG islands." Figure 1 shows a plot of CpG sites across a typical CpG island. Apart from the obvious local concentration of CpGs, Fig. 1 also illustrates two other typical features: the island is about 1 kb long (between 0.5 and 2 kb is usual); and it includes two exons and an intron (the first one to three exons are normally included). CpG clusters of this kind are normally nonmethylated at all times, while the flanking regions are methylated.

In choosing restriction enzymes that specifically cleave at CpG islands, the most important requirements are that the enzyme recognition site should contain a CpG, and that cleavage should be blocked by methylation of that CpG[3,4] (although the latter is not important when cloned DNA is being tested[5]). There are, however, other aspects of CpG enzymes that make discrimination for or against islands more stringent. For example,

[1] A. Bird, M. Taggart, M. Frommer, O. J. Miller, and D. Macleod, *Cell (Cambridge, Mass.)* **40,** 91 (1985).

[2] A. P. Bird, *Nature (London)* **321,** 209 (1986).

[3] A. P. Bird, *Trends Genet.* **3,** 342 (1987).

[4] W. R. A. Brown and A. P. Bird, *Nature (London)* **322,** 477 (1986).

[5] S. Lindsay and A. P. Bird, *Nature (London)* **327,** 336 (1987).

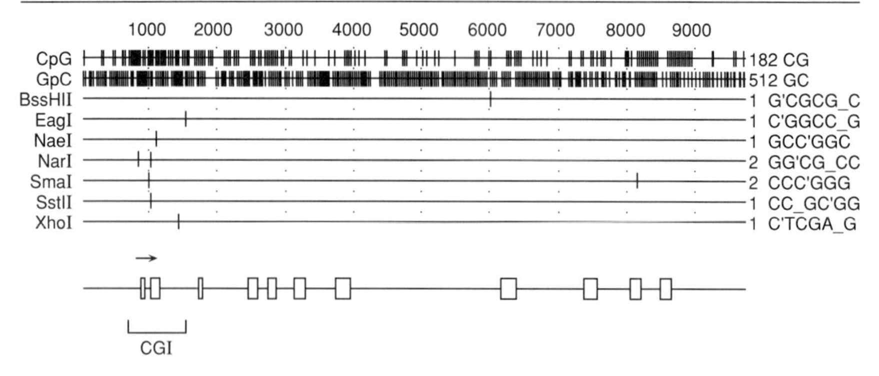

FIG. 1. Concentration of rare-cutter sites in a typical CpG island gene, showing the distribution of CpGs and, for comparison, GpCs in a 9722-nucleotide sequence of the mouse nicotinic acetylcholine receptor (β subunit) gene [as determined by A. Buonanno, J. Mudd, and J.-P. Merlie, *J. Biol. Chem.* **264,** 7611 (1989)]. CpG frequency is depressed relative to GpC over most of the sequence, but not in the region between nucleotide 700 and 1600. This region corresponds to the CpG island (see bracket labeled CGI) and is expected to be nonmethylated in all cells of the animal. A diagnostic feature of CpG islands is that nonmethylated sites for rare cutters are concentrated within them. Note that because the *Bss*HII site (about nucleotide 6000) and the rightmost *Sma*I site (about nucleotide 8200) are outside the CpG island region, they are likely to be methylated and therefore uncuttable in uncloned genomic DNA.

the number of CpGs per site, the presence of A/T pairs, and the number of base pairs in the recognition site can all have dramatic effects on patterns of cleavage.[6] In this chapter, we discuss the factors affecting choice of rare-cutting restriction enzymes for mapping and cloning experiments, and we present examples of their use.

Principles of Method

The usefulness of CpG islands as gene markers depends on the idea that most CpG islands signify genes. How reliable is this idea? Examination of the DNA databases shows that all "housekeeping genes" examined so far have islands,[7] together with a significant fraction of genes that are expressed in a tissue-specific fashion. It is unlikely that the databases reflect the true ratio of tissue-specific and housekeeping genes, because genes that are tissue specific have historically commanded more interest than genes that are expressed ubiquitously. In spite of the bias, we have looked for islands in genes for which long tracts of genomic sequence are available. Out of 29 mouse genes whose coding regions and flanks had been

[6] A. P. Bird, *Nucleic Acids Res.* **17,** 9485 (1989).

[7] M. Gardiner-Gardner and M. Frommer, *J. Mol. Biol.* **196,** 261 (1987).

sequenced, 12 had CpG island-like clusters according to the criteria in Ref. 1. A similar calculation for human genes in the database showed that 58% (38 out of 65) were associated with CpG islands. Given the likely underrepresentation of ubiquitously expressed genes, these figures strongly suggest that at least 50% of genes in these two mammals are CpG island associated.

The foregoing discussion makes it clear that a large fraction of genes have 5′ domains that are marked by CpG islands. Are all CpG islands found at 5′ domains of this kind, or are there islands located in nongenic DNA? The best evidence that islands signify genes comes from studies in which islands selected at random have been tested for the presence of an associated gene. These studies show that once a CpG island has been found, it is very likely to mark a gene.[8-10] Particularly impressive are the analyses of 170 kb of DNA in the mouse major histocompatibility complex (MHC) region,[11] and of 540 kb in the human MHC class III region.[12] Identification of CpG islands led to the discovery of 5 new genes in the mouse and 12 in the human MHC regions.

If most CpG islands correspond to genes, and a large fraction of genes have CpG islands, then it follows that the number of islands should resemble the number of genes. The average frequency of CpG islands in mammalian DNA is calculated to be about 1/150 kb of DNA, making an expected total of about 20,000 islands in the whole genome, assuming that CpG islands comprise 1% of the genome and have an average size of 1.5 kb.[4] A survey of the frequency of clustered rare-cutter sites in mapped regions of the human genome gives a somewhat lower figure of approximately 1/300 kb, which, if representative, would give a total of 10,000 islands per genome. Neither of these figures is wholly reliable, because both make important assumptions. In particular the second estimate assumes that one cluster of rare-cutter sites detected on pulsed-field gels corresponds to one CpG island, which will be an underestimate in regions where several islands are clustered together (see, e.g., Ref. 13). A further reservation is that several genes are known to be arranged head to head with two

[8] P. Lavia, D. Macleod, and A. P. Bird, *EMBO J.* **6,** 2773 (1987).

[9] X. Estivill, M. Farrall, P. J. Scambler, G. M. Bell, K. M. F. Hawley, N. J. Lench, G. Bates, H. C. Kruyer, P. A. Frederick, P. Stanier, E. K. Watson, R. Williamson, and B. J. Wainwright, *Nature (London)* **326,** 840 (1987).

[10] G. A. Rappold, L. Stubbs, S. Labeit, R. B. Crkvenjakov, and H. Lehrach, *EMBO J.* **6,** 1975 (1987).

[11] K. Abe, J.-F. Wei, F.-S. Wei, Y.-C. Hsu, H. Uehara, K. Artzt, and D. Bennett, *EMBO J.* **7,** 3441 (1988).

[12] C. A. Sargent, I. Dunham, and R. D. Campbell, *EMBO J.* **8,** 2305 (1989).

[13] C. Huxley and M. Fried, *Mol. Cell. Biol.* **10,** 605 (1990).

transcription start sites diverging from a single CpG island. Thus one island can signify two genes (e.g., see Ref. 8). In spite of these limitations, the numbers fall within the wide range of estimates for the number of genes in the mammalian genome.

As shown in Table I, CpG enzymes can be classified into groups based on two parameters: (1) the proportion of all genomic sites that are within CpG islands, and (2) the average number of sites per island. The first parameter measures how effectively the enzyme discriminates between island and nonisland DNA, and the second, the likelihood that any particular island will contain a site. The various groups of enzyme have distinct properties, and therefore distinct uses in the analysis of mammalian DNA. Care is needed in choosing the right enzyme for the job, and in interpreting the results obtained with a particular enzyme. We will briefly discuss the uses of these enzymes here, but will illustrate them with examples in later sections. The results shown in Table I were derived from a study of 37 human CpG islands.

To identify the position of CpG islands in genomic DNA, group II enzymes are the most useful because 75% of their sites are within islands, the remaining sites probably being methylated and therefore not cuttable. All the enzymes in this group have cutting sites that contain 2 CpGs and are 100% G + C in base composition. The high proportion of these sites in islands is due to the rarity of such sequences in the (A + T)-rich, CpG-deficient majority of the genome. Clusters of group II sites are probably the best means of identifying an island, because every island will on average have a site for at least one of these enzymes. Group I sites, although nearly exclusive to islands, are not present in every island. Thus the presence of a genomic *Not*I site is an extremely strong indicator of the presence of a CpG island, but cannot be used to identify the sites of all islands. Because of the rarity of sites in the genome, *Not*I is one of the most useful enzymes for making very long-range restriction maps and for the detection of chromosome breakpoints from a distance. The newly available group I enzyme *Asc*I appears to share the same properties as *Not*I.

Only a small fraction of sites for group III enzymes are associated with CpG islands. These enzymes have only a single CpG in their recognition sequences. Thus while CpG islands have the same frequency of sites for these enzymes as for those of group II, there are many more interisland (often methylated) sites. Therefore the presence of these sites in genomic DNA can be used to confirm the presence of an island, but is not in itself adequate diagnosis of the presence of an island. This is especially true in cloned DNA, where methylation of nonisland sites is removed.

For unknown reasons the percentage of sites for group IV enzymes in

CpG islands is much (up to 10-fold) lower than expected. In fact, well over 90% of sites are outside of CpG islands. In part this is probably a reflection of the presence of A and T residues in the recognition sites of these enzymes. It follows that the presence of such sites within cloned (i.e., unmethylated) DNA cannot be considered significant evidence for the presence of CpG islands. Because most genomic group IV sites are in interisland DNA they are therefore subject to variable methylation. As a result these enzymes are not reliable for the detection of altered restriction fragments resulting from chromosome breakpoints. Group IV enzymes are, however, useful for the establishment of physical linkages because large restriction fragments can be generated that avoid CpG islands.

The number of sites for a particular enzyme per island and the proportion of cutting sites within islands are sensitive to the G + C content of the islands and of the genome as a whole and therefore vary even between different mammalian species. In plants the depletion of CpG dinucleotide in the genome as a whole is not so extreme, and in fish CpG islands are in fact quite (A + T)-rich.[14] The distribution of sites for rare cutters in these organisms is therefore different from that in mammals. In this chapter we will address ourselves only to the detection of mammalian genes.

Recently, several new rare-cutter restriction enzymes, that do not fall within the groups described in Table I, have become available.

1. *Srf*I (GCCCGGGC) and *Fse*I (GGCCGGCC). These, like group I enzymes, have an 8 bp G + C recognition site and thus show the same number of cutting sites per island as *Not*I and *Asc*I. However, the sites for *Srf*I and *Fse*I contains only 1 CpG dinucleotide and so there will be more interisland sites for these enzymes than is the case for the group I enzymes.

2. *Csp*I/*Cpo*I (CGGA/TCCG), with a 7 bp recognition sequence that contains 2 CpG dinucleotides, but the site also contains 1 A or T residue. This means that there will be more interisland sites than found for either group I or II enzymes.

Materials and Methods

Source of Biological Material

The choice of biological starting material is dictated by the availability of fresh blood or other tissue, or of transformed cell lines. In general, circulating lymphocytes, lymphoblastoid cell lines, or fibroblast cell lines give more consistent results than solid tissue. This is probably due to the

[14] S. Cross, P. Kovarik, J. Schmidtke, and A. Bird, *Nucleic Acids Res.* **19,** 1469 (1991).

TABLE I
DIFFERENT GROUPS OF RESTRICTION ENZYMES USED FOR MAPPING HUMAN GENOME

Group	Number of CpGs	G + C content	Enzyme[a]	Site	Number of sites per island[b]		Estimated percentage of all sites in islands[c]	Uses and comments
					Expected	Observed		
I	2	8/8	*Not*I	GCGGCCGC	0.14	0.35	93	Nearly all genomic sites are in islands, but only a minority of islands have a site. Large size of fragments is useful for extending PFGE maps and detection of chromosome breakpoints
			*Asc*I	GGCGCGCC	0.14	0.27		
II	2	6/6	*Bss*HII	GCGCGC	1.68	2.11	76	Most genomic sites are in islands; nonisland sites are often methylated in genomic DNA. Nearly every island should have a site for one of these enzymes. Clusters of these sites are the best indication of the presence of a CpG island
			*Eag*I	CGGCCG	1.68	1.70		
			*Sst*II/*Sac*II	CCGCGG	1.68	1.89		
III	1	6/6	*Nae*I	GCCGGC	1.68	1.14	41	As for group II, nearly every island will have a site for one of these, but there are many more nonisland genomic sites that can be subject to variable methylation. Therefore, these enzymes are more useful for mapping islands in genomic DNA than in cloned DNA
			*Nar*I	GGCGCC	1.68	1.65		
			*Sma*I	CCCGGG	1.68	2.14		

IV	2	4/6	*Mlu*I	ACGCGT	0.42	0.03	4	Frequency of sites in islands is lower than expected. Most sites are not in islands and are therefore subject to variable methylation. Thus these enzymes are not so useful for defining the position of islands or chromosome breakpoints, but are very useful for establishing physical linkage of markers
			*Nru*I	TCGCGA	0.42	0.11		
			*Pvu*I	CGATCG	0.42	0.08		
			*Spl*I	CGTACG	0.42	0.03		
V[d]	1	4/6	*Sal*I	GTCGAC	0.42	0.14	5	Uses as for group IV
			*Xho*I	CTCGAG	0.42	0.43		
VI	1	4/4	*Hha*I	GCGC	46.2	21.86	26	Fragment sizes too small to be useful for PFGE analysis. Many interisland sites. Used for fine mapping of CpG islands
			*Hpa*II	CCGG	46.2	20.35		

[a] All enzymes listed are blocked by methylation.

[b] Assumes that the average size of islands is 1.4 kb, based on a survey of 37 human CpG islands. This may be an underestimate because many of the islands in Genbank may not be complete.

[c] Based on the observed average frequency of island sites for the group. Assumptions about base composition of mammalian genomes are those laid out in W. R. A. Brown and A. P. Bird, *Nature* (*London*) **322,** 477 (1986).

[d] There are many enzymes in this group; only the two most commonly used are listed.

difficulty in preparing dispersed but intact cells from tissues with a large connective tissue content. However, successful results have been obtained with a variety of tissues and even tumor tissue.[15] An important consideration is the observation that, in cell lines that have been established in culture for a long time, methylation of islands occurs at genes not required for growth in culture conditions. In some lines more than half of all the islands are methylated.[16] This phenomenon is less extreme in relatively newly established lymphoblastoid cell lines. Methylation of CpG islands in culture means that some CpG islands will now be resistant to cleavage by rare cutters. This would allow long-range maps to be extended past CpG islands that are consistently cut in tissue DNA, in a manner analogous to partial digestion.

Pulsed-Field Gel Electrophoresis

Because the frequency of CpG islands in the genome is on the order of one per several hundred kilobases of DNA, conventional gel electrophoresis is unable to resolve fragments generated by the cleavage of mammalian DNA with enzymes shown in Table I. Pulsed-field gel electrophoresis (PFGE) circumvents this problem, allowing restriction fragments of up to several megabases to be resolved. There are many PFGE systems available commercially, but all rely on the underlying principle of subjecting the DNA molecules to electric fields whose direction alters, causing reorientation of the DNA molecules within the agarose. Commonly used systems include the CHEF-DR II (Bio-Rad, Richmond, CA) and Pulsaphor (LKB, Rockville, MD) systems. For most purposes the same batches of agarose used for normal gel electrophoresis can be used; however, there are specialized agarose products available for PFGE that, for example, facilitate the resolution of large DNA molecules (such as FastLane; FMC Bioproducts, Rockland, ME).

Embedding DNA in Agarose

To generate intact restriction fragments of up to several megabases from mammalian DNA, the DNA must be protected from shearing forces during its preparation. This is achieved by embedding whole cells in a solid matrix of agarose prior to DNA preparation. This agarose must be of a low melting point (LMP) variety and of a high level of purity. We

[15] B. Royer-Pokora, S. Ragg, B. Heckl-Ostreicher, M. Held, U. Loos, K. Call, T. Glaser, D. Housman, G. Saunders, B. Zabel, B. Williams, and A. Poustka, *Genes, Chromosomes, Cancer* **3,** 89 (1991).

[16] F. Antequera, J. Boyes, and A. Bird, *Cell (Cambridge, Mass.)* **62,** 503 (1990).

routinely use ultrapure LMP agarose (Bethesda Research Laboratories, Gaithersburg, MD) although other brands are satisfactory; e.g., InCert agarose (FMC Bioproducts).

The following is a protocol for the embedding of mammalian cells in agarose blocks.

1. Harvest cells from culture, white blood cells from circulating lymphocytes, or homogenize tissue (filtering through muslin to remove connective tissue). Wash cells three times in cold phosphate-buffered saline (PBS).
2. Count cells, repellet, and resuspend in PBS to 2.5×10^7 cells/ml.
3. Add an equal volume of molten (but cooled to 42°) 1% (w/v) LMP ultrapure agarose (Bethesda Research Laboratories) made up in PBS. Mix well and aliquot into molds.
4. Leave in the cold room until set, then push plugs gently out into NDS (sufficient to cover the plugs well).

NDS: 0.5 *M* ethylenediamine tetraacetic acid (EDTA), 10 m*M* Tris-HCl, 1% (w/v) laurylsarcosine, pH 9.5

(Use NaOH pellets to maintain pH while dissolving the EDTA. Autoclave before adding laurylsarcosine.)

5. Add proteinase K to 0.5 mg/ml. Incubate at 50° for 24 hr.
6. Replace with fresh NDS and proteinase K and continue incubation for a further 48 hr.
7. Store plugs at 4° in fresh NDS. They will keep for several years like this.

Restriction Enzyme Digestion of Plugs

Rare-cutting restriction enzymes are available from all enzyme manufacturers. The group I enzyme *Asc*I has recently become available from New England BioLabs (Beverly, MA). *Spl*I is available from Amersham (Arlington Heights, IL), and *Srf*I is available from Stratagene. *Csp*I/*Cpo*I are available from Promega and Amersham, respectively. Note that *Nae*I, *Nar*I and *Sac*II show site preferences, i.e. they cleave some sites better than others depending on the surrounding DNA sequence.[17] Restriction digests are carried out as described below and according to the specifications of the manufacturer, although we have found that addition of Triton X-100 at 0.01% (v/v) aids digestion by *Not*I.

1. Incubate each plug in 5 ml TE + 5 μl fresh phenylmethylsulfonyl fluoride (PMSF) (20 mg/ml in 2-propanol). Leave at 50° for 30 min.

[17] New England BioLabs catalog, p. 135 (1991).

2. Repeat step 1.

3. Incubate each plug in 5 ml TE at 4° for 30 min.

4. Repeat step 3.

5. Incubate in 1 ml restriction buffer at 4° for 30 min [no 2-mercaptoethanol or bovine serum albumin (BSA) is necessary at this stage].

6. Repeat step 5.

7. Add 100 μl of restriction buffer (plus BSA, 2-mercaptoethanol, and spermidine as specified by the supplier). Add 10–20 units of enzyme and incubate between 4 hr and overnight.

Cloning Vectors

Several cloning vectors have been designed with polylinkers that allow the cloning of group I or II enzyme sites (see Table I). These provide a means of directly isolating potential coding sequences from complex mammalian genomes. Vectors commonly used include plasmids (e.g., Bluescript KS and SK series), phage (e.g., EMBL5 and 6), cosmids (e.g., scos2 and Lorist 6), and YACs (e.g., pYAC55 and pJS97/pJS98). *Escherichia coli* contains restriction systems McrA and McrBC that cleave DNA when it is methylated at specific cytosines. This is therefore a consideration when choosing a host strain for cloning mammalian DNA, because the Mcr restriction systems have been shown to have both quantitative and qualitative effects on the recovery of cloned DNA.[18,19]

Applications

Introducing 11p13 Wilms' Tumor Locus

Our description of the use of rare-cutter restriction enzymes in genome mapping will center on studies of the human 11p13 region, involved in the contiguous gene syndrome, the WAGR (Wilms' tumor, aniridia, genitourinary anomalies, and mental retardation) syndrome. The region of DNA involved in these anomalies has been extensively studied by several groups, generating many cloned DNA markers and genes. The construction of a consensus PFGE map was an important landmark along the road

[18] P. A. Whittaker, A. J. B. Campbell, E. M. Southern, and N. E. Murray, *Nucleic Acids Res.* **16,** 6725 (1988).

[19] J. P. Doherty, M. W. Graham, M. E. Linsenmeyer, P. J. Crowther, M. Williamson, and D. M. Woodcock, *Gene* **98,** 77 (1991).

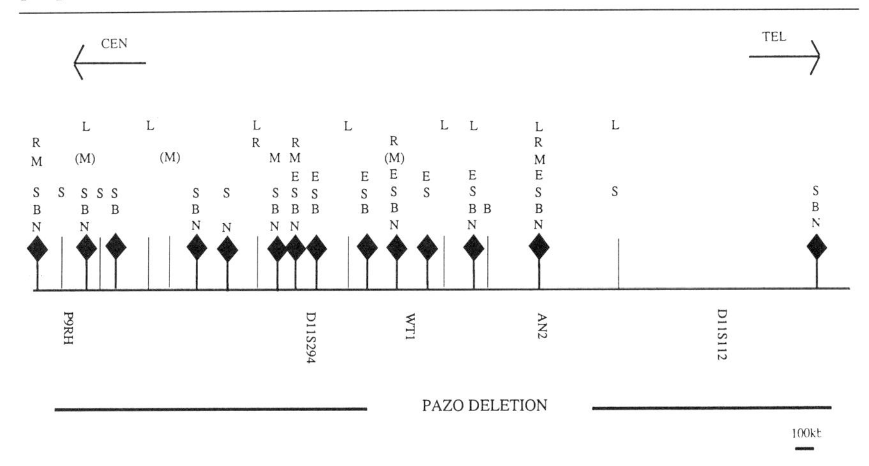

FIG. 2. Long-range restriction map of the region of human chromosome 11p13 involved in the WAGR syndrome. The positions of the candidate genes for Wilms' tumor (*WT1*) and aniridia (*AN2*) are shown. The location of probes D11S294, D11S112, and P9RH are also shown. Genomic cutting sites detected by PFGE are indicated: B, *Bss*HII; E, *Eag*I; L, *Sal*I; M, *Mlu*I; N, *Not*I; R, *Nru*I; S, *Sac*II. (M) indicates *Mlu*I sites that are cut only in some cases, presumably due to varying states of methylation. CpG islands (◆) are defined by the coincidence of two or more sites for group I and II enzymes. The extent of the deletion in the individual PAZO is also shown.

to the isolation of the Wilms' tumor (*WT1*) gene.[20–23] CpG islands were mapped over several megabases of DNA, and many of them cloned and associated genes identified. Establishing the position of chromosome breakpoints in relation to the physical map has been an important step in correlating particular genes with specific phenotypes.[24] Figure 2 shows the consensus long-range map of 11p13 with the position of CpG islands and known genes indicated.

[20] W. A. Bickmore, D. J. Porteous, S. Christie, A. Seawright, J. M. Fletcher, J. C. Maule, P. Couillin, C. Junien, N. D. Hastie, and V. van Heyningen, *Genomics* **5,** 685 (1989).

[21] D. A. Compton, M. M. Weil, C. Jones, V. M. Riccardi, L. C. Strong, and G. F. Saunders, *Cell (Cambridge, Mass.)* **55,** 827 (1988).

[22] M. Gessler and G. A. P. Bruns, *Genomics* **5,** 43 (1989).

[23] E. A. Rose, T. Glaser, C. Jones, C. L. Smith, W. H. Lewis, K. M. Call, M. Minden, E. Champagne, L. Bonetta, H. Yeger, and D. E. Houseman, *Cell (Cambridge, Mass.)* **60,** 495 (1990).

[24] V. van Heyningen, W. A. Bickmore, A. Seawright, J. M. Fletcher, J. Maule, G. Fekete, M. Gessler, G. A. P. Bruns, C. Huerre-Jeanpierre, C. Junien, B. R. G. Williams, and N. D. Hastie, *Proc. Natl. Acad. Sci. U.S.A.* **87,** 5383 (1990).

Mapping CpG Islands in Genomic DNA

To map CpG islands in genomic DNA, clustering of sites for group II enzymes is the most rigorous criterion. The additional presence of sites for group I enzymes is strong confirmatory evidence in favor of a CpG island, but these will be present in only approximately one in five islands. Southern hybridization of genomic DNA digested with group II enzymes often results in hybridizing fragments that appear to be the same size, within the limits of resolution of PFGE (± 5–10 kb). This could result from *bona fide* clustering of cutting sites at CpG islands or be due to fortuitous production of similar sized restriction fragments. These two possibilities can be resolved and the diagnosis of a CpG island strengthened by double digestion with these enzymes, because if the cutting sites are actually clustered on the DNA then digestion with several enzymes in combination should not reduce the size of the hybridizing fragments. An example of this type of analysis is shown in Fig. 3. Probe D11S294 is close to the *WT1* locus, being deleted in all of our WAGR individuals with 11p13 constitutional deletions.[20] We therefore wished to ascertain whether there were neighboring CpG islands because the genes associated with such islands might be important in relationship to the disease phenotype. Hybridization of D11S294 to *Bss*HII and *Sst*II genomic digests generated very similar sized (90 kb) fragments (Fig. 3). The presence of CpG islands 90 kb apart, flanking D11S294, is confirmed by *Bss*HII plus *Sst*II double digestion, which does not decrease the size of the hybridizing fragment, confirming that the sites are indeed clustered together within a few kilobases. In addition it can be seen that these CpG islands appear to contain sites for the group III enzyme *Nar*I. Hybridization to *Not*I digests produces a 500-kb fragment, thus *Not*I sites are not present in both of the CpG islands flanking D11S294; indeed, completion of the long-range restriction map (Fig. 2) shows that only the more proximal of the two CpG islands contains a *Not*I site.

Identification of Chromosome Breakpoints

Because *Not*I sites do not occur in every CpG island, fragments generated by this enzyme (and also presumably by *Asc*I) are generally larger than those generated by enzymes of group II. This is of particular use in the identification of chromosome breakpoints arising from deletions or translocations, as illustrated in Fig. 4. The individual PAZO has aniridia and cryptorchidism/hypospadias. We suspected an 11p13 deletion, although none was visible cytogenetically. Using D11S294 as a probe, the proximal deletion breakpoint in PAZO could not be detected with group II enzymes because the fragment sizes in this region were small (on the

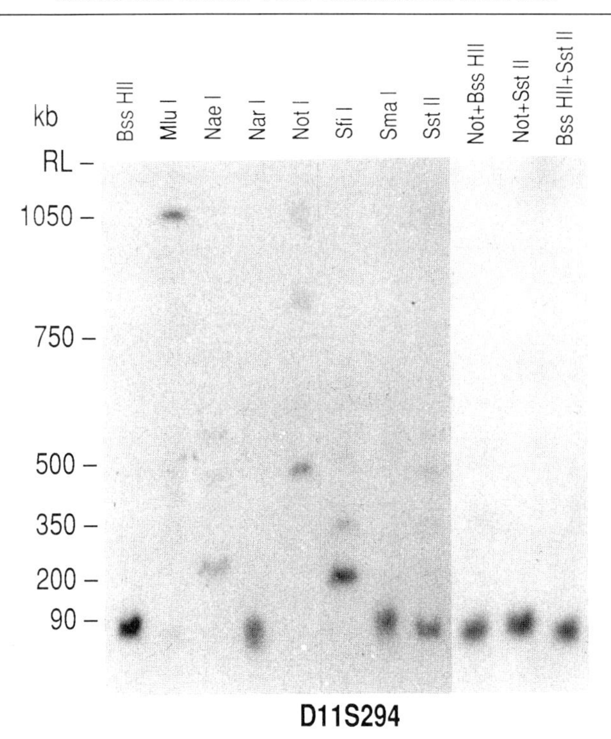

FIG. 3. Mapping CpG islands in genomic DNA. The probe D11S294 was used to probe PFGE blots of DNA from a normal lymphoblastoid cell line digested with a range of rare-cutter enzymes. Hybridizing fragments obtained with *Bss*HII, *Nar*I, *Sma*I, and *Sst*II appear to be the same size (90 kb). This suggests that CpG islands may flank this probe in the genome. The coincidence of the *Bss*HII and *Sst*II cutting sites is confirmed by double digestion with these enzymes, which does not result in a decrease in size of the hybridizing fragment.

order of 100 kb). However, using *Not*I, which generates a 500-kb fragment in normal individuals, two hybridizing fragments are apparent in PAZO: one of normal size from the unaltered chromosome 11 homolog and one of 1.5 Mb from the deleted chromosome where the normal 500-kb D11S294 hybridizing *Not*I fragment has been broken and joined onto the 1.4-Mb *Not*I fragment from the distal deletion breakpoint (Figs. 2 and 4b). The distal deletion breakpoint in PAZO could be detected with the group II enzyme *Bss*HII (Fig. 4c), because fragments generated in that particular region of the genome by such enzymes are quite large. The novel fragment observed in PAZO is in this case smaller than the normal fragment (1.2 as opposed to 1.4 Mb). The extent of the deletion in PAZO, as deduced from this analysis by PFGE, is 1 Mb. The overlap of this deletion with that of

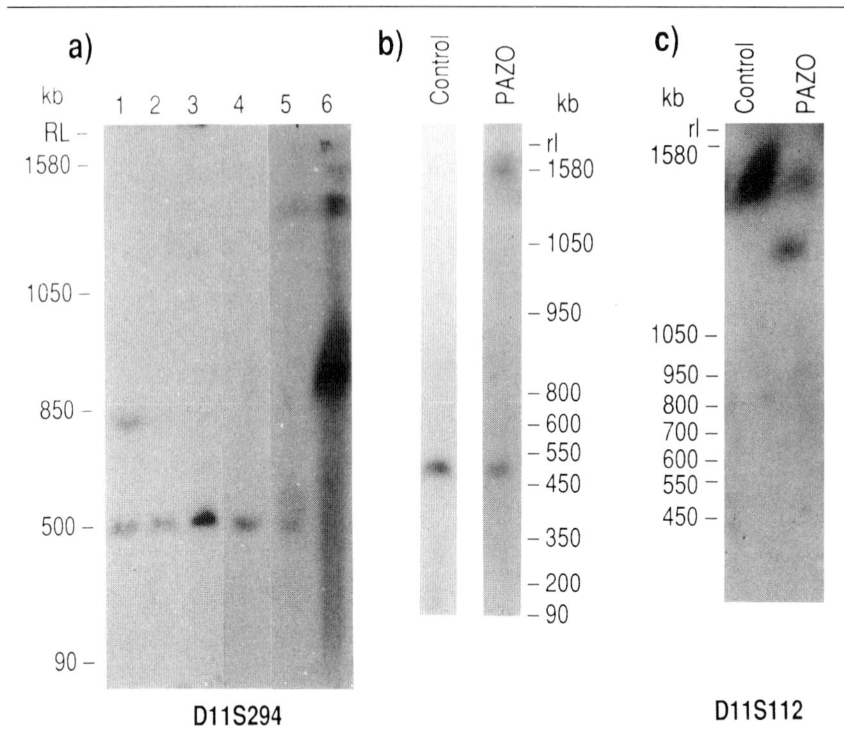

FIG. 4. Detection of chromosome breakpoints and methylation of islands in culture. (a) *Not*I digests of six cell lines probed with D11S294. Lane 1, normal lymphoblastoid cell line; lane 2, GM4613 fibroblast cell line carrying an 11p13 translocation; lane 3, SIMO lymphoblastoid cell line carrying an 11p13 translocation; lane 4, ANX somatic cell hybrid carrying a human 11p13 deletion chromosome on a rodent background; lane 5, 1W1 somatic cell hybrid carrying a normal human chromosome 11 on a rodent background; lane 6, EJ188D human bladder carcinoma cell line. (b) *Not*I digests of lymphoblastoid DNAs from a normal individual and from individual PAZO probed with D11S294. (c) *Bss*HII digests of lymphoblastoid DNAs from a normal individual and from PAZO, probed with D11S112.

an individual with cryptorchidism/hypospadias but no aniridia (DAR), led us to propose that the 11p13 Wilms' tumor gene (*WT1*) plays a role in correct genital development,[24] later supported by studies showing expression of the *WT1* gene in the early gonadal ridge.

Chromosome Banding and Distribution of CpG Islands

Analysis of different regions of the human and mouse genomes has shown that CpG islands are not located at random; rather, their distribution is biphasic. In some parts of the genome CpG islands occur frequently over many megabases of DNA while elsewhere they occur very rarely. It has been suggested that these two compartments of the genome are equivalent to the two types of chromosome band that can be distinguished morphologically on mammalian chromosomes by a variety of techniques. According to this hypothesis a majority of mammalian genes (and thus CpG islands) are clustered in regions of the chromosome that correspond cytologically to Giemsa-negative staining regions (R bands) and that Giemsa-positive bands contain few if any genes.[25] Indeed, it may be that this uneven distribution of genes is directly responsible for chromosome banding patterns. Thus knowing the chromosomal location of a particular DNA marker can give an indication of the size range of fragments that might be generated by group I and II enzymes.

Uses and Abuses of Group IV Enzymes

The choice of group I and II enzymes for analysis of chromosome rearrangements, rather than enzymes of group IV, is important because the high percentage of sites for the former in CpG islands (which are constitutively unmethylated) ensures that any altered fragment seen is likely to be due to chromosome rearrangement rather than variable methylation of a site. Indeed this is demonstrated by the common *Not*I fragment size among several different chromosomes 11 in Fig. 4a. In contrast, because most group IV cutting sites are outside of CpG islands they are subject to variable levels of methylation. As a result a new variant seen in an individual with such an enzyme may well be due to a methylation variation rather than a chromosome rearrangement. Indeed, there are cases where *Mlu*I has been used in the analysis of chromosome breakpoints where it has been shown subsequently that the altered restriction fragments seen are most likely due to such a methylation change, rather than to a chromosome breakpoint.[20,21] Figure 5 illustrates the variation in methylation status of *Mlu*I sites between individuals. With two DNA markers from the 11p13 region many hybridizing fragments of differing sizes can be seen on chromosome 11s from different sources (lymphoblastoid cell lines, fibroblast cell lines, and somatic cell hybrids). Some of these chromosome 11s have translocations or deletions in the WAGR region that would be detected by the probes used, but the high degree of methylation variation in normal chromosome 11s makes it impossible to distinguish between the real chromosome breakpoints and the methylation variants. This extreme level of polymorphism, can, however, be put to good use. Because group II enzymes cleave principally at CpG islands,

[25] W. A. Bickmore and A. T. Sumner, *Trends Genet.* **5,** 144 (1989).

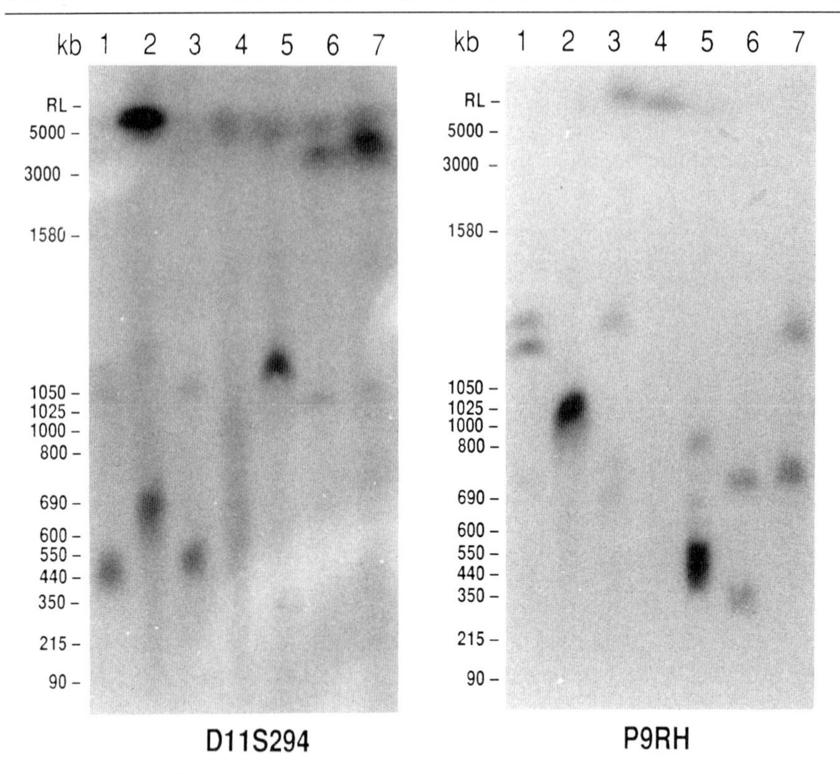

FIG. 5. Variability of methylation of *Mlu*I sites in genomic DNA. *Mlu*I digests of DNAs from different sources probed with D11S294 and P9RH. Lane 1, DG fibroblast cell line carrying an 11p13 deletional translocation; lane 2, SIMO lymphoblastoid cell line carrying an 11p13 translocation; lane 3, ANNA lymphoblastoid cell line from an individual with an 11p13 deletion; lane 4, ANX somatic cell hybrid carrying the deleted chromosome 11 from ANNA on a rodent cell background; lane 5, TRAKE lymphoblastoid cell line carrying an 11p13 deletion; lane 6, PAZO-lymphoblastoid cell line carrying an 11p13 deletion; lane 7, ASTY lymphoblastoid cell line with no detectable 11p13 deletion.

and therefore produce fragments of very similar sizes by PFGE, it can be difficult to distinguish physical linkage of two DNA probes from comigration of hybridizing fragments on PFGE gels. However, if two DNA probes show the same pattern of *Mlu*I fragments across several individuals, this is a very strong indicator of physical linkage. This is analogous to the use of restriction fragment length polymorphisms (RFLPs) in genetic linkage analysis and indeed methylation variation can be considered as a form of RFLP that is often inherited in a Mendelian manner.[26]

[26] A. J. Silva and R. White, *Cell (Cambridge, Mass.)* **54,** 145 (1988).

De Novo Methylation of CpG Islands in Cell Lines

In contrast to the variations in methylation state of *Mlu*I sites in 11p13 we have found little evidence for methylation of the CpG islands themselves in a variety of fresh bloods, lymphoblastoid cell lines, fibroblast cell lines, and somatic cell hybrids. This is in line with the observation that CpG islands remain unmethylated regardless of the expression of the associated gene. One exception is illustrated in Fig. 3. In the EJ188D cell line we observe methylation of the telomeric *Not*I site of the D11S294 hybridizing fragment, resulting in an 850-kb fragment rather than a 500-kb one. The EJ188D cell line was established at least 20 years ago. It may therefore exhibit methylation of CpG islands associated with genes superfluous to growth in culture in common with other well-established cell lines.[16] It is of interest to note that the CpG island that becomes methylated in this case is that associated with the *WT1* gene (Fig. 2), a tissue-specific transcription factor not expressed in EJ188D and that is presumably not required for growth in culture.[27] This same site has also been seen to be methylated somatically, in some Wilms' tumors, leading to the speculation that this methylation may have played a role in inactivation of this tumor suppressor gene and hence in tumorigenesis.[15]

Cloning of CpG Islands

One of the first steps along the road to isolating a particular gene based on its location in the genome is to obtain DNA markers from the desired region, to enable the construction of genetic and physical maps. If the cloning of these DNA markers is targeted to transcribed regions in the first place, you can kill two birds with one stone. This can be achieved by cloning CpG islands from an appropriate somatic cell hybrid or from flow-sorted chromosomes. It is then possible to travel from one CpG island to another by chromosome jumping.[28] This approach has been applied to the 11p13 region. We have been able to isolate CpG islands from the WAGR region by cloning DNA from a somatic cell hybrid containing 11p into λEMBL6, which will clone *Not*I or *Eag*I sites. We have shown that one of these islands is conserved in evolution, detects a transcript in Northern blots, and encodes an open reading frame (W. A. Bickmore, unpublished observations, 1991). Gessler *et al.* have shown the power of this approach,

[27] K. M. Call, T. Glaser, C. Y. Ito, A. J. Buckler, J. Pelletier, D. A. Haber, E. A. Rose, A. Kral, H. Yeger, W. H. Lewis, C. Jones, and D. E. Houseman, *Cell (Cambridge, Mass.)* **60,** 509 (1990).

[28] A. Poustka, T. M. Pohl, D. P. Barlow, A.-M. Frischauf, and H. Lehrach, *Nature (London)* **325,** 353 (1987).

isolating four CpG islands from the WAGR region using a *Bss*HII jumping library.[29] One of these included the CpG island associated with the gene predisposing to Wilms' tumor.

Detection of CpG Islands in Cloned DNA

In many cases regions of the genome have already been cloned in vectors that were not designed specifically for the isolation of CpG islands. Such cloned DNA can be screened for the presence of CpG islands. This requires careful consideration of the enzymes used and definition of CpG islands as the blockage of enzyme sites by methylation is now removed. Particular attention must therefore be paid to the properties of the enzymes set out in Table I. Clusters of sites for group I and II enzymes in cloned DNA are indicative of the presence of CpG islands. However, sites for enzymes other than those of groups I and II are not significant in cloned DNA—for example, only a few percent of *Mlu*I sites are within CpG islands and therefore the vast majority of *Mlu*I sites apparent in cloned DNA are not in CpG islands. This type of analysis has been carried out on yeast artificial chromosome (YAC) clones from the WAGR region.[30] The choice of enzymes used in this study means that the authenticity of some CpG islands, defined by the coincidence of sites for only one group II enzyme with, for example, *Mlu*I, *Nru*I, and *Sal*I sites, remains in some doubt. An example of a well thought out choice of restriction enzymes for the study of cloned DNA is found in the analysis of overlapping cosmid clones from the human MHC region.[31] Analysis of >500 kb of this DNA for clusters of group I and II cutting sites has been used to define the position of CpG islands and led to the identification of 12 novel genes in this region.

Caveats and Pitfalls

We hope that we have illustrated that by judicious choice of restriction enzymes, the ability to isolate genes via CpG islands and to identify the position of genes in a particular genomic region is a powerful tool in molecular genetics. However, it is important to stress some drawbacks that call for cautious interpretation of data.

[29] M. Gessler, A. Poustka, W. Cavenee, R. L. Neve, S. H. Orkin, and G. A. P. Bruns, *Nature (London)* **343,** 774 (1990).

[30] L. Bonetta, S. E. Kuehn, A. Huang, D. J. Law, L. M. Kalikin, M. Koi, A. E. Reeve, B. H. Brownstein, H. Yeger, B. R. G. Williams, and A. P. Feinberg, *Science* **250,** 994 (1990).

[31] N. Fischel-Ghodsian, R. D. Nicholls, and D. R. Higgs, *Nucleic Acids Res.* **15,** 9215 (1987).

1. First, although clustering of sites for group I and II enzymes is the most reliable method for identifying the position of CpG islands in cloned or genomic DNA, in the latter case it should be borne in mind that the limit of resolution of PFGE (plus or minus 5–10 kb) is considerably larger than the average size of CpG islands. Therefore what appear to be clustered cutting sites in PFGE maps may in fact prove to be dispersed over many kilobases of DNA and thus may not represent islands.

2. While to date virtually all CpG islands have been found in association with genes, the converse is not true. Many tissue-specific genes lack CpG islands and some CpG islands will not have cutting sites for the restriction enzymes usually used in their identification (e.g., the CpG island of the hamster adenine phosphoribosyltransferase (APRT) gene). Restriction enzymes can therefore identify only a proportion (albeit probably a large one) of genes. The only example so far of an unmethylated CpG island that is not associated with a functional gene is the $\psi\alpha 2$ pseudogene in the human α-globin locus.[31] Other nonprocessed pseudogenes have lost their CpG islands through methylation and mutation.

3. In general CpG islands are unmethylated. The exceptions seem to be the inactive X chromosome,[32] tumor cells,[15] fragile sites,[33] and cell lines that have been in culture for many years.[16] In the latter case, the islands that are methylated seem to be those associated with genes whose transcription is superfluous to growth in culture. Methylation differences in cell lines may be used to advantage to effectively create partial restriction digests enabling maps to be extended beyond a particular CpG island normally unmethylated in somatic cells. Obviously when searching for CpG islands in a particular region of DNA, the use of well-established cell lines should be avoided. Conversely, treatment of cell lines with 5-azacytidine may enable methylated CpG islands to be unmasked.[34]

To date the only example of methylation of gene-associated CpG islands in uncultured somatic cells has been in the cases discussed above. We cannot, however, rule out the possibility that some other CpG islands become methylated in somatic tissues. There have been several reports in the literature of the identification of clusters of CpG cutting sites in cloned DNA, particularly in YACs, which are not detectable (i.e., are methylated) in genomic DNA. These sites have been termed CpG islands, yet many have been defined as cutting sites for enzymes other than those of groups I and II, so that we consider that their authenticity as CpG islands is in

[32] D. H. Keith, J. Singer-Sam, and A. D. Riggs, *Mol. Cell. Biol.* **6,** 4122 (1986).

[33] A. Vincent, D. Heitz, C. Petit, C. Kretz, I. Oberle, and J.-L. Mandel, *Nature (London)* **349,** 624 (1991).

[34] C. Dobkin, C. Ferrando, and W. T. Brown, *Nucleic Acids Res.* **15,** 3183 (1987).

doubt.[30,35] There are others that do indeed seem to be *bona fide* islands in that they have clusters of cutting sites for group II enzymes.[36] It is not yet established if such sites represent (1) fortuitous clustering of sites in a highly (G + C)-rich stretch of DNA, (2) a rare class of CpG island that is methylated in most cells (but presumably not in the germ line), or (3) "dead" islands perhaps associated with recently inactivated genes or pseudogenes. Only detailed analysis of the sequences concerned will answer this question.

Acknowledgments

W.A.B. is a Lister Institute Research Fellow. We thank the MRC Human Genetics Unit Photography Department for preparation of figures.

[35] D. Wilkes, J. Shaw, R. Anand, J. Riley, P. Winter, J. Wallis, A. G. Driesel, R. Williamson, and S. Chamberlain, *Genomics* **9,** 90 (1991).
[36] R. Anand, D. J. Ogilvie, R. Butler, J. H. Riley, R. S. Finniear, S. J. Powell, J. C. Smith, and A. F. Markham, *Genomics* **9,** 124 (1991).

[23] Characterization of Type II DNA-Methyltransferases

By David Landry, Janet M. Barsomian, George R. Feehery, and Geoffrey G. Wilson

Introduction

DNA-methyltransferases (MTases) catalyze the postreplicative addition of a methyl group to the N-6 position of adenine,[1-5] or to the N-4 or 5 position of cytosine[6-10] in DNA. These are the commonest DNA

[1] U. Kuhnlein, S. Linn, and W. Arber, *Proc. Natl. Acad. Sci. U.S.A.* **63,** 556 (1969).
[2] S. Hattman, *J. Virol.* **7,** 690 (1971).
[3] J. P. Brockes, P. R. Brown, and K. Murray, *Biochem. J.* **127,** 1 (1972).
[4] P. H. Roy and H. O. Smith, *J. Mol. Biol.* **81,** 427 (1973).
[5] A. Dugaiczyk, J. Hedgpeth, H. W. Boyer, and H. M. Goodman, *Biochemistry* **13,** 503 (1974).
[6] A. A. Janulaitis, S. Klimasauskas, M. Petrusyte, and V. Butkus, *FEBS Lett.* **161,** 131 (1983).
[7] V. V. Butkus, S. J. Klimasauskas, and A. A. Janulaitis, *Anal. Biochem.* **148,** 194 (1985).
[8] M. Ehrlich, M. A. Gama-Sosa, L. H. Carreira, L. G. Ljungdahl, K. C. Kuo, and C. W. Gehrke, *Nucleic Acids Res.* **13,** 1399 (1985).
[9] H. W. Boyer, L. T. Chow, A. Dugaiczyk, J. Hedgpeth, and H. M. Goodman, *Nature (London) New Biol.* **244,** 40 (1973).
[10] S. Hattman, E. Gold, and A. Plotnik, *Proc. Natl. Acad. Sci. U.S.A.* **69,** 187 (1972).

modifications; more esoteric kinds occur,[11,12] but they will not be discussed here. *S*-Adenosylmethionine (AdoMet) is the universal donor of the methyl group[13,14]; it is required for catalysis and perhaps also for DNA binding.[15,16] The methyl group appears to be transferred directly from AdoMet to DNA without the formation of covalent enzyme-AdoMet intermediates.[17,18] Simple kinetics suggest that MTases act as monomers.[16,19]

Bacterial MTases frequently occur as part of restriction–modification (R–M) systems.[20–22] These are simple immune systems that protect bacteria from infections by foreign DNA molecules.[23–27] One component of R–M systems, the endonuclease (ENase), recognizes specific nucleotide (nt) sequences in DNA and catalyzes double-stranded (ds) cleavage of the DNA (restriction) if the sequence is unmodified. The other component, the MTase, methylates the DNA of the cell at the same sequences (modification), rendering the DNA resistant to such cleavage.[28,29]

[11] S. Hattman, *J. Virol.* **32,** 468 (1979).
[12] H. R. Revel, *in* "Bacteriophage T4" (C. K. Mathews, E. M. Kutter, G. Mosig, and P. Berget, eds.), p. 156. American Society for Microbiology, Washington, D.C., 1983.
[13] W. Arber, *J. Mol. Biol.* **11,** 247 (1965).
[14] W. Arber, *Prog. Nucleic Acid Res. Mol. Biol.* **14,** 1 (1974).
[15] P. Modrich, *Q. Rev. Biophys.* **12,** 315 (1979).
[16] W. E. Jack, R. A. Rubin, A. Newman, and P. Modrich, *in* "Gene Amplification and Analysis" (J. G. Chirikjian, ed.), Vol. 1, p. 165. Elsevier/North Holland, 1981.
[17] J. C. Wu and D. V. Santi, *J. Biol. Chem.* **262,** 4778 (1987).
[18] A. L. Pogolotti, A. Ono, R. Subramaniam, and D. V. Santi, *J. Biol. Chem.* **263,** 7461 (1988).
[19] R. A. Rubin and P. Modrich, *J. Biol. Chem.* **252,** 7265 (1977).
[20] B. Suri, V. Nagaraja, and T. A. Bickle, *Curr. Top. Microbiol. Immunol.* **108,** 1 (1984).
[21] H. O. Smith and S. V. Kelly, *in* "DNA Methylation: Biochemistry and Biological Significance" (A. Razin, H. Cedar, and A. D. Riggs, eds.), p. 39. Springer-Verlag, New York, 1984.
[22] R. Yuan and H. O. Smith, *in* "DNA Methylation: Biochemistry and Biological Significance" (A. Razin, H. Cedar, and A. D. Riggs, eds.), p. 73. Springer-Verlag, New York, 1984.
[23] B. Endlich and S. Linn, "The Enzymes" (P. D. Boyer, ed.), Vol. 14, p. 137. Academic Press, New York, 1981.
[24] D. H. Krüger and T. A. Bickle, *Microbiol. Rev.* **47,** 345 (1983).
[25] R. Yuan and D. L. Hamilton, *in* "DNA Methylation: Biochemistry and Biological Significance" (A. Razin, H. Cedar, and A. D. Riggs, eds.), p. 11. Springer-Verlag, New York, 1984.
[26] T. A. Bickle, *in* "*Escherichia coli* and *Salmonella typhimurium:* Cellular and Molecular Biology" (F. C. Neidhardt, J. L. Ingraham, K. B. Low, B. Magasanik, M. Schaechter, and H. E. Umbarger, eds.), p. 692. American Society for Microbiology, Washington, D.C., 1987.
[27] G. G. Wilson and N. E. Murray, *Annu. Rev. Genet.* **25,** 585 (1991).
[28] R. J. Roberts, *Nucleic Acids Res.* **18** (Suppl.), 2331 (1990).
[29] C. Kessler and V. Manta, *Gene* **92,** 1 (1990).

The best characterized MTases are those from type II and type IIs R–M systems.[21,30] The ENases and MTases from these systems function as independent proteins encoded by separate genes, and they are widely used in the laboratory for recombinant DNA manipulation. The ENases require Mg^{2+} and generally act only on unmodified DNA. The MTases require AdoMet and act on unmodified DNA or on DNA that is already modified on one strand (hemimethylated). The enzymes recognize specific 4- to 8-nt sequences within dsDNA; for type II systems the sequences are symmetric, and for type IIs systems the sequences are usually asymmetric. Modification affects nucleotides within the recognition sequence and probably prevents endonucleolytic cleavage by steric hindrance.

We describe here procedures to characterize the modification activity of type II and IIs MTases. We discuss purification of the MTase, the design of oligodeoxynucleotide substrates, the methylation of natural and synthetic substrates, fidelity of sequence recognition, and the identification of the modified base and the determination of its position within the recognition sequence.

Materials and Methods

Reagents

Nonradioactive AdoMet (Sigma, St. Louis, MO) is prepared as a 32 m*M* stock solution in H_2SO_4–ethanol (9 : 1), pH 2.0, and stored at −20°. [^{3}H]AdoMet (New England Nuclear, Boston, MA) is used at low specific activity (~10 Ci/mmol, 0.5 mCi/ml). Circles (2.4 cm in diameter) of DE-81 paper (Whatman, Clifton, NJ) are used for sample collection, and are counted in Optifluor (Packard, Meriden, CT).

A Heat Systems-Ultrasonics (Farmingdale, NY) sonicator is used to disrupt cells. The following columns (Pharmacia, Piscataway, NJ) are used for protein purification: Sephadex G-50, 10 ml; 1.5 × 5 cm heparin-Sepharose; 4 × 20 cm DEAE-Sepharose; 1 ml Mono Q; 1 ml Mono S high-performance liquid chromatography (HPLC); and 100 ml Superose 6 column materials. Phosphocellulose is obtained from Whatman. Poly Cat A (Custom LC, Inc.), heparin TSK-GEL (TosoHaas, Philadelphia, PA), and Poros Q (Perseptive Biosystems) are also used for HPLC.

The Waters (Milford, MA) model 650E protein purification system or Pharmacia (Piscataway, NJ) fast protein liquid chromatography (FPLC) system (equipped with a GP25 gradient programmer, two p500 pumps, a

[30] W. Szybalski, S. C. Kim, N. Hasan, and A. J. Podhajska, *Gene* **100,** 13 (1991).

UV-1 detector, 5MPA mixer, and an LKB REC102 recorder) are used in several purification steps.

T4 polynucleotide kinase, restriction enzymes, and DNA methylases are from New England Biolabs (Beverly, MA). *Crotalus adamanteus* venom phosphodiesterase I is from Worthington Biochemicals (Freehold, NJ). N^4-Methyldeoxycytidine (m^4C) is synthesized by thiation of deoxyuridine followed by amination with methylamine.[31] 5-Methyldeoxycytidine (m^5C), N^6-methyldeoxyadenosine (m^6A), and the other standard deoxynucleosides (Table I) are from Sigma. Fluorescent cellulose thin-layer chromatography (TLC) plates (DC-Fertigplatten cellulose F) are from Merck (Rahway, NJ). Nucleoside cyanoethylphosphoramidites are from Biosearch. The chemical phosphorylation reagent, 2-[2-4,4′-dimethoxytrityloxyethylsulfonyl]ethyl(2 - cyanoethyl) - (*N,N* - diisopropyl)phosphoramidite, used for the 5′ phosphorylation, is from Glen Research (Herndon, VA).

Synthesis of Oligonucleotide Substrates

The synthesis of 5′-phosphorylated and unphosphorylated oligonucleotides is performed at the 1 μM scale using β-cyanoethylphosphoramidite chemistry (Biosearch 8600 automated DNA synthesizer).[32] The oligonucleotides are deprotected with ammonia, and the ammonia is removed by evaporation. The crude oligonucleotides (1 μmol) are purified by polyacrylamide gel electrophoresis (PAGE) and further purified by HPLC with an RCM C_8 cartridge (Waters). Gel extracts are loaded directly onto the column and washed with five column volumes of HPLC buffer (0.1 *M* aqueous triethylammonium bicarbonate, pH 7.0). A binary elution system is used, starting with HPLC buffer and ending in 15% (v/v) acetonitrile in the same buffer (30-min linear gradient at 2 ml/min). Salts are removed by coevaporation with water in a vacuum centrifuge (Speed-Vac).

Thin-Layer Chromatography of Modified and Unmodified Deoxynucleosides

Two TLC solvents are used in parallel to differentiate between m^6A, m^4C, and m^5C. In solvent D (80 : 20, ethanol–water),[33] m^6A and m^4C comigrated, separate from m^5C. In solvent G (66 : 33 : 1, isobutyric

[31] I. Wempen, R. Dushinsky, L. Kaplan, and J. Fox, *J. Am. Chem. Soc.* **83,** 686 (1961).
[32] N. D. Sinha, J. Biernat, J. McManus, and H. Koster, *Nucleic Acids Res.* **12,** 4539 (1984).
[33] H. G. Khorana, A. F. Turner, and J. P. Vizsolyi, *J. Am. Chem. Soc.* **83,** 686 (1961).

TABLE I
CELLULOSE THIN-LAYER CHROMATOGRAPHY OF DEOXYNUCLEOSIDES[a]

Compound	Symbol	Relative migration rate	
		Solvent D[b]	Solvent G[c]
Thymidine	T	1.00	1.00
Deoxycytidine	C	0.83	1.05
Deoxyadenosine	A	0.79	1.27
Deoxyguanosine	G	0.67	0.91
Deoxyinosine	I	0.61	0.72
5-Methyldeoxycytidine	m^5C	0.87	1.10
N^4-Methyldeoxycytidine	m^4C	1.00	1.08
N^6-Methyldeoxyadenosine	m^6A	1.00	1.30
S-Adenosylmethionine	AdoMet	0	1.13

[a] Migration of deoxynucleosides and methylated deoxynucleosides on cellulose thin-layer chromatography. Solvent D separates m^5C from m^4C + m^6A; solvent G separates m^6A from m^4C + m^5C. Deoxynucleotides do not migrate in either solvent. Values given are relative distances, measured with respect to the migration of thymidine.

[b] Solvent D is 80 : 20 ethanol–water (v/v).

[c] Solvent G is 66 : 33 : 1 isobutyric acid–water–ammonium hydroxide (v/v).

acid–water–ammonium hydroxide), m^4C and m^5C migrate closely, separate from m^6A (Table I).

Polyacrylamide Gel Electrophoresis of Oligonucleotides[34]

Reagents

Gel running buffer (10×): 0.5 *M* Tris–borate, 10 m*M* ethylenediaminetetraacetic acid (EDTA), pH 8.3

Acrylamide stock solution: 20% (w/v) acrylamide, 7 *M* urea, 5% (w/v) bisacrylamide in 1× gel running buffer

Ammonium persulfate stock solution: 10% (w/v) ammonium persulfate in deionized water. One milliliter is added to 150 ml degassed acrylamide stock solution. Polymerization is initiated with 70 μl *N,N,N′,N′*-tetramethylethylenediamine (TEMED). The gels are set at room temperature for 1 hr before use

Procedure. Strand separation of oligonucleotides and DNA fragments from the partial degradation with phosphodiesterase I are separated on a 20 cm × 20 cm × 3 mm 7 *M* urea polyacrylamide gel. Before loading the samples, the polyacrylamide gels are electrophoresed at constant 20 mA

[34] T. Maniatis and A. Efstratiadis, this series, Vol. 65, p. 299.

until the bromphenol blue migrates 17 cm from the origin. Each reaction is loaded into a 2-cm wide by 1.75-cm deep well. The samples are electrophoresed at 20 mA until the bromphenol blue migrates 12 cm.

After synthesis, each crude oligonucleotide (1 μmol) is dissolved in 300 μl of 7 *M* urea in 1× PAGE running buffer and loaded directly onto a 7 *M* urea, 20% (w/v) polyacrylamide gel (20 cm × 40 cm × 3 mm, containing one 15-cm wide lane). Before loading the sample, the gel is electrophoresed until the bromphenol blue migrates 30 cm. The sample is electrophoresed until the bromphenol blue migrates at least 35 cm.

Oligonucleotide bands are detected by UV quenching after placing the gel on a fluorescent cellulose TLC plate covered with plastic wrap and illuminated with a 254-nm UV lamp. Gel slices containing the oligonucleotides are extracted in 0.5 *M* ammonium acetate, 1 m*M* EDTA, pH 7.0.[35]

Methyltransferase Purification

Source of Methylase

DNA-methyltransferases can be purified from natural bacteria strains or from recombinants. The former sources are the most direct, but cloning is often helpful if the natural strain is difficult to handle or produces insufficient MTase for characterization. Numerous MTase genes have been cloned into *Escherichia coli*.[36] The clones are usually isolated selectively by their ability to modify themselves and become resistant to restriction enzyme digestion.[37–39]

The cloning of MTase genes is performed as follows: bacterial DNA fragments, from convenient digests of total cellular DNA, are ligated to suitably cleaved preparations of a plasmid vector and transformed into *E. coli*. The vector must contain sites for the MTase being cloned. The transformants are grown overnight to allow recombinants that have acquired the MTase gene to become modified, then the DNA of the entire plasmid population is purified. The DNA is incubated with a restriction enzyme that digests only the unmodified DNA (usually, the cognate endonuclease), and any plasmids resistant to digestion are recovered by transformation. The transformants are examined individually; usually, many

[35] H. O. Smith, this series, Vol. 65, p. 371.

[36] G. G. Wilson, *Nucleic Acids Res*. **19,** 2539 (1991).

[37] E. Szomolányi, A. Kiss, and P. Venetianer, *Gene* **10,** 219 (1980).

[38] A. Janulaitis, P. Povilionis, and K. Sasnauskas, *Gene* **20,** 197 (1982).

[39] J. E. Brooks, R. M. Blumenthal, and T. R. Gingeras, *Nucleic Acids Res*. **11,** 837 (1983).

are found to carry the MTase gene. See Ref. 40 for a detailed discussion of this method. Frequently, cloned MTase genes express poorly at first, and it is necessary to engineer higher levels of expression to purify and characterize the enzyme.

Methyltransferases that recognize symmetric sequences (type II) sequentially modify both strands of the sequence; those that recognize asymmetric sequences (type IIs) usually methylate only one strand. Consequently, in type II systems, modification is the result of a single MTase, whereas in type IIs systems it is often the result of a pair of MTases, one for each strand.[41–43] When cloning type IIs modification genes, care must be taken to isolate both MTases because each activity, alone, is usually sufficient to protect the recognition sequence from digestion during selection. The original clones for the *Fok*I MTase, for example, were incomplete.[44]

Column Chromatography

The techniques for purifying methylases are similar to those used for purifying many enzymes.[45,46] Bacterial cells expressing the methyltransferase are grown to saturation, harvested by centrifugation, and resuspended in lysis buffer. The cells are disrupted by sonication,[47] or by passage through a Gaulin press,[48] and the debris is removed by centrifugation. The supernatant is chromatographed on a column of derivatized beaded agarose. The bound protein is eluted and fractions containing methylase activity are identified, and then chromatographed by HPLC. For the initial characterization of most MTases, 1 liter of cell culture and two chromatography runs are sufficient.

The cells are suspended in 2 to 3 vol of lysis buffer (10 m*M* KPO_4, pH 7.4, 10 m*M* EDTA, 10 m*M* 2-mercaptoethanol). Sonication is carried out on ice in a stainless steel 30-ml conical tube using a small tip (3 mm). Cell

[40] K. D. Lunnen, J. M. Barsomian, R. R. Camp, C. O. Card, S. Z. Chen, R. Croft, M. C. Looney, M. M. Meda, L. S. Moran, D. O. Nwankwo, B. E. Slatko, E. M. Van Cott, and G. G. Wilson, *Gene* **74,** 25 (1988).

[41] D. Landry, M. C. Looney, G. R. Feehery, B. E. Slatko, W. E. Jack, I. Schildkraut, and G. G. Wilson, *Gene* **77,** 1 (1989).

[42] H. Sugisaki, K. Kita, and M. Takanami, *J. Biol. Chem.* **264,** 5757 (1989).

[43] H. Sugisaki, K. Yamamoto, and M. Takanami, *J. Biol. Chem.* **266,** 13952 (1991).

[44] D. Nwankwo and G. Wilson, *Mol. Gen. Genet.* **209,** 570 (1987).

[45] M. Dixon, *in* "The Enzymes" (P. D. Boyer, ed.), p. 24. Academic Press, New York, 1979.

[46] E. Rossomando, E. Stellwagen, M. Gorbunoff, R. Kennedy, S. Ostrove, S. Weiss, L. Giri, R. Chicz, and F. Regner, this series, Vol. 182, p. 309.

[47] M. Cull and C. McHenry, this series, Vol. 182, p. 147.

[48] F. Wyers, A. Sentenac, and P. Fromageot, *Eur. J. Biochem.* **270,** 270 (1973).

breakage is achieved by repeated 1-min pulses at a 50% duty cycle. The sample should be kept below 10° to prevent degradation. The liberated protein (Bradford assay[49]) or nucleic acid (A_{260}) is monitored after each sonication step. Cell disruption is complete when the concentration of liberated protein or nucleic acid reaches a plateau.

The disrupted cell suspension is centrifuged (10,000 rpm) for 1 hr at 4° and the supernatant is collected. (The supernatant can be frozen at −20° at this point without loss of MTase activity.) The supernatant is loaded onto a chromatography column. The column of choice for small-scale (~10 ml lysate) purification is a 1.5 × 5 cm heparin-Sepharose column. (Many MTases and ENases bind to heparin.) The column is washed with 5 column volumes (50 ml) of lysis buffer containing 0.05 *M* NaCl, and then eluted with 100 ml of lysis buffer containing a linear gradient of 0.05 to 1 *M* NaCl. Alternatively, the column can be eluted sequentially with 5-ml steps of lysis buffer containing 0.1, 0.3, and 0.5 *M* NaCl. For larger lysates (10–1000 ml) an initial 4 × 20 cm DEAE-Sepharose or phosphocellulose column can be used, followed by a 1.5 × 20 cm heparin-Sepharose column.

Several alternative 1-ml columns using the Pharmacia FPLC or Waters HPLC systems are employed for the second purification step. These include Mono Q, Mono S, Poly Cat A, heparin TSK-GEL, and Poros Q. Typical runs involve loads of 50 ml or under, followed by a 5-ml wash, and an 80-ml linear gradient from 0 to 1 *M* NaCl. The flow rate is 1 ml/min at all stages. One-milliliter fractions are collected, and the protein peaks are automatically recorded. Modification activity can often be correlated with an individual protein peak from the HPLC profile. Molecular mass is measured by sodium dodecyl sulfate (SDS)-PAGE, and Superose gel filtration.[50,51]

Characterization of Methyltransferase Activity

Protection Assays

Qualitative, sequence-specific modification activity is assayed by protection against restriction enzyme digestion. Quantitative modification activity, specific and otherwise, is assayed by radioactive incorporation of [^{3}H]AdoMet. Generally, we use natural DNA molecules (plasmids, phages, etc.) for the former assays, and synthetic double-stranded oligonucleotides for the latter.

[49] M. M. Bradford, *Anal. Biochem.* **72,** 248 (1976).
[50] R. Chicz and F. Regnier, this series, Vol. 182, p. 392.
[51] D. Garfin, this series, Vol. 182, p. 425.

Protection assays involve incubating the DNA with AdoMet and the MTase fraction, reincubating the DNA with the cognate restriction enzyme, then electrophoresing the products to observe the resistance to digestion that results from modification. The substrate DNA must contain one or more recognition sites for the methylase and its cognate endonuclease. These sites should be readily cleaved by the endonuclease and cleavage should elicit a well-defined pattern. When the substrate has only one or two sites, it may be necessary to predigest it with an unrelated enzyme so that cleavage by the principal enzyme is unmistakable.

Crude or partially purified cell extract (1–10 μl) is added to methylation buffer (50 m*M* Tris-HCl, pH 7.5, 10 m*M* EDTA, 5 m*M* 2-mercaptoethanol, 80 μM AdoMet) containing 50 μg/ml DNA in a 50-μl volume. (EDTA is present to inhibit contaminating nucleases.) The reaction is incubated for 1 hr at the growth temperature of the natural host for the methylase (e.g., 37°), then stopped by incubation at 70° for 15 min. If the MTase is heat stable it can be inactivated by phenol–chloroform extraction, and the modified DNA can be recovered by precipitation. The reaction is stopped to inactivate contaminating nucleases as well as to inactivate the MTase. If the nucleases are not inactivated, degradation of the substrate can occur during the next step when Mg^{2+} is added and the DNA is digested with the corresponding ENase.

Following inactivation, an equal volume of digestion buffer (50 m*M* Tris-HCl, pH 8.0, 40 m*M* $MgCl_2$, 5 m*M* 2-mercaptoethanol, and salt according to the requirements) is added together with 20 U (units) of the corresponding ENase. The mixture is incubated for an additional hour at a suitable temperature. Digestion is terminated by the addition of 0.1 vol of 10× stop dye [20% (w/v) Ficoll, 0.15% (w/v) bromphenol blue, 10 m*M* Tris-HCl, pH 8.0, 10 m*M* EDTA, 0.1% (w/v) SDS]. The SDS prevents binding of protein to the DNA, which may otherwise retard the mobility of the DNA fragments on the agarose gel.

The samples are electrophoresed in an agarose gel alongside digested and undigested control DNAs. Modification blocks endonucleolytic cleavage (Fig. 1). A dilution series of the MTase on the DNA is used to titer modification activity. One unit is the minimum amount required to completely prevent cleavage of 1 μg of DNA in 1 hr at the optimum reaction temperature. It is important to titer the MTase activity on substrates that are intended for future experiments because undermethylation or overmethylation can both hinder characterization. The *Hga*I methylases, for example, exhibit star activity (methylation of noncanonical sequences) when as little as 10-fold excess enzyme is used (J. M. Barsomian, and G. G. Wilson, unpublished observations, 1990).

FIG. 1. Sequence-specific methylation. Successive column chromatography fractions, collected during the purification of the $M_I \cdot$ *Hga*I MTase (reaction: GACGC/GCGTC → GACGC/GCGTC; the methylated nucleotide is underlined), were incubated with phage λ DNA (>100 sites) and AdoMet. The reactions were terminated by heat inactivation, then they were incubated with the *Hga*I endonuclease in the presence of Mg^{2+}. The fractions were finally electrophoresed in an agarose gel, stained with ethidium bromide, and viewed under ultraviolet light. In fractions containing MTase activity (E and F), the *Hga*I sites become modified, and thus become resistant to *Hga*I digestion.

[^{3}H]AdoMet

For efficient methylation, AdoMet must be present in excess. Usually this means a concentration of 10 μM or more.[52,53] Test for limiting AdoMet concentrations by titrating across the range of 0–20 μM AdoMet in 2 μM increments, under constant MTase and DNA concentrations in a standard protection assay. Endonuclease cleavage becomes apparent as AdoMet becomes limiting.

[^{3}H]AdoMet is available in a high (75 Ci/mmol) or low (10 Ci/mmol) specific activity at a radioactive concentrations of 0.5 mCi/ml. The low

[52] G. E. Herman and P. Modrich, *J. Biol. Chem.* **257,** 2605 (1982).

[53] N. O. Reich and E. A. Everett, *J. Biol. Chem.* **265,** 8929 (1990).

specific activity is preferred because the AdoMet concentration is higher (50 μM). Typical methylation assays use 2 μl of low specific activity [^{3}H]AdoMet in a reaction volume of 10–20 μl.

Comparative Stoichiometry

To assess the fidelity of sequence recognition by a methyltransferase, the number of methyl groups incorporated per recognition sequence is compared to the number incorporated by a characterized methylase such as M · *Hae*III or M · *Eco*RI.

A reaction mixture is prepared containing 50 μg/ml of a standard DNA (e.g., pBR322, phage λ) in methylation buffer and 10 μM [^{3}H]AdoMet. The mixture is divided in two. The test MTase is added to one part and the control MTase to the other. At 15-min intervals, 20-μl samples are withdrawn from each tube and heated to 70° for 10 min to stop the reactions. Ten microliters from each stopped sample is spotted onto a DE-81 filter. The filters are washed three times in 10 ml/filter of 50 m*M* Tris-HCl, pH 7.5, 1 m*M* EDTA, and 0.2 *M* NaCl, twice in 2-propanol, and (optional) once in ether (in a well-ventilated hood). The filters are air dried and then counted in a scintillation vial containing 5 ml scintillant. The remaining 10 μl of each stopped sample is incubated with the cognate restriction enzyme, then electrophoresed to determine the degree of modification. When the DNA has been fully modified, the ratio of ^{3}H incorporated in the control and test reactions should be equal to the ratio of the recognition sequences (Fig. 2). If the test MTase incorporates more counts than expected, suspect infidelity. If it incorporates fewer counts, suspect hemimethylation.

Synthetic Oligonucleotide Substrate

Synthetic double-stranded oligonucleotides are the substrate of choice for the characterization of MTase activity.[41,54] The oligonucleotides are designed to contain a single recognition sequence for the MTase, and occasionally recognition sequences for additional, control, MTases. The oligonucleotide is methylated *in vitro* by the MTase in the presence of [^{3}H]AdoMet and the nature of the methylated base, and its position within the recognition sequence, are determined. To identify the methylated base (m^{6}A, m^{5}C, or m^{4}C), the oligonucleotide is completely degraded, the products are separated, and the labeled base is identified. Degradation can be achieved using phosphodiesterase I and alkaline phosphatase, or by

[54] P. J. Greene, M. S. Poonian, A. L. Nussbaum, L. Tobias, D. E. Garfin, H. W. Boyer, and H. M. Goodman, *J. Mol. Biol.* **99,** 237 (1975).

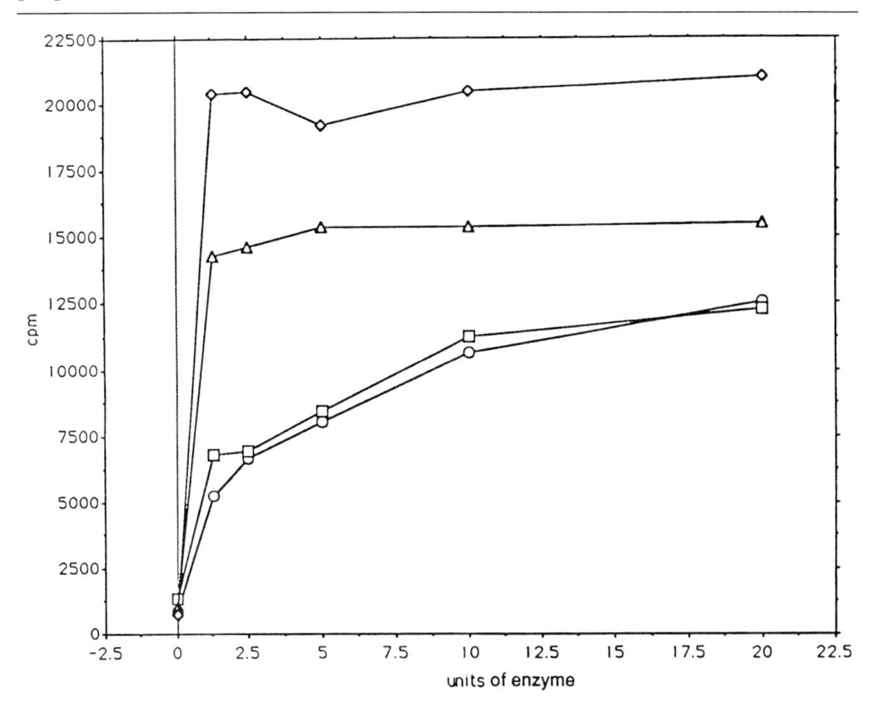

FIG. 2. Stoichiometry of methylation. Increasing quantities of various methyltransferases were incubated with phage λ DNA and [^{3}H]AdoMet. The DNA was collected on filters, and the incorporated radioactivity was measured. Sufficient enzyme (1 unit) was added to the first tube in each series to achieve full modification as judged by resistance to endonuclease digestion; in subsequent tubes, enzyme was present in excess. The M · *Hae*III (triangles; 149 sites in λ DNA) and M · *Hha*I (diamonds; 215 sites) MTases each transfer two methyl groups per recognition sequence (*Hae*III: GGCC/GGCC → GGCC/GGCC; *Hha*I: GCGC/GCGC → GCGC/GCGC; the methylated nucleotides are underlined). The M_1 · *Hga*I and M_2 · *Hga*I MTases (squares and circles respectively; 102 sites) each transfer only one methyl group per recognition sequence (M_1: GACGC/GCGTC → GACGC/GCGTC: M_2: GACGC/GCGTC → GACGC/GCGTC). At 10-fold excess enzyme–substrate levels, both *Hga*I MTases exhibit "star" activity by methylating GACGT in addition to GACGC.

hydrolysis with perchloric acid.[7,8] To avoid destruction of minor bases, we use the former method.[55,56] Separation of the degradation products can be accomplished by HPLC or by TLC.[57–59] We use TLC for its simplicity and reliability. To determine the position of the methylated base, the individual strands of the modified oligos are partially digested and the

[55] Y. I. Buryanov, L. V. Andreev, and N. V. Eroshina, *Bokhimia* **39,** 31 (1968).

[56] Y. Iwanami and G. M. Brown, *Arch. Biochem. Biophys.* **126,** 8 (1968).

[57] C. W. Gehrke, R. A. McCune, M. A. Gama-Sosa, M. Ehrlich, and K. C. Kuo, *J. Chromatogr.* **301,** 199 (1984).

fragments are separated by PAGE. The fragments are excised from the gel and counted for radioactivity.

The oligonucleotide is designed to have the recognition sequence of the MTase in the center. One strand is made 22 nt long and the other is 25 nt long. The difference in strand lengths facilitates later strand separation. The melting temperature of oligonucleotides of this length is well above the customary MTase reaction temperature. The sequence of the oligonucleotide is designed to be nonpalindromic, except within the recognition sequence if necessary. (Nonpalindromic oligonucleotides yield a more even distribution of fragments on partial degradations because phosphodiesterase I digests ssDNA at a 20-fold greater rate than ds DNA.) The sequence external to the recognition sequence should be ~70% A–T and should not include hairpins. Some MTases have been reported to methylate ssDNA in addition to dsDNA.[60] One strand of the oligonucleotide can be used as a substrate to determine if the MTase has any ssDNA activity.

Methylation of Oligonucleotide

Each 5′-phosphorylated oligonucleotide strand (2.5 μg; 0.05 OD_{254}) is annealed in 30 μl of reaction buffer (50 m*M* Tris-HCl, pH 7.5, 50 m*M* EDTA, 25 m*M* 2-mercaptoethanol) by incubating at 72° for 5 min, then at 25° for 30 min. One microliter of the annealed mixture (~10 pmol of recognition sites) is diluted to 50 μl with reaction buffer, and incubated at 37° for 1 hr with 5 μl (10–15 units) of methyltransferase and 2 μl (100 pmol; 1 μCi) of [^{3}H]AdoMet. To enable the oligonucleotide to be visualized on a gel 2.5 μg of each unlabeled single-stranded oligonucleotide is added, followed by 30 μl of formamide and 1 μl of 1× stop dye solution [10× stop dye solution: 0.3% (w/v) xylene cyanol, 0.3% (w/v) bromphenol blue, 0.3% (w/v) EDTA, pH 8.0 in deionized formamide]. The strands are denatured by heating to 95° for 5 min and then separated by running the mixture on a gel containing 7 *M* urea, 20% (w/v) polyacrylamide. The bands are excised and the separated strands are desalted by gel filtration through 10-ml Sephadex G-50 columns. The oligomers are usually labeled to a specific activity of 5×10^4 disintegrations per minute (dpm) per μg.

Identification of the Modified Base

Each of the purified, modified single-stranded oligonucleotides [~1×10^4 counts per minute (cpm)] is individually digested with excess bacterial

---

[58] H. K. Mangold, *in* "Thin-Layer Chromatography" (E. Stahl, ed.), p. 786. Springer-Verlag, New York, 1969.

[59] K. Randerath, *Biochem. Biophys. Res. Commun.* **6,** 452 (1961/62).

[60] S. Cerritelli, S. S. Springhorn, and S. A. Lacks, *Proc. Natl. Acad. Sci. U.S.A.* **86,** 9223 (1989).

alkaline phosphatase (1 μl, 1000 U/ml) and phosphodiesterase I (1 μl, 100 U/ml) in 50 μl of 20 mM Tris-HCl, pH 8.4, 10 mM $MgCl_2$ at 37° for 2 hr. The digests are desalted using a 10-ml Sephadex G-10 gel-filtration column that was first equilibrated with deionized water. Fractions (0.5 ml) are eluted with water and the fraction containing the ^{3}H peak is concentrated by vacuum centrifugation. The concentrated fraction is divided in half, and each half is spotted onto a 20 × 20 cm fluorescent cellulose TLC plate. The unlabeled methylated deoxynucleosides (0.1 A_{260} units of each; Table I) are spotted on top of the sample to act as internal standards, and again spotted on separate lanes to act as external standards. One of each pair of plates is developed in solvent D [80 : 20 (v/v) ethanol–water] and the other is developed in solvent G. [66 : 33 : 1 (v/v) isobutyric acid–water–ammonium hydroxide]. The plates are illuminated at 254 nm and the spots in the sample lane formed by the m^{5}C, m^{4}C, and m^{6}A are scraped and collected using a 1-ml Pipetman tip (plugged at the narrow end with siliconized glass wool) attached to an aspirator pump. The sample is eluted from the cellulose with 1 ml of water and added to 10 ml of scintillation cocktail. The amount of tritium in each sample is measured (10-min periods). The modified base is the sample containing ^{3}H.

^{32}P End-Labeling and Modification of Oligonucleotides

Single-stranded oligonucleotide substrate (0.015 μg) is dissolved in 5 μl of reaction buffer [75 mM Tris-HCl, pH 7.5, 10 mM dithiothreitol (DTT), and 10 mM $MgCl_2$] and incubated at 37° for 0.5 hr with 1 μl of [γ-^{32}P]ATP (3000 Ci/mmol) and 1 μl (10,000 U/ml) T4 polynucleotide kinase. The reaction is stopped by heating to 95° for 5 min.

After heating, the mixture is microfuged for 15 sec and 3 μl (0.015 μg) of complementary oligonucleotide and 3 μl of 0.5 M EDTA, pH 8.0 are added. The mixture is annealed at 25° for 15 min. After annealing, 1 μl of AdoMet (3.2 mM, prepared fresh) and 2–5 units of MTase are added. The reaction is incubated at 37° for 1 hr. The reaction is stopped by heating to 95° for 5 min. The mixture is stored at −20°; aliquots are diluted with water before adding the appropriate amounts to the phosphodiesterase I partial digestion reaction mixture as described in the next section.

Identification of Position of Methylation

Approximately 5 × 10^4 dpm of each strand of 5′-phosphorylated ^{3}H-methylated oligonucleotide is mixed with 75 μg of the unmethylated, nonradiolabeled phosphorylated oligonucleotide, and sufficient 5′-^{32}P end-labeled (modified with cold AdoMet) oligonucleotide to achieve approximately the same levels of ^{3}H and ^{32}P activity (Fig. 3). Each mixture is heated to 95° for 5 min, microfuged for 15 sec, and immediately placed on

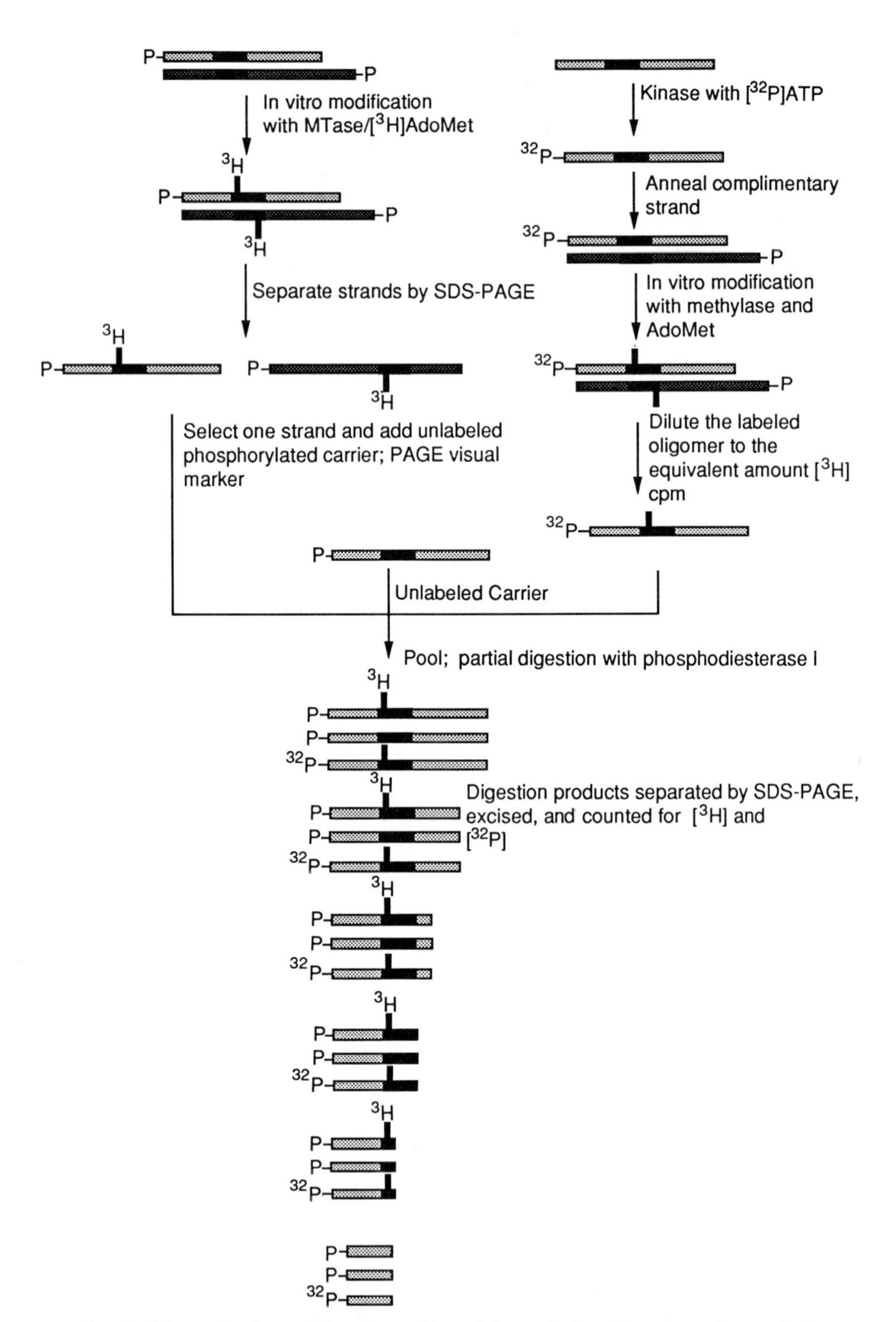

FIG. 3. Scheme for determining the position of the methylated base in each strand of the recognition sequence.

ice. Each oligonucleotide mixture is incubated with 1 μl (100 U/ml) of phosphodiesterase I in 50 μl of 20 m*M* Tris-HCl, pH 8.4, 10 m*M* $MgCl_2$ at 37°. Aliquots (1 μl) are withdrawn from each reaction, every 2 min for the first 10 min followed by 2 μl every 2 min thereafter. The aliquots are immediately pooled into two stop solutions containing 20 μl of 100 m*M* EDTA, 30 μl formamide, and 1 μl of 1× stop dye solution and placed at 4°. The digests are separated by electrophoresis on a 20% (w/v) polyacrylamide, 7 *M* urea gel. Gel slices containing the partially digested fragments are crushed in microfuge tubes and incubated overnight in 300 μl of 0.5 *M* ammonium acetate, 1 m*M* EDTA, pH 7.0. The slices are microfuged and the supernatants collected. The slices are washed twice with 300 μl of water. The ammonium acetate supernatants are pooled with the water washes, added to 5–10 ml scintillation cocktail, and measured for ^{32}P and ^{3}H radioactivity. The measured values are corrected for background and for channel cross-talk. The ^{3}H : ^{32}P ratio normalizes the data for differences in the quantities of DNA extracted from the gel bands. For both strands, the last high ^{3}H : ^{32}P ratio coincides with the base that is methylated.

Acknowledgments

We wish to thank Dr. Donald Comb for support and encouragement, and our colleagues for helpful discussions.

[24] Amino Acid Sequence Arrangements of DNA-Methyltransferases

By GEOFFREY G. WILSON

Introduction

DNA-methyltransferases (MTases) catalyze the addition of methyl groups to adenine and cytosine residues in DNA. The methyl group is transferred from *S*-adenosylmethionine[1] (AdoMet). One methyl group is transferred per residue, producing N^4-methylcytosine[2,3] (m^4C),

[1] W. Arber, *J. Mol. Biol.* **11,** 247 (1965).

[2] A. A. Janulaitis, S. Klimasauskas, M. Petrusyte, and V. Butkus, *FEBS Lett.* **161,** 131 (1983).

[3] M. Erhlich, M. A. Gama-Sosa, L. H. Carreira, L. G. Ljungdahl, K. C. Kuo, and C. W. Gehrke, *Nucleic Acids Res.* **13,** 1399 (1985).

5-methylcytosine[4] (m^5C), and N^6-methyladenine[5–8] (m^6A). The enzymes interact with particular nucleotide (nt) sequences in DNA—usually comprising four to eight specific bases—and methylate an A or C residue within only these sequences.[9,10] Individual MTases are specific for a particular nucleotide sequence, a particular base within the sequence, and a particular atom within the base.[11] With rare exceptions,[12,13] the enzymes recognize only double-stranded (ds) substrates and act as monomers, methylating one strand at a time.[14–20]

DNA methyltransferases from bacteria and viruses display a wide variety of sequence specificities[11,21–27] (over 200 have been demonstrated

[4] S. Hattman, E. Gold, and A. Plotnik, *Proc. Natl. Acad. Sci. U.S.A.* **69,** 187 (1972).

[5] S. Hattman, *J. Virol.* **7,** 690 (1971).

[6] J. P. Brockes, P. R. Brown, and K. Murray, *Biochem. J.* **127,** 1 (1972).

[7] U. Kuhnlein and W. Arber, *J. Mol. Biol.* **63,** 9 (1972).

[8] P. H. Roy and H. O. Smith, *J. Mol. Biol.* **81,** 427 (1973).

[9] H. W. Boyer, L. T. Chow, A. Dugaiczyk, J. Hedgpeth, and H. M. Goodman, *Nature (London) New Biol.* **244,** 40 (1973).

[10] A. Dugaiczyk, J. Hedgpeth, H. W. Boyer, and H. M. Goodman, *Biochemistry* **13,** 503 (1974).

[11] M. Nelson and M. McClelland, *Nucleic Acids Res.* **19,** 2045 (1991).

[12] A. G. de la Campa, P. Kale, S. S. Springhorn, and S. A. Lacks, *J. Mol. Biol.* **196,** 457 (1987).

[13] S. Cerritelli, S. S. Springhorn, and S. A. Lacks, *Proc. Natl. Acad. Sci. U.S.A.* **86,** 9223 (1989).

[14] R. A. Rubin and P. Modrich, *J. Biol. Chem.* **252,** 7265 (1977).

[15] G. E. Herman and P. Modrich, *J. Biol. Chem.* **257,** 2605 (1982).

[16] U. Günthert, R. Lauster, and L. Reiners, *Eur. J. Biochem.* **159,** 485 (1986).

[17] S. Guha, *Gene* **74,** 77 (1988).

[18] V. U. Nwosu, B. A. Connolly, S. E. Halford, and J. Garnett, *Nucleic Acids Res.* **16,** 3705 (1988).

[19] K. Kita, H. Kotani, H. Sugisaki, and M. Takanami, *J. Biol. Chem.* **264,** 5751 (1989).

[20] Y. Maekawa, H. Yasukawa, and B. Kawakami, *J. Biochem. (Tokyo)* **107,** 645 (1990).

[21] H. O. Smith and S. V. Kelly, *in* "DNA Methylation: Biochemistry and Biological Significance" (A. Razin, H. Cedar, and A. D. Riggs, eds.), p. 39. Springer-Verlag, New York, 1984.

[22] B. Suri, V. Nagaraja, and T. A. Bickle, *Curr. Top. Microbiol. Immunol.* **108,** 1 (1984).

[23] R. Yuan and H. O. Smith, *in* "DNA Methylation: Biochemistry and Biological Significance" (A. Razin, H. Cedar, and A. D. Riggs, eds.), p. 73. Springer-Verlag, New York, 1984.

[24] S. Hattman, J. Wilkinson, D. Swinton, S. Schlagman, P. M. Macdonald, and G. Mosig, *J. Bacteriol.* **164,** 932 (1985).

[25] M. Noyer-Weidner, S. Jentsch, J. Kupsch, M. Bergbauer, and T. A. Trautner, *Gene* **35,** 143 (1985).

[26] J. L. Van Etten, Y. Xia, D. E. Burbank, and K. E. Narva, *Gene* **74,** 113 (1988).

[27] D. H. Krüger, T. A. Bickle, M. Reuter, C.-D. Pein, and C. Schroeder, *in* "Nucleic Acid Methylation" (G. A. Clawson, ed.), p. 113. Alan R. Liss, New York, 1990.

or inferred[28]) and they are becoming attractive subjects for investigating protein–DNA interactions.[29–33] They function in bacteria as components of DNA repair systems,[34] and as components of restriction–modification (R–M) systems, in which role they protect DNA from endonucleolytic digestion.[21,23,35] The amino acid sequences of ~100 MTases have been determined, most derived from bacterial R–M systems.[36] The sequences are quite diverse, but they can be organized into distinct classes, the members of which resemble one another to varying degrees.[37–41] We summarize here the classes of MTases that have been discerned, and the principal amino acid sequence motifs and arrangements that differentiate them.

Methyltransferase Classes

Based on the amino acid sequence, DNA methyltransferases divide naturally into those that methylate the endocyclic carbon-5 of cytosine (m^5C-MTases) and those that methylate the exocyclic nitrogen of adenine and cytosine (m^4C- and m^6A-MTases). The former group constitutes a single, fairly homogeneous class,[38,40] but the latter group is heterogeneous, comprising enzymes of three general amino acid sequence architectures, referred to here as α, β, and γ. These MTases share common amino acid sequence motifs, but the positions, orders, and characteristics of the motifs vary according to architecture.[39,41] Enzymes of the γ architecture appear all to be m^6A-MTases; they constitute the m^6A_γ class. Both m^4C- and m^6A-

[28] R. J. Roberts, *Nucleic Acids Res.* **18** (Suppl.), 2331 (1990).

[29] M. Hümbelin, B. Suri, D. N. Rao, D. P. Hornby, H. Eberle, T. Pripfl, S. Kenel, and T. A. Bickle, *J. Mol. Biol.* **200,** 23 (1988).

[30] T. A. Trautner, T. S. Balganesh, and B. Pawlek, *Nucleic Acids Res.* **16,** 6649 (1988).

[31] K. Wilke, E. Rauhut, M. Noyer-Weidner, R. Lauster, B. Pawlek, B. Behrens, and T. A. Trautner, *EMBO J.* **7,** 2601 (1988).

[32] Z. Miner, S. L. Schlagman, and S. Hattman, *Nucleic Acids Res.* **17,** 8149 (1989).

[33] S. Som, L. F. Yang, and S. Friedman, *ASM Abstr.* **5,** A436 (1991).

[34] F. Barras and M. G. Marinus, *Trends Genet.* **5,** 139 (1989).

[35] T. A. Bickle, *in* "*Escherichia coli* and *Salmonella typhimurium:* Cellular and Molecular Biology" (F. C. Neidhardt, J. L. Ingraham, K. B. Low, B. Magasanik, M. Schaechter, and H. E. Umbarger, eds.), p. 692. American Society for Microbiology, Washington, D.C., 1987.

[36] G. G. Wilson, *Nucleic Acids Res.* **19,** 2539 (1991).

[37] S. Chandrasegaran and H. O. Smith, *in* "Structure and Expression" (M. H. Sarma and R. H. Sarma, eds.), Vol. 1, p. 149. Adenine Press, Guiderland, New York, 1988.

[38] R. Lauster, T. A. Trautner, and M. Noyer-Weidner, *J. Mol. Biol.* **206,** 305 (1989).

[39] R. Lauster, *J. Mol. Biol.* **206,** 313 (1989).

[40] J. Pósfai, A. S. Bhagwat, G. Pósfai, and R. J. Roberts, *Nucleic Acids Res.* **17,** 2421 (1989).

[41] S. Klimasauskas, A. Timinskas, S. Menkevicius, D. Butkienè, V. Butkus, and A. Janulaitis, *Nucleic Acids Res.* **17,** 9823 (1989).

MTases occur with the α architecture, and these constitute the m^4C_α and m^6A_α classes. Other m^4C- and m^6A-MTases occur with the β architecture, and these constitute the m^4C_β and m^6A_β classes (see Appendix).

In some respects, the m^4C-MTases resemble one another more closely than they resemble m^6A-MTases, and vice versa, hence it is reasonable to divide the α and β groups into classes, one for each kind of enzyme. It might be better, as an alternative, to recognize m^4C- and m^6A-MTases as separate groups, and to divide each of these into α and β classes. Either way, the result is the same; the former method is used here. The specificities of MTases are often inferred indirectly, by a process of elimination,[11] rather than by direct chemical analysis, and the arguments used are occasionally somewhat circular. MTases are frequently classified by their similarity to previously characterized enzymes, but notwithstanding, the classification scheme is consistent. All enzymes that produce m^5C, for example, do resemble m^5C-MTases; and only those that resemble m^5C-MTases do, in fact, produce m^5C. Thus, while incompletely proven, the scheme is generally considered to be correct.

m^5C-MTases. This class comprises nearly 50 enzymes that range in length between ~300 and 500 amino acids. The enzymes are moderately similar to one another (typically, 30–35% overall identity). The similarities occur as 10- to 15-amino acid motifs, scattered throughout their length. Ten such motifs have been noted.[38,40,42] The motifs occur in invariant order (Fig. 1a[43–46]), and the six most conserved can be readily distinguished in all but one enzyme (Fig. 2a).

A conserved sequence resembling motif I of the m^5C-MTases (LF-G-GG) can be recognized in most of the other methyltransferases, too. Its ubiquity has led to the speculation that it forms the AdoMet-binding site.[39,41,43] The remaining motifs are unique to m^5C-MTases. Motif IV (D----G-PCQ--S--G) is highly conserved and is thought to be the active site; the cysteine residue forms a transient covalent bond with carbon-6 of cytosine, facilitating the subsequent methylation of carbon-5.[44,45] An extended region of variable length and sequence separates motifs VIII and IX; it is thought to be the DNA sequence recognition domain.[30,31,46] The motifs characteristic of the bacterial m^5C-MTases also

[42] A. Düsterhöft, D. Erdmann, and M. Kröger, *Nucleic Acids Res.* **19,** 1049 (1991).
[43] H. O. Smith, T. M. Annau, and S. Chandrasegaran, *Proc. Natl. Acad. Sci. U.S.A.* **87,** 826 (1990).
[44] J. C. Wu and D. V. Santi, *J. Biol. Chem.* **262,** 4778 (1987).
[45] S. Friedman, *ASM Abstr.* **5,** A436 (1991).
[46] J. Walter, M. Noyer-Weidner, and T. A. Trautner, *EMBO J.* **9,** 1007 (1990).

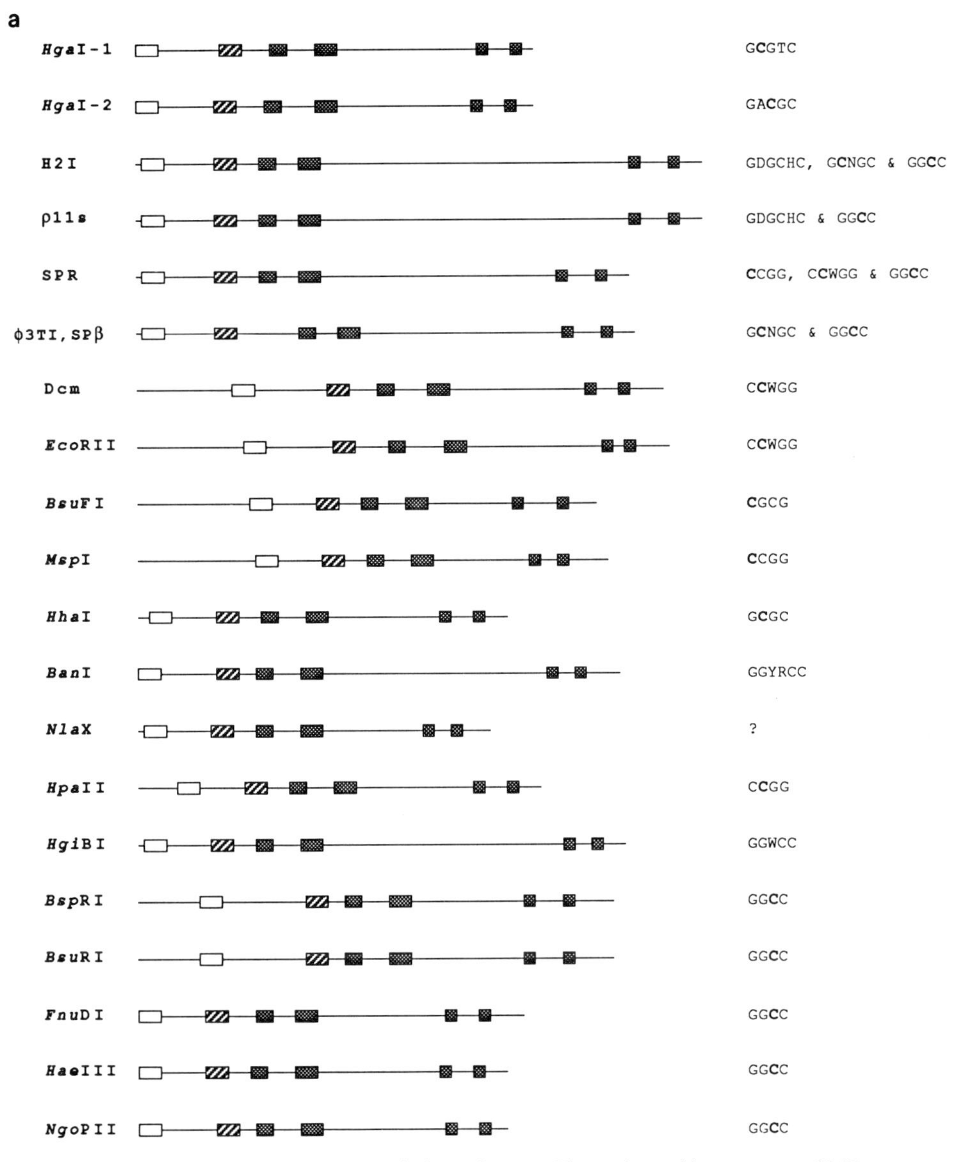

FIG. 1. Architecture of DNA methyltransferases. The amino acid sequences of MTases are depicted as lines with amino termini to the left. The principal conserved sequence motifs are shown as boxes. The relative positions and sizes of the motifs, and the lengths of the proteins, are drawn to scale. The open box depicts conserved motif I, found in most of the MTases, and hypothesized to form the AdoMet-binding site.[39,41,43] The enzymes are arranged

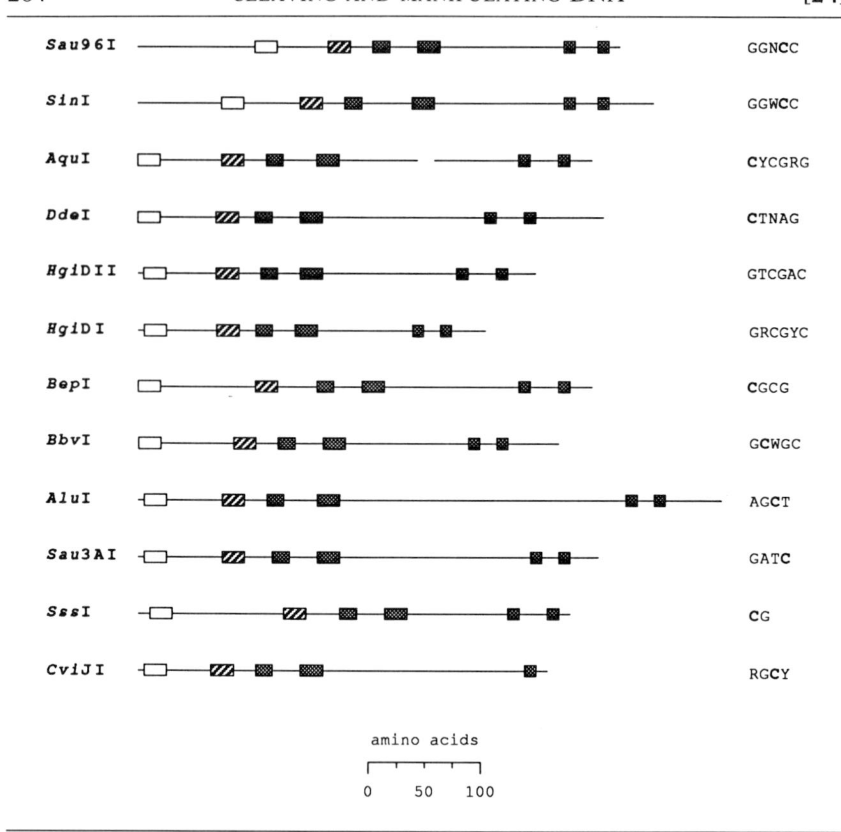

approximately according to amino acid sequence similarity, neighboring proteins sharing greatest overall similarity. Identical, or nearly identical, proteins have been combined. The names of the enzymes are given on the left, and their recognition sequences are shown on the right; the bases that probably become methylated are printed in bold. The individual amino acid sequences that constitute the motifs are shown in Fig. 2. (a) m^5C-Methyltransferases. Six conserved motifs are shown. Motif I is drawn as an open box. Motif IV, thought to encompass the active site,[44,45] is represented by the striped boxes. The remaining motifs are stippled. The nonconserved section between motifs VIII and IX is thought to form the DNA sequence-recognition domain.[30,31,46] The motifs are numbered according to Pósfai *et al.*[45] (b) m^4C_α-MTases and m^6A_α-MTases. Two conserved motifs are shown. Motif I is drawn as an open box, motif IV as a filled box. The motifs are numbered according to Hattman *et al.*[24] The m^4C-MTases are drawn above the m^6A-MTases. (c) m^4C_β-MTases and m^6A_β-MTases. Motif I is drawn as an open box, motif IV as a filled box. Note that the order of the motifs is reversed from that in the α group. The m^4C-MTases are drawn above the m^6A-MTases. (d) m^6A_γ-MTases. Motif I is drawn as an open box, motif IV as a filled box. The type II enzymes are drawn above the type I enzymes.

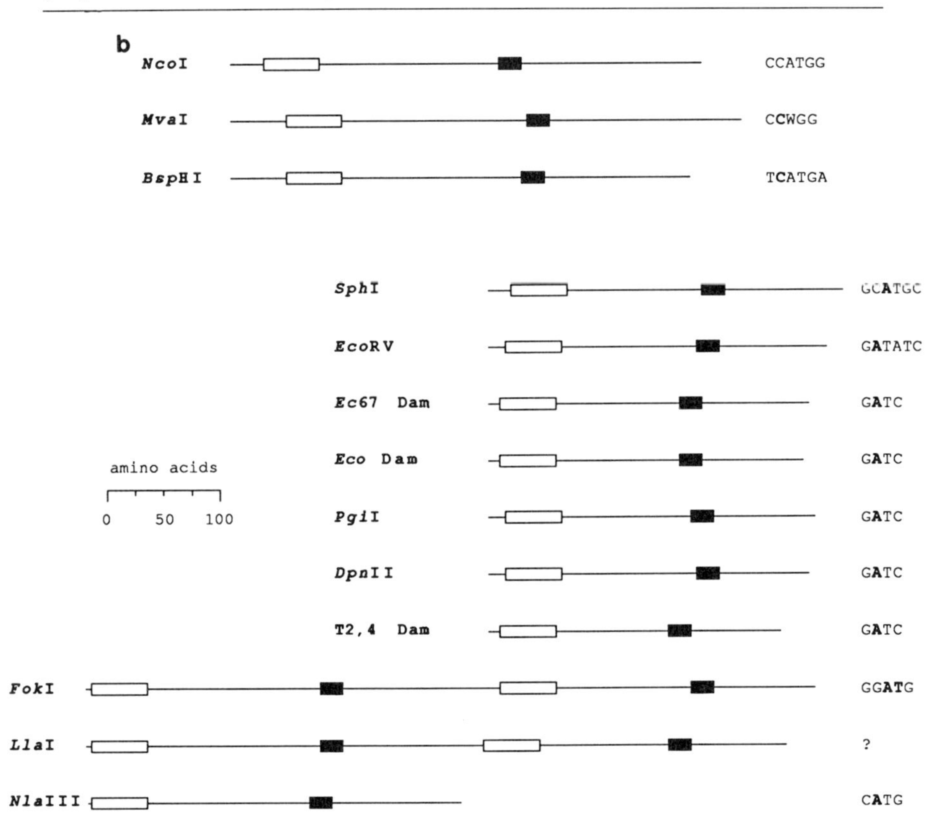

FIG. 1. (*continued*)

occur in the carboxy-terminal 500 or so amino acids of mammalian CpG MTases.[47,48]

α-MTases. m^4C- and m^6A-MTases contain two principal conserved motifs. Motif I corresponds to the LF-G-GG motif of the m^5C-MTases. In the α-MTases this motif occurs near the amino terminus, as it does in the m^5C-MTases (Fig. 1b). The second motif, IV, contains the tetrapeptide XPPY (Fig. 2b). In the m^6A_α- and m^6A_β-MTases the first residue is aspartate: DPPY. The two motifs are generally separated by 100–150 amino acids (Fig. 1b and c). Beside these two motifs, other regions of sequence similarity between MTases in different architectural groups, and different mechanistic classes, can sometimes be discerned.[24,49] These secondary

[47] T. Bestor, A. Laudano, R. Mattaliano, and V. Ingram, *J. Mol. Biol.* **203,** 971 (1988).

[48] P. D. Andrews, C. Taylor, and D. P. Hornby, Second NEB workshop on biological DNA modification, Berlin, September 2–7, 1990.

[49] R. Lauster, A. Kriebardis, and W. Guschlbauer, *FEBS Lett.* **220,** 167 (1987).

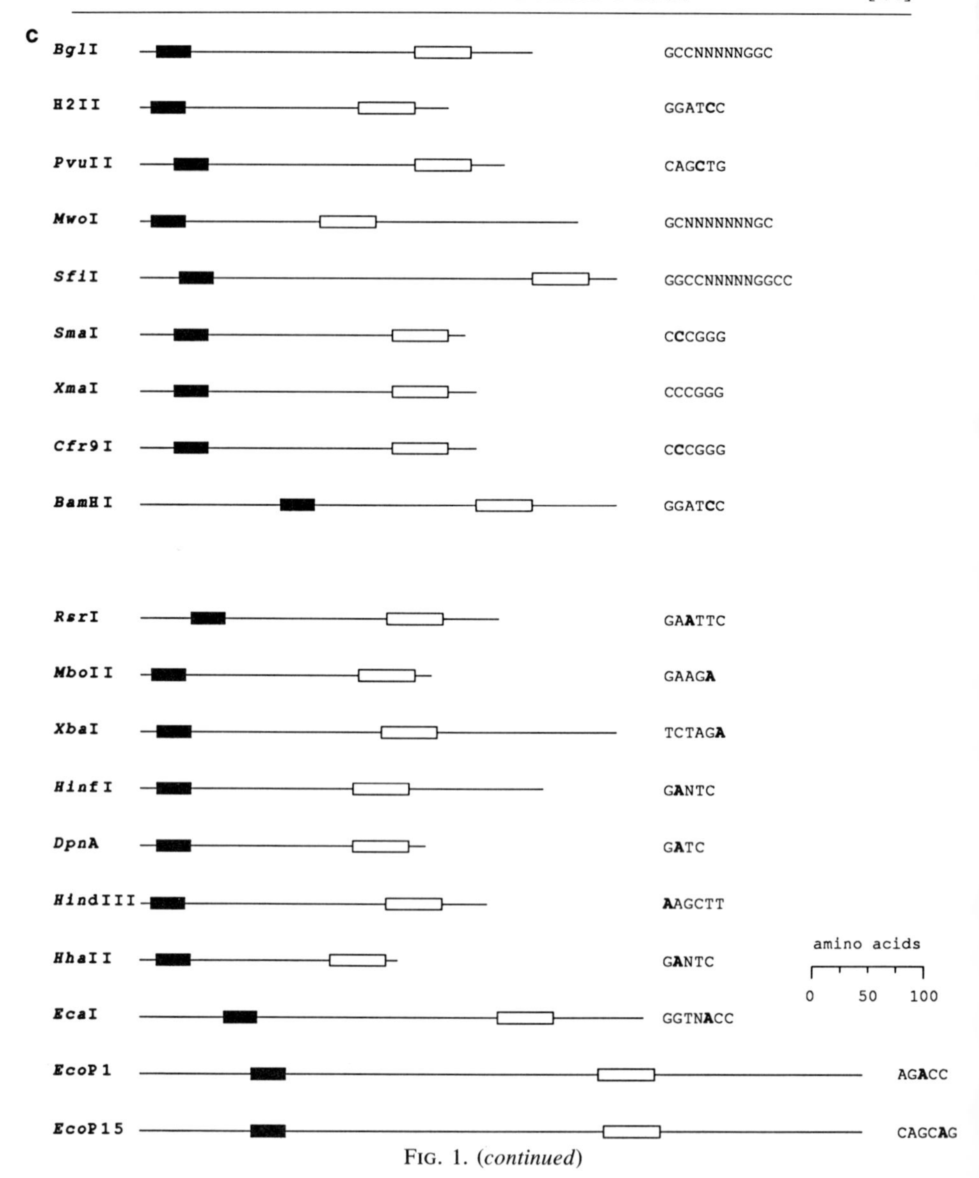

FIG. 1. (*continued*)

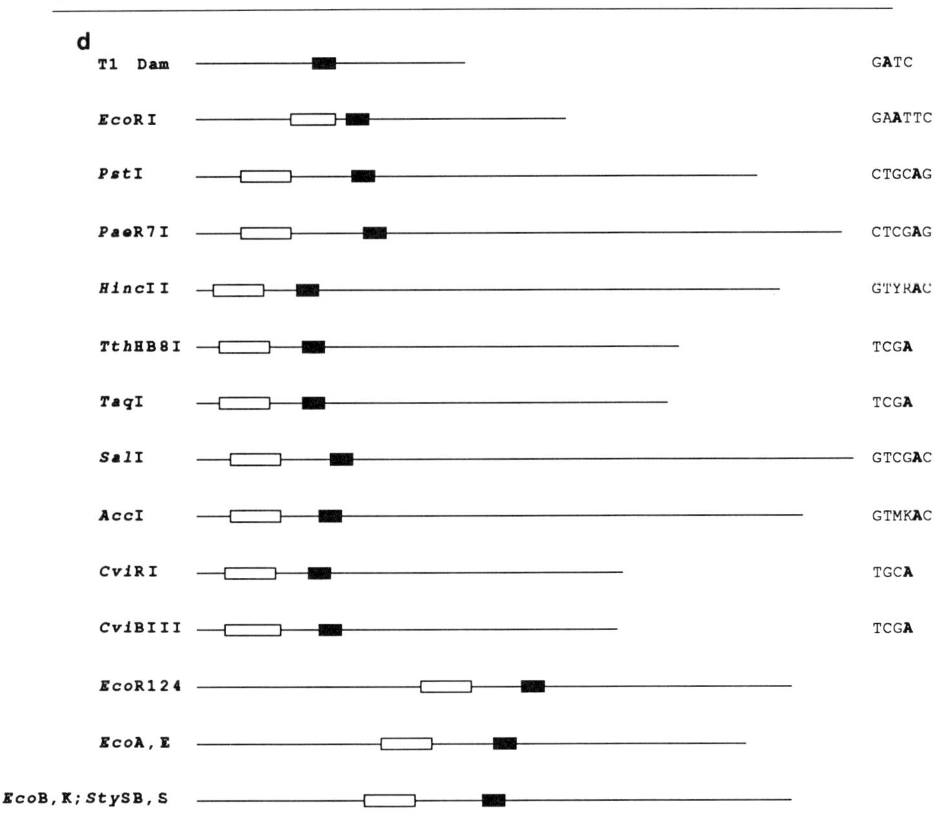

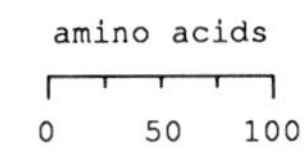

FIG. 1. (*continued*)

motifs are not well conserved, however, and defining them presents a challenge; we do not attempt to define them here.

M · *Mva*I produces m^4C[41]; M · *Nco*I is presumed to do so as well, because it does not resemble an m^5C-MTase, and R · *Nco*I is not blocked by adenine methylation. M · *Bsp*HI may also produce m^4C because it resembles M · *Mva*I and M · *Nco*I more closely than it does m^6A-MTases. These three enzymes currently constitute the m^4C_α class. In the m^4C_α-MTases and the m^4C_β-MTases, the first residue of motif IV is usually serine: SPPY[41] (Fig. 2b and c). The remaining α enzymes, known or assumed to be m^6A-MTases, include most of the characterized Dam MTases,[24,49] and two enzymes that have a double structure: M · *Fok*I and a

	Motif I	Motif IV	Motif VI	Motif VIII	Motif IX	Motif X
*Hga*I-1	GLSLFSSAGIGEYFL	DFLIASPPCQGMSVAGKNRD	PAYILIENVPFL	LDAADYGTPQRRKRAIIRLNK	RVLSVLELMRLTGLP	RQIIGECIPP
*Hga*I-2	AMSLFSSAGIGELDL	KFLLATPPCQGLSSVGKNKH	LDFILIENVPRF	LNAKDYGICQSRPRAIIKMYK	RVLSLLETFIVSSID	RTIIGEAIPP
H2I	VMSLFSGIGAFEAAL	DLLVGGSPCQSFSVRGYRKG	PRYFVFENVKGL	LNSKFFNVPQNRERIYIIGVR	RKLTPLECWRLQAFD	YKEAGNSITV
rho11s	VMSLFSGIGAFEAAL	DLLVGGSPCQSFSVAGYRKG	PRYFVFENVKGL	LNSKFFNVPQNRERIYIIGVR	RKLTPLECWRLQAFD	YKEAGNSITV
SPR	VMSLFSGIGAFEAAL	DLLVGGSPCQSFSVAGHRKG	PKFFVFENVKGL	LNSKFFNVPQNRERLYIIGIR	RRLTPLECFRLQAFD	YKQAGNSITV
Phi3TI	VMSLFSGIGAFEAAL	DLLTSGFPCPTFSVAGGRDG	PKFVILENVKGL	LNSKFFNVPQNRERVYIIGIR	RKLSPLECWRLQAFD	YKQAGNSITV
Dcm	FIDLFAGIGGIRRGF	DVLLAGFPCQPFSLAGVSKK	PAMFVLENVKNL	IIDGKHFLPQHRERIVLVGFR	RRLTPRECARLMGFE	YRQFGNSVVV
*Eco*RII	FIDLFAGIGGIRKGF	DVLLAGFPCQPFSLAGVSKK	PAIFVLENVKNL	VIDGKHFLPQHRERIVLVGFR	RRLTPRECARLMGFE	YRQFGNSVVV
*Bsu*FI	FIDLFAGIGGIRLGF	DVLLAGFPCQPFSNIGKREG	PKMFLLENVKGL	MDAQNFGLPQRRERIVIVGFH	RLFSELELKRLMGFP	YRQFGNSVAV
*Msp*I	FIDLFSGIGGIRQSF	DILCAGFPCQPFSHIGKREG	TPVLFLENVPGL	LDASHFGIPQKRKRFYLVAFL	RLLTTNECKAIMGFP	YRQMGNSVVV
*Hha*I	FIDLFAGLGGFRLAL	DILCAGFPCQAFSISGKQKG	PKVVFMENVKNF	LNALDYGIPQKRERIYMICFR	RKLHPRECARVMGYP	YKQFGNSVVI
*Ban*I	FVDLFAGIGGIRIGF	DFLLAGFPCQPFSYAGKQQG	PKAFLLENVRGL	LNSSTFGVPQNRVRIYILGIL	RRITPRECARLQGFP	YKQLGNSVTV
*Nla*X	IIDLFAGIGGIRLGF	DILSAGFPCQPFSQAGLKKG	PQAFLLENVKQL	LKARDFGIPQNRERIYLVGFL	RKITPPEAARLQGFP	YRQFGNSVCV
*Hpa*II	FIDLFAGIGGFRIAM	DILCAGFPCQAFSIAGKRGG	PKAFFLENVKGL	VNAKNFGVPQNRERIYIVGFH	RKMTPREWARLQGFP	YKQFGNSVAV
*Hgi*BI	FIDLFAGIGGFRLGL	DVLVGGVPCQPWSIAGKNQA	PKAFIFENVKGL	LNSFDFGVAQNRDRVFIVGIQ	RKITVSEAARLQGFP	FRLIGNSVAP
*Bsp*RI	VLSLFCGAGGLDLGF	NLVIGGFPCPGFSEAGPRLV	PEIFVAENVKGM	LNARDYGVPQIRERVIIVGVR	RRLSVKEIKRIQTFP	YKQIGNAVPV
*Bsu*RI	VLSLFSGCGGLDLGF	NLILGGFPCPGFSEAGPRLI	PEIFVAENVKGM	LNARDYGVPQLRERVIIEGVR	RRLSVKEIARVQTFP	YKQIGNAVPV
*Fnu*DI	LLSLFSGAGGLDLGF	DGIIGGPPCQSWSEAGSLRG	PKFFLAENVKGM	LNAFDYGVAQDRERVFYVGFR	RRLSIRECARIQGFP	YKMVGNAVPV
*Hae*III	LISLFSGAGGLDLGF	DGIIGGPPCQSWSEGGSLRG	PIFFLAENVKGM	LNANDYGVAQDRKRVFYIGFR	RRLTVRECARVQGFP	YKMIGNAVPV
*Ngo*PII	IISLFSGCGGLDLGF	DGIIGGPPCQSWSEAGALRG	PKFFLAENVSGM	ANAKDYGVAQERKRVFYIGFR	RRMTVREVARIQGFP	YKMIGNAVPV
*Sau*96I	VFETTAGAGGLALGL	DILSGGYPCQTFSYAGKRNG	PKAFIAENVRGL	LNSWNYDVAQKRERIVIIGIR	RPFSIREYARIQSFP	YRQIGNAVPV
*Sin*I	ALSFTSGAMGLDLGI	DLIMGGPPCQAFSTAGKRLG	PKYIVIENVRGL	YNSANFGVPQIRERVIIICSR	RPLSVQEYKVIQQFP	YRQLGNAVPI
*Aqu*I	LISLFSGAGGMDIGF	DLVIGGPPCQSFSLAGKRMG	PKCFVMENVKGM	LNAADFGVPQFRERVFIVGNR	RMLTVRELACLQTFP	FGQVGNAVPP
*Dde*I	IIDLFAGCGGFSHGF	DGIIGGPPCQGFSLSGNRDQ	PKFFVMENVLGI	LNACDYGVPQSRQRVFFIGLK	RNFTAREGARIQSFP	YQQIGNAVPP
*Hgi*DI	TIDLFAGCGGMSLGF	ELIIGGPPCQDFSSAGKRDE	PAWVIMENVERA	LDASLCGVPQLRKRTFVIGHR	RPLTTKERSLIQTFP	EQMIGNAVPV
*Bep*I	VLSLFSGCGGMDLGL	DVVTGGFPCQDFSFAGKRKG	PKVFIAENVKGL	LNAKNYGVAQNRERVIFIGIS	RRLTVRECALIQSFP	YKIIGNAVPP
*Bbv*I	KGELFCGPGGLALGA	DAFTFGFPCNDYSIVGEHKG	PLVFIAENVGGL	YKFEEYGVPQRRHRIIIVGIR	RALTNRERARLQTFP	RKQIGMAVPP
*Sau*3AI	VVELFAGVGGFRLGL	DMIVGGFPCQDYSVARSLNG	PKYLLLENVDRL	INAADYGNAQRRRRVFIFGYK	RTLTPIEAERLNGFP	YFCMGNALVV
*Sss*I	VFEAFAGIGAQRKAL	DLLTYSFPCQDLSQQGIQKG	PKYLLMENVGAL	LNAADFGSSQARRRVFMISTL	RKMNSDETFLYIGFD	IFVCGNSISV
*Cvi*JI	TLELFAGIAGISHGL	DMITAGFPCTGFSIAGSRTG	PKIVFLENSHML	CRASIIGAHHQRHRWFCLAIR	RHLSGIWCAWLMGYD	?
Consensus	---LF-G-GG---GF	D-L-GGFPCQ-FS-AG-R-G	PK-F--ENVKGL	LNA---GVPQ-RER--I-G-R	R-LT-RE-ARLQ-FP	Y-Q-GN-V-V

b

	Motif I		Motif IV
*Nco*I	GFSAQWAQSVIAESGA	QRVLDPYSGSGTTVIAAEESGAAGLGVD	WADLILTSPPYAN
*Mva*I	MMIPQLAKEFIELTQQVKPEI	KKLYDPFMGSGTSLVEGLAHGLEVYGTD	MFDIVVTSPPYGD
*Bsp*HI	KFPSVLAGKIIDLFPPKDKN	SYILDNFCGSGTTLVEAKLRGINSVGLD	NIYLVISHPPYLN
Consensus	----QLA---I-L---	----DPF-GSGTTLVEA---G---YG-D	--DLV-TSPPY-N
*Sph*I	GSKRALASQILSLFPHGGV	PRLVEPFAGSAAISVAAR	PDELVYMDPPYQG
*Eco*RV	GIKTKLVPCIKRIVPKNFN	GVWVEPFMGTGVVAFNVA	RDDVVYCDPPYIG
*Ec*67 Dam	GNKTAIMSELKKHLPAG	PRLVEPFAGSCAVMMATD	VGDVVYCDPPYDG
Eco Dam	GGKYPLLDDIKRHLPKG	ECLVEPFVGAGSVFLNTD	DASVVYCDPPYAP
*Pgi*I	GGKRQLIPEIKNLLPKGILS	HPYYEPFIGGGALLFELQ	RSSFVYLDPPYHP
*Dpn*II	GGKRQLLPVIRELIPKTY	NRYFEPFVGGGALFFDLA	TGDFVYFDPPYIP
T2,4 Dam	GNKQSLLPELKPHFPKY	DRFVDLFCGGLSVSLNVN	DGDFVYVDPPYLI
*Fok*I-C	GGKHKLLNQIVPLFPDKI	DTFVDLFSGGFNVGINVN	QNDLVYCDPPYLI
*Lla*I-C	GSKDDVIPRIFKLLPKHV	TTFVDAMGGAFNVGANRT	DDTVFYFDPPYLV
*Fok*I-N	GSKVNLLDNIQEVIEENVKDDA	HVFMDLFSGTGIVGENFK	YGDILYIDPPYNG
*Lla*I-N	GNKTNLLNFIQQVIKKHDIQG	QTFADLFAGTGSVGDYFK	SGDIAYIDPPYTI
*Nla*III	GSKLKLSNWLETEISNVAGHSLSD	KVFCDLFAGTGIVGRKFK	SGDILYLDPPYNA
Consensus	G-K--LL--I----PK-	--FVDPF-G-G-V-----	-GD-VY-DPPY--

FIG. 2. Conserved amino acid sequence motifs. The MTases are arranged in the same orders as in Fig. 1, but some sequences have been omitted. The motifs are printed, left to right, in the order in which they occur in the proteins. The boundaries of the motifs are debatable. Alignments were made using the University of Wisconsin Genetics Computer Group (UWGCG) programs Gap, Lineup, Pileup, and Profile [J. Devereux, P. Haeberli, and O. Smithies, *Nucleic Acids Res.* **12,** 387 (1984)]. Consensus sequences were calculated manually, with help from the UWGCG program Pretty. An amino acid symbol in the consensus signifies that that amino acid occurs in at least 50% (regular type) or at least 75% (boldface type) of the aligned proteins. A hyphen signifies that no amino acid predominates at that position. The consensuses are based on all the sequences that we have access to, including some not shown here. Highly conserved amino acids are also shown in boldface type in the individual sequences. (a) Conserved motifs of m^5C-methyltransferases; (b) conserved motifs of m^4C_α- and m^6A_α-MTases; (c) conserved motifs of m^4C_β- and m^6A_β-MTases; and (d) conserved motifs of m^6A_γ-MTases.

C

```
                Motif IV                                          Motif I

BglI       GDARELLK   CIEEESIALSVWSPPYHVGK  KY    HEAKFPLLLPQRLI KLLTQKGDTVLDCFMGSGTTAVAALSESRNFIGIE
H2II       NDCVQFMKE  NIGDCTIDLTVTSPPYDDLR  NY    HPAIFPEKLAEDHI LSWSNEGDIVFDPFMGSGTTAKMAALNNRKYIGTE
PvuII      GDSLELLE   SFPEESISLVMTSPPFALQRKKEY    HPARFPAKLPEFFI RMLTEPDDLVVDIFGGSNTTGLVAERESRKWISFE
MwoI       GDVFSALR   CLEDNSISVALTSPPYWRQR  DY    HTAVFPEKLVSSLLSRCNLKDGDYILDPFAGTGTTGAVVKKMKYQLYPKD
SfiI       GDSLDCLA   KLPDESINTVVTSPPYWAVR  DY    HFAVMPRKLAHFAL KATLPMNGSCLDPFMGSGTTGRVVRELGGRFVGVD
SmaI       GDAREAVQ   GLDSEIFDCVVTSPPYWGLR  DY    HFAVFPRAMARLCV LAGSRPGGKVLDPFFGSGTTGVVCQELDRECVGIE
XmaI       GDALTVLR   RLPSGSVRCVVTSPPYWGLR  DY    HFATFPPELIRPCI HASTEPGDYVLDPFFGSGTVGLVCQDENRQYVGIE
Cfr9I      GDALSVLR   RLPSGSVRCIVTSPPYWGLR  DY    HFATFPTELIRPCI LASTKPGDYVLDPFFGSGTVGVVCQQEDRQYVGIE
BamHI      GDCLELFK   QVPDENVDTIFADPPFNLDK  EY    KFNELSVKLLDRII TMSTNEGDVVLDPFGGSGTTFAVSEMLGRKWIGFE

Consensus  GD-L--L-   -LP-ES----VTSPPY---R  DY    HFA-FP--L----I ---T--GD-VLDPF-GSGTTG-V-----R-Y-G-E

RsrI       CDCLDTLA   KLPDDSVQLIICDPPYNIMLAD      HPTQKPAAVIERLV RALSHPGSTVLDFFAGSGVTARVAIQEGRNSICTD
MboII      MNCFDFLD   QVENKSVQLAVIDPPYNLSKAD      HITPKPRDLIERII RASSNPNDLVLDCFMGSGTTAIVAKKLGRNFIGCD
XbaI       GDCLVELA   RLEECEADLVYLDPPFFTNRRH      YPTQKPILLLERII EISTDPGDFIVDPFCGSGTTLVAAAILGRRAFGID
HinfI      GDCIEKLK   TIPNESIDLIFADPPYFMQTEG      HSTQKPESLLYKVI LSSSKPNDVVLDPFFGTGTTGAVAKALGRNYIGIE
DpnA       SDTFKFLS   KMKPESMDMIFADPPYFLSNGG      HPTQKPEYLLERII LASTKEGDYILDPFVGSGTTGVVAKRLGRRFIGID
HindIII    SDSIFEIK   KLDSNSIHAIISDIPYGIDYDD      HVAQKPLNLMKLLI DLVTKEEQIVLDPFAGSGTTLLAAKELNRHFIGYE
HhaII      KMNGIDLFK  LIPDKAVKIAFFDPQYRGVLDK      HTHSKPIEMQKQLI LATTQEGDLILDPASGGYSVFECCKQTNRNFIGCD
EcaI       SDNSLALNRL MVEGKKAKLIYLDPPYATGMGF      YPTEKNFNMMKLIV GASSNPGDLVIDPFCGSGSTLHAASLLQRKWIGID
EcoP1      GDNLEVLKHMVNAYAEKVNMIYIDPPYNTGKDG      FEGPKPVPLITDLV KIGTKKDSLVLDFFAGSGTTAEAVAYLNEKDSGCR
EcoP15     GDNLEVLKHMVNAYAEKVKMIYIDPPYNTGKDG      FTNAKTIKLVEDLI SFACDGEGIVLDFFAGSGTTAHTVFNLNNKNKTSY

Consensus  -D----L-   -----SV-LI--DPPY------      H-TQKP--L---LI -------D-VLD-F-GSGTT---A--L-R--IG-D
```

d

	Motif I		Motif IV
T1 Dam	?		GEGDVIFCNPPYEP
*Eco*RI	GFSSSEAAKNGFYYEYHKENGKKLVFDD	ISVSSFCGDGDFR	KKSDIVVTNPPFSL
*Pst*I	GQFMSSSAVSELMANLFESYVGEH	EILDAGAGVGSLT	PKYNKAILNPPYLK
*Pae*R7I	GAIFTRSEVVDFILDLAGYTEDQPLHEKRL	LEPSFGGGDFL	GQFDFVVGNPPYVR
*Hinc*II	GQFFTPTHIVKYMIGLMTKNKNA	SILEPSSGNGVFL	PQYDSIIGNPPYVR
*Tth*HB8I	GRVETPPGLVRFMVGLAEARKGV	RVLEPACADGPFL	EAFDLILGNPPYGI
*Taq*I	GRVETPPEVVDFMVSLAEAPRGG	RVLEPACAHGPFL	EAFDLILGNPPYGI
*Sal*I	GQFATPPSLAGEIMRYTIDLHEETRI	NFLEPSCGSGSFF	PVASLLVANPPYVR
*Acc*I	AQFFTPFPIAYAMAKWILGNKQLK	TVLEPAFGLGVFS	NKYDGIICNPPYFK
*Cvi*RI	GVFFTPKDIRDIVFEELGDFEPT	NILEPTCGTGEFI	EKFDLIIGNPPYFT
*Cvi*BIII	GIFFTPKTVREKLFGFTEHFQNTPGF	SILEPSCGTGEII	GKFDFIVGNPPYVV
*Eco*R124	GEFFTPQHVSKLIAQLAMHGQTHVN	KIYDPAAGSGSLL	KPFDAIVSNPPYSV
*Eco*K	GQYFTPRPLIKTIIHLLKPQPRE	VVQDPAAGTAGFL	PKAHIVATNPPFGS
Consensus	G-FFTP---------L-------	--LEP-CG-G-FL	---D-I--NPPY--

FIG. 2. *(continued)*

M · *Lla*I.[19,50,51] The carboxy-terminal halves of these proteins are closely similar to the T4 Dam MTase, and the amino-terminal halves are closely similar to the M · *Nla*III MTase.[50,52] M · *Fok*I appears to be a joint protein, formed by the fusion of somewhat dissimilar, ancestral MTases of identical sequence specificity but opposite strand specificity. The amino-terminal portion of M · *Fok*I methylates only the top strand of the asymmetric recognition sequence, while the carboxy-terminal portion of M · *Fok*I methylates only the bottom strand.[50,53] M · *Lla*I is probably similar.[51] The recognition sequences of the m^6A_α-MTases often include the dinucleotide AT, the adenine of which is the target for methylation (Fig. 1b).

β-MTases. β-MTases resemble α-MTases except, remarkably, they are reversed.[39] The amino acid sequences of the two principal motifs are very similar between the two groups (Fig. 2b and c), but they occur in opposite orders. In the β-MTases, motif IV occurs at the amino terminus, and motif I occurs toward the carboxy terminus; in the α-MTases, they occur the other way around (Fig. 1b and c). The catalytic consequences of this motif reversal, if any, are unknown.

Many of the β-MTases are m^4C-MTases. M · *Bam*HI, M · *Cfr*9I, M · *Pvu*II, and M · *Sma*I are known to produce m^4C,[41,54–56] and M · *Xma*I, M · *Mwo*I, M · *Bgl*I, and M · *Sfi*I probably do as well, because their recognition sequences contain only G · C base pairs, and they resemble these enzymes rather than m^5C-MTases. Among the m^6A_β-MTases, the two type III enzymes M · *Eco*P1 and M · *Eco*P15 are somewhat different, and M · *Eca*I is intermediate between them and the other members of the class.

γ-MTases. The two principal motifs are separated by a relatively short distance in the γ-MTases—40 to 80 amino acid (Fig. 1d). Asparagine occupies the first position of motif IV: NPPY (Fig. 2d). This class could reasonably be divided into two subclasses, one for the type I modification subunits (*Eco*A, *Eco*K, *Sty*SB, etc.) and the other for the type II enzymes. Type I and type II enzymes behave differently and their architectures are somewhat different (Fig. 1d). Type I MTases lack recognition sequence specificity; they act as heteromultimers, deriving specificity from accom-

[50] M. C. Looney, L. S. Moran, W. E. Jack, G. R. Feehery, J. S. Benner, B. E. Slatko, and G. G. Wilson, *Gene* **80,** 193 (1989).

[51] C. Hill, L. A. Miller, and T. R. Klaenhammer, *J. Bacteriol.* **173,** 4363 (1991).

[52] D. Labbé, H. J. Höltke, and P. C. K. Lau, *Mol. Gen. Genet.* **224,** 101 (1990).

[53] H. Sugisaki, K. Kita, and M. Takanami, *J. Biol. Chem.* **264,** 5757 (1989).

[54] J. E. Brooks, P. D. Nathan, D. Landry, L. A. Sznyter, P. A. Waite-Rees, C. L. Ives, L. S. Moran, B. E. Slatko, and J. S. Benner, *Nucleic Acids Res.* **19,** 841 (1991).

[55] T. Tao, J. Walter, K. J. Brennan, M. M. Cotterman, and R. M. Blumenthal, *Nucleic Acids Res.* **17,** 4161 (1989).

[56] S. Klimasauskas, D. Steponaviciene, Z. Maneliene, M. Petrusyte, V. Butkus, and A. Janulaitis, *Nucleic Acids Res.* **18,** 6607 (1990).

panying specificity polypeptides.[57,58] Phage T1 Dam lacks a readily recognizable sequence corresponding to motif I; it is presumed to belong to the m^6A_γ class because it possesses asparagine in motif IV (Fig. 2d).

Appendix: Classification of DNA-Methyltransferases

Enzyme[a]	Specificity[b]	Class[c]	Refs.[d]	GenEMBL number[e]
*Acc*I	GTMK**A**C	m^6A_γ	59, 60	
*Alu*I	AG**C**T	m^5C	61	
*Aqu*I	**C**YCGRG	m^5C	62	M28051
*Bam*HI	GGAT**C**C	m^4C_β	54, 63	X55285
				X54303
*Ban*I	GGYRCC	m^5C	20	D00704
*Ban*III	ATCGAT	m^6A_γ	64	
*Bbv*I	G**C**WGC	m^5C	65	
*Bcn*I	C**C**SGG	m^4C_α	66	
*Bep*I	**C**GCG	m^5C	67	X13555
*Bgl*I	GCCN$_5$GGC	m^4C_β	68	
*Bsp*HI	T**C**ATGA	m^4C_α	69	
*Bsp*RI	GG**C**C	m^5C	70	X15758
*Bsp*6I	GCN$_2$GC	m^5C	71	
*Bsu*BI	CTGC**A**G	m^6A_γ	72	
*Bsu*FI	**C**CGG	m^5C	46	X51515

(continued)

[57] P. Kannan, G. M. Cowan, A. S. Daniel, A. A. F. Gann, and N. E. Murray, *J. Mol. Biol.* **209,** 335 (1989).
[58] C. Price, J. Lingner, T. A. Bickle, K. Firman, and S. W. Glover, *J. Mol. Biol.* **205,** 115 (1989).
[59] S.-Z. Chen, L. S. Moran, B. E. Slatko, and G. G. Wilson, unpublished (1989).
[60] B. Kawakami, H. Christophe, M. Nagatomo, and M. Oka, *Agric. Biol. Chem.* **55,** 1553 (1991).
[61] B.-H. Zhang, T. Tao, G. G. Wilson, and R. Blumenthal, in preparation.
[62] C. Karreman and A. de Waard, *J. Bacteriol.* **172,** 266 (1990).
[63] P. G. Vanek, J. F. Connaughton, W. D. Kaloss, and J. G. Chirikjian, *Nucleic Acids Res.* **18,** 6145 (1990).
[64] B. Kawakami, A. Sasaki, M. Oka, and Y. Maekawa, *Agric. Biol. Chem. Tokyo* **54,** 3227 (1990).
[65] J. M. Barsomian and G. G. Wilson, unpublished (1990).
[66] S. Menkevicius, V. Butkus, and A. Janulaitis, personal communication (1991).
[67] D. Kupper, J.-G. Zhou, P. Venetianer, and A. Kiss, *Nucleic Acids Res.* **17,** 1077 (1989).
[68] K. D. Lunnen and G. G. Wilson, unpublished (1990).
[69] R. D. Morgan, personal communication (1991).
[70] G. Pósfai, A. Kiss, S. Erdei, J. Pósfai, and P. Venetianer, *J. Mol. Biol.* **170,** 597 (1983).
[71] A. Lubys and A. Janulaitis, personal communication (1991).
[72] G. Xu, W. Kapfer, J. Walter, and T. A. Trautner, Second NEB workshop on biological DNA modification, Berlin, September 2–7, 1990, p. 85.

Appendix (*continued*)

Enzyme[a]	Specificity[b]	Class[c]	Refs.[d]	GenEMBL number[e]
*Bsu*RI	GG**C**C	m^{5}C	73	X02988
*Cfr*9I	C**C**CGGG	m^{4}C$_\beta$	41	X17022
*Cfr*10I	R**C**CGGY	m^{5}C	74	
CpG MTase	**C**G	m^{5}C	47, 75	X14805
*C*viAII	C**A**TG	m^{6}A	76	M86639
*C*viBIII	TCG**A**	m^{6}A$_\gamma$	77	X06618
*C*viJI	RG**C**Y	m^{5}C	78	M27265
*C*viRI	TGC**A**	m^{6}A$_\gamma$	79	M38173
Dcm	C**C**WGG	m^{5}C	80, 81	X13330
*Dde*I	**C**TNAG	m^{5}C	82	Y00449
*Dpn*A	G**A**TC	m^{6}A$_\beta$	83	
*Dpn*II	G**A**TC	m^{6}A$_\alpha$	83	M11226
*Eag*I	CGGCCG	m^{5}C	84	
*Eca*I	GGTN**A**CC	m^{6}A$_\beta$	85	X17111
*Eco*A	G**A**GN$_7$G**T**CA	m^{6}A$_\gamma$	86	
*Eco*B	TG**A**N$_8$**T**GCT	m^{6}A$_\gamma$	87	
Eco Dam	G**A**TC	m^{6}A$_\gamma$	88	J01600
*Eco*E	G**A**GN$_7$ATGC	m^{6}A$_\gamma$	86	
*Eco*K	A**A**CN$_6$G**T**GC	m^{6}A$_\gamma$	89	X06545
*Eco*P1	AG**A**CC	m^{6}A$_\beta$	29	
*Eco*P15	CAGCAG	m^{6}A$_\beta$	29	

[73] A. Kiss, G. Pósfai, C. C. Keller, P. Venetianer, and R. J. Roberts, *Nucleic Acids Res.* **13,** 6403 (1985).

[74] P. Povilonis, S. Menkevicius, D. Butkiene, R. Vaisvila, V. Butkus, and A. Janulaitis, personal communication (1991).

[75] T. Bestor, A. W. Page, and L. L. Carlson, Second NEB workshop on biological DNA modification, Berlin, September 2–7, 1990, p. 47.

[76] Y. Zhang, M. Nelson, J. W. Nietfeldt, D. E. Burbank, and J. L. Van Etten, *Nucleic Acids Res.*, in press (1992).

[77] K. E. Narva, D. L. Wendell, M. P. Skrdla, and J. L. Van Etten, *Nucleic Acids Res.* **15,** 9807 (1987).

[78] S. L. Shields, D. E. Burbank, R. Grabherr, and J. L. Van Etten, *Virology* **176,** 16 (1990).

[79] C. Stefan, Y. Xia, and J. L. Van Etten, *Nucleic Acids Res.* **19,** 307 (1991).

[80] T. Hanck, N. Gerwin, and H.-J. Fritz, *Nucleic Acids Res.* **17,** 5844 (1989).

[81] A. Sohail, M. Lieb, M. Dar, and A. S. Bhagwat, *J. Bacteriol.* **172,** 4214 (1990).

[82] L. A. Sznyter, B. Slatko, L. Moran, K. H. O'Donnell, and J. E. Brooks, *Nucleic Acids Res.* **15,** 8249 (1987).

[83] S. A. Lacks, B. M. Mannarelli, S. S. Springhorn, and B. Greenberg, *Cell (Cambridge, Mass.)* **46,** 993 (1986).

[84] L. Sznyter, L. Moran, B. E. Slatko, and J. E. Brooks, personal communication (1989).

[85] V. Brenner, P. Venetianer, and A. Kiss, *Nucleic Acids Res.* **18,** 355 (1990).

[86] A. S. Daniel, G. Cowan, and N. E. Murray, personal communication (1991).

[87] J. Kelleher, A. S. Daniel, and N. E. Murray, personal communication (1991).

[88] J. E. Brooks, R. M. Blumenthal, and T. R. Gingeras, *Nucleic Acids Res.* **11,** 837 (1983).

[89] W. A. M. Loenen, A. S. Daniel, H. D. Braymer, and N. E. Murray, *J. Mol. Biol.* **198,** 159 (1987).

Appendix (*continued*)

Enzyme[a]	Specificity[b]	Class[c]	Refs.[d]	GenEMBL number[e]
*Eco*RI	GA**A**TTC	m^6A	90, 91	J01675
*Eco*RII	C**C**WGG	m^5C	92	X05050
*Eco*RV	G**A**TATC	m^6A_α	93	M19941
*Eco*R124	GAAN_6RTCG	m^6A_γ	58	X13145
*Eco*57I	C**T**GA**A**G	m^6A_γ	94	M74821
*Eco*72I	CA**C**GTG	m^5C	66	
EC_{67} Dam	G**A**TC	m^6A_α	95	M55249
*Fnu*DI	GG**C**C	m^5C	96	
*Fok*I	GG**AT**G	m^6A_α	19, 50	J04623 M28828
ϕ3TI	G**C**NGC GG**C**C	m^5C	97	M13488
ϕ3TII	?T**C**GA?	m^5C	98	
*Hae*III	GG**C**C	m^5C	99	
*Hga*I-1	G**C**GTC	m^5C	100, 101	D90363
*Hga*I-2	GA**C**GC	m^5C	100, 101	D90363
*Hgi*BI	GGWCC	m^5C	102	X55137
*Hgi*CI	GGYRCC	m^5C	103	
HgiCII	GGWCC	m^5C	103	
HgiDI	GRCGYC	m^5C	42	X55140
*Hgi*DII	GTCGAC	m^5C	104	X55141
*Hgi*EI	GGWCC	m^5C	103	
*Hha*I	G**C**GC	m^5C	105	J02677

(*continued*)

[90] P. J. Greene, M. Gupta, H. W. Boyer, W. E. Brown, and J. M. Rosenberg, *J. Biol. Chem.* **256,** 2143 (1981).
[91] A. K. Newman, R. A. Rubin, S.-H. Kim, and P. Modrich, *J. Biol. Chem.* **256,** 2131 (1981).
[92] S. Som, A. S. Bhagwat, and S. Friedman, *Nucleic Acids Res.* **15,** 313 (1987).
[93] L. Bougueleret, M. Schwarzstein, A. Tsugita, and M. Zabeau, *Nucleic Acids Res.* **12,** 3659 (1984).
[94] R. Vaisvila and A. Janulaitis, personal communication (1991).
[95] M. Y. Hsu, M. Inouye, and S. Inouye, *Proc. Natl. Acad. Sci. U.S.A.* **87,** 9454 (1990).
[96] B.-H. Zhang, E. M. Van Cott, J. S. Benner, and G. G. Wilson, unpublished (1990).
[97] A. Tran-Betcke, B. Behrens, M. Noyer-Weidner, and T. A. Trautner, *Gene* **42,** 89 (1986).
[98] P. A. Terschüren and T. A. Trautner, Second NEB workshop on biological DNA modification, Berlin, September 2–7, 1990, p. 96.
[99] B. E. Slatko, R. Croft, L. S. Moran, and G. G. Wilson, *Gene* **74,** 45 (1988).
[100] J. M. Barsomian, D. Landry, L. S. Moran, B. E. Slatko, G. R. Feehery, and G. G. Wilson, unpublished (1989).
[101] H. Sugisaki, K. Yamamoto, and M. Takanami, *J. Biol. Chem.* **266,** 13952 (1991).
[102] A. Düsterhöft, D. Erdmann, and M. Kröger, *Nucleic Acids Res.* **19,** 3207 (1991).
[103] D. Erdmann, Ph.D. Thesis, Justus-Liebig-Universität, Giessen (1991).
[104] A. Düsterhöft and M. Kröger, *Gene* **106,** 87 (1991).
[105] M. Caserta, W. Zacharias, D. Nwankwo, G. G. Wilson, and R. D. Wells, *J. Biol. Chem.* **262,** 4770 (1987).

Appendix (*continued*)

Enzyme[a]	Specificity[b]	Class[c]	Refs.[d]	GenEMBL number[e]
*Hha*II	G**A**NTC	m^6A_β	106	K00508
*Hinc*II	GTYR**A**C	m^6A_γ	107	X52124
			108	X54197
*Hind*III	**A**AGCTT	m^6A_β	109	
*Hinf*I	G**A**NTC	m^6A_β	110	
HpaI	GTTA**A**C	m^5C	111	
*Hpa*II	C**C**GG	m^5C	112	X51322
H2I	GDGCHC	m^5C	114	
	G**C**NGC			
	GG**C**C			
H2II	GGAT**C**C	m^4C_β	113	X53032
*Msp*I	**C**CGG	m^5C	115	X14191
*Lla*I	?	m^6A_α	51	
MboII	GAAG**A**	m^6A_β	116	X56977
*Mun*I	CAATTG	m^6A_β	117	
*Mva*I	C**C**WGG	m^4C_α	41	X16985
*Mwo*I	GCN$_7$GC	m^4C_β	118	
*Nco*I	CCATGG	m^4C_α	119	
*Nde*I	CAT**A**TG	m^6A_α	120	
*Ngo*MI	G**C**CGGC	m^5C	121	

[106] B. Schoner, S. Kelley, and H. O. Smith, *Gene* **24,** 227 (1983).

[107] H. Ito, A. Sadaoka, H. Kotani, N. Hiraoka, and T. Nakamura, *Nucleic Acids Res.* **18,** 3903 (1990).

[108] P. A. Waite-Rees, D. O. Nwankwo, L. S. Moran, B. E. Slatko, G. G. Wilson, and J. S. Benner, unpublished (1989).

[109] D. O. Nwankwo, L. S. Moran, B. E. Slatko, and G. G. Wilson, unpublished (1990).

[110] S. Chandrasegaran, K. D. Lunnen, H. O. Smith, and G. G. Wilson, *Gene* **70,** 387 (1988).

[111] P. A. Waite-Rees, L. S. Moran, B. E. Slatko, L. Hornstra-Coe, and J. S. Benner, *Gene*, in press (1992).

[112] C. O. Card, G. G. Wilson, K. Weule, J. Hasapes, A. Kiss, and R. J. Roberts, *Nucleic Acids Res.* **18,** 1377 (1990).

[113] J. F. Connaughton, W. D. Kaloss, P. G. Vanek, G. A. Nardone, and J. G. Chirikjian, *Nucleic Acids Res.* **18,** 4002 (1990).

[114] C. Lange, M. Noyer-Weidner, T. A. Trautner, M. Weiner, and S. A. Zahler, *Gene* **100,** 213 (1991).

[115] P. M. Linn, C. H. Lee, and R. J. Roberts, *Nucleic Acids Res.* **17,** 3001 (1989).

[116] H. Bocklage, K. Heeger, and B. Müller-Hill, *Nucleic Acids Res.* **19,** 1007 (1991).

[117] N. Zareckaja, R. Vaisvila, and A. Janulaitis, personal communication (1991).

[118] K. D. Lunnen, L. S. Moran, B. E. Slatko, and G. G. Wilson, unpublished (1991).

[119] B. Zhang, E. M. Van Cott, J. S. Benner, L. S. Moran, B. E. Slatko, and G. G. Wilson, unpublished (1991).

[120] P. A. Waite-Rees, C. Polisson, K. R. Silber, L. S. Moran, B. E. Slatko, and J. S. Benner, *Nucleic Acids Res.* (in press).

[121] R. Chien and D. Stein, personal communication (1991).

Appendix (*continued*)

Enzyme[a]	Specificity[b]	Class[c]	Refs.[d]	GenEMBL number[e]
*Ngo*MIV	GGN$_2$**C**C	m^{5}C	122	
*Ngo*PII	GG**C**C	m^{5}C	123	X06965
*Nla*III	C**A**TG	m^{6}A$_\alpha$	52, 124	X54485
*Nla*X	?	m^{5}C	52	X54485
*Pae*R7I	CTCG**A**G	m^{6}A$_\gamma$	125	X03274
*Pgi*I	G**A**TC	m^{6}A$_\alpha$	126	M63469
*Pst*I	CTGC**A**G	m^{6}A$_\gamma$	127	K02081
*Pvu*II	CAG**C**TG	m^{4}C$_\beta$	55	X13778
P1 Dam	G**A**TC	m^{6}A$_\alpha$	128	
*Rsr*I	GA**A**TTC	m^{6}A$_\beta$	129, 130	X16456
$\rho 11_s$	GDG**C**HC	m^{5}C	131	X05242
	GG**C**C			
*Sal*I	GTCG**A**C	m^{6}A$_\gamma$	132	
*Sau*3AI	GAT**C**	m^{5}C	133	M32470
*Sau*96I	GGN**C**C	m^{5}C	134, 135	X53096
*Sfi*I	GGCCN$_5$GGCC	m^{4}C$_\beta$	136	
*Sin*I	GGW**C**C	m^{5}C	137	J03391
*Sma*I	C**C**CGGG	m^{4}C$_\alpha$	138, 139	X16458
SPβ	G**C**NGC	m^{5}C	97	M19513
	GG**C**C			M19514

(continued)

[122] D. Stein, personal communication (1991).
[123] K. M. Sullivan and J. R. Saunders, *Nucleic Acids Res.* **16,** 4369 (1988).
[124] R. D. Morgan, R. R. Camp, and G. G. Wilson, unpublished (1989).
[125] G. Theriault, P. H. Roy, K. A. Howard, J. S. Benner, J. E. Brooks, A. F. Waters, and T. R. Gingeras, *Nucleic Acids Res.* **13,** 8441 (1985).
[126] J. A. Banas, J. J. Ferretti, and A. Progulske-Fox, *Nucleic Acids Res.* **19,** 4189 (1991).
[127] R. Y. Walder, J. A. Walder, and J. E. Donelson, *J. Biol. Chem.* **259,** 8015 (1984).
[128] J. M. Coulby and N. L. Sternberg, *Gene* **74,** 191 (1988).
[129] W. Kaszubska, C. Aiken, C. D. O'Connor, and R. I. Gumport, *Nucleic Acids Res.* **17,** 10403 (1989).
[130] F. H. Stephenson and P. J. Greene, *Nucleic Acids Res.* **17,** 10503 (1989).
[131] B. Behrens, M. Noyer-Weidner, B. Pawlek, R. Lauster, T. S. Balganesh, and T. A. Trautner, *EMBO J.* **6,** 1137 (1987).
[132] M. R. Rodicio, B. E. Slatko, L. S. Moran, and G. G. Wilson, unpublished (1990).
[133] S. Seeber, C. Kessler, and F. Götz, *Gene* **94,** 37 (1990).
[134] L. Sziláк, P. Venetianer, and A. Kiss, *Nucleic Acids Res.* **18,** 4659 (1990).
[135] D. O. Nwankwo, L. S. Moran, B. E. Slatko, and G. G. Wilson, Second NEB workshop on biological DNA modification, Berlin, September 2–7, 1990, p. 76.
[136] E. M. Van Cott, L. S. Moran, B. E. Slatko, and G. G. Wilson, unpublished (1990).
[137] C. Karreman and A. de Waard, *J. Bacteriol.* **170,** 2527 (1988).
[138] S. Heidmann, W. Seifert, C. Kessler, and H. Domdey, *Nucleic Acids Res.* **17,** 9783 (1989).
[139] J. C. Dunbar, B. E. Withers, L. A. Fox, L. R. Carlock, K. D. Lunnen, and G. G. Wilson, in preparation.

Appendix (*continued*)

Enzyme[a]	Specificity[b]	Class[c]	Refs.[d]	GenEMBL number[e]
*Sph*I	GC**A**TGC	m^6A_α	140	
SPR	C**C**WGG **C**CGG GG**C**C	m^5C	141, 142	K02124
*Sss*I	**C**G	m^5C	143	X17195
*Sty*SB	GA**G**N$_6$**RT**A**Y**G	m^6A_γ	87	
*Sty*SP	AA**C**N$_6$**GT**RC	m^6A_γ	87	
*Taq*I	TCG**A**	m^6A_γ	144, 145	
*Tth*HB8I	TCG**A**	m^6A_γ	146	
T1 Dam	G**A**TC	m^6A	147	J05393
T2 Dam	G**A**TC	m^6A_α	148	M22342
T4 Dam	G**A**TC	m^6A_α	24, 149	K03113
*Xba*I	TCT**A**GA	m^6A_β	96	
*Xma*I	C**C**CGGG	m^4C_β	139	

[a] Sequenced DNA methyltransferases are listed in alphabetical order in the first column. The nomenclature is based on the taxonomic names of the organisms in which the enzymes occur.[150]

[b] Recognition sequences of the MTases[28] are listed; only one strand of the duplex is shown, printed in the 5′ to 3′ orientation. The base that becomes methylated, if it is known or can be inferred,[11] is printed in boldface type (e.g., **A** or **C**). For symmetric sequences, the same base is methylated on the complementary strand. For asymmetric sequences, the base that is methylated on the complementary strand, if any, is indicated by printing its partner in bold type (**T** or **G**).

[c] The likely methylation products, and the architectural classes to which the enzymes belong, are listed in the third column. The designations α, β, and γ are personal preferences; a nomenclature has not been agreed on.

[d] References reporting the sequences of the enzymes are given.

[e] If the sequence is available from the GenBank/EMBL database, its accession number is given.

[140] J. E. Brooks, L. S. Moran, B. E. Slatko, and G. G. Wilson, unpublished (1990).

[141] H.-J. Buhk, B. Behrens, R. Tailor, K. Wilke, J. J. Prada, U. Günthert, M. Noyer-Weidner, S. Jentsch, and T. A. Trautner, *Gene* **29,** 51 (1984).

[142] G. Pósfai, F. Baldauf, S. Erdei, J. Pósfai, P. Venetianer, and A. Kiss, *Nucleic Acids Res.* **12,** 9039 (1984).

[143] P. Renbaum, D. Abrahamove, A. Fainsod, G. G. Wilson, S. Rottem, and A. Razin, *Nucleic Acids Res.* **18,** 1145 (1990).

[144] B. E. Slatko, J. S. Benner, T. Jager-Quinton, L. S. Moran, T. G. Simcox, E. M. Van Cott, and G. G. Wilson, *Nucleic Acids Res.* **15,** 9781 (1987).

[145] F. Barany, B. Slatko, M. Danzitz, D. Cowburn, I. Schildkraut, and G. G. Wilson, *Gene* **112,** 91 (1992).

[146] F. Barany, M. Danzitz, J. Zebala, and A. Mayer, *Gene* **112,** 3 (1992).

[147] E. Schneider-Scherzer, B. Auer, E. J. de Groot, and M. Schweiger, *J. Biol. Chem.* **265,** 6086 (1990).

[148] Z. Miner and S. Hattman, *J. Bacteriol.* **170,** 5177 (1988).

[149] P. M. Macdonald and G. Mosig, *EMBO J.* **3,** 2863 (1984).

[150] H. O. Smith and D. Nathans, *J. Mol. Biol.* **81,** 419 (1973).

Acknowledgments

We thank colleagues and co-workers for helpful discussions and for generously permitting access to unpublished sequences. We give particular thanks to Drs. Donald Comb, Richard Roberts, and Ira Schildkraut for encouragement.

[25] Use of DNA Methyltransferase/Endonuclease Enzyme Combinations for Megabase Mapping of Chromosomes

By MICHAEL NELSON and MICHAEL MCCLELLAND

Introduction

Based on a knowledge of the methylation sensitivities of restriction–modification enzymes, and the specificities of bacterial DNA methyltransferases, it is possible to mix and match site-specific DNA methylation and restriction endonuclease cleavage reactions to produce rare or novel DNA cleavages (Dobritsa and Dobritsa, 1979; Nelson *et al.*, 1984; McClelland *et al.*, 1985). Detailed procedures for the purification, assay, and use of methylase/endonuclease combinations in sequential two-step reactions were described in a previous volume of this series (McClelland, this series; Nelson and McClelland, this series; Nelson and Schildkraut, this series). Three-step procedures have also been described, based on methylase/endonuclease combinations *and* sequence-specific masking by bacterial repressor proteins (Koob *et al.*, 1988), polypyrimidine triplexes (Maher *et al.*, 1989; Hanvey *et al.*, 1989), or other DNA methyltransferases (McClelland and Nelson, 1987; McClelland and Nelson, 1988a; Posfai and Szybalski, 1988). This chapter describes four selected DNA methylase/endonuclease combinations that may be used for megabase mapping of chromosome fragments in the size range from 100 to 2000 kilobases (kb), paying special attention to features of these reactions that have sometimes proven problematic.

As a practical matter, it is important to recognize that *sequential multistep* DNA methylation and cleavage reactions can be used in conjunction with pulsed-field gel electrophoresis (PFGE) (Schwartz and Cantor, 1984; Gardiner *et al.*, 1986; Cantor *et al.*, 1988) only when adequate precautions are taken to ensure enzyme purity, substrate DNA integrity, and reaction yield. Accordingly, it has proven necessary to (1) critically control the purity of methyltransferase and endonuclease enzymes, (2) minimize the number of steps in multistep enzymatic treatment protocols, (3) use as substrates high molecular weight (bacterial chromosome) controls in parallel with eukaryotic DNA during methylation and cleavage reactions, and (4) define reaction conditions that are compatible with pulsed field electrophoretic separation. DNA methylation and subsequent cleavage reactions are normally carried out *in situ* on unsheared chromosomes embedded in agarose plugs (McClelland, this series; Smith and Cantor, this series).

Molecular Basis for Sensitivity of Restriction Enzymes to Methylation

Three chemically distinct classes of bacterial DNA modification methyl transfer reactions have been identified: unsaturated 5,6-double bond methylation at ^{5m}C and exocyclic aminomethylations at either ^{4m}C or ^{6m}A (Klimasauskas *et al.*, 1990). When some bases within a restriction endonuclease DNA recognition sequence are methylated, the nuclease does not cleave, or inefficiently cleaves its target.

At the molecular level, this inability of restriction endonucleases to cut methylated DNA can be explained using *Eco*RI and *Eco*RV endonucleases as instructive models. Adenine aminomethylation at *canonical* GA^{6m}ATTC modification sites, or at G^{6m}AATTC, prevents *Eco*RI from cutting (Brennan *et al.*, 1987). Based on the *Eco*RI endonuclease–DNA cocrystal X-ray structure (McClarin *et al.*, 1986; Rosenberg, 1991), methylation of either adenine perturbs essential hydrogen bond contacts to Glu-144 and Arg-145, resulting in reduced binding affinity to methylated substrate DNA. In contrast, cytosine ring methylation at GAATT^{5m}C would not be expected to interfere with critical DNA : protein contacts inferred from X-ray diffraction data (McClarin *et al.*, 1986). Therefore, the reduced rate of *Eco*RI cleavage on cytosine-modified substrates (Brennan *et al.*, 1987) can be attributed to steric distortions of the enzyme–substrate complex during catalysis (Heitman and Model, 1990).

Similarly, based on the X-ray structure of *Eco*RV endonuclease (Winkler, *et al.*, 1991), hydrogen bonding of Asn-185 to the first adenine of GATATC is perturbed by G^{6m}ATATC methylation. However, the failure of *Eco*RV to cleave canonically modified sites results from nonproductive enzyme–substrate complexes, rather than reduced substrate-binding affinity (Halford, 1991; Taylor *et al.*, 1991; Newman *et al.*, 1990). At *noncanonically* modified *Eco*RV sites such as GATAT^{5m}C or GAT^{6m}ATC, interpretation of X-ray crystal structural data is less clear. Although the major groove cytosine N-4 group of GATATC contacts a glycine in the *Eco*RV polypeptide, cleavage at GATAT^{5m}C is not grossly impaired. In contrast, methylation at GAT^{6m}ATC apparently adversely affects conformational changes of the *Eco*RV endonuclease–DNA reaction intermediate (Mazzerelli *et al.*, 1989), because X-ray structural data imply no protein contacts to the second adenine of the *Eco*RV recognition sequence (Halford, 1991).

More generally, all three common types of bacterial DNA methylation, ^{5m}C, ^{4m}C, and ^{6m}A, are bulky modifications of the DNA major groove, and can be expected to sterically hinder the binding and/or catalytic reactions of many DNA-binding proteins (Riggs, 1975; Adams, 1990), including restriction–modification enzymes.

Over 1000 different restriction–modification enzymes have been identified (Roberts, 1990, 1991); and 240 restriction endonucleases have been characterized with respect to their sensitivities to site-specific ^{5m}C, ^{4m}C, and ^{6m}A modifications (Nelson and McClelland, 1991; McClelland and Nelson, 1988b; Kessler and Manta, 1990). In general, when site-specific DNA methylation results in a reduced rate of cleavage by a restriction endonuclease, the level of this inhibition is two or more orders of magnitude (Nelson and McClelland, 1991). For use in DNA blocking reactions, a list of over 130 characterized DNA methyltransferases and their modification specificities has been compiled; therefore a large number of possible "designer" methylase/endonuclease combinations are possible (Nelson and McClelland, 1991).

DNA Methyltransferases Used as Generalized "Blocking Reagents"

Theoretical Considerations: DNA Fragment Size, Reaction Yield, Comparison to Other Methods

Overlapping methylase/endonuclease sites are defined by methylase recognition sequences of 2 to 6 bp, flanking one or both sides of a 6- to 8-bp restriction target. Such methylase/endonuclease sites are therefore 6.5 to 16 bp long, occurring at random once every $4^{6.5}$ to 4^{16} bp. Within this size range, fragments in the 4^7 to 4^{11} (~100 to 2000 kb) range are most useful for large-scale electrophoretic mapping, because smaller fragments (<100 kb) may be obtained using conventional restriction digests, whereas larger fragments (>2000 kb) are not separable by PFGE (Schwartz and Cantor, 1984; Gardiner *et al.*, 1986; Cantor *et al.*, 1988). Depending on the nonrandom G + C content, dinucleotide frequencies, and trinucleotide frequencies of the DNA to be analyzed (McClelland and Nelson, 1987; Arnold *et al.*, 1988), methylase/endonuclease combinations may be chosen that produce PFGE-separable fragments in the 100- to 2000-kb range (McClelland *et al.*, 1986). While it is relatively easy to carry out three-step liquid reactions in solution on 40- to 50-kb viral DNA substrates (bacteriophage λ or adenovirus-2), it has proven considerably more difficult to carry out sequential multistep methylase/endonuclease cleavages reliably when chromosomes are embedded in agarose plugs (McClelland and Nelson, 1988a; Koob and Szybalski, 1990).

When adequate precautions are taken, high-yield (>99%) cleavages of whole *bacterial* chromosomes embedded in agarose plugs are possible using methylase/endonuclease combinations, resulting in as few as 2 to 14 fragments (Weil and McClelland, 1989; Qiang *et al.*, 1990; Hanish and McClelland, 1991). Applications to the megabase mapping of eukaryotic chromosomes by pulsed-field electrophoresis are technically feasible (Waterbury *et al.*, 1990; Wilson and Hoffman, 1991; Hanish *et al.*, 1991; Shukla *et al.*, 1991) but are not always easy to carry out.

Although elegant in principle, DNA cleavage methods based on polypyrimidine triplexes (Moser and Dervan, 1987; Strobel *et al.*, 1988; LeDoan *et al.*, 1987; Perroualt *et al.*, 1990) or semisynthetic oligonucleotide/enzyme hybrids (Chen and Sigman, 1987; Ebright *et al.*, 1990; Pei *et al.*, 1990) have practical limitations in terms of reaction yield (Moser and Dervan, 1987; Ebright *et al.*, 1990; Sigman and Chen, 1990) and specificity (Chen and Sigman, 1987; Strobel and Dervan, 1990; Francois *et al.*, 1989) when applied to DNA fragments in the megabase size range. "Rare-cutting" strategies that rely on oligonucleotide base pairing are theoretically limited by (1) sequence mismatching (Moser and Dervan, 1987; Griffin and Dervan, 1989; Roberts and Crothers, 1991), (2) the thermal instability of short DNA duplexes (Bolton and McCarthy, 1962; Craig *et al.*, 1990) or triplexes (Howard *et al.*, 1966; Plum *et al.*, 1991) shorter than 13 to 15 bp, and (3) the inseparability of DNA fragments greater than 3–5 Mb in size using pulsed-field gel electrophoresis. Bacterial repressor proteins define sequences of more than 12 bp (Brennan and Mathews, 1989), and while it is possible to engineer rare methylase/repressor sites into transposons, retroviruses, or yeast artificial chromosomes (Koob and Szybalski, 1990; Strobel and Dervan, 1990) it is also possible to engineer methylase/endonuclease sites into retroposons that can be more efficiently cleaved using one-step enzymatic protocols (Hanish and McClelland, 1991; Wong and McClelland, 1992). As F. Sanger (1988) stated, "Throughout the work on proteins and RNA it had usually been found that enzymatic methods were more successful than chemical ones because of their greater specificity. I was thus rather prejudiced towards the use of enzymes. . . ."

In summary, as compared to methods based on polypyrimidine triplexes (Moser and Dervan, 1987; Strobel *et al.*, 1988) or bacterial repressor proteins (Koob *et al.*, 1988; Ebright *et al.*, 1990), sequential enzymatic treatments *efficiently* produce chromosome fragments that are directly separable by PFGE. However, when using sequential methylase/endonuclease DNA cleavage schemes, enzyme-catalyzed reaction parameters must be carefully controlled at each stage of a multistep process.

Technical Considerations in Methyltransferase Reactions: S-Adenosylmethionine, S-Adenosylhomocysteine Diffusion Limitations, Temperature, Enzyme Source, Buffer Conditions, Enzyme Concentration, and Streamlining Multistep Protocols

In general, DNA methyltransferases catalyze the sequence-specific transfer of the methyl group of *S*-Adenosylmethionine (SAM) to adenine or cytosine residues, producing methylated DNA and *S*-Adenosylhomocysteine (SAH) reaction products.

$$\text{SAM} + \text{DNA} \xrightarrow{\text{MTase}} \text{methyl-DNA} + \text{SAH}$$

This reaction has been studied by detailed kinetic analysis in the cases of M · *Eco*RI adenine methyltransferase (Pogolotti *et al.*, 1988; Reich and Mashoon, 1991) and M · *Hha*I cytosine methyltransferase (Wu and Santi, 1987). In optimizing DNA methylation reactions, it is important to recognize that SAH is both a product inhibitor of enzyme-catalyzed SAM-dependent DNA methylation ($K_i \sim 10\ \mu M$), and is produced nonenzymatically by hydrolysis of SAM. In the mild alkaline (pH 7.0–9.0) conditions required for DNA methyltransferase reactions, relatively short reaction times (2–4 hr) must be used, because inhibitory SAH levels accumulate over longer reaction times.

$$\text{SAM} + H_2O \xrightarrow{\text{MTase}} CH_3OH + \text{SAH}$$

As a practical matter, DNA methylation reactions *in situ* should be carried out using SAM concentrations about 10- to 20-fold over the K_m(SAM), about 100–200 μM; and precautions should be taken to minimize nonenzymatic SAH buildup. Three specific recommendations are useful in this regard: (1) SAM should be added to reactions of agarose-embedded DNA from freshly thawed acidic concentrated stocks in 10 m*M* H_2SO_4–10% ethanol. (2) Incubation temperatures between 23 and 30° stabilize SAM in aqueous alkaline reactions *in situ*, although these temperatures are suboptimal for [^{3}H]SAM incorporation in DNA methylase reactions *in vitro*. (3) To avoid buildup of inhibitory SAH levels, agarose-embedded DNA samples can be briefly washed after 1- to 2-hr reactions, using fresh reaction buffer. *S*-Adenosylhomocysteine readily washes out of agarose gels (<15 min), whereas higher molecular weight methylases (30,000–70,000) diffuse slowly into and out of agarose gels.

Although DNA methyltransferases have been purified from a number of sources, the usual source of these enzymes is restriction endonuclease-producing bacteria (Nelson and McClelland, this series, 1991). For use in megabase mapping experiments, two points are cogent: (1) A large number of restriction–modification genes have been cloned (Wilson, 1990, 1991), and these recombinant clones are excellent sources of starting material for methylase purifications (Qiang *et al.,* 1990). (2) On the other hand, whereas not all restriction enzyme-producing strains produce detectable quantities of DNA methylases, other strains produce large quantities of methylases, but little restriction enzyme activity (Nelson and McClelland, this series).

DNA methylases from bacteria are usually inhibited by sodium or potassium chloride concentrations higher than 50 m*M* (Hanish and McClelland, 1988), and in some cases Tris inhibits enzyme activity (Hulsmann *et al.,* 1990). Accordingly, chloride-free buffers are generally preferred in sequential methylase/endonuclease protocols. DNA methyltransferases from bacteria, bacteriophages, blue-green algae, and eukaryotic algal viruses are stable and active in either of two chloride-free buffer systems: (1) "0.5× KGB" [10 m*M* Tris–acetate, 50 m*M* potassium glutamate, 0.5 m*M* 2-mercaptoethanol, 50 μg/ml bovine serum albumin (BSA), pH 7.5], supplemented with 1–2 m*M* ethylenediaminetetraacetic acid (EDTA) and 100 μM SAM (Hanish and McClelland, 1988), or (2) an "all-natural buffer" (50 m*M* glycylglycine, 5 m*M* potassium citrate, 5 m*M* glutathione, 50 μg/ml BSA, 100 μM SAM, pH 8.2) (Nelson, 1991).

Reactions in agarose plugs are diffusion limited. Therefore to achieve complete methylation of DNA embedded at the core of 1-mm-thick 0.8–1.0% (w/v) agarose plugs, it has proven necessary to use 10–30 units of DNA methyltransferase per microgram of DNA (Qiang *et al.,* 1990; Weil and McClelland, 1989; Wilson and Hoffman, 1991). Because 1 unit of DNA methyltransferase is defined as the amount of enzyme required to protect 1 μg of λ phage DNA from cleavage by cognate restriction enzyme for 1 hr in 50 μl *liquid* reaction mixture (Nelson and McClelland, this series), enzyme purity becomes a serious consideration at high enzyme-to-substrate ratios. For use in megabase mapping experiments, DNA methylases must not degrade chromosomal DNA nonspecifically more than once per million base pairs. In order to use a 30-fold excess of enzyme without nonspecific DNA cleavage, purified enzymes must not have endonuclease contaminants above 1 part per 30 million bp. While this level of purity is technically feasible, commercially prepared DNA methylases are usually not pure enough for megabase mapping experiments, and PFGE distortions are superimposed on any DNA degradation due to enzyme impurities.

Two promising methods have been developed that minimize some diffusion-limited yield and purity problems of enzymatic DNA manipulations *in situ*. These methods are based on either agarose bead emulsions (McClelland, this series; Kauc *et al.,* 1989; Koob and Szybalski, 1990) or thin (0.1 m*M*) DNA-agarose films supported on polyvinyl diflouride (PVDF) membranes (Millipore Corp., Bedford, MA) (Nelson, 1991). Less enzyme is required in emulsions or membrane-supported films, due to their increased surface area-to-volume ratios relative to agarose plugs. However, these new procedures have not been adapted for the multistep procedures described below.

Finally, many of the above complications can be avoided if DNA modifications are carried out *in vivo*. For example, when *E. coli* or *Salmonella* strains carry appropriate DNA modification genes on multicopy plasmids, it is possible to achieve complete methylation of plasmid and chromosomal DNA. Technically demanding two-step methylase/endonuclease protocols can thus be streamlined to simple one-step restriction digests (Qiang *et al.*, 1990; Hanish and McClelland, 1991; Wong and McClelland, 1991).

Control Substrates for Debugging Multistep Methylase/Endonuclease Reactions

Whereas restriction mapping of phage and plasmid-sized molecules is often carried out in the absence of control DNA digestion, it has proven necessary to be more rigorous when designing adequate controls for methylase/endonuclease cleavages in megabase mapping experiments using pulsed field gels. This problem is partly heuristic: multistep procedures are inherently less reliable than one-step procedures. As a practical matter, it has proven necessary to prepare small (3–5 Mb) bacterial chromosomes of selected G + C contents, embedded in agarose, to monitor DNA methylation independently and cleavage reactions *in situ*. Depending on the G + C content and nonrandom sequence character of the DNA to be analyzed, different bacterial chromosomal DNAs may be selected as suitable high molecular weight control substrates for pulsed-field gel mapping experiments (McClelland *et al.*, this series). These bacterial chromosomes can be cleaved into a small number of DNA fragments and serve as positive controls to monitor enzyme purity, reaction yield, and the quality of PFG separations. Agarose-embedded bacterial chromosomes are commercially available (New England BioLabs, Beverly, MA; Promega Corp., Madison, WI) or can be prepared as agarose plug "DNA inserts" (McClelland *et al.*, 1987; Smith and Cantor, this series; Hung and Bandziulis, 1990).

Technical Considerations for Endonuclease Reactions in Agarose Plugs: Enzyme Concentration, Time, Temperature, and G + C Content

As with methyltransferase reactions, endonuclease digestion is diffusion limited when DNA is embedded in agarose plugs. In addition, restriction endonucleases require divalent magnesium ion, and often require different salt, pH, and temperature conditions than are optimal for methyltransferase activity. Furthermore, many endonucleases are inhibited by agarose or by-products of its manufacture, and not all restriction enzymes are stable over the course of long *in situ* reactions. Two technical improvements have simplified these problems: (1) A single reaction buffer based on potassium glutamate can be used for all methylases and restriction endonucleases used in megabase mapping experiments (McClelland *et al.*, 1987; Hanish and McClelland, 1988). (2) Many commercially available endonucleases used for PFGE mapping experiments are titered and tested for functional purity by digesting bacterial chromosomes embedded in agarose plugs (Hung and Bandziulis, 1990).

As a guideline, Table I gives the number of units of restriction enzyme required to give complete digestion, in the absence of detectable nonspecific degradation, of ~1–2 μg bacterial chromosomal DNA embedded in agarose plugs.

Four General Strategies Using Methylase/Endonuclease Combinations for Megabase Mapping of Chromosomes

Due to the large number of different restriction and modification enzymes that may be used for rare-cutting and megabase mapping, it is beyond the scope of this chapter to list all possible combinations. In general, however, we may distinguish four types of strategies based on methylase/endonuclease combinations. These megabase mapping strategies have the general formulas: (1) methylase 1/endonuclease 1 "competition reactions," (2) methylase 1/endonuclease 2 "cross-protection reactions," (3) adenine methylase/*Dpn*I endonuclease cleavages, and (4) methylase 1/methylase 2/endonuclease 2 "double blocking reactions."

Partial Digest Mapping Using Simultaneous Methylase/Endonuclease Competition (Methylase 1/Endonuclease 1)

Example: M·BspRI/NotI Competition. Due to the diffusion-limited nature of enzymatic reactions *in situ,* it is frequently difficult to achieve partial digests of high molecular weight DNA embedded in agarose plugs. Digestion is often complete at the outer edges of agarose plugs, but DNA remains uncut at the core. However, because DNA methylases and endonucleases are moderately sized proteins (Wilson, 1991), both enzymes are diffusion limited in modifying or cutting DNA, respectively. Accordingly, Hanish and McClelland (1990) used simultaneous M·*Bsp*RI methylase/

*Not*I endonuclease *competing* reactions as a means of producing uniform partial digests:

$$\begin{array}{ccc} \text{—GC}\underline{\text{GGCC}}\text{GC— + SAM} & \xrightarrow{\text{GG}^{5m}\text{CC MTase}} & \text{—GCGG}^{5m}\text{CCGC—} \\ \text{Mg}^{2+} \downarrow \textit{Not}\text{I} & & \text{Mg}^{2+} \not\downarrow \textit{Not}\text{I} \\ \text{—GC—OH + pGGCCGC—} & & \end{array}$$

Whereas 15–30 units of M · *Bsp*RI methylase is required for complete protection of 1 μg of embedded *Salmonella* chromosomal DNA against excess (8 units) endonuclease cleavage, 2 to 5 units of methylase per microgram is required for partial digestion.

High molecular weight DNA (~2 μg) embedded in agarose plugs (~1 × 1 × 3 mm; low-melt LE agarose, FMC Bioproducts, Rockland, ME), prepared according to Weil and McClelland (1989) is equilibrated for 30 min at room temperature in 200 to 300 μl modified 1.5× KGB buffer (15 m*M* Tris–acetate, 150 m*M* potassium glutamate, 15 m*M* magnesium acetate, 75 μg/ml bovine serum albumin, pH 7.5) (McClelland *et al.*, 1987; Hanish and McClelland, 1988) in 1.5-ml plastic microfuge tubes. After equilibration, samples are placed on ice, buffer is removed with an Eppendorf pipette, and 100 μl of fresh 1.5× KGB reaction buffer, supplemented with 160 μ*M* SAM, is added. *S*-Adenosylmethionine is freshly thawed from a frozen 32 m*M* SAM stock solution in 10 m*M* H_2SO_4–10% ethanol.

To different 100 μl reaction mixtures kept on ice, 2, 4, 6, or 8 units of M · *Bsp*RI DNA methyltransferase (Koncz *et al.*, 1978; Hanish and McClelland, 1990) is added. Prior to mixing, 8 units of *Not*I (Boehringer-Mannheim, Indianapolis, IN; Promega Corp., Madison, WI; or Stratagene, Inc., La Jolla, CA) is added, and reactions are incubated for 8 hr at 23°. Salt conditions, temperature, and SAM concentrations are chosen according to limitations of enzyme purity and SAM stability, as described above in the section Technical Considerations in Methyltransferase Reactions.

Agarose plugs are then placed in loading wells of a 0.8% (w/v) agarose gel, placed in 10 m*M* Tris–acetate–0.25 m*M* EDTA running buffer in a Gene-Line transverse alternating field electrophoresis unit (Beckman Instruments, Palo Alto, CA) as described by Gardiner *et al.* (1986). Pulsed-field gels are run at 10 V/cm using a pulse time of 20 sec for 12 hr.

Other Methylase/Endonuclease Competitions for Megabase Mapping. More generally, if a restriction endonuclease is sensitive to a given site-specific modification, then an appropriate methyltransferase may be chosen for use in methylase/endonuclease competition. A large number of methylase/endonuclease competition reactions are possible, based on

TABLE I
PARAMETERS FOR *in Situ* DIGESTION OF GENOMIC DNA BY RESTRICTION ENDONUCLEASES[a]

		Genome			Conditions for digestion			
Restriction endonuclease	Recognition sequence	Source	Size (Mb)	G + C (%)	Enzyme/DNA (U/μg)	Temperature (°C)	Time (hr)	Number of fragments
*Apa*I	GGGCCC	*S. mutans*	2.2	37	20/2	37	4	15–20
		H. influenzae	2.0	38	20/2	37	8	20
*Ase*I	ATTAAT	*R. sphaeroides*	4.4	65	20/2	37	8	11
		C. crescentus	4.0	60	20/2	37	8	13
*Avr*II	CCTAGG	*E. coli* K	4.7	50	1/2	37	8	13
*Bln*I	CCTAGG	*E. coli* K	4.7	50	10/2	37	8	13
		S. typhimurium	5.0	53	10/2	37	8	~15
*Bgl*I	$GCCN_5GGC$	*S. aureus*	3.0	34	30/2	37	3	20–25
*Bss*HII	GCGCGC	*C. jejuni*	1.7	35	50/2	50	6	~15
		S. aureus	3.0	34	14/2	50	4	>25
*Bst*BI	TTCGAA	*R. sphaeroides*	4.4	65	5/2	50	16	>25
*Cla*I	ATCGAT	*M. bovis*	2.9	45	12/1	37	4	>25
*Csp*I	CGGWCCG	*M. bovis*	2.9	45	8/1	30	5	8
*Dpn*I(M · *Cla*I)	ATCGATCGAT	*S. aureus*	3.0	34	10/2	37	6	2
		P. aeruginosa	5.9	67	10/2	37	6	15
*Eag*I	CGGCCG	*S. aureus*	3.0	34	10/2	37	5	16
		H. influenzae	2.0	38	10/2	37	4	20–25
*Ecl*XI	CGGCCG	*S. aureus*	3.0	34	10/2	37	4	~10
*Eco*47III	AGCGCT	*S. aureus*	3.0	34	8/2	37	4	>25
*Fsp*I	TGCGCA	*S. aureus*	3.0	34	20/2	37	16	>25
*Ksp*I	CCGCGG	*S. aureus*	3.0	34	10/2	37	4	~20
*Mlu*I	ACGCGT	*S. aureus*	3.0	34	10/2	37	3	25–30

*Nae*I	GCCGGC	*H. influenzae*	2.0	38	30/2	37	>16	~28
*Nhe*I	GCTAGC	*T. celer*	1.9	—	10/2	37	6	5
		M. bovis	2.9	45	5/1	37	3	20–25
*Not*I	GCGGCCGC	*S. aureus*	3.0	34	10/2	37	4	1
		M. bovis	2.9	45	5/1	37	3	7
		S. mutans	2.2	37	20/2	37	4	10
		B. cereus	5.7	36	20/2	37	4	11
		E. coli K	4.7	50	10/2	37	5	~22
*Nru*I	TCGCGA	*S. aureus*	3.0	34	10/2	37	16	25–30
*Pac*I	TTAATTAA	*E. coli* K	4.7	50	20/2	37	16	>25
*Rsr*II	CGGWCCG	*H. influenzae*	2.0	38	30/2	37	>16	4
		S. aureus	3.0	34	15/2	37	16	7
*Sac*II	CCGCGG	*S. aureus*	3.0	34	10/2	37	4	~20
		S. mutans	2.2	37	20/2	37	4	>30
*Sal*I	GTCGAC	*C. jejuni*	1.7	35	50/1	37	6	6
		M. bovis	2.9	45	16/1	37	4	~15
*Sgr*AI	CRCCGGYG	*C. trachomatis*	1.0	45	10/1	37	20	4
*Sfi*I	GGCCN$_5$GGCC	*S. aureus*	3.0	34	10/2	37	16	1
		E. coli K	4.7	50	10/2	50	16	~20
*Sma*I	CCCGGG	*S. aureus*	3.0	34	20/2	22	3	18
		H. influenzae	2.0	38	20/2	37	8	16–22
*Spe*I	ACTAGT	*T. celer*	1.9	—	10/2	37	6	5
		C. crescentus	4.0	60	20/2	37	8	17–23
		M. bovis	2.9	45	5/1	37	3	20–25
		P. aeruginosa	5.9	67	5/1	37	6	36
*Spl*I	CGTACG	*S. aureus*	3.0	34	20/2	37	3	25–30
*Spo*I	TCGCGA	*S. aureus*	3.0	34	10/2	37	4	25–30
*Srf*I	GCCCGGGC	*B. subtilis*	4.7	45	10/2	37	16	4
*Sse*8387I	CCTGCAGG	*M. bovis*	2.9	45	10/2	37	16	~10
		C. trachomatis	1.0	45	25/1	37	20	17
*Ssp*I	AATATT	*M. bovis*	2.9	45	2/1	37	4	~10
		R. sphaeroides	4.4	65	4/2	37	5	~20

(*continued*)

TABLE I (*continued*)

Restriction endonuclease	Recognition sequence	Genome: Source	Genome: Size (Mb)	Genome: G + C (%)	Conditions for digestion: Enzyme/DNA (U/μg)	Conditions for digestion: Temperature (°C)	Conditions for digestion: Time (hr)	Conditions for digestion: Number of fragments
*Xba*I	TCTAGA	*M. bovis*	2.9	45	10/1	37	3	20–25
*Xho*I	CTCGAG	*M. bovis*	2.9	45	16/1	37	4	~18

[a] Modified from the data of Hung and Banziulis (1990) and Davis (1990/1991), based on the original experiments of McClelland *et al.* (1987). Additions have been made for *Apa*I, *Rsr*II, and *Sma*I digestion of *Haemophilus influenzae* Rd DNA (Lee and Smith, 1988), *Ase*I digestion of *Rhodobacter sphaeroides* 2.4.1 DNA (Suwanto and Kaplan, 1989), *Ase*I and *Spe*I digestion of *Caulobacter crescentus* ATCC 19089 DNA (Ely *et al.*, 1990), *Avr*II digestion of *E. coli* K DNA (Daniels, 1991), *Bln*I digestion of *Salmonella typhimurium* LT2 DNA (Wong and McClelland, 1992), *Bss*HII and *Sal*I digestion of *Campylobacter jejuni* DNA (Nuitjen *et al.*, 1990), *Dpn*I digestion of *Staphylococcus aureus* 3A (Weil and McClelland, 1989), *Dpn*I and *Spe*I digestion of *Pseudomonas aeruginosa* PAO1 DNA (Romling *et al.*, 1989), *Eag*I, *Nae*I, and *Rsr*II digestion of *Haemophilus influenzae* b DNA (Butler and Moxon, 1990), *Ecl*XI and *Ksp*I digestion of *Staphylococcus aureus* 3A DNA (Poddar and McClelland, 1992), *Nhe*I and *Spe*I digestion of *Thermococcus celer* DNA (Noll, 1989), *Not*I digestion of *Bacillus cereus* ATCC 10987 DNA (Kolsto *et al.*, 1990), *Not*I, *Apa*I, and *Sac*II digestion of *Streptococcus mutans* MT8148 DNA (Okahashi *et al.*, 1990), *Sgr*AI and *Sse*8387I digestion of *Chlamydia trachomatis* DNA (Birkelund and Stephens, 1992), *Srf*I digestion of *Bacillus subtilis* 168 DNA (Simcox *et al.*, 1991), and *Sse*8387I digestion of *Moraxella bovis* DNA (Kotani *et al.*, 1991; Nelson, 1991).

the known specificities of methyltransferases and the sensitivity of restriction endonucleases to site-specific methylation (Nelson and McClelland, 1991). It is likely that many methylase/endonuclease competitions can be carried out in 0.5× KGB, supplemented with 160 μM SAM, as described by Hanish and McClelland for liquid reactions. However, investigators should prepare desired DNA methylases to adequate purity levels, as described above in the section Theoretical Considerations. For example, due to inadequate purity of commercial M · *Hae*III (GG^{5m}CC) methylase (New England BioLabs) and its poor activity in agarose plugs, we have been unable to use this methylase in place of M · *Bsp*RI (GG^{5m}CC). However, when adequate precautions are taken, highly purified M · *Fnu*DII can be used in methylase competition reactions with *Asc*I (GGCGCGCC), *Bss*HII (GCGCGC), *Mlu*I (ACGCGT), *Nru*I (TCGCGA), *Spo*I (TCG-CGA), or *Sac*II (CCGCGG).

Blocking Subset of Rare-Cutting Endonuclease Sites at Overlapping Methylase/Endonuclease Targets: Sequential Two-Step DNA "Cross-Protection" (Methylase 1/Endonuclease 2)

Example: M · FnuDII/NotI Cleavage of Escherichia coli Chromosome into Large Pieces. When methylase and endonuclease target sequences overlap, it is sometimes possible to limit DNA cleavage to those restriction targets that are not methylated by overlapping methylation (Nelson *et al.*, 1984; Nelson and Schildkraut, this series). This "cross-protection" procedure has been adapted to pulsed-field gel mapping of whole chromosomes, using M · *Fnu*DII or M · *Bep*I modification (^{5m}CGCG) followed by *Not*I cleavage. Because *Not*I is sensitive to methylation at GCGGC^{5m}CGC (McClelland and Nelson, 1988b; Nelson and McClelland, 1991), those *Not*I sites that are preceded by C (CGCGGCCGC) and/or followed by G (GCGGCCGCG) can be selectively protected from cleavage.

$$\text{GCGGCCGCG} + \text{SAM} \xrightarrow{^{5m}\text{CGCG MTase}} \text{GCGGC}^{5m}\text{CGCG} \underset{\text{Mg}^{2+}}{\overset{NotI}{\not\longrightarrow}}$$

$$\text{(A/G/T)GCGGCCGC(A/C/T)} \underset{\text{Mg}^{2+}}{\xrightarrow{NotI}} \text{(A/G/T)GC—OH} + \text{pGGCCGC(A/C/T)}$$

Sequential M · *Fnu*DII/*Not*I cleavage therefore defines DNA cleavage at (A/G/T)GGCGGCCGC(A/C/T), and limits cleavage to 14 of 22 *Not*I sites in the *E. coli* RR1 chromosome (Qiang *et al.*, 1990). These 14 large *Not*I fragments give identical PFG banding patterns, whether produced by *in vivo* modification from high-copy M · *Bep* or M · *Fnu*DII modifying plasmids, or by *in situ* methylation using highly purified M · *Fnu*DII methyltransferase. Because many investigators may wish to use cross-protection reactions on eukaryotic chromosomal DNAs, the *in situ* procedure using purified methylase in a two-step methylase/endonuclease procedure is described below. However, especially in pulsed-field gel mapping of bacterial chromosomes, the *in vivo* plasmid modification approach should be considered as the method of choice.

Source of ^{5m}*CGCG Methylase.* M · *Fnu*DII methyltransferase (^{5m}C-GCG) is prepared according to Qiang *et al.* (1990) and titered against *Bst*UI (CGCG) endonuclease (New England BioLabs). One unit of methylase is defined as the amount of enzyme required to protect 1 μg of λ DNA in a 30-μl liquid reaction mixture in 1 hr at 30° from cleavage by a 10-fold excess of *Bst*UI. M · *Fnu*DII methylase may be purchased from New England BioLabs, but in our hands the purity of this commercial enzyme has not been adequate for megabase mapping experiments (see the section Technical Considerations in Methyltransferase Reactions, above). We have been unable to prepare adequate quantities of pure ^{5m}CGCG methylase from *Fusobacterium nucleatum* 4H, *Acinobacter calcoaceticus, Thermoplasma acidophilum* ATCC (American Type Culture Collection, Rockville, MD) 25905, and *Bacillus subtilis* ATCC 6633. However, M · *Bsu*E (Gaido *et al.,* 1988; Shukla *et al.,* 1991) or M · *Bep*I (Kupper *et al.,* 1988) methyltransferases are suitable alternatives to M · *Fnu*DII.

Testing ^{5m}*CGCG Methylase against BstUI, MluI, and NotI Cleavage in Liquid Reactions.* Due to the time and effort required to prepare substrates and to run pulsed-field gels, it is generally useful to check reaction components, methylases, and endonucleases in liquid reactions against adenovirus-2 DNA prior to attempting *in situ* reactions and pulsed-field gel separations. DNA methylations are carried out at 30° on 1 μg of adenovirus-2 DNA in 30-μl reaction mixtures in 0.5× KGB, supplemented with 100 μM SAM and 2 mM EDTA (Qiang *et al.,* 1990; Hanish and McClelland, 1988). Following DNA methylation (3–6 hr) using approximately 2–5 units of methylase, magnesium acetate is added to a 10 mM final concentration, and 5 units *Bst*UI (New England BioLabs), *Mlu*I (New England BioLabs), or *Not*I (Boehringer-Mannheim or Stratagene) is added. DNA that has been mock-modified (no methylase) or ^{5m}CGCG modified is then digested with *Bst*UI (CGCG) at 65°, *Mlu*I (ACGCGT) or *Not*I (GCGGCCGC) at 37° for 2–3 hr (10- to 15-fold enzyme excess), and reaction samples are loaded onto 1.0% (w/v) agarose gels. After electrophoresis and staining, complete adenovirus-2 DNA protection against *Bst*UI and *Mlu*I endonucleases should be seen in ^{5m}CGCG-methylated DNA samples. In contrast, a distinct subset of *Not*I fragments is seen, corresponding to GCGGC^{5m}CGCG cross-protection.

In situ ^{5m}*CGCG Methylation of Chromosomal DNA Embedded in Agarose Plugs. Escherichia coli* K or *Bacillus subtilis* 168 chromosomal

DNA (~2 μg) embedded in agarose plugs (Weil and McClelland, 1989; Smith and Cantor, this series) is equilibrated in magnesium-free 0.5× KGB, supplemented with 2 m*M* EDTA (Hanish and McClelland, 1988). Agarose plugs (~1 × 1 × 3 mm) are incubated for 30 min in 200–300 μl of buffer (0.5× KGB + 2 m*M* EDTA) at room temperature in 1.5-ml microfuge tubes. Samples are then placed on ice, buffer is removed using an Eppendorf pipette, and 100 μl of fresh methylase reaction buffer (0.5× KGB + 2 m*M* EDTA + 100 μ*M* SAM) is added. *S*-Adenosylmethionine is freshly thawed from a frozen 32 m*M* stock solution in 10 m*M* H_2SO_4–10% ethanol.

Purified M · *Fnu*DII methylase is added (20 units) to each agarose plug in 100 μl of reaction buffer, followed by incubation at 30° for 2 hr. Reaction buffer is then removed using an Eppendorf pipette, and 100 μl of fresh methylase reaction buffer and 20 units more methylase are added, followed by incubation at 30° for 2–4 hr. Two rounds of *in situ* DNA methylation in EDTA buffer are carried out to ensure complete methylation. Reactions are terminated by heating at 50° for 30 min.

If desired, agarose plugs containing *E. coli* RR1 chromosomal DNA, with or without M · *Fnu*DII (or M · *Bep*I) modifying plasmid, can be used as controls for *in situ* methylase reactions on eukaryotic chromosomes, because this bacterial strain shows an easily detectable cross-protection pattern after *Not*I cleavage (8 of 22 sites blocked; Qiang *et al.*, 1990). However, readers should be cautioned that many mammalian DNAs are partially modified at ^{5m}CG (Lindsay and Bird, 1987), and *Not*I is sensitive to GCGGC^{5m}<u>CG</u>C modification (McClelland and Nelson, 1988b; Nelson and McClelland, 1991). Some partial digestion is to be expected due to *in vivo* ^{5m}CG methylation of *Not*I sites, but this complication has not limited the utility of *in situ* ^{5m}CGCG cross-protection in mapping *Not*I sites in human DNA (Shukla *et al.*, 1991).

In Situ NotI Cleavage of Chromosomal DNA Embedded in Agarose Plugs. After two methylase treatments in EDTA buffer, and heat treatment to inactivate trace nuclease contaminants in the DNA methylase, agarose plugs are equilibrated at 23° in 200–300 μl of 1× KGB (McClelland *et al.*, 1987) containing 10 m*M* magnesium acetate and 0.1% (v/v) Triton X-100. Equilibration buffer is removed using an Eppendorf pipette, and 100 μl of fresh 1× KGB, containing 10 m*M* magnesium acetate and 0.1% (v/v) Triton X-100, is added. Twelve units of pulsed-field mapping grade *Not*I (Promega, Stratagene, Boehringer-Mannheim, or New England BioLabs) is added, followed by a 4-hr incubation at 37°. *Not*I sites cleaved after ^{5m}CGCG cross-protection are a subset of unmodified *Not*I sites, a relationship that helps to identify adjacent *Not*I fragments. On average, these modified DNA fragments are twice as large as unmodified *Not*I fragments, which should be considered when adjusting pulse intervals for PFG electrophoretic separations.

After *Not*I digestion, agarose plugs are placed in loading wells of a 0.8% (w/v) agarose gel, placed in 10 m*M* Tris–acetate–0.25 m*M* EDTA running buffer in a Gene-Line transverse alternating field electrophoresis unit (Beckman Instruments) as described by Gardiner *et al.* (1986). Pulsed-field gels are run at 10 V/cm using a pulse time of 22 sec for 18 hr.

Other High-Frequency Cross-Protection Cleavages for Megabase Mapping. Several additional cross-protection reactions may be useful for megabase mapping and ordering large chromosome fragments produced by rare-cutting restriction enzymes (McClelland *et al.,* 1987). Several of these methylase/endonuclease combinations potentially useful for human chromosome mapping are outlined in Table II.

Methyl-Dependent Cleavage Using Adenine Methylases and DpnI

Principle of Adenine Methylase/DpnI Cleavage Strategy. McClelland *et al.* (1984) described a two-step method for creating specific 8- and 10-bp cleavages, using selected adenine methyltransferases in combination with the methyl-dependent endonuclease *Dpn*I.

$$\text{DNA} + \text{SAM} \xrightarrow{-\text{G}^{6\text{m}}\text{A MTase}} \begin{matrix}\text{—G}^{6\text{m}}\text{ATC—} \\ \text{—CT}_{6\text{m}}\text{AG—}\end{matrix} \xrightarrow[\text{Mg}^{2+}]{DpnI} \begin{matrix}\text{—G}^{6\text{m}}\text{A—OH} \\ \text{—CTp}\end{matrix} + \begin{matrix}\text{pTC—} \\ \text{HO—}_{6\text{m}}\text{AG—}\end{matrix}$$

This *Dpn*I methyl-dependent cleavage strategy has been demonstrated with M · *Taq*I (TCG^{6m}A) and M · *Cla*I (ATCG^{6m}AT) adenine methylases to cleave chromosomal DNAs into a few large pieces (McClelland, this series; Weil and McClelland, 1989; Waterbury *et al.,* 1990). M · *Bsp*106I (ATCG^{6m}AT) and M · *Mbo*II (GAAG^{6m}A) methylases have been used to cut plasmid DNA substrates containing ATCGATCGAT and GAAGATC-TTC sites, respectively (McClelland *et al.,* 1985; Nelson, 1991). *In vivo* modification by M · *Cvi*BIII (TCG^{6m}A), M · *Bsp*106I (ATCG^{6m}AT), and M · *Xba*I (TCTAG^{6m}A) methylases have also been used to create the respective bimethylated G^{6m}ATC *Dpn*I sites:

$$\begin{matrix}\text{TCG}^{6\text{m}}\text{ATCGA,} & \text{ATCG}^{6\text{m}}\text{ATCGAT,} & \text{and} & \text{TCTAG}^{6\text{m}}\text{ATCTAGA,} \\ \text{AGCT}_{6\text{m}}\text{AGCT} & \text{ATGCT}_{6\text{m}}\text{AGCTA} & & \text{AGATCT}_{6\text{m}}\text{AGATCT}\end{matrix}$$

Adenine methyltransferases that have been tested are mostly insensitive to ^{5m}C methylation, except for M · *Taq*I. Furthermore, *Dpn*I is insensitive to cytosine modification and cuts G^{6m}AT^{5m}C sites efficiently (Nelson and McClelland, 1991). In contrast, most rare-cutting restriction endonucleases suitable for mammalian genome mapping are sensitive to ^{5m}CG methylation. The 8- to 14-bp sequences defined by adenine methylase/*Dpn*I

TABLE II
HIGH-FREQUENCY CROSS-PROTECTIONS FOR MEGABASE MAPPING OF MAMMALIAN DNA

Restriction endonuclease[a]	Recognition sequence[b]	Blocking MTase	Specificity[b,c]	Cross-protection overlap[c,d]
*Csp*I (*Rsr*II)	CGGWCCG	M · *Hpa*II	G^{5m}CGG	(C)CGGWCCG(G)[d]
		M · *Bce*FI	ACGGC	(A)CGGWCCG(T)
		M · *Aci*I	GCGG	(G)CGGWCCG(G)
*Ecl*XI (*Eag*I)	CGGCCG	M · *Hpa*II	C^{5m}CGG	(C)CGGCCG(G)[d]
		M · *Fnu*4HI	G^{5m}CNGC	(G)CGGCCG(C)[d]
		M · *Bce*FI	ACGGC	(A)CGGCCG(T)
		M · *Aci*I	GCGG	(G)CGGCCG(C)
*Fse*I	GGCCGGCC	M · *Sau*96I	GGN^{5m}CC	(G)GGCCGGCC(C)
		M · *Eae*I	YGG^{5m}CCR	(Y)GGCCGGCC(R)
*Ksp*I (*Sac*II, *Sst*II)	CCGCGG	M · *Fnu*4HI	G^{5m}CNGC	(G)CCGCGG(C)[d]
*Mlu*I	ACGCGT	M · *Hga*I	GACGC	(G)ACGCGT(C)
*Nae*I	GCCGGC	M · *Bsp*RI	GG^{5m}CC	(G)GCCGGC(C)[d]
		M · *Cvi*JI	RG^{5m}CB	(R)GCCGGC(B)
*Not*I	GCGGCCGC	M · *Fnu*DII	^{5m}CGCG	(C)GCGGCCGC(G)[d]
		M · *Bep*I	^{5m}CGCG	(C)GCGGCCGC(G)[d]
		M · *Bsu*EI	^{5m}CGCG	(C)GCGGCCGC(G)[d]
*Sfi*I	GGCCN$_5$GGCC	M · *Sau*96I	GGN^{5m}CC	GGCC(C)N$_3$(G)GGCC
		M · *Sss*I	^{5m}CG	GGCC(G)N$_3$(C)GGCC[d]
*Srf*I	GCCCGGGC	M · *Bsp*RI	GG^{5m}CC	(G)GCCCGGGC[d]
		M · *Cvi*JI	RG^{5m}CB	(R)GCCCGGGC(B)
*Sna*BI	TACGTA	M · *Cvi*QI	GT^{6m}AC	(G)TACGTA(C)
*Spl*I (*Bsi*WI)	CGTACG	M · *Mae*II	ACGT	(A)CGTACG(T)

[a] Cross-protections of endonucleases that are potentially useful for megabase mapping in mammalian genomes are listed. Alternative commercially available isoschizomers are given in parentheses.

[b] W = A or T; Y = C or T; R = A or G; B = C or G or T; ^{5m}C or ^{6m}A denote methylated bases.

[c] See Nelson and McClelland (1991) for specificities of modification methylases.

[d] Cross-protections that have been demonstrated (Nelson and McClelland, 1987; Qiang *et al.*, 1990; Simcox and Simcox, 1991; Shukla *et al.*, 1991; Nelson, 1991).

combinations, and their general insensitivity to cytosine methylation, make methylase/*Dpn*I cleavage schemes suitable for megabase mapping of whole chromosomes. Several adenine methylase/*Dpn*I cleavages that have been demonstrated are listed in Table III.

For example, the *Staphylococcus aureus* chromosome has been cut into two pieces, about 1 and 2 megabases in size, using M · *Cla*I and *Dpn*I enzymatic reactions *in situ* (Weil and McClelland, 1989), and yeast chromosomes have been cut into megabase-sized pieces using M · *Cla*I and *Dpn*I (Waterbury *et al.*, 1990). The M · *Cla*I/*Dpn*I cleavage scheme is described in detail below.

TABLE III
ADENINE METHYLASE/*Dpn*I CLEAVAGES SUITABLE FOR MEGABASE MAPPING

Methylase	Modification specificity	MTase/*Dpn*I rare cleavage specificity	Size (bp)	Sensitivity to ^{5m}C	Chromosomes cut into megabase pieces
M · *Taq*I[a,b]	TCG^{6m}A	TCGATCGA	8	+	*E. coli* K[c]
M · *Cvi*BIII	TCG^{6m}A	TCGATCGA	8	—	*E. coli* K[c]
M · *Cla*I[a]	ATCG^{6m}AT	ATCGATCGAT	10	—	*S. aureus*[c] *S. cerevisiae*[c]
M · *Bsp*106I[b]	ATCG^{6m}AT	ATCGATCGAT	10	—	*B. sphaericus*[c,d]
M · *Mbo*II	GAAG^{6m}A	GAAGATCTTC	10	—	—
M · *Ncu*I	GAAG^{6m}A	GAAGATCTTC	10	?	—
M · *Bsa*BI	GATN$_4$ATC	GATN$_3$GATCN$_3$ATC	10	?	—
M · *Mam*I	GATN$_4$ATC	GATN$_3$GATCN$_3$ATC	10	?	—
M · *Xba*I	TCTAG^{6m}A	TCTAGATCTAGA	12	—	*E. coli* K,[d] *Salmonella*[d]
M · *Eco*R124I	CG^{6m}AYN$_6$TTC	GAAN$_4$CGATCGN$_4$TTC	12	?	—
M · *Eco*R124/3I	CG^{6m}AYN$_7$TTC	GAAN$_5$CGATCGN$_5$TTC	12	?	—
M · *Eco*B	TG^{6m}AN$_8$TGCT	AGCAN$_5$TGATCAN$_5$TGCT	14	?	—
M · *Pae*PAOI	?	?	~9	?	*P. aeruginosa*[d,e]

[a] Commercially available (New England BioLabs).
[b] Commercially available (Stratagene, Inc.).
[c] Two-step methylase/endonuclease reaction *in vitro*.
[d] Modification *in vivo;* one-step endonuclease cleavage *in vitro*.
[e] See Romling *et al.* (1989).

M · ClaI/DpnI Cleavage Reaction: Dependent on High-Purity Methylase

$$\begin{matrix}\text{—ATCGATCGAT—}\\ \text{—TAGCTAGCTA—}\end{matrix} + \text{SAM} \xrightarrow[\text{EDTA}]{\text{M}\cdot Cla\text{I MTase}}$$

$$\begin{matrix}\text{—ATCG}^{6m}\text{ATCG}^{6m}\text{AT—}\\ \text{—T}_{6m}\text{AGCT}_{6m}\text{AGCTA—}\end{matrix} + \text{SAH} \xrightarrow[\text{Mg}^{2+}]{Dpn\text{I}} \begin{matrix}\text{—ATCG}^{6m}\text{A—OH}\\ \text{—TAGCTp}\end{matrix} + \begin{matrix}\text{pTCGAT—}\\ \text{HO—AGCTA—}\end{matrix}$$

The two-step M · *Cla*I/*Dpn*I megabase cleavage scheme has been used reliably in several laboratories, with the caveat that adequately pure M · *Cla*I methylase is essential for megabase mapping applications. Highly purified M · *Cla*I or M · *Bsp*106I methylases (ATCG^{6m}AT) are prepared to purity levels suitable for megabase mapping according to previously described methods (McClelland, 1981; Nelson and McClelland, this series). Suitable high-purity M · *Bsp*106I is available from Stratagene, but commercial M · *Cla*I methylase should be used with *extreme* caution. Methylase purity must permit >20 units of DNA methylase to be used per microgram of high molecular weight DNA, embedded in agarose plugs (1 unit is defined as the amount of methylase required to protect 1 μg of phage λ DNA from cleavage by 10-fold excess *Cla*I endonuclease for 1 hr in a 50-μl *liquid* reaction containing 0.5× KGB (10 m*M* Tris–acetate, 50 m*M* potassium acetate, 2 m*M* EDTA, 160 μ*M* SAM, 50 μg/ml BSA, pH 7.5 at 30°). Methylase purity is checked by incubating 10 to 20 units of M · *Cla*I (or M · *Bsp*106I) and 1–2 μg *S. aureus* 3A DNA (Promega) embedded in agarose in a 100-μl reaction of the above buffer, 0.5× magnesium-free KGB (Hanish and McClelland, 1988) containing 100 μ*M* SAM and 2 m*M* EDTA, at 30° for 4 hr. Agarose plugs are then heated for 30 min at 55° to inactivate trace endonuclease contaminants, and then equilibrated in 300 μl 1× KGB (10 m*M* Tris–acetate, 50 m*M* potassium glutamate, 10 m*M* magnesium acetate, 1 m*M* 2-mercaptoethanol, 100 μg/ml BSA, pH 7.5). Equilibration buffer is removed and replaced by 100 μl fresh 1× KGB and 10 units *Cla*I endonuclease. Digestion is carried out at 37° for 4–6 hr, and samples are analyzed by pulsed-field electrophoresis using short (10–15 sec) pulse times. Methylase preparations that fail to protect high molecular weight bacterial DNA from *Cla*I endonuclease completely, or result in detectable substrate degradation, should not be used in *Dpn*I cleavage experiments.

In Situ ATCG^{6m}AT Methylation of Chromosomal DNA Embedded in Agarose. Staphylococcus aureus ISP8 DNA embedded in ~1 × 2 mm agarose plugs (as described by Weil and McClelland, 1989) is equilibrated in 100 μl 0.5× magnesium-free KGB containing 2 m*M* EDTA and 160 μ*M* SAM (Hanish and McClelland, 1988) at room temperature. M · *Cla*I or M · *Bsp*106I methylase (6 units) is added, followed by incubation at 30° for 8–12 hr. Reactions are terminated by heating to 55° for 30 min.

In Situ Cleavage of a Bacterial Chromosome into Two Large Pieces. Methylated chromosomal DNA, embedded in agarose, is equilibrated in 1× KGB for 20–30 min at room temperature. Buffer is removed and replaced by 100 μl fresh 1× KGB. Five units *Dpn*I endonuclease (Boehringer-Mannheim or New England BioLabs) is added, followed by incubation for 6 hr at 37°. *Dpn*I purity can be tested by doubly digesting *S. aureus* ISP8 or *S. aureus* 3A with 5 units *Dpn*I and 5 units *Csp*I (Promega) for 6–8 hr at 30°. *Csp*I reliably produces 10 fragments in the chromosome of *S. aureus* ISP8, the sizes of which are approximately 750, 600, 500, 300, 275, 200, 125, 80, 70, and 50 kbp (Weil and McClelland, 1989). Degradation of this simple pattern by *Dpn*I (or M · *Cla*I methylase) is indicative of enzyme purity problems.

Initially it is advantageous to carry out simultaneous (*Csp*I + *Dpn*I) cleavage of the *S. aureus* ISP8 chromosome, because the very high molecular weight reaction products of M · *Cla*I/*Dpn*I cleavage are about 1.0 and 1.9 million bp in size. Enzyme purity and substrate integrity are more readily detected in this double digest, because *Csp*I reliably gives complete, undegraded digestion of control *S. aureus* DNA, and requisite PFG pulse times (10–20 sec) and running times (12–18 hr) are shorter. M · *Cla*I/*Dpn*I or M · *Bsp*106I/*Dpn*I digestion of *S. aureus* ISP8 DNA results in cleavage of the 750- and 275-kb *Csp*I fragments (Weil and McClelland, 1989). When clean (M · *Cla*I/*Dpn*I + *Csp*I) double-digest patterns are obtained using short pulse times, separation of larger (1–2 Mb) fragments after M · *Cla*I/*Dpn*I digestion can be attempted. Agarose plugs containing methylated and *Dpn*I-digested DNA are loaded onto a transverse alternating field electrophoresis unit (Beckman Instruments), and separated using 300- to 500-sec pulse times at 5 V/cm for 48 hr.

Streamlining Two-Step Adenine Methylase/DpnI Cleavages. Whereas the feasibility of *in situ* two-step methylase/*Dpn*I protocols for megabase mapping of chromosomes has been demonstrated (Weil and McClelland, 1989; Waterbury *et al.,* 1990), it is possible to streamline these reactions to functional one-step protocols. First, when a suitable high-purity adenine methylase is available, methylation and *Dpn*I cleavage reactions can be carried out in 1× KGB buffer containing both 10 m*M* magnesium acetate and 100–160 μ*M* SAM. However, it is our experience that trace contaminant nucleases in even highly purified DNA methylase preparations usually preclude this coupling of reactions for megabase mapping.

On the other hand, it is possible to carry out certain adenine methylations *in vivo,* as described above for plasmid-coded cytosine modification or in certain modification-proficient cells (Romling *et al.,* 1989; Cummings *et al.,* 1974). Hanish and McClelland (1991) have used plasmid-encoded M · *Xba*I modification (TCTAG^{6m}A) and transposons containing rare 12-bp (TCTAGATCAGA) M · *Xba*I/*Dpn*I cleavage sites to map insertions into a *dam*$^-$ *Salmonella typhimurium* chromosome. In this application, it is not necessary to purify M · *Xba*I adenine methyltransferase. Highly selective chromosome cleavage is easily carried out by one-step *Dpn*I digestion of chromosomal DNA from bacteria containing the modifying plasmid and integrated transposon.

Secondary Reactions in Selected Methylase/DpnI Cleavages. In some methylase/*Dpn*I combinations, secondary DNA cleavages have been observed at unexpected sites. For example, M · *Taq*I/*Dpn*I or M · *Cvi*BIII/*Dpn*I cleavage of bacteriophage λ *dam*$^-$ DNA, which contains no TCGATCGA sites, gives cleavage at two TCGATCCG sites. A simple explanation for this phenomenon is not apparent because (1) both strands of G^{6m}ATC sequences must be methylated for efficient double-stranded *Dpn*I cleavage to occur (Geier and Modrich, 1979; Lacks and Greenberg, 1980); (2) hemimethylated (TCG^{6m}ATC) *Dpn*I sites in λ *dam*$^-$ DNA and several other plasmid and phage DNA substrates are not cut by excess *Dpn*I (McClelland, 1981; McClelland *et al.,* 1984, 1985; Nelson, 1991); (3) *Dpn*I cleavage at hemimethylated M · *Cla*I (ATCG^{6m}ATC) or M · *Xba*I (TCTAG^{6m}A) sites in whole bacterial chromosomes is not observed, although such sites are present (Weil and McClelland, 1989; Hanish and McClelland, 1991). It is likely that, at a reduced rate, *Dpn*I cuts the methylated strand of certain hemimethylated G^{6m}ATC sites in a sequence context-dependent fashion.

Three-Step "Double-Blocking" Reactions (Methylase 1/Methylase 2/ Endonuclease 2)

M · HpaII/M · BamHI/BamHI Cleavage in Vitro. Some DNA methylases, like their endonuclease partners, cannot modify DNA target sites that are sterically hindered. By selectively protecting certain rare methylase sites that overlap sequences defined by other DNA methylases (Posfai and Szybalski, 1988; McClelland and Nelson, 1988a), bacterial operators (Koob and Szybalski, 1990), or polypyrimidine DNA triplexes (Maher *et al.*, 1989; Hanvey *et al.*, 1989), it is possible to limit DNA cleavage to defined target sites. However, such procedures are necessarily three-step sequential treatments: (1) an initial "masking" reaction is carried out using a DNA methylase, site-specific DNA-binding protein, or an oligonucleotide triplex; (2) a blocking reaction using a DNA methylase is carried out at all unoccluded restriction-modification target sites; (3) endonuclease cleavage is limited to selected *unmodified* restriction sites.

For example, like other DNA-binding proteins, including restriction endonucleases, some DNA methylases cannot bind to target sites that carry modifications at noncanonical sites. In some cases, a DNA modification methylase and its endonuclease partner may differ, such that the methylase cannot methylate certain modified sites, whereas the cognate endonuclease cuts this noncanonically modified sequence (McClelland and Nelson, 1988a; Nelson and McClelland, 1991). Such is the case of M · *Bam*HI methylase and *Bam*HI endonuclease with respect to noncanonical GGATC^{5m}C modification.

—CCGGATCCGG— / —GGCCTAGGCC— + SAM —[M · *Hpa*II MTase (1)]→ —C^{5m}CGGATC^{5m}CGG— / —GG$_{5m}$CCTAGG$_{5m}$CC— + SAH

(2) ✗ M · *Bam*HI MTase ↓

—CCG—OH / —GGCCTAGp + pGATCCGG— / HO—GCC— ←[(3) *Bam*HI, Mg^{2+}]— —C^{5m}CGGATC^{5m}CGG— / —GG$_{5m}$CCTAGG$_{5m}$CC—

Based on this knowledge of the differential sensitivity of restriction and modification enzymes to site-specific methylation, three-step reaction sequences (McClelland and Nelson, 1988a) such as M · *Hpa*II/M · *Bam*HI/ *Bam*HI are possible. Using this sequential double-blocking scheme, it is possible to selectively cut the single CCGGATCCGG site in phage λ and adenovirus-2 DNAs. Nonoverlapping *Bam*HI sites are methylated by M · *Bam*HI, and therefore canonically methylated against *Bam*HI cleavage. However, we have been unable to apply this strategy to large-scale genomic mapping due to technical limitations inherent to multistep procedures.

Two-Step M · BamHI/BamHI, M · TaqI/TaqI, and M · TaqI/AsuII Cleavages. On the other hand, streamlined versions of this sequential double-methylation strategy may be more technically feasible. Because some DNA methylations can be carried out *in vivo,* it is sometimes possible to reduce sequential three-step protocols to manageable two-step DNA cleavage protocols. For example, it is possible to use *in vivo* bacterial *dcm* modification ($C^{5m}CWGG$) to define M · *dcm*/M · *Bam*HI/*Bam*HI cleavage at rare, modified $C^{5m}CWGGATC^{5m}CWGG$ sites in *E. coli;* and mammalian cell (^{5m}CG) methylation limits M · *Bam*HI/*Bam*HI cleavage to bimethylated $^{5m}CGGATC^{5m}CG$ sites (Nelson, 1991). M · *Taq*I cannot methylate $T^{5m}CGA$ sequences, whereas *Taq*I endonuclease cuts this noncanonically modified sequence. Therefore, it is possible to selectively cut ^{5m}CG-methylated DNAs at $T^{5m}CGA/T^{5m}CGA$ sequences, by sequential treatment with M · *Taq*I methylase and *Taq*I endonuclease. M · *Taq*I/*Taq*I cleavage ($T^{5m}CGA$) or M · *Taq*I/*Asu*II cleavage ($TT^{5m}CGAA$), based on sequential double methylation reactions (McClelland and Nelson, 1988a; Posfai and Szybalski, 1988),

$$\begin{matrix}\text{—T}^{5m}\text{CGA—}\\ \text{—AG}_{5m}\text{CT—}\end{matrix} + \text{SAM} \xrightarrow[\quad]{\text{M}\cdot Taq\text{I MTase}} \!\!\!\!\!\!\!\!\!\!\!\!\!\!\times\;\;\;\;\;\;\;\;\; \xrightarrow[\text{Mg}^{2+}]{Taq\text{I}} \begin{matrix}\text{—T—OH}\\ \text{—AG}_{5m}\text{Cp}\end{matrix} + \begin{matrix}\text{p}^{5m}\text{CGA—}\\ \text{HO—T—}\end{matrix}$$

$$\begin{matrix}\text{—TCGA—}\\ \text{—AGCT—}\end{matrix} + \text{SAM} \xrightarrow{\text{M}\cdot Taq\text{I MTase}} \begin{matrix}\text{—TCG}^{6m}\text{A—}\\ \text{—}_{6m}\text{AGCT—}\end{matrix} + \text{SAH} \xrightarrow[\text{Mg}^{2+}]{Taq\text{I}}\!\!\!\!\!\!\!\times$$

may be useful in cutting mammalian chromosomes selectively in G + C rich, ^{5m}CG-methylated islands (Bird and Southern, 1978; Lindsay and Bird, 1987). Such two-step procedures rely on *in vivo* cellular ^{5m}CG methylation, preventing M · *Taq*I from blocking (via adenine methylation) endonucleolytic cleavage at cytosine-modified *Taq*I (or *Asu*II) sites.

References

Adams, R. L. P., *Biochem J.* **265,** 309 (1990).

Arnold, J., Cuticchia, A. J., Newsome, D. A., Jennings, W. W., and Ivarie, R., *Nucleic Acids Res.* **16,** 7145 (1988).

Bird, A. P., and Southern, E. M., *J. Mol. Biol.* **118,** 27 (1978).

Birkelund, S., and Stephens, R. S., *J. Bacteriol.* **174,** 2742 (1992).

Bolton, E. T., and McCarthy, B. J., *Proc. Natl. Acad. Sci. U.S.A.* **48,** 1390 (1962).

Brennan, C. A., Van Cleve, M. D., and Gumport, R. I., *J. Biol. Chem.* **261,** 7270 (1987).
Brennan, R. G., and Mathews, B. W., *J. Biol. Chem.* **264,** 1903 (1989).
Butler, P. D., and Moxon, E. R., *J. Gen. Microbiol.* **136,** 2333 (1990).
Cantor, C. R., Smith, C. L., and Matthew, M. K., *Annu. Rev. Biophys. Chem.* **17,** 287 (1988).
Chen, C. B., and Sigman, D. S., *Science* **237,** 1197 (1987).
Craig, A. G., Nizetic, D., Hoheisel, J. D., Zehetner, G., and Lehrach, H., *Nucleic Acids Res.* **18,** 2653 (1990).
Cummings, D. J., Tait, A., and Goddard, J. M., *Biochim. Biophys. Acta* **374,** 1 (1974).
Daniels, D. L., *Nucleic Acids Res.* **18,** 2649 (1991).
Davis, T., *New England BioLabs catalog,* p. 64 (1990–1991).
Dobritsa, A. P., and Dobritsa, S. V., *Gene* **10,** 105 (1979).
Ebright, R. H., Ebright, Y. W., Pendergrast, P. S., and Gunasekera, A., *Proc. Natl. Acad. Sci. U.S.A.* **87,** 2882 (1990).
Francois, J.-C., Saison-Behmoaras, T., Barbier, C., Chassignol, M., Thuong, N. T., and Helene, C., *Proc. Natl. Acad. Sci. U.S.A.* **86,** 9702 (1989).
Gaido, M., Prostko, C. R., and Strobl, J. S., *J. Biol. Chem.* **263,** 4832 (1988).
Gardiner, K., Laas, W., and Patterson, D., *Somatic Cell. Mol. Genet.* **12,** 185 (1986).
Geier, G. E., and Modrich, P., *J. Biol. Chem.* **254,** 1408 (1979).
Griffin, L. C., and Dervan, P. B., *Science* **245,** 967 (1989).
Halford, S. E., *personal communication* (1991).
Hanish, J., and McClelland, M., *Gene Anal. Tech.* **5,** 105 (1988).
Hanish, J., and McClelland, M., *Nucleic Acids Res.* **18,** 3287 (1990).
Hanish, J., and McClelland, M., *Nucleic Acids Res.* **19,** 829 (1991).
Hanish, J., Rebelsky, D., McClelland, M., and Westbrook, C., *Genomics* **10,** 681 (1991).
Hanvey, J. C., Shimuzu, M., and Wells, R. D., *Nucleic Acids Res.* **18,** 157 (1989).
Heitman, J., and Model, P., *EMBO J.* **9,** 3369 (1990).
Howard, F. B., Frazier, J., Singer, M. F., and Miles, H. T., *J. Mol. Biol.* **16,** 415 (1966).
Hulsmann, K.-H., Bergeret-Coulaaud, A., and Hahn, U., *Nucleic Acids Res.* **18,** 7189 (1990).
Hung, L., and Bandziulis, R., *Promega Notes* **24,** 1 (1990).
Kauc, L., Mitchell, M., and Goodgall, S., *J. Bacteriol.* **171,** 2474 (1989).
Kessler, C., and Manta, V., *Gene* **92,** 1 (1990).
Klimasauskas, S., Timinskas, A., Menkevicius, S., Butkiene, D., Butkus, V., and Janulaitis, A., *Nucleic Acids Res.* **17,** 9823 (1990).
Kolsto, A.-B., Gronstad, A., and Oppegaarad, H., *J. Bacteriol.* **172,** 3821 (1990).
Koncz, C., Kiss, A., and Venetianer, P., *Eur. J. Biochem.* **89,** 523 (1978).
Koob, M., and Szybalski, W., *Science* **250,** 270 (1990).
Koob, M., Grimes, E., and Szybalski, W., *Science* **241,** 1084 (1988).
Kotani, H., Nomura, Y., Kawashima, Y., Sagawa, H., Kita, A., Ito, H., and Kato, I., *Nucleic Acids Res.* **18,** 5637 (1991).
Kupper, D., Jian-Guang, Z., Kiss, A., and Venetianer, P., *Gene* **74,** 33 (1988).
Lacks, S., and Greenberg, B., *J. Mol. Biol.* **114,** 153 (1980).
LeDoan, T., Perroualt, L., Praseuth, D., Habhoub, N., Decout, J.-L., Thuong, N. T., Lhomme, J., and Helene, C., *Nucleic Acids Res.* **15,** 7749 (1987).
Lee, J. L., and Smith, H. O., *J. Bacteriol.* **170,** 4402 (1988).
Lindsay, S., and Bird, A. P., *Nature* (*London*) **327,** 336 (1987).
McClarin, J. A., Frederick, C. A., Wang, B.-C., Greene, P., Boyer, H. W., Grable, J., and Rosenberg, J. M., *Science* **234,** 1526 (1986).
McClelland, M., *Nucleic Acids Res.* **9,** 6795 (1981).

McClelland, M., this series, Vol. 155, p. 22.
McClelland, M., and Nelson, M., *in* "Gene Amplification and Analysis" (J. Chirikjian, ed.), Vol. 5, p. 257. Elsevier Science, New York, 1987.
McClelland, M., and Nelson, M., *Gene* **74,** 169 (1988a).
McClelland, M., and Nelson, M., *Gene* **74,** 291 (1988b).
McClelland, M., Kessler, L., and Bittner, M., *Proc. Natl. Acad. Sci. U.S.A.* **81,** 983 (1984).
McClelland, M., Nelson, M., and Cantor, C. R., *Nucleic Acids Res.* **13,** 7171 (1985).
McClelland, M., Jones, R., Patel, Y., and Nelson, M., *Nucleic Acids Res.* **15,** 5985 (1986).
McClelland, M., Hanish, J., Nelson, M., and Patel, Y., *Nucleic Acids Res.* **16,** 364 (1987).
Maher, L. J., Wold, B., and Dervan, P. B., *Science* **245,** 725 (1989).
Mazzerelli, J., Scholtissek, S., and McLaughlin, L. W., *Biochemistry* **28,** 4616 (1989).
Moser, H. E., and Dervan, P. B., *Science* **238,** 645 (1987).
Nelson, M., unpublished observations (1991).
Nelson, M., and McClelland, M., this series, Vol. 155, p. 32.
Nelson, M., and McClelland, M., *Nucleic Acids Res.* **19**(Suppl), 2045 (1991).
Nelson, M., and Schildkraut, I., this series, Vol. 155, p. 41.
Nelson, M., Christ, C., and Schildkraut, I., *Nucleic Acids Res.* **12,** 5165 (1984).
Newman, P. C., Williams, D. M., Cosstick, R., Seela, F., and Connolly, B. A., *Biochemistry* **29,** 9902 (1990).
Noll, K. M., *J. Bacteriol.* **171,** 6270 (1989).
Nuitjen, P. J. M., Bartels, C., Beumink-Pluym, N. M. C., Gaastra, W., and Van der Zeijst, A. M., *Nucleic Acids Res.* **18,** 6211 (1990).
Okahashi, N., Sasakawa, C., Okada, N., Yamada, M., Yoshikawa, M., Tokuda, M., Takahashi, I., and Koga, T., *J. Gen. Microbiol.* **136,** 2217 (1990).
Qiang, B.-Q., McClelland, M., Poddar, S., Spoukaskas, A., and Nelson, M., *Gene* **88,** 101 (1990).
Pei, D., Corey, D. R., and Schultz, P. G., *Proc. Natl. Acad. Sci. U.S.A.* **87,** 9858 (1990).
Perrouault, L., Asseltine, U., Rivalle, C., Thuong, N. T., Bisagni, E., Giovangelli, C., LeDoan, T., and Helene, C., *Nature (London)* **344,** 358 (1990).
Plum, G. E., Park, Y.-W., Singleton, S. F., Dervan, P. B., and Breslauer, K. J., *Proc. Natl. Acad. Sci. U.S.A.* **87,** 9436 (1991).
Poddar, S., and McClelland, M., manuscript in preparation (1992).
Pogolotti, A. L., Ono, A., Subramaniam, R., and Santi, D. V., *J. Biol. Chem.* **263,** 7461 (1988).
Posfai, G., and Szybalski, W., *Nucleic Acids Res.* **16,** 6245 (1988).
Reich, N. O., and Mashoon, N., *Biochemistry* in press (1991).
Riggs, A. D., *Cytogenet. Cell. Genet.* **14,** 9 (1975).
Roberts, R. J., *Nucleic Acids Res.* **18**(Suppl), 2331 (1990).
Roberts, R. J., *Nucleic Acids Res.* **19**(Suppl), 2077 (1991).
Roberts, R. W., and Crothers, D. M., *Proc. Natl. Acad. Sci. U.S.A.* **88,** 9397 (1991).
Romling, U., Grothues, U., Bautsch, W., and Tummler, B., *EMBO J.* **8,** 4081 (1989).
Rosenberg, J. M., *Curr. Opin. Struct. Biol.* **1,** 104 (1991).
Sanger, F., *Annu. Rev. Biochem.* **57,** 1 (1988).
Schwartz, D. C., and Cantor, C. R., *Cell (Cambridge, Mass.)* **37,** 67 (1984).
Shukla, H., Kobayashi, Y., Arenstorf, H., Yasukochi, Y., and Weissman, S., *Nucleic Acids Res.* **19,** 4233 (1991).
Sigman, D. S., and Chen, C. B., *Annu. Rev. Biochem.* **59,** 207 (1990).
Simcox, T., and Simcox, M. L., *Gene* in press (1991).
Smith, C. L., and Cantor, C. R., this series, Vol. 155, p. 449.
Strobel, S. A., and Dervan, P. B., *Science* **249,** 73 (1990).

Strobel, S. A., Moser, H. E., and Dervan, P. B., *J. Am. Chem. Soc.* **110,** 7927 (1988).
Suwanto, A., and Kaplan, S., *J. Bacteriol.* **171,** 5850 (1989).
Taylor, J. D., Badcoe, I. G., Clarke, A. R., and Halford, S. E., *Biochemistry* **30,** 8743 (1991).
Waterbury, P. G., Rehfuss, R. P., Carroll, W. T., Smardon, A. M., Faldasz, B. D., Huckaby, C. S., and Lane, M. J., *Nucleic Acids Res.* **17,** 9493 (1990).
Weil, M. D., and McClelland, M., *Proc. Natl. Acad. Sci. U.S.A.* **86,** 51 (1989).
Wilson, G. G., *Gene* **74,** 281 (1990).
Wilson, G. G., *Nucleic Acids Res.* **19,** 2539 (1991).
Wilson, W. W., and Hoffman, R. M., *Anal. Biochem.* **191,** 370 (1991).
Winkler, F. K., D'Arcy, A., Blocker, H., Frank, R., and van Boom, J. H., *J. Mol. Biol.* **217,** 235 (1991).
Wong, K. K., and McClelland, M., *J. Bacteriol.* **174,** 1656 (1992).
Wu, J. C., and Santi, D. V., *J. Biol. Chem.* **262,** 4786 (1987).

[26] Conferring New Specificities on Restriction Enzymes: Cleavage at Any Predetermined Site by Combining Adapter Oligodeoxynucleotide and Class-IIS Enzyme

By ANNA J. PODHAJSKA, SUN CHANG KIM, and WACLAW SZYBALSKI

Introduction

Class-II restriction endonucleases (ENases), the most widely used tools in genetic engineering today, cleave double-stranded (ds) DNA at preexisting recognition sites, which usually are 4–8 bp long.[1] Because the number of such recognition sites is quite limited, it would be convenient to have a means for cutting DNA at any predetermined site. Such a method, as adopted for single-stranded (ss) DNA, is described here and depends on the use of a specially designed oligodeoxyribonucleotide (oligo) adapter carrying the class-IIS ENase recognition site, which directs the cognant class-IIS ENase to a specific predetermined cleavage site.[2–4]

Principle of Method

The method exploits the mechanism of cleavage employed by the class-IIS ENase, as exemplified here by the enzyme, *Fok*I. *Fok*I recognizes a

[1] C. Kessler and V. Manta, *Gene* **92,** 1 (1990); R. J. Roberts, *Nucleic Acids Res.* **17,** r347 (1989).
[2] W. Szybalski, *Gene* **40,** 169 (1985).
[3] A. J. Podhajska and W. Szybalski, *Gene* **40,** 175 (1985).
[4] S. C. Kim, A. J. Podhajska, and W. Szybalski, *Science* **240,** 504 (1988).

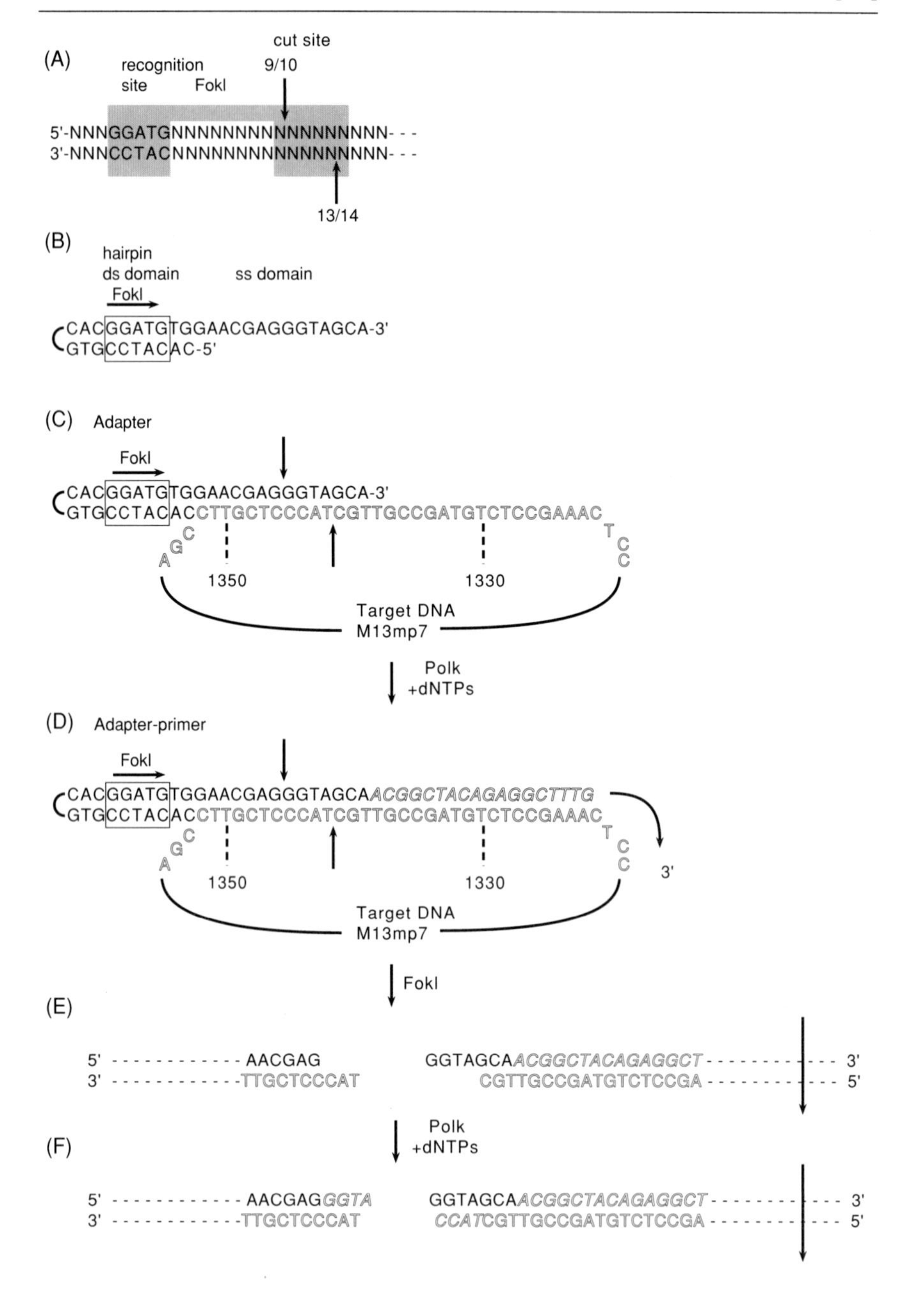
(A)
recognition site
FokI
cut site 9/10
5'-NNNGGATGNNNNNNNNNNNNNNNNN- - -
3'-NNNCCTACNNNNNNNNNNNNNNNNN- - -
13/14
(B)
hairpin ds domain
ss domain
FokI
CACGGATGTGGAACGAGGGTAGCA-3'
GTGCCTACAC-5'
(C) Adapter
FokI
CACGGATGTGGAACGAGGGTAGCA-3'
GTGCCTACACCTTGCTCCCATCGTTGCCGATGTCTCCGAAAC
1350
1330
Target DNA M13mp7
PolK +dNTPs
(D) Adapter-primer
FokI
CACGGATGTGGAACGAGGGTAGCAACGGCTACAGAGGCTTTG
GTGCCTACACCTTGCTCCCATCGTTGCCGATGTCTCCGAAAC
3'
1350
1330
Target DNA M13mp7
FokI
(E)
5' - - - - - - - - - - - - AACGAG GGTAGCAACGGCTACAGAGGCT - - - - - - - - - - - 3'
3' - - - - - - - - - - - -TTGCTCCCAT CGTTGCCGATGTCTCCGA - - - - - - - - - - - 5'
PolK +dNTPs
(F)
5' - - - - - - - - - - - - AACGAGGGTA GGTAGCAACGGCTACAGAGGCT - - - - - - - - - - 3'
3' - - - - - - - - - - - -TTGCTCCCAT CCATCGTTGCCGATGTCTCCGA - - - - - - - - - - 5'

5-bp sequence, 5′-GGATG, and cuts 9 and 13 nucleotides (nt) away
CCTAC
from this site (see Fig. 1A). In native DNA, both the 5-bp recognition site and the cleavage site are on the same molecule. However, it is possible to place the *Fok*I recognition site on a special hairpin oligo adapter, shown in Fig. 1B.[2–4] *Fok*I binds to the ds domain of such an adapter, but cannot cleave the ss domain unless it is paired with a complementary region of the target ssDNA, as represented in Fig. 1C.[3] By specifying the proper complementary sequence for the ss domain in the adapter, one could cut the target ssDNA at any predetermined site between any two bases (e.g., between T and C, as shown in Fig. 1C).

Experimental Procedures

Solutions

*Fok*I buffer: 20 m*M* Tris-HCl, pH 7.5, 10 m*M* $MgCl_2$, and 1 m*M* dithiothreitol (DTT)

TE: 10 m*M* Tris-HCl, 1 m*M* ethylenediaminetetraacetic acid (EDTA), pH 8.0/24°

Alkaline loading buffer (ALB): 30 m*M* NaOH, 2 m*M* EDTA, 7% (w/v) Ficoll, 0.1% (w/v) sodium dodecyl sulfate (SDS), 0.01% (w/v) bromphenol blue

---

FIG. 1. Design of an adapter oligodeoxynucleotide (oligo) for instructing the *Fok*I enzyme to cleave M13 ssDNA at a predetermined point. (A) Both the *Fok*I recognition and cutting domains for dsDNA are schematically outlined. The staggered cleavage points are represented by vertical arrows. (B) The 34-mer adapter oligo is designed to bind to *Fok*I and to cut M13 ssDNA between nt 1341 and 1342 [P. M. G. F. Van Wezenbeek, T. J. M. Hulsebos, and J. G. G. Schoenmakers, *Gene* **11,** 129 (1980)]. (C) The adapter is a 34-mer oligo with a 10-bp hairpin ds domain carrying the *Fok*I recognition site (boxed) and a 14-nt ss domain complementary to nt 1339–1352 of phage M13mp7 ss target DNA. This 34-mer adapter is directing the *Fok*I-mediated cleavages, as shown by the vertical arrows. (D) Alternatively, the 3′ end of the 34-mer adapter–primer is elongated by PolIk and all four dNTPs, thus converting the M13mp7 ss target to dsDNA. (E) Addition of *Fok*I results in staggered cleavages creating a predetermined end, the position of which depends solely on the design and the sequence of the adapter-primer. (F) Because of the presence of PolIk and four dNTPs, the cohesive ends are filled in. The bold letters represent the adapter and the outlined letters indicate the target ssDNA. Only one kind of adapter with the 5′ hairpin ds domain and with the 3′ ss target recognition domain is shown here. The high precision of cleavage was assessed by sequence analysis.[4]

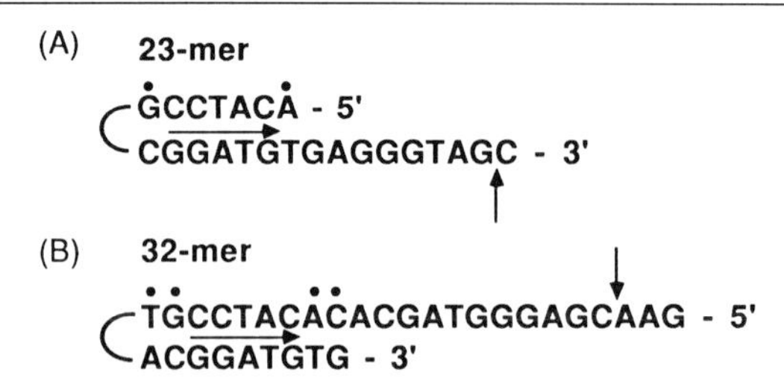

FIG. 2. The 23- and 32-nt oligo adapters. The 5-bp *Fok*I recognition sequences are marked by horizontal arrows within the ds hairpin domains. The additional base-paired nucleotide flanking both ends of the *Fok*I recognition sequence in the ds domain are indicated by heavy dots. Vertical arrows indicate the predicted *Fok*I cut sites in the adapters, after being annealed to the target ssDNA. The 23-mer adapter showed very poor binding to the *Fok*I enzyme and did not direct any cleavage.[5]

Design of Adapter

The adapter shown in Fig. 1B will serve as an example. Its hairpin ds domain is 10 bp long and consists of a 5-bp *Fok*I recognition site and 3- and 2-bp ds flanking sequences. The ss domain is 14 nt long and its sequence, which is complementary to the target ssDNA, determines the exact point where the target will be cleaved (Fig. 1C).[3,4]

Notes on Adapter Design

a. The hairpin ds domain shown in Fig. 1B (composed of a 5-bp *Fok*I recognition site and two flanking sequences, 2 and 3 bp long) was found to be quite efficient both in binding to *Fok*I and in subsequent cleavage. A ds domain, shorter by 1 bp and with flanking sequences of only 2 bp (see Fig. 2B),[3,5] was found to be as efficient as that shown in Fig. 1B. However, a ds domain with each flanking sequence only 1 bp in length was a poor *Fok*I binder and unsuitable as an adapter (see Fig. 2A).

b. Adapters can dimerize by hybridization between two open hairpin domains, but they still remain active in the present procedure. Heating to 70° (5 min) and rapid cooling in ice produces predominantly monomeric adapters, as confirmed by 10% (w/v) polyacrylamide gel electrophoresis.[5]

c. The ss domain of the adapter in Fig. 1B is 14 nt long, but a longer ss sequence would assure higher hybridization specificity for a given site on the target DNA. The design of the adapter could allow some mispairing

[5] S. C. Kim and W. Szybalski, unpublished.

in the region between the recognition site and cutting domain (around nt 1350 in Fig. 1C), while still directing rather efficient and precise cleavage, providing that 14 or more nucleotides remain base paired.[5] The latter property could be exploited when designing specific adapters.

d. The 3′ end in the ss domain of the adapter (shown in Fig. 1C) permits its use as a primer, as described below (and shown in Fig. 1D and E).[4] Adapters with a 5′-end ss domain were also constructed and found to direct precise cutting just as efficiently[3,5]; however, they cannot be used as primers.

e. If required, the recognition specificity of *Fok*I could be increased to 7 bp, 5′-CCGGATG, by double methylation.[6,7]
GGCCTAC

f. Analogous adapters could be designed for several other class-IIS ENases.[8]

Enzymes

The class-IIS ENases must be free of endonucleolytic activity toward ssDNA.[3,4] *Fok*I was selected, because of our previous experience with this enzyme, and because the M · *Fok*I methyltransferase has become commercially available, permitting protection of dsDNA from *Fok*I digestion at preexisting restriction sites. The experimental approach we describe here is not limited to *Fok*I. The procedure probably could be adapted to many of the over 35 other class-IIS ENases and their isoschizomers.[8]

Cleaving ssDNA to Produce Single-Stranded Products

1. Suspend M13mp7 ssDNA (2 μg) in 20 μl of *Fok*I buffer, and add a 7.5-fold molar excess of the 34-mer or 32-mer adapters shown in Figs. 1B and 2 (see note *a* below).
2. Heat the sample for 10 min at 70°, cool it rapidly in ice, and incubate for 30 min at 37° for an effective hybridization (see note *c* below).
3. Add *Fok*I enzyme (12 units) and digest for 2 hr at 37°.
4. Precipitate the sample with 4 vol of 95% ethanol in the presence of 0.3 *M* sodium acetate, place the tube in dry ice for 10 min, spin in a microfuge for 10 min, and then wash with 70% ethanol.
5. Dry the sample under vacuum, and then resuspend in 20 μl of ALB.

[6] G. Pósfai and W. Szybalski, *Nucleic Acids Res.* **16,** 6245 (1988).
[7] S. C. Kim, G. Pósfai, and W. Szybalski, *Gene* **100,** 45 (1991).
[8] W. Szybalski, S. C. Kim, N. Hasan, and A. J. Podhajska, *Gene* **100,** 13 (1991).

6. Load the resuspended sample onto a 1% (w/v) agarose gel, run for 2 hr at 150 V, and then stain the bands with ethidium bromide.

Notes on Cleaving ssDNA

a. The M13mp7 ssDNA and adapter can be stored for several months at 4° in TE buffer. Any range between a 5- and 15-fold molar excess of adapter to M13mp7 ssDNA works well.

b. Agarose gels (0.8–1%, w/v) are recommended for ssDNA of the M13 size.

c. The annealing time can be reduced to 5 min with hardly any change in the digestion pattern.

d. As the target, one could use dsDNA with a ss gap at the adapter-annealing and -cutting region.[2,8]

Cleaving ssDNA to Produce Double-Stranded Fragment(s)

1. Suspend M13mp7 ssDNA (2 μg) in 100 μl of *Fok*I buffer, together with a 7.5-fold molar excess of the 34-mer adapter–primer (Fig. 1B).
2. Heat the sample for 10 min at 70°, cool it rapidly in ice, and then anneal at 37° for 30 min (see note *a* below).
3. Add *Escherichia coli* polymerase I, Klenow fragment (PolIk, 10 units; New England BioLabs, Beverly, MA), and all four dNTPs (8 μl of each dNTP at 2.5 m*M*) to the reaction sample; add *Fok*I (8 units) 5 min after the addition of the PolIk, and take samples after 15 min, 30 min, 1 hr, and 2 hr.
4. Stop the reaction in each sample by adding 100 μl of 5 *M* ammonium acetate and 400 μl of 95% ethanol (see note *b* below).
5. Place the samples in dry ice, spin the sample in a microfuge for 10 min, wash twice with 70% ethanol, dry under vacuum, and then redissolve in *Fok*I buffer.
6. Load the resuspended samples on a 1.5% (w/v) agarose gel (see note *c* below), run for 2 hr at 150 V, and then stain with ethidium bromide. Further details and photographs of the resulting gels are as in Ref. 4.

Notes on Cleaving to Produce Double-Stranded Fragment(s)

a. To avoid the nonspecific binding of the adapter, a 37° hybridization temperature is recommended. (Also, see note *c*, above, in the *Notes on Adapter Design* section.) The annealing time can be reduced to 5 min with hardly any change in the digestion pattern.

b. To remove the remaining dNTPs after the reaction, use ammonium acetate (final concentration, 2.5 *M*) instead of sodium acetate.

c. A 1.5% (w/v) agarose gel is suitable for dsDNA fragments.

Conclusions

A specially designed adapter–primer oligo permits one to produce DNA fragments with a predetermined end located between any specified two base pairs of the target DNA. This approach is especially useful for DNA fragments cloned in an ssDNA-generating vector, and it could also complement our other preprogrammed DNA-trimming method.[9] The present experiment should also be important for dissecting the mechanisms of enzyme recognition, endonucleolytic cleavage, and methylation.

Cleavage of ssDNA vectors is helpful for the alternative method of sequencing in the case of sequence ambiguities encountered during the dideoxy sequencing method. Only one or two hairpin adapters, complementary to the universal primer site(s), permit precise cleavage of ssDNA and its ^{32}P end-labeling by PolIk (for the already prepared ss plasmids, which were used for dideoxy sequencing), for Maxam–Gilbert sequencing of the strand opposite to that already sequenced by the dideoxy technique.[10]

[9] N. Hasan, S. C. Kim, A. J. Podhajska, and W. Szybalski, *Gene* **50,** 55 (1986).
[10] B. Goszczynski and J. D. McGhee, *Gene* **104,** 71 (1991).

[27] Triple Helix-Mediated Single-Site Enzymatic Cleavage of Megabase Genomic DNA

By SCOTT A. STROBEL and PETER B. DERVAN

Oligonucleotide-directed triple helix formation is a generalizable chemical approach for the recognition and cleavage of a single target site within several megabase pairs of duplex genomic DNA.[1-4] Pyrimidine oligodeoxyribonucleotides 15 to 25 bases in length form a highly specific triple helix structure with purine tracts in double-stranded DNA of high complexity.[1-4] The pyrimidine oligonucleotide binds by Hoogsteen hydrogen bonding in the major groove of the DNA duplex parallel to the Watson–Crick purine strand.[1] The recognition motif is generalizable to homopurine target sites utilizing thymine binding to adenine–thymine base pairs (T · A-T

[1] H. E. Moser and P. B. Dervan, *Science* **238,** 645 (1987).
[2] S. A. Strobel, H. E. Moser, and P. B. Dervan, *J. Am. Chem. Soc.* **110,** 7927 (1988).
[3] S. A. Strobel and P. B. Dervan, *Science* **249,** 73 (1990).
[4] S. A. Strobel and P. B. Dervan, *Nature (London)* **350,** 172 (1991).

triplet)[5] and N-3 protonated cytosine binding to guanine–cytosine base pairs (C + G-C triplet).[6–8] 5-Bromouracil (^{Br}U) and 5-methylcytosine (^{Me}C) can be substituted in the third strand for T and C, respectively, to generate oligonucleotides with higher binding affinities at slightly basic pH values.[9] The triple helix motif is partially extended to mixed purine-pyrimidine sequences utilizing guanine binding to thymine–adenine base pairs (G · T-A triplet)[10] and alternate strand triple helix formation.[11] More recently, a G-rich oligodeoxyribonucleotide third strand has been shown to bind antiparallel to the Watson–Crick purine strand via G · G-C, A · A-T, T · A-T base triplets.[12] The binding site size of triple helix recognition is sufficiently large to statistically identify a unique site in the human genome (>16 bp).[13,14]

Koob *et al.* demonstrated the feasibility of using a DNA-binding protein to uniquely block the action of a methylase at a single overlapping recognition site while methylating all other sites.[15,16] Using this Achilles' heel cleavage procedure, they demonstrated single-site cleavage of yeast and *Escherichia coli* genomes on digestion with a restriction enzyme able to cut only at unmethylated sites.[16] Protein-mediated Achilles' heel cleavage may not, however, be readily generalizable to unique cleavage at selected genetic markers without artificial insertion of the target sequence due to the paucity of applicable DNA-binding proteins. A more general approach for recognition of endogenous DNA sequences might be a chemical approach based on oligonucleotide-directed triple helix formation. This strategy offers a generalizable DNA-binding motif that is capable of locally protecting the DNA from methylation at an overlapping target site while not affecting the activity of the methylase at its other recognition sites.[17,18] Triplex formation, global methylation, and triple helix disruption result in

[5] G. Felsenfeld, D. R. Davies, and A. Rich, *J. Am. Chem. Soc.* **79,** 2023 (1957).
[6] M. N. Lipsett, *J. Biol. Chem.* **239,** 1256 (1964).
[7] P. Rajagopal and J. Feigon, *Biochemistry* **28,** 7859 (1989).
[8] C. de los Santos, M. Rosen, and D. Patel, *Biochemistry* **28,** 7282 (1989).
[9] T. J. Povsic and P. B. Dervan, *J. Am. Chem. Soc.* **111,** 3059 (1989).
[10] L. C.Griffin and P. B. Dervan, *Science* **245,** 967 (1989).
[11] D. A. Horne and P. B. Dervan, *J. Am. Chem. Soc.* **112,** 2435 (1990).
[12] P. A. Beal and P. B. Dervan, *Science* **251,** 1360 (1991).
[13] P. B. Dervan, *in* "Nucleic Acids and Molecular Biology" (F. Eckstein and D. M. J. Lilley, eds.), Vol. 2, p. 49. Springer-Verlag, Heidelberg, 1988.
[14] P. B. Dervan, *in* "Oligonucleotides as Inhibitors of Gene Expression" (J. S. Cohen, ed.), p. 197. MacMillan, London, 1990.
[15] M. Koob, E. Grimes, and W. Szybalski, *Science* **241,** 1084 (1988).
[16] M. Koob and W. Szybalski, *Science* **250,** 271 (1990).
[17] L. J. Maher, B. Wold, and P. B. Dervan, *Science* **245,** 725 (1989).
[18] J. C. Hanvey, M. Shimizu, and D. Wells, *Nucleic Acids Res.* **18,** 157 (1990).

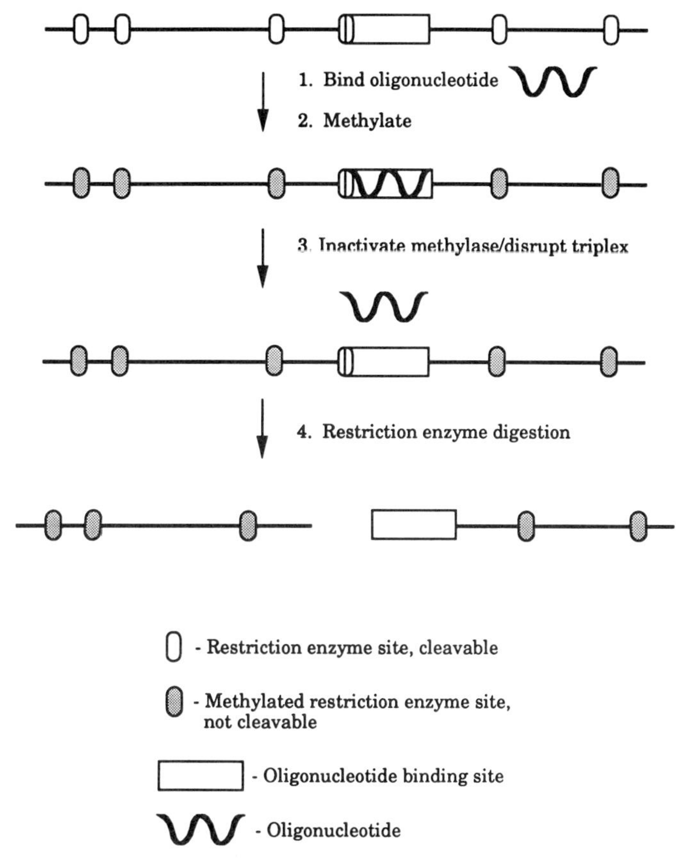

FIG. 1. General scheme for single-site enzymatic cleavage of genomic DNA by oligonucleotide-directed triple helix formation (4). Chromosomal DNA is equilibrated with an oligodeoxyribonucleotide in a methylase-compatible buffer containing polycation. *Eco*RI methylase, which methylates the central adenines of the sequence 5′-GAATTC-3′ and renders the sequence resistant to cleavage by *Eco*RI restriction endonuclease, is added and allowed to methylate to completion. The methylase is inactivated and the triple helix is disrupted at 55° in a high-pH buffer containing detergent. After washing extensively, the chromosomal DNA is reequilibrated in restriction enzyme buffer and cut to completion with *Eco*RI restriction endonuclease. The cleavage products are separated by pulsed-field gel electrophoresis and efficiencies quantitated by Southern blotting. (From Ref. 4.)

DNA that is resistant to endonuclease digestion at all sites except the one previously bound by the oligonucleotide[17,18] (Fig. 1). Subsequent restriction enzyme digestion produces nearly quantitative single-site cleavage of large genomic DNA at endogenous sequences.[4] In this chapter we fully describe the procedure for triple helix-mediated enzymatic cleavage of the *Saccharomyces cerevisiae* genome at a single predetermined site on chromosome III (340 kb) using a 24-base oligonucleotide and *Eco*RI methylation and restriction enzymes.

Materials and Methods

Media, Solutions, and Reagents

YPD medium: 1% (w/v) Bacto yeast extract, 2% (w/v) Bacto-peptone, 2% (w/v) glucose. Autoclave yeast extract and peptone for 20 min, cool, and add 1/25 vol of sterile 50% (w/v) glucose

Spheroplasting solution: 0.9 *M* sorbitol, 20 m*M* Tris, pH 7.5, 50 m*M* ethylenediaminetetraacetic acid (EDTA), pH 8.0, 7.5% (v/v) 2-mercaptoethanol, and 20 μg/ml zymolyase 100T (ICN, Costa Mesa, CA). Zymolyase is prepared as a stock solution at 2 mg/ml in 10 m*M* phosphate buffer, pH 7.0, and 50% (w/v) glycerol

Sucrose solution: Same as above, without zymolyase

NDS solution: 0.5 *M* EDTA, pH 8.0, 10 m*M* Tris, pH 7.5, 1% (w/v) laurylsarcosine, 1.0 or 0.5 mg/ml proteinase K (Bethesda Research Laboratories, Gaithersburg, MD)

Triple helix/*Eco*RI methylase solution (THEM): 100 m*M* NaCl, 100 m*M* Tris-HCl, 10 m*M* EDTA, 2 m*M* spermine tetrahydrochloride (nuclease-free grade; Sigma, St. Louis, MO). Adjust final solution to pH 6.6, 7.0, 7.4, or 7.8

Inactivation buffer: 1% (w/v) laurylsarcosine, 100 m*M* Tris-HCl, 10 m*M* EDTA. Adjust final solution to pH 9.0

Wash buffer: 10 m*M* Tris-HCl, 10 m*M* EDTA at final pH 9.0

2× KGB buffer: 200 m*M* potassium glutamate, 50 m*M* Tris–acetate, 20 m*M* magnesium acetate. Adjust to pH 7.6 with acetic acid or potassium hydroxide. Add bovine serum albumin (BSA) and 2-mercaptoethanol to a final concentration of 100 μg/ml and 1 m*M*, respectively

Preparation of Yeast Chromosomal DNA

1. Add 50 μl of saturated overnight cultures of haploid yeast strain SEY6210 or recombinant derivatives[3,4] to 100 ml of YPD medium. Grow overnight to OD_{600} between 4 and 6.

2. Harvest the cells by centrifugation at 5000 rpm for 5 min at 5°. Discard the supernatant and wash with 20 ml of 50 m*M* EDTA, pH 8.0. Harvest and drain the pellet well.

3. Resuspend the yeast in 6–8 ml of spheroplasting solution and incubate at 37° for 60 min. Gently swirl the solution every 15 min to resuspend

the cells. After spheroplasting is complete, centrifuge the cells as before, decant, and resuspend in 6–8 ml of sucrose solution. Add an equal volume of liquefied 1.4% LMP Incert (FMC, Rockland, ME) agarose in sucrose solution preheated to 42°, mix, and cast in 6 × 6 × 100 mm molds. Solidify the agarose on ice to prevent leakage.

4. Carefully remove the plugs from the molds and transfer two plugs (~3.5 ml/plug) to 25 ml of NDS solution in a 50-ml conical centrifuge tube. Incubate with gentle shaking at 50° for 24 hr. The plugs, which were originally pale white, become transparent on lysis. Decant and add a second 25-ml aliquot of NDS solution with 0.5 mg/ml proteinase K and incubate at 50° for an additional 3 hr.

5. Wash the plugs three times (with 30 ml for 20–30 min each time) with gentle shaking using 0.5 *M* EDTA, pH 8.0. Repeat the three washes with 10 m*M* EDTA. Add 1/100 vol of 100 m*M* phenylmethylsulfonyl fluoride (PMSF) dissolved in 100% ethanol and incubate on shaker for at least 1 hr to inactivate any residual proteinase K. Complete five 20- to 30-min washes with 30 ml of 10 m*M* EDTA and store the plugs at 4° until use.

Comments. DNA prepared from SEY6210 was more susceptible to degradation during preparation than other yeast strains tested. For this reason the standard procedure for DNA preparation from *S. cerevisiae*[19] was modified to maximize chromosomal integrity. In addition to the preparation in agarose blocks described above, DNA was prepared in agarose beads.[16] Little difference was found in either the quality of the DNA or its reactivity in subsequent enzymatic treatments.

Preparation of Oligodeoxynucleotides

1. Synthesize oligonucleotides on a 1-μmol scale using commercially available *N,N*-diisopropyl-β-cyanoethyl phosphoramidites (BrdU and 5-MedC phosphoramidites are availble through Cruachem, Herndon, VA). Deprotect oligonucleotides in concentrated ammonium hydroxide at 55° overnight and rotoevaporate to dryness.

2. Purify oligonucleotides by 15% (w/v) polyacrylamide gel electrophoresis. Briefly visualize bands under a hand-held short-wavelength UV lamp and cut oligonucleotides from gel. Care should be taken with oligonucleotides containing BrdU due to UV-induced strand scission. Crush the polyacrylamide slice to a powder and extract the oligonucleotide in 8.0 ml of 200 m*M* NaCl, 1 m*M* EDTA, pH 8.0 at 37° overnight. Pass extracted

[19] C. L. Smith, S. R. Klco, and C. R. Cantor, *in* "Genome Analysis: A Practical Approach" (K. Davies, ed.), p. 41. IRL Press, Oxford, 1988.

oligonucleotides through a 0.45-μm cellulose acetate Centrex filter (Schleicher & Schuell, Keene, NH) to remove polyacrylamide fragments.

3. Desalt the oligonucleotides by slowly loading eluent on disposable C_{18} reversed-phase column (Waters, Milford, MA). Remove salt by washing extensively with 20 ml of distilled water, and eluting in 2.0 ml of 40% (v/v) acetonitrile in water. Determine the concentration by OD_{260} absorbance. Transfer oligonucleotides to Eppendorf tubes in 20-nmol aliquots and dry on a Speed-Vac. Store at $-20°$ in the dark until use.

Triple Helix-Mediated Enzymatic Cleavage

1. Cut 1.5-mm slices of yeast plugs (~50-μl volume) and transfer to 1.8-ml flat-bottom microcentrifuge tubes. Wash twice with 1.0 ml of THEM buffer for 15 min. Decant and overlay the plug with 120 μl of THEM buffer.

2. Add 2 μl of 100 μM oligodeoxynucleotide to the overlay and incubate at room temperature on shaker (150 rpm) for 8 hr to facilitate oligonucleotide diffusion and triple helix formation.

3. Add 1 μl of 20 mg/ml acylated BSA (nuclease-free grade; Sigma), 1 μl of 32 m*M* *S*-adenosylmethionine in 5 m*M* H_2SO_4, 10% ethanol, and 2 μl of 40 units/μl *Eco*RI methylase (New England BioLabs, Beverly, MA). Incubate an additional 3–5 hr at room temperature with shaking.

4. Simultaneously disrupt the triplex and inactivate the methylase by removing the supernatant above the plug and adding 1.0 ml of inactivation buffer. Heat plugs to 55° for 30 min. Place tubes on ice following inactivation to harden the agarose plug before decanting.

5. Remove laurylsarcosine, oligonucleotide, and methylase from plug by washing four times, 15 min each time, with 1.0 ml of wash buffer.

6. Equilibrate the DNA in restriction enzyme buffer by washing twice for 15 min with 1.0 ml of 2× KGB. Decant and overlay with 120 μl of buffer. Add 1 μl of 20 units/μl *Eco*RI restriction enzyme (New England BioLabs) and digest DNA for 6 hr at room temperature on a shaker.

7. Decant plugs and add 1.0 ml of 0.5× TBE. Disrupt any DNA–protein interactions by heating to 55° for 10 min and cooling quickly on ice.

8. Resolve cleavage products by pulsed-field gel electrophoresis[20,21] using a 0.5× TBE, 1% (w/v) agarose gel at 14°, 200 V on a CHEF II system (Bio-Rad, Richmond, CA). Ramp switch times from 10 to 40 sec for 18 hr and 60 to 90 sec for 6 hr. Visualize products by ethidium bromide

[20] D. C. Schwartz and C. R. Cantor, *Cell (Cambridge, Mass.)* **37,** 67 (1984).

[21] G. F. Carle and M. V. Olson, *Nucleic Acids Res.* **12,** 5647 (1984).

staining (500 ng/ml) and Southern blotting with 0.45-μm Nytran membranes according to manufacturer protocols (Schleicher & Schuell).

Comments. The incubation time necessary for complete methylation is dependent on the pH of the buffer. The pH optimum is 8.0 for *Eco*RI methylase, but triple helix formation is more effective at neutral pH. Methylation is complete within 3 hr at pH 7.4, but 4.5 hr is necessary at pH 6.6. Thus, the methylation time must be optimized for the conditions chosen.

Commercial preparations of *Eco*RI methylase contain endonuclease impurities that are active in the presence of magnesium. Fortunately, methylases do not require a metal cofactor for activity. To prevent nonspecific degradation of the chromosomal DNA, the methylation reactions were performed without magnesium in the presence of the metal chelator EDTA. After methylation is complete, the methylase impurities must be inactivated in warm detergent before addition of the restriction enzyme buffer.

Because triple helix formation can specifically block the activity of both methylases and restriction enzymes[17,18] it is necessary to remove the oligonucleotide from the target site. Utilizing the observation that triple helix formation is temperature, polycation, and pH dependent[1,3] the complex can be readily disrupted in the absence of spermine at moderate temperatures and high pH. After triplex disruption and before the addition of magnesium (a dication necessary for the restriction enzyme that also stabilizes triple helix formation) the plug must be thoroughly washed to remove the oligonucleotide.

Two-Dimensional Pulsed-Field Gel Electrophoresis

1. Follow steps 1–3 as described above. After complete methylation of the DNA but before restriction enzyme digestion, separate the yeast chromosomes by pulsed-field gel electrophoresis in a 0.5× TBE, 0.8% (w/v) Incert LMP agarose gel at 14° and 200 V. Set switch times at 60 sec for 18 hr and 90 sec for 6 hr to resolve all the yeast chromosomes fully. Include an untreated control lane to serve as a size standard.

2. Cut the size standard from the geland stain with ethidium bromide. Do not stain the methylated DNA. Based on the location of the chromosomal bands in the standard lane, caerfully cut a full-length strip of methylated DNA and place it in a 15-ml centrifuge tube containing 4 ml of 2× KGB buffer. Equilibrate for 10 min at room temperature.

3. Decant and add 4 ml of 2× KGB buffer and 10 μl of *Eco*RI restriction enzyme (200 units). Digest at room temperature with shaking (60 rpm) for 6 hr.

4. Load strip onto a second 1% (w/v) agarose pulsed-field gel containing 0.5× TBE at 14° and 200 V. Set switch times as in step 1. Visualize products by ethidium bromide staining.

Results and Discussion

Studies with plasmid DNA demonstrate that oligonucleotide-directed triple helix formation in combination with appropriate methylases and restriction enzymes can limit the operational specificity of restriction enzymes to endonuclease recognition sequences that overlap oligonucleotide-binding sites.[17,18] To demonstrate the feasibility of single-site enzymatic cleavage of genomic DNA by triple helix-mediated Achilles' heel cleavage, a sequence containing an overlapping 24-bp purine tract and 6-bp *Eco*RI site was inserted proximal to the *LEU2* gene[22] on the short arm of chromosome III (340 kb) by homologous recombination (Fig. 2.)[3] Oligodeoxynucleotides designed to form triple helix complexes that overlap half of the *Eco*RI recognition site were synthesized with CT, ^{Me}CT, and ^{Me}C^{Br}U deoxynucleotides (Fig. 2).[9] The genetic map of yeast chromosome III[23,24] and affinity cleaving data[3] indicate that cleavage at the target site should produce two fragments 110 ± 10 and 230 ± 10 kb in size (Fig. 2).

Use of the ^{Me}CT oligodeoxynucleotide for triple helix formation at pH 7.4 followed by *Eco*RI methylase and endonuclease treatments resulted in exclusive cleavage of chromosome III at the target site (Fig. 3, lane 1). No cleavage was detected on any other chromosomes under these conditions. Cleavage was not observed in the absence of oligonucleotide (lane 2, Fig. 3) or in a yeast strain lacking the target sequence (lane 3, Fig. 3). The expected 110-kb product was visualized with ethidium bromide staining and confirmed by Southern blotting with a *HIS4* marker (Fig. 3C).[25] The 230-kb product comigrated with chromosome I, but was detected by Southern blotting with a *LEU2* marker (Fig. 3B).[22] The cleavage efficiency was 94 ± 2% with a 3-hr methylase incubation. Longer methylation times resulted in a gradual reduction in the observed cleavage efficiency, suggesting that the oligonucleotide dissociation rate might be the limiting factor for efficiency in this system.[26]

The specificity of triple helix formation has been shown to be pH dependent.[1-3] At lower pH, sequences of near but imperfect similarity can

[22] A. Andreadis, Y. P. Hsu, G. B. Kohlhow, and P. Shimmel, *Cell (Cambridge, Mass.)* **31,** 319 (1982).

[23] R. K. Mortimer and D. Schild, *Microbiol. Rev.* **49,** 181 (1985).

[24] G. F. Carle and M. V. Olson, *Proc. Natl. Acad. Sci. U.S.A.* **82,** 3756 (1985).

[25] J. K. Keesey, R. Bigelis, and G. R. Fink, *J. Biol. Chem.* **254,** 7427 (1979).

[26] L. J. Maher, P. B. Dervan, and B. Wold, *Biochemistry* **29,** 8820 (1990).

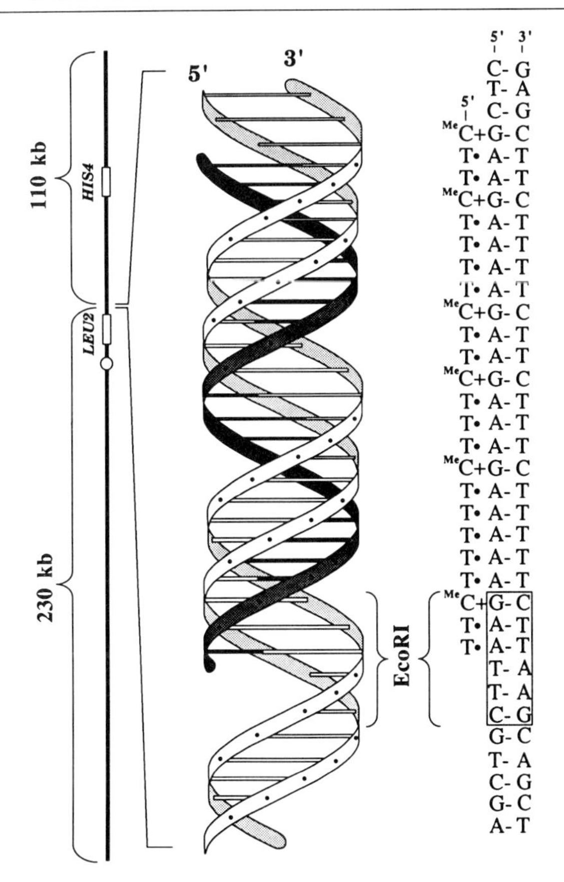

FIG. 2. *Left:* Genetic map of *S. cerevisiae* chromosome III. The locations of the *HIS4* and *LEU2* loci (boxes), the centromere (circle), and the triple helix-*Eco*RI target site are indicated. The expected sizes of the cleavage products are shown. *Right:* Schematic diagram of the triple helix complex overlapping the *Eco*RI methylase–endonuclease site. The pyrimidine oligonucleotide is bound in the major groove parallel to the purine strand of the DNA duplex, and covers half of the *Eco*RI site. (From Ref. 4.)

be bound and subsequently cleaved. In agreement with this observation, secondary cleavage sites were revealed as a function of oligonucleotide composition and pH (Fig. 4). Using either the ^{Me}CT or ^{Me}C^{Br}U oligodeoxynucleotides at or below pH 7.4, 170 secondary cleavage product was observed (Fig. 4, lanes 5 and 9). The cleavage site can be assigned to chromosome II (820 kb) by two-dimensional pulsed-field gel electrophoresis (Fig. 5). Additional secondary cleavage sites were observed with the ^{Me}C^{Br}U oligodeoxynucleotide at pH 6.6 (Fig. 4, lane 9). The 180- and

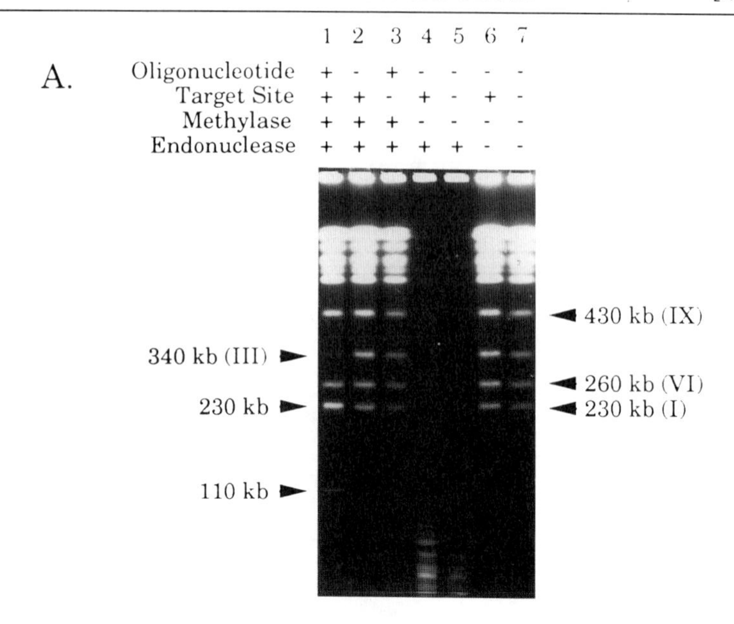

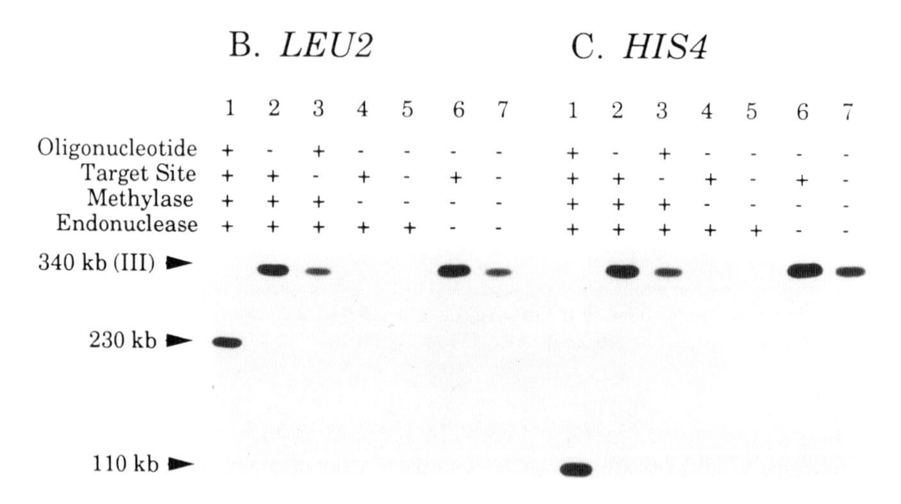

FIG. 3. Single-site enzymatic cleavage of yeast genomic DNA with reagents indicated above the lanes. Products were visualized by ethidium bromide staining (A) and Southern blotting with *LEU2* (B) and *HIS4* (C) chromosome markers. (From Ref. 4.)

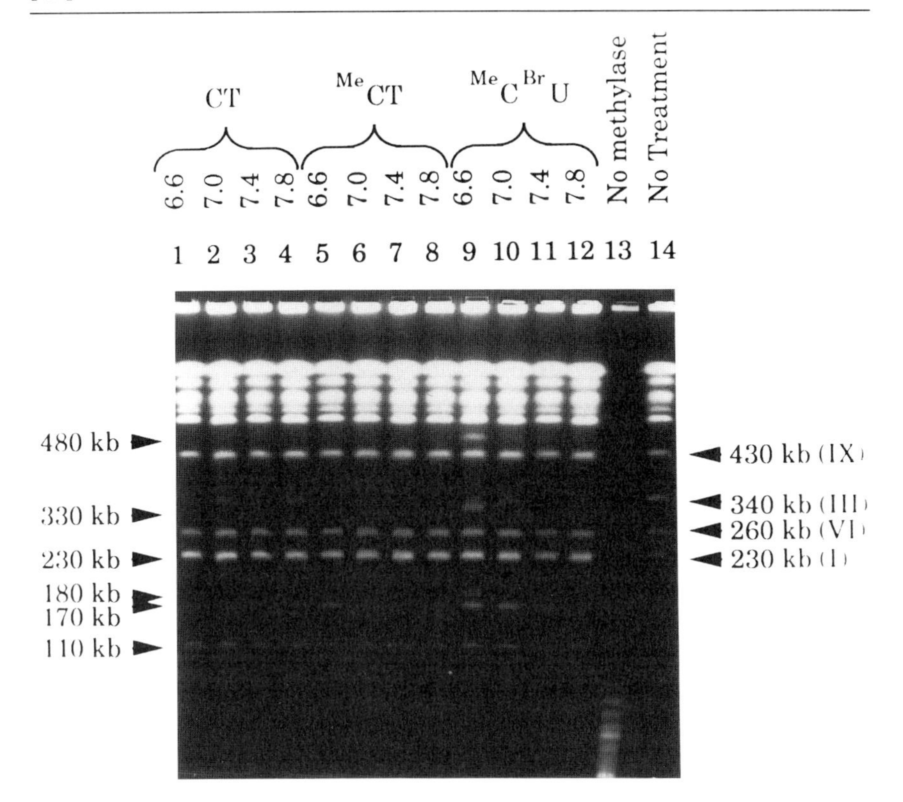

FIG. 4. Triple helix-mediated enzymatic cleavage of the yeast genome as a function of oligonucleotide composition and pH. Lanes 1–12: Reaction on a yeast strain containing the triple helix target site with oligonucleotides (CT, ^{Me}CT, and ^{Me}C^{Br}U) and pH (6.6, 7.0, 7.4, and 7.8) as indicated above the lanes. (From. Ref. 4.)

480-kb products were assigned to chromosome XI (660 kb), and the 330- and 780-kb products were assigned to the VII/XV doublet (1100 kb) (Fig. 5). Cleavage at all secondary sites could be eliminated at a threshold pH for each oligonucleotide (Fig. 4, lanes 2, 8, and 12). Because of the greater triple helix stability at pH values optimal for the methylase (pH 8.0), ^{Me}C-substituted oligonucleotides are preferred over C substitutions in this assay. ^{Br}U substitutions offered little additional pH stability and increased the number of secondary cleavage sites observed. This property of ^{Br}U-substituted oligonucleotides is useful in searching for endogenous homopurine-methylase sites in unsequenced DNA by using degenerate oligodeoxynucleotides containing ^{Me}C and ^{Br}U.

Many methylases, including AluI and Dam, exhibit optimal activity in low salt buffers. Unfortunately, spermine rapidly precipitates DNA at low sodium chloride concentrations. To prevent precipitation during triple

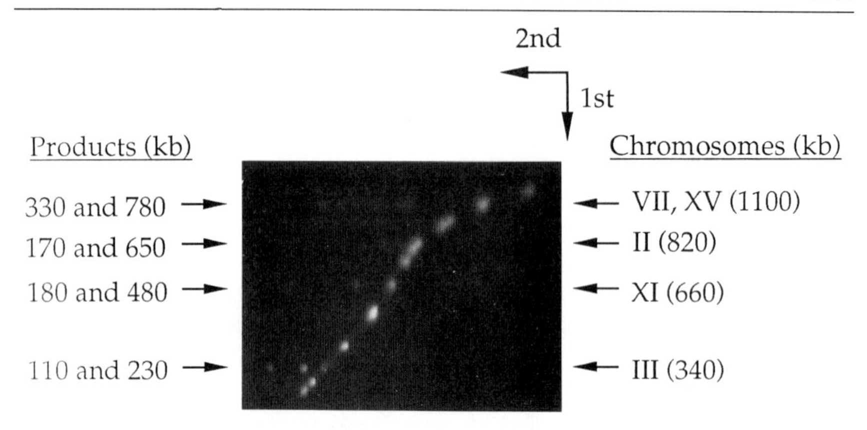

FIG. 5. Two-dimensional pulsed-field gel analysis of $^{Me}C^{Br}U$ oligonucleotide incubated at pH 6.6. Intact chromosomes form a diagonal with cleavage products located immediately to the left of the chromosome of origin.

helix formation and methylation reactions, the salt concentration must be adjusted to 25–50 m*M*. The enzyme activity retained under these conditions is sufficient to yield complete methylation of genomic DNA if adequate time and enzyme concentration are employed.

Not all methylases are compatible with this technique. *Bam*HI methylase is significantly inhibited by spermine. *Hae*III restriction enzyme can cut hemimethylated sites, making it much more difficult to achieve complete methylation.[27] An additional concern when working with genomic DNA, particularly with enzymes sensitive to CpG methylation such as *Hpa*II and *Hha*I, is the prior methylation state of the target DNA. A large yet variable percentage of CpG sequences in mammalian cells are methylated *in vivo*.[28,29] Achilles' heel cleavage at these sites would not be possible unless methylation-free DNA could be prepared.

Affinity cleaving using oligodeoxynucleotide–EDTA–Fe, a technique that generates cleavage at all sites of oligonucleotide binding but in low yield, demonstrated that triplex formation is occurring at several positions on the yeast chromosomes besides the single target site.[3] Triple helix-mediated Achilles' heel cleavage exposes only those sites that are tightly bound and partially overlap a complete methylase/endonuclease site.[4,17,18] The probability of finding an endogenous homopurine sequence that overlaps an *Eco*RI methylase site is quite low, possibly one every 500 kb or

[27] C. Kessler and H. J. Holtke, *Gene* **47,** 1 (1986).

[28] A. Bird, *Nature (London)* **321,** 209 (1986).

[29] A. Bird, M. Taggart, M. Frommer, O. J. Miller, and D. Macleod, *Cell (Cambridge, Mass.)* **40,** 91 (1985).

less.[3] The frequency of targetable sequences is increased, however, when other methylase/endonuclease pairs are considered. In addition to *Taq*I and *Eco*RI,[4,17,18] single-site protection of plasmid DNA has been achieved with *Msp*I, *Hpa*II, and *Alu*I methylase/endonucleases pairs. Several other methylases could be used, but remain untested. To date, complete methylation of yeast genomic DNA under conditions compatible with triple helix formation has been achieved with *Eco*RI, *Alu*I, and *dam* methylases. The list will expand as target sites of interest are identified. Although all enzyme sets tested might not be effective, the use of several different methylase/endonuclease combinations could increase the frequency of cleavable sites in endogenous sequences to as high as one every 10–30 kb.[3,27]

The generalizability of triple helix-mediated enzymatic cleavage affords high specificity that can be readily customized to unique genetic markers without artificial insertion of a target sequence. Extensive sequencing to identify target sites could be avoided by using degenerate oligonucleotides to screen genetic markers for overlapping triple helix/endonuclease sites. The potential generalizability of triple helix-mediated Achilles' heel cleavage, a technique capable of near-quantitative cleavage at a single site in at least 14 megabase pairs of DNA, could assist in physical mapping of chromosomal DNA and expedite isolation of DNA segments linked to disease.

Acknowledgment

We are grateful to Dr. John Hanish for helpful discussions, the National Institutes of Health for grant support, and the Howard Hughes Medical Institute for a predoctoral fellowship to S.A.S.

[28] Conferring New Cleavage Specificities of Restriction Endonucleases

By MICHAEL KOOB

Introduction

Achilles' cleavage (AC) is a simple, general method for combining the cleavage specificities of restriction endonucleases with the binding specificities of other DNA-binding molecules[1,2] (Fig. 1). With this technique, any restriction site that can be protected from methylation by its

[1] M. Koob, E. Grimes, and W. Szybalski, *Science* **241,** 1084 (1988).
[2] M. Koob, E. Grimes, and W. Szybalski, *Gene* **74,** 165 (1988).

cognate methyltransferase (MTase) can be made unique. Previous studies have demonstrated that AC allows both plasmid and intact chromosomal DNA to be efficiently and specifically cleaved at restriction sites protected from methylation by repressor proteins,[3,4] eukaryotic transcription factors,[5] synthetic oligodeoxynucleotides capable of forming a triple helix structure,[6,7] and by RecA–oligodeoxynucleotide nucleoprotein filaments.[5] Although most of these MTase-blocking agents have only a limited range of DNA-binding specificities, the binding specificity of RecA nucleoprotein filaments is derived entirely from the oligonucleotide moiety. RecA-mediated AC (RecA-AC), therefore, allows the recognition site for any restriction endonuclease on DNA of any size to be converted into a unique cleavage site.

The first section of this chapter describes a set of protocols for completely methylating both plasmid and intact chromosomal DNA; the second section discusses general aspects of the AC process; and the last section contains a protocol for RecA-AC, which is at this time the most general and therefore the most useful of the AC reactions.

Methylation of Plasmid and Genomic DNA

Obtaining complete methylation is the most critical aspect of AC. The high level of impurities present in many of the commercially available MTase preparations makes methylation reactions less straightforward than enzymatic digestion. Although partial methylation or a small amount of degradation by nucleases has little noticeable effect on plasmid DNA, incomplete methylation or nuclease activity is disastrous when working with chromosomal DNA. The following protocol is designed to avoid these problems. Precautions are taken to ensure that the MTase as well as the methyl donor *S*-adenosylmethionine (SAM), which readily undergoes alkaline hydrolysis, remain active. In addition, nonspecific degradation by contaminating nucleases is avoided by performing the methylation reaction in the absence of divalent cations, which are required for most nucleases but not by MTases, and then inactivating the nucleases before Mg^{2+} is added for the subsequent cleavage reaction.

[3] M. Koob and W. Szybalski, *Science* **250,** 271 (1990).

[4] E. Grimes, M. Koob, and W. Szybalski, *Gene* **90,** 1 (1990).

[5] M. Koob, unpublished data, 1991.

[6] J. C. Hanvey, M. Shimizu, and R. D. Wells, *Nucleic Acids Res.* **18,** 157 (1989).

[7] S. A. Strobel and P. B. Dervan, *Nature (London)* **350,** 172 (1991).

Buffers and Reagents

Standard 10× MTase buffer: Prepare a buffer of 0.5 *M* Tris-HCl (pH 7.5/25°) and 0.1 *M* ethylenediaminetetraacetic acid (EDTA) (pH 8.0/25°) from sterile stock solutions and store at room temperature (see note *a* below). Dilute to 1× with pure, sterile H_2O and adjust to 10 m*M* dithiothreitol (DTT) [see note (b) below] just before use

S-Adenosylmethionine (SAM): SAM is typically supplied with the MTase. Store this stock solution in 1-μl portions at −20°. Just before use dilute with sufficient ice-cold 5 m*M* H_2SO_4 to make a 1.6 m*M* (20×) solution

MTases: Divide the sample in small portions when first received from the supplier and store at −20°. Use each aliquot only a few times to avoid inactivation associated with repeated temperature shifts

TE: 10 m*M* Tris-HCl, 1 m*M* EDTA (pH 8.0/25°)

TEX: Add 0.01% (v/v) Triton X-100 (Sigma, St. Louis, MO) to TE and store at room temperature

ES: Add 1% (w/v) sodium *N*-lauroylsarcosine to sterile 0.5 *M* EDTA (pH 8.0/25°) and store at room temperature

Notes on Buffers and Reagents

a. Although most MTases will work well in this buffer, some require NaCl or potassium glutamate for full activity. Follow the recommendations of the supplier.

b. An active reducing agent is critical for optimal methylation. Use 1 *M* DTT stock stored in small aliquots at −20°.

Methylation of Plasmid DNA

Methylation reactions on plasmid DNA are performed in much the same way as are typical digestion reactions with restriction endonucleases. Set up a reaction with sterile doubly distilled H_2O, 0.1 vol 10× MTase buffer (with fresh DTT), 0.1 vol bovine serum albumin (BSA; 1 mg/ml stock), and plasmid DNA and then add the MTase and freshly diluted SAM. Incubate at the recommended temperature (typically 37°) for 10 to 30 min (depending on the activity of the MTase). Stop the reaction by heating to 65° for 10 min or by deproteinization (e.g., column purification). Adjust the buffer to that suitable for the restriction enzyme (add salt and sufficient Mg^{2+} to compensate for the EDTA) or precipitate the DNA and resuspend it in the appropriate restriction buffer. Digest the modified DNA with a restriction endonuclease.

Methylation of Genomic DNA Embedded in Agarose Microbeads

1. Pipette 20–30 μl of microbeads [see note (a) below] into a clear or light-colored Eppendorf microfuge tube.
2. Wash the beads [see note (a) below] once with 200 μl of TEX and then twice with 200 μl of 1× methylation buffer (fresh DTT) without BSA.
3. Add 0.1 vol BSA (2–3 μl of sterile 1 mg/ml stock solution), MTase [see note (b) below], and 1–1.5 μl of freshly prepared 20× SAM to the buffer-equilibrated pellet and mix thoroughly with the end of the pipette tip. Place the microfuge tube in the appropriate water bath and incubate for 1 hr (shaking is not necessary).
4. To inactivate the MTase and any contaminating nucleases, add 25 μl of ES to the methylation reaction and incubate at 52° for 15–30 min.
5. Wash the beads once with 200 μl of TE, once with 200 μl of TEX, and then twice with 200 μl of the supplier's recommended digestion buffer without BSA.
6. Digest and analyze the DNA by pulsed-field gel electrophoresis (PFGE) as described [see note (a) below].

Notes on Methylating Genomic DNA

a. The preparation and enzymatic manipulation of genomic DNA in agarose microbeads is described in [2] in this volume.[8] Refer to [2] for more details concerning the physical handling, digestion, and electrophoresis of DNA in microbeads.

b. The amount of MTase needed to completely methylate the DNA should be determined empirically for each MTase preparation.

Achilles' Cleavage

The major steps in the AC procedure are outlined in Fig. 1. To perform an AC reaction, a methylation reaction is performed as described in the preceding section. However, the MTase-blocking agent is added to the DNA after the addition of BSA but before the addition of MTase and SAM. Inactivation of the methyltransferase, contaminating nucleases, and the MTase-blocking agent and the subsequent digestion with the restriction endonuclease are performed as described.

Methylation protection can be mediated by any DNA-binding molecule or group of molecules that form sequence-specific complexes capable of excluding an MTase. In practice, however, conditions must be found in which both the blocking agent binds DNA and the appropriate MTase are

[8] M. Koob and W. Szybalski, this volume, [2].

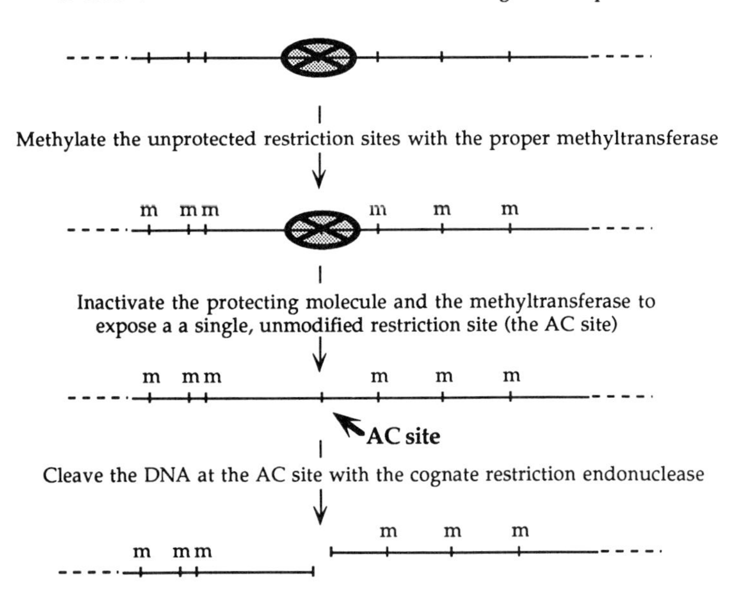

FIG. 1. The Achilles' cleavage procedure for creating new cleavage specificities.

active. In some cases, such as LacI-mediated AC (LacI-AC), the binding conditions and the conditions for methylation coincide exactly. Often, however, the optimal buffer or reaction temperatures are not the same and a compromise must be found between binding and methylation conditions. In general, it is better to use conditions more favorable for binding and simply use more MTase than would otherwise be required. If the AC reaction will eventually be applied to chromosomal DNA, conditions should first be optimized on a model plasmid template containing a known AC site. If at all possible, Mg^{2+} should not be included in the reaction buffer (refer to the preceding section and to the note below concerning nucleases in MTases).

A protocol for RecA-AC, which allows DNA of any size to be efficiently cleaved at any predetermined restriction site, is given in the next section.

RecA-Mediated Achilles' Cleavage

The RecA protein from *Escherichia coli* promotes the strand exchange of single-stranded DNA (ssDNA) fragments with homologous duplex DNA (dsDNA) in an ATP-dependent process. If the nonhydrolyzable ATP analog adenosine 5′-(γ-thio)triphosphate (ATPγS) is used in place of ATP, the RecA–ssDNA presynaptic nucleoprotein filament remains in a stable complex with the homologous dsDNA.[9–11] These nucleoprotein filaments can act as the MTase-blocking agent in an AC reaction. Since the binding specificity of the nucleoprotein filament is derived entirely from the ssDNA moiety, any restriction site can be made unique with the RecA-AC procedure.

In this procedure, the RecA nucleoprotein filament is first formed by polymerizing the RecA on an oligonucleotide of any sequence in the presence of ATPγS. The target dsDNA is then added. After a half-hour incubation, the concentration of Mg^{2+} is increased to improve the efficiency of methylation protection (presumably by stabilizing the filament–dsDNA complex) and a methylation reaction is performed with the appropriate MTase.

Buffers and Reagents

10× RecA/methylation buffer: Prepare a buffer of 250 m*M* Tris–acetate (pH 7.5/25°) and 40 m*M* magnesium acetate from sterile stock solutions and store at room temperature. Dilute to 1× with sterile doubly-distilled H_2O and adjust to 10 m*M* DTT just before use

Oligodeoxynucleotide (oligo): Synthesize an oligo with a sequence identical to that of the targeted sequence. Any unique sequence containing a restriction site for which an MTase is available can be used. For plasmid DNA, the oligo should be at least 24 nucleotides (nt) in length. However, longer oligos (30–36 nt) result in more efficient RecA-AC of plasmid DNA. The oligo should be at least 60 nt for AC of chromosomal DNA embedded in agarose. Resuspend the purified oligo in H_2O

RecA: This can be purchased commercially (Pharmacia, Piscataway, NJ). Calculate the molarity of the RecA stock (M_r 37,800) from the concentration (mg/ml) provided by the manufacturer

ATPγS: Several companies sell this nonhydrolyzable ATP analog (e.g., Sigma). Make a solution in water so that its concentration is two to four times that of the RecA stock (e.g., 0.4 m*M* for 0.165 m*M* RecA) and store in small aliquots at −70°.

MTases: Unlike most MTase-blocking agents used in AC reactions, RecA DNA-binding activity is strictly dependent on the presence of

[9] M. M. Cox and I. R. Lehman, *Proc. Natl. Acad. Sci. U.S.A.* **78,** 3433 (1981).

[10] P. W. Riddles and I. R. Lehman, *J. Biol. Chem.* **260,** 170 (1985).

[11] S. M. Honigberg, D. K. Gonda, J. Flory, and C. M. Radding, *J. Biol. Chem.* **260,** 11845 (1985).

Mg^{2+}. The MTases used for RecA-AC of chromosomal DNA must therefore be free of contaminating nucleases. To test for nucleases, incubate yeast or λDNA markers with the MTase in 1× RecA buffer for 1 hr at 37° and then analyze by PFGE (no digestion). Degradation will be obvious if nucleases are present

Titrating RecA Stock with Oligodeoxynucleotide

Experimentally, the challenge in applying RecA-AC is to optimize the sequence-specific methylation protection afforded by RecA–oligo filaments and to eliminate the nonspecific protection that results from RecA monomers interacting with the dsDNA. To do this, it is vital to form the RecA–ssDNA complex with the proper molar ratio of RecA monomers to nucleotides. A nucleotide: RecA monomer ratio of about 3 : 1 works best (e.g., ~0.165 μg oligo and ~6.25 μg RecA), but the working ratio should always be determined experimentally for your reagents with a titration reaction. Make a series of dilutions of the oligo in water and perform the RecA-AC reaction on linearized plasmid DNA as described below. Whenever possible, work with plasmid DNA that contains the sequence to be cleaved and at least one other (preferably more) recognition site for the enzyme used. In this case, choose the ratio that optimizes protection of the targeted restriction site and eliminates the nonspecific protection of the other site(s). If a clone of the target sequence is not available, work with a frequent-cutting (4 bp) restriction/MTase pair and any plasmid DNA to assay nonspecific protection and choose the ratio at which nonspecific protection is just eliminated.

RecA-Mediated Achilles' Cleavage of Plasmid DNA

1. To form a RecA–oligo nucleoprotein filament, combine 1 μl of 10× RecA buffer, 6 μl of doubly distilled H_2O, 1 μl of oligo (at the proper concentration; see above), and 1 μl of RecA stock. Mix well with the pipette tip. Warm to 37° for 30–60 sec [see note (a) below], add 1 μl of ATPγS, and incubate at 37° for another 10 min [see note (b) below].

2. Combine the plasmid DNA, 2 μl of 10× RecA buffer (40 m*M* magnesium acetate, fresh DTT), 3 μl of BSA (1 mg/ml) and sufficient double-distilled H_2O to bring the volume to 17 μl. Add this mixture to the RecA–oligo filament. Incubate for 30 min at 37°.

3. Adjust the concentration of magnesium acetate to 8.0 m*M* [see note (c) below]. Add the MTase, 1 μl of the freshly diluted SAM stock (2.4 m*M*), and incubate at 37° for 10–30 min.

4. Inactivate the MTase and the RecA (e.g., 65°, 15 min). Adjust the buffer to that recommended for the appropriate restriction enzyme and digest the modified plasmid DNA in the usual way.

Notes on RecA-Mediated Achilles' Cleavage of Plasmid DNA

a. This is important for forming complete nucleoprotein filaments. The reaction temperature should not be allowed to fall significantly below 37° when the ATPγS is added.

b. As an alternative to working at precisely the proper ratio of RecA to oligo, excess RecA can be used if 1 μl of oligo(dT) (24–60 nt; at the same concentration as the other oligo) is added at this point to absorb the RecA monomers still in solution.

c. Magnesium acetate concentrations ranging from 10 to 20 m*M* have about the same effect.

RecA-Mediated Achilles' Cleavage of Genomic DNA Embedded in Agarose Microbeads

1. Wash 20–30 μl of microbeads once with 200 μl of TEX and twice with 200 μl of 1× RecA (4 m*M* magnesium acetate and fresh DTT). Add 2–3 μl of BSA (1 mg/ml) to the microbead pellet and warm to 37° [see note (a) below].

2. Prepare the RecA filament by setting up a reaction mixture with 1 μl of 10× RecA, 4 μl of doubly distilled H_2O, 2 μl of oligo (see above), and 2 μl of RecA stock. Warm to 37°. After 30–60 sec add 1 μl of ATPγS. Incubate at 37° for 10 min. If excess RecA is used [see note (b) above], add 1 μl of oligo(dT) at this point.

3. Add the RecA–oligo filament (5–10 μl) to the microbeads and add sufficient magnesium acetate to maintain the concentration of magnesium acetate at 4 m*M*. Place the reaction on ice and diffuse overnight [see note (b) below] for an alternative to steps 2 and 3.

4. Warm the reaction mixture to 37° and incubate for 30 min. Adjust the concentration of magnesium acetate to 8.0 m*M* and add fresh DTT to 10 m*M*. Add the MTase and 1–1.5 μl of SAM and incubate at 37° for 1 hr.

5. Inactivate the RecA and the MTase by adding 25 μl of ES to the reaction and incubate at 52° for 30 min. Wash the microbeads once with 200 μl of TE, once with 200 μl of TEX, and twice with 1× restriction buffer (no BSA). Add 0.1 vol of BSA and the restriction enzyme (0.5–1 μl) and digest for 1 hr.

Notes on RecA-Mediated Achilles' Cleavage of Genomic DNA

a. Details concerning the preparation and manipulation of genomic DNA in agarose microbeads are given in the first section of this chapter and in [2] in this volume.[8]

b. To form the RecA filament in the presence of genomic DNA, add 1–2 μl of the oligo to the microbead pellet and incubate at 37° for 15 min. Add 1–2 μl (to correspond with the amount of oligo added) of the RecA stock and continue incubating at 37° for 10 min. Add 1 μl of ATPγS and proceed as in step 4.

Concluding Remarks

Restriction enzymes that efficiently cleave DNA at short but specific sequences have played a central role in the development of modern molecular genetics. By dramatically increasing the specificity of these enzymes, AC amplifies the power of conventional recombinant DNA technology, particularly when it is applied to the physical mapping and precise molecular dissection of multimegabase genomes.

[29] Modified T7 DNA Polymerase for DNA Sequencing

By CARL W. FULLER

When bacteriophage T7 infects its *Escherichia coli* host, it directs the production of a phage-specific DNA polymerase.[1,1a] This enzyme is a two-subunit protein.[2–4] The smaller subunit, thioredoxin (M_r 12,000), is the product of the host *trxA* gene. The larger subunit (M_r 80,000) is the product of the T7 gene *5*. The two subunits are tightly associated, purifying in a 1 : 1 stoichiometry from infected cells. This enzyme has two known catalytic activities, DNA-dependent DNA polymerase and a 3′ → 5′-exonuclease activity that is active on both single-stranded and double-stranded DNA. Both of these activities are also present in purified gene *5* protein alone, although the DNA polymerase activity and the double-stranded exonuclease are both greatly increased by the binding of thioredoxin.[5] This increase has been attributed to the high processivity conferred by the thioredoxin subunit.[6]

[1] P. Grippo and C. C. Richardson, *J. Biol. Chem.* **246,** 6867 (1971).
[1a] S. L. Oley, W. Stratling, and R. Knippers, *Eur. J. Biochem.* **23,** 497 (1971).
[2] P. Modrich and C. C. Richardson, *J. Biol. Chem.* **250,** 5508 (1975).
[3] P. Modrich and C. C. Richardson, *J. Biol. Chem.* **250,** 5515 (1975).
[4] K. Hori, D. F. Mark, and C. C. Richardson, *J. Biol. Chem.* **254,** 11598 (1979).
[5] S. Tabor, H. E. Huber, and C. C. Richardson, *J. Biol. Chem.* **262,** 16212 (1987).
[6] H. E. Huber, S. Tabor, and C. C. Richardson, *J. Biol. Chem.* **262,** 16224 (1987).

Exonuclease-Deficient Forms of T7 DNA Polymerase

The 3′ → 5′-exonuclease activity of T7 DNA polymerase is quite potent, much more so than the exonuclease activity of *E. coli* DNA polymerase I. In fact, when assayed on single-stranded substrate in the absence of dNTPs it hydrolyzes DNA about as fast as the polymerase can synthesize DNA under optimal polymerization conditions.[7] Because chain-termination DNA sequencing relies on the stable incorporation of dideoxynucleotides at the 3′ ends of DNA,[8] it is not surprising that T7 DNA polymerase is not suitable for sequencing. T7 DNA polymerase catalyzes the incorporation of dideoxynucleotides into DNA, and also catalyzes their removal, resulting in little useful sequence information.

As early as 1979 it was recognized that the active sites for polymerase and exonuclease may be different because the enzyme had been isolated in two different forms, one with high exonuclease specific activity and another with low exonuclease activity.[8a,9] Prolonged incubation (in dialysis) of high exonuclease preparations of T7 DNA polymerase in the absence of ethylenediaminetetraacetic acid (EDTA) resulted in a reduction of the exonuclease activity without a reduction in polymerase activity, suggesting that the low exonuclease preparations were the result of modification during purification.[9]

Tabor and Richardson[10,11] reported an oxidation procedure that readily inactivated the exonuclease activity with little effect on the polymerase activity. This procedure requires oxygen, a reducing agent (dithiothreitol), and stoichiometric amounts of iron. They postulated that the iron binds to the enzyme, where cycles of reduction and oxidation generate reactive oxygen species that oxidize amino acids within the exonuclease active site.

The specific amino acids that, when oxidized, reduce exonuclease activity have not been identified, but the oxidized or modified form of T7 DNA polymerase was found to have several advantages for use in DNA sequencing using dideoxynucleotides.[12,13] These will be described in more detail below. This chemically modified form of T7 DNA polymerase is known commercially as Sequenase Version 1.0 T7 DNA polymerase (United States Biochemical Corp., Cleveland, OH).

[7] S. Adler and P. Modrich, *J. Biol. Chem.* **254,** 11605 (1979).
[8] F. Sanger, S. Nicklen, and A. R. Coulson, *Proc. Natl. Acad. Sci. U.S.A.* **74,** 5463 (1977).
[8a] H. Fischer and D. C. Hinkle, *J. Biol. Chem.* **255,** 7956 (1980).
[9] M. J. Engler, R. L. Lechner, and C. C. Richardson, *J. Biol. Chem.* **258,** 11165 (1983).
[10] S. Tabor and C. C. Richardson, *J. Biol. Chem.* **262,** 15330 (1987).
[11] S. Tabor and C. C. Richardson, U.S. Patent 4,946,786 (1990).
[12] S. Tabor and C. C. Richardson, *Proc. Natl. Acad. Sci. U.S.A.* **84,** 4767 (1987).
[13] S. Tabor and C. C. Richardson, U.S. Patent 4,795,699 (1989).

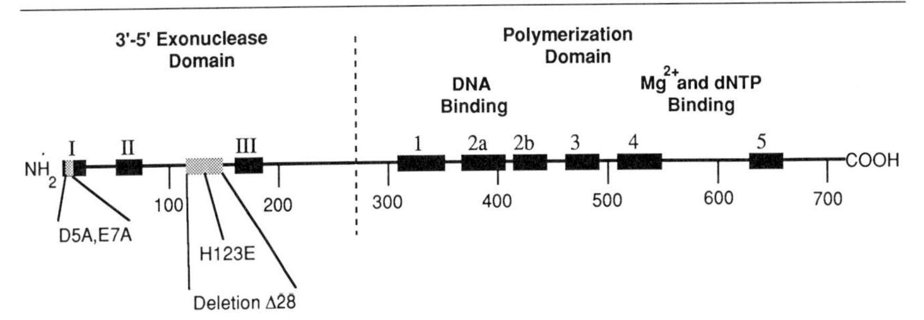

FIG. 1. Map of the amino acid sequence of T7 DNA polymerase. The product of gene 5 of bacteriophage T7 is a protein composed of 704 amino acids [J. J. Dunn and F. W. Studier, *J. Mol. Biol.* **8,** 452 (1983)] with both exonuclease and polymerase activities. Black bars labeled I–III are conserved amino acid sequence motifs in the exonuclease domain [L. Blanco, A. Bernad, M. A. Blasco, and M. Salas, *Gene* **100,** 27 (1991)]. Similar motifs in the polymerase domain are numbered 1–5. Site-directed mutations known to affect exonuclease activity are marked in gray. These include the Δ28 deletion mutant used for DNA sequencing [S. Tabor and C. C. Richardson, *J. Biol. Chem.* **264,** 6447 (1989)] and a double point mutation (D5A, E7A) that has been kinetically characterized [S. S. Patel, I. Wong, and K. A. Johnson, *Biochemistry* **30,** 511 (1991)].

The three-dimensional structure determination of the large fragment of *E. coli* DNA polymerase I (the Klenow enzyme) and site-directed mutagenesis have revealed distinct domains for polymerase and exonuclease activities.[14–16] There is considerable amino acid sequence homology between T7 DNA polymerase and the *E. coli* enzyme. Indeed, at least nine highly conserved sequence motifs have been identified in known DNA-dependent DNA polymerases.[17] These are outlined in Fig. 1, which shows the alterations of the exonuclease domain in a site-directed mutagenesis experiment that resulted in varying degrees of inactivation of the exonuclease without effecting polymerase activity.[18,19] A more recently characterized mutant[20] is also indicated in Fig. 1.

These mutagenesis studies confirm that the exonuclease domain of T7 DNA polymerase resides in the N-terminal portion of the molecule, as

[14] D. L. Ollis, P. Brick, R. Hamlin, N. G. Xoung, and T. A. Steitz, *Nature (London)* **313,** 762 (1985).

[15] P. S. Fremont, J. M. Freidman, L. S. Beese, M. R. Sanderson, and T. A. Steitz, *Proc. Natl. Acad. Sci. U.S.A.* **85,** 8924 (1988).

[16] V. Derbyshire, P. S. Freemont, M. R. Sanderson, L. Beese, J. M. Freidman, C. M. Joyce, and T. A. Steitz, *Science* **240,** 199 (1988).

[17] L. Blanco, A. Bernad, M. A. Blasco, and M. Salas, *Gene* **100,** 27 (1991).

[18] S. Tabor and C. C. Richardson, *J. Biol. Chem.* **264,** 6447 (1989).

[19] S. Tabor and C. C. Richardson, U.S. Patent 4,942,130 (1990).

[20] S. S. Patel, I. Wong, and K. A. Johnson, *Biochemistry* **30,** 511 (1991).

predicted by sequence homology. The D5A, E7A double mutation[20] is within exonuclease sequence motif I. The point mutation H123E[18] lies between exonuclease motifs II and III. This latter mutation reduces the exonuclease activity by more than 98% when measured using uniformly labeled DNA substrate but less than 50% when measured using 3′ end-labeled substrate DNA. Deletion mutations within this region, including a 28-amino acid deletion extending from Lys118 to Arg145 (called Δ28), further reduce the exonuclease-specific activity of the enzyme without reducing polymerase activity. Exonuclease activity of the enzyme produced from the Δ28 deletion is reduced to unmeasurable levels even when 3′-^{32}P-labeled DNA is used.[18] This enzyme also has excellent properties for DNA sequencing and has the commercial name Sequenase Version 2.0 T7 DNA polymerase.

T7 DNA polymerase Δ28 has a specific polymerase activity that is about sixfold greater than that of either native or chemically modified enzyme when activity is measured using primed M13 DNA as template.[18] Chemically modified T7 DNA polymerase, which has a low but detectable level of exonuclease (about 0.5%, when measured using uniformly labeled substrate), has a specific activity somewhat greater than that of the native enzyme. Since DNA sequencing depends on both the molarity and activity of polymerase, different unit definitions have been adopted for the commercial preparations of these enzymes.

Both the chemically modified and the Δ28 enzymes are effective for DNA sequence analysis.[12,13,18,19,21] They have similar properties and are used in similar ways. Thus only protocols and results for the genetically altered Δ28 T7 DNA polymerase will be discussed here, but similar results can be obtained using chemically modified T7 DNA polymerase.

The exonuclease activity of T7 DNA polymerase is partially responsible for the fidelity of replication by this enzyme.[9] While this may be of concern for some applications of DNA polymerases *in vitro,* reduced fidelity of exonuclease-free forms of DNA polymerase will not affect the accuracy of DNA sequencing experiments. This is because the bands observed on the sequencing gel are the result of synthesis on a large population of template molecules, most of which are accurately replicated. Errors of misincorporation would result in very slight increases in background at levels too small to be observed in ordinary sequencing experiments.

Properties of T7 DNA Polymerase

The chain-termination method of DNA sequencing involves the synthesis of a DNA strand by a DNA polymerase.[8] Synthesis is initiated at

[21] C. W. Fuller, *Comments* **15**(2), 1 (1988).

only one site, where a primer anneals to the template. Chain growth is terminated by using a 2′,3′-dideoxynucleoside triphosphate (ddNTP) lacking the 3′-hydroxyl group required for continued DNA synthesis (hence the name chain termination).

All known DNA polymerases initiate synthesis only at the 3′ end of a polynucleotide primer base paired to a DNA template. Extension of a unique primer by DNA polymerase in the presence of all four deoxy- and one dideoxynucleoside triphosphate yields a population of molecules with common 5′ ends, but different 3′ ends, depending on the site at which a ddNMP was incorporated. In a typical sequencing experiment, four separate reactions are performed, each with a different ddNTP, thus the size of each fragment is determined by the sequence of the template. The products of the four reactions are analyzed by electrophoresis using a denaturing polyacrylamide gel, which separates the products by size.

The quality of the results of a DNA sequencing experiment depend on the capabilities of the DNA polymerase. It must initiate DNA synthesis at the 3′ end of a synthetic oligonucleotide primer and terminate with the incorporation of a chain-terminating dideoxynucleotide.[8] It must also readily utilize any of several analogs of deoxynucleoside triphosphates, including dideoxynucleotides, α-thionucleotides (for labeling with ^{35}S), and analogs of dGTP (e.g., dITP and 7-deaza-dGTP), which are used to eliminate secondary structure problems (compressions) that sometimes occur during gel electrophoresis.[12,22]

Modified T7 DNA polymerase is able to incorporate the nucleotide analogs used for DNA sequencing. The rate of incorporation of α-thio-dNMPs for labeling sequences is essentially equal to that of normal dNMPs.[12,20] Similarly, the dGTP analogs used to eliminate compression artifacts in sequencing gels (including 7-deaza-dGTP and dITP) are readily used by modified T7 DNA polymerase.[12,22] While dITP is more effective in completely eliminating compression artifacts, 7-deaza-dGTP is a better substrate for the polymerase.[12] (See the section on DNA sequencing methods below for detailed descriptions of reagent concentrations.) Most important is the capacity of T7 DNA polymerase to use the dideoxynucleoside triphosphates. A convenient measure of the ability of a polymerase to use dideoxynucleotides is the ratio of deoxynucleotide to dideoxynucleotide used in typical sequencing reactions. These ratios, which reflect the k_{cat}/K_m for the deoxynucleotide divided by the k_{cat}/K_m value for the dideoxynucleotide, are chosen so that termination by incorporation of a dideoxynucleotide occurs on average about once for every 50 deoxynucleotides incorporated. The large fragment of *E. coli* DNA polymerase I (Klenow), which incorporates dideoxynucleotides relatively poorly, re-

[22] S. Mizusawa, S. Nishimura, and F. Seela, *Nucleic Acids Res.* **14,** 1319 (1986).

quires a ratio of ddTTP to dTTP of about 100 : 1.[8] The ratios for ddATP, ddGTP, and ddCTP differ, but are also high, varying from 20 : 1 to 40 : 1. This indicates that the Klenow enzyme discriminates strongly (up to about 5000-fold) against the incorporation of dideoxynucleotides. The concentration ratios used with modified T7 DNA polymerase are about 0.1 : 1 and identical for all four ddNTPs.[12] When Mn^{2+} is used as cofactor for polymerization, the ratios decrease to about 0.02 : 1.[23–25] Under these conditions, modified T7 DNA polymerase does not kinetically distinguish between dNTPs and ddNTPs. Running sequencing reactions in the presence of Mn^{2+} is discussed in detail below.

The DNA polymerase used for sequencing must faithfully synthesize DNA using any template, even those with strong secondary structures. It is best if the synthesis terminates *only* with the incorporation of a chain-terminating dideoxynucleotide so that false, background bands are not seen on the sequencing gel. Furthermore, the frequency of dideoxynucleotide-associated terminations should be independent of surrounding sequence so that uniform band intensities are produced. Low exonuclease forms of T7 DNA polymerase, unlike the native, high exonuclease activity form, have potent strand displacement activity.[9,18] In fact, it appears that the lack of strand displacement activity in native T7 DNA polymerase is a direct result of the presence of an exonuclease activity.[18] The strand displacement activity gives modified T7 DNA polymerase the ability to synthesize on (and thus sequence) templates with secondary structures. It is highly processive, remaining bound to the primer–template for the polymerization of hundreds or thousands of nucleotides without dissociating.[10] This property helps eliminate background terminations that might interfere with reading sequence. Finally, it incorporates dideoxynucleotides at nearly the same relative frequency, regardless of neighboring sequence. This provides uniform band intensities as discussed later.

Method for Sequencing with Modified T7 DNA Polymerase

The most commonly used method for sequencing with modified T7 DNA polymerase was introduced by Tabor and Richardson to take advantage of some of its useful properties.[12,26] The DNA synthesis is carried out on primer–template in two steps (see Fig. 2). The first is the labeling step, in which only deoxynucleoside triphosphates (including an α-labeled

[23] S. Tabor and C. C. Richardson, *Proc. Natl. Acad. Sci. U.S.A.* **86,** 4076 (1989).
[24] S. Tabor and C. C. Richardson, U.S. Patent 4,962,020 (1990).
[25] C. W. Fuller, *Comments* **16**(3), 1 (1989).
[26] C. W. Fuller, *Comments* **14**(4), 1 (1988).

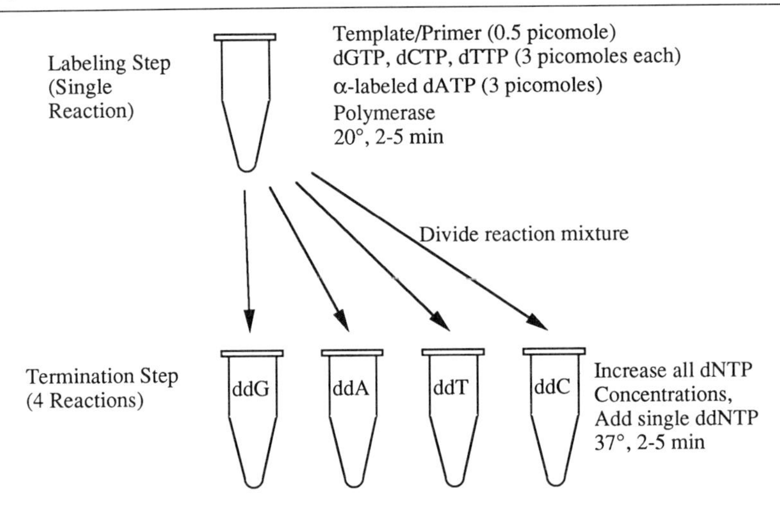

FIG. 2. Scheme of the two-step protocol used for sequencing with modified T7 DNA polymerase. A single labeling reaction containing primed template DNA is followed by four separate termination reactions, one for each ddNTP. The reaction conditions are optimized for each step independently.

nucleotide) are present, and the second is the chain-termination step, using both deoxy- and dideoxynucleotides.

In the first step, the primer is extended using limiting amounts of the deoxynucleoside triphosphates, including one radioactively labeled dNTP. This reaction is allowed to continue long enough for most of the nucleotide present, including the labeled nucleotide, to be incorporated into DNA chains. The lengths of these chains vary randomly from several nucleotides to hundreds of nucleotides, depending on the amount of nucleotide and primed template present initially. Ideally, about 0.5 pmol of primed template is present along with about 3 pmol of each dNTP. This step is carried out with an excess of polymerase (usually about 2–3 pmol) and at a temperature and nucleotide concentration at which polymerization is not highly processive. On average, each primer is extended by some 25 nucleotides, but because synthesis is not synchronous, extensions range from just a few nucleotides to about 100 nucleotides. All of the extended products are labeled by incorporation of the α-labeled dNMP. Because the lengths of the extensions are determined largely by the relative amounts of template and nucleotide, the reaction time and temperature are not critical. Thus, while the reaction may be essentially complete in just a few seconds, it can be continued for several minutes without difficulty and the average length of extension can be varied by varying the amount of nucleotide used.

To begin the second step, four equal aliquots of the labeling step reaction mixture are transferred to termination reaction vials. These are prefilled with a supply of all the deoxynucleoside triphosphates and one of the four dideoxynucleoside triphosphates. The temperature and concentrations of nucleotides are chosen so that polymerization will be rapid and processive. DNA synthesis begins and continues until all growing chains are terminated by a dideoxynucleotide. During this step, the average number of nucleotides by which the primer–template molecules are extended depends on the ratio of concentration of deoxy- to dideoxynucleoside triphosphates. On average, this can range from only a few nucleotides to hundreds of nucleotides. The reactions are stopped by the addition of EDTA and formamide, denatured by heating, and loaded onto electrophoresis gels.

The protocol originally introduced for chain-termination sequencing with Klenow enzyme consisted of running four parallel reactions.[8] Each of these had a carefully balanced, complex mixture of dNTPs, and one ddNTP. Both sequence-specific chain termination and labeling of products had to be achieved in this single set of reactions. Thus the conditions used are a compromise of the optimal conditions for labeling and termination. The two-step protocol is used so that reaction conditions can be chosen individually for the goals of efficient labeling of product DNA and of efficient termination by dideoxynucleotides. Few compromises must be made in choosing reaction conditions. This also allows changing the standard conditions if the focus of a particular experiment demands it. Labeling is very efficient, requiring less radioactive nucleotide than older procedures.[12,26] Termination reactions can be run at very high concentrations of dNTPs if necessary and a higher temperature where the polymerization is more processive, resulting in lower backgrounds and better strand displacement synthesis. Finally, because labeling for all four lanes occurs in a single reaction, lane-to-lane variations in resulting band intensities are minimized, making gels easier to interpret. Detailed protocols are provided in the section on DNA sequencing methods.

Results

The results of a sequencing experiment run on a high-resolution denaturing electrophoresis gel are shown in Fig. 3. Note that the general intensities of the bands are the same in each of the lanes, that there is little background between the bands representing chain terminations, and that the bands are uniformly dark. Together these factors contribute to the "readability" of the sequence. It is relatively easy to interpret these sequencing experiments with confidence even when band-to-band spacing is small at the resolution limit of the gel.

Variations in Band Intensities

The variations in band intensity on a sequencing gel can only be studied when the bands on the radiogram are quantified. Figure 4 shows the results of optical scanning of a DNA sequencing gel autoradiogram. Scans of all four lanes have been superimposed. Shown are the results of three sequencing experiments using the same template DNA. The top set of scans (Fig. 4A) is from a sequence generated using Klenow enzyme with the two-step protocol. In the center (Fig. 4B) is the same sequence generated using Δ28 modified T7 DNA polymerase using the two-step protocol in the presence of Mg^{2+}. The bottom set of scans (Fig. 4C) is from an experiment using modified T7 DNA polymerase in the presence of both Mg^{2+} and Mn^{2+}. Lines have been drawn at the levels of the highest and lowest peaks in this small region of the sequence, and the area between the lines is shown in gray. This area represents the variability in band intensities (peak heights) intrinsic in each sequencing method. This is important because variability in band intensities created by the properties of the enzyme is indistinguishable from other kinds of variability (noise) in the sequencing experiment.[23,24,25,27]

The variability in band intensity is the consequence of variations in the rates of incorporation of dideoxynucleotides caused by the local sequence of the template. The mechanistic details that govern sequence context sensitivity are not yet understood, but context-related events can be noticed (and perhaps used to advantage) by anyone attempting to read a sequence generated by the Klenow enzyme. Over larger sequences than those displayed in Fig. 4, the band intensities in sequencing gels generated using Klenow enzyme have been observed to vary over a 14-fold range (data not shown; see also Refs. 12, 23, 25, and 27). If a particularly strong band is close to a weak band, it is easy to imagine that the weak band might be missed or misinterpreted as background noise. Band intensities obtained with modified T7 DNA polymerase vary over a much smaller range of about fourfold under the same conditions.

Tabor and Richardson[23,24] reported running sequencing reactions in the presence of Mn^{2+} as well as Mg^{2+}. Both modified T7 DNA

[27] J. Z. Sanders, S. L. MacKellar, B. J. Otto, C. T. Dodd, C. Heiner, L. E. Hood, and L. M. Smith, *in* "Procedures of the Sixth Conversation in Biomolecular Stereodynamics," (R. H. Sarma and M. H. Sarma, eds.), Academic Press, New York, 1989.

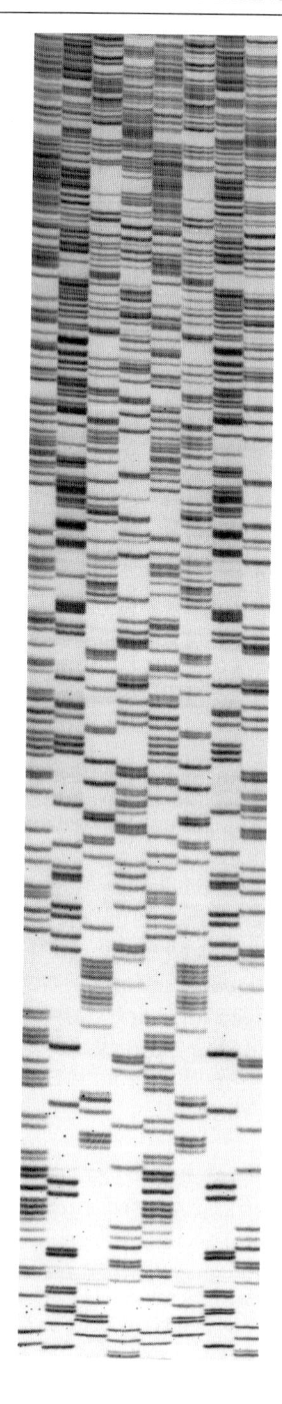

polymerase and the large fragment of DNA polymerase I (Klenow enzyme) need greatly reduced concentrations of dideoxynucleotides (relative to deoxynucleotides) for sequencing in the presence of Mn^{2+}. In the case of modified T7 DNA polymerase, the ratio of dideoxynucleotide to deoxynucleotide that generates usable sequencing band patterns is decreased from about 0.1 : 1 in the presence of Mg^{2+} to 0.02 : 1 in the presence of Mn^{2+}.

The bands generated by T7 DNA polymerase using Mn^{2+} are considerably more uniform than those generated with Mg^{2+} [23,25] (this is shown in Fig. 4C). In the presence of Mn^{2+}, over 95% of the bands fall within 10% of the mean corrected band intensity. While the human visual system filters out much of the band intensity variation seen in autoradiograms, the use of Mn^{2+} with modified T7 DNA polymerase improves the confidence in the interpretation of a sequencing gel. The increased uniformity also should improve the interpretation of sequences read by machine.[23]

Use of Pyrophosphatase in Sequencing Reactions

In the initial experimentation with chemically modified T7 DNA polymerase, it was noticed that under certain conditions the intensities of some of the bands on resulting sequencing gels can be weak.[12] This phenomenon is particularly apparent when the termination reactions are run for 30 min or longer, and when dITP is used in place of dGTP. This observation holds true even for Δ28 T7 DNA polymerase[28,29] and thus cannot be attributed to exonuclease activity, because Δ28 T7 DNA polymerase has no measurable exonuclease activity.[18] As shown in Fig. 5, some bands become less intense when the termination reactions are allowed to incubate for extended periods of time, especially when using dITP. When the reactions are incubated for 2 min, the bands have normal intensities; however, after a 10- to 60-min incubation, approximately 1 in 10 bands is reduced in intensity. Note that the rates at which individual bands decrease in intensity vary widely,

[28] S. Tabor and C. C. Richardson, *J. Biol. Chem.* **265,** 8322 (1990).
[29] C. C. Ruan, S. B. Samols, and C. W. Fuller, *Comments* **17**(1), 1 (1990).

FIG. 3. Sequencing gel generated using modified T7 DNA polymerase. This gel, which resolves bases 250–500 bases from the priming site, is the result of a sequencing experiment using M13mp18 template and Mg^{2+}, following the protocol described in text using Mg^{2+}. Note the absence of background between the bands and the uniform intensities of the bands both between lanes and within lanes. These factors contribute to the accurate interpretation of sequencing experiments.

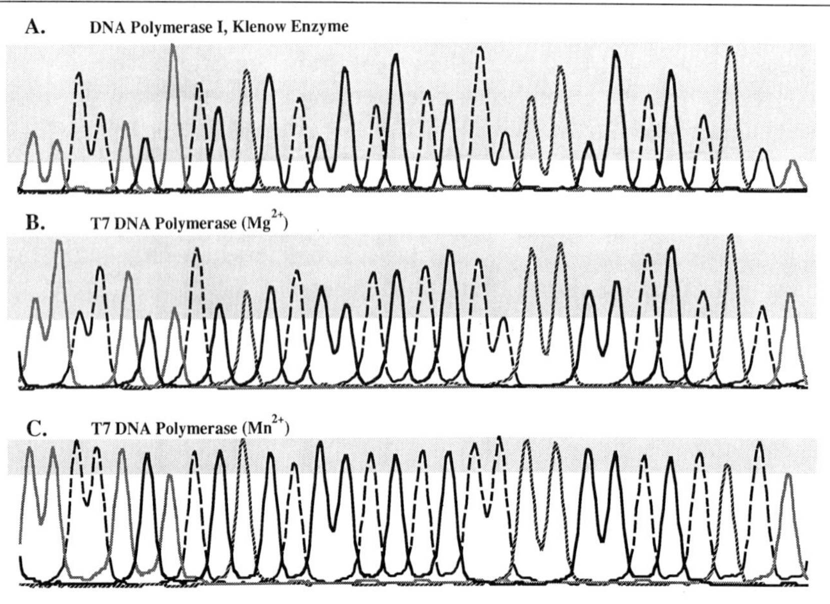

FIG. 4. Densitometer scans of DNA sequencing gel autoradiograms. An autoradiogram like the one shown in Fig. 3 was scanned using a computer-interfaced densitometric scanner (SciScan 5000, United States Biochemical Corp.). Scans of the four lanes (G, A, T, and C) are superimposed and the area between the heights of the tallest and shortest peaks shaded. *Top:* Scans of a sequence generated using DNA polymerase I (Klenow enzyme) and the two-step procedure (Fig. 2). *Center:* Scans from a sequence made using modified T7 DNA polymerase in the presence of Mg^{2+}. *Bottom:* Scans from a sequence made using modified T7 DNA polymerase in the presence of Mn^{2+}. Band intensities are even more uniform when Mn^{2+} is present in the reaction mixture. When band intensities are uniform, they can be interpreted even in sequence regions where the resolution of the gel is marginal. Using fluorescent-labeled primers and a laser-scanning instrument (Model 373A, Applied Biosystems, Foster City, CA), sequences of 450–500 nucleotides can be obtained routinely [B. F. McArdle and C. W. Fuller, *Editorial Comments* **17**(4), 1 (1991)].

depending on the DNA sequence. A similar result has been observed when sequencing using avian myeloblastosis virus (AMV) reverse transcriptase.[29]

The explanation for band intensity loss during prolonged incubation is that DNA polymerase also catalyzes pyrophosphorolysis.[28,29] Pyrophosphorolysis is the reversal of the polymerization wherein DNA and pyrophosphate react to form a deoxynucleoside triphosphate, leaving the DNA one base shorter. The polymerization, pyrophosphorolysis, and exonuclease reactions are summarized as follows:

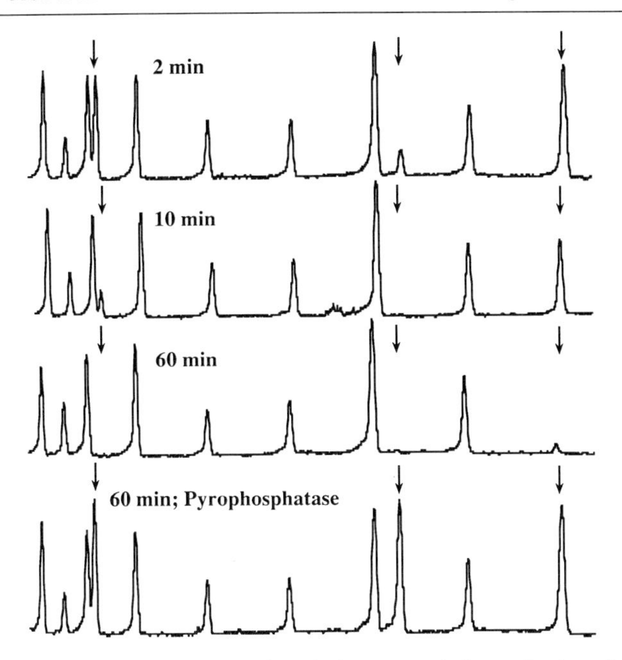

FIG. 5. Sequence-specific pyrophosphorolysis can result in weakening of some bands with prolonged reaction times. Shown are densitometer scans of ddTTP lanes obtained as in Fig. 4. The termination reactions used to generate these results were run for the indicated times at 37° using dITP in place of dGTP. The bottom scan is the result of a sequencing reaction containing 0.005 units of inorganic pyrophosphatase incubated for 60 min. Some of the bands (arrows) decrease in intensity with longer reaction times. The decreases occur at different rates but none occur when pyrophosphate is removed from the reaction mixtures by the action of pyrophosphatase.

Polymerase: $$DNA_{(n)} + dNTP \xrightarrow{M^{2+}} DNA_{(n+1)} + PP_i$$

Pyrophosphorylase: $$DNA_{(n)} + dNTP \xleftarrow{M^{2+}} DNA_{(n+1)} + PP_i$$

Exonuclease: $$DNA_{(n)} + H_2O \xleftarrow{M^{2+}} DNA_{(n-1)} + dNMP$$

where M^{2+} is Mg^{2+} or Mn^{2+}.

Under the conditions normally used for DNA synthesis (i.e., high dNTP concentrations and low pyrophosphate concentration), the forward reaction (polymerization) is greatly favored over the reverse reaction (pyrophosphorolysis). The equilibrium constant favors polymerization by about 10^4-fold.[20] However, pyrophosphorolysis does occur given enough time. The rate of pyrophosphorolysis varies with the neighboring sequence,[20,28,29] reducing the intensity of some bands while leaving others unchanged.

Inorganic pyrophosphatase catalyzes the hydrolysis of pyrophosphate to two molecules of orthophosphate.

$$P_2O_7^{4-} + H_2O \xrightarrow{Mg^{2+}} 2HPO_4^{2-}$$

This enzyme plays an important role in DNA synthesis *in vivo,* rendering DNA polymerization essentially irreversible by removing pyrophosphate. The addition of inorganic pyrophosphatase (from yeast) to DNA sequencing reactions completely stabilizes all band intensities, even when incubating the reactions for 60 min and when dITP is used in place of dGTP (Fig. 5). Pyrophosphatase can be used for sequencing with dGTP or its analogs (dITP, 7-deaza-dGTP, etc.) and is effective in the presnce of Mg^{2+} or Mn^{2+}. The use of pyrophosphatase with modified T7 DNA polymerase assures the stability of all DNA generated in the DNA-sequencing reaction, providing additional assurance that the sequence determined is accurate. Pyrophosphorolysis also can be reduced by using high concentrations of dNTPs.[28]

DNA Sequencing Methods

Materials and Reagents

Reaction buffer (5× concentrate): 200 m*M* Tris-HCl, pH 7.5, 100 m*M* $MgCl_2$, 250 m*M* NaCl

Primer: 0.5 pmol/μl

Dithiothreitol (DTT) solution: 100 m*M*

Modified T7 DNA polymerase, either chemically modified or genetically modified deletion Δ28[12,18]: These should be adjusted to a concentration of 13 units/μl or approximately 1.0 mg/ml

Yeast inorganic pyrophosphatase: 5 units/ml in 10 m*M* Tris-HCl, pH 7.5, 0.1 m*M* EDTA, 50% (v/v) glycerol. *Note:* Pyrophosphatase can be premixed with modified T7 DNA polymerase at a ratio of about 0.001 units pyrophosphatase for each 3 units polymerase. (One unit of pyrophosphatase hydrolyzes 1 μmol of pyrophosphate/min at 25°)

Enzyme dilution buffer: 10 m*M* Tris-HCl, pH 7.5, 5 m*M* DTT, 0.5 mg/ml bovine serum albumin (BSA)

Manganese buffer (for dGTP): 150 m*M* sodium isocitrate, pH 7.0, 100 m*M* $MnCl_2$

Stop solution: 95% formamide, 20 m*M* EDTA, 0.05% (w/v) bromophenol blue, 0.05% (w/v) xylene cyanol FF

Nucleotide mixtures: Concentrations of nucleotides are listed in Table I. They are prepared in water or 50 m*M* NaCl as indicated

Labeled dATP: Either [α-^{32}P]dATP or [α-^{35}S]dATP is required for autoradiographic detection of the sequence. Nucleotide labeled with ^{35}S has the advantages of high resolution[30] and operator safety. The specific activity for either ^{35}S or ^{32}P should be 1000–1500 Ci/mmol; ^{32}P should be less than 2 weeks old while ^{35}S is usable for 4–6 weeks

Water: Only deionized, distilled should be used for the sequencing reactions

TE buffer: 10 m*M* Tris-HCl, 1 m*M* EDTA, pH 7.5. It is used for template preparation

TBE buffer (10 ×): 0.9 *M* Tris, 0.9 *M* boric acid, 25 m*M* disodium EDTA for sequencing gels

Gel monomers: Acrylamide and *N,N'*-methylenebisacrylamide should be freshly dissolved for preparing gels. Other reagents (urea, Tris, boric acid, and EDTA) should be electrophoresis grade

All nucleotide mixtures should be stored frozen at −20° and for longest life be kept on ice when thawed for use. The buffer, primer, and stop solutions can be stored for 4–8 weeks at 4°. Store modified T7 DNA polymerase at −20°. Dilute only the amount of enzyme needed in ice-cold buffer and use it immediately. All of these reagents are available from United States Biochemical Corp. (Cleveland, OH).

Necessary Equipment

Constant temperature bath: Sequencing requires incubations at room temperature, 37, 65, and 75°. The annealing step requires slow cooling from 65° to room temperature

Electrophoresis equipment: While a standard, nongradient sequencing gel apparatus is sufficient for much sequencing work, the use of field-gradient ("wedge") gels allow much greater reading capacity on the gel.[31] A power supply offering constant voltage operation at 2000 V or greater is essential

Gel handling: If ^{35}S sequencing is desired, a large tray for washing the gel (to remove urea) and a gel-drying apparatus are necessary. Gels containing ^{35}S must be dry and in direct contact with the film at room temperature for fast, sharp exposures

Autoradiography: Any large-format autoradiography film and film holder can be used. Development is done according to the instructions of the film manufacturer

[30] M. D. Biggin, T. J. Gibson, and G. F. Hong, *Proc. Natl. Acad. Sci. U.S.A.* **80,** 3963 (1983).

[31] W. Ansorge and S. Labeit, *J. Biochem. Biophys. Methods* **10,** 237 (1984).

TABLE I
NUCLEOTIDE MIXTURES USED FOR DNA SEQUENCING WITH MODIFIED T7 DNA POLYMERASE

Mixes	Concentration (μM)		
	dGTP	dITP	7-Deaza-dGTP
Labeling			
For use with α-labeled dATP	dGTP, 1.5	dITP, 3.0	7-Deaza-dGTP, 1.5
	dTTP, 1.5	dTTP, 1.5	dTTP, 1.5
	dCTP, 1.5	dCTP, 1.5	dCTP, 1.5
For use with α-labeled dCTP	dGTP, 1.5	dITP, 3.0	7-Deaza-dGTP, 1.5
	dTTP, 1.5	dTTP, 1.5	dTTP, 1.5
	dATP, 1.5	dATP, 1.5	dATP, 1.5
Sequence extending[a]	dGTP, 180	dITP, 360	
	dATP, 180	dATP, 180	
	dTTP, 180	dTTP, 180	
	dCTP, 180	dCTP, 180	

	Concentration (μM)														
	dGTP					dITP					7-Deaza-dGTP				
Termination[a]	Mix	G	A	T	C	Mix	G	A	T	C	Mix	G	A	T	C
For use with Mg^{2+}	dGTP	80	80	80	80	dITP	160	160	160	160	7-Deaza-dGTP	80	80	80	80
	dATP	80	80	80	80	dATP	80	80	80	80	dATP	80	80	80	80
	dTTP	80	80	80	80	dTTP	80	80	80	80	dTTP	80	80	80	80
	dCTP	80	80	80	80	dCTP	80	80	80	80	dCTP	80	80	80	80
	ddGTP	8	—	—	—	ddGTP	1.6	—	—	—	ddGTP	8	—	—	—
	ddATP	—	8	—	—	ddATP	—	8	—	—	ddATP	—	8	—	—
	ddTTP	—	—	8	—	ddTTP	—	—	8	—	ddTTP	—	—	8	—
	ddCTP	—	—	—	8	ddCTP	—	—	—	8	ddCTP	—	—	—	8
For use with Mn^{2+}	dGTP	80	80	80	80										
	dATP	80	80	80	80										
	dTTP	80	80	80	80										
	dCTP	80	80	80	80										
	ddGTP	1.6	—	—	—										
	ddATP	—	1.6	—	—										
	ddTTP	—	—	1.6	—										
	ddCTP	—	—	—	1.6										

[a] Mixes contain 50 mM NaCl.

Protocols

All sequencing reactions are run in 0.5- or 1.5-ml plastic centrifuge tubes. These should be kept capped to minimize evaporation of the small volumes employed. Additions should be made with disposable-tip micropipettes and care should be taken not to contaminate stock solutions. The solutions must be thoroughly mixed after each addition, typically by "pumping" the solution two or three times with the micropipette, avoiding the creation of air bubbles. At any stage where the possibility exists for some solution to cling to the walls of the tube, it should be centrifuged. With care and experience these reactions can be completed in 10–15 min.

Annealing Template and Primer

1. For each set of four sequencing lanes, a single annealing (and subsequent labeling) reaction is used. In a centrifuge tube combine the following:

Primer, 1 μl
Reaction buffer, 2 μl
DNA (1 μg, single stranded), 7 μl

The total volume should be 10 μl; if a smaller volume of DNA solution is used, the balance should be made up with distilled water. This is a 1 : 1 (primer–template) molar stoichiometry. The use of too little template will narrow the effective sequencing range, resulting in faint bands near the bottom of the gel. *Note:* Double-stranded DNA must be denatured prior to annealing; see below.

2. Warm the capped tube to 65° for 2 min, then allow the temperature of the tube to cool slowly to room temperature over a period of about 30 min. Cooling can be done by using a small heating block or a small beaker of 65° water as a temperature bath, placed at room temperature on the bench to cool slowly. After annealing, *place the tube on ice.*

It is necessary to denature double-stranded DNA templates prior to performing the sequencing reactions. It is recommended that double-stranded supercoiled plasmid DNA be denatured by the alkaline denaturation method,[32–36] while linear DNA can be effectively denatured by boiling.

[32] E. J. Chen and P. H. Seeburg, *DNA* **4,** 165 (1985).

[33] M. Hattori and Y. Sakaki, *Anal. Biochem.* **152,** 232 (1986).

[34] M. Haltiner, T. Kempe, and R. Tijan, *Nucleic Acids Res.* **13,** 1015 (1985).

[35] H. M. Lim and J. J. Pene, *Gene Anal. Tech.* **5,** 32 (1988).

[36] F. Tonneguzzo, S. Glynn, E. Levi, S. Mjolsness, and A. Hayday, *BioTechniques* **6,** 460 (1988).

1. Add 0.1 vol of 2 *M* NaOH, 2 m*M* EDTA to the DNA sample (2–4 pmol, typically 3–5 μg) to a final concentration of 0.2 *M* NaOH, 0.2 m*M* EDTA. Incubate 30 min at 37°.[35]

2. Add 0.1 vol of 3 *M* sodium acetate (pH 4.5–5.5) and 2–4 vol of cold 95% ethanol to this mixture. Incubate 15 min at −70°.

3. Centrifuge 15 min and carefully remove the supernatant.

4. Wash the pellet by adding 500 μl of cold 70% ethanol, inverting the tube once, and centrifuging for 1 min. Draw off the ethanol, taking care not to disturb the pellet.

5. Dry the pellet. Resuspend the pellet in 7 μl sterile distilled water and proceed with the annealing exactly as for single-stranded DNA. If desired, the dried pellet can be stored at −20° for up to 1 week.

Linear, double-stranded DNA can be denatured for sequencing by mixing it with an excess of primer and boiling for about 2 min. The tube is removed from the boiling water bath and immediately plunged into ice water to cool quickly. Further heating for annealing is avoided, and the primed template is used directly for the sequencing reactions.

Labeling Step

1. Dilute modified T7 DNA polymerase to working concentration (1.6 units/μl). Dilution from a storage concentration of 13 units/μl by a factor of 1 : 8 is recommended so that glycerol used to stabilize the enzyme for long-term storage is not added in large quantity to the sequencing reactions. Excess glycerol will cause a distortion on the sequencing gel. Pyrophosphatase can be added to the diluted enzyme as well. Add 0.5 μl of 5 units/ml pyrophosphatase to 1 μl (13 units) of modified T7 DNA polymerase and 6.5 μl of dilution buffer.

2. To the annealed template–primer add the following (on ice):

Template–primer (above), 10 μl
DTT (0.1 *M*), 1.0 μl
Labeling mix, 2.0 μl
[α-^{35}S]dATP, 0.5 μl (see note below)
Diluted enzyme, 2.0 μl (always add enzyme last)

Mix thoroughly (avoiding bubbles) and incubate for 2–5 min at room temperature. *Note:* The amount of labeled nucleotide can be adjusted according to the needs of the experiment. Either [α-^{32}P]dATP or [α-^{35}S]dATP can be used. Nominally, 0.5 μl of 10 μCi/μl and 10 μ*M* (1000 Ci/mmol) should be used.

Termination Reactions

During this step, the concentration of dNTPs is increased and ddNTPs are added, resulting in the further elongation of chains. The termination reactions are largely complete within a few seconds. Because it is critical that the mixture is at 37° (or warmer) during this period, it is very important to prewarm the termination reaction tubes for at least 1 min prior to beginning the reaction. Prewarming the reaction tubes will assure the reaction is run at the appropriate temperature. Running the termination reaction cooler than 37° can result in artifact bands on the sequencing gel.

1. Have on hand four tubes labeled G, A, T, and C.
2. Place 2.5 μl of the ddGTP termination mix in the tube labeled G. Similarly fill the A, T, and C tubes with 2.5 μl of the ddATP, ddTTP, and ddCTP termination mixes, respectively. Cap the tubes to prevent evaporation. (This is best done before beginning the labeling reaction.)
3. Prewarm the tubes at 37° at least 1 min.
4. When the labeling incubation is complete, remove 3.5 μl and transfer it to the tube labeled G. Mix, centrifuge, and continue incubation of the G tube at 37°. Similarly transfer 3.5 μl of the labeling reaction to the A, T, and C tubes, mixing and returning them to the 37° bath.
5. Continue the incubations for a total of 3–5 min.
6. Add 4 μl of stop solution to each of the termination reactions, mix thoroughly, and store on ice until ready to load the sequencing gel. Samples labeled with ^{35}S can be stored at −20° for 1 week with little degradation. Samples labeled with ^{32}P should be run the same day.
7. When the gel is ready for loading, heat the samples to 75–80° for 2 min and load immediately on the gel. Use 2–3 μl in each lane.

Extending Sequences Beyond 400 Bases from the Primer

To increase the ratio of deoxy- to dideoxynucleotides in the termination step, thereby increasing the average length of extensions in the termination step, a sequence extending mix can be used. This method can extend the reactions thousands of bases and does not necessarily sacrifice information closer to the primer.

Table II shows the approximate relative extensions that can be achieved using the sequence extending mixes. For example, if sequences are visible to 400 bases without the extending mix, a relative extension of 4.0 would give bands to approximately 400 × 4, or 1600 bases. It is important to remember that although it is possible to extend the primers in sequencing reactions to lengths of thousands of bases, the current gel technology is incapable of resolving DNA molecules greater than about 700 bases in length.

TABLE II
USE OF SEQUENCE EXTENDING MIXES

Volume (μl)			Approximate relative extension
Termination mix	Extension mix	Total	
2.5	0.0	2.5	1.0
2.0	0.5	2.5	1.5
1.5	1.0	2.5	2.5
1.0	1.5	2.5	4.0

Sequence extending mix is also useful for increasing the nucleotide concentration where one nucleotide is depleted due to an unusual base composition of the template [e.g., add sequence extending mix to the T termination reactions when sequencing through poly(A) tails].

Improving Band Intensities Close to Primer

The addition of Mn^{2+} to the reaction buffer changes the ratio of reactivities of dNTPs to ddNTPs by approximately fivefold, even when both Mg^{2+} and Mn^{2+} are present.[23] Thus, simply adding Mn^{2+} to the sequencing reaction buffer has the same effect as increasing the relative concentrations of all four ddNTPs. This makes terminations occur closer to the priming site, emphasizing sequences close to the primer. This is useful because reactions that contain less than the normal amount of template DNA result in weak bands near the primer. It is often difficult to prepare or measure the correct amount of template DNA and reactions carried out with small amounts of DNA. Running the sequencing reactions in the presence of Mn^{2+} will usually restore the sequences close to the primer even when less template is available. To use Mn^{2+}, simply add 1 μl of manganese buffer (above) to the labeling reaction.

Running Sequencing Reactions in 96-Well Microassay Plates

Plates with 96 U-bottom wells are ideal for running several sets of termination reactions at the same time. Care should be taken not to allow reactions to evaporate to dryness. Plates can be covered with plastic film if necessary. Labeling reactions are performed in small centrifuge vials while termination reactions are run in the plate as follows.

1. Prepare the plate by adding 2.5 μl of termination mixes to appropriate wells in the plate and keep the plate at room temperature or on ice until the labeling reactions are finished.

2. When the labeling reactions are nearly finished, float the 96-well plate in a 37° water bath to prewarm the termination reactions. It is helpful to use a clip or weight to prevent the plate from slipping in the bath. A 37° heating block may also be used.

3. When the labeling reactions are done, transfer 3.5 μl of the labeling reaction product into each of the four termination wells, mixing as usual.

4. When all of the labeling reactions have been distributed to the appropriate wells of the plate (12–24 sets will fit on a single plate), continue incubation of the plate at 37° until all of the termination reactions have run at least 2 min and no more than 10 min.

5. Remove the plate and add stop solution (4 μl) to each reaction.

6. Immediately prior to loading the gel, denature the reaction products by heating the plate in a water bath at 75° for 2 min. Avoid placing the plate in the oven for this incubation. The plate temperature will not reach 75° in the 2-min time period in an oven.

Compressions

The final analysis step in DNA sequencing, whether done using chain-termination (Sanger, dideoxy) or chain-cleavage (Maxam–Gilbert) methods, involves the use of a denaturing polyacrylamide electrophoresis gel to separate DNA molecules by size. Resolution of one nucleotide in several hundred can be achieved using relatively simple equipment but there can be difficulties in resolving some sequences.

Electrophoretic separation based solely on size requires complete elimination of secondary structure from the DNA. This is typically accomplished by using high concentrations of urea in the polyacrylamide matrix and running the gels at elevated temperatures. For most DNAs, this is sufficient to give a regularly spaced pattern of bands in a sequencing experiment. Compression artifacts are recognized by anomalous band-to-band spacing on the gel. Several bands run closer together than normal, which may be unresolved. This is usually followed by several bands running farther apart than normal. Compression artifacts are caused whenever stable secondary structures exist in the DNA under the conditions prevailing in the gel matrix during electrophoresis. The folded DNA structure runs faster through the gel matrix than an equivalent unfolded DNA, catching up with the smaller DNAs in the sequence.

Typical gel conditions (7 *M* urea and 50°) denature A/T secondary structures, but not all structures with three or more G/C pairs in succession. In the case of dideoxy sequencing, these base-paired structures will not exist until synthesis extends around the loop and along the 3′ side of the stem. This is the sequence region, which migrates anomalously on gels. If this same strand were being sequenced using Maxam–Gilbert methods with a 5′ end label, the same compression would be observed. However, sequencing of the opposite strand would have the compression displaced to the other side of the inverted repeat.

One way to eliminate gel compression artifacts is to use an analog of dGTP during the sequencing reactions. This is simple and usually quite effective. Recommended nucleotide mixtures for sequencing with dITP[12,37] or 7-deaza-dGTP[22] are shown above. When these analogs replace dGTP in the sequencing experiment, the dG nucleotides in the product DNA are replaced by dI or 7-deaza-dG. These form weaker base pairs with dC, which are more readily denatured for gel electrophoresis. With the most difficult compression artifacts dITP fully disrupts secondary structure while 7-deaza-dG only partially resolves the sequence. It is always a good practice to run dGTP reactions alongside dITP or 7-deaza-dGTP reactions because these analogs may potentiate other problems with the sequence.

An alternative to using dGTP analogs is to use a sequencing gel formulation that more fully denatures the DNA. This can be accomplished by adding 20–40% (v/v) formamide to gels containing 7 *M* urea. These can be as effective as using dITP and their preparation is not really more difficult than normal sequencing gels.

1. Place 40 ml formamide (or 20 ml for a 20% gel) in a beaker with a magnetic stir bar.
2. Add the following:

Acrylamide, 7.6 g
N,N′-Methylenebisacrylamide, 0.4 g
Urea, 42.0 g
TBE buffer (10×), 10 ml

3. Cover with Parafilm and warm in a 65° water bath (tray) with stirring until dissolved. This takes 30–60 min and the mixture reaches a temperature of about 50°. When completely dissolved add water to a total volume of 100 ml.
4. Cover and continue stirring at room temperature (no water bath) to cool to ~25°. Vacuum filter with a paper or nitrocellulose filter.
5. Add 1 ml of 10% (w/v) ammonium persulfate and 0.15 ml of TEMED. Pour immediately. Pouring this viscous mixture requires a nearly vertical angle for the glass plates. The gel should be polymerized within 30 min.

[37] J. A. Gough and M. E. Murray, *J. Mol. Biol.* **166,** 1 (1983).

Note: Running these gels requires a higher voltage (30–60% higher) than normal gels, and DNA migrates about half as fast. Soaking the gel in 5% acetic acid, 20% methanol helps prevent swelling. Soak 15 min for a 0.4-mm-thick gel and dry normally.

Troubleshooting

Film Blank or Nearly Blank

1. If using single-sided film, the emulsion side must be placed facing the dried gel.
2. DNA preparation may be bad; repurify the DNA.
3. Labeled nucleotide may be too old.
4. Some component may be missing.
5. Enzyme may have lost activity.

Bands Smeared

1. DNA preparation may be contaminated; repurify the DNA.
2. Gel may be bad. Gels should be cast with freshly made acrylamide solutions and should polymerize rapidly, within 15 min of pouring. Try running a second gel with the same samples.
3. Gel may have been run too cold. Sequencing gels should be run with a surface temperature of 50–55°.
4. Gel may have been dried too hot or not flat enough to be evenly exposed to film.
5. Samples may not have denatured. Make sure samples are always heated to 75° for at least 2 min (longer in a heat block) immediately prior to loading on gel.

Sequence Faint near Primer

1. Insufficient DNA in the sequencing reaction; a minimum of 0.5 pmol DNA is required for sequencing close to the primer, this usually corresponds to about 1 μg of single-stranded M13 DNA and 3–4 μg of plasmid DNA. Try adding 1 μl of manganese buffer (above) to the labeling reaction or increasing the amount of DNA.
2. Insufficient primer; use a minimum of 0.5 pmol. Primer-to-template mole ratio should be 1 : 1 to 5 : 1.

Bands Appear across All Four Lanes

1. DNA preparation may be bad; repurify the template DNA.
2. Reagents may not be mixed thoroughly during the reactions; mix

carefully after each addition, avoiding bubbles and centrifuging to bring all solution to the tip of the tube.

3. Be sure that the annealing step is not run too long or too hot; it is usually sufficient to heat the mixture to 65° and cool to room temperature within 15–30 min.

4. The labeling step should not be run warmer than 20° or longer than 5 min. Doing so will often result in many "pause" sites in the first 100 bases from the primer. The termination step should not be run cooler than 37°. Room-temperature termination reactions (even ones in which the tubes are not prewarmed) will cause pausing at sites further than 100 bases from the primer. Termination reactions (but not labeling reactions) can be run up to 50°, which may improve results for some templates.

5. Sequences may have strong secondary structure. Modified T7 DNA polymerase will pause at sites of exceptional secondary structure, especially when dITP is used. Try reducing the concentration of nucleotides in the labeling step to keep extensions during this step from reaching the pause site or using more polymerase on difficult templates. If the problem persists, the addition of 0.5 μg of *E. coli* single-strand binding protein (SSB; United States Biochemical Corp.) during the labeling reaction usually eliminates the problem. SSB must be inactivated prior to running the gel. Add 0.1 μg of proteinase K (United States Biochemical Corp.) and incubate at 65° for 20 min after adding the stop solution.

Bands in Two or Three Lanes

1. Heterogeneous template DNA may possibly be the result of spontaneous deletions arising during M13 phage growth. Plaque purify the clone and limit phage growth to less than 6–8 hr.

2. Reaction mixtures may be insufficiently mixed.

3. The sequence may be prone to compression artifacts in the gel. Compressions occur when the DNA [usually (G + C)-rich] synthesized by the DNA polymerase does not remain fully denatured during electrophoresis. Try using dITP or 7-deaza-dGTP in the reaction mixtures to eliminate gel compressions or use a gel prepared with 40% formamide along with 7 *M* urea.

Some Bands Faint

Termination reaction time may be too long and pyrophosphatase omitted. Try reducing the termination reaction time (1 min is usually sufficient) or adding 0.001 units pyrophosphatase to the labeling step.

Sequence Fades Early in One Lane

Template DNA may have a biased nucleotide composition. This is common for cDNA templates that have poly(A) sequences. In this case, the T lane does not extend as far as the others. This is caused by early exhaustion of dTTP and ddTTP in the reactions. Try adding sequence extending mix to the T reaction only (use 2 μl sequence extending mix and 1 μl T termination mix). This situation may also be improved by adding extra dTTP to the labeling reaction (1 μl of 500 μl*M* dTTP).

Section III

Reporter Genes

[30] Nondestructive Assay for β-Glucuronidase in Culture Media of Plant Tissue Cultures

By JEAN GOULD

Introduction

The Standard β-Glucuronidase Assay. The development of the β-glucuronidase (GUS) gene fusion system[1] and the proliferation of many GUS gene (*uidA*) constructions has made the screening for transformed plant tissues easy and rapid. Although milligram amounts of tissue are required, the method is destructive and there are times when valuable tissue must be sacrificed for the assay. This destructive aspect of the assay can be a drawback for some applications of the GUS reporter system.[2]

Derivation of Nondestructive β-Glucuronidase Assay. The nondestructive method is a simple derivation of the standard GUS fluorimetric assay using methylumbelliferylglucuronide (MUG) described by Jefferson *et al.*[1,3] In this derivation, the substrate is placed in the culture medium for the assay rather than in a tissue extract. This procedure was developed to avoid extraction of tissues in screening progeny.[4] The reagents and solutions are the same as described by Jefferson; however, MUG can be used in the phosphate buffer described for the GUS colorimetric assay, as well as in the standard GUS lysis buffer. This variation of the GUS assay is easily performed on the spent culture medium after routine reculture of the tissue.

Principle of Method

The method is based on the common observation that plant tissues in culture usually lose or release many cytosolic proteins into the culture medium.[5] Some proteins are actively exported from cultured plant cells; other proteins can leak from injured or dying cultured cells. The proteins that are actively exported usually have a function outside of the cell, such as the proteins of the extracellular matrix and cell wall. The exported or

[1] R. A. Jefferson, T. Kavanagh, and M. W. Bevan, *EMBO J.* **6,** 3901 (1987).
[2] M. R. Hanson, *Plant Cell* **1,** 169 (1990).
[3] R. A. Jefferson, *Plant Mol. Biol. Rep.* **5,** 387 (1988).
[4] J. H. Gould and R. H. Smith, *Plant Mol. Biol. Rep.* **7,** 209 (1989).
[5] J. H. Gould and T. Murashige, *Plant Cell Tissue Organ Cult.* **4,** 29 (1985).

lost proteins accumulate in the culture medium and many remain active in this environment for weeks. The nondestructive assay method for GUS is based on this natural release of proteins from cultured plant tissues and the stability of GUS in the culture medium.

As a general rule, if GUS activity can be detected in transformed plant tissues, it can probably be detected in the culture medium as well. There are of course exceptions depending on the promoter and other sequences used with the GUS gene fusion. The assay of the culture medium is straightforward when a constitutive promoter is used. If the promoter sequences are regulated, GUS will not be detected in the medium unless the appropriate induction signal has been used and the enzyme has time to accumulate. β-Glucuronidase can also be synthesized by the transforming *Agrobacterium* and released into the culture medium. If the promoter sequences are specific for eukaryotic transcription, such as in "Kozaks GUS" described by Jefferson,[3] synthesis by the *Agrobacterium* can be negligible.

Materials and Reagents

Materials and Equipment. Refrigerator (4°) and freezer (−20°) space, pH meter, balance, distilled or deionized water, autoclave, and sterile transfer facilities should be available.

The material of the culture vessels should transmit enough ultraviolet (UV) light for the reaction to be detected. Most commercial polystyrene plates absorb in the UV and fluoresce in the blue range, effectively shielding the medium; however, some polystyrene, such as Falcon (Becton Dickinson, Oxnard, CA) "Optilux," can be used.

Equipment and Reagents

Ultraviolet mineral lamp or UV light box
Pipette, 10–100 μl
Nitrocellulose hydrophilic membrane filters, 0.2 μm
Plastic syringe, 10–15 ml
Methylumbelliferylglucuronide (MUG; Sigma, St. Louis, MO)
Monobasic and dibasic sodium phosphate

Preparation of Solutions

Phosphate buffer: Combine 15.8 ml 1 *M* Na_2HPO_4 and 9.8 ml 1 *M* NaH_2PO_4; adjust pH to 7.00; dilute to 500 ml

Divide into 50-ml aliquots, dispense into containers, and sterilize by autoclaving. Store at room temperature.

MUG assay solution: Combine 7 mg MUG, 10 ml phosphate assay buffer, pH 7, 2 mg/10 ml chloramphenicol (optional if sterile tissue is assayed)

After MUG has dissolved, store frozen (−20°) until just before use; *mix well* when thawed. Dispense to the culture medium through a 0.2-μm membrane filter to sterilize. Refreeze immediately after use.

Stop buffer (optional): 15.7 g Disodium carbonate

Dilute to 500 ml with deionized water. Store at 4°.

Protocol

Assay of Media. Assays may be made in agar or liquid media. After the plant tissues have been transferred from the culture medium, dispense 10–20 μl sterile MUG assay buffer to the agar surface where the tissue was in direct contact, or directly into the liquid medium. This procedure is referred to as "MUGing." The "MUGed" plates are then sealed with Parafilm and incubated overnight at 25–37°. Following incubation, examine the plates using a UV light box or a hand-held mineral lamp. Ten microliters of stop buffer can be added to each "MUGed" area and allowed to diffuse for 10 min prior to examination; however, this is usually not necessary. The positive reaction product 4-methylumbelliferone (4-MU) will appear as a blue fluorescent spot in the medium (Fig. 1). When a significant amount of GUS has accumulated and activity is high, the medium can become an intense fluorescent blue throughout. On the other hand, a small amount of GUS or low activity produces an ephemeral 4-MU spot that can be observed briefly until the compound is diluted by diffusion and cannot be seen.

Documentation of Assay. The equipment used to photograph ethidium bromide-stained gels can be used to photograph these assays. The normal camera settings used to photograph gels are good settings with which to start. If the orange filter is used, a 1-sec or longer exposure may be required. We used a UV light box (Fotodyne), Polaroid camera, CRT hood, and Polaroid 107C or 667 black-and-white instant pack film to photograph the GUS assay shown in Fig. 1. If this equipment is not available, documentation can be frustrating but possible using a mineral lamp, 35-mm camera, a broad range of exposures, and a fine grain variable ASA black and white film (Panatonic; Kodak, Rochester, NY).

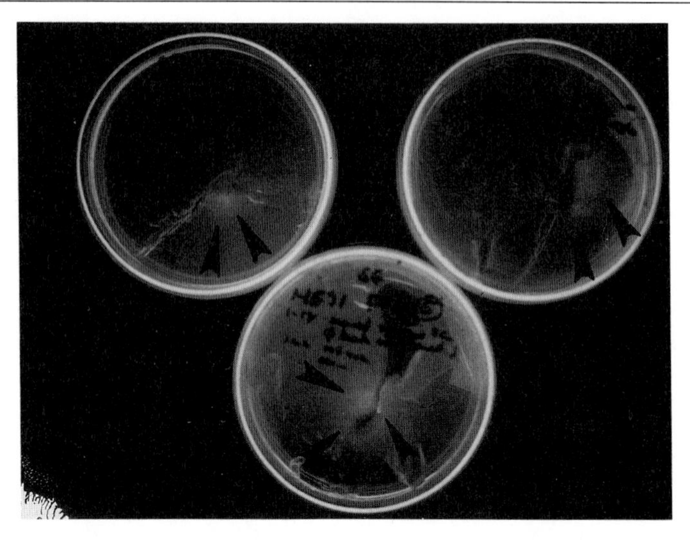

FIG. 1. An example of a positive response for GUS in agar-solidified culture medium of germinating F_1 seedlings from regenerated and putatively transformed cotton plants. The light areas in the agar (arrowheads) are the areas of the fluorescent blue product 4-MU from the hydrolysis of MUG. Conditions for the photograph: Polaroid high-speed, black and white film #667; orange filter for ethidium bromide/DNA photography; exposure, 1 sec; aperture setting, 4.5; UV light source, Foto/Prep I (Fotodyne, New Berlin, WI) on the analytical, or high, setting.

Applications

Assay of Spent Culture Medium. This assay is best performed just after transfer of potentially transformed tissues to fresh medium, as described above.

Assay of Germination Medium. For us, the most useful application of this assay has been in screening aseptically germinated F_1 embryos and seed. Mature and developing embryos are frequently removed from the seed and germinated in culture to release them from dormancy, to hasten development, and improve seedling survival. For isolated embryos, protein leakage can be substantial. However, enzyme loss from normally germinating seedlings can be low due to the presence of an intact cuticle. The assay of these seedlings can be improved by introducing a wound in the root to produce leakage during germination.

We aseptically excised and germinated the mature embryos of corn and wheat as a simple means to break dormancy. The medium used was a hormone-free Murashige and Skoog-based medium, described previously.[6] With cotton, mature seeds were germinated[7]; however, cotton embryos were sometimes "rescued" from bolls taken from dying plants or broken branches. Figure 1 is an example of a positive reaction in the germination medium of rescued F_1 cotton embryos.

A precaution in this application is that the seed coat of some species harbor and release fluorescent compounds; for example, sunflower contains scopoletin. The presence of these compounds in the germination medium can make this assay impossible to interpret. In these cases, the seed coat should be removed prior to germination or the seedling should be transferred to a fresh medium.

Assay for Endogenous β-Glucuronidase-Like Enzymes. The assay can also be used in the media of fungi, bacteria, and untransformed cultured plant tissues, to determine the presence or activity of native GUS-like proteins.

Alternate Procedures

Medium plus Methylumbelliferylglucuronide. Methylumbelliferylglucuronide, to a final concentration of 1 m*M*, can be added directly during the preparation of the medium. The compound is dissolved in 5 ml water, filter sterilized, and added to a cooled, autoclaved medium. Tissues to be assayed are transferred to this medium and cultured for several days, then transferred back to a regular medium. Neither the substrate, MUG, nor the accumulation of 4-MU (the fluorescent product) in the medium was toxic to petunia shoots during 1 to 3 days of culture. This variation is as effective as addition of MUG after culture. Accumulation of the fluorescent product in the culture medium can be observed directly without transferring the tissues; however, MUG is inefficiently used and excessive UV radiation can damage the plant tissues.

β-Glucuronidase Antibody. Purified GUS antibody currently available (Clontech, Palo Alto, CA) may be substituted in this assay to screen specifically for the GUS protein rather than activity. This derivation may be helpful when the synthesized GUS protein is not active or if the plant tissues in question have an endogenous enzyme with GUS-like activity that does not cross-react with the antibody. The procedure using the antibody should be identical to that described above, replacing MUG with appropriately diluted antibody and protein-stabilizing compounds. The

[6] J. H. Gould, M. E. Devey, O. Hasegawa, E. Ulian, G. Peterson, and R. H. Smith, *Plant Physiol.* **95,** 426 (1991).

[7] J. H. Gould, unpublished results (1991).

antibody-containing solution (50–100 μl) should be placed on the medium, to one side of the tissue site, and the antibody allowed to diffuse outward. The protein/antibody reaction should be detectable as a white precipitate.

Other Antibodies. In theory, the assay can be modified for use in screening for other reporter genes by using the antibodies specific for the encoded proteins, such as heat shock, chloramphenicol acyltransferase, or any other stable solution protein that can be induced and synthesized in the cytosol under the conditions of the culture.

Concluding Remarks

Precautions. A positive reaction is not necessarily indicative of transformation. As with the standard GUS assay, the fluorescent product, 4-MU, can result from endogenous β-glucuronidases of bacterial and fungal contaminants as well as enzymes endogenous to the plant tissue being assayed.[8] Fluorescent spots in culture medium can indicate contamination of the tissues with the *Agrobacterium* vector. Presence of a substrate such as MUG can induce synthesis of β-glucuronidases in many saprophytic microorganisms in overnight assays. When this type of contamination is possible, the MUG assay buffer should contain an inhibitor of prokaryotic protein synthesis, such as chloramphenicol (170–200 μg/ml) or kanamycin (50–60 μg/ml).[9]

[8] Cy Hu, P. Chee, R. Chesney, J. P. Miller, and W. O'Brien, *Plant Cell Rep.* **9,** 1 (1990).
[9] J. Sambrook, E. Fritsch, and T. Maniatis, "Molecular Cloning," 2nd Ed., Cold Spring Harbor Laboratory, Cold Spring Harbor, New York, 1989.

[31] Secreted Placental Alkaline Phosphatase as a Eukaryotic Reporter Gene

By BRYAN R. CULLEN and MICHAEL H. MALIM

Introduction

A major focus of current molecular biological research is the identification and functional dissection of the cis-acting sequences and trans-acting factors that regulate eukaryotic gene expression *in vivo*. A common method of addressing this question is to link the regulatory sequence of interest, e.g., an inducible promoter/enhancer element, to a reporter gene. After transfection into the appropriate cells, the reporter gene product can then provide an accurate, quantitative measure of the level of gene expression directed by the promoter/enhancer element, and mutated derivatives thereof, in response to wild-type or mutant forms of the appropriate trans-activator.

The most common indicator gene assay relies on the prokaryotic gene for chloramphenicol acetyltransferase (CAT).[1] Chloramphenicol acetyltransferase has many advantages as a reporter gene, including the absence of interfering activities in eukaryotic cells, high protein stability, and high sensitivity. However, the most commonly utilized assay for CAT activity requires the use of an expensive radioactive substrate and is relatively time consuming.[1] In addition, cell cultures expressing the intracellular CAT enzyme must be harvested, lysed, and extracted prior to analysis. This precludes any kinetic analysis of gene expression within one culture over time and renders more difficult the simultaneous analysis of CAT RNA expression. These considerations have led to a search for alternate reporter genes that, ideally, would combine the sensitivity of the CAT assay with a more convenient, nonradioactive assay and that would produce a secreted indicator gene product. This quest has been given further impetus by the general perception that experiments designed to quantitate the level of gene expression *in vivo* require the use of an accurate internal control. Such a control must, by definition, rely on a second, distinct reporter gene construct that would not interfere with the accurate analysis of CAT activity.

Several alternate or complementary reporter gene systems have been proposed. Among the more prevalent are the genes for the intracellular enzymes β-galactosidase[2] and luciferase[3] and the secreted protein human growth hormone.[4] In this chapter, we will describe the use of a novel indicator gene, producing a secreted form of the human enzyme placental alkaline phosphase,[5] that we believe offers a number of advantages. These include very high stability, efficient secretion by all cells tested, and the availability of a simple, inexpensive, and highly quantitative assay that does not require any unusual equipment or reagents.

Principle of Method

The gene for human placental alkaline phosphatase is a member of a multigene family that, in humans, includes at least three distinct iso-

[1] C. M. Gorman, L. F. Moffat, and B. H. Howard, *Mol. Cell. Biol.* **2,** 1044 (1982).

[2] G. An, K. Hidaka, and L. Siminovitch, *Mol. Cell. Biol.* **2,** 1628 (1982).

[3] S. J. Gould and S. Subramani, *Anal. Biochem.* **7,** 5 (1988).

[4] R. F. Selden, K. Burke-Howie, M. E. Rowe, H. M. Goodman, and D. D. Moore, *Mol. Cell. Biol.* **6,** 3173 (1986).

[5] J. Berger, J. Hauber, R. Hauber, R. Geiger, and B. R. Cullen, *Gene* **66,** 1 (1988).

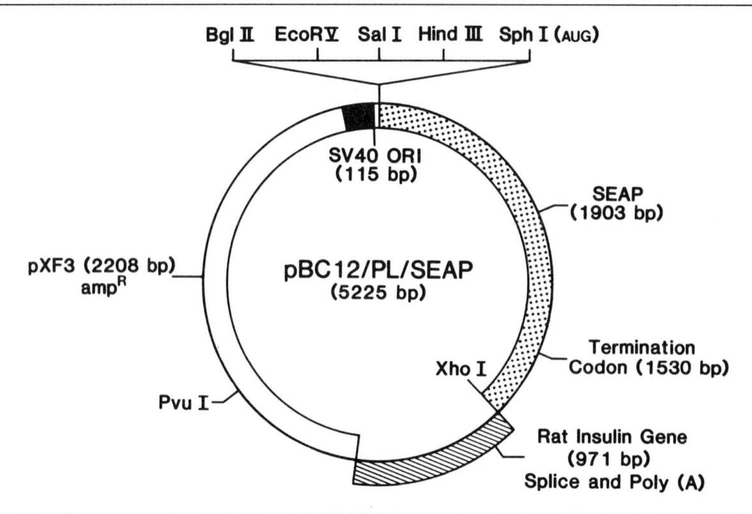

FIG. 1. Structure of the plasmid pBC12/PL/SEAP. The plasmid is designed to facilitate insertion of promoter sequences upstream of the *SEAP* coding sequences.[5] The polylinker *Sph*I site also encodes the translation initiation codon of *SEAP* (GCAUGC).[6] Unique restriction enzyme sites are indicated.

zymes.[6,7] Although all alkaline phosphatases are able to hydrolyze phosphate esters, most optimally at high pH, placental alkaline phosphatase is unusual in at least two respects. The first is the high temperature stability of this enzyme relative to other isozymes detected in humans or in other animals. The second is the highly restricted species and tissue expression of the placental isozyme, which is confined to the placenta of higher primates and, occasionally, transformed cells derived from these same species.[6,7] Although alkaline phosphatases are normally membrane-bound cell surface proteins, they can be inefficiently released into supernatant media in an active form.[6–8] To convert the 513-amino acid cell surface form of placental alkaline phosphatase into a fully active, constitutively secreted enzyme, we introduced a termination codon after codon 489 to produce an efficiently secreted gene product that we term SEAP (secreted alkaline phosphatase).[5,8] The gene encoding SEAP was then introduced into a convenient shuttle vector termed pBC12/PL/*SEAP*. This plasmid (Fig. 1) contains the *SEAP* gene flanked 5′ by a polylinker sequence and

[6] W. Kam, E. Clauser, Y. S. Kim, Y. Wai Kan, and W. J. Rutter, *Proc. Natl. Acad. Sci. U.S.A.* **82,** 8715 (1985).

[7] J. L. Millan, *J. Biol. Chem.* **261,** 3112 (1986).

[8] J. Berger, A. D. Howard, L. Brink, L. Gerber, J. Hauber, B. R. Cullen, and S. Udenfriend, *J. Biol. Chem.* **263,** 10016 (1988).

3′ by an intron and polyadenylation signals derived from the genomic rat preproinsulin II gene.[5] The ability to propagate the vector in bacteria is conferred by a selectable marker (Amp^r) and origin of replication derived from the pBR322 derivative pXF3. In addition, pBC12/PL/SEAP contains a functional but transcriptionally inert origin of replication from simian virus 40 (SV40), thus permitting plasmid replication in cells that express functional SV40 T antigen. The polylinker sequence is convenient for the introduction of regulatory sequences of interest and derivatives containing the long terminal repeat (LTR) promoters of Rous sarcoma virus (RSV), human immunodeficiency virus (HIV), and human T cell leukemia virus (HTLV) have been described, as has a construct bearing the immediate early promoter of human cytomegalovirus.[5]

After insertion of the promoter of interest into pBC12/PL/*SEAP*, the expression plasmid can be introduced into tissue culture cell lines using an appropriate transfection protocol (the SV40 T antigen-expressing cell line COS is particularly ideal as this cell line is readily transfectable and can replicate the pBC12 series plasmids to high copy number). Vectors expressing trans-acting factors that modulate the activity of the introduced promoter can be cotransfected if desired. The SEAP enzyme is efficiently secreted into the supernatant medium in all cell lines tested thus far. Maximal levels are generally secreted between 48 and 72 hr after transfection. In our hands, SEAP activity is essentially stable in tissue culture media at 37°.

Materials and Reagents

Diethanolamine (Cat. No. D45-500; Fisher Scientific, Springfield, NJ)
L-Homoarginine and *p*-nitrophenol phosphate disodium (Cat. Nos. H-1007 and 104-0, respectively; Sigma Chemical Co., St. Louis, MO)
2× SEAP buffer (for 50 ml):

Amount	Stock	Final concentration
10.51 g	Diethanolamine (100% solution)	2 *M*
50 μl	1 *M* $MgCl_2$	1 m*M*
226 mg	L-Homoarginine	20 m*M*

Make up in water and store at 4° (no need to autoclave)
120 m*M* *p*-Nitrophenol phosphate: Dissolve 158 mg in 5 ml of 1× SEAP buffer (make fresh)

Assay for Secreted Alkaline Phosphatase

1. Transfect cells as normal and include a negative control (e.g., a transfection cocktail with no SEAP expression vector in it).

2. Culture at 37° for 48 to 72 hr.

3. Remove 250 μl of each culture supernatant and transfer to Eppendorf tubes. To be on the safe side, maintain the cultures at 37° until "good" SEAP data have been obtained.

4. Heat tubes at 65° for 5 min to inactivate endogenous phosphatases (SEAP is relatively heat stable).

5. Spin for 2 min at room temperature in an Eppendorf microfuge at full speed.

6. Transfer supernatants to fresh Eppendorf tubes. These may be stored at −20° indefinitely.

7. In Eppendorf tubes add 100 μl of 2× SEAP buffer to 100 μl of the supernatant medium (or empirically determined dilutions thereof). As a zero standard make one sample substituting the medium with water.

8. Mix by vortexing.

9. Transfer the contents of each tube to a well of a flat-bottomed microtiter plate. Avoid creating air bubbles.

10. Incubate the plate at 37° for 10 min.

11. During this incubation make the *p*-nitrophenol phosphate (substrate) solution and prewarm to 37°.

12. Add 20 μl of the substrate solution to each well, preferably using a multipipetter.

13. Using an enzyme-linked immunosorbent assay (ELISA) plate reader, measure the light absorbance at 405 nm (A_{405}) at regular intervals (e.g., every 5 min) over the next 30 min while continuing to incubate the plate at 37°.

14. Calculate the levels of SEAP activity using data points that yield a rate of change in light absorbance that is linear with respect to time. (If available, a computer-linked kinetic ELISA plate reader is most convenient, as this will precisely calculate the linear change in A_{405} in all 96 wells of a microtiter plate over time.)

15. Secreted alkaline phosphatase activity can be expressed simply as the change in light absorbance per minute per sample. More formally, SEAP activity can be given in milliunits (mU) per milliliter, where 1 mU is defined as the amount of SEAP that will hydrolyze 1.0 pmol of *p*-nitrophenyl phosphate per minute.[5] This equals an increase of 0.04 A_{405} units/min. The specific activity of SEAP is 2000 mU/μg protein.[9]

[9] E. Ezra, R. Blacher, and S. Udenfriend, *Biochem. Biophys. Res. Commun.* **116,** 1076 (1983).

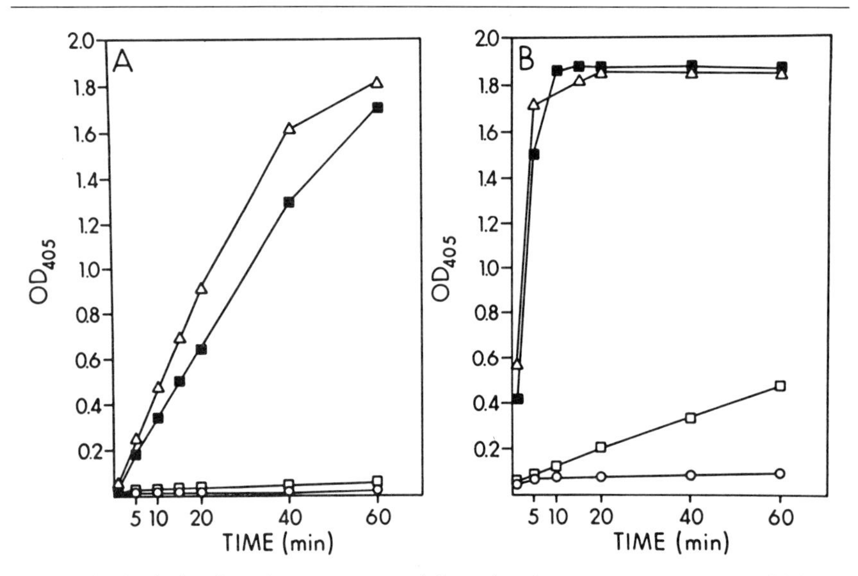

FIG. 2. Analysis of *in vivo* promoter activity using the *SEAP* reporter gene. COS cell cultures were transfected[5] with the indicated expression plasmids and supernatant medium sampled 60 hr later. The change in A_{405} with time is shown for SEAP assays using either 10 μl (A) or 100 μl (B) of medium from transfected cultures. The assay was performed as described in text. The actual levels of SEAP activity determined in this assay were as follow: pBC12/RSV/*SEAP* (△) = 275 mU/ml; pBC12/HIV/*SEAP* (Tat induced) (■) = 247 mU/ml; pBC12/HIV/*SEAP* (basal) (□) = 2.2 mU/ml; negative control (○) ≈ 0.1 mU/ml. (Reproduced from Berger *et al.*[5] by permission.)

An example of an experiment in which the SEAP assay was used to quantitate relative promoter activities in transfected COS cell cultures is given in Fig. 2. In this experiment, we used plasmids that contained *SEAP* linked to the RSV LTR, a highly active constitutive promoter, or the HIV LTR, an inducible promoter that displays a relatively low basal activity. These constructs were transfected into COS cell cultures either alone or, in the case of the pBC12/HIV/*SEAP* construct, together with an expression vector that encodes the HIV Tat trans-activator.[5] Supernatant media were sampled 60 hr after transfection and SEAP activity quantitated as described above. The figure presents the rate of change in light absorbance at 405 nm in SEAP assays that used either 10 μl (Fig. 2A) or 100 μl (Fig. 2B) of supernatant medium. The former level may be more suitable for accurate quantitative analysis of SEAP activity when very active promoter constructs are tested, in this case the RSV LTR and the trans-activated HIV LTR, while the latter may be more suitable for weak promoter constructs, such as the basal HIV LTR promoter. As demonstrated by this example, the SEAP assay is not only rapid and convenient but also sensitive and highly quantitative. In addition, this experiment demonstrates that endogenous alkaline phosphatases found either in tissue culture cells or fetal calf serum are not normally detected by this assay. This results both from the presence of L-homoarginine (an inhibitor of other alkaline phosphatases) in the assay buffer and from the preincubation at 65°, which eliminates heat-sensitive isozymes.[10]

Concluding Remarks

In this chapter, we have described *SEAP*, a novel indicator gene producing secreted alkaline phosphatase. The *SEAP* gene has several advantages when compared to *CAT* or to other prevalent reporter genes. Notably, SEAP is a highly stable, secreted enzyme that can be accurately quantitated using an inexpensive and rapid assay that is similar to the alkaline phosphatase-based ELISA run routinely by many laboratories.[11] The fact that SEAP is secreted means that several assays can be performed on a single culture over time, thus permitting analysis of temporal changes in gene expression that occur in response to different stimuli. The SEAP assay can be used with the majority of cell types and has been shown to function extremely well in injected *Xenopus* oocytes.[12] However, the SEAP assay does have one disadvantage relative to CAT, i.e., lower sensitivity. The assay described in this chapter cannot reliably quantitate levels of SEAP enzyme expression that fall significantly below 50 pg of protein per milliliter. This level of sensitivity is between 10- and 50-fold lower than the sensitivity of radioactivity-based assays for CAT expression.[1] In part, this lower sensitivity may reflect a requirement for protein dimerization in the production of enzymatically active alkaline phosphatase. Therefore, in cells that are poorly transfectable or in other experimental settings where expression of the transfected plasmid is likely to be low, SEAP may not be suitable. Nevertheless, it is clear that SEAP will present considerable advantages in some settings and SEAP may well form an ideal internal control for CAT assays in many experimental protocols. The *SEAP* vectors described in this chapter are available from the authors on request.

[10] J. Berger, A. D. Howard, L. Gerber, B. R. Cullen, and S. Udenfriend, *Proc. Natl. Acad. Sci. U.S.A.* **84,** 4885 (1987).

[11] R. B. McComb and G. N. Bowers, Jr., *Clin. Chem.* **18,** 97 (1972).

[12] S. S. Tate, R. Urade, R. Micanovic, L. Gerber, and S. Udenfriend, *FASEB J.* **4,** 227 (1990).

[32] Use of Fluorescent Chloramphenicol Derivative as a Substrate for Chloramphenicol Acetyltransferase Assays

By DENNIS E. HRUBY and ELIZABETH M. WILSON

The DNA sequences encoding bacterial chloramphenicol acetyltransferase (CAT) enzyme are commonly fused to heterologous transcriptional regulatory signals and used as a "reporter gene" for measuring the rate of transcription of chimeric genes, as well as the translation and/or stability of the chimeric transcripts within the context of transformed bacterial cells, transfected tissue culture cells, or genetically engineered recombinant viruses. The *cat* gene is a particularly powerful and sensitive tool when used in tissue culture systems because mammalian cells lack this enzyme, and hence contain no endogenous CAT activity. Furthermore, the activity of the CAT enzyme can be easily measured and quantitated.[1] Two general types of CAT assay procedures (with various modifications) are currently available. The first involves using [^{14}C]chloramphenicol as the substrate for acetylation. The acetylated derivatives are resolved from the substrate by thin-layer chromatography and quantitated by either autoradiography and subsequent densitometric scanning or by eluting the acetylated derivatives from the plate followed by scintillation counting.[2] The second method utilizes [^{14}C]acetyl-CoA as the substrate. In this case, the acetylated reaction products are collected by organic extraction and quantitated directly by scintillation counting.[3] Both methods, although quite effective, use radioactive substrates that are expensive, potentially hazardous, and require scintillation counting for quantitation. As an alternative procedure, an assay that is based on the use of a fluorescent chloramphenicol derivative not subject to similar limitations has been developed.[4] Using a fluorescent chloramphenicol substrate to assay CAT activity has a number of advantages over the traditional methods: (1) No radioactive materials are required. This dispenses with the need for film, fluors, scintillation counters, and disposal of radioactive wastes; (2) the assay is rapid and easily quantified; (3) the results can be evaluated immediately. Taken together, these results would suggest that the use of a fluorescent chloramphenicol substrate provides an attractive alternative

[1] C. M. Gorman, L. F. Moffat, and B. H. Howard, *Mol. Cell. Biol.* **2,** 1044 (1982).
[2] J. N. Miner, S. L. Weinrich, and D. E. Hruby, *J. Virol.* **62,** 297 (1988).
[3] M. J. Sleigh, *Anal. Biochem.* **156,** 251 (1986).
[4] D. E. Hruby, J. M. Brinkley, H. C. Yang, R. P. Haugland, S. L. Young, and M. H. Melner, *BioTechniques* **8,** 3 (1990).

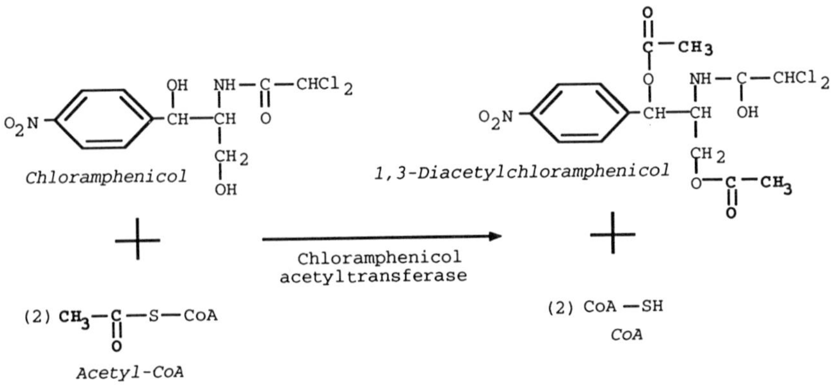

FIG. 1. Acetylation of chloramphenicol by bacterial CAT enzyme.

method for measuring CAT activity in the extracts of cells which express this enzyme.

Principle of Method

Cells expressing the *cat* gene are broken open and a low-speed cytoplasmic extract is prepared. The extract is then incubated with acetyl-CoA and a fluorescent chloramphenicol substrate at 37°. If there is active CAT enzyme present, the substrate will become acetylated at the 1- and/or 3-positions of the molecule (Fig. 1).[5] The reaction is terminated by the addition of ice-cold ethyl acetate, which extracts the fluorescent chloramphenicol substrate and any acetylated derivatives into the organic phase. Following drying and being redissolved in a small volume of ethyl acetate, the reaction products are resolved by ascending thin-layer chromatography on silica plates using chloroform-methanol (9 : 1) as the solvent.[6] After chromatography the plates are air dried. At this point, the results of the assay can be qualitatively assessed by visual inspection using either visible or ultraviolet light. This represents one of the convenient features of the fluorescent chloramphenicol substrate as it, and its acetylated derivatives, are visible during chromatography, which facilitates the excision of the spots as well as allowing the investigator to monitor the CAT assay results quickly. To quantitate the reaction, the spots corresponding to the acetylated CAT derivatives are excised, eluted in methanol, and the fluorescence measured using a dual-beam fluorimeter. The compound, which is

[5] W. V. Shaw, this series, Vol. 43, p. 737.

[6] W. M. Hodges and D. E. Hruby, *Anal. Biochem.* **160,** 65 (1987).

yellow in visible light, has an extinction coefficient of 73,200 cm^{-1} M^{-1} in methanol, absorbs light maximally at 505 nm, and emits at 512 nm.[7]

Materials and Reagents

Bodipy chloramphenicol (FAST CAT), 1.0 m*M*, in 0.1 *M* Tris-HCl (pH 7.8)–methanol (9 : 1): Molecular Probes (Eugene, OR)
Thin-layer chromatography (TLC) reference standard (0.2 m*M* mixture of 1-,3-, and 1,3-acetyl derivatives of Bodipy chloramphenicol in ethyl acetate): Molecular Probes)
Chloramphenicol acetyltransferase (from *Escherichia coli*): Sigma Chemical Company (St. Lous, MO)
Acetyl-CoA: Sigma Chemical Company
Ethyl acetate: EM Science (Cherry Hill, NJ)
Silica thin-layer chromatography plates with flexible backs: J. T. Baker Chemical Company (Phillipsburg, NJ)
Chloroform: J.T. Baker Chemical Company
Methanol: J. T. Baker Chemical Company

Equipment Needed

Variable microliter pipetting device
Polypropylene 1.5-ml microfuge tubes
Vortex mixer
Water bath
Microcentrifuge
Vacuum desiccator
Thin-layer chromatography chamber
Glass cuvettes
Fluorimeter
Water pump (aspirator) with trap for collecting liquid waste

Assay Procedure

Preparation of Enzyme Extracts

Bacterial Cells. Inoculate transformed bacterial cells into the appropriate liquid medium and grow until the suspension has an optical density at 600 nm of approximately 0.6. Transfer 0.2 ml of the culture into a 1.5-ml microfuge tube and pellet the cells by centrifugation at 15,000 rpm for 2 min in a microcentrifuge. Aspirate off the medium and resuspend the

[7] P. L. Moore, A. P. Guzikowski, H. C. Kang, and R. P. Haugland, unpublished observations (1989).

pelleted cells in 0.5 ml of 100 m*M* Tris-HCl (pH 8.0). Add 20 μl of lysis medium [100 m*M* ethylenediamine tetraacetic acid (EDTA), 100 m*M* dithiothreitol (DTT), 50 m*M* Tris-HCl (pH 8.0)]. Add a small drop of toluene from a fine-tip Pasteur pipette. Incubate at 30° for 30 min. The extract can then be assayed immediately for CAT activity or stored at −20°.[8]

Mammalian Cells. Scrape transfected cells, or recombinant virus-infected cells, from a confluent 100 × 20 mm tissue culture plate and transfer to a 15-ml conical tube. Centrifuge at 2000 rpm at 4° for 10 min. Remove the medium, resuspend the cells in 5 ml of phosphate-buffered saline, and centrifuge as above. Aspirate off the supernatant and resuspend the pellet in 1 ml 250 m*M* Tris-HCl (pH 7.8). Sonicate the cell suspension six times for 10-sec periods. Subject the suspension to three cycles of freeze-thawing. Centrifuge at 12,000 rpm for 5 min at 4° in a microcentrifuge. The supernatant can then be assayed for CAT activity or frozen at −20° until later.[9]

Enzyme Assay

Both positive and negative controls should be included with each set of assays. The reaction buffer alone serves as a convenient negative control. Purified CAT enzyme (0.1 unit), which is commercially available, is an appropriate positive control. Mix 59 μl extract with 1 μl of 1.0 m*M* Bodipy chloramphenicol in 0.1 *M* Tris-HCl (pH 7.8)–methanol (9 : 1) and incubate at 37° for 5 min. To inactivate endogenous acetylating enzymes, the extracts can be heated at 65° for 10 min prior to assay. The bacterial CAT enzyme is stable to this treatment whereas most other acetylating activities are not.[10] Add 10 μl of 4 m*M* acetyl-CoA and continue the incubation for a fixed period of time, between 15 and 60 min. The acetyl-CoA should be freshly prepared before each set of assays. Dissolve 0.5 mg of acetyl-CoA in 135 μl of water to obtain a 4 m*M* solution. The concentration of substrate in the reaction mixture is approximately 142 μ*M*, which is well above its K_m of 7.4 μ*M*. Because the rate of enzymatic conversion is dependent on the concentration of CAT enzyme present, longer incubation periods will be necessary when assaying cell extracts expressing very low levels of enzyme. Stop the reaction by adding 1 ml ice-cold ethyl acetate and mixing vigorously. If vortexing is not possible, shake each sample vigorously for at least 30 sec to ensure complete extractions. Centrifuge briefly to separate the phases. Remove the top 900 μl of ethyl acetate and transfer to a fresh tube. Dry the reaction products down by evaporating off the ethyl

[8] J. Tomizawa, *Cell* (*Cambridge, Mass.*) **40,** 527 (1985).
[9] D. Hruby, unpublished observations (1988).
[10] D. W. Crabb and J. E. Dixon, *Anal. Biochem.* **163,** 88 (1987).

acetate. The drying procedure is conveniently carried out using a Savant (Hicksville, NY) Speed Vac concentrator or other similar device. Resuspend the yellow residue left in the bottom of the tube in a small volume (20–30 μl) of ethyl acetate. The reaction products are ready for TLC or may be stored at $-20°$ for later analysis. The ethyl acetate extracts are stable indefinitely, as long as they are stored in a cool place away from light. If the samples are to be stored for an extended length of time before assaying, the containers should be tightly closed to prevent contamination and excessive evaporation of the solvent.

Thin-Layer Chromatography

Draw a light pencil line at 2 cm along the long axis of a 10×20 cm silica TLC plate. Use an automatic pipetting device to deliver 5 μl of the reaction products on the plate at the 2-cm level. Best results will be obtained by repeatedly applying portions of the sample, allowing the solvent spot to partially dry between applications (this technique will minimize spreading of the spot and give optimal resolution). Multiple reactions may be analyzed on the same plate provided the samples are separated by at least 1.5 cm. After the spots have dried, put the plate into a sealed chromatography chamber that is filled to a depth of 1 cm with chloroform–methanol (9 : 1). It is essential that the chromatography chamber be well sealed to prevent evaporation, which will alter the eluent composition. Placing one or more filter paper strips or pads in the tank along with the plate will help keep the atmosphere in the tank saturated with eluent and improve the reproducibility of the separations. Close the chamber and allow the buffer to ascend the plate. The solvent front should be allowed to advance to near the top of the plate for optimum separation of the products. However, do not allow the plates to remain too long in the chamber after the solvent front reaches the top, as the separated bands will diffuse and be more difficult to visualize. Remove the plate and air dry. The results of the chromatography can be monitored by visual inspection of the yellow compounds. The visualization can be enhanced by viewing the plate under ultraviolet light. Plates can be photographed using a variety of cameras and film. Good results have been obtained with a 35-mm camera using Ektachrome 400 film and a Polaroid camera (Fotodyne Co., New Berlin, WI) with type 55 positive/negative black and white print film.

Analysis of Results

There are two approaches to quantitating the results. The first and simplest, as it requires no additional equipment, is to carry out the CAT

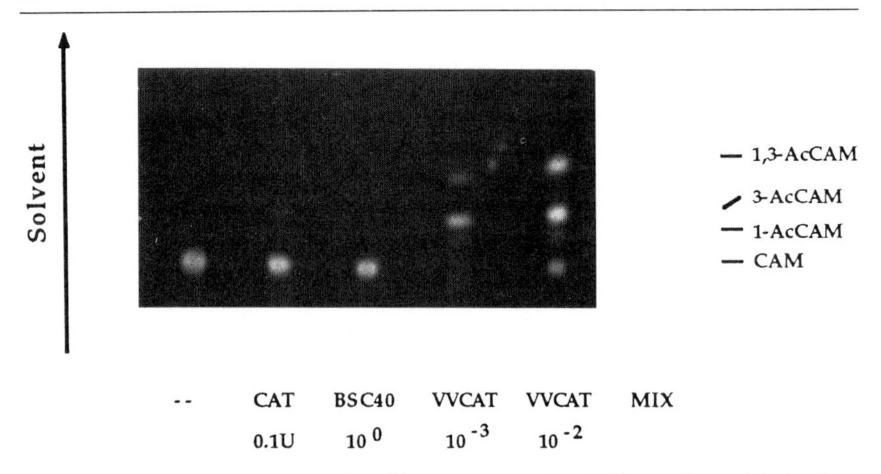

FIG. 2. Thin-layer chromatography of fluorescent acetylated chloramphenicol derivatives. The CAT assays and chromatography were carried out exactly as described in text. Extracts tested were as follow: --, water; CAT (0.1 U), purified chloramphenicol acetyltransferase; BSC40, a cytoplasmic extract of uninfected monkey kidney tissue culture cells; and VVCAT, a cytoplasmic extract of BSC40 cells infected with a recombinant vaccinia virus expressing the bacterial *cat* gene. The numbers below the BSC40 and VVCAT extracts indicate the dilutions that were used. The identities of the chloramphenicol substrate and acetylated derivatives are indicated at the right.

assay on serial dilutions of the extract and use visual inspection to determine the approximate end point of enzyme activity. An example of this type of analysis is presented in Fig. 2. Alternatively, following the TLC lightly circle the bands corresponding to the unreacted substrate and all of the acetylated products for each sample, using a soft lead pencil. This is best accomplished using a UV illuminator to ensure that all of the products are visualized. Scrape and combine the acetylated bands for each sample in clean tubes. Scrape the unreacted substrate band for each sample into separate tubes. Add a precise and constant volume of spectrophotometric-grade methanol to each tube and vortex for approximately 1 min to extract the compounds. After centrifugation, withdraw a measured aliquot from each tube, taking care not to disturb the compacted silica gel. Any convenient volume that can be accurately measured will suffice, but the smaller the volume, the more concentrated the solution and, hence, the greater the signal. Standard 3-ml cuvettes usually require at least 2 ml of solution for measurement. Smaller cuvettes of 0.5 ml are readily available and are preferable, because a smaller volume can be measured. Measure the fluorescence at 512 nm either by exciting the samples at 490 nm or by using fluoroscein filters. A fluorometer with a variable-wave-

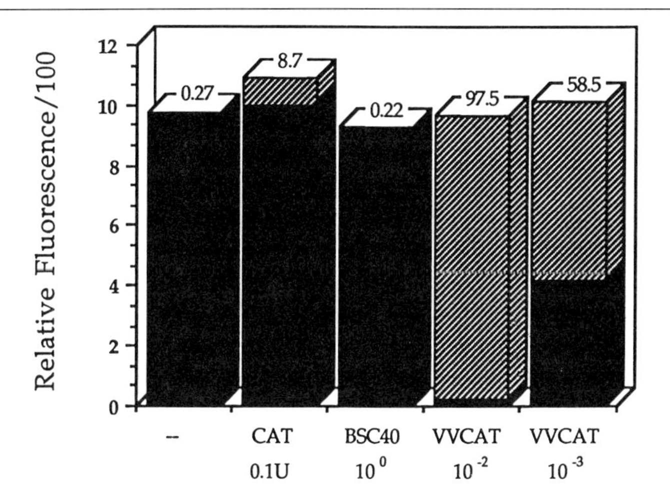

FIG. 3. Quantitative analysis of CAT enzyme assays. The spots corresponding to the fluorescent chloramphenicol substrate and the acetylated derivatives (1-AcCAM, 3-AcCAM, and 1,3-AcCam) were excised from the thin-layer plate shown in Fig. 2. After elution with methanol, the relative fluorescence of each sample was measured in a dual-beam fluorimeter. The samples were excited at 490 nm and the emitted light measured at 512 nm. The results are shown in a layered bar graph, in which the unreacted substrate is shown in black and the acetylated products are indicated by the diagonal stripes. The numbers shown above each bar indicate the percentage substrate conversion measured for each reaction.

length diffraction grating will allow use of optimal excitation and emission wavelengths. However, the fluorescent substrate and products give strong fluorescence with fluorescein filters if a fluorimeter is not available. Determine the conversion of substrate into products from the relative intensities of the fluorescent signals, using the following formulas:

$$\text{Conversion (\%)} = (\text{FI}_{\text{products}}/\text{FI}_{\text{substrate}}) \times 100$$

$$\text{Products } (\mu M) = \text{conversion (\%)} \times \text{substrate added to the reaction } (\mu M)$$

where FI is the fluorescence intensity. As an example, the TLC plate shown in Fig. 2 was subjected to this type of analysis and the results are presented in Fig. 3. Alternatively, the conversion can be measured by the relative absorption at 505 nm of the products and the unreacted substrate. However, the fluorometric method is inherently much more sensitive and, hence, the method of choice.

Final Comments

The gene producing the bacterial CAT enzyme has been widely used in a variety of systems as a reporter gene because of the ease and rapidity of the enzyme assay and the absence of any endogenous background activity. The development of a modified CAT enzyme assay that uses fluorescent chloramphenicol substrates promises to enhance the utility of this system by eliminating the use of radioactive materials. This is not only a matter of convenience and cost saving but should also allow this reporter gene system to be used by investigators who are not able to work with radioactive materials and by instructors in classroom situations.

[33] *pac* Gene as Efficient Dominant Marker and Reporter Gene in Mammalian Cells

By SUSANA DE LA LUNA and JUAN ORTÍN

Introduction

DNA-mediated gene transfer has become a standard technique to introduce genes into animal cells. It has enabled the functional studies of alterations produced *in vitro,* as well as the overproduction of the corresponding gene products. However, because the vast majority of the genes are not selectable, the development of selectable markers for cotransformation experiments has been a crucial step in the generalization of this approach.

Since the recovery of hypoxanthine–guanine phosphoribosyltransferase (HPRT) activity mediated by DNA,[1] biochemical markers have been extensively used. Some of them, like the genes coding for HPRT[1], adenine phosphoribosyltransferase (APRT),[2] thymidine kinase (TK),[3] dihydrofolate reductase (DHFR),[4] or aspartate transcarbamylase (aspartate carbamoyltransferase; ATCase)[5] require the use of cell lines that are deficient in the corresponding enzymes. Hence, they are selectable but not dominant markers and their usefulness is limited. Other genes, like those encoding xanthine–guanine phosphoribosyltransferase (GPT), asparagine synthase (AS), or adenosine deaminase (ADA) can serve

[1] E. H. Szybalska and W. Szybalski, *Proc. Natl. Acad. Sci. U.S.A.* **48,** 2026 (1962).
[2] M. Wigler, A. Pellicer, S. Silverstein, R. Axel, G. Urlaub, and L. Chasin, *Proc. Natl. Acad. Sci. U.S.A.* **76,** 1373 (1979).
[3] M. Perucho, D. Hanahan, L. Lipsich, and M. Wigler, *Nature* (*London*) **285,** 207 (1980).
[4] R. J. Kaufman and P. A. Sharp, *J. Mol. Biol.* **159,** 601 (1982).
[5] J. C. Ruiz and G. M. Wahl, *Mol. Cell. Biol.* **6,** 3050 (1986).

as dominant markers when used under the appropriate experimental conditions.[6–8]

However, antibiotic resistance markers constitute the paradigm among dominant ones. They have been constructed from bacterial genes by engineering the control elements for expression in mammalian cells. The pioneer and most widely used among them is the one encoding aminoglycoside phosphoribosyltransferase (*neo*),[9,10] the gene able to determine resistance to neomycin in bacteria and to G418 in mammalian cells. The list has grown to include those encoding hygromycin phosphotransferase (*hyg*)[11,12] and puromycin acetyltransferase (*pac*),[13,14] which determine resistance to hygromycin and puromycin, respectively. The latter encodes puromycin acetyltransferase from *Streptomyces alboniger*. Its development as a dominant marker and its practical use are the subjects of this chapter.

Principle of Method

Puromycin is an aminonucleoside antibiotic produced by *S. alboniger*.[15] It blocks protein synthesis by interacting with the A site of the large ribosomal subunit of both prokaryotic and eukaryotic ribosomes,[16] because it resembles an amino acid attached to the terminal adenosine of tRNA. This interaction is followed by peptide bond formation, which releases the nascent polypeptide in the form of polypeptidylpuromycin. The antibiotic inhibits the growth of gram-positive bacteria and various animal and insect cells.[13,17,18] Fungi and gram-negative bacteria are resistant due to the low permeability of the antibiotic.[16]

[6] R. Mulligan and P. Berg, *Proc. Natl. Acad. Sci. U.S.A.* **78,** 2072 (1981).
[7] M. Cartier, M. W.-M. Chang, and C. P. Stanners, *Mol. Cell. Biol.* **7,** 1623 (1987).
[8] R. J. Kaufman, P. Murtha, D. E. Ingolia, C.-Y. Yeung, and R. E. Kellems, *Proc. Natl. Acad. Sci. U.S.A.* **83,** 3136 (1986).
[9] F. Colbere-Garapin, F. Horodniceanu, P. Khourilsky, and A.-C. Garapin, *J. Mol. Biol.* **150,** 1 (1981).
[10] P. J. Southern and P. Berg, *J. Mol. Appl. Genet.* **1,** 327 (1982).
[11] R. F. Santerre, N. E. Allen, J. N. Hobbs, Jr., N. Rao, and R. J. Schmidt, *Gene* **30,** 147 (1984).
[12] T. J. Giordano and W. T. McAllister, *Gene* **88,** 285 (1990).
[13] J. Vara, A. Portela, J. Ortín, and A. Jiménez, *Nucleic Acids Res.* **14,** 4617 (1986).
[14] S. de la Luna, I. Soria, D. Pulido, J. Ortín, and A. Jiménez, *Gene* **62,** 121 (1988).
[15] J. N. Porter, R. I. Hewitt, C. W. Hesseltine, C. W. Krupka, J. A. Lowery, W. S. Wallace, N. Bohonos, and J. H. Williams, *Antibiot. Chemother.* **2,** 409 (1952).
[16] D. Vázquez, "Inhibitors of Protein Biosynthesis." Springer-Verlag, Berlin, Heidelberg, and New York, 1979.
[17] M. Kuwano, K. Takenaka, and M. Ono, *Biochim. Biophys. Acta* **563,** 479 (1979).
[18] A. M. Fallon and V. Stollar, *Somatic Cell Genet.* **8,** 521 (1982).

Like other antibiotic-producing *Streptomyces*, *S. alboniger* ribosomes are sensitive to their own antibiotic.[19] Puromycin resistance is mediated by a puromycin acetyltransferase (PAC) that inactivates the antibiotic by N-acetylation of the tyrosinyl group, rendering a product that cannot accept peptidyl moieties.[20] The gene encoding this activity (*pac* gene) has been cloned and its sequence has been determined; it confers puromycin resistance on introduction into *Escherichia coli* and *Streptomyces lividans*.[21,22] Furthermore, the *pac* gene has been cloned under the control of the simian virus 40 (SV40) early promoter in a series of shuttle vectors[13,14] and used as a dominant marker for the selection of transformed mammalian cells by allowing them to grow in the presence of the antibiotic.

The *pac*-containing constructs could be grouped in two classes: (1) pSV_2*pac* and pSV_2*pac*ΔP, in which the *pac* gene is substituted for β-globin in $pSV_2\beta$-globin, and (2) pBS*pac* and pBS*pac*ΔP, which lack the SV40 small-t intron and the pBR322 toxic sequences[23]; the first member of each class contains the original prokaryotic promoter of the *pac* gene. All these recombinants have been used for the selection of transformed mammalian cells by their growth in medium containing puromycin. Moreover, the fact that PAC activity can be measured in an easy, rapid, and reproducible assay (note that this enzymatic activity is the counterpart of chloramphenicol acetyltransferase (CAT), when puromycin is used as substrate) makes the *pac* gene useful as a reporter gene to analyze eukaryotic promoter functions.

Methods

Calcium Phosphate Transfection

Many methods have been developed to introduce cloned DNA into mammalian cells. These techniques include calcium phosphate transfection,[2,24] DEAE-dextran-mediated transfection,[25] protoplast fusion,[26] lipo-

[19] J. Pérez-González, J. Vara, and A. Jiménez, *J. Gen. Microbiol.* **131,** 2877 (1985).

[20] J. Pérez-González, J. Vara, and A. Jiménez, *Biochem. Biophys. Res. Commun.* **113,** 772 (1983).

[21] J. Vara, F. Malpartida, D. Hopwood, and A. Jiménez, *Gene* **33,** 197 (1985).

[22] R. A. Lacalle, D. Pulido, J. Vara, M. Zalacaín, and A. Jiménez, *Gene* **79,** 375 (1989).

[23] M. Lusky and M. Botchan, *Nature* (*London*) **293,** 79 (1981).

[24] F. L. Graham and A. J. van der Eb, *Virology* **52,** 456 (1973).

[25] J. H. McCutchan and J. S. Pagano, *J. Natl. Cancer Inst.* **41,** 351 (1968).

[26] W. Schaffner, *Proc. Natl. Acad. Sci. U.S.A.* **77,** 2163 (1980).

some-mediated fusion,[27] electroporation,[28,29] and microinjection.[30] We normally use calcium phosphate transfection to transform cell lines. The protocol presented below works well for every cell line tested for puromycin selection. Descriptions for optimizing the different parameters affecting transfection efficiencies or detailed protocols for the other procedures can be found in general manuals.[31-33]

Solutions

HBS (2×) (280 m*M* NaCl, 50 m*M* HEPES): Dissolve 1.6 g of NaCl and 1 g of *N*-2-hydroxyethylpiperazine-*N'*-2-ethanesulfonic acid (HEPES; Sigma, St. Louis, MO) in a total volume of 90 ml of doubly distilled H_2O. Adjust the pH to 7.11 ± 0.05 with 1 *N* NaOH, add 2 ml of 100× phosphate solution, and then adjust the volume to 100 ml. Sterilize by filtration (0.22-μm filter) and store in 10-ml aliquots at −20°.

Phosphate (100×): Mix 1 vol 70 m*M* H_2NaPO_4 and 1 vol 70 m*M* HNa_2PO_4

$CaCl_2$ (2 *M*): Dissolve 5.92 g of $CaCl_2 \cdot 2H_2O$ (Mallinckrodt, St. Louis, MO) in 20 ml of doubly distilled H_2O. Sterilize by filtration (0.22-μm filter) and store in 1-ml aliquots at −20°

DNA: High molecular weight (HMW) DNA from calf thymus (Boehringer Mannheim, Indianapolis, IN) at a concentration of 2.5 mg/ml. Plasmid DNA should be extracted with phenol–chloroform, precipitated with ethanol, and resuspended in TE [10 m*M* Tris-HCl, pH 7.5, 1 m*M* ethylenediaminetetraacetic acid (EDTA)] at a concentration of 1 mg/ml

Procedure

1. Twenty-four hours before transfection, seed 100-mm tissue culture plates with 1×10^6 growing cells. Change the medium 4 hr before the addition of the precipitate (10 ml/dish).

[27] R. J. Mannino and S. Gould-Fogerite, *BioTechniques* **6,** 682 (1988).

[28] E. Neumann, M. Schaefer-Ridder, Y. Wang, and P. H. Hofschneider, *EMBO J.* **1,** 841 (1982).

[29] U. Zimmermann, *Biochim. Biophys. Acta* **694,** 227 (1982).

[30] M. R. Capecchi, *Cell (Cambridge, Mass.)* **22,** 479 (1980).

[31] B. R. Cullen, this series, Vol. 152, p. 684.

[32] J. Sambrook, E. F. Fritsch, and T. Maniatis, "Molecular Cloning: A Laboratory Manual," 2nd Ed., Vol. 3, Sect. 16.30. Cold Spring Harbor Laboratory, Cold Spring Harbor, New York, 1989.

2. Prepare the calcium–DNA solution (per dish) as follows:
   a. Mix 429 μl sterile doubly distilled H_2O, 6 μl of HMW DNA, and 5 μl of plasmid DNA in a 5-ml plastic tube (e.g., Falcon 2054; Becton Dickinson, Oxnard, CA). Slowly add 60 μl of 2 *M* $CaCl_2$. Mix by slowly pipetting up and down.
   b. Place 500μl of 2 × HBS in a sterile 15-ml polystyrene tube (e.g., Falcon 2051).
   c. Add the calcium–DNA solution dropwise with a Pasteur pipette while bubbling through a plugged 1-ml sterile plastic pipette inserted into the mixing tube. Use a mechanical pipettor attached to the plastic pipette or blow through a rubber tubing attached to the pipette.
   d. Allow the precipitate to form for 30 min at room temperature without agitation.

3. Resuspend the precipitate by gentle pipetting and add it directly on the medium that covers the cells. Agitate the plate to mix the precipitate and medium.

4. Incubate the transfected cells for 16–24 hr, wash the cultures thoroughly, and add fresh medium.

Comments

1. The initial cell density can vary for each cell type, but the optimal would be that which renders a subconfluent monolayer the day for application of the selection or splitting the transfected cells.

2. Transfection efficiency can vary between different batches of 2× HBS. The HEPES buffer must be pH 7.11 ± 0.05.[24] The pH of the solution can change during storage.

3. To achieve the highest transformation efficiencies, plasmid DNAs should be purified by equilibrium centrifugation in CsCl–ethidium bromide density gradients,[34] but DNA obtained by the alkaline lysis procedure[35] also works well if impurities are removed (e.g., using Geneclean, BIO 101, La Jolla, CA).

4. Carrier DNA must also be of high purity; some authors recommend

[33] F. M. Ausubel, R. Brent, R. E. Kingston, D. D. Moore, J. G. Seidman, J. A. Smith, and K. Struhl (eds.), "Current Protocols in Molecular Biology," Chap. 9. Greene Publ. and Wiley (Interscience) New York, 1987.

[34] J. Sambrook, E. F. Fritsch, and T. Maniatis, "Molecular Cloning: A Laboratory Manual," 2nd Ed., Vol. 1, Sect. 1.42. Cold Spring Harbor Laboratory, Cold Spring Harbor, New York, 1989.

[35] H. C. Birnboim and J. Doly, *Nucleic Acids Res.* **7,** 1513 (1979).

using mammalian HMW DNA prepared in the laboratory,[36,37] but we routinely use commercial DNA with satisfactory results.

5. In cotransfections, we use a 5–10 : 1 ratio of nonselectable gene to *pac* recombinant; remember that final DNA concentration in the mix should be 20 μg/ml.

6. Because the pH of the culture medium turns acidic after the addition of the calcium–DNA precipitate, be careful to maintain a CO_2 concentration in the incubator not higher than 5–7%.

Puromycin Selection

Solutions

Puromycin stock solution: Dissolve 10 mg of puromycin dihydrochloride (Sigma) in 1 ml of distilled H_2O. Sterilize by filtration (0.22-μm filter) and store in 100-μl aliquots at −20°. We have not noticed differences in potency between different lots of puromycin, as has been observed with G418.

Procedure. After transfection, selection is applied to allow isolation of stably transformed clones. This can be done in two ways, depending on the efficiency of stable transformation and the purpose of the experiment. Thus, 200–300 colonies are easily counted in a 100-mm dish, but it is better to limit the number to 20–30 if individual clones are to be picked up and expanded.

1. If selection is going to be applied directly to the transfected plate, allow cells to duplicate in nonselective medium before the addition of the antibiotic.

2. If the culture is going to be split, recover the cells by trypsinization and replate them at a proper dilution; then allow the cells to grow just until subconfluency before applying selection.

3. In any case, wash the cultures and feed them with fresh selective medium every 3–4 days.

The selection conditions will be defined by the minimum level of puromycin that prevents cell growth, the strength of the promoter that controls *pac* gene expression, and the number of integrated copies required. For the culture cell lines we have used (Vero, COS-1, ts-COS, BHK-21, L,

[36] J. Sambrook, E. F. Fritsch, and T. Maniatis, "Molecular Cloning: A Laboratory Manual," 2nd Ed., Vol. 2, Sect. 9.14. Cold Spring Harbor Laboratory, Cold Spring Harbor, New York, 1989.

[37] B. G. Herrmann and A.-M. Frischauf, this series, Vol. 152, p. 180.

HeLa, MDCK) the toxic concentration of puromycin is about 2.5 μg/ml; if other cell lines are to be used, the toxic concentration must be previously determined. With our *pac* recombinants,[13,14] in which the *pac* gene is under the control of the SV40 early promoter, we use 10 μg/ml puromycin in the selective medium (except for HeLa cells, for which we use 2.5–5 μg/ml). Cell death is quite fast and within 16–24 hr most cells are detached from the plates. For rapidly growing cells such as BHK-21 or MDCK, resistant colonies are easily distinguished in 1 week. For other cell lines it is necessary to wait 1 week longer. When colonies are apparent, they can either be picked up by trypsinization or fixed and stained if only the number of transformed clones is required.

Comments. It is recommended that a mock transfection be carried out in parallel in order to determine the background of puromycin-resistant colonies. The frequency of spontaneous mutation to puromycin resistance ranged between 0.1 and 5 colonies/10^6 transfected cells, depending on the cell line.[13,14]

If the cloning efficiency of the recipient cell line is low, it will be very difficult to obtain isolated colonies after selection, so we recommend a less drastic selection method: (1) Use the minimum level of puromycin that kills nontransformed cells; (2) allow cells to duplicate twice before applying selection; (3) add the selective medium and, when 50% of the cells have died, wash off the antibiotic; (4) incubate the culture for 24–48 hr in the absence of puromycin; (5) add the selective medium again; (6) repeat this cycle until the colonies are apparent.

Analysis of Puromycin Acetyltransferase Activity

Puromycin acetyltransferase activity can be measured by a modification of the procedure described by Shaw to assay chloramphenicol acetyltransferase (CAT) activity,[38,39] based on the differential distribution of acetyl-CoA and *N*-acetylpuromycin between an organic and an aqueous phase.

Solutions

Phosphate-buffered saline (PBS): 137 m*M* NaCl, 2.7 m*M* KCl, 6.4 m*M* Na_2HPO_4, 1.4 m*M* KH_2PO_4 Lysis buffer: 50 m*M* Tris-HCl, pH 8.5, 2 m*M* EDTA, 10% (v/v) glycerol, and 0.5% (v/v) Nonidet P-40 (NP-40)

Incubation buffer: 0.1 *M* Tris-HCl, pH 8.0, 0.2 m*M* [^{3}H]acetyl-CoA (5–20 cpm/pmol), and 0.2 m*M* puromycin

[38] W. V. Shaw, this series, Vol. 43, p. 737.

[39] J. Vara, J. A. Pérez-Gónzalez, and A. Jiménez, *Biochemistry* **24,** 8074 (1985).

Procedure

1. Wash cell monolayers twice with cold PBS, add 1 ml of PBS, and scrape off the cells into microfuge tubes with a rubber policeman. Keep the tubes on ice while other plates are processed.

2. Pellet the cells by low-speed centrifugation at 4° and remove PBS carefully.

3. Resuspend the cells (up to 10^6) in 100 μl of lysis buffer and keep on ice for 5 min.

4. Centrifuge for 5 min at 12,000 *g* and 4°, and recover the supernatant to assay activity. Store at −20°.

5. Carry out the reactions in a final volume of 50 μl, containing incubation buffer and 5–10 μl of cell extract, for 30 min at 30°.

6. Stop the reactions with 200 μl of 5 *M* NaCl–0.1 *M* borate, pH 9.0.

7. Add 1 ml of toluene-based scintillator and shake the mix vigorously. Wait for 5 min to allow phase separation to occur and count.

Comments

1. One unit of PAC activity corresponds to 1 μmol of acetyl group transferred to puromycin per minute. Specific activity is defined as milliunits per microgram of protein.

2. For high background or low PAC activity cell extracts it is convenient to make a double set of reactions, with and without puromycin.

3. Synthesis of *N*-acetylpuromycin can be detected with higher sensitivity using the classic thin-layer chromatography method used to detect CAT activity.[39]

4. To overcome the presence of acetyl-CoA hydrolases in the extract, use 0.3 units of acetyl-CoA synthetase and 1 m*M* ATP in the assay.[40]

Properties of pac as Selective Marker

The efficiency of transformation of several cultured cell lines with *pac*-containing constructs is indicated in Table I. As a comparative reference, the results of some parallel experiments in which the widely used construct pSV$_2$*neo* was used are also included. It is clear that *pac* gene is more efficient in certain cell lines, like BHK21, and slightly less efficient in others, as in the L cell line.

The efficiency of cotransformation is difficult to determine, because it depends on several factors: the cell type, the state of DNA, the ratio between selectable and nonselectable gene, and, of course, the nature of

[40] A. Nieto and P. Esponda, 1991, personal communication.

TABLE I
TRANSFORMATION EFFICIENCY WITH DIFFERENT *pac*-CONTAINING CONSTRUCTS

	Cell line			
Construct	BHK-21	Vero	L	ts-COS
pSV_2*neo*	1000/1200[a]	66/400	162/800	24/74
pSV_2*pac*	240/668	2/10	19/37	ND[b]
pSV_2*pac*ΔP	2760/2950	45/90	28/48	ND
pBS*pac*	6255/6990	72/115	82/118	102/254
pBS*pac*ΔP	9780/10275	190/225	400/426	ND

[a] The numbers refer to resistant colonies obtained from 10^6 recipient cells in two independent experiments. Selection was done in medium containing 10 μg/ml puromycin.

[b] ND, Not determined.

the nonselectable gene. Use of the *pac* gene has allowed the establishment of cell lines that express genes such as the influenza virus nucleoprotein,[41] the adenovirus VA-1 RNA,[42] the RNA polymerase of phage T7[43] or various proteins of the foot-and-mouth disease virus,[44] with cotransformation efficiencies ranging between 5 and 70%.

There is no way to predict the stability of the marker in stably transformed cell lines in the absence of selection, because it is probably affected by the specific characteristics of the parental cell line, the integration site in chromosomal DNA, and the possible toxicity of the cotransformed gene. We have PAC-expressing cell lines that maintain puromycin resistance in the absence of the antibiotic for more than 20 passages, and others that lose the resistant phenotype in 50% of the culture by the fifth passage. Therefore, we recommend growing transformed cell lines in medium containing 2.5 μg/ml puromycin, a concentration that is toxic for most of the culture cell lines not expressing PAC.

Concluding Remarks

The usefulness of a new dominant marker, such as *pac* gene, must be evaluated in the frame of other available ones. In view of its versatility, the only sensible comparison should include other antibiotic resistance markers such as *neo* or *hyg*.

[41] S. de la Luna, A. Portela, C. Martínez, and J. Ortín, *Nucleic Acids Res.* **15,** 6117 (1987).

[42] C. Martínez, S de la Luna, F. Peláez, J. A. López-Turiso, J. Valcárcel, C. de Haro, and J. Ortín, *J. Virol.* **63,** 5445 (1989).

[43] S. de la Luna, A. Beloso, and J. Ortín, 1991, unpublished data.

[44] E. Martínez-Salas, 1991, unpublished data.

Efficiency comparisons are always difficult to make because they depend on (1) intrinsic properties of the gene, (2) the characteristics of the particular construct used, (3) the recipient cell line, and (4) the experimental protocol used for transformation. In our hands, the *pac* gene provides similar efficiency as *neo*, but we have no direct data with regard to *hyg* efficiency. In any case, PAC expression does not interfere with G418 selection and hence it is possible to perform double-selection experiments[45] or to transform cell lines that already express one of the resistant genes.[42] It is worth mentioning in this context that *neo* has been shown to decrease the expression of cotransforming genes, while *pac* did not influence adjacent promoters.[46]

Some of the advantages of *pac* over other available selectable markers derive from its ease of use and low cost. Selection with puromycin is fast, because sensitive cells die in the first few days of treatment, and the development of resistant colonies is easy to follow. The cost of the drug required for the selection is about 10 times less than that for G418 and about the same as that for hygromycin.

However, the main advantage of *pac* is the possibility of using it both as a selectable marker and as a reporter gene: it is possible to predict the efficiency of a given construct for transformation by a fast transient expression assay and determination of PAC activity.[14] Likewise, it is possible to test PAC activity in the different tissues of animals made transgenic for the *pac* gene.[40] Moreover, proper optimization of the *pac* gene should make it useful as a marker for gene targeting.

Acknowledgments

The authors thank C. Martínez and A. Beloso for excellent technical assistance and P. Esponda, R. A. Lacalle, E. Martínez-Salas, and A. Nieto for the communication of unpublished data. Work in the author's laboratory was supported by Grants BIO88-0191-C02 and BIO89-0545-C02 from the Comisión Interministerial de Ciencia y Tecnología and by an institutional grant from Fundación Ramón Areces to the CBM.

[45] J. C. de la Torre, S. de la Luna, J. Díez, and E. Domingo, *J. Virol.* **63,** 2385 (1989).

[46] P. Artelt, R. Grannemann, C. Stocking, J. Friel, J. Bartsch, and H. Hauser, *Gene* **99,** 249 (1991).

[34] Luciferase Reporter Gene Assay in Mammalian Cells

By ALLAN R. BRASIER and DAVID RON

The firefly luciferase reporter gene has been widely applied in both transient and stable transfection protocols in eukaryotic cells to measure transcriptional regulation of DNA elements.[1-6] Firefly luciferase (EC 1.13.12.7, luciferin 4-monooxygenase) produces light by the ATP-dependent oxidation of luciferin. In the presence of excess substrate, light output from samples containing firefly luciferase becomes proportional to enzyme concentration in the assay, thus allowing for its quantitation. Characteristics of the luciferase reporter system that make it especially suitable for gene regulation studies include the speed and sensitivity of the assay, compatibility of measurement with other transfected reporter genes such as alkaline phosphatase,[7] and the use of nonradioactive reagents. Furthermore, the rapid turnover of the luciferase mRNA and protein allows for interrupted cycloheximide incubations to quantitate transcriptional induction mediated by preformed transcriptional activators.[8,9] Accumulation of transfected reporter activity can be measured subsequent to a stimulus delivered in the presence of protein synthesis inhibition. Luciferase enzymatic activity correlates with changes in reporter mRNA abundance, thus obviating tedious RNA analysis.[6,8,10]

We describe in this chapter conventional harvest and assay of transfected luciferase reporter activity, the use of cotransfected reporter gene alkaline phosphatase to monitor changes in plate-to-plate transfection efficiency and, finally, a modified "minilysate" protocol that allows for simultaneous measurement of both luciferase reporter activity and assay of nuclear DNA-binding activity within the same plate of transfected cells.

[1] S. J. Gould and S. Subramani, *Anal. Biochem.* **175,** 5 (1988).
[2] A. R. Brasier, J. E. Tate, and J. F. Habener, *BioTechniques* **7,** 1116 (1989).
[3] J. Alam and J. L. Cook, *Anal. Biochem.* **188,** 245 (1990).
[4] T. M. Williams, J. E. Buerlein, S. Ogden, L. J. Kricka, and J. A. Kant, *Anal. Biochem.* **176,** 28 (1989).
[5] S. K. Nordeen, *BioTechniques* **6,** 454 (1988).
[6] D. Ron, A. R. Brasier, K. A. Wright, J. E. Tate, and J. F. Habener, *Mol. Cell. Biol.* **10,** 1023 (1990).
[7] P. Henthorn, P. Zervos, M. Raducha, H. Harris, and T. Kadesch, *Proc. Natl. Acad. Sci. U.S.A.* **85,** 6342 (1988).
[8] A. R. Brasier and D. Ron, *Methods Neurosci.* **5,** 181 (1991).
[9] M. Subramanian, L. J. Schmidt, C. E. Crutchfield, and M. J. Getz, *Nature (London)* **340,** 64 (1989).
[10] D. Ron, A. R. Brasier, K. A. Wright, and J. F. Habener, *Mol. Cell. Biol.* **10,** 4389 (1990).

Luciferase Reporter Vectors

deWet and colleagues reported the sequence of the luciferase gene from the firefly *Photinus pyralis* and constructed pSV2-based reporter vectors that directed immunologic and enzymatic luciferase expression in transfected CV-1 cells.[11] These workers identified two potential translation initiation sites and observed that deletion of the 5′-most initiation codon increased luciferase reporter activity by twofold and reduced luciferase activity in CV-1 cells transfected with promoterless plasmids. Nordeen introduced bidirectional multiple cloning sites into the pSV232AL-AΔ5′ plasmid containing the simian virus 40 (SV40) polyadenylation signal upstream of the multiple cloning site to lower background from cryptic promoters within the pBR322 plasmid.[5] We inserted the LΔ5′ cDNA into a modified pGEM3 vector, which has the advantage of being a high-copy plasmid that yields greater amounts of plasmid DNA.[2] Although this vector lacks the polyadenylation signal sequences upstream of the multiple cloning site, primer extension analysis of poly(A)$^+$-selected RNA from transfected hepatoma cells has demonstrated correct transcript initiation.[6] We have also introduced an SV40 polyadenylation cassette into the pGEM3-based pOLUC plasmid to generate the plasmid designated pOALUC. This was done to minimize any spurious effect of cryptic promoter activity arising from within the plasmid using an identical strategy as described by Maxwell *et al.*[12] The map of this vector is given in Fig. 1A, and the sequence of the multiple cloning site in Fig. 1B. Unique restriction sites are indicated by an asterisk for use in cloning promoters to test transcriptional regulatory elements.

Cellular Transfections with Luciferase Reporters

Luciferase reporter plasmids are prepared by cesium chloride density ultracentrifugation and are introduced into cultured cells using conventional techniques, with the optimal method of transfection being determined empirically for individual cell lines. We currently transfect human hepatoma HepG2 cell lines with calcium phosphate coprecipitation. Routinely, precipitates are prepared in triplicate (for statistical purposes) containing 20 μg of supercoiled plasmid DNA (total) consisting of 10 μg of luciferase reporter, 3.5 μg pSV2APAP (alkaline phosphatase internal

[11] J. R. de Wet, K. V. Wood, M. DeLuca, D. R. Helinski, and S. Subramani, *Mol. Cell. Biol.* **7,** 725 (1987).

[12] I. H. Maxwell, G. S. Harrison, W. M. Wood, and F. Maxwell, *BioTechniques* **7,** 276 (1989).

A.

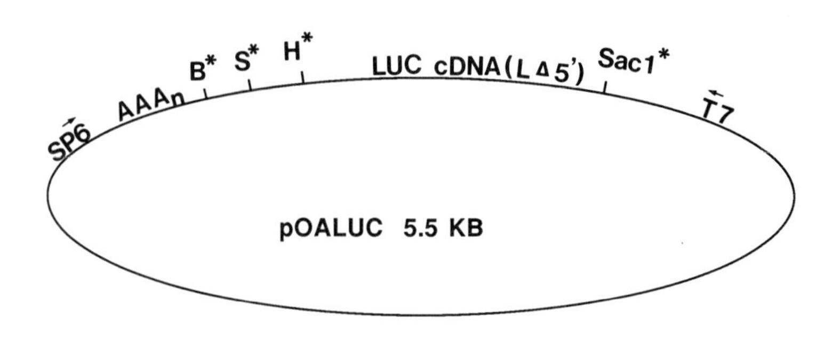

B.

FIG. 1. (A) Restriction map of pOALUC, a promoterless luciferase reporter vector. The vector contains a trimerized SV40 polyadenylation cassette[12] ligated as a *Bam*HI/*Bgl*II fragment into the *Bam*HI site of pOLUC.[2] The SV40 cassette upstream of the multiple cloning site reduces background from plasmid-initiated transcripts.[12] The vector is 5.5 kbp in size. (B) Sequence of the multiple cloning site 5′ to the luciferase LΔ5′ cDNA from *Photinus pyralis*. Restriction endonuclease sites are indicated as B, *Bam*HI; S, *Sma*I; H, *Hin*dIII; Sc, *Sac*I. Nucleotide positions over the luciferase cDNA (LΔ5′) refer to coordinates as described by deWet.[11]

control reporter[7]), and 6.5 μg of carrier pGEM plasmid DNA. Thus, each 60-mm plate is transfected with a total of 6.7 μg of total DNA, of which 3.3 μg is luciferase reporter, 1.25 μg is pSV2APAP alkaline phosphatase reporter plasmid, and 2.2 μg is carrier pGEM3 plasmid. The triplicate precipitates are prepared in 5-ml snap-cap Falcon (Becton Dickinson, Oxnard, CA) tubes and the DNA mix is brought to a final volume of 225 μl with 0.45-μm membrane filtered distilled water. Twenty-five microliters of 2.5 *M* $CaCl_2$ is added, followed by 250 μl of 2× HBS [HEPES-buffered saline; 275 m*M* NaCl, 10 m*M* KCl, 1.4 m*M* sodium phosphate monobasic, 11 m*M* dextrose, 42 m*M* *N*-2-hydroxytehylpiperazine-*N*′-2-ethanesulfonic acid (HEPES), pH 7.5]. The precipitate is allowed to stand at room temperature for 20 min after mixing and 167 μl of this precipitate is added dropwise to a monolayer of 1× HBS-washed cells. The precipitate is allowed to incubate on the cell monolayer for 15 min, followed by the addition of 3 ml of culture medium. The glycerol shock after 4 hr yields the same transfection efficiency as that obtained by allowing the precipitates to stand on the plate overnight. Cos-1 cells (used in the subsequent section) are transfected by the DEAE-dextran technique,[13] shocked in phosphate-buffered saline (PBS)/10% (v/v) dimethyl sulfoxide (DMSO) and harvested 48 hr after transfection.

Cell Lysis

Reporter enzymatic activity accumulates within transfected cells in a time-dependent fashion for 48–72 hr. The optimal time from transfection until harvest for luciferase reporter activity may vary among different cell types. HepG2 cells are harvested 48–60 hr after transfection, a time at which the luciferase expression directed by the viral SV40 early region promoter or angiotensinogen promoter is maximal.[2] For cellular extraction of luciferase reporter activity, cells are washed in PBS and then lysed *in situ* by the addition of Triton lysis buffer [1% (v/v) Triton X-100, 25 m*M* glycylglycine, pH 7.8, 15 m*M* $MgSO_4$, 4 m*M* ethylene glycolbis(β-aminoethyl ether)-*N,N,N′,N′*-tetraacetic acid (EGTA), 1 m*M* dithiothreitol (DTT, added immediately before use)]. The presence of reducing agent is critical to maintain luciferase enzymatic activity. The cells are scraped, transferred to 1.5-ml microfuge tubes, and spun at room temperature in a microcentrifuge at maximum speed (10,000 rpm) for 5 min. One hundred microliters of each supernatant is immediately assayed for luciferase activity.

Luciferase Reporter Assay

The ATP concentration is an important variable in the luciferase assay, as ATP is both a substrate and a regulator of luciferase enzymatic activity.[14] ATP (1–2 m*M*) is optimal for assay of transfected luciferase activity in HepG2 cells.[2] We assay 100 μl of each lysate by adding 360 μl of luciferase assay buffer (25 m*M* glycylglycine, pH 7.8, 15 m*M* $MgSO_4$, 4 m*M* EGTA, 15 m*M* potassium phosphate, pH 7.8, 2 m*M* ATP, 1 m*M* DTT; concentrated phosphate buffer and DTT are added immediately before

[13] M. A. Lopata, D. W. Cleveland, and B. Sollner-Webb, *Nucleic Acids Res.* **12,** 5707 (1984).

[14] M. DeLuca and W. D. McElroy, *Biochem. Biophys. Res. Commun.* **123,** 764 (1984).

use). The assay is initiated by injection of 200 μl of 0.2 m*M* luciferin solution (0.2 m*M* synthetic crystalline luciferin in 25 m*M* glycylglycine, pH 7.8, 15 m*M* $MgSO_4$, 4 m*M* EGTA, 2 m*M* DTT) diluted from concentrated aqueous 1 m*M* luciferin stock solution in the same buffer frozen at −70° in the dark. Unused diluted luciferin is discarded.

We quantitate luciferase activity using an automated luminometer [we use LKB (Rockville, MD) 1251 luminometer driven by an IBM-compatible PC with term emulator program and a dot-matrix printer]. A program written in BASIC programming language that we have designed is reproduced in Fig. 2. The data output is continuously printed each second, as well as peak and integrated luminescence values between 1 and 16 sec after luciferin injection into the sample. Thus, this program allows for collection of data without the additional cost of purchasing a chart recorder. With the addition of luciferin in the presence of ATP, magnesium ions, and oxygen, luciferase produces a peak flash of light followed by a plateau that decays to 80% of the initial peak values within 16 sec.[2,13] Both peak values and integrated light output values between 1 and 16 sec are proportional to luciferase enzyme concentrations.[1,2] We have found integrated values to have a lower intraassay variation than peak values and therefore prefer to use the former measurement in the evaluation of our data. The luciferase assay, in our hands, is linear over a wide range of proteins (up to 200 μg total cellular protein from detergent lysates), thus avoiding the need for sample dilutions.[2] A more complete discussion of interfering substances in the luciferase assay is given by Thore.[15]

Measurement of Cotransfected Reporter Activity

Provided that the transfected cells do not express endogenous alkaline phosphatase enzyme, aliquots of cellular lysate from the above-mentioned Triton lysis-prepared extracts can be assayed for phosphatase activity exactly as described by Henthorn and Kadesch.[7] Assay blanks are reconstituted by addition of the same amount of Triton lysis buffer to the alkaline phosphatase (PAP) assay buffer. Test luciferase reporter activity can then be normalized for plate-to-plate variations in transfection efficiency. Chloramphenicol acetyltransferase (CAT[16]) has also been used as a cotransfected reporter to monitor transfection efficiency[2]; however, the presence of Triton in the cytoplasmic extract is inhibitory in the CAT assay and thus lowers its sensitivity.

[15] A. Thore, *Sci. Tools* **26,** 30 (1979).

[16] C. M. Gorman, L. F. Moffat, and B. H. Howard, *Mol. Cell. Biol.* **2,** 1044 (1982).

LUCIFERASE PROGRAM FOR PC-DRIVEN LUMINOMETER

```
10   MIXER PULSE
20   I DELAY 1
30   I TIME 15
31   C FACTOR 0.1
32   P FACTOR 0.1
33   I FACTOR 0.1
40   FIRST IN
50   LOOP 25
60   PRINT POSITION
70   DISPENSER 3.1
80   START I
90   START P
100  LOOP 16
110  PRINT C;
120  WAIT 1
130  END LOOP
140  WRITE
150  WRITE ' NO. INTEGRAL PEAK'
155  PRINT POSITION;
160  PRINT I;
170  PRINT P
180  P RESET
190  I RESET
191  WAIT 5
200  NEXT SAMPLE
210  END LOOP IF NO SAMPLE
```

FIG. 2. BASIC program for operation of the LKB term emulator program. As entered, the program will direct the LKB luminometer to inject luciferin, record continuous, peak, and integrated luminescence between 1 and 16 sec after injection of the luciferin, and advance to the next sample.

Assay for DNA-Binding Proteins and Luciferase Reporter Activity in Transfected Cells

Changes in gene transcription can be correlated with alterations in the abundance or structure of nuclear DNA-binding proteins that interact with specific cis-regulatory sequences in the promoter region of the gene.[17,18] These alterations can be detected by assaying *in vitro* for the sequence-specific DNA-binding activity of proteins isolated from nuclei of cells that have undergone a controlled physiological perturbation. It is often advantageous to be able to correlate, in the same cell, changes in reporter gene activity with the simultaneously occurring alterations in those DNA-binding proteins that are hypothesized to play a role in the regulatory event being studied. We have applied an existing protocol for the rapid preparation of nuclear extract from cultured cells[19] to allow the simultaneous measurement of luciferase reporter gene activity in transfected cells.

In the example presented in Fig. 3, this combined assay was used to evaluate DNA-binding activity and transcriptional activation potential of chimeric proteins generated by in-frame ligation of a truncated form of the activation domain of the cloned human transcription factor CREB and the DNA-binding domain of the yeast transcription factor Gal-4 [designated Gal(1–74); Ref. 20]. DNA (1–2 μg) prepared by alkaline lysis of *Escherichia coli* transformed with the ligation product CREB-GAL(1–74) in a replicating eukaryotic expression plasmid was used to transfect Cos-1 cells. The cells were cotransfected with 100 ng of reporter gene plasmid that contains the Gal-4 binding site (UAS) ligated upstream of a minimal promoter–luciferase reporter construct. Seventy-two hours following transfection, we simultaneously harvested cells for nuclear UAS-binding activity by electrophoretic gel mobility shift assay (EMSA), and cytosol for luciferase activity directed by the cotransfected reporter gene. The strategy is depicted in Fig. 3A.

To harvest the transfected Cos-1 cells for simultaneous assay of DNA-binding activity and reporter activity, transfected 60-mm plates are washed twice with 3 ml of cold phosphate-buffered saline (PBS) and scraped in 0.5 ml of PBS/0.5 m*M* ethylenediaminetetraacetic acid (EDTA) on ice after a 5-min equilibration. The cells are scraped with a rubber policeman and then transferred to a microfuge tube and pelleted by spinning for 2 min at 4500 *g* at 4° in a tabletop microcentrifuge. The cell pellet is resuspended

[17] A. R. Brasier, D. Ron, J. E. Tate, and J. F. Habener, *EMBO J.* **9,** 3933 (1990).

[18] C. Q. Lee, Y. Yun, J. P. Hoeffler, and J. F. Habener, *EMBO J.* **9,** 4455 (1990).

[19] E. Schreiber, P. Matthias, M. Muller, and W. Schaffner, *Nucleic Acids Res.* **17,** 6419 (1989).

[20] J. Ma and M. Ptashne, *Cell (Cambridge, Mass.)* **51,** 113 (1987).

in 0.4 ml of buffer A [10 m*M* HEPES, pH 7.8, 10 m*M* KCl, 0.1 m*M* EGTA, 0.1 m*M* EDTA, 1 m*M* phenylmethylsulfonyl fluoride (PMSF), 1 m*M* DTT] and cells are allowed to swell on ice for 15 min. Lysis of the cells is effected by adding 20 μl of 10% (v/v) Nonidet P-40 (NP-40) in H_2O, rapidly vortexing the cells for 10 sec, and immediately pelleting the nuclei by spinning for 30 sec at maximal speed (10,000 *g*, 4°) in a tabletop microcentrifuge. The supernatant containing the cytosolic fraction with the luciferase enzymatic activity is transferred to a different microcentrifuge tube and stored on ice for further analysis. In a typical transfection experiment as little as 25–50 μl of this lysate is sufficient to generate a reliable signal.

To the nuclear pellet, 2 vol of buffer C [10 m*M* HEPES, pH 7.8, 400 m*M* KCl, 0.1 m*M* EGTA, 0.1 m*M* EDTA, 1 m*M* PMSF, 1 m*M* DTT, 0.1% (v/v) NP-40] is added, resuspended by vigorous vortexing for 15–30 sec, followed by 15 min of gentle agitation at 4° to allow for lysis of the nuclei and extraction of the DNA-binding proteins. The insoluble chromatin and membranes are removed by centrifugation for 5 min in a microfuge at maximal speed and the supernatant is saved for analysis of DNA-binding activity. Typically for a confluent 60-mm plate of cells, we obtain 30–50 μl of nuclear extract at a concentration of 3–10 μg/μl that is suitable for EMSA, Western/Southern blot analysis, or immunoprecipitation.

The autoradiogram of the EMSA and data from the luciferase assay are shown in Fig. 3B. Cells transfected with reporter gene only contain no nuclear proteins that bind the UAS (lane 1, Fig. 3B, consistent with the fact that mammalian cells do not have a Gal-4 homolog). Luciferase activity from the UAS-containing reporter gene is consistently low. Transfection of cells with a positive control Gal-4 chimera that contains the full-length activating domain of CREB (lane 7, Fig. 3B) induces a novel UAS-binding species and activates the reporter gene. The ligations used to transfect the Cos-1 cells represented in lanes 2 and 4 (Fig. 3B) were successful and gave rise to a chimeric UAS-binding protein that activates the reporter gene. Consistent with the smaller size of the encoded protein, the gel-shifted bands generated by the ligations being tested are of greater mobility than the positive control shown in lane 7 (Fig. 3B). The minipreparation DNA transfected and assayed in lanes 3, 5, and 6 (Fig. 3B) is from unsuccessful ligations and failed to encode a DNA-binding protein.

Using this assay we have detected DNA-binding proteins from the b-ZIP class (CREB,C/EBP), zinc finger class (Gal-4), and Rel-like proteins (NFκB). Furthermore, comparison of luciferase reporter gene activity from transfected cells harvested with this combined technique with that from plates harvested by the plate lysis method demonstrates comparable levels of enzymatic activity. There are three major limitations of this technique. The first lies in the method of preparation of the nuclear extract,

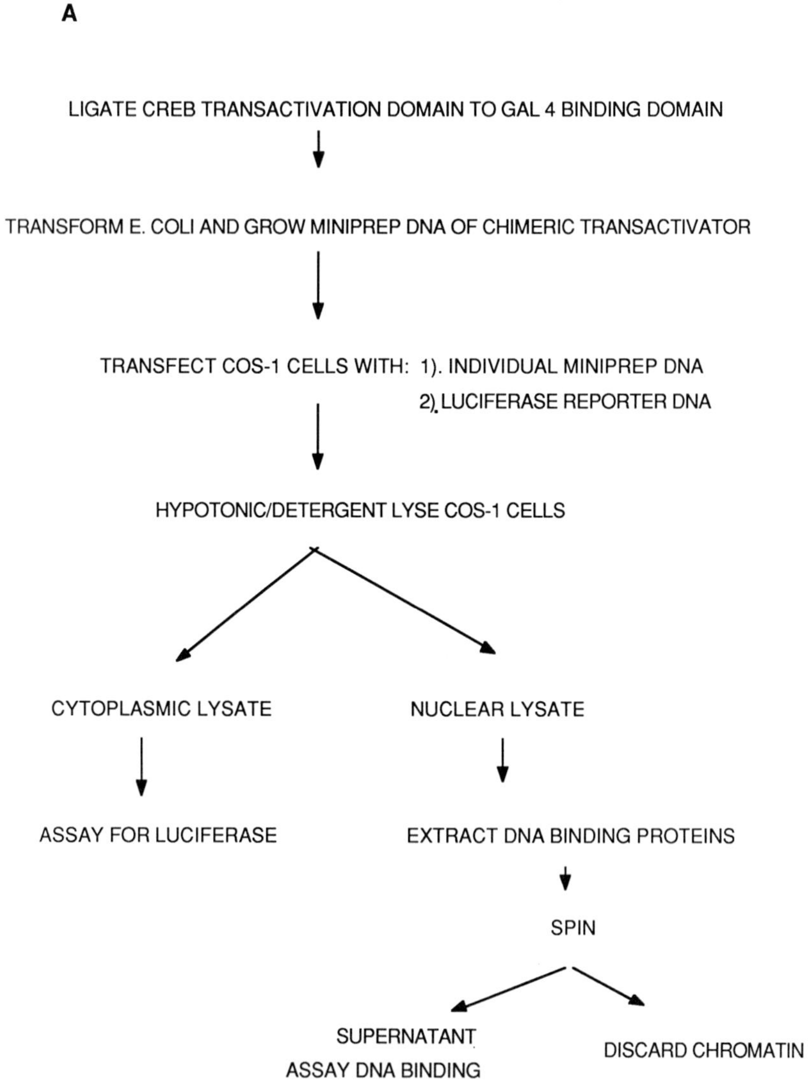

A

LIGATE CREB TRANSACTIVATION DOMAIN TO GAL 4 BINDING DOMAIN

TRANSFORM E. COLI AND GROW MINIPREP DNA OF CHIMERIC TRANSACTIVATOR

TRANSFECT COS-1 CELLS WITH: 1). INDIVIDUAL MINIPREP DNA
2). LUCIFERASE REPORTER DNA

HYPOTONIC/DETERGENT LYSE COS-1 CELLS

CYTOPLASMIC LYSATE

ASSAY FOR LUCIFERASE

NUCLEAR LYSATE

EXTRACT DNA BINDING PROTEINS

SPIN

SUPERNATANT
ASSAY DNA BINDING

DISCARD CHROMATIN

FIG. 3. (A) Cos-1 cells were transfected by the DEAE-dextran technique with 1–2 μg of plasmid DNA encoding chimeric DNA-binding proteins that contain the Gal-4-binding domain and the CREB activation domain, along with 100 ng of a reporter plasmid that contains the Gal-4-binding site (UAS) upstream of an angiotensinogen minimal promoter–luciferase reporter plasmid (UASp59RLG). Seventy-two hours later, plates were processed for nuclear

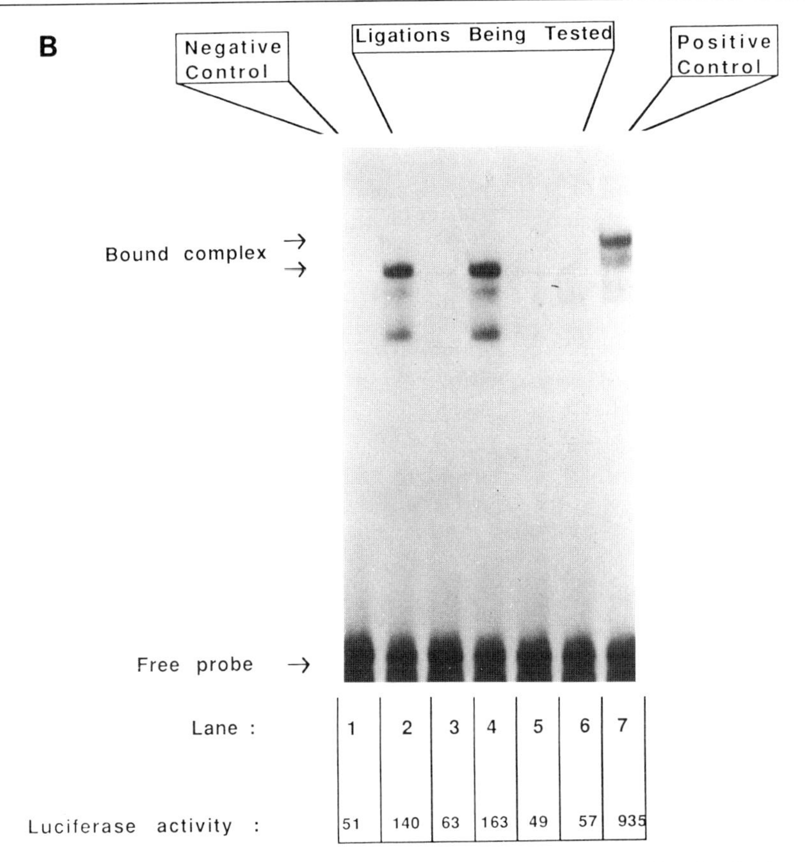

protein and luciferase enzymatic activity. (B) Autoradiogram after EMSA using 5–10 μg of nuclear protein (contained in 2 μl of nuclear extract) using as a probe the radioactively labeled oligonucleotide containing the UAS sequence. Complexes were resolved by 4.5% PAGE as described.[18] Luciferase activity was measured in 50 μl of cytosolic extract and results are expressed in integrated light units after subtracting machine background. Lane 1, cells transfected with the reporter plasmid only; lanes 2–6, transfected with plasmids from the ligation product; lane 7, transfected with a positive control Gal-4-CREB chimera.[19]

i.e., reliance on detergent lysis to disrupt the cellular membrane and leave the nuclei intact. The concentration of detergent used during the lysis step needs to be calibrated for the specific cell type and growth conditions. Second, the presence of detergent and the high salt concentration in the nuclear extract limits the amount of extract that can be used in a solution-phase binding assay to 10% of the reaction volume. Otherwise the final salt and detergent concentrations become inhibitory to DNA binding. Finally, a critical assumption in this combined luciferase reporter/DNA-binding assay is that one can measure the changes in DNA-binding proteins that occur in the transfected cell population. Transiently transfected cells, on which the analysis of reporter gene activity is being carried out, represent only a small minority of the total cell population of the plate (0.001–1% of cells contain transfected DNA[3]). Although reporter enzymes are not expressed in nontransfected cells, binding proteins that interact with DNA regulatory elements may be present in both transfected and nontransfected cells. Thus, the DNA-binding activity derived from an expression plasmid may be too faint to be distinguished from the background produced by nontransfected cells. The advantage of a system that relies on the SV40 large T antigen-expressing Cos-1 cells to replicate eukaryotic expression vectors containing the SV40 origin of replication to high levels is that plasmid-encoded proteins may be detected. Thus, not all transiently transfectable cell lines can be used for this combined reporter/DNA-binding assay strategy.

Conclusions

The firefly luciferase reporter gene is a highly versatile tool for studies of transcriptional regulation. The reporter assay is extremely sensitive, reproducible, and compatible with internal control alkaline phosphatase reporters. These features allow for measurements of low levels of transcription and correction for variations in transfection efficiency. Further, the use of nonradioactive reagents for detection is a significant advantage in an era where radioactive waste disposal is an issue. The rapid mRNA and protein turnover of luciferase in mammalian cells are properties that can be utilized to measure inducible inhancer activity by preformed DNA-binding proteins. The use of the previously described interrupted cycloheximide pulse–chase analysis[6,8] allows the investigator to measure reporter gene induction by stimulation with hormones in the presence of protein synthesis inhibition. This is followed by removal of hormone and protein synthesis inhibitor and subsequent measurement of reporter activity. Under these conditions, transcripts synthesized by preformed proteins correlate with luciferase enzymatic activity. This simplified strategy avoids the necessity of time-consuming mRNA purification and transcript analysis. We show here that the Triton lysis strategy can be easily adapted to assay both overexpressed (transfected) DNA-binding proteins or endogenous inducible DNA-binding activity. Thus, the cells from the same plate can be used to monitor reporter gene activity and DNA-binding activity. This allows a more precise correlation of the two activities.

Acknowledgment

We thank Joel F. Habener, in whose laboratory these experiments were conducted, for support.

[35] Transient Expression Analysis in Plants Using Firefly Luciferase Reporter Gene

By KENNETH R. LUEHRSEN, JEFFREY R. DE WET, and VIRGINIA WALBOT

Introduction

Reporter genes have been extensively used to study gene expression. The hallmark of an excellent reporter gene is the ease with which its product can be assayed. Widely used reporter genes typically include those encoding chloramphenicol acetyltransferase (CAT), β-galactosidase (*lacZ*), and neomycin phosphotransferase (*neo*); however, each of these suffers from one or more disadvantages including high backgrounds, costly and tedious assay procedures, and low signal-to-noise ratio. A new generation of reporter genes has been developed, including those encoding *Escherichia coli* β-glucuronidase (*uidA* gene) GUS activity[1] and bacterial (EC 1.14.14.3, alkanal monooxygenase)[2] and firefly luciferases (EC 1.13.12.7, luciferin 4-monooxygenase).[3,4] Here we describe the use of the firefly luciferase gene as a reporter in plant transient expression assays.

The enzymatic properties of firefly luciferase have been well studied; the first luciferase gene cloned was from the North American firefly, *Photinus pyralis*.[3,4] The enzyme has a molecular mass of 62 kDa and requires only luciferin, ATP, Mg^{2+}, and molecular oxygen for the production of yellow-green (560 nm) light.[5] The enzyme kinetics show a linear

[1] R. A. Jefferson, *Plant Mol. Biol. Rep.* **5,** 387 (1987).
[2] E. A. Meighen, *Microbiol. Rev.* **55,** 123 (1991).
[3] J. R. de Wet, K. V. Wood, D. R. Helinski, and M. DeLuca, *Proc. Natl. Acad. Sci. U.S.A.* **82,** 787 (1985).
[4] J. R. de Wet, K. V. Wood, M. DeLuca, D. R. Helinski, and S. Subramani, *Mol. Cell. Biol.* **7,** 725 (1987).
[5] S. J. Gould and S. Subramani, *Anal. Biochem.* **175,** 5 (1988).

response over eight decades of dilution. Although firefly luciferase is normally transported to and expressed in peroxisomes,[6] this localization is not an absolute requirement as the enzyme can be assayed *in vivo* in bacteria and in cell-free extracts. There are several advantages of luciferase over the commonly used CAT: detection is 10–1000 times more sensitive[4,7] (as little as 10^{-20} mol can be detected), the assays are much cheaper and quicker, assays do not require hazardous radioactive chemicals, and there is no inherent background light production in cells. A disadvantage of using firefly luciferase is the cost of a luminometer, although some assay methods permit the use of a standard scintillation counter to monitor activity.

Firefly luciferase has been widely used in many organisms to monitor gene expression. The luciferase cDNA has been successfully expressed in bacteria,[8] yeast,[9] *Dictyostelium,*[10] monocot and dicot plants,[11] and mammalian tissue culture cells.[4] Most applications are based on the assay of luciferase enzyme in cell extracts. It is also possible to analyze luciferase expression *in vivo* because the substrate luciferin will diffuse across biological membranes. This approach has allowed identification of *E. coli* cultures expressing luciferase[8] and direct video imaging of tobacco protoplasts.[12] Transgenic tobacco plants have been constructed in which luciferase expression has been localized and quantitated in individual organs and tissues.[11,13,14] Thus, the firefly luciferase gene is a sensitive and versatile reporter gene.

Plasmids Incorporating Firefly Luciferase Reporter Gene

To investigate the effects of changing various components of a transcriptional unit on the level of transient gene expression in plant protoplasts, we designed and constructed modular expression vectors. These vectors have at least one unique restriction site between each functional component of the transcriptional unit so that individual segments can

[6] S. J. Gould, G. A. Keller, M. Schneider, S. H. Howell, L. J. Garrard, J. M. Goodman, B. Distel, H. Tabak, and S. Subramani, *EMBO J.* **9,** 85 (1990).

[7] T. M. Williams, J. E. Burlein, S. Ogden, L. J. Kricka, and J. A. Kant, *Anal. Biochem.* **176,** 28 (1989).

[8] K. V. Wood and M. DeLuca, *Anal. Biochem.* **161,** 501 (1987).

[9] H. Tatsumi, T. Masuda, and E. Nakano, *Agric. Biol. Chem.* **52,** 1123 (1988).

[10] P. K. Howard, K. G. Ahern, and R. A. Firtel, *Nucleic Acids Res.* **16,** 2613 (1988).

[11] D. W. Ow, K. V. Wood, M. DeLuca, J. R. de Wet, D. R. Helinski, and S. H. Howell, *Science* **234,** 856 (1986).

[12] D. R. Gallie, W. J. Lucas, and V. Walbot, *Plant Cell* **1,** 301 (1989).

[13] W. M. Barnes, *Proc. Natl. Acad. Sci. U.S.A.* **87,** 9183 (1990).

[14] M. Schneider, D. W. Ow, and S. H. Howell, *Plant Mol. Biol.* **14,** 935 (1990).

be easily exchanged. To facilitate the exchange of reporter genes, we introduced an *Nco*I restriction site at the initiation codon of the firefly luciferase gene using oligonucleotide directed mutagenesis. The sequence at the initiation codon was changed from 5′-GGTAAAATGG-3′ to 5′-GGTACCATGG-3′. This maintained the consensus sequence as defined by Kozak for a eukaryotic translational start site[15,16] and introduced an *Nco*I site (5′-CCATGG-3′) and an overlapping *Kpn*I site (5′-GGTACC-3′). The altered sequence is present in the luciferase cassettes carried in pJD250 and pJD251 (Fig. 1) and does not affect expression levels in maize protoplasts (J. R. de Wet, unpublished observations, 1990). The luciferase cassettes in pJD261, pJD293, and pJD294 (Fig. 1) carry an alternative sequence upstream of the luciferase initiation codon. This alternative sequence (5′-GTCGACCATGG-3′) was obtained from the β-glucuronidase vector pRAJ275[1] and consists of a *Sal*I site overlapping the *Nco*I site; this sequence maintains a consensus translational start site. The firefly luciferase cassette in pJD293 also features a different arrangement of restriction sites at its 3′ end.

In constructing our plant expression vectors we chose to use the cauliflower mosaic virus (CaMV) *35S* promoter; this is a strong promoter in both dicots and monocots.[17–19] The CaMV *35S* promoter fragment we used extends from −363 to +1 relative to the transcriptional start site[20]; this is a region responsible for the activity of the promoter.[21] A restriction fragment carrying the polyadenylation signal of the nopaline synthase (*nos*) gene from the Ti plasmid of *Agrobacterium tumefaciens*[18,22] forms the 3′ end of the transcriptional unit. Restriction maps of the basic CaMV *35S* promoter–*nos* 3′ vectors, pJD288 and pJD290, are shown in Fig. 2.

The CaMV *35S* promoter vector pJD290 was constructed as follows: the restriction fragment from the *Dde*I site at base 7069 to the *Hph*I site at base 7444 from CaMV strain 1841[23] was isolated and treated with T4 DNA polymerase in the presence of all four deoxyribonucleoside triphosphates (dNTPs). The resultant blunt-ended fragment was 426 bases in

[15] M. Kozak, *Cell (Cambridge, Mass.)* **44,** 283 (1986).
[16] M. Kozak, *J. Cell Biol.* **108,** 229 (1989).
[17] J. Callis, M. Fromm, and V. Walbot, *Genes Dev.* **1,** 1183 (1987).
[18] M. Fromm, L. P. Taylor, and V. Walbot, *Nature (London)* **319,** 791 (1986).
[19] J. T. Odell, S. Knowlton, W. Lin, and C. J. Mauvais, *Plant Mol. Biol.* **10,** 263 (1988).
[20] H. Guilley, R. Dudley, K. G. Jonard, E. Balazs, and K. E. Richards, *Cell (Cambridge, Mass.)* **30,** 763 (1982).
[21] R.-X. Fang, F. Nagy, S. Sivasubramanian, and N.-H. Chua, *Plant Cell* **1,** 141 (1989).
[22] H.-J. Fritz, *in* "DNA Cloning: A Practical Approach" (D. M. Glover, ed.), p. 151. IRL Press, Oxford, 1985.
[23] R. C. Gardner, A. J. Howarth, P. Hahn, M. Brown-Luedi, R. J. Shepherd, and J. Messing, *Nucleic Acids Res.* **9,** 2871 (1981).

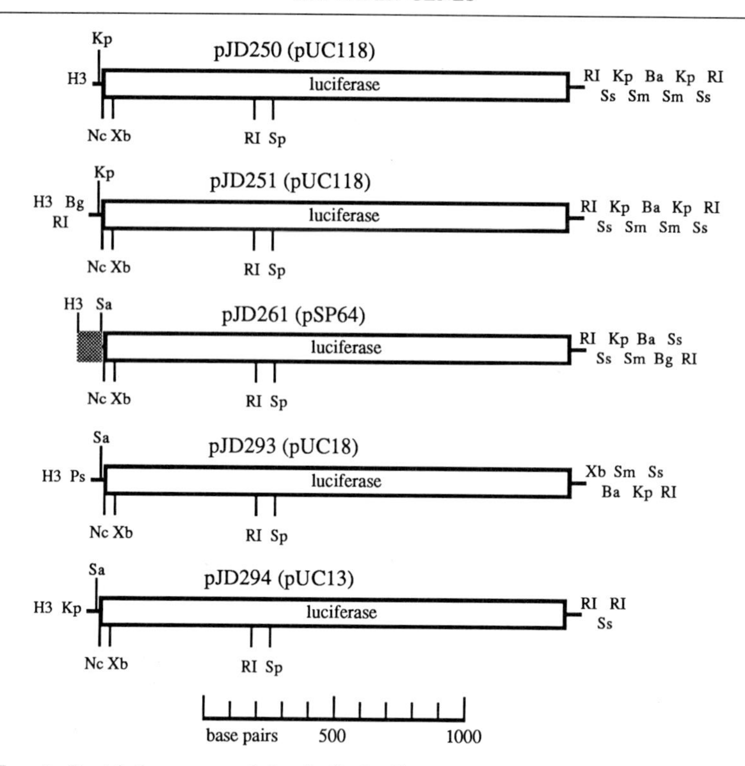

FIG. 1. Restriction maps of the firefly luciferase cassettes. Only the firefly luciferase gene and flanking restriction sites are shown. The cloning vector carrying the diagrammed fragments is in parentheses following the name of each of the luciferase plasmids. The *Nco*I restriction site that was introduced into the luciferase gene by mutagenesis occurs at -2 relative to the translational start site. The shaded box in pJD261 represents the tobacco mosaic virus (TMV) Ω sequence. pSP64 is a product of Promega Corporation. The restriction sites are abbreviated as follows: *Bam*HI, Ba; *Bgl*II, Bg; *Eco*RI, RI; *Hin*dIII, H3; *Kpn*I, Kp; *Nco*I, Nc; *Pst*I, Ps; *Sal*I, Sa; *Sst*I, Ss; *Sma*I, Sm; *Xba*I, Xb.

length and extended from -363 to $+3$ relative to the transcriptional start site[20] of the *35S* promoter. This fragment was inserted into the *Sma*I site of pUC18[24] to produce pJD264. The plasmid pJD264 was digested with *Eco*RI and the 5′ overhangs were filled in. A 260-bp *Pst*I to *Xba*I restriction fragment containing the poly(A) addition region of the *nos* gene[25] was isolated from the plasmid pCaMV*neo*[18] and made blunt ended by treatment

[24] C. Yanisch-Perron, J. Vieira, and J. Messing, *Gene* **33,** 103 (1985).

[25] R. T. Fraley, S. G. Rogers, R. B. Horsch, P. R. Sanders, J. S. Flick, S. P. Adams, M. L. Bittner, L. A. Brand, C. L. Fink, J. S. Fry, G. R. Galluppi, S. B. Goldberg, N. J. Hoffman, and S. C. Woo, *Proc. Natl. Acad. Sci. U.S.A.* **80,** 4803 (1983).

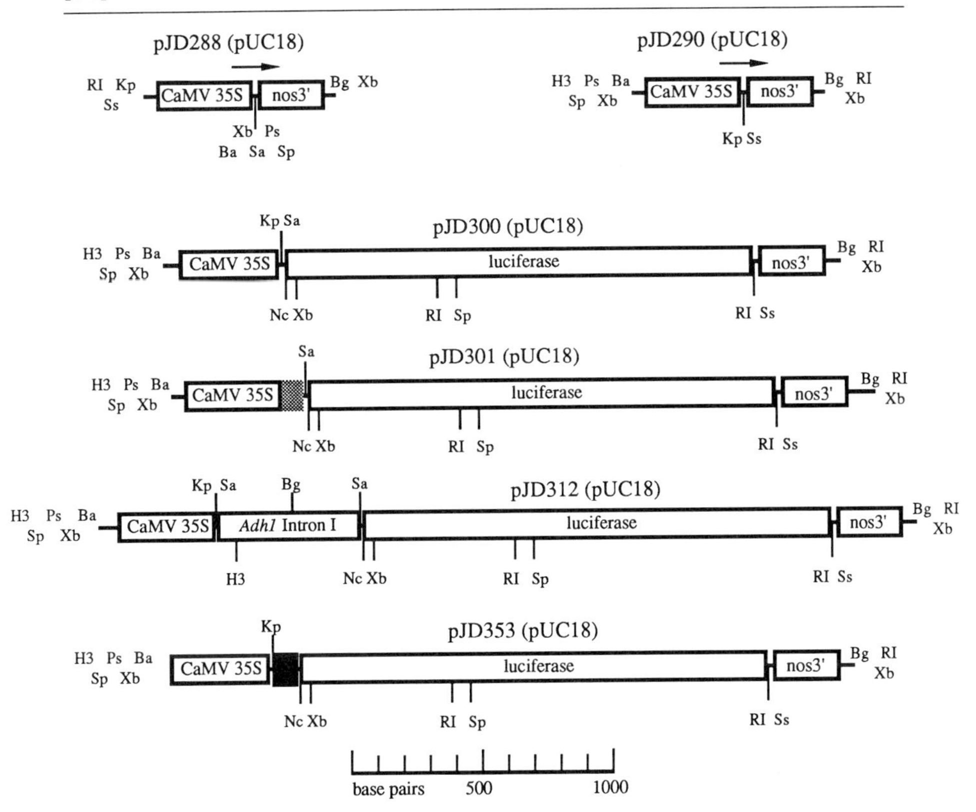

FIG. 2. Restriction maps of the CaMV *35S* promoter luciferase expression vectors. The name of each expression vector is followed in parentheses by the name of the plasmid vector carrying the diagrammed transcription units. The shaded box in pJD301 represents the TMV Ω sequence. The black box in pJD353 is the *Adh1* untranslated leader region. The restriction sites are abbreviated as follows: *Bam*HI, Ba; *Bgl*II, Bg; *Eco*RI, RI; *Hin*dIII, H3; *Kpn*I, Kp; *Nco*I, Nc; *Pst*I, Ps; *Sal*I, Sa; *Ssp*I, Sp; *Sst*I, Ss; *Sma*I, Sm; *Xba*I, Xb.

with T4 DNA polymerase in the presence of all four dNTPs. This fragment was then inserted into the blunt-ended *Eco*RI site of pJD264 to produce pJD287. The *Sal*I site in the pUC18 polylinker region of pJD287 was eliminated by digesting the plasmid with *Sal*I, removing the overhangs by digestion with mung bean nuclease, and recircularizing the plasmid with T4 DNA ligase to produce pJD290.

The plasmid pJD288 was constructed using the same CaMV *35S* promoter and *nos* poly(A)-containing restriction fragments as were used in the construction of pJD290. The CaMV *35S* promoter fragment was inserted into the *Hin*cII site of pUC18 to produce pJD265. The *nos* 3′

fragment was then inserted into the *Hin*dIII site of pJD265, after the *Hin*dIII staggered ends had been filled in to produce pJD288.

An *Nco*I site was introduced into the luciferase gene at the transcriptional start site using the gapped duplex method of oligonucleotide mutagenesis.[22] Briefly, the *Hin*dIII to *Bam*HI restriction fragment containing the luciferase gene was isolated from pJD204[4] and inserted into pUC118[26] that had been digested with *Hin*dIII and *Bam*HI to produce pJD243. The single-stranded DNA form of pJD243 was propagated in the *E. coli* strain BW313 (*dut ung thi-1 relA spoT1 F' lsyA*).[27] The uracil-substituted single-stranded template was hybridized to pJD243 that had been digested with *Hin*dIII and *Xba*I to form the gapped duplex and to the phosphorylated oligonucleotide 5'-pCCGGTACTGTTGGTA<u>CC</u>ATGGAAGACGCCA-3'. The two underlined C's were A's in the original luciferase sequence. The circles were repaired with T4 DNA polymerase and T4 DNA ligase and transformed into a dut^+ ung^+ strain of *E. coli*. The mutagenesis should introduce a *Kpn*I site overlapping a following *Nco*I site. A successfully mutagenized plasmid clone was identified by digestion with *Kpn*I and *Nco*I and named pJD250. pJD251 was constructed in the same manner as pJD250 except that the *Hin*dIII–*Bam*HI luciferase restriction fragment was obtained from pJD205.[4]

The tobacco mosaic virus (TMV) Ω leader, shown previously to enhance expression in a number of plant species,[28] was added to the luciferase gene as follows. The *Eco*RI site at the 3' end of the β-glucuronidase gene in pRAJ275[1] was cut and filled. The plasmid was then digested with *Hin*dIII and the β-glucuronidase-containing restriction fragment was inserted into pUC18 that had been digested with *Hin*dIII and *Hin*cII. The resultant plasmid was digested with *Sal*I and *Bam*HI, and the β-glucuronidase restriction fragment was inserted into the TMV Ω-containing plasmid pJII101[29] that had been digested with *Sal*I and *Bam*HI. This resulted in the introduction of an *Nco*I site immediately downstream from the *Sal*I site at the 3' end of the Ω sequence in pJII101. The β-glucuronidase gene was then removed by digestion with *Nco*I and *Bam*HI and replaced by an *Nco*I–*Bam*HI fragment obtained from pJD250 to produce the Ω–luciferase plasmid pJD261. The *Sal*I–*Bam*HI luciferase gene fragment from pJD261 was inserted into pUC13[30] that had been digested with *Sal*I and *Bam*HI to

[26] J. Vieira and J. Messing, this series, Vol. 153, p. 3.

[27] T. A. Kunkel, *Proc. Natl. Acad. Sci. U.S.A.* **82,** 488 (1985).

[28] D. R. Gallie, D. E. Sleat, J. W. Watts, P. C. Turner, and T. M. A. Wilson, *Nucleic Acids Res.* **15,** 3257 (1987).

[29] D. R. Gallie, D. E. Sleat, J. W. Watts, P. C. Turner, and T. M. A. Wilson, *Nucleic Acids Res.* **15,** 8693 (1987).

[30] J. Messing, this series, Vol. 101, p. 20.

produce pJD291. A *Kpn*I site was inserted 5′ of the luciferase gene by digesting pJD291 with *Pst*I, using T4 DNA polymerase in the presence of all four dNTPs to remove the 3′ overhangs, ligating *Kpn*I linkers (pGGG-TACCC) to the blunt ends of the plasmid, digesting the DNA with *Kpn*I, and recircularizing the plasmid. The resultant plasmid was named pJD294.

The CaMV *35S* luciferase expression vector pJD300 was constructed by inserting the *Kpn*I–*Sst*I luciferase gene fragment from pJD294 into pJD290 that had been digested with *Kpn*I and *Sst*I. The CaMV*35S*–Ω–luciferase expression vectors were constructed as follows: pJD290 was digested with *Kpn*I, the 3′ overhangs were removed, and the DNA was then digested with *Sst*I. The plasmid pJD261 was digested with *Hin*dIII, the 5′ overhangs were removed by digestion with mung bean nuclease, and the Ω–luciferase fragment was released from the vector by digestion with *Sst*I. This fragment was inserted into the prepared pJD290 DNA producing the vector pJD301.

The plasmid pMANC1,[31] which has an *Nco*I site introduced at the translational start site, was used as a source of the untranslated leader from the *Zea mays* alcohol dehydrogenase-1 (*Adh1*) gene. A *Kpn*I linker (pGGGTACCC) was inserted into the *Ban*II site base 29 bp upstream of the *Adh1* transcriptional start site after the *Ban*II 3′ overhangs had been removed. The *Adh1* untranslated leader was then inserted into pJD300 as a *Kpn*I to *Nco*I restriction fragment to produce pJD353. pJD353 is expressed approximately fourfold higher in maize than pJD300.[32]

The presence of *Adh1* intron 1 in a transcription unit has been shown to increase reporter gene expression in maize[17,33] and other monocots.[34,35] The luciferase expression vector containing the first intron of the maize *Adh1* gene, pJD312, was constructed as follows: the 557-bp *Bcl*I–*Bam*HI *Adh1* intron 1-containing restriction fragment[17] was inserted into the *Bam*HI site of pUC9 such that the *Bcl*I site of the intron 1 fragment was proximal to the *Sal*I site of pUC9. The resultant plasmid was subsequently digested with *Bam*HI, the 5′ overhangs were filled in, and phosphorylated *Sal*I linkers were ligated to the blunt ends. The plasmid DNA/*Sal*I linker ligation mix was then digested with *Sal*I, and the *Adh1* intron 1 fragment was isolated and inserted into the *Sal*I site of pJD300 to produce pJD312.

Other luciferase cassette vectors useful for creating translational fusions have also been reported.[36] In addition to the DNA-based vectors

[31] L. Lee, C. Fenoll, and J. L. Bennetzen, *Plant Physiol.* **85,** 327 (1987).

[32] J. R. deWet and V. Walbot, unpublished data (1990).

[33] K. R. Luehrsen and V. Walbot, *Mol. Gen. Genet.* **225,** 81 (1991).

[34] J. Kyozuka, T. Izawa, M. Nakajima, and K. Shimamoto, *Maydica* **35,** 353 (1990).

[35] J. H. Oard, D. Paige, and J. Dvorak, *Plant Cell Rep.* **8,** 156 (1989).

[36] C. D. Riggs and M. J. Chrispeels, *Nucleic Acids Res.* **15,** 8115 (1987).

described above, this laboratory has previously described a series of plasmids from which luciferase mRNA can be transcribed *in vitro*.[12] The mRNAs from these vectors can be directly introduced into plant protoplasts. These vectors have been useful in assessing posttranscriptional controls such as the TMV Ω translational enhancer element and the effect of capping and poly(A) tails on translation and mRNA stability.

Transient Expression Assay

The transient assay protocol described below utilizes electroporation[37] for gene transfer and was developed for maize suspension cells. With minor modifications we have adapted it to maize callus tissue as well as rice, carrot, tobacco, and bean suspension cultures[12,38,39]; firefly luciferase has been expressed in all these cell types.

Instrumentation Required

The protocols described below require specialized instrumentation. We list the devices we utilize and suggest equivalent instruments or alternative sources. We have extensively used the Promega Corp. (Madison, WI) X-Cell 450 electroporation apparatus; however, this is no longer being marketed; an apparatus with similar electrical parameters is available from Hoefer (San Francisco, CA) (model PG101). The most convenient method to detect luciferase expression is with a luminometer; we have used models 2001 and 2010 from Analytical Luminescence Laboratory (San Diego, CA) and model 3010 from Analytical Scientific Instruments (Alameda, CA) and find these machines to be adequate. However, the high price might preclude the purchase of a luminometer. A scintillation counter can substitute[40] if expression is high enough. β-Glucuronidase enzyme assays can be done spectrophotometrically but the more sensitive fluorescence assay is preferred. A full-featured fluorometer works well but is costly; we have found the low cost model TKO-100 minifluorometer from Hoefer performs well for GUS assays and, in a separate procedure, can be used to quantitate DNA concentration.

Reagents

Black Mexican sweet (BMS) medium: One package Murashige and Skoog salt (GIBCO, Grand Island, NY)/liter, 1 ml 1000× vitamins

[37] M. Fromm, J. Callis, L. P. Taylor, and V. Walbot, this series, Vol. 153, p. 351.
[38] P. Leon, F. Planckaert, and V. Walbot, *Plant Physiol.* **95,** 968 (1991).
[39] F. Planckaert and V. Walbot, *Plant Cell Rep.* **8,** 144 (1989).
[40] V. T. Nguyen, M. Morange, and O. Bensaude, *Anal. Biochem.* **171,** 404 (1988).

(1000× vitamin stock contains 1.3 mg/ml niacin, 250 μg/ml thiamin, 250 μg/ml pyroxidine, 250 μg/ml pantothenate; store at −20°), 130 mg/liter asparagine, 20 g/liter sucrose, 200 mg/liter inositol, 2 mg/liter 2,4-D (2,4-dichlorophenoxyacetic acid). The final solution is adjusted to pH 5.8 with 0.1 *N* NaOH and autoclaved to sterilize. This is a modification of the MS medium previously described[41]

Protopalst isolation medium (PIM): 90 g/liter mannitol, 0.73 g/liter $CaCl_2 \cdot 2H_2O$, 0.97 g/liter MES [2-(*N*-morpholino)ethanesulfonic acid], final solution adjusted to pH 5.8 with 0.1 *N* NaOH

PIM + enzymes: To 100 ml PIM add 0.3 g cellulase (CELF/Worthington, Freehold, NJ), 1 g cytolyase (Genencor, South San Francisco, CA), 20 mg pectolyase Y23 (Seishin Pharmaceuticals, Tokyo, Japan), 0.5 g bovine serum albumin (BSA) (Sigma, St. Louis, MO), 50 μl 2-mercaptoethanol. Centrifuge the mixture for 5 min at 3000 *g* to pellet any insoluble material and sterilize by passage through a 0.45-μm Nalgene filter (Nadge, Rochester, NY)

Electroporation solution (EPH): 36.4 g/liter mannitol, 8.95 g/liter KCl, 0.58 g/liter NaCl, 0.59 g/liter $CaCl_2 \cdot 2H_2O$, 2.38 g/liter *N*-2-hydroxyethylpiperazine-*N'*-2-ethanesulfonic acid (HEPES), final solution adjusted to pH 7.2 with 0.1 *N* NaOH

Plating out medium for transients (POMT): To 800 ml of BMS medium add 54.6 g mannitol and add 200 ml of filter-sterilized conditioned BMS medium (below)

Conditioned BMS medium: Centrifuge and recover the supernatant from a 2- to 4-day-old BMS culture. The medium is filter sterilized through a 0.45-μm Nalgene filter

Routine Maintenance of Suspension Cultures

The BMS suspension line we use has been in culture for several years, and the cells are densely cytoplasmic and homogeneous in size. Cultures are split 1 : 1 into fresh medium every 3–4 days. We use 40 ml of culture medium in a 100-ml flask and shake at 110 rpm at 25°.

Protoplast Preparation and Electroporation

The BMS suspension is routinely maintained as described above. For the highest levels of transient gene expression, we find it best if the cells are in a logarithmic growth phase. A 1 : 1 transfer on each of 2 days before electroporation satisfies that requirement and results in a sufficient density of cells for several (about eight) gene transfers per flask. Cultured cells (~100 ml) are transferred to sterile 50-ml screw-cap plastic tubes (Corning,

[41] T. Murashige and F. Skoog, *Physiol. Plant.* **15,** 473 (1962).

Corning, NY) and centrifuged at 150 *g* for 2 min at room temperature, and the supernatant is discarded or used for POMT (above). The cells are washed once with 50 ml PIM and centrifuged as before. The cell pellet is dispersed in 50 ml of PIM + enzymes and 10 ml is aliquoted to each of 5 petri plates (100 × 15 mm); these are incubated at 25° on a rotary shaker at 50 rpm for 3–5 hr. The optimal time of digestion is variable and is terminated when the protoplasts are close to spherical yet still retain some cell shape (e.g., some cell wall) as monitored by light microscopy. The digested protoplasts are transferred to a 50-ml screw-cap tube and centrifuged as before. The pellet is washed twice with 50 ml PIM to remove the protoplasting enzymes and once with 50 ml EPH. Finally, the protoplasts are suspended in a total volume of 8–12 ml EPH (~2–4 × 10^6 protoplasts/ml). For the highest levels of expression, an optional 10-min heat shock at 45° increases transient gene expression of luciferase or GUS approximately two- to fourfold. With or without the optional heat shock, the protoplasts are chilled on ice for >30 min before electroporation.

Each transfection is set up by dispensing 0.5 ml of EPH containing 5–50 μg of luciferase expression plasmid, 10–20 μg of a GUS plasmid (optional; see below), and 75 μg of sheared, nondenatured salmon sperm DNA (as a carrier) into a 1-ml plastic semimicrocuvette (Cat. No. 58017-847; VWR Scientific, Philadelphia, PA). BMS protoplasts (0.5 ml) are added to the cuvette and gently mixed. Within 2 min of protoplast addition, a flame-sterilized stainless steel electrode (0.4-cm plate separation; model E4 obtained from Prototype Design Services, Madison, WI) is inserted into the cuvette and a 12-msec pulse of 450 V/cm (capacitors charged to 1550 μF) is applied with an electroporation apparatus. For our laboratory cultures, the optimal conditions for tobacco protoplasts are 450 V/cm (1550 μF) with a 5-msec pulse; carrot protoplasts are electroporated at 700 V/cm (1550 μF) with a 6-msec pulse. A prior report describes the optimization procedure for a new cell source.[37]

Immediately after discharge, the protoplasts are removed with a Pasteur pipette and gently dispensed into 7.5 ml POMT in a 100 × 15 mm petri plate. The transfected protoplasts are incubated statically at 25° for 20–40 hr. Generally, expression can be detected as early as 2 hr and is linear for the first 30–40 hr; expression decreases thereafter as a result of degradation of the input plasmid DNA and turnover of mRNA and luciferase protein.

Luciferase Assays

For assessing luciferase expression in plant cells, we have traditionally used an adaptation of the luciferase assay reported in de Wet *et al.*[4]

Promega Corporation has described an improved assay system,[42] composed of the cell culture lysis reagent (CCLR) and luciferase assay reagent (LAR) buffers described below; this is available in kit form (Cat. No. E1500). We will describe each assay system and compare the efficacy of each.

Reagents

Luciferase extraction buffers: Standard buffer is 100 m*M* potassium phosphate, pH 7.8, 1 m*M* ethylenediaminetetraacetic acid (EDTA), 7 m*M* 2-mercaptoethanol, 10% (v/v) glycerol; autoclave to sterilize. The alternative CCLR buffer is 25 m*M* Tris–phosphate, pH 7.8, 2 m*M* dithiothreitol (DTT), 2 m*M* 1,2-diaminocyclohexane-*N,N,N′,N′*-tetraacetic acid, 10% (v/v) glycerol, 1% (v/v) Triton X-100

Luciferase assay buffers: The standard buffer is 25 m*M* Tricine, pH 7.8, 15 m*M* $MgCl_2$, 5 m*M* ATP, 7 m*M* 2-mercaptoethanol, 0.5 mg/ml BSA. We make this fresh from stock solutions of 0.1 *M* Tricine, pH 7.8, 1 *M* $MgCl_2$, 0.1 *M* ATP, pH 7.5, and BSA at 10 mg/ml. The alternative buffer LAR is 20 m*M* Tricine, pH 7.8, 1.07 m*M* $(MgCO_3)_4Mg(OH)_2 \cdot 5H_2O$, 2.67 m*M* $MgSO_4$, 0.1 m*M* EDTA, 33.3 m*M* DTT, 270 μM coenzyme A, 470 μM luciferin, 530 μM ATP. Store the LAR buffer in small aliquots at $-80°$

Potassium D-luciferin (10 m*M*): Potassium luciferin (Cat. No. 1600) from Analytical Luminescence Laboratory has a molecular weight of 318.41. To prepare, dissolve 32 mg of potassium luciferin in 10 ml water and store at 4° in the dark. The working solution is 250 μM and is made by dilution with distilled water

Cell extracts are prepared by transferring the transfected protoplasts to a 15-ml conical screw-cap tube; a rubber policeman can be used to dislodge any protoplasts adhering to the petri dish. The protoplasts are spun at 150 *g* for 2 min at 4° and the supernatant is removed by aspiration. The protoplasts are resuspended in 0.4 ml extraction buffer and then transferred to a 1.5-ml microfuge tube. To disrupt the cells, we sonicate on ice for 10 sec at 75 W with an Ultrasonics (Long Island, NY) sonifier (model W185D) using the microtip; alternatively, using the CCLR buffer [containing 1% (v/v) Triton X-100], sonication can be omitted and a 10-sec vortexing at high speed will solubilize the enzyme. The disrupted protoplasts are centrifuged at 12,000 *g* for 5 min at 4° to pellet the cell debris. The luciferase activity remains stable for at least 1 month in extracts stored at $-80°$.

[42] K. V. Wood, *in* "Bioluminescence & Chemiluminescence: Current Status" (P. E. Stanley and J. Kricka, eds.), Wiley, Chichester, 1991.

Our standard assay is as follows: add 5–50 μl of extract to 200 μl assay buffer in a luminometer cuvette and allow the mixture to equilibrate to room temperature (about 15 min). The cuvette is placed in the counting chamber of a luminometer and 100 μl of 250 μM luciferin is injected into the cuvette to start the reaction. The photons emitted are integrated over a 10-sec period and are expressed as light units (lu)/10 sec.

Few cell extracts contain a source of chemiluminescence, hence background can be ascribed to the luminometer. The machines we have tested have a background of 50–200 events/10-sec counting period. This is a very low background considering that 20 μl of extract from cells electroporated with 10 μg plasmid pJD300 and incubated for 20–40 hr routinely yields >20,000 lu/10 sec (see below and Table I).

The quantity of light emitted per unit time falls off rapidly after an initial burst about 0.3 sec[5] after luciferin is added. Consequently, longer assay times do not accurately measure luciferase activity. Mechanistically, luciferase activity is inhibited by oxyluciferin, the reaction product, because this molecule remains in the active site. An alternative procedure extending the sensitivity of the assay takes advantage of a buffer system promoting a more rapid turnover of the luciferase enzyme. Using the LAR buffer described above, the rate of light emission is essentially constant for several minutes ($t_{1/2}$ 5 min), allowing longer assay times and hence increased sensitivity. The improved assay is as follows: add 20 μl of cell extract to 100 μl of the LAR buffer (equilibrated to 22°), place in the luminometer, and measure the light produced for the desired time. Alternatively, using the luminometers described above, 100 μl of LAR buffer can be injected directly into the cell-free extract to initiate light production.

The LAR assay is also well suited for use with a scintillation counter. Using the standard assay, the time required to initiate the reaction and to insert the vial in the counting chamber will miss the initial burst of light (t = 0.3 sec). The extended emission of light using the LAR buffer allows sufficient time for sample manipulation and detection using a scintillation counter. Maximal sensitivity requires turning off the coincidence counter to ensure that all photons are counted as events.

β-Glucuronidase

Important in comparing the expression of several luciferase constructs is the significant variation caused by differences in protoplast viability, protoplast recovery, and purity among luciferase plasmid preparations. A simple correction for protoplast recovery can be made by standardizing the luciferase activity with the total protein in each extract and representing expression as a specific activity (lu/10 sec/μg protein). To correct each

TABLE I
TRANSIENT EXPRESSION ASSAY USING FIREFLY LUCIFERASE REPORTER GENE[a]

Assay and plasmid	Extraction buffer	Sonication	Protein (mg/ml)	GUS (pmol/min/20 μl)	Luciferase activity			
					Standard assay (lu/10 sec)	LAR assay (lu/30 sec)	Protein corrected (lu/30 sec/μg protein)	GUS corrected (lu/30 sec/pmol MU/min)
Assay A								
20-hr control	Std	Yes	1.5	0	42	184	6	0
(pJD300)	Std	Yes	1.5	12.7	33,369	185,624	6164 ± 356	14691 ± 1314
40-hr control	Std	Yes	1.1	0	64	156	7	0
(pJD300)	Std	Yes	1.3	23.7	61,508	320,629	12171 ± 1123	13628 ± 2261
Assay B								
Control	Std	Yes	1.3	0	61	182	8	0
(pAL61)	Std	Yes	1.3	22.9	84,585	497,185	19695 ± 1596	21683 ± 1287
Control	CCLR	No	1.1	0	191	167	7	0
(pAL61)	CCLR	No	1.2	19.2	62,311	602,496	25346 ± 1311	31585 ± 3668
Control	CCLR	Yes	1.9	0	62	162	4	0
(pAL61)	CCLR	Yes	1.7	20.8	61,592	561,524	16198 ± 462	27394 ± 5470

[a] Electroporations using BMS protoplasts and the plasmids indicated were done as described in text. In assay A, the protoplasts were incubated for either 20 or 40 hr after electroporation; in assay B, incubation was for 40 hr. The compositions of the standard and CCLR extraction buffers are given in the section on luciferase assays. Sonication was done as described in the text. Protein determinations were with the Bradford reagent (Bio-Rad, Richmond, CA) compared with BSA standards. β-Glucuronidase assays were done as described in the text. The 30-sec luciferase readings for 20 μl of extract using the LAR buffer were corrected with either the total protein or the GUS activity in 20 μl of extract; the results from two or four replicates were averaged and a standard deviation is given as a ± value. Std, Standard.

transfection assay for expression variation, we generally include an equivalent mass of a GUS-expressing plasmid in each cuvette. Each luciferase reading is then corrected by the amount of GUS activity in each extract [lu/10 sec/pmol 4-methylumbelliferyl-β-D-glucuronide (MUG) hydrolyzed/min]. The GUS enzyme is fully active in each of the luciferase extraction buffers described above and is stable when stored at $-80°$. The GUS assay is a modification of that described by Jefferson.[1]

Reagents

GUS assay buffer (2×): 100 mM $NaPO_4$, pH 7, 20 mM 2-mercaptoethanol, 20 mM Na_2EDTA, 0.2% (w/v) sodium laurylsarcosine, 0.2% (v/v) Triton X-100

GUS assay buffer (2×) + MUG: Add 3.15 mg MUG (Sigma) to 5 ml 2× GUS assay buffer; make fresh for each use

Methylumbelliferone (MU; Sigma) standard (100 nM): Add 1.98 mg MU to 10 ml water and make two successive 100× dilutions with 0.2 M $NaCO_3$

The quantities of extract used and the incubation times are given as guidelines and should be altered according to the amount of GUS activity in each extract. For each assay, dispense 75 μl of 2× GUS assay buffer + MUG into a 1.5-ml microfuge tube. Add 75 μl of extract and place the tube in a 37° water bath for 5 min to allow the GUS enzyme to reach V_{max}. Remove 40 μl and dispense into 960 μl of 0.2 M Na_2CO_3 (t = 0 min) and repeat this at t = 30 and t = 60 min. Read the fluorescence (excitation at 365 nm and emission at 455 nm) in a TKO-100 (Hoefer) or comparable fluorometer and compare with a 100 nM MU standard; MU is the product of the β-glucuronidase reaction. The slope of the GUS activity curve is expressed in picomoles MU/per minute or in picomoles MU/per minute per milligram protein if a specific activity measurement is desired.

Results. Table I shows the expression results from two transient assays. In assay A (Table I), 10 μg of pJD300 and 10 μg of the GUS expression plasmid $pCal_1Gc$[33] per transfection were electroporated into BMS protoplasts. After incubation for 20 or 40 hr, the protoplasts were harvested and extracts were prepared and assayed for luciferase, GUS, and protein concentration. As shown, there is about twice the luciferase activity present at 40 hr than at 20 hr. Also, using the LAR buffer results in an approximately fivefold increase in light units detected when compared with the standard assay buffer. In assay B (Table I), 10 μg of the luciferase expression plasmid pAL61[33] and 10 μg of the GUS expression plasmid $pCal_1Gc$ per cuvette were electroporated into BMS protoplasts. After incubation for 40 hr, extracts were prepared using the extraction buffer and treatment indicated and assays were performed using the parameters shown. The inclusion of Triton X-100 in the extraction buffer did not significantly affect luciferase or GUS expression but did obviate the protoplast sonication step, thus simplifying the procedure. Overall, the best results were obtained using the CCLR extraction and the LAR assay buffers. We estimate that sensitivities ~10-fold above the standard assay can be routinely accomplished using the improved buffer system and the 30-sec assay time; an additional ~4-fold gain can be achieved using a 120-sec assay time [maximum setting on the Monolight 2010 (Analytical Luminescence Laboratory)].

Factors Affecting Luciferase Expression

The physiological state of the BMS suspension cells before and after gene transfer is the primary determinant of expression level. We have noted as much as a 50-fold variation in the absolute level of luciferase expression between batches of protoplasts. As alluded to above, it is important that the cells are logarithmically growing when harvested for protoplasting. The extent of protoplasting is also critical; the removal of most but not all of the cell wall is the most effective compromise between efficient gene transfer and protoplast viability. As most of the commercially available protoplasting enzymes are crude preparations, there is considerable variation in the digestion rates and the viability of protoplasts after treatment; we recommend testing specific lots from different vendors and then purchasing amounts sufficient to last several years. The optional heat shock treatment *before* gene transfer results in enhanced expression, probably by inducing a stress response that renders the protoplasts better able to withstand electroporation. Stress *after* gene transfer can also affect expression levels; when maize protoplasts were subjected to a 42° heat shock 5 hr after electroporation, luciferase expression was inhibited while GUS expression was not.[43] Presumably this is a function of luciferase mRNA and/or enzyme instability at the heat shock temperature.

It is also possible that constructing translational fusions with luciferase might alter the specific activity of the enzyme. Such fusions have been constructed and it was found that substantial luciferase activity could be detected (although the specific activities of the native enzyme and the fusions have not been directly compared). For example, we have added 26 and 31 amino acids to the N terminus of luciferase and noted little difference in activity compared with native luciferase (pAL61 and pAL74).[33] Howard *et al.*[10] have added from 14 to 127 amino acids for 3

[43] L. Pitto, D. Gallie, and V. Walbot, unpublished data (1990).

different N-terminal translational fusions and detected activity for each. The entire *neo* gene was added to the C terminus of luciferase, and this fusion also retained high activity.[13] Thus, the luciferase enzyme is able to accommodate translational fusions at both its N and C termini.

Other Methods of Detecting Luciferase Expression

Apart from the enzyme assays described above, there are additional methods available to detect luciferase gene transfer and expression.

RNA Analysis

Often the need arises to determine the structure and abundance of RNA transcribed from an expression plasmid. We have found that the luciferase mRNA can be detected in transfected maize protoplasts by either Northern blot or RNase protection analyses. The procedures we use for analyzing transcript RNA have been described in detail.[33]

Whole-Cell and Tissue Analysis

In addition to assaying luciferase activity in cell-free extracts, cells, tissues, and organs can be tested by diffusing the substrate luciferin across biological membranes. At neutral pH, luciferin is negatively charged and thus does not freely diffuse across membranes; generally luciferin is protonated in a mild acidic buffer to facilitate its passage. Gallie *et al.*[12] used video imaging (VIM) technology to detect luciferase activity in transiently transfected tobacco protoplasts. Tobacco plants stably transformed with CaMV *35S*–luciferase constructs have been used to assess tissue-specific expression patterns using standard photographic film and/or light-enhancing hardware.[11,13,14] The luciferase protein has also been detected immunocytochemically.[6]

Future Applications

Multiple Luciferase Reporter Genes

Because light is so readily detected and quantified, luciferase expression is at present among the most sensitive reporter gene activities. Additional improvements are possible, however, to increase the flexibility of luciferase assays. For example, rather than including as an internal transfection and viability control a second reporter gene activity that requires a separate assay, such as GUS, it may be possible to use two different luciferase proteins, each utilizing the same luciferin substrate.

All of the insect-derived luciferases utilize the same substrate, but the enzymes are distinct.[44] The enzyme derived from the firefly *P. pyralis* emits light maximally at 560 nm while the four luciferases of the click beetle (*Pyrophorus plagiophthalamus*) have maxima at 548, 560, 578, and 590 nm. Analysis of overlapping spectra is a routine problem in photochemistry (i.e., assay of the two forms of phytochrome), and should pose no problem in deriving equations for quantifying the levels of green (548 nm) and orange (590 nm) forms of click beetle luciferase once the quantal efficiency of each enzyme is determined. Analysis of spectral quality will require a spectrofluorometer.

An alternative approach would be to capitalize on the variety of bioluminescent reactions, including luciferases with different substrate luciferins than the insect enzymes. The Monolight 2010 luminometer (Analytical Luminescence Laboratory) has two injection devices, allowing sequential injection of two different luciferins and measurement of light. Alternatively, for machines with just one injection device, two samples could be prepared from the same cell extract and each assayed with its luciferin substrate. To date, three different enzymes with unique substrate requirements have been described: (1) The heterodimeric bacterial luciferase[2] from the *lux* operon of *Vibrio* species has been widely used; (2) luciferase from the marine crustacean *Vargula hilgendorfii*[45] has been reported. It has not yet found wide application, in part because the substrate is still in limited supply. The *V. hilgendorfii* enzyme is secreted from its host and from mammalian cells, allowing assay of the medium without sacrificing the cells, hence this enzyme may prove invaluable in assessing stable transformants or other materials of limited quantity; (3) aequorin from coelenterates[46] has been used as a reporter activity in mammalian cells and utilizes a different chemical mechanism than the beetle luciferases.

Targeting to Cellular Compartments

In vivo firefly luciferase is targeted to the peroxisomes; ATP and oxygen are both available in this compartment and luciferin can diffuse into the organelle. The peroxisome/lysosome targeting signal is thought to reside in the few terminal amino acids of luciferase.[47] With appropriate transit sequences it may also be possible to direct nuclear-encoded luciferase protein into the mitochondria, chloroplasts, or endomembrane system

[44] K. V. Wood, Y. Amy Lam, H. H. Seliger, and W. D. McElroy, *Science* **244,** 700 (1989).

[45] E. M. Thompson, S. Nagata, and F. I. Tsuji, *Gene* **96,** 257 (1990).

[46] H. Tanahashi, T. Ito, S. Inouye, F. I. Tsuji, and Y. Sakaki, *Gene* **96,** 249 (1990).

[47] S. J. Gould, G. A. Keller, N. Hosken, J. Wilkinson, and S. Subramani, *J. Cell Biol.* **108,** 1657 (1989).

of plant cells. Provided all of the required substrates are available, luciferase expression should be possible from each compartment. With additional study, it may be possible to design luciferase translational fusion proteins that are inactive until appropriately targeted, that is, removal of the transit sequence and/or acquisition of other posttranslational modifications. If feasible, such reporter gene activities would provide very sensitive indicators for analyzing the kinetics of protein translocation into the various organelles both *in vivo* and *in vitro*.

Altering Half-Life of Luciferase

In mammalian cells the firefly luciferase protein has a short half-life of about 3 hr, compared to about 50 hr for CAT.[48] Although protein turnover data are not available for plants, peak enzyme levels are achieved within 3–5 hr after introduction of luciferase mRNA into protoplasts[12] and can be readily decreased by a brief heat shock or incubation of cells for >12 hr.[43] These observations suggest that luciferase protein is not long lived in plant cells. Protein engineering could increase the enzyme half-life, but an alternative approach is now available. Inclusion of luciferin analogs in *in vitro* assay buffers or in cell culture media serves to increase the luciferase half-life severalfold.[48] One advantage of a short enzymatic half-life that should not be forgotten is that it allows repetitive induction of luciferase activity from a regulatable promoter, without building up a high background of reporter gene expression. Consequently, experimental design may favor a luciferase reporter activity with a short or long half-life. Similarly, the luciferase mRNA can be engineered to give a short or long half-life by manipulating the composition of the 5′ and 3′ untranslated regions as required for particular experiments.

Acknowledgments

We thank Keith Wood and Peter Christie for helpful comments. Development of luciferase reporter genes for maize cells was aided in part by support of K.R.L. by an American Cancer Society postdoctoral fellowship (PF2943), of J.R.D. by a National Institutes of Health (NIH) Postdoctoral Fellowship (GM11767-02), and by a grant from the NIH (GM 32422).

[48] J. F. Thompson, L. S. Hayes, and D. B. Lloyd, *Gene* **103,** 171 (1991).

[36] The *bar* Gene as Selectable and Screenable Marker in Plant Engineering

By KATHLEEN D'HALLUIN, MARC DE BLOCK, JÜRGEN DENECKE, JAN JANSSENS, JAN LEEMANS, ARLETTE REYNAERTS, and JOHAN BOTTERMAN

Introduction

Plants can be transformed by using *Agrobacterium*-based T-DNA vectors or by direct uptake of DNA. However, for the selection of transformed cells or tissues, selectable marker genes are required. Antibiotic resistance genes, such as those encoding neomycin phosphotransferase II (*neo*) and hygromycin phosphotransferase (*hpt*), have frequently been used for this purpose.

Genes conferring resistance to herbicidal compounds are potential alternatives for selection. Herbicide resistance genes generally code for a modified target protein insensitive to the herbicide or for an enzyme that degrades or detoxifies the herbicide in the plant before it can act.[1] Resistance to glyphosate or sulfonylurea herbicides has been obtained by using genes coding for the mutant target enzymes, 5-enolpyruvylshikimate-3-phosphate synthase (EPSPS) and acetolactate synthase (ALS), respectively, which have been identified in amino acid biosynthesis pathways. Resistance to glufosinate ammonium, bromoxynil, and 2,4-dichlorophenoxyacetate (2,4-D) have been obtained by using bacterial genes encoding a phosphinothricin acetyltransferase, a nitrilase, or a 2,4-dichlorophenoxyacetate monooxygenase, respectively, which detoxify the respective herbicides.

Although several herbicide resistance genes have been successfully transferred to plants, their use as selectable marker genes has only rarely been described. A gene encoding a mutant ALS enzyme has been used as a selectable marker in transformation of corn,[2] tobacco,[3a] cotton, and sugar beet.[3b] In other cases transformants were selected on the basis of kanamycin and herbicide resistance genes used for screening the trans-

[1] J. Botterman and J. Leemans, *Trends Genet.* **4,** 219 (1988).

[2] M. E. Fromm, F. Morrish, C. Armstrong, R. Williams, J. Thomas, and T. M. Klein, *Bio/Technology* **8,** 833 (1990).

[3a] A. Reynaerts, unpublished results (1992).

[3b] K. D'Halluin, M. Bossut, E. Bonne, B. Mazur, J. Leemans, and J. Botterman, *Bio/Technology* **10,** 78 (1992).

formed shoots.[4,5] The success of using herbicide resistance genes for selection purposes is also correlated with its mode of action. Because bromoxynil is an inhibitor of photosystem II, a photosynthetically active tissue is required to make the bromoxynil-specific nitrilase gene useful for *in vitro* selection. The 2,4-D-degrading enzyme could not be used as a selectable marker either, presumably because transformed cells in a chimeric tissue cannot develop into shoots and they are overgrown or inhibited by untransformed callus that is rapidly developing in the presence of 2,4-D.

In this chapter we report the use of the *bar* gene as a selectable marker in plant transformation, as a screenable marker in tissue culture and plant breeding, and as a reporter gene in plant molecular biology.

Bialaphos Resistance Gene (*bar*)

The bialaphos resistance gene (*bar*) is part of the biosynthesis pathway of bialaphos.[6] The strains *Streptomyces hygroscopicus* and *Streptomyces viridochromogenes* produce bialaphos, a tripeptide antibiotic. It consists of phosphinothricin (PPT), which is an irreversible inhibitor of glutamine synthetase (glutamate–ammonia ligase), and two L-alanine residues. Glufosinate ammonium, the chemically synthesized PPT, and bialaphos are used as nonselective herbicides. The *bar* gene codes for phosphinothricin acetyltransferase (PAT), which converts PPT with high affinity into a nonherbicidal acetylated form by transferring the acetyl group from acetyl-CoA on the free amino group of PPT (Fig. 1).

The *bar* gene has served as a useful assayable marker gene in plant molecular biology. To guarantee correct translation initiation in plants, an ATG initiation codon was introduced instead of the GTG codon used in the *Streptomyces* strain and the second codon was changed to introduce an *Nco*I site at the 5′ end of the coding region (Fig. 2). Chimeric gene constructs containing the *bar* coding region under control of different promoters have been transferred to several crops. The *bar* gene has also been successfully used as a selectable marker in some plant species using PPT or bialaphos as a selective agent (Table I). Transgenic plants were resistant to herbicide applications in the greenhouse and in field conditions.[7,8]

[4] B. L. Miki, H. Labbé, J. Hattori, T. Ouellet, J. Gabard, G. Sunohara, P. J. Charest, and V. N. Iyer, *Theor. Appl. Genet.* **80,** 449 (1990).

[5] A. McHughen, *Plant Cell Rep.* **8,** 445 (1989).

[6] C. J. Thompson, N. R. Movva, R. Tizard, R. Crameri, J. E. Davies, M. Lauwereys, and J. Botterman, *EMBO J.* **6,** 2519 (1987).

[7] W. De Greef, R. Delon, M. De Block, J. Leemans, and J. Botterman, *Bio/Technology* **7,** 61 (1989).

[8] K. D'Halluin, J. Botterman, and W. De Greef, *Crop Sci.* **30,** 866 (1990).

$$CH_3-P(=O)(OH)-CH_2-CH_2-CH(NH_2)-COOH \xrightarrow[\text{PAT}]{\text{Ac-CoA}} CH_3-P(=O)(OH)-CH_2-CH_2-CH(NH-Ac)-COOH$$

PHOSPHINOTHRICIN ACETYLPHOSPHINOTHRICIN

FIG. 1. The reaction mechanism catalyzed by phosphinothricin acetyltransferase (PAT). In the presence of acetyl-CoA PPT is converted to acetylphosphinothricin (Ac-PPT).

The *bar* Gene as Selectable Marker in Plant Transformation

Although the *bar* gene can be used as a selectable marker in transformation experiments of several plant species using PPT or bialaphos as a selective agent (Table I), it is not possible to give a general transformation procedure for different plant species. It is generally known that the selection conditions in transformation/regeneration procedures must be adapted for each plant species, variety, or line. Concentrations of PPT for selection purposes can vary between 0.5 and 100 mg/liter. To this end, some general comments based on published data or personal communications are given.

The use of the *bar* gene as a selectable marker gene is strongly related with the mode of action of PPT. Phosphinothricin is an inhibitor of glutamine synthetase, which plays a central role in the assimilation of ammonia

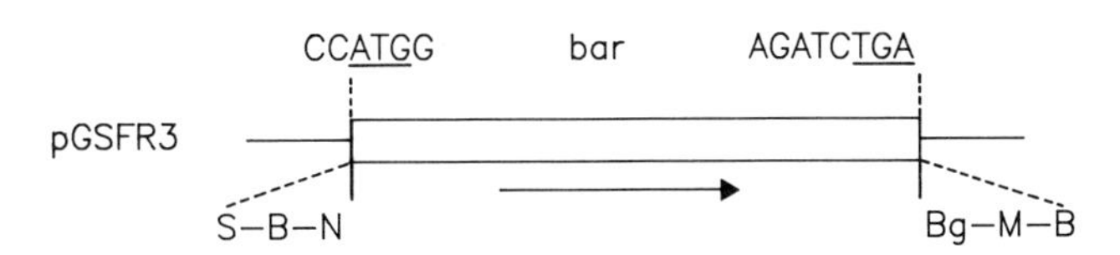

FIG. 2. Schematic representation of the plasmid pGSFR3 carrying a *bar* coding region cassette. The *bar* coding region is represented by an open box with the nucleotide sequence flanking the translation initiation and stop codon above. Suitable restriction sites (S, *Sma*I; B, *Bam*HI; N, *Nco*I; Bg, *Bgl*II; M, *Mlu*I) allowing the retrieval of the *bar* gene cassette are indicated. The nucleotide sequence is available from the EMBL database (accession number X05822).

TABLE I

PLANT SPECIES TRANSFORMED WITH A CHIMERIC *bar* GENE CONSTRUCT USED AS SELECTABLE MARKER AND/OR REPORTER GENE[a]

Plant species	Selection	Reporter gene	*Agrobacterium*	Ref.
Arabidopsis thaliana		+	+	*b*
Beta vulgaris	+	+	+	*c*
Brassica oleracea	+	+	+	*d*
Brassica napus	+	+	+	*d*
Cichorium intybus		+	+	*e*
Cucumis melo		+	+	*f*
Daucus carota		+	+	*g*
Fragaria vesca		+	+	*g*
Gossypium hirsutum	+	+	+	*h*
Lactuca sativa	+	+	+	*i*
Lycopersicum esculentum	+	+	+	*f, j*
Nicotiana tabacum	+	+	+	*j*
Medicago sativa	+	+	+	*k, l*
Populus species	+	+	+	*m*
Solanum tuberosum	+	+	+	*n, o*
Zea mays	+	+	−	*p, q*

[a] The use of the *bar* gene as marker for selection or as reporter gene is represented by + signs. Transformations based on *Agrobacterium* are indicated in a separate column.

[b] D. Valvekens, M. Van Montagu, and M. Van Lijsebettens, *Proc. Natl. Acad. Sci. U.S.A.* **85,** 5536 (1988).

[c] K. D'Halluin, M. Bossut, E. Bonne, B. Mazur, J. Leemans, and J. Botterman, *Bio/Technology* **10,** 78 (1992).

[d] M. De Block, D. De Brouwer, and P. Tenning, *Plant Physiol.* **91,** 694 (1989).

[e] J. Janssens, unpublished results (1990).

[f] G. Donn, M. Nilges and S. Morocz, "Abstracts VII International Congress on Plant Tissue Cell Culture," p. 53. Amsterdam, 1990.

[g] G. Donn, R. Dirks, P. Eckes, and B. Uijtewaal, "Abstracts VII International Congress on Plant Tissue Cell Culture," p. 176. Amsterdam, 1990.

[h] A. Reynaerts, unpublished results (1990).

[i] J. Janssens and A. Reynaerts, unpublished results (1990).

[j] M. De Block, J. Botterman, M. Vandewiele, J. Dockx, C. Thoen, V. Gosselé, N. Rao Movva, C. Thompson, M. Van Montagu, and J. Leemans, *EMBO J.* **6,** 2513 (1987).

[k] P. Eckes, B. Uijtewaal, and G. Donn, *Abstr. UCLA Symp.*, 334 (1989).

[l] K. D'Halluin, J. Botterman, and W. De Greef, *Crop Sci.* **30,** 866 (1990).

[m] M. De Block, *Plant Physiol.* **93,** 1110 (1990).

[n] M. De Block, *Theor. Appl. Genet.* **76,** 767 (1988).

[o] J. Kraus, J. Dettendorfer, and R. Nehls, "Abstracts VII International Congress on Plant Tissue Cell Culture," p. 52. Amsterdam, 1990.

[p] W. J. Gordon-Kamm, T. M. Spencer, M. L. Mangano, T. R. Adams, R. J. Daines, W. G. Start, J. V. O'Brien, S. A. Chambers, W. R. Adams, N. G. Willets, T. B. Rice, C. J. Mackey, R. W. Krueger, A. P. Kausch, and P. G. Lemaux, *Plant Cell* **2,** 603 (1990).

[q] M. E. Fromm, F. Morrish, C. Armstrong, R. Williams, J. Thomas, and T. M. Klein, *Bio/Technology* **8,** 833 (1990).

and in the regulation of the nitrogen metabolism in plants.[9,10] It is the only enzyme that detoxifies ammonia produced during nitrate reduction, photorespiration, and amino acid degradation in plant cells. Following an application of PPT, the ammonia metabolism in the plant is disturbed and NH_3 accumulates to phytotoxic levels in the cells. To stabilize the pH of the plant substrate it is important to add a buffer. We frequently use 2-(*N*-morpholino)ethanesulfonic acid (MES; Sigma, St. Louis, MO) at a concentration of 0.5 g/liter. Because ammonia is mainly produced during the reaction linked with photosynthetic electron transport, its accumulation is higher in plants exposed to light than in those kept in the dark. Exposure of leaf disks to light increases the sensitivity to PPT, which can lead to rapid cell death, even before transformed cells can be selected. The incubation of leaf disks in the dark or under low light can reduce the sensitivity for PPT but also reduces the efficiency of the selection and a higher frequency of nontransformed shoots might be obtained. Leaf disks of tomato, potato, and alfalfa became necrotic on exposure to PPT in the light.[8,11,12] Consequently, there is a significant lower recovery of transformants compared to other selective agents such as kanamycin. In *Brassica* species, shoot formation occurs faster after selection on kanamycin.[13] On the other hand, with cotton, selection on PPT or on kanamycin provided comparable recovery of transformed embryogenic callus.[14]

In general, selection is recommended as soon as possible after cocultivation. Applying the selection pressure in alfalfa only after the 2,4-D pulse increased the recovery of somatic embryos, but nearly all germinated plantlets were nontransformed.[8] At a certain stage during development alfalfa embryos are probably less sensitive to the herbicide. The duration of selection pressure is also very important. With corn, all plants derived from calli that were not immediately under selection pressure prior to plant regeneration were shown to be untransformed. Maintaining the selection during the regeneration protocol was required to recover transgenic plants.[2]

In metabolically active tissue the effect of PPT can be so detrimental that transformed cells within a chimeric tissue might not develop further due to the ammonia accumulation in nontransformed cells. Ammonia

[9] B. J. Miflin and P. J. Lea, *Annu. Rev. Plant Physiol.* **28,** 299 (1977).

[10] T. A. Skokut, C. P. Wolk, J. Thomas, J. C. Meeks, and P. W. Shaffer, *Plant Physiol.* **62,** 299 (1978).

[11] M. De Block, J. Botterman, M. Vandewiele, J. Dockx, C. Thoen, V. Gosselé, N. Rao Movva, C. Thompson, M. Van Montagu, and J. Leemans, *EMBO J.* **6,** 2513 (1987).

[12] M. De Block, *Theor. Appl. Genet.* **76,** 767 (1988).

[13] M. De Block, D. De Brouwer, and P. Tenning, *Plant Physiol.* **91,** 694 (1989).

[14] A. Reynaerts, unpublished results (1990).

could diffuse across the cell membrane into neighboring transformed cells, thereby causing cell death. Transformation of protoplasts instead of explant tissues might overcome this problem. The *bar* gene was efficiently used as a selectable marker in tobacco protoplasts.[11,15] In electroporated rice protoplasts, the transformation frequencies on kanamycin and PPT were comparable. Kanamycin could be used to select for transformed cells early after protoplast regeneration. At the callus stage, kanamycin appeared to be less efficient than PPT in inhibiting the growth of untransformed calli. For PPT selection in rice it proved essential to adapt the culture substrate and omit the amino acids from the substrate.[16]

In cereal transformation a general difficulty has been the lack of an effective selective agent for totipotent cultures.[17] Graminaceous plants show a high natural resistance to some aminoglycosides, such as kanamycin, making it difficult to distinguish transformed from nontransformed tissue.[18,19] Gordon-Kamm *et al.*[20] found bialaphos an effective selective agent for embryogenic suspension cultures. On the other hand, we found that the success rate of transformation in corn was higher with kanamycin than with PPT.[21]

In conclusion, the successful use of selective agents in plant transformation experiments requires an optimization of different parameters such as the choice of cell type, the stage of cell development, the rate of cell growth, the time of application, the concentration and the duration of exposure to the selective agent, the composition of growth substrate, and the growth conditions.

The *bar* Gene as Screenable Marker in Tissue Culture and Plant Breeding

Putative transformants can be easily screened by *in vitro* or *in vivo* spraying with the herbicide. Shoots *in vitro* are sprayed with a 1%

[15] J. Denecke, J. Botterman, and R. Deblaere, *Plant Cell* **2,** 51 (1990).

[16] R. Dekeyser, B. Claes, M. Marichal, M. Van Montagu, and A. Caplan, *Plant Physiol.* **90,** 217 (1989).

[17] I. Potrykus, *Trends Biotechnol.* **7,** 269 (1989).

[18] I. Potrykus, M. W. Saul, J. Petruska, J. Paszkowski, and R. D. Shillito, *Mol. Gen. Genet.* **199,** 183 (1985).

[19] R. M. Hauptmann, V. Vasil, P. Ozias-Akins, Z. Tabaeizadeh, S. G. Rogers, R. T. Fraley, R. B. Horsch, and I. K. Vasil, *Plant Physiol.* **86,** 602 (1988).

[20] W. J. Gordon-Kamm, T. M. Spencer, M. L. Mangano, T. R. Adams, R. J. Daines, W. G. Start, J. V. O'Brien, S. A. Chambers, W. R. Adams, N. G. Willets, T. B. Rice, C. J. Mackey, R. W. Krueger, A. P. Kausch, and P. G. Lemaux, *Plant Cell* **2,** 603 (1990).

[21] K. D'Halluin, unpublished results (1992).

(v/v) aqueous solution of the formulated glufosinate ammonium[21a] Plants in the greenhouse are sprayed with a 2% (v/v) aqueous solution of Basta. Spraying is done from four sides in a 1-m^2 surface using an air-brush line. Due to its non-systemic action, the whole plant or a part of the leaf tissue can be treated. Resistance can be tested in a callus induction test where tissue explants of putative transformants are transferred to callus induction medium containing different PPT concentrations. The rooting of tobacco shoots on medium containing 2 μg/ml PPT also allowed the distinguishing of transformed from untransformed shoots.

The herbicide resistance trait can be used for screening the progeny of transgenic lines in plant breeding. In contrast to antibiotic resistance markers, it offers a great advantage for selection of progeny grown in soil. Herbicide resistance can be monitored by spraying the seedlings in the soil or by a localized application of Basta to the leaves with a brush. The resistance gene can be physically linked to other genes conferring agronomically useful traits that will not provide selectable or directly screenable phenotypes among the initial products of transformation. Subsequently, in a plant breeding program the trait can be transferred to other lines through crossing by following the easily assayable herbicide resistance phenotype linked to other genes. The spraying of the seedlings can be used to follow segregation. For example, hybrid breeding is facilitated by coupling a genetic male sterility gene to a dominant selectable marker such as the *bar* gene.[22]

The *bar* Gene as Reporter Gene in Plant Molecular Biology

The *bar* gene has been transferred under control of different promoters in plant cells both for transient assays[23] and for stable plant transformation.[11] The *bar* coding region has been hooked to several promoter fragments for the analysis of expresion patterns in different organ systems and at different stages of development.[24,25] The *bar* coding region has also been

[21a] Glufosinate ammonium can be purified from Basta (Hoechst AG, Frankfurt, Germany) (200 g/liter glufosinate ammonium) by extraction of the surfactant twice with ethyl acetate; it is also commercially available as Pestanal (Riedel-deHaën, Seelze, Germany).

[22] C. Mariani, M. De Beuckeleer, J. Truettner, J. Leemans, and R. B. Goldberg, *Nature (London)* **347,** 737 (1990).

[23] J. Denecke, V. Gosselé, J. Botterman, and M. Cornelissen, *Methods Mol. Cell. Biol.* **1,** 19 (1989).

[24] E. De Almeida, V. Gosselé, C. Muller, J. Dockx, A. Reynaerts, J. Botterman, E. Krebbers, and M. Timko, *Mol. Gen. Genet.* **218,** 78 (1989).

[25] A. De Clercq, M. Vandewiele, R. De Rycke, J. Van Damme, M. Van Montagu, E. Krebbers, and J. Vandekerckhove, *Plant Physiol.* **92,** 899 (1990).

used to analyze the expression of dicistronic transcriptional units and to study the mechanism of antisense control.[26–28] The *bar* gene has also been successfully used in protein targeting studies. It was shown that the enzyme is secreted when targeted to the lumen of the endoplasmic reticulum by signal peptide-mediated translocation.[15]

It was also shown that the presence of foreign proteins at the carboxy terminus of PAT does not disturb the acetyltransferase activity.[29] The binding constants for PPT of the native acetyltransferase and of C-terminal fusion proteins did not differ significantly. This means that the substrate-binding site geometry in the fusion protein is not fundamentally changed relative to the PPT-binding site in the native enzyme. The presence of a *Bgl*II restriction site immediately in front of the stop codon allows the creation of in-frame gene fusions at the 3′ end (Fig. 2). The *bar* gene allowing 3′-terminal gene fusions was applied for the analysis of suppressor tRNAs and anticodon interactions in different sequence contexts in transgenic plants.[30] For the functional analysis of C-terminal retention signals in reticuloplasmins in plant cells, PAT was shown to be a suitable protein as a carrier molecule. The presence of these signals at the PAT C terminus was shown to have no influence on the PAT enzyme but causes a reduction of the rate of PAT secretion.[31]

Methods to Detect the *bar* Gene Product

The gene product is easily monitored by enzymatic or immunological assay. With a polyclonal antiserum raised against the enzyme purified from an *Escherichia coli* overproducing strain,[29] amounts of 2 ng PAT in 100 μg protein of total cell extracts are detectable. Its enzymatic activity can be analyzed by chromatographic detection of ^{14}C-labeled acetylated PPT. Enzymatic activities are clearly observed after an overnight exposure. More accurate values of enzyme activity are obtained if the enzyme kinetics are analyzed spectrophotometrically. However, background activities due to acetylation of other substrates present in crude plant cell extracts might interfere with this assay. It was observed that background activities are strongly dependent on the origin of the plant tissues. For

[26] M. Cornelissen, *Nucleic Acids Res.* **18,** 7203 (1989).

[27] G. Angenon, J. Uotila, A. Kurkela, T. Teeri, J. Botterman, M. Van Montagu, and A. Depicker, *Mol. Cell. Biol.* **9,** 5676 (1989).

[28] M. Cornelissen and M. Vandewiele, *Nucleic Acids Res.* **17,** 833 (1989).

[29] J. Botterman, V. Gosselé, C. Thoen, and M. Lauwereys, *Gene* **102,** 33 (1991).

[30] G. Angenon *et al.*, unpublished results (1991).

[31] J. Denecke, R. De Ryche, and J. Botterman, *EMBO J.* **11,** 2345 (1992).

example, the background activity is much higher in potato than in tobacco extracts. The determination of the ammonia level is a very sensitive indicator of glutamine synthetase activity. The ammonia level did not change significantly in transgenic plants 24 hr after treatment with PPT, while there is a rapid increase in control plants treated with the herbicide. Finally, enzymatic activities can be correlated with protein levels using Western blotting.

Protocol 1: Thin-Layer Chromatography

The PAT assay is based on the detection of ^{14}C labeled acetylated PPT after separation by thin-layer chromatography. As a labeled substrate, [1-^{14}C]acetyl-CoA or [3,4-^{14}C]PPT can be used.[6,32] Because only ^{14}C-labeled acetyl-CoA is commercially available we describe the procedure with nonradioactive PPT as substrate.

1. Grind 100 mg tissue in Eppendorf tubes, on ice, with 100 μl extraction buffer [25 m*M* Tris-HCl, pH 7.5, 1 m*M* Na_2EDTA, 0.15 mg/ml phenylmethylsulfonyl fluoride (PMSF; Sigma] to which 5 mg polyvinyl polypyrrolidone (PVPP; Sigma) was added.
2. Centrifuge for 10 min in an Eppendorf centrifuge at 4° at 13,000 rpm; this yields a crude extract.
3. Measure the protein concentration in the crude extract following the Bio-Rad (Richmond, CA) assay relative to bovine serum albumin (BSA; Sigma) as standard.
4. Dilute the crude extract in extraction buffer to a final concentration of 0.1 mg/ml protein.
5. Prepare a reaction mixture containing 1 μl of a 3 m*M* PPT stock, 1 μl of a 2 m*M* acetyl-CoA (Sigma) stock, 2 μl of [^{14}C]acetyl-CoA (58.1 mCi/mmol; Du Pont–New England Nuclear, Boston, MA) and 12 μl diluted extract.
6. Incubate for 30 min at 37°.
7. Spot 5 μl reaction mixture on a silica gel thin-layer chromatography (TLC) plate (Kieselgel 60; Merck, Rahway, NJ).
8. Carry out chromatography in a 3 : 2 mixture of 1-propanol and NH_4OH [25% (v/v) NH_3] for 9 hr.
9. Visualize the ^{14}C-labeled acetylated PPT by exposure to X-ray film overnight.

A slightly modified procedure can be used for a more quantitative analysis. For this procedure, larger amounts of plant tissue are required.

[32] E. Strauch, W. Wohlleben, and A. Puhler, *Gene* **63,** 65 (1988).

1. Grind 0.5 g of tissue in 500 μl extraction buffer.
2. Measure the protein concentration as in step 3 above and dilute to a final concentration of 1 mg/ml in extraction buffer.
3. Purify partially the crude cell extract by differential ammonium sulfate precipitation (30 to 60%). To a 500-μl extract, appropriate volumes of a 100% ammonium sulfate solution are added and the mixture is incubated for 1 hr on ice. The pellet is spun down at 4° for 10 min at 13,000 rpm.
4. Resuspend protein pellets in 200 μl of 100 m*M* Tris-HCl, pH 7.5, and keep on ice.
5. Prepare a reaction mixture consisting of 100 m*M* Tris-HCl, pH 7.5, a 1 : 10 volume dilution of [^{14}C]acetyl-CoA (58.1 mCi/mmol), a 10-fold molar excess of cold acetyl-CoA, and appropriate dilutions of the protein samples.
6. Preincubate at 37° for 2 min.
7. Reaction is started by the addition of PPT to a final concentration of 1 m*M*. Usually the final reaction mixture is 50 μl.
8. Samples of 5 μl are taken after 2, 5, 10, and 20 min and spotted directly on the silica gel plate.
9. After separation by chromatography (as above), the reaction product and acetyl-CoA are quantified by scintillation counting.
10. Dilutions are such that less than 10% of the acetyl-CoA reacts with PPT after 20 min for all samples, and the slope of the curve is taken as a measure of the relative activity. Because the assay is performed at suboptimal acetyl-CoA concentrations, the quantification is based on a standard curve. For this purpose, a dilution series using purified PAT as standard is made in extracts from an untransformed plant processed in the same manner as the samples to be tested.

Protocol 2: Spectrophotometric Assay

PAT activity is quantified by measuring enzyme kinetics. The method is based on the generation of free CoA sulfhydryl groups during the transfer of the acetyl group to PPT. The reaction of the reduced CoA with 5,5′-dithiobis(2-nitrobenzoic acid) (DTNB; Aldrich, Milwaukee, WI) yields a molar equivalent of free 5-thio-2-nitrobenzoic acid with a molar extinction coefficient of 13,600 at 412 nm.

1. Prepare a crude cell extract as described in protocol 1.
2. Measure the protein concentration and dilute to 1 mg/ml.
3. Perform reactions at 35° in 1 ml reaction buffer (100 m*M* Tris-HCl, pH 7.5, 0.5 m*M* acetyl-CoA, 0.4 mg/ml DTNB, and 1 m*M* PPT) using a recording spectrophotometer equipped with a temperature-controlled cuvette chamber.

4. After establishing a base line rate of DTNB reduction in the presence of plant extract and acetyl-CoA, start the reaction by the addition of substrate. The difference in the slope of the two curves (OD/min) divided by 13.6 represents the micromoles DTNB reduced per minute at 35°.

Protocol 3: Detection of Ammonia in Plant Extracts

1. Extract 250 mg leaf material in 1 ml water containing 50 mg PVPP.
2. Centrifuge for 5 min in an Eppendorf centrifuge.
3. Dilute 200 μl supernatant with 800 μl water.
4. Add 20 μl of the diluted plant extract to 1.5 ml reagent A [5 g phenol, 25 mg sodium nitroprusside (Sigma), 500 ml water], followed by the addition of 1.5 ml reagent B (2.5 g NaOH, 1.6 ml NaOCl (with 13% available chlorine), 500 ml water).
5. Incubate the reaction mixture for 15 min at 37°.
6. Measure the absorbance at room temperature at 625 nm.
7. Determine NH_4^+-N on a standard curve (μg NH_4^+-N/g fresh weight = μg determined NH_4^+-N $\times$ 450).

The standard curve is made using NH_4Cl in concentrations ranging from 0.1 to 2 μg NH_4^+-N (3.82 g NH_4Cl = 1 g NH_4^+-N); 20 μl of a nontreated control plant is added to the reaction mixtures.

Protocol 4: Detection of Phosphinothricin Acetyltransferase by Immunoblotting

1. Prepare a crude cell extract as described in protocol 1. Extraction buffer includes 1% (v/v) 2-mercaptoethanol, 50 m*M* Tris-HCl, pH 6.8, 0.13 mg/ml leupeptin.
2. Measure the protein concentration and dilute to 1 μg/ml protein.
3. Boils 50 μl extract for 10 min with 50 μl loading buffer [200 m*M* Tris-HCl, pH 8.8, 5 m*M* EDTA, 1 *M* sucrose, 0.1% (v/v) bromphenol blue, 3% (w/v) sodium dodecyl sulfate (SDS), 7.5 m*M* dithiothreitol (DTT; Sigma)].
4. Separate by electrophoresis on a 12.5% (w/v) SDS–polyacrylamide gel.
5. Soak the gel for 10 min in blotting solution (25 m*M* Tris, 192 m*M* glycine, pH 8.3, 10% methanol) while gently rocking.
6. In parallel, prewet a sheet of nitrocellulose in blotting buffer.
7. Place gel carefully on nitrocellulose sheet. Remove all air bubbles.
8. Sandwich the gel and filter between two layers consisting of a Whatman (Clifton, NJ) filter and a sponge presoaked in blotting solution.

9. Transblot for 3–4 hr at 40 V; make sure that the nitrocellulose is toward the anode.

10. Incubate the filter for 2–4hr at room temperature or overnight at 4° in PBS-ovalbumin buffer [150 mM NaCl, 12.6 mM $NaH_2PO_4 \cdot 2H_2O$, 14 mM KH_2PO_4, pH 7.4, 0.5% (w/v) ovalbumin] with rocking.

11. Soak the filter at room temperature for 2 hr with gentle shaking in anti-PAT antibody solution.

12. Wash three times in PBS containing 0.1% (v/v) Triton X-100 for 10 min.

13. Incubate for 2 hr at room temperature in a second antibody solution (alkaline phosphatase-conjugated anti-rabbit IgG; Sigma).

14. Wash once with PBS [with 0.1% (v/v) Triton X-100] for 30 min and three times with PBS for 30 min.

15. Add freshly made staining solution [30 mg *p*-nitro blue tetrazolium chloride (NBT; Bio-Rad) in 1 ml 70% (v/v) *N,N*-dimethylformamide, 15 mg 5-bromo-4-chloro-3-indolyl phosphate toluidine salt (BCIP; Sigma) in 1 ml 100% *N,N*-dimethylformamide, 98 ml carbonate buffer (0.1 M $NaHCO_3$ + 1.0 mM $MgCl_2$) (pH 9.8)] and shake until the reaction product is visible.

16. Rinse the filter in distilled water and dry between two sheets of Whatman 3MM paper in the dark.

[37] Selectable Markers for Rice Transformation

By ALLAN CAPLAN, RUDY DEKEYSER, and MARC VAN MONTAGU

Introduction

Choosing Convenient Method to Introduce DNA

A number of techniques have been developed to transfer DNA into rice. Some techniques, such as polyethylene glycol (PEG)-mediated or electroporation-mediated gene transfer, have been used repeatedly by several groups,[1–9] and are becoming well-established routines for genetic

[1] K. Toriyama, Y. Arimoto, H. Uchimiya, and K. Hinata, *Bio/Technology* **6,** 1072 (1988).

[2] H. M. Zhang, H. Yang, E. L. Rech, T. J. Golds, A. S. Davis, B. J. Mulligan, E. C. Cocking, and M. R. Davey, *Plant Cell Rep.* **7,** 379 (1988).

[3] K. Shimamoto, R. Terada, T. Izawa, and H. Fujimoto, *Nature* (*London*) **338,** 274 (1989).

[4] R. Matsuki, H. Onodera, T. Yamauchi, and H. Uchimiya, *Mol. Gen. Genet.* **220,** 12 (1989).

[5] S. K. Datta, A. Peterhans, K. Datta, and I. Potrykus, *Bio/Technology* **8,** 736 (1990).

TABLE I
REPRESENTATIVE TRANSFORMATION EFFICIENCIES OF RICE PROTOPLASTS

Selectable marker[a]	DNA concentration (μg/ml)	Method of introduction[b]	Selective agent (μg/ml)[c]	Transformation frequency (10^{-6})	Estimated copy number	Ref.
*P35S:npt*II	16	E	G418 (20)	4	1–2	1
*P35S:npt*II	20	E	Kan (100)	280	1–2	2
P35S:hpt	10	E	Hyg (20)	Up to 6500	2–10	3
P35S:hpt	1	E	Hyg (50)	120–200	≥3	4
P35S:hpt	1.3	P	Hyg (25)	5–44[d]	50–100	5
Pnos:hpt	40	P	Hyg (50, 100)	76; 59	—	6
P35S:hpt	40	P	Hyg (50, 100)	197; 176	1–10	6
*P35S:npt*II	50	P	Kan (100)	20	2–10	7
*P35S:npt*II	75	P	Kan (200)	3250–6500[d]	20–30	8
P35S:hpt	20	E	Hyg (50)	134–137	1–10	9

[a] *P35S*, cauliflower mosaic virus (CaMV) promoter; *Pnos*, nopaline synthase promoter; *npt*II, neomycin phosphotransferase II gene; *hpt*, hygromycin phosphotransferase gene.
[b] E, Electroporation; P, PEG.
[c] Kan, Kanamycin; Hyg, hygromycin.
[d] Indica variety of rice; all other experiments use japonica varieties.

engineering projects. Others including pollen tube transformation,[10] particle guns,[11] and *Agrobacterium tumefaciens*,[12] which have been introduced more recently, involve more specialized skills, and hence seem more difficult to apply repeatedly. Genes transferred by any of the first four methods mentioned rely on host DNA repair systems to link the foreign genes to the chromosomes. The integration sites may be distributed at random along the sequences of the introduced DNA, and thus may disrupt the gene being transferred. However, *Agrobacterium* and its tumor-inducing (Ti) plasmid-mediated system of cell transformation should be able to transfer DNA in a defined manner producing predictable end points for integration.

Electroporation and PEG-mediated gene transfer do not differ greatly either in terms of efficiency or ease of application (Table I). Each functions

[6] A. Hayashimoto, Z. Li, and N. Murai, *Plant Physiol.* **93,** 857 (1990).
[7] A. Peterhans, S. K. Datta, K. Datta, G. J. Goodall, I. Potrykus, and J. Paszkowski, *Mol. Gen. Genet.* **222,** 361 (1990).
[8] J. Peng, L. J. Lyznik, L. Lee, and T. K. Hodges, *Plant Cell Rep.* **9,** 168 (1990).
[9] Y. Tada, M. Sakamoto, and T. Fujimura, *Theor. Appl. Genet.* **80,** 475 (1990).
[10] Z.-X. Luo and R. Wu, *Plant Mol. Biol. Rep.* **6,** 165 (1988).
[11] M. E. Fromm, F. Morrish, C. Armstrong, R. Williams, J. Thomas, and T. M. Klein, *Bio/Technology* **8,** 833 (1990).
[12] D. M. Raineri, P. Bottino, M. P. Gordon, and E. W. Nester, *Bio/Technology* **8,** 33 (1990).

by opening pores through which DNA can diffuse into the cell, and eventually into the nucleus, where transcription is possible. Both of these methods require similar amounts of purified DNA (generally ranging from 10 to 40 μg/sample). For the sake of convenience, it is advisable to maintain the genes that are to be transferred on high-copy number, stably segregating *Escherichia coli* vectors such as those derived from the modified ColE1 origin present in plasmids such as pUC18.[13] Whereas it has been demonstrated that plasmids as large as 200 kilobases (kb) can be electroporated into protoplasts,[14] it is prudent to keep in mind that larger plasmids often yield less DNA during isolation from bacteria.

One way of circumventing size constraints is to separate the sequences to be transferred into two sets. For example, the selected marker and the gene under investigation may be cloned on individual vectors and mixed prior to DNA transfer. Molar ratios of 1 : 2, 1 : 1, or 2 : 1 (selected–unselected) yield similar frequencies of cotransformation that differ as much between replicates as between planned variations (14–85%).[4,7,9] For some experiments such two-component systems may offer several advantages. First, the cotransformed vector may contain the same restriction sites or expression sequences present in the selected marker without complicating the cloning strategy or increasing the risk of homology-dependent deletion formation that can occur between repeated sequences in the same bacteria. Second, the selected marker and the nonselected gene can integrate at independent loci and thus segregate during meiosis. In this way, primary transformants can produce descendants that are no longer drug resistant, yet retain the unselected gene of interest. Such plants thus have very little extraneous DNA, and can be retransformed with the same selectable marker, should later experiments require the introduction of other genes.

Materials

Protoplast and Plant Culture Media and Solutions

Linsmaier and Skoog (LS) medium[15]: 1 liter of deionized water, 4.4 g of Murashige and Skoog salts (Sigma Chemical Co., St. Louis, MO), 1 mg of thiamin hydrochloride, 100 mg of *myo*-inositol, 30 g of sucrose, 4 g of agarose, pH 5.7. Autoclave, cool to 55°, and add 2 mg of 2,4-dichlorophenoxyacetic acid (2,4-D) dissolved in dimethyl sulfoxide (DMSO; from a stock of 10 mg/ml)

[13] C. Yanisch-Perron, J. Vieira, and J. Messing, *Gene* **33,** 103 (1985).
[14] W. H. R. Langridge, B. J. Li, and A. A. Szalay, *Plant Cell Rep.* **4,** 355 (1985).
[15] E. M. Linsmaier and F. Skoog, *Physiol. Plant.* **18,** 100 (1965).

CPW13M[16]: 1 liter of deionized water, 101 mg of KNO_3, 27.2 mg of KH_2PO_4, 1480 mg of $CaCl_2 \cdot 2H_2O$, 0.025 mg of $MgSO_4 \cdot 7H_2O$, 0.16 mg of KI, 0.025 mg of $CuSO_4 \cdot 5H_2O$, 975 mg of 2-(*N*-morpholino) ethanesulfonic acid, 130 g of mannitol, pH 5.8. Autoclave

KpR (per liter of deionized H_2O)[17]: As this medium is quite complex, it is advisable to make several stock solutions. Macronutrients can be added as dry powder: 1900 mg of KNO_3, 600 mg of NH_4NO_3, 600 mg of $CaCl_2 \cdot 2H_2O$, 300 mg of $MgSO_4 \cdot 7H_2O$, 300 mg of KCl, 170 mg of KH_2PO_4. Micronutrients are added as 10 ml of a 100× stock solution containing (per liter) 1000 mg of $MnSO_4 \cdot H_2O$, 300 mg of H_3BO_3, 200 mg of $ZnSO_4 \cdot 7H_2O$, 75 mg of KI, 25 mg of $Na_2MoO_4 \cdot 2H_2O$, 2.5 mg of $CuSO_4 \cdot 5H_2O$, and 2.5 mg of $CoCl_2 \cdot 6H_2O$. Organic acids are added as 10 ml of a 100× stock solution containing (per liter) 1 g of citric acid, 1 g of malic acid, 1 g of fumaric acid, and 0.5 g of sodium pyruvate. Adjust pH of stock solution to 5.5 with NH_4OH. Several vitamins are added as 10 ml of a 100× stock containing (per liter) 100 mg of nicotinamide, 100 mg of ascorbic acid, 100 mg of pyridoxinhydrochloride, 100 mg of thiamin hydrochloride, 50 mg of calcium pantothenate, 50 mg of choline chloride, 20 mg of folic acid, and 10 mg of riboflavin. KpR medium is completed by addition of 100 mg of *myo*-inositol (as dry powder), 10 μl of *p*-aminobenzoic acid (1 mg/ml) stock in H_2O), 10 μl of vitamin A (0.5 mg/ml stock in ethanol), 10 μl of vitamin D_3 (0.5 mg/ml stock in ethanol), 10 μl of vitamin B_{12} (1 mg/ml stock in ethanol), and 1.0 ml of biotin (0.5 mg/ml stock in water). Add 1.25 g of from a dry mix consisting of D-sucrose (10 g), D-fructose (5 g), D-ribose (5 g), D-xylose (5 g), D-mannose (5 g), D-cellobiose (5 g), L-rhamnose (5 g), D-mannitol (5 g), D-sorbitol (5 g). Iron is supplied to the medium as $FeSO_4 \cdot 7H_2O$ (28.75 mg) and Na_2 EDTA (37.25 mg). In our experiments we use 10% (w/v) glucose as osmoticum, omit the casamino acids and the coconut water, and use 0.5 mg of 2,4-D (10 mg/ml stock in DMSO), 1 mg of naphthaleneacetic acid (10 mg/ml stock in DMSO), and 0.5 mg of zeatin (5 mg/ml stock in DMSO) as hormones. The final pH of the solution should be 5.7. In general, the medium is filter sterilized using a 0.2-μm filter unit (Nalgene, Nalge, Rochester, NY), and stored at 4°. To make agarose for embedding protoplasts, use deionized water to prepare a 2% Seaplaque agarose (FMC, Rockland, ME) solution, autoclave, cool to 45°, and mix with an equal volume of 2× KpR solution prior to use

[16] E. M. Frearson, J. B. Power, and E. C. Cocking, *Dev. Biol.* **33,** 130 (1973).
[17] K. N. Kao, *Mol. Gen. Genet.* **150,** 225 (1977).

Electroporation buffer (EPR): 10 mM N-(2-hydroxyethyl)piperazine-N'-(2-ethanesulfonic acid) (HEPES), pH 7.2, 4 mM $CaCl_2$, 10% (w/v) glucose

Solutions for NPTII Assay

Use distilled water for all solutions and buffers.

Extraction buffer (4×): 100 mM Tris-HCl (pH 6.8), 0.3 mg/ml leupeptin, 2% (v/v) 2-mercaptoethanol
Loading buffer (10×): 50% (v/v) glycerol, 10% (v/v) 2-mercaptoethanol, 0.5% (v/v) sodium dodecyl sulfate (SDS), 0.005% (v/v) bromphenol blue
Separation gel: For 30 ml, 10 ml of 30% (w/v) acrylamide, 3.9 ml of 2% (w/v) bisacrylamide, 11.2 ml of 1 M Tris-HCl (pH 8.7), 4.5 ml of water. When ready to pour gel, add 0.2 ml of 10% (w/v) ammonium persulfate (APS) and 0.04 ml of N,N,N',N'-tetramethylethylenediamine) (TEMED)
Stacking gel: For 10 ml, 1.67 ml of 30% (w/v) acrylamide, 1.3 ml of 2% (w/v) bisacrylamide, 1.25 ml of 100 M Tris-HCl (pH 6.8), 5.6 ml of water. When ready to pour, add 0.1 ml of 10% (w/v) APS and 0.01 ml of TEMED
Running buffer: 6 g/liter Tris, 14.9 g/liter glycine (do not adjust pH)
Reaction buffer: 50 mM Tris-HCl (pH 7.5), 200 mM NH_4Cl, 25 mM $MgCl_2$, 0.5 mM dithiothreitol (DTT)
Washing buffer: 100 mM Na_2HPO_4 (pH 7.0; adjusted with HCl)

Methods and Discussion

Possible Selective Agents

The efficiency of any selection technique is determined both by the ability of the selective agent to inhibit untransformed cells, and by the ability of the detoxifying gene to protect transformed ones. The first criterion can be assayed prior to any transformation experiment by analyzing the growth rate (in our case, as a function of the increase in fresh weight) of the target cells on medium supplemented with increasing concentrations of the desired selective agents. An increasing number of selective agents (Table II[18–22]) are now available, but the most commonly used products

[18] L. Herrera-Estrella, M. De Block, E. Messens, J.-P. Hernalsteens, M. Van Montagu, and J. Schell, *EMBO J.* **2,** 987 (1983).
[19] C. Waldron, E. B. Murphy, J. L. Roberts, G. D. Gustafson, S. L. Armour, and S. K. Malcolm, *Plant Mol. Biol.* **5,** 103 (1985).
[20] R. M. Hauptmann, V. Vasil, P. Ozias-Akins, Z. Tabaeizadeh, S. G. Rogers, R. T. Fraley, R. B. Horsch, and I. K. Vasil, *Plant Physiol.* **86,** 602 (1988).

TABLE II
REPRESENTATIVE SELECTION AGENTS FOR TISSUE CULTURE

Selective agent (abbreviation)[a]	Mode of action	Detoxifying gene	Detoxification reaction
Kanamycin (Kan)	Inhibition of organellar ribosomes	*npt*II[18]	Phosphorylation
G418	Primarily inhibits 80S ribosomes	*npt*II[18]	Phosphorylation
Hygromycin (Hyg)	Primarily inhibits 80S ribosomes	*hpt*[19]	Phosphorylation
Methotrexate (Mtx)	Inhibits dihydrofolate reductase	*dhrf*[20]	Mtx-resistant dihydrofolate reductase
Bleomycin (Blm)	Causes double-stranded breaks in DNA	*ble*[21]	Unknown
Phosphinothricin (Ppt)	Inhibits glutamine synthase	*bar*[22]	Acetylation

[a] All chemicals are available from Sigma Chemical Company, except Ppt, which is available as Basta (Hoechst, Somerville, NJ) and contains ±200 mg/ml Ppt.

are kanamycin, hygromycin, and G418. We have evaluated the power of these as well as of phosphinothricin, bleomycin, and methotrexate to inhibit rice callus growth. Other agents that could be useful, and should be tested in similar experiments, include streptomycin, glyphosate (an herbicide that inhibits aromatic amino acid biosynthesis), and chlorsulfuron (which inhibits branched amino acid synthesis).

Autoclave basic agar medium and allow it to cool to approximately 60°. Prepare 50 mg/ml aqueous stocks of phosphinothricin, G418, kanamycin, hygromycin, and bleomycin. Filter sterilize these through 0.22-μm disposable filters and store frozen in small aliquots (they may be kept at least 1 month). Methotrexate is dissolved in 0.1 *M* NaOH at a concentration of 10 mg/ml and immediately diluted into agar. Dilute each of the other chemicals into cooled medium, add 2,4-D (10 mg/ml in dimethyl sulfoxide; may be stored 6 months at −20°) to 2 μg/ml, and pour approximately 30 ml into sterile petri dishes. Using an alcohol-sterilized forceps, transfer approximately 70–100 mg of callus (we routinely use 15 individually grown callus lines) onto each dilution of every chemical tested. The growth rates of independently induced calli may differ by a factor of two to three,

[21] J. Hille, F. Verheggen, P. Roelvink, H. Franssen, A. van Kammen, and P. Zabel, *Plant Mol. Biol.* **7,** 171 (1986).

[22] M. De Block, J. Botterman, M. Vandewiele, J. Dockx, C. Thoen, V. Gosselé, R. Movva, C. Thompson, M. Van Montagu, and J. Leemans, *EMBO J.* **6,** 2513 (1987).

even on nonselective medium. Thus, it is best to place a portion of each individual callus line on each medium to be compared, including one on permissive medium.

Weigh calli after 5–7 days, and at 10 ± 2-day intervals. This may be done by transferring material to sterile, disposable, preweighted petri dishes. Each of the subpopulations derived from a callus lineage may be weighed independently. Cover and weigh the dish to determine the new fresh weight. The calli can then be transferred onto freshly prepared plates with the corresponding concentration of selective agent.

To simplify weighing, calli may also be placed onto sterile filter paper laid flat onto the solid medium so they can be weighed *en masse*. If calli are maintained on filter paper, use two forceps to lift the paper into the empty petri dish.

As seen in Fig. 1,[23] methotrexate (2 μg/ml), phosphinothricin (10 μg/ml), and hygromycin (25 μg/ml) inhibited growth within 2 weeks. G418 ($\leq$50 μg/ml) and bleomycin (20 μg/ml) required 3 weeks to do the same. On the other hand, kanamycin inhibited rice callus growth only partially, even at concentrations 5- to 10-fold above those used to select transgenic tissue from many other plants. However, when smaller calli were used ($<$500 μm), lower concentrations of kanamycin (300 μg/ml) were capable of retarding growth efficiently, and protoplast division could be completely arrested at 200 μg of kanamycin/ml. This indicates that, when using kanamycin, it is important to start the selection for transformed cells soon (less than 2 weeks) after introduction of DNA into protoplasts.

It should be noted that several additives including 10 m*M* glutamic acid, 25 m*M* proline, and 10 m*M* arginine reduce the inhibition of callus growth by 20 μg of phosphinothricin/ml from 91% to 69, 51, and 35%, respectively. It is therefore necessary to omit them, or other amino acid sources such as casein hydrolysate or coconut water, from the medium when stringent selection is desired. In our experience (Table III[24]), omitting these supplements does not markedly decrease callus formation from protoplasts.

Evaluating Effectiveness of Selectable Markers

The effectiveness with which a selectable marker protects transformants depends on at least the three following factors: first, the ability of the promoter and 3′ end sequences to provide adequate amounts of translatable RNA in those cells most susceptible to the selective agent

[23] R. Dekeyser, B. Claes, M. Marichal, M. Van Montagu, and A. Caplan, *Plant. Physiol.* **90,** 217 (1989).

[24] J. A. Thompson, R. Abdullah, and E. C. Cocking, *Plant Sci.* **47,** 123 (1986).

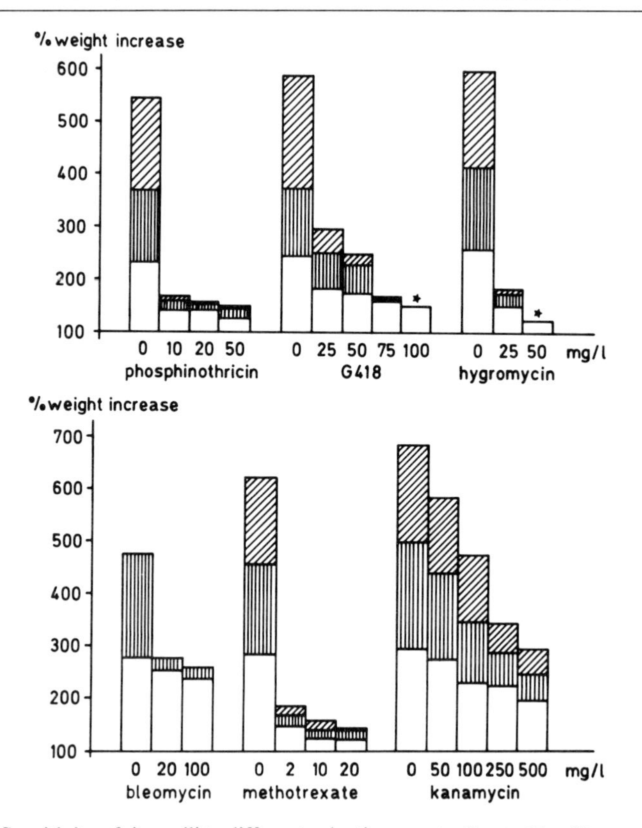

FIG. 1. Sensitivity of rice calli to different selective agents. Ten to 15 calli were transferred to media containing different concentrations of selective agent and the initial fresh weight was determined. The calli were reweighed after 10 to 12 days (white), 20 to 24 days (vertical stripes), and 28 to 31 days (slanted stripes) of incubation. The growth rate is defined as 100 times the ratio of fresh weight at day x to the initial fresh weight. The values are the averages of three independent experiments. An asterisk means there was no further increase in weight. (Reproduced from Dekeyser *et al.*[23] with permission by the American Society of Plant Physiologists.)

(mRNA levels can be boosted in monocots by inclusion of an intron in the primary transcript[25]); second, the rate with which the foreign enzyme inactivates the toxic substrate; third, the use of culture conditions that ensure that transformants are not adversely effected by metabolites from dying, untransformed cells.

The capacities of the expression system (that is, the promoter, 3′ end, and intron, if desired) can be assessed rapidly by assaying enzyme activity

[25] J. Callis, M. Fromm, and V. Walbot, *Genes Dev.* **1,** 1183 (1987).

TABLE III
NUMBER OF PROTOPLASTS REGENERATING TO CALLI[a]

Treatment	Medium[b]			
	KpR	KpR − cw − ca	KpR + Kan_{200}	KpR − cw − ca + Ppt_{10}
Unelectroporated	3.0×10^{-3}	3.0×10^{-3}	0	0
Electroporated				
Without DNA	1.8×10^{-3}	—	0	0
With pLD1	1.8×10^{-3}	—	3.0×10^{-5}	0
With pGSFR280	—	1.6×10^{-3}	—	0.7×10^{-5}

[a] The frequencies are the average of two independent experiments. (Reproduced from Dekeyser *et al.*[23] with permission by the American Society of Plant Physiologists.)

[b] KpR, Protoplast culture medium containing coconut water and casamino acids[24]; − cw, without coconut water; − ca, without casamino acids; + Kan_{200}, supplemented with 200 mg kanamycin/liter; + Ppt_{10}, supplemented with 10 mg phosphinothricin/liter.

produced by different constructs transiently expressed in protoplasts. The second point can be functionally investigated by comparing the growth rates of transformed and untransformed lines on selective media. The last factor generally requires frequent refreshment of the selection medium, and elimination of any plant material that does not appear healthy.

Protoplast Preparation, Electroporation, and Culture

Following the method of Thompson *et al.*,[24] we isolated protoplasts from established cell suspension cultures at the third to fifth day after subculture. To enrich for small calli, the cultures are first sieved through a 250-μm nylon mesh. One gram of drained calli is mixed with 20 ml of CPW13M medium containing 1% (w/v) cellulase RS (Yakult Honsha Co., Ltd., Nishinomiya, Japan), and 0.1% (w/v) pectolyase Y23 (Seishin Pharmaceutical Co., Ltd., Nihonbashii, Japan). The calli are incubated at room temperature in the dark on a rotary platform shaker (40 rpm) for 3 hr (until protoplasts begin to appear), followed by a further incubation at 25° for 2–3 hr without shaking. The digested mixture is then passed through a series of sterilized nylon sieves of 64-, 45-, and 30-μm diameter mesh to separate undigested cell clumps from protoplasts. The protoplasts are pelleted by centrifugation at 80 *g* for 5 min, and washed once in CPW13M and twice in protoplast electroporation buffer. At this point, protoplasts are pooled, counted using a hemacytometer, and adjusted to a final concentration of 5 to 7.5 × 10^6/ml.

Two hundred microliters of this suspension is pipetted into disposable spectrophotometer cuvettes and mixed with 10 μg pLD1[23] plasmid DNA (or an equimolar equivalent of the other plasmids), and 11 μl of a 3 *M*

NaCl stock. After 10 min on ice, an electrical shock of 375 V/cm from a 200-μF capacitor (the pulse time is approximately 54 msec) is delivered, and the protoplasts are left on ice for 15 min.

For transient gene expression, protoplasts are gently dispersed into 5 ml KpR liquid and cultured for 48 hr in dim light at 25°. In most successful experiments (as seen in Table III) the electrical shock reduces cell viability by no more than 40–50%. The number of viable protoplasts may be estimated by diluting them into KpR medium containing 5 μg/ml fluorescein diacetate (from a stock of 500 μg/ml acetone). Viable cells fluoresce when viewed at 450–490 nm with a fluorescent light microscope through a 520-nm cutoff filter.

To select transformants resuspend protoplasts in 3 ml KpR liquid, heat to 45° (5 min), and immediately chill 10 sec on ice. Then spin 4 min, 80 *g* and resuspend protoplasts to a density of approximately 3–5 $\times$ 10^5 viable protoplasts/ml with KpR medium containing 1% Seaplaque agarose. Quickly pipette 3.5 ml into sterile, 5-cm petri dishes, and allow medium to solidify. Cultivate the protoplasts in the dark at 25°. After 1 week, transfer each of four quarters of the agarose disks to separate 9-cm petri dishes containing 5 ml KpR medium with or without selective agent. Refresh the medium each week. Microcalli may be isolated and transferred to LS medium after 3 weeks.

Electroporation of Leaf Bases

We have developed an alternative method to analyze transient gene expression in intact rice tissue instead of in protoplasts.[26] To use this technique, dehusked rice seeds are sterilized by soaking for 45 min in a solution of 0.4% (v/v) commercial bleach, 0.1% (w/v) Na_2CO_3, 3% (w/v) NaCl, 0.15% (w/v) NaOH, and 0.01% (w/v) sodium dodecyl sulfate. They are rinsed five times with sterile water and germinated in the dark at 25° on a medium with MS salts, 1% (w/v) sucrose, and 0.8% (w/v) agar (pH 5.6). Seven days later (when the etiolated seedlings are 4 to 6 cm high) the lower leaf segment, commonly called the leaf base region, is isolated. The seedling is transferred to a large, disposable petri dish and cut with a clean, sterilized scalpel blade immediately above the site where the secondary roots pierce through the leaves. After removing the coleoptile, a second transverse cut is made approximately 3 mm above the first one. These leaf bases are then subdivided into segments with a length of 1 to 2 mm, and incubated for 3 hr in electroporation buffer (EPR) containing 0.2 m*M* spermidine. Next, the EPR medium is removed, the explants are washed

[26] R. A. Dekeyser, B. Claes, R. M. U. De Rycke, M. E. Habets, M. Van Montagu, and A. B. Caplan, *Plant Cell* **2,** 591 (1990).

twice with EPR, and the segments from 25 leaf bases (about 60 pieces in total) are transferred to a disposable spectrophotometer cuvette containing 0.2 ml EPR supplemented with 0.2 m*M* spermidine. Twenty micrograms of supercoiled plasmid DNA is then added to the buffer. After incubation of the DNA/explant mixture for 1 hr at room temperature (with regular, gentle shaking), add 11 μl of a 3 *M* NaCl stock, mix, and put the cuvette on ice for 10 min. Parallel stainless steel electrodes, 2 mm thick with inner surfaces 6 mm apart, are inserted in such a way that the leaf bases are gathered between them. Then one pulse with an electrical field strength of 375 V/cm is discharged from a 900-μF capacitor. The homemade electroporation unit consists of an ISCO power supply connected to an array of capacitors arranged in a circuit as described.[27] The explants are left on ice for 15 min and washed by three successive additions and removals of 0.3 ml KpR medium. Then the explants contained in the liquid medium are pipetted into 4.5 ml of liquid KpR medium with 2 mg 2,4-D/liter. The 5-cm diameter petri dishes are then incubated in the dark at 25°. Experiments have demonstrated that using a chloride-free EPR buffer[9] containing 75 m*M* aspartic acid, 75 m*M* glutamic acid, and 4 m*M* calcium gluconate instead of 150 m*M* NaCl and 4 m*M* $CaCl_2$ increases the efficiency of transient expression approximately twofold.

Analyzing Gene Expression in Protoplasts or Cells

A number of reporter systems are available, varying in convenience and sensitivity. Most of our work is first done using the neomycin phosphotransferase II (*npt*II) gene. No plant species investigated to date has been found to have a phosphotransferase activity comigrating with NPTII, so that even weak expression systems can be monitored reliably.

NPTII Assay

Preparation of Samples from Transformed Tissues. For assaying callus or plants, transfer approximately 0.1 to 0.2 g of tissue to a 1.6-ml Eppendorf centrifuge tube. Crush tissue in 0.1 to 0.2 ml of 2× extraction buffer using a metal or plastic rod. The tissue is generally homogenized within 2–3 min. At that time, spin 3 min at 4°, ≥9000 *g*, to sediment particulate debris and pipette supernatant to second tube. Chill on ice until needed.[28]

To assay leaf bases, the liquid medium is carefully removed and all leaf bases are crushed within one Eppendorf tube together with 0.15 ml 2× extraction buffer. To assay protoplasts, the medium containing the

[27] M. Fromm, L. P. Taylor, and V. Walbot, *Proc. Natl. Acad. Sci. U.S.A.* **82,** 5824 (1985).
[28] B. Reiss, R. Sprengel, H. Will, and H. Schaller, *Gene* **30,** 211 (1984).

cells is gently pipetted into a 10-ml centrifuge tube, and centrifuged for 3 min at 80 *g*. This pellets the protoplasts at the bottom of the tube. The overlaying KpR medium is slowly removed with a capillary tube connected to a peristaltic pump, leaving behind 75 to 100 μl. Next, the protoplasts are gently resuspended by moving the tube between thumb and fingers, and transferred to an Eppendorf tube by using a wide-mouthed 200- or 1000-μl tip pipettor. After addition of 0.25 vol (about 30 μl) of a 4× extraction buffer, the protoplasts are immediately chilled on ice. The protoplasts are broken by sonication for 5 to 6 sec. The sonicator probe should be set just below the surface of the liquid, and the amplitude set at 10 to 15 μm from peak to peak. To remove the cell debris, centrifuge the Eppendorf tubes at 4° for 1 min at ≥9000 *g* in an Eppendorf centrifuge and withdraw the supernatant (containing the NPTII protein) to a precooled Eppendorf tube; store on ice.

To compare the NPTII activity in different samples, equal amounts of cellular proteins are used in the NPTII assay. The protein content may be measured by the Bio-Rad (Richmond, CA) protein assay. Five microliters of each sample is mixed with 800 μl distilled water and 200 μl Bio-Rad dye, and transferred to a spectrophotometer cuvette. Before measuring the absorbance at 595 nm the reaction is allowed to proceed for 5 to 10 min. Absorption of the sample should be compared to a blank containing 5 μl of extraction buffer. If desired, a standard curve can be obtained using different amounts of bovine serum albumin (1 to 20 μg). For each NPTII determination, transfer a similar amount of protein to a precooled Eppendorf tube (we routinely load 50 μg of protein), and mix with 0.1 vol of 10× loading buffer before loading on the nondenaturing protein gel.

Preparation of Nondenaturing Gel. To prepare the protein gels, glass plates (15 × 17 cm) are thoroughly cleaned with detergent, then rinsed successively with tap water and ethanol, and dried (avoid contact between plates and skin). The plates are assembled with appropriate spacers (thickness, 1 mm) and clamps and, if necessary, sealed at the outside with a 1% (w/v) agarose solution. Next, the separation gel is poured between the glass plates up to 5 cm from their top. After carefully covering the gel solution with 0.5 cm of distilled water, the gel is allowed to polymerize (approximately 30 min at room temperature). Then the water is removed and the stacking solution is poured between the glass plates. A comb (thickness, 1 mm) with wells that can contain 100 μl of sample is inserted (avoid trapping air bubbles under the comb) and the gel is allowed to polymerize. After gently removing the comb and the bottom spacer, the gel is placed in an electrophoresis tank filled with running buffer and cooled for 1 hr before loading. The samples are loaded into the wells and run at 4–10° at 15 V/cm until the bromphenol blue marker has migrated 10 cm into the separation gel (approximately 5 hr). Then the two glass plates are laid horizontally and gently separated. Using a scalpel blade, the upper 10 cm of the separation gel is isolated and given three 15-min washes (with shaking) in 300 ml freshly made reaction buffer (chilled at 4°).

Phosphotransferase Assay. Meanwhile, 500 mg of agarose (Sigma type II) is added to 50 ml reaction buffer, rapidly heated in a microwave oven to dissolve the agarose, and cooled to 45° in a water bath. Subsequently, in a disposable beaker or tube, add 10 μl [γ-^{32}P]ATP (100 μCi, 3000 Ci/mmol) and 100 μl of a freshly prepared aqueous kanamycin sulfate stock (50 mg/ml) to 30 ml of the agarose and mix thoroughly. Lay the rinsed separation gel into a matching tray or box, and let it equilibrate to room temperature. Next, pour the agarose solution evenly over the separation gel. Wait for the gel to solidify and cover first with one sheet of Whatman (Clifton, NJ) P81 paper and then with two sheets of Whatman 3MM paper (all cut to size and prewet in reaction buffer), a 5-cm stack of paper towels (cut to size), and a 1-kg weight evenly distributed over the top of the towels. After 3 to 12 hr of blotting, discard the paper towels and remove the Whatman P81 paper. Wash this four times with 200 ml of washing solution for 15 min each, at 80°. The Whatman P81 paper is then covered with thin plastic wrap, and exposed to X-ray film (XAR; Kodak, Rochester, NY) using an intensifying screen at −70°. Exposure time varies from 10 min to 72 hr, depending on the strength of the NPTII signal. As controls for the NPTII assay, we used protein extracts from an untransformed tobacco plant (blank) and a tobacco plant transformed with the P2′-*npt*II construct (positive control). NPTII activity is detected by production and binding of phosphorylated kanamycin to the Whatman P81 paper. One often sees two slower migrating bands above this that are also detected in untransformed material and probably correspond to phosphorylated proteins.

To quantitate the NPTII activity, the autoradiogram is matched with the Whatman P81 paper, the NPTII spots are located, cut out with a scalpel blade, air dried, submerged in scintillation liquid, and counted in a Beckman (Palo Alto, CA) scintillation counter.

Conclusions

Of the six promoters tested in protoplasts, and the four in leaf bases (Table IV), the two strongest and least influenced by the source of protoplast or explant are derived from the mannopine synthase *2′* gene and the cauliflower mosaic virus (CaMV) 35S transcript. It is useful to note that the ratio of activities of the P*35S* and P*nos* constructs measured in transient expression assays[23] (approximately 3.6) is similar to the ratio of transfor-

TABLE IV
RELATIVE NPTII ACTIVITY (%) PRODUCED BY DIFFERENT CHIMERIC GENES IN PROTOPLASTS AND LEAF BASES[a]

Promoter fused to *npt*II gene (reference)	Protoplasts from		Explants from	
	Suspension cultures	Leaves	Leaf bases (L/D)[b]	Leaves (L/D)
None (23, 26)	ND	ND	ND/ND	ND/ND
P*2′* (23, 26)	100	100	93/100	20/100
P*35S* (23, 26)	40	—	95/98	26/96
P*nos* (23)	11	8	—	—
P*1′* (23)	10	—	—	—
P*z4* (23)	0.4	0.4	—	—
P*4.7* (23)	0.2	0.2	—	—
Pextensin (26)	—	—	—/6	—
P*cab* (26)	—	—	ND/ND	83/ND

[a] In each column, all experiments have been normalized to P*2′*-driven NPTII activity. (Reproduced from Dekeyser *et al.*[26] with permission by the American Society of Plant Physiologists.)

[b] Leaf bases or leaves have been pregrown and maintained in light (L) or dark (D) for 4 days after electroporation. ND, Not detected.

mation frequencies obtained with each promoter under selective conditions (2.6 and 3.0).[6] This indicates transient assays measure parameters that are useful for predicting transformation efficiencies.

We have measured the growth rates of calli containing each of four different selectable markers (P*nos* : *npt*II, P*2′* : *npt*II, P*35S* : *bar,* and P*1′* : *dhfr*) on different concentrations of the corresponding selective agents. These calli contained similar numbers of inserts,[23] and as seen in Table III were isolated at comparable frequencies, although some were selected on 200 μg kanamycin/ml, and some with 10 μg phosphinothricin/ml. Similarly, there were comparable levels of NPTII activity in the calli analyzed, irrespective of how they were initially selected.[23]

The first relevant observation made on the regenerate material was that there was no significant difference in the growth rates of transformed and untransformed calli on 0, 100, or 500 μg kanamycin/ml. On the other hand, the same transformed lines grew 8–10 times faster than untransformed calli on both 100 and 500 μg G418/ml, for which rice has no endogenous resistance. Thus, to minimize the frequency of escapes from selection, one should use G418 instead of kanamycin whenever possible.

Calli containing the P*35S* : *bar* construct also grew well under selection, in fact, 10–15 times faster than untransformed calli on 10 or 50 μg of phosphinothricin/ml. Calli containing P*1'* : *dhfr* did not grow rapidly on either 2 or 10 μg methotrexate/ml. All untransformed material died within 1 week. Either of these selection systems could, therefore, be used in place of the aminoglycoside detoxification system. However, it must be noted that the majority of fertile, transgenic plants obtained to date have been isolated using P*35S* : *hpt*[3,5,6,29]

Variables in Gene Expression in Transformed Material

The expression of transferred genes often varies from transformant to transformant. Sometimes the new genes are poorly expressed in some lines, or even silent.[29,30] In other cases, apparently homogeneous calli do not express β-glucuronidase activity uniformly,[30] or NPTII-deficient plants regenerate from NPTII-positive material.[2] *A priori,* these phenomena might indicate that the originally selected colonies were in fact polyclonal; however, other factors beyond the control of the investigator might produce similar results. For example, gene expression can be reduced by changes in DNA methylation or organization occurring in response to tissue culture conditions.[31,32] In other cases, activation or inactivation of nearby chromosomal promoters could silence gene transcription or generate antisense transcripts to reduce the expression of the foreign gene. On the other hand, foreign gene expression in most calli stabilizes within 3 to 5 months of culture,[8] and can remain constant in the absence of selection.[2,3,5,6,30]

Although the presence of the selectable marker can sometimes be verified by assaying for the corresponding enzyme activity, the presence of the cotransferred genes that have no overt phenotype can be ascertained only by performing Southern hybridization.[33] To answer questions about the organization of the new genes, one should always use restriction enzymes with targets that are generally not methylated in eukaryotes such as *Bgl*II, *Eco*RI, or *Hin*dIII. Preferably, at least one (or a mixture of two) of the enzymes should be chosen because it cuts twice within the gene under investivation and generates a fragment that must be intact for the gene to function. In appropriate reconstruction experiments, one can compare the intensity of hybridization of probes to genomic DNA and to

[29] R. Terada and K. Shimamoto, *Mol. Gen. Genet.* **220,** 389 (1990).

[30] W. Zhang and R. Wu, *Theor. Appl. Genet.* **76,** 835 (1988).

[31] P. T. H. Brown, J. Kyozuka, Y. Sukekiyo, Y. Kimura, K. Shimamoto, and H. Lörz, *Mol. Gen. Genet.* **223,** 324 (1990).

[32] E. Müller, P. T. H. Brown, S. Hartke, and H. Lörz, *Theor. Appl. Genet.* **80,** 673 (1990).

[33] T. Maniatis, E. F. Fritsch, and J. Sambrook, "Molecular Cloning: A Laboratory Manual." Cold Spring Harbor Laboratory, Cold Spring Harbor, New York, 1980.

similarly cut plasmid to estimate the copy number of the foreign gene. A single-copy equivalent of a 10-kb sequence in rice, assuming a diploid genome size of 4×10^8 bp, is approximately 2.5×10^{-5} μg/μg genomic DNA. Similar analyses with a second restriction enzyme that cuts once, or not at all, in the transferred DNA can clarify the number of independent integration sites of the gene in the host sequence.

We have analyzed seven transformed rice callus lines using *Bgl*II, *Eco*RI, and a mixture of *Eco*RV and *Bst*EII. Forty percent had single copies of the selected markers. Neither these transformants, nor others listed in Table IV, showed a consistent correlation between copy number and either gene expression or levels of resistance. Because deletions and mutations may be common during integration of foreign sequences, it is possible many of these genes are not expressed.

Acknowledgments

This work was supported by grants from the Rockefeller Foundation (RF 86058 #59) and the Services of the Prime Minister (U.I.A.P. #120CO187). R.D. was a Research Assistant of the National Fund for Scientific Research (Belgium).

[38] Thaumatin II: A Sweet Marker Gene for Use in Plants

By MICHAEL WITTY

Introduction

Thaumatin II is a natural product of the West African plant *Thaumatococcus daniellii*. The fruit of this plant produces a family of five or more extremely sweet proteins[1] called thaumatins that are traditionally used by West Africans to sweeten food and beverages.[2] In 1839 Daniell, the first European to write of thaumatin, described it as having ". . . an indescribable yet intense degree of dulcidity. . . ."[2] Thaumatin has long been a substance well known for the intensity of its taste and the small amount needed to produce an obvious sensation. Thaumatin solutions taste sweet at concentrations as low as 10^{-8} *M*.[3] Thaumatins also lower the taste

[1] H. van der Wel and K. Loeve, *Eur. J. Biochem.* **31,** 221 (1972).

[2] W. F. Daniell, *Pharm. J.* **14,** 158 (1855).

[3] J. D. Higginbotham, *in* "Developments in Sweeteners I," p. 87. Applied Science Publ., London, 1979.

threshold of many sweet and savory compounds.[4] It is therefore not surprising that the food processing industry has found several uses for the sweet-tasting *T. danielli* fruit extracts.[5] The two most abundant thaumatins, thaumatin I and thaumatin II, have been well characterized.[6,7] Thaumatin II cDNA has been cloned and expressed in *Solanum tuberosum*. Biologically active recombinant (r)-thaumatin II is produced at quantities above the sweet taste threshold.[8]

Thaumatin protein has potential as a food additive[4] and the thaumatin gene may have potential as a palatability gene for the improvement of crop cultivars.[9] The thaumatin gene may have a role in biotechnology other than as a palatability gene. It has potential as a marker gene.[10]

Principle of Method

The function of marker genes is to allow easy identification of transgenic plants. The most commonly used plant marker genes, coding for NPTII, CAT, OCS, GUS,[11,12] and BSR,[13] marker genes,[14] all require sophisticated tissue culture facilities, trained workers, or expensive reagents for selection procedures. Transgenic plants propagated on a large scale in the field should have a simple marker gene for the easy identification of transgenic plants and their progeny. A sensitive and extremely convenient candidate is the CaMV *35S*-5′ (cauliflower mosaic virus *35S* 5′ region)–preprothaumatin II–*tml*-3′ [3′ region of the tumor size (*tml*) locus] chimera. In the absence of such a convenient marker gene identification of transgenic plant progeny would have to be carried out using expensive laboratory techniques. This chapter describes how the CaMV-*35S*-5′–preprothaumatin II–*tml*-3′ chimera may be used as a marker gene.

[4] J. D. Higginbotham, M. Lindley, and P. Stephens, *in* "The Quality of Foods and Beverages" (G. Charalambous and G. Inglett, eds.), Vol. 1, p. 91. Academic Press, New York, 1981.

[5] P. Stephens, *Food (London)* **March** (1983).

[6] R. B. Iyengar, P. Smits, F. van der Ouderaa, H. van der Wel, J. van Brouwershaven, P. Ravenstein, G. Richters, and P. D. Wassenaar, *Eur. J. Biochem.* **96,** 193 (1979).

[7] A. M. de Vos, M. Hatada, H. van der Wel, H. Krabbendam, A. F. Peerdeman, and S. H. Kim, *Proc. Natl. Acad. Sci. U.S.A.* **82,** 1406 (1985).

[8] M. Witty and W. J. Harvey, *N. Z. J. Crop Hortic. Sci.* **18,** 77 (1990).

[9] M. Witty, *Trends Biotechnol.* **8,** 113 (1990).

[10] M. Witty, *Nucleic Acids Res.* **17,** 3312 (1989).

[11] K. Weising, J. Schell, and G. Kahl, *Annu. Rev. Genet.* **22,** 421 (1988).

[12] L. Herrera-Estrella and J. Simpson, *in* "Plant Molecular Biology: A Practical Approach" (C. H. Shaw, ed.), p. 131. IRL Press, 1988.

[13] NPTII, Neomycin phosphotransferase II; CAT, chloramphenicol acetyltransferase; OCS, octopine synthase; GUS, β-glucuronidase; BSR, blasticidin-S deaminase.

[14] T. Kamakura, K. Yoneyama, and I. Yamaguchi, *Mol. Gen. Genet.* **223,** 332 (1990).

Materials and Reagents

Permission to use thaumatin II cDNA was granted by Unilever Research Laboratories (Vlaardingen, The Netherlands). A mixture of thaumatin I and II was kindly given by Dr. Ledeboer. Mixed thaumatin I and thaumatin II was found suitable for producing antibodies for use in this work, presumably because thaumatin I and thaumatin II are immunologically similar.[15,16] The method used was essentially that of van der Wel and Bel. Antiserum produced using mixed thaumatin I and thaumatin II antigen does not cross-react with wild-type *S. tuberosum* plant extracts. A purified thaumatin I and II mixture can be purchased from the Sigma Chemical Company (St. Louis, MO). Food-grade thaumatin was a kind gift of J. D. Higginbotham. Food-grade thaumatin or Talin can be purchased from Tate and Lyle, Ltd. U.K. All other materials were obtained from standard sources.

Methods

The binary plasmid vector pWIT2 (Fig. 1), which contains a CaMV *35S*–preprothaumatin II–*tml*-3′ chimera, is engineered using conventional published methods[17] and transferred to *Agrobacterium rhizogenes* A4T by triparental mating. Axenic *S. tuberosum* cv. '*Iwa*' plants are decapitated and then inoculated with *A. rhizogenes*:pWIT2. After about 10 days, hairy roots approximately 1 cm long are excised and incubated on MS (Murashige and Skoog) medium plus 3% (w/v) sucrose and 250 mg/liter cefotaxime for 2 days. The hairy roots are then transferred to the same medium plus 25 mg/liter kanamycin for selection and propagation of independent hairy root lines.

Transformed hairy roots are incubated for 1 month on MS medium plus 1% (w/v) sucrose, 250 mg/liter casamino acids, 1 mg/liter indole-3-acetic acid (1AA), 1 mg/liter N^6-benzyladenine (BAP), 10 mg/liter gibberellin A_3 (GA_3), and 25 mg/liter kanamycin.[18] After 1 month shoots appear, which are then rooted on MS medium plus 3% (w/v) sucrose, 250 mg/liter casamino acids, and 0.5 mg/liter kinetin. Whole plants are transferred to soil and grown to maturity in glasshouses.

Detection of transgenic tissue lines producing thaumatin is simple. A small amount, approximately 0.1 g, is tasted. The taste of thaumatin-producing tissues is obvious: a characteristic slowly developing and in-

[15] H. van der Wel and W. J. Bel, *Chem. Senses Flavor* **3,** 99 (1978).
[16] C. A. M. Hough and J. A. Edwardson, *Nature* (*London*) **271,** 381 (1978).
[17] M. Witty, *Biotechnol. Lett.* **12,** 131 (1990).
[18] N. O. Espinoza and J. H. Dodds, *Plant Sci.* **41,** 121 (1985).

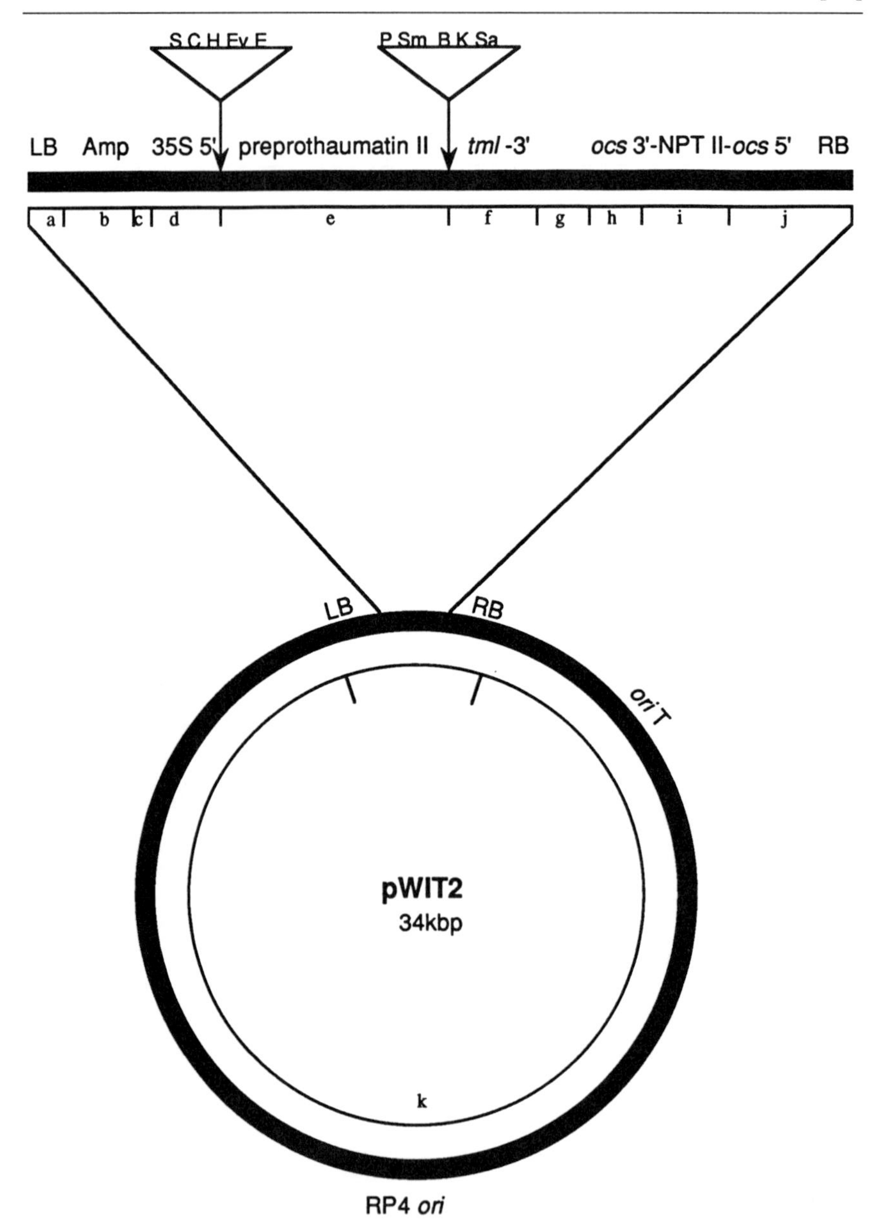

FIG. 1. pWIT2. *Amp,* Ampicillin resistance gene; LB, T-DNA, left border; RB, T-DNA, right border; C, *Cla*I; B, *Bam*HI; E, *Eco*RI; Ev, *Eco*RV; H, *Hin*dIII; K, *Kpn*I; P, *Pst*I; S, *Sal*I; Sa, *Sac*I; Sm, SmaI. (a) Left T-DNA border, 625 to 2212. From R. F. Barker, K. B. Idler, D. V. Thompson, and J. D. Kemp, *Plant Mol. Biol.* **2,** 335 (1983). (b) and (g) are the result of homologous recombination between the pUC-*cat* region of pCGN587 and the pUC19

tensely sweet taste and aftertaste. Some experience of the taste of thaumatin should be gained using dilute solutions of Talin before transgenic tissue is tasted. Because of the persistence of the taste of thaumatin this practice tasting is best carried out the day before screening. Control tissue should be tasted before putative transformants to gain experience of the taste of ordinary tissue. With some practice, small pieces of thaumatin-producing tissues can be identified. Usually this simple procedure is sufficient to identify transformed lines. However, statistical significance for taste observations, which can be used as published evidence of thaumatin production, can be gained using sensory analysis, if this is thought necessary.

The example presented below more precisely defines sensory analysis of transgenic *S. tuberosum* hairy root lines expressing the CaMV *35S*-5′–preprothaumatin II–*tml*-3′ chimera. A triangle-type difference test[19] was used to compare 0.1-g samples of hairy root lines 1B, 1C, 1D, 2A, 2B, 2D, and 1Y with ordinary hairy root tissue as controls. The tissue was ground in liquid nitrogen, weighed as a dry powder, thawed, centrifuged by accelerating to 12,000 rpm, and quickly frozen again to form a compact pellet that could be easily shaken out of an Eppendorf tube. This procedure was necessary to produce samples uniform in composition and appearance. Twelve panelists, known to be able to detect 100 μl of 10^{-7} M Talin in distilled water, were asked to identify the uniquely different 0.1-g sample among three and to describe how the like samples differed from the rest. The number of panelists correctly identifying thaumatin-producing tissue lines was as follows:

Line:	1B	1C	1D	2A	2B	2D	1Y
Number:	10	10	12	11	11	11	12

[19] A. Kramer and B. A. Twigg, "Quality Control for the Food Industry," Vol. 1, 3rd Ed. AVI Publishing, 1970.

region of pWIT10. (c) Tn*5* DNA, *Hin*dIII (1196) to *Bgl*II (1516). (d) CaMV *35S* promoter region, 7144 to 7544. R. C. Gardner, A. J. Howarth, P. Hahn, M. Brown-Luedi, R. J. Shepherd, and J. Messing, *Nucleic Acids Res.* **9,** 2871 (1981). (e) Preprothaumatin II cDNA. (f) *tml*-3′, 9062 to 11207. (h) Octopine synthase (*ocs*) 3′ region, 11207 to 12823. From J. J. Fillatti, J. Sellmer, B. McCown, B. Haissig, and L. Comai, *Mol. Gen. Genet.* **206,** 192 (1987). (i) Neomycin phosphotransferase II coding region, 1196 to 1516 of Tn*5*. (j) Octopine synthase 5′ region and right T-DNA border, 13643–15208. (k) pRK290 *Sal*I to *Eco*RI region, including the RP4 origin of replication and *ori* T functions.

By consulting the published statistical tables of Kramer and Twigg it is seen that these results show ($p < 0.1\%$) that the characteristic taste of thaumatin, which panelists had been trained to recognize, was detected in all the plant cell lines.

It should be noted that sensory analysis is notoriously sensitive to bias. Perception is proportional to expectation. For this reason experiments should be performed carefully, under the supervision of experienced food scientists whenever possible. A particular problem in conducting taste tests using thaumatin is the persistance of its taste. This causes carryover of taste sensation from one sample to the next. A compromise must be found between lengthening the time between tastings and making sure that not too much time has elapsed or panelists will forget the tastes they have just experienced.

The sweet phenotype has been observed in plants grown to maturity. The taste of r-thaumatin had been observed in every tissue type examined by the author, including root, hairy root, tuber, stolon, stem, and leaf. Southern blots have shown the presence of DNA homologous to the thaumatin chimera; Western blots confirmed the presence of thaumatin antigenic matter; protein slot blots have been used to determine that r-thaumatin production is on the order of 10^{-7} *M*.[17]

Concluding Remarks

When tasting cultured plant material toxicity of the growth medium should be considered. For example, the herbicide 2,4-D (2,4-dichlorophenoxyacetic acid) is a common growth factor used in plant tissue.[20] An advantage of hairy root culture is that growth factor-free medium can be used to generate samples for tasting. Some plants produce toxins that should be treated with caution. Green tissues of *S. tuberosum* produce poisonous alkaloids, including solanine and chaconine,[21] and *Nicotiana* tissues produce nicotine in culture.[22] *Agrobacterium* is not considered to be hazardous to human health.[23,24] However, the hairy roots used in sensory analysis had been cultivated for several weeks on antibiotic-free medium and were considered to be *Agrobacterium* free. Thaumatin protein

[20] S. S. Bhojwani, "Plant Tissue Culture," p. 32. Elsevier, Amsterdam, 1983.

[21] M. R. Cooper and A. W. Johnson, "Poisonous Plants in Britain and Their Effects of Animals and Man," p. 20. Ministry of Agriculture, Fisheries, and Food, Reference Book 161, 1984.

[22] M. J. C. Rhodes, M. Hilton, A. J. Parr, J. D. Hamill, and R. J. Robins, *Biotechnol. Lett.* **8,** 415 (1986).

[23] H. Lautrop, *Acta Pathol. Microbiol. Scand. Suppl.* **187,** 63 (1967).

[24] P. S. Riely and R. E. Weaver, *J. Clin. Microbiol.* **5,** 172 (1977).

has been rigorously examined and found to be safe for human consumption.[25]

Panelists were able to detect thaumatin in 0.1-g samples despite the astringent background taste of the hairy root tissue, and the very small sample size. The transgenic r-thaumatin samples could survive the rigorous treatment involved in taste testing without the sweet taste factor being destroyed. Expression of the thaumatin II gene under control of the CaMV *35S*-5′ and *tml*-3′ elements causes accumulation of thaumatin protein sufficient to cause an obvious change in phenotype in a higher plant. This change of phenotype is sufficiently marked for it to be used to detect transgenic tissue lines or whole plants. It is believed that this marker gene would be of particular use in the agricultural application of new plants generated by genetic manipulation.

[25] J. D. Higginbotham, D. J. Snodin, K. K. Eaton, and J. W. Daniel, *Food Chem. Toxicol.* **21,** 815 (1983).

[39] Transformation of Filamentous Fungi Based on Hygromycin B and Phleomycin Resistance Markers

By PETER J. PUNT and CEES A. M. J. J. VAN DEN HONDEL

Introduction

Filamentous fungi are attractive organisms in which to study biological processes at a molecular level. Many filamentous fungi display a considerable metabolic diversity. Especially for *Aspergillus nidulans* and *Neurospora crassa,* extensive biochemical and genetic information is available concerning various metabolic pathways. Therefore these organisms are well suited for a molecular biological study of cellular metabolism. Filamentous fungi are also suitable for the study of cellular differentiation, because they have a complex life cycle with distinct cellular differentiation.

One of the first requirements to carry out modern molecular biological research is the availability of a DNA-mediated transformation system to generate genetically modified cells that have taken up DNA of interest. Since the first description of such a system for the filamentous fungus *N. crassa*[1] many other transformation systems have been developed.[2,3]

To permit selection of transformed cells, markers are used that are capable of complementing a mutation (auxotrophic markers) or that provide a new property to the host cell, such as antibiotic resistance (dominant selectable markers). The latter type of selection has the advantage that mutant strains are not required. This implies that genetically uncharacterized species [among which many (plant) pathogenic and industrially used species] can also be transformed using dominant selectable markers.

Two of these markers, which are widely used for fungal transformation, provide resistance against hygromycin B (HmB) and phleomycin (Phle). Hygromycin B is an aminoglycosidic antibiotic that disturbs protein synthesis by interfering with peptidyl-tRNA translocation and causing misreading.[4] Hygromycin B resistance genes, isolated from *Streptomyces hygroscopicus*[5] and *Escherichia coli*,[6] encode HmB phosphotransferases that inactivate the antibiotic by phosphorylation.

Phleomycin is a metalloglycopeptide antibiotic causing DNA strand scission. Phleomycin resistances genes, isolated from *Streptoalloteichus hindustanus* and *E. coli*, encode proteins that inactivate the antibiotic by binding to it.[7] In this chapter we describe fungal transformation based on the HmB resistance gene of *E. coli* (*hph*) and the Phle resistance gene of *S. hindustanus* (*ble*).

Materials

Strains and Plasmids

As recipients for transformation experiments the following strains (or derivatives thereof) are used: *A. nidulans* FGSC4, *Aspergillus niger*

[1] M. E. Case, M. Schweizer, S. R. Kushner, and N. H. Giles, *Proc. Natl. Acad. Sci. U.S.A.* **76,** 5259 (1979).

[2] T. Goosen, C. J. Bos, and H. W. J. van den Broek, *in* "Handbook of Applied Mycology, Fungal Biotechnology" (D. K. Arora, K. G. Mukerji, and R. P. Elander, eds.), Vol. 4, p. 151. Dekker, New York, 1992.

[3] C. A. M. J. J. van den Hondel, and P. J. Punt, *in* "British Mycological Society Symposium Volume 18: Applied Molecular Genetics of Fungi" (J. F. Peberdy, C. E. Caten, J. E. Ogden, and J. W. Bennett, eds.), p. 1. University Press, Cambridge, 1992.

[4] A. Gonzalez, A. Jimenez, D. Vasquez, J. E. Davies, and D. Schindler, *Biochim. Biophys. Acta* **251,** 459 (1978).

[5] F. Malpartida, M. Zalacain, A. Jimenez, and J. Davies, *Biochem. Biophys. Res. Commun.* **117,** 6 (1983).

[6] L. Gritz and J. Davies, *Gene* **25,** 179 (1983).

[7] A. Gatignol, H. Durand, and G. Tiraby, *FEBS Lett.* **230,** 171 (1990).

N402,[8] *A. niger* CBS513.88,[9] *A. niger* ATCC1015,[10] *Aspergillus ficuum* NRRL3135,[11] *Aspergillus oryzae* IMI44242,[12] and *Penicillium chrysogenum* ATCC48271[13] (Table I).

In most transformation experiments vectors pAN7-1[8] and pAN8-1[14] are used.

Antibiotics

Hygromycin B is purchased from Calbiochem (La Jolla, CA), Phleomycin from Cayla (Toulouse, France). Stock solutions (50–200 mg/ml) of both antibiotics are prepared in water. These stocks can be stored for several months at 4° without any noticeable change in activity.

Other Materials

Protoplasting enzyme (NovoZym 234) is obtained from NOVO biolaboratories (Bagsvaerd, Denmark) and polyethylene glycol from either Serva (Heidelberg, Germany) (PEG 4000) or Merck (Rahway, NJ) (PEG 4000–6000). All other reagents used are analytical grade.

Method

For transformation of a fungal species with the aid of an antibiotic resistance marker the following aspects are of importance: (1) sensitivity of the host strain to the antibiotic, (2) preparation of protoplasts that are competent to take up vector DNA and are capable of subsequent regeneration, (3) expression of the antibiotic resistance gene in the fungal host, by fungal expression signals, and (4) selection procedures for optimal transformation frequencies.

[8] P. J. Punt, R. P. Oliver, M. A. Dingemanse, P. H. Pouwels, and C. A. M. J. J. van den Hondel, *Gene* **56,** 117 (1987).

[9] W. van Hartingsveldt, C. M. J. van Zeijl, A. E. Veenstra, J. A. van den Berg, P. H. Pouwels, R. F. M. van Gorcom, and C. A. M. J. J. van den Hondel, "Proceedings of the 6th International Symposium on Genetics of Industrial Microorganisms," p. 107. Strasbourg, 1990.

[10] R. F. M. van Gorcom, J. G. Boschloo, A. Kuyvenhoven, J. Lange, A. J. van Vark, C. J. Bos, J. A. M. van Balken, and C. A. M. J. J. van den Hondel, *Mol. Gen. Genet.* **223,** 192 (1990).

[11] E. J. Mullaney, P. J. Punt, and C. A. M. J. J. van den Hondel, *Appl. Microbiol. Biotechnol.* **28,** 451 (1988).

[12] I. E. Mattern, S. Unkles, J. R. Kinghorn, P. H. Pouwels, and C. A. M. J. J. van den Hondel, *Mol. Gen. Genet.* **210,** 460 (1987).

[13] M. Kolar, P. J. Punt, C. A. M. J. J. van den Hondel, and H. Schwab, *Gene* **62,** 127 (1988).

[14] I. E. Mattern, P. J. Punt, and C. A. M. J. J. van den Hondel, *Fungal Gen. Newslett.* **35,** 25 (1988).

TABLE I
ANTIBIOTIC RESISTANCE OF VARIOUS FUNGAL STRAINS[a]

	Hygromycin B concentration (μg/ml)						Phleomycin concentration (μg/ml)					
Strain	0	10	50	100	500	1000	0	2	5	10	50	100
A. nidulans FGSC4	+	+	+	+	+	+	+	+	+	−	−	−
A. niger N402	+	+	−	−	nt	nt	+	+	−	−	−	−
A. oryzae IMI44242 *nia*							+	+	+	+	±	−
P. chrysogenum ATCC48271	+	+	+	+	+	+*	+	+	+	−	nt	nt

[a] Diluted spore suspensions are plated on solidified minimal media containing various concentrations of the antibiotic: +, significant growth; +*, significant growth but no sporulation; ±, sparse growth; −, no (or very little) growth; nt, not tested.

Antibiotic Sensitivity

Although most filamentous fungi are relatively resistant to many antibiotics, preliminary experiments in our laboratory indicated that *A. niger* is relatively sensitive to HmB and Phle. To test the resistance level of this and various other fungal strains, diluted spore suspensions are plated on solidified growth media containing 0–1000 μg/ml HmB or 0–100 μg/ml Phle. In this way, differences in resistance level are observed between species (Table I). It should be noted that the composition of the growth medium used for resistance tests may influence the results of these tests. Resistance levels on complete medium [containing 0.5% (w/v) yeast extract and 0.2% (w/v) casamino acids] are higher (up to twofold) than on minimal media. Furthermore, Phle has a reduced antibiotic activity in acidic media (pH < 5) and in hypertonic media (as used for protoplast regeneration).[7,15] The antibiotic activity of HmB is decreased in hypertonic media with ionic osmotic stabilization.[16]

In general, to prevent outgrowth of protoplasts on hypertonic media we use a twofold higher antibiotic concentration than required to prevent growth of spores on isotonic media (Table I).

Transformation Procedure

Fungal transformation procedures are generally based on preparation of protoplasts and subsequent PEG/$CaCl_2$-mediated DNA uptake. The

[15] B. Austin, R. M. Hall, and B. M. Tyler, *Gene* **93,** 157 (1990).
[16] D. Cullen, S. A. Leong, L. J. Wilson, and D. J. Henner, *Gene* **57,** 21 (1987).

following procedure, based on the methods described by Goosen *et al.*[17] and Yelton *et al.*[18] is routinely used in our laboratory for transformation of various fungal species (e.g., *Aspergillus, Penicillium*).

1. Conidiospores are inoculated (to a final concentration of $1-2 \times 10^6$/ml) in complete medium [70 m*M* $NaNO_3$, 7 m*M* KCl, 11 m*M* KH_2PO_4, 2 m*M* Mg_2SO_4, 1% (w/v) glucose, 0.5% (w/v) yeast extract, 0.2% (w/v) casamino acids], containing trace elements [1000 × stock: 76 m*M* $ZnSO_4$, 178 m*M* H_3BO_3, 25 m*M* $MnCl_2$, 18 m*M* $FeSO_4$, 7.1 m*M* $CoCl_2$, 6.4 m*M* $CuSO_4$, 6.2 m*M* Na_2MoO_4, 174 m*M* ethylenediaminetetraacetic acid (EDTA)] and vitamins (1000 × stocks: 100 mg/liter thiamin, 100 mg/liter riboflavin, 100 mg/liter nicotinamide, 50 mg/liter pyridoxine, 10 mg/liter pantothenic acid, 0.2 mg/liter biotin). Auxotrophic *Aspergillus* strains (especially Arg^-, Pyr^-, Trp^-) require additional supplementation (with arginine, uridine, and tryptophan, respectively). The cultures are incubated 16–20 hr at 30° with vigorous shaking [300–400 rpm in a New Brunswick (Edison, NJ) incubator], yielding 1–5 g mycelium (wet weight) per 100 ml.
2. Mycelium is collected by filtration through sterile Myracloth (Calbiochem, La Jolla, CA), rinsed with and resuspended in 0.27 *M* $CaCl_2$, 0.6 *M* NaCl, 1.7 Osm (1 g mycelium/20 ml).
3. Protoplasts are prepared by incubation with NovoZym 234 at a final concentration of 1–5 mg/ml at 30° with slow agitation (50–100 rpm). At 30-min intervals protoplast formation is checked on a microscope. When many free protoplasts are observed and the mycelium is not yet completely broken down into small fragments the suspension is put on ice (for *A. niger* usually after 60–120 min). Protoplast yields of 10^7-10^8 protoplasts/g mycelium are routinely obtained in this way. In other experiments Caylase C3 (Cayla) was successfully used as protoplasting enzyme, resulting in similar protoplast yields as observed for NovoZym 234.
4. Protoplasts are separated from the mycelium by filtration through sterile Myracloth or gauze. The protoplast suspension is diluted (1 : 1) in 1.2 *M* sorbitol, 10 m*M* Tris-HCl (pH 7.5), 50 m*M* $CaCl_2$, 35 m*M* NaCl, 1.7 Osm (STCl700), and incubated on ice for 5–10 min.
5. Protoplasts are collected by centrifugation (3000 rpm, 10 min, 0°) and washed twice with STC1700. Finally, the protoplasts are resuspended (10^7-10^8/ml) in STC1700.
6. Protoplast suspension (100–150 μl) is mixed with 1–20 μl of transforming DNA (5–10 μg) and incubated at 20–25° for 20–30 min. The

[17] T. Goosen, G. Bloemheuvel, C. Gysler, D. A. de Bie, H. W. J. van den Broek, and K. Swart, *Curr. Genet.* **11,** 499 (1987).

[18] M. M. Yelton, J. E. Hamer, and W. E. Timberlake, *Proc. Natl. Acad. Sci. U.S.A.* **81,** 1470 (1984).

addition of 10–20 mM aurin tricarboxylic acid (a DNase inhibitor) during this incubation results in a two- to fourfold increase of the final transformation frequency. Subsequently, in three steps 250, 250, and 850 μl of 60% PEG 4000 (or PEG 6000) in 10 mM Tris-HCl (pH 7.5), 50 mM $CaCl_2$ are carefully mixed with the DNA–protoplast mixture. For transformation with high molecular weight DNA (>50 kb) 25% PEG instead of 60% is used in the first step; in this way aggregation of the DNA is avoided. Subsequently, the suspension is incubated for 20 min at 20–25°.

7. Polyethylene glycol-treated protoplast suspensions are diluted by addition of 5–10 ml of STC1700 and the protoplasts are collected by centrifugation (10 min, 3000 rpm, 0°) and resuspended in 500 μl of STC1700. About 100–200 μl of this suspension is plated onto osmotically stabilized (1.2 M sorbitol) (selective) agar medium.

8. Protoplast viability is determined by plating serial dilutions (10^{-3}–10^{-6}) of the protoplast suspension before and after PEG treatment onto solidified nonselective agar media, with and without 1.2 M sorbitol.

Transformation Vectors

To accomplish expression in fungal hosts, (prokaryotic) HmB and Phle resistance genes are provided with fungal expression signals. In our transformation vectors the 5′-flanking sequences of the efficiently expressed *gpdA* gene of *A. nidulans* are used to drive expression of the resistance genes. Furthermore, the 3′-flanking sequences of the *trpC* gene of *A. nidulans* are used as terminator sequences (Fig. 1).

Selection Procedures

Selection of cells that have taken up the marker gene may be carried out by plating protoplasts directly onto agar media containing the selective agent at concentrations high enough to prevent background outgrowth of nontransformed protoplasts. However, this way of (direct) selection may also prevent outgrowth of transformed cells in which the marker gene is not yet expressed at a sufficiently high level at the time of selection, as was described for HmB selection in *Saccharomyces cerevisiae*. To overcome this problem, cells treated with transforming DNA are plated onto nonselective media and after several hours of incubation a top layer of selective medium is applied.[19]

Hygromycin B Selection. Initially, an indirect selection procedure was used in transformation experiments with *A. niger*.[8] In subsequent experiments we have noticed that direct selection is also possible and results in

[19] K. R. Kaster, S. G. Burgett, and T. D. Ingolia, *Curr. Genet.* **8,** 353 (1984).

higher transformation frequencies with *A. niger* as well as with a related species (Table II).

We find that, in contrast to *A. niger*, several fungal species show high natural resistance to HmB (Table I). For example, *A. nidulans* still grows in the presence of 1 mg/ml HmB. Therefore, direct selection with this (high) concentration of HmB is not possible. However, indirect selection provides a way to isolate HmB^r transformants of this species (Table II). For this type of selection DNA-treated cells are plated onto a nonselective bottom layer medium and are overlayed with 2 vol of selective medium (1 mg/ml HmB) after 12–16 hr of incubation at 30–37°. After 2–4 days, transformants grow through the top layer and start to sporulate. After prolonged incubation (>5 days) untransformed mycelium also grows through and sporulates. Similar indirect procedures may also be used for other fungal species with high natural resistance.

Phleomycin Selection. For both *A. niger* and *A. nidulans*, $Phle^r$ transformants are obtained by direct selection (Table II). However, for *A. oryzae* direct selection does not result in $Phle^r$ transformants. Although we succeeded in isolating $Phle^r$ *A. oryzae* strains with the aid of cotransformation,[12] various indirect Phle selection protocols in *A. oryzae* have also been unsuccessful.

Discussion

Further Applications of Method

In this chapter we have limited ourselves to a description of the results obtained with the vectors pAN7-1 and pAN8-1 in several *Aspergillus* species and *Penicillium chrysogenum*. However, these vectors have also been used successfully in numerous other fungal species, both ascomycetes and basidiomycetes.[3] Especially in genetically uncharacterized species, for which no other transformation markers are available as yet, pAN7-1 and pAN8-1 have been useful to establish genetic transformation. Gene cloning strategies have also been developed for some of these species using a cosmid vector (pAN7-2) containing the *hph* cassette.[20]

Besides the vectors described in this chapter several other fungal HmB or Phle selection vectors have been developed.[2] In general, these vectors give similar results as obtained with pAN7-1 and pAN8-1.

Further applications of the transformation systems, based on HmB and Phle selection, emerge from the data collected for different fungal species.

[20] R. N. Cooley, R. F. M. van Gorcom, C. A. M. J. J. van den Hondel, and C. E. Caten, *J. Gen. Microbiol.* **137,** 2085 (1991).

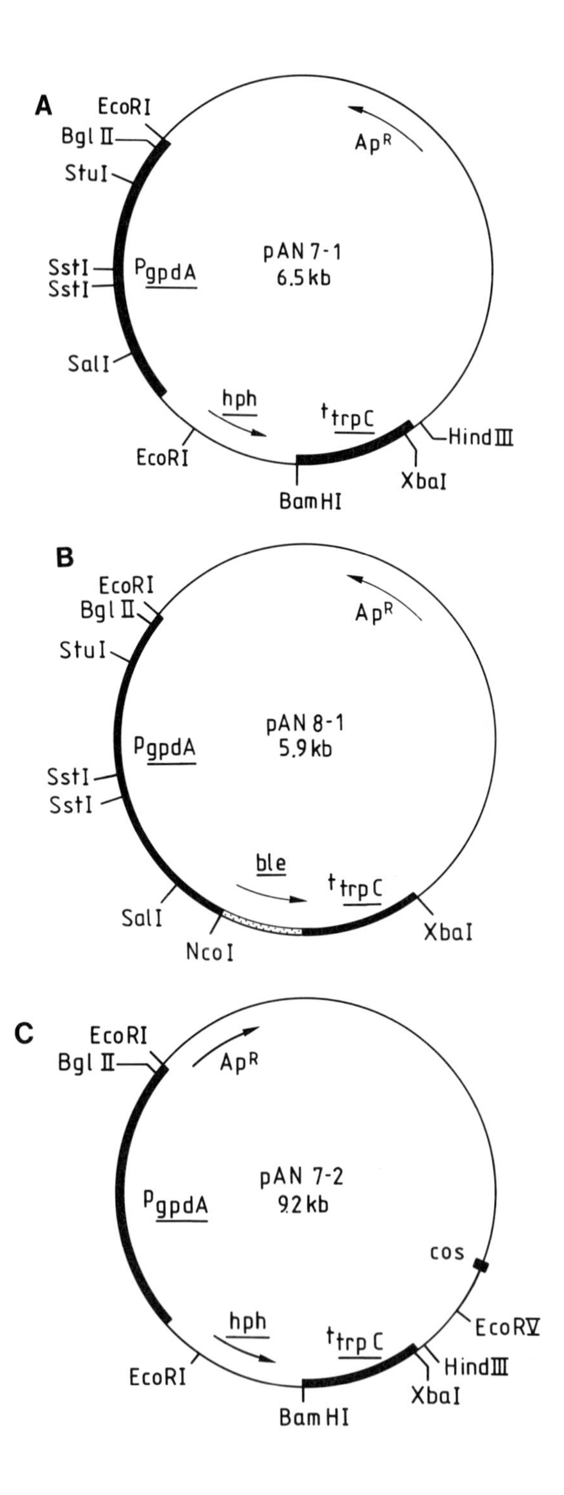

A
EcoRI
Bgl II
StuI
SstI
SstI
SalI
EcoRI
ApR
pAN 7-1
6.5 kb
PgpdA
hph
ttrpC
HindIII
XbaI
BamHI
B
EcoRI
Bgl II
StuI
SstI
SstI
SalI
NcoI
ApR
pAN 8-1
5.9 kb
PgpdA
ble
ttrpC
XbaI
C
EcoRI
Bgl II
ApR
pAN 7-2
9.2 kb
PgpdA
cos
EcoRV
hph
ttrpC
HindIII
XbaI
EcoRI
BamHI

1. Both selection markers can be efficiently used for cotransformation experiments in various fungi. Cotransformation frequencies (of the unselected vector) of 50% or more are found.[3]

2. In virtually all fungal species one or more vector molecules integrate into the host genome as is indicated by Southern analysis of HmB^r or Phle^r transformants. For a few species the data suggest that the HmB resistance level of transformants is a measure of the number of integrated gene copies of pAN7-1,[8,21] which may provide a way to select for multicopy transformants. The question of whether the Phle resistance level is also correlated to the number of integrated pAN8-1 copies has not yet been tested.

Comparison with Other Dominant Selectable Markers

The potential of dominant selectable markers for transformation of filamentous fungi obviously was recognized by many research groups. Therefore, besides the markers described in this chapter, vectors with other dominant selectable markers have been developed for filamentous fungi. In general, two types of marker genes can be discriminated[3]: (1) Genes encoding antibiotic-inactivating proteins and (2) genes encoding antibiotic-resistant proteins that can take over an essential antibiotic-sensitive function in the fungal cell. Both HmB resistance and Phle resistance genes belong to the first category. In general, this type of marker results in a fully dominant phenotype of the transformants and shows a broad host range. Dominant markers based on antibiotic-resistant proteins (e.g., benomyl resistance genes and oligomycin resistance genes) often result in a semidominant phenotype and their use may be limited to one or a few species.

A special dominant selectable marker is the *A. nidulans amdS* gene.

[21] A. Herrera-Estrella, G. H. Goldman, and M. van Montagu, *Mol. Microbiol.* **4,** 839 (1990).

FIG. 1. Schematic representation of relevant vectors. (A) Vector pAN7-1; (B) vector pAN8-1; (C) cosmid vector pAN7-2 is a derivative of cosmid pJB8 [D. Ish-Horowicz and J. R. Burke, *Nucleic Acids Res.* **9,** 2989 (1981)] containing the *hph* cassette of pAN7-1. To construct this vector, a 3.8-kb *Eco*RI (partial)–*Hin*dIII fragment from pAN7-1 was cloned into pJB8 which was digested with *Eco*RI and *Hin*dIII. For each vector a number of convenient cloning sites are indicated. Vector pAN7-2 contains a unique *Eco*RV and *Bgl*II site to facilitate cloning of partially *Hae*III/*Alu*I- and *Sau*3AI-digested genomic DNA, respectively. p_{gpdA}, promoter region of the *A. nidulans gpdA* gene; t_{trpC}, terminator region of the *A. nidulans trpC* gene; *hph,* HmB phosphotransferase gene from *E. coli*; ble, *S. hindustanus* gene encoding Phle-binding protein; Ap^R, ampicillin resistance; *cos,* cos site of bacteriophage λ. Arrows indicate the direction of transcription. The complete nucleotide sequence of pAN7-1 and pAN8-1 is not known.

TABLE II

TRANSFORMATION FREQUENCIES[a] OBTAINED WITH pAN7-1 AND pAN8-1 IN VARIOUS FUNGAL SPECIES

Species	Hygromycin B selection			Phleomycin selection		
	Indirect (200 μg/ml)	Indirect (1000 μg/ml)	Direct (100–200 μg/ml)	Direct (5 μg/ml)	Direct (10–20 μg/ml)	Direct (50 μg/ml)
A. nidulans FGSC4[b]	—	5–15	—	—	1–3	—
A. niger N402[b]	5–15	—	30–70	10–20	10–20	—
A. niger CBS513.88	—	—	20	—	—	—
A. niger ATCC1015[b]	3–10	—	5–40	—	—	—
A. ficuum NRRL 3135	5	—	10–15	—	—	30
P. chrysogenum ATCC48271				—	20	—

[a] Number of transformants per microgram DNA is given.
[b] Or derivatives from these strains.

This nutritional marker confers the ability to use acetamide and acrylamide as a nitrogen (and carbon) source. Most filamentous fungi do not readily use these compounds as such. Although several species can be transformed with the *amdS* gene,[2] a prerequisite for the use of this marker is that the fungal species is able to grow on defined synthetic medium with NH_4^+ as the sole nitrogen source. Many (especially plant pathogenic) fungi, however, require complex media to support growth. Such species cannot be transformed with the *amdS* gene as a selectable marker.

Acknowledgments

We wish to acknowledge all colleagues at the Medical Biological Laboratory who contributed unpublished results to us. Wim van Hartingsveldt and Peter Pouwels are gratefully acknowledged for stimulating discussions and critical reading of the manuscript.

[40] Positive Selection Vectors Based on Palindromic DNA Sequences

By J. ALTENBUCHNER, P. VIELL, and I. PELLETIER

Insertion of DNA fragments into a bacterial cloning vector is an inefficient process, whereby transformed colonies often contain only the vector DNA. The self-ligation of the vector can be reduced by removing the 5′-phosphate groups at the ends of the insertion site by alkaline phosphatase, but this reduces the ligation efficiency. Another way to overcome this problem is to use positive selection vectors. Many sophisticated vectors have been constructed, which allow growth or survival of only the transformed colony. This happens when an insert inactivates a gene, the product of which is lethal under certain conditions or in certain strains.[1-3] Despite the obvious advantage of these vectors they are rarely used because of limitations, such as the need for special strains, plasmid instability, low transformation frequency, and lack of useful restriction sites.

[1] B. Nilsson, M. Uhlen, S. Josefson, S. Gatenbeck, and L. Philipson, *Nucleic Acids Res.* **11,** 8019 (1983).
[2] D. Dean, *Gene* **15,** 99 (1981).
[3] L. S. Ozaki, S. Maeda, K. Shimada, and Y. Takagi, *Gene* **8,** 301 (1980).

Principle of Method

It was observed that long palindromic DNA sequences inhibit the propagation of plasmids in bacteria.[4] The reason for this lethal effect might be the formation of cruciform structures, which strongly reduce the replication rate[5] or which are mistaken as recombination intermediates and cleaved by nucleases.[6] This effect is overcome in *sbcC* mutants, which tolerate palindromic DNA for unknown reasons.[7] The viability is also restored by separating the inverted repeats by a nonpalindromic sequence.

A series of useful cloning vectors was constructed on the basis of this principle. In various replicons long inverted repeats from random DNA were generated containing symmetrical polylinker sequences separated by nonpalindromic DNA. Removal of this stuffer fragment by cutting the plasmid at the symmetrical polylinker sequences and ligation of the remaining vector fragment restore the efficacy of the palindromic sequences. This treatment leads to a reduction of the transformation frequency by several orders of magnitude unless DNA fragments are provided to replace the stuffer fragment.[4,8]

In this chapter we describe two new versatile vectors for which we applied this principle of positive selection. We used the *lac* promoter and the T4 transcription terminator sequence instead of random DNA to generate the palindromic DNA and combined it with the advantage of direct screening of recombinant DNA using the α-complementation system of *lacZ*.[9]

Material and Methods

The bacterial strains and plasmids used are listed in Tables I and II. Restriction enzymes, T4 DNA ligase, and Klenow polymerase are obtained from Boehringer Mannheim (Indianapolis, IN) and used as recommended by the manufacturer. For transformation of *Escherichia coli* the TSS method of Chung *et al.*[10] is used. Routinely a transformation rate of 2×10^6 transformants/μg vector (pIC19H) is obtained. Preparation and transformation of *Streptomyces lividans* protoplasts, small-scale plasmid preparations, and isolation of chromosomal DNA from *S. lividans* are

[4] C. E. Hagan and G. E. Warren, *Gene* **24,** 317 (1983).

[5] J. C. Lindsey and D. R. F. Leach, *J. Mol. Biol.* **206,** 779 (1989).

[6] K. Mizuuchi, M. Mizuuchi, and M. Gellert, *J. Mol. Biol.* **156,** 229 (1982).

[7] A. F. Chalker, D. R. F. Leach, and R. G. Lloids, *Gene* **71,** 201 (1988).

[8] J. Elhai and C. P. Wolk, *Gene* **68,** 119 (1988).

[9] C. Yanish-Perron, J. Vieira, and J. Messing, *Gene* **33,** 103 (1985).

[10] C. T. Chung, S. L. Niemela, and R. H. Miller, *Proc. Natl. Acad. Sci. U.S.A.* **86,** 2172 (1989).

TABLE I
BACTERIAL STRAINS

Strain	Genotype	Ref.[a]
E. coli JM109	*recA endA gyrA96 thi hsdR17 supE44* Δ(*lac-proAB*) (F′ *traD36 proAB lacI*^{q}*Z*Δ*M15*)	*1*
E. coli JC7623	*recB21 recC22 sbcB15 sbcC201 thi hisG4* Δ(*gpt-proAB*)*62 argE3 thr-1 leuB6 kdgK51 rfbD1 ara-14 lacY1 galK2 mtl-1 rpsL31*	*2*
S lividans TK64	*pro-2 str-6*	*3*

[a] Key to references: (*1*) C. Yanisch-Perron, J. Vieira, and J. Messing, *Gene* **33,** 103 (1985); (*2*) S. J. Morton, D. R. F. Leach, and R. G. Lloyd, *Nucleic Acids Res.* **17,** 8033 (1989); (*3*) D. A. Hoopwood, T. Kieser, H. M. Wright, and J. M. Bibb, *J. Gen. Microbiol.* **129,** 2257 (1983).

TABLE II
PLASMIDS

Plasmid	Marker[a]	Ref.[b]
pIC19H, pIC20H, pIC20R	Ap^r, *lacZ′*	(*1*)
pHP45Ω	Sm^r, Sp^r, Ap^r	(*2*)
pIJ350	Ts^r	(*3*)
pIJ4083	Ts^r, *xylE*	(*4*)
pIJ702	Ts^r, *mel*	(*5*)
pJOE773, pJOE930, pIC20HE	Ap^r, *lacZ′*	This work
pJOE875	Ap^r, Ts^r, *mel*	This work
pJOE814.1	Ap^r, *xylE*	This work

[a] Ap^r, Ampicillin resistance; Sm^r, streptomycin resistance; Sp^r, spectinomycin resistance; Ts^r, thiostrepton resistance.

[b] Key to references: (*1*) J. L. Marsh, M. Erfle, and E. J. Wykes, *Gene* **32,** 481 (1984); (*2*) P. Prentki and H. M. Krisch, *Gene* **29,** 303 (1984); (*3*) T. Kieser, D. A. Hopwood, H. M. Wright and C. J. Thompson, *Mol. Gen. Genet.* **185,** 223 (1982); (*4*) T. M. Clayton and M. J. Bibb, *Nucleic Acids. Res.* **18,** 1077 (1992); (*5*) E. Katz, C. H. Thompson and D. A. Hopwood, *J. Gen. Microbiol.* **129,** 2703 (1983).

carried out as described by Hopwood *et al.*[11] From agarose gels DNA fragments are purified by dissolving the agarose in NaI and binding the DNA to glassmilk (Clean-A-Gene kit; Andes Scientific, Inc.) To determine the catechol 2,3-dioxygenase activity *E. coli* cells are grown in L-broth at 37° to an optical density (OD_{600}) of 0.3. The *lac* regulatory system is induced by adding isopropyl-β-thio-galactoside (IPTG) to a final concentration of 1 m*M*. After further incubation for 1 hr the cells are harvested and lysed

[11] D. A. Hopwood, M. J. Bibb, K. F. Chater, T. Kieser, C. J. Bruton, H. M. Kieser, D. J. Lydiate, H. Schrempf, C. P. Smith, and J. M. Ward, "Genetic Manipulation of Streptomyces. A Laboratory Manual." The John Innes Foundation, Norwich, 1985.

by ultrasonication. Catechol 2,3-dioxygenase activity is measured by following the formation of 2-hydroxymuconic semialdehyde from catechol at 370 nm as described by Zukowski *et al.*[12] The protein concentration of the crude extract is determined by the method of Bradford[13] using bovine serum albumin (BSA) as reference. For other methods, such as agarose gel electrophoresis, standard conditions as described by Sambrook *et al.*[14] are applied.

Construction and Properties of Vector pJOE930

The plasmid pIC19H is a derivative of the well-known pUC series with just a modified polylinker sequence. A 342-bp *Pvu*II fragment of pIC19H containing the *lac* promoter and the N-terminal end of *lacZ*, including the polylinker sequence, was cloned into the *Nde*I site of the same plasmid after the 5′ protruding ends of the *Nde*I site were filled in by Klenow polymerase. Hereby a 684-bp-long palindrome was obtained in the resulting plasmid pJOE930 (Fig. 1). The two inverted repeats are separated by a 125-bp nonpalindromic sequence, which allows the propagation of pJOE930 in any *E. coli* strain. Transformants of *E. coli* JM109 give blue-colored colonies on agar plates containing 5-bromo-4-chloro-3-indolyl-β-D-galactopyranoside (X-Gal). This shows that the *lacZ* α fragment still complements the *lacZ*ΔM15 mutation on the F′ plasmid in JM109. When the DNA between the inverted repeats is removed by cutting the vector at the symmetrical polylinker sequences and the remaining vector fragment ligated, the transformation frequency in JM109 drastically drops more than 100-fold (Table IV). This demonstrates that pJOE930 fulfills the criterion for a positive selection vector.

The plasmid pJOE930 should also be an excellent expression vector due to the inducible *lac* promoters on both sides of the multiple cloning site. To see if the two equally strong promotors in pJOE930 influence each other, for example, by formation of RNA–RNA hybrids, which might lower the expression of inserted genes, we cloned a promoterless *xylE* gene as a *Hin*dIII fragment in both orientations into pJOE930 (see below). The specific activity of the catechol 2,3-dioxygenase encoded by *xylE* in pJOE930 was determined in cells of JM109 induced and noninduced by IPTG and compared with pIC19H, in which the *xylE* gene was inserted in the same manner. As shown in Table III, the IPTG-induced expression of the *xylE* gene in pJOE930 in both orientations is as high as that observed

[12] M. M. Zukowski, D. F. Gaffney, D. Speck, M. Kaufmann, A. Findeli, A. Wisecup, and J.-P. Lecocq, *Proc. Natl. Acad. Sci. U.S.A.* **80,** 1101 (1983).

[13] M. Bradford, *Anal. Biochem.* **72,** 248 (1976).

[14] J. Sambrook, E. F. Fritsch, and T. Maniatis, "Molecular Cloning: A Laboratory Manual." Cold Spring Harbor Laboratory, Cold Spring Harbor, New York, 1989.

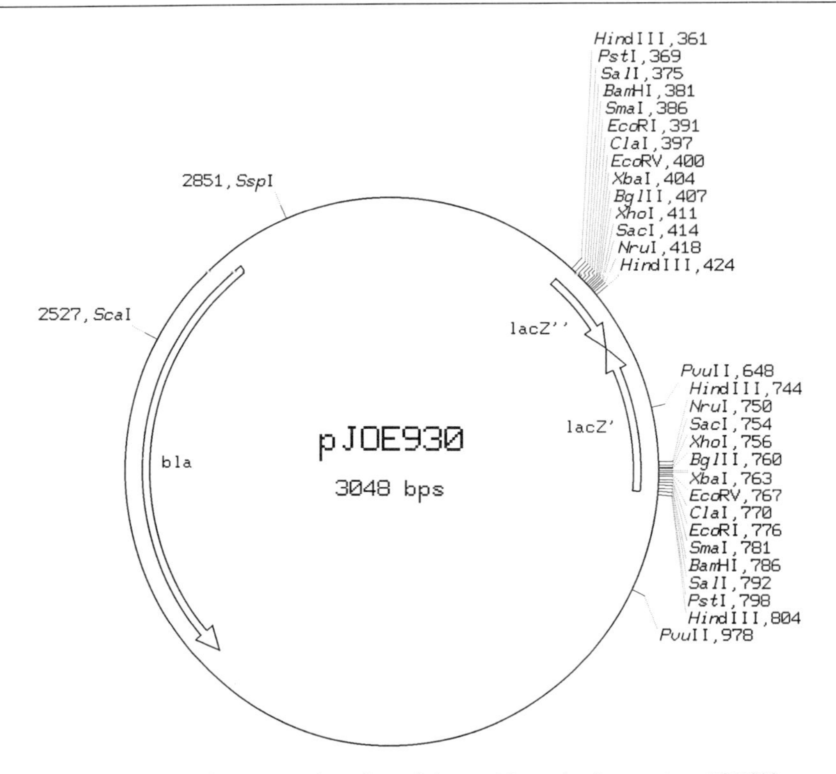

FIG. 1. Restriction map and marker of the positive selection vector pJOE930.

in pIC19H. Only the basal level of *xylE* expression in pJOE930 is slightly higher than that obtained with pIC19H.

Construction of pJOE773 and Two Derivatives

In some cloning experiments one would like to avoid expression of the cloned gene by a nearby promoter of the vector. For this purpose we constructed the positive selection vector pJOE773 with transcription terminators as inverted repeats. The plasmid pJOE773 contains the 2362-bp *Pvu*II vector fragment of pIC19H with the ampicillin resistance gene and the origin of replication. To both *Pvu*II ends a 124-bp *Sau*3A/*Hin*dIII fragment containing the T4 transcription terminator sequence from the Ω element was ligated, followed by the polylinker sequence of pIC19H. The two inverted DNA sequences are separated by the *lacZ* α fragment isolated as an *Hae*II fragment from pIC20R, where we deleted the polylinker sequence with *Eco*RI (Fig. 2). Transformation of pJOE773 into JM109 and growth on LB agar plates containing X-Gal and IPTG results in blue

TABLE III
CATECHOL 2,3-DIOXYGENASE ACTIVITY IN *Escherichia coli* JM109 WITH PROMOTERLESS *xylE* GENE IN pJOE930, pIC19H, AND pJOE773

Vector	Orientation of *xylE*	Induction with IPTG	nmol/(min · mg)
pJOE930	a	−	1122
		+	4493
	b	−	849
		+	4429
pIC19H	a	−	840
		+	4789
	b	−	170
		+	169
pJOE773	a	−	10
	b	−	12

colonies as observed with pJOE930. Removal of the *lacZ* α fragment without replacing it by another fragment also drastically reduces the transformation frequency (see below).

The host range of pJOE773 can easily be extended by converting it in a shuttle vector as shown in Fig. 2. Inserting the *Streptomyces* high-copy plasmid pIJ350 as an *Nde*I fragment into the *Nde*I site of pJOE773 generated the shuttle vector pJOE875. [Actually a derivative of pIJ350 was used with a mutation in the *Bam*HI site[15] and the *Kpn*I site deleted by S1 nuclease.] Instead of the *lacZ* α complementation, which is of course not functional in *Streptomyces,* the melanin production genes, obtained from pIJ702 as a *Bcl*I fragment, were placed in the center of the palindromic sequence. Transformation of pJOE875 into *S. lividans* leads to dark colonies with black halos on R2 agar plates supplemented with tyrosine and $CuSO_4$. Therefore they can easily be distinguished from transformants without the melanin-encoding fragment. The transformation rate of pJOE875 in *S. lividans* without a stuffer fragment between the inverted repeats is reduced 50- to 100-fold similar to that observed in *E. coli.* This has already been shown by Kieser and Melton[15] with a similar plasmid.

Furthermore, the plasmid pJOE773 can easily be converted into a promoter probe vector (Fig. 2). The promotor probe vector pJOE814.1 with *xylE* as reporter gene was constructed in the following way: First, the polylinker sequence of the plasmid pIC20H was improved by replacing the *Kpn*I/*Sac*I region with a newly synthesized polylinker sequence to give pIC20HE (Fig. 3). The promoterless *xylE* gene from the *Streptomyces*

[15] T. Kieser and R. E. Melton, *Gene* **65,** 83 (1988).

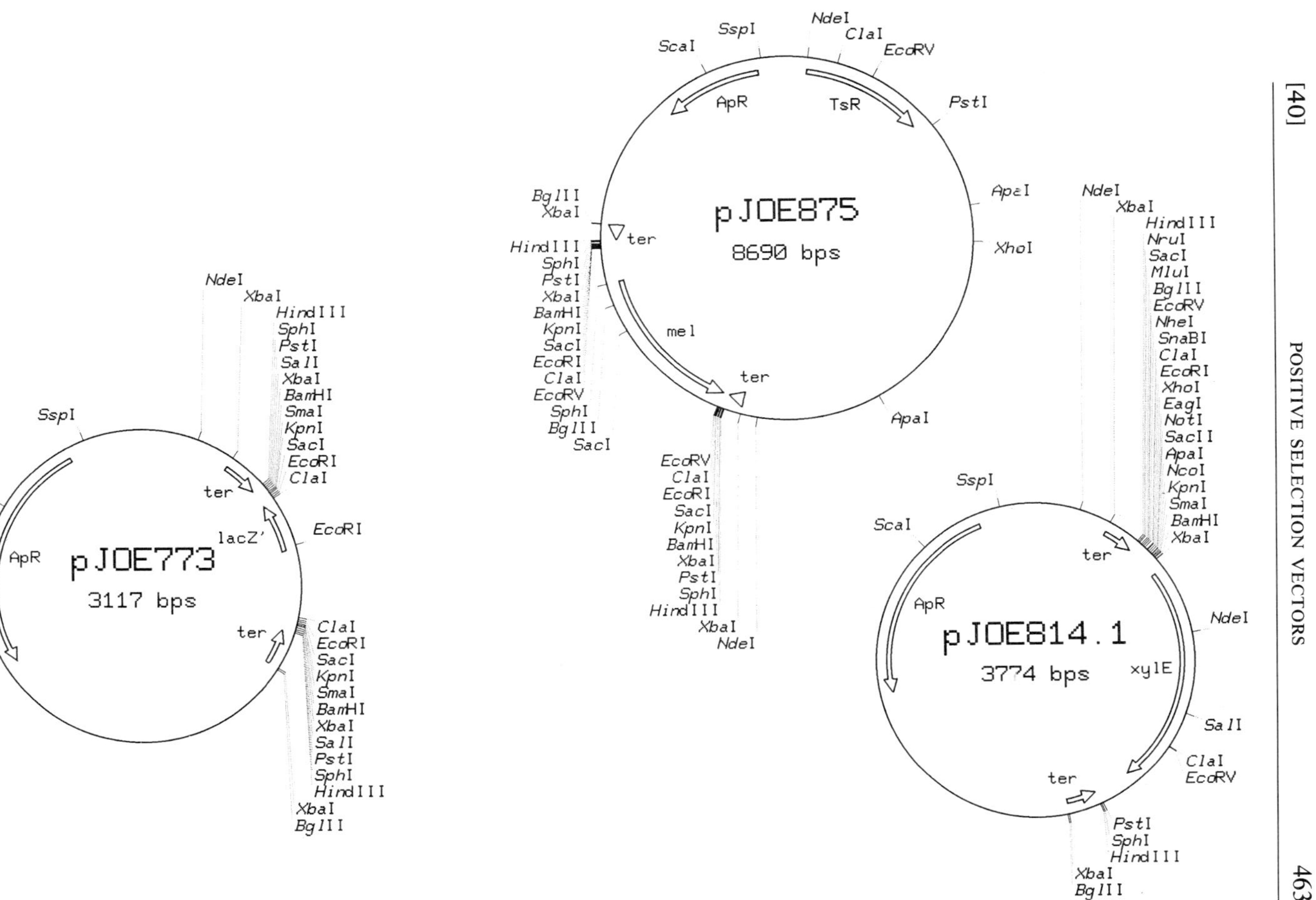

FIG. 2. Restriction maps of the positive selection vector pJOE773 and the two derivatives, the shuttle vector pJOE875 and the promoter probe vector pJOE814.1 with *xylE* as reporter.

```
                                                                  SacII
                                                                  -----
                    SphI         SalI          BamHI      KpnI       ApaI    EagI
                    ----         ----          -----      ----       ----    ----
               HindIII      PstI        XbaI        SmaI       NcoI          NotI
               -------      ----        ----        ----       ----          ----
atgaccatgattacgccAAGCTTGCATGCCTGCAGGTCGACTCTAGAGGATCCCCGGGTACCATGGCCCGCGGCCGC
.         .         .         .         .         .         .         .
tactggtactaatgcggTTCGAACGTACGGACGTCCAGCTGAGATCTCCTAGGGGCCCATGGTACCCGGGCGCCGGCG


      EcoRI        SnaBI      NheI         BglII            NruI
      -----        -----      ----         -----            ----
  XhoI        ClaI        NaeI       EcoRV          MluI      SacI      HindIII
  ----        ----        ----       -----          ----      ----      -------
CTCGAGAATTCATCGATACGTAGCCGGCTAGCGATATCAGATCTACGCGTACGAGCTCGCGAAAGCTTggcactg
 .         .         .         .         .         .         .         .
GACGTCTTAAGTAGCTATGCATCGGCCGATCGCTATAGTCTAGATGCGCATGCTCGAGCGCTTTCGAAccgtgac
```

FIG. 3. DNA sequence and restriction sites of the polylinker sequence of the plasmid pIC20HE. Lower-case letters indicate *lacZ* DNA.

promoter probe vector pIJ4083 was inserted as a *Bam*HI/*Pst*I fragment into a *Bam*HI/*Pst*I-linearized pIC20HE. From the resulting plasmid the *xylE* gene was integrated into pJOE773 as a *Hin*dIII fragment by replacing the *lacZ* α fragment. The catechol 2,3-dioxygenase activity expressed by this plasmid in JM109 is sufficiently low (Table III) to be used as a sensitive system for detecting and measuring promoter activities.

How to Use Positive Selection Vectors

There are several ways to use the positive selection vectors pJOE773 and pJOE930 for cloning DNA fragments. This was demonstrated by cloning *Bam*HI-digested chromosomal DNA of *S. lividans* into the vector pJOE930.

In the simplest case the chromosomal DNA and the vector are digested by the same restriction enzyme and, after inactivation of the enzyme by phenol extraction and ethanol precipitation, ligated and transformed. As seen in Table IV this gives about 70% transformed colonies, which are white on X-Gal agar plates and should contain *S. lividans* DNA. Isolation of plasmid DNA from 48 colonies confirmed the assumption. Under identical conditions about the same number of transformants with DNA inserted in the vector pIC19H are obtained but here these plasmids comprise only 10% of the total number of transformants.

For many cloning experiments an insertion rate of 70% is already sufficient. On the other hand, it is very easy to get nearly 100% insertions by cutting the vector with a second restriction enzyme further inside the polylinker sequence. Two examples are shown in Table IV, where the vector was cleaved with *Sma*I and *Eco*RI, respectively, in addition to

TABLE IV
TRANSFORMATION FREQUENCIES OF pJOE930, pIC19H, AND pJOE773 DEPENDENT ON STRAIN, DIGESTION, PURIFICATION OF VECTOR FRAGMENT, AND PRESENCE OF INSERT DNA

Vector	Digestion/ purification	Strain	Ratio of vector to insert[a]	Number of transformants[b]	White colonies (%)
pJOE930	*Bam*HI	JM109	1:2	10,400	69
pIC19H	*Bam*HI	JM109	1:2	63,300	11
pJOE930	*Bam*HI/*Sma*I	JM109	1:2	5,800	98
pJOE930	*Bam*HI/ *Eco*RI	JM109	1:2	7,300	99
pJOE930	*Bam*HI/gel	JM109	1:2	4,800	>99
pJOE930	*Bam*HI/gel	JM109	1:0	45	>99
pJOE773	*Bam*HI/gel	JM109	1:0	15	>99
pJOE930	*Bam*HI/gel	JC7623	1:0	52,000	—
pJOE773	*Bam*HI/gel	JC7623	1:0	76,000	—

[a] Ligation was carried out overnight at 10° in a volume of 20 μl containing 200 ng vector DNA and, when indicated, 600 ng *Bam*HI-digested chromosomal DNA of *S. lividans*. Assuming an average fragment size of 4 kb, this corresponds to a ratio of 1:2.
[b] The number of transformants is given per 1 μg vector DNA.

*Bam*HI. These experiments reproducibly gave an insertion rate of more than 98%.

Another, slightly more laborious way is to purify the vector fragment on an agarose gel. The *lacZ* α fragment is sufficiently small not to contaminate the vector fragment and therefore again more than 98% insertions is obtained. The purified vector fragments of pJOE773 and pJOE930 were also used to see the reduction of the transformation frequency when no additional fragments were added. In both cases a more than 100-fold lower transformation frequency was obtained. The plasmids isolated from these transformants showed deletions of the whole palindromic sequence.

The most convenient way to use these vectors would be to bring the plasmids without a stuffer fragment into an *sbcC* strain, which tolerates the palindromes, and then transform them into an *sbcC* wild type after insertion of foreign DNA between the inverted repeats. Transformation of ligated vector fragments of pJOE930 and pJOE773 without the *lacZ* α fragment into the *sbcC* mutant JC7623 gave transformation frequencies similar to pIC19H cleaved by *Bam*HI and ligated again. Surprisingly, plasmid DNA isolated from these transformants could not be cleaved by enzymes with the restriction sites in the palindromic sequence. This was first considered to be caused by an extrusion of the palindrome under nonphysiological conditions occurring at the plasmid DNA preparation.

But it turned out to be mainly the result of deletions of the palindromic sequence.

Concluding Remarks

Long palindromic DNA sequences lead to instability and, when exceeding a certain threshold of about 150 bp, to nonviability of the carrier replicon in bacteria. The viability is restored by placing any nonpalindromic DNA sequence at least 50 bp in length between the inverted repeats.[16] Like others before us, we used this principle to construct positive selection vectors and improved them by screening recombinants directly and using them for other applications.

The plasmid pJOE930 was constructed in a single cloning step from pIC19H and in the same way positive selection vectors can be obtained from any plasmid of the pUC series. The advantage of having a strong promoter on both sides of the integration site is to make the expression of an inserted gene orientation independent. This should halve the number of clones needed in cloning experiments in which the gene must be identified by expression via an *E. coli* promotor.

The vector pJOE773 with the transcription terminators flanking the insertion site is useful for verification of the functionality of a promoter in an inserted DNA fragment. When the expression of the cloned gene by a vector promoter is harmful to the cell, the transcription terminators protect the cells from this adverse effect. This allows successful construction of genomic libraries. Furthermore, the vector is ideal for constructing promoter probe vectors. If the promoterless gene is first cloned into the polylinker sequence of pIC20HE as shown above, it can be brought into pJOE773 in one step as a *Hin*dIII fragment together with many useful cloning sites.

Due to the symmetry of the polylinker sequences, fragments can easily be exchanged between pJOE773 and pJOE930 as well as between shuttle vectors such as pJOE875 and pIC19H, if there are no *Hin*dIII sites in the cloned fragments. Insertion into pIC19H allows the double-stranded DNA sequencing of the cloned fragments with the universal and reverse primer. This is not possible in pJOE773 or pJOE930. Thus, these procedures further augment the versatility of the vector system.

Acknowledgments

We thank everyone who sent us strains and plasmids, Hilde Watzlawick and Gisela Kwiatkowski, who synthesized the oligonucleotides, and Professor Ralf Mattes for support and encouragement.

[16] G. J. Warren and R. L. Green, *J. Bacteriol.* **161,** 1103 (1985).

Section IV

Vectors for Cloning Genes

[41] Compilation of Superlinker Vectors

By JÜRGEN BROSIUS

There are 64 possibilities for uninterrupted palindromic hexameric restriction enzyme recognition sequences. Enzymes for 52 of those sequences are commercially available.[1] Many of the remaining 12 will follow over the next years. A large polylinker has been synthesized that accommodates all the above 64 restriction sites as well as 2 octameric (*Not*I and *Sfi*I) and a few interrupted palindromic and nonpalindromic hexamer sites.[2] Several variants of the linker have been inserted into a number of popular multifunctional cloning and expression vectors.[2] These vectors will facilitate many cloning and subcloning applications because of their potential to accommodate restriction fragments irrespective of the enzymes by which they have been generated. In most cases these restriction sites will be preserved.

Multifunctional Cloning Vectors

Characteristics of some of the more frequently requested as well as four further developed cloning and expression vectors are summarized in Table I. pSL180, pSL190, pSL1180, and pSL1190[2] correspond to commonly used vectors pUC18, pUC19, pUC118, and pUC119.[3] The polylinkers of the latter group have been replaced with the SL1 version [317 base pairs (bp), flanked by *Eco*RI and *Hin*dIII] of the superlinker (Fig. 1B). Because of the extended interruption of the Lac$Z\alpha'$ peptide derivative, a color discrimination between nonrecombinant and insert-containing vectors is not possible anymore. Therefore, bacterial colonies harboring those plasmids are white on indicator plates containing 5-bromo-4-chloro-3-indolyl-β-D-galactopyranoside (X-Gal) and isopropyl-β-D-thiogalactopyranoside (IPTG). pSL1180/pSL1190 differ from pSL180/pSL190 by the presence of the intergenic region (*IR*) from filamentous single-stranded phage M13, which contains the origin of replication as well as signals for phage packaging. This permits the generation of single-stranded DNA following superinfection with helper phage KO7 using a protocol described in an earlier volume of this series.[4] The single-stranded DNA corresponds, as

[1] R. J. Roberts and D. Macelis, *Nucleic Acids Res.* **20**(Suppl.), 2167 (1992).
[2] J. Brosius, *DNA* **8,** 759 (1989).
[3] J. Messing, *Gene* **100,** 3 (1991).
[4] J. Vieira and J. Messing, this series, Vol. 153, p. 3.

TABLE I
PLASMIDS WITH SUPERLINKERS[a]

Plasmid	Parent	Size (bp)	Linker	Relevant features
pSL180[b]	pUC18[c]	2946	SL1	Cloning vector; no color screening for inserts
pSL190[b]	pUC19[c]	2946	SL1	As pSL180; linker in opposite orientation
pSL1180[b,d]	pUC118[e]	3422	SL1	Cloning vector; no color screening for inserts; M13 origin of replication for generation of single-stranded DNA (the coding strand with respect to the *lacZ* region)
pSL1190[b,d]	pUC119[e]	3422	SL1	As pSL1180; linker in opposite orientation
pSE1200[b]	pUC120[f]	3442	SL2	Cloning and expression vector; no color selection for inserts; M13 origin; *lac* promoter/operator; *lac* RBS; ATG start codon in *Nco*I site
pSL300[b]	pBS KS(+)[g]	3264	SL2	SL2 replaces linker from pBluescript KS M13(+); cloning vector; T3 and T7 transcription promoters flanking superlinker; f1 origin of replication for generation of single-stranded DNA; translation restart distal to linker into *lacZα'*; subtle color discrimination (midblue to light blue) for insert screening
pSL301[b,h]	pSL300[b]	3267	SL2	As pSL300; different translation restart from linker into *lacZα'*; subtle color discrimination (midblue to light blue) for insert screening
pSL350[i]	pSE1200[b]	3594	SL2	Cloning and expression vector; T3 and T7 transcription promoters flanking superlinker; M13 origin of replication; *lac* promoter/operator; T7 gene *10* translational enhancer; minicistron upstream from ATG-containing *Nco*I site for translational restart; open reading frame throughout superlinker; distal to linker translational restart into wild-type *lacZα*; subtle color discrimination (blue to midblue) for insert screening
pSL351[i]	pSL350[i]/ pUC181[j]	3594	SL2	As pSL350; M13 origin in opposite orientation; single-stranded DNA (yield significantly lower as with pSL350) representing the anticoding strand with respect to the *lacZ* region

TABLE I (*continued*)

Plasmid	Parent	Size (bp)	Linker	Relevant features
pKK489-7[i]	pSL350[i]	3303		As pSL350; shorter polylinker (62 bp) from pTrc99A[l]; useful for future constructions requiring sites not present in the linker
pSE220[b]	pKK233-2[k]	4926	SL2	Expression vector; *trc* promoter; *lac* operator/RBS; inducible with IPTG, ATG in *Nco*I site; *rrnB* transcription terminators
pSE280[b,h]	pSE220[b]	3865	SL2	As pSE220; shorter, remnants of *tet* gene removed
pSE380[b,h]	pTrc99A[l]	4467	SL2	As pSE280; harbors in addition *lac* repressor (*lac*I) gene
pSE420[i,h]	pSE380[b]	4617	SL2	As pSE380; T7 gene *10* translational enhancer; A/T-rich regions flanking RBS; minicistron upstream from CCATGG (=start codon in *Nco*I cloning site) for efficient translational restart; open reading frame throughout superlinker
pKK480-3[i]	pSE420[i]	4760		As pSE420; rat calmodulin gene instead of linker; model system for very high level of recombinant protein expression
pKK520-3[i]	pKK232-8[m]	5417	SL2	Promoter selection vector; linker upstream from promoter-deficient chloramphenicol acetyltransferase gene

[a] All vectors confer resistance to ampicillin.
[b] J. Brosius, *DNA* **8,** 759 (1989).
[c] J. Norrander, T. Kempe, and J. Messing, *Gene* **26,** 101 (1983).
[d] Plasmid commercially available from Pharacia-LKB Biotechnology, Inc. (Piscataway, NJ).
[e] J. Vieira and J. Messing, this series, Vol. 153, p. 3.
[f] J. Norrander, J. Vieira, I. Rubinstein, and J. Messing, *J. Biotechnol.* **2,** 157 (1985).
[g] J. M. Short, J. M. Fernandez, J. A. Sorge, and W. D. Huse, *Nucleic Acids Res.* **16,** 7583 (1988).
[h] Plasmid commercially available from Invitrogen Corp. (San Diego, CA).
[i] This paper.
[j] J. Vieira and J. Messing, *Gene* **100,** 189 (1991)
[k] E. Amann and J. Brosius, *Gene* **40,** 183 (1985).
[l] E. Amann, B. Ochs, and K.-J. Abel, *Gene* **69,** 301 (1988).
[m] J. Brosius, *Gene* **27,** 151 (1984).

A.

```
                DraIII
             PflMI
           Tth111I         XmnI   BstBI    AlwNI BsgI        AatII
NcoI     BstEII            BsmI       EcoRI      BspMI          SalI Bsu36I 1D
              .          .          .          .          .          .          .
CCATGGCTGGTGACCACGTCGTGGAATGCCTTCGAATTCAGCACCTGCACATGGGACGTCGACCTGAGGT  70

                                                                      (ClaI)
             Eco0109I                                                 BsaBI
           ApaI    13A          MunI                           10B PvuI
  13D    SmaI       4D      BamHI    15D   (BclI) BspHI  BglII BssHII      EcoRV
              .          .          .          .          .          .          .
AATTATAACCCGGGCCCTATATATGGATCCAATTGCAATGATCATCATGACAGATCTGCGCGCGATCGAT 140

         DraI
             1A        SphI  3D                  AgeI       2A
  Eco47III         FspI    NheI 13B     XbaI  KpnI        HpaI        NaeI   SnaBI
              .          .          .          .          .          .          .
ATCAGCGCTTTAAATTTGCGCATGCTAGCTATAGTTCTAGAGGTACCGGTTGTTAACGTTAGCCGGCTAC 210

                                                                Narl
      BspEI    AseI      AvrII     NdeI    EagI       PvuII                  BsiWI
Bst1107I   SspI      StuI      NsiI     NotI     PstI      KasI  ClaI   MluI
              .          .          .          .          .          .          .
GTATACTCCGGAATATTAATAGGCCTAGGATGCATATGGCGGCCGCCTGCAGCTGGCGCCATCGATACGC 280

                                                   BglI
                                                   SfiI
                                XhoI      SpeI MscI                  AflII
     NruI   SacII 5A 12D   SacI      ScaI       PmlI       ApaLI    HindIII
              .          .          .          .          .          .          .
GTACGTCGCGACCGCGGACATGTACAGAGCTCGAGAAGTACTAGTGGCCACGTGGGCCGTGCACCTTAAGCTT
```

B.

```
   BstBI MscI     AatII
EcoRI       NcoI      SalI
              .          .
GAATTCGAATGGCCATGGGACGTCGAC...
```

FIG. 1. Sequences of superlinkers SL1 and SL2. (A) The coding strand of the 353-bp-long linker SL2 is shown in its entirety. The ATG start codon, which is proximal to an open reading frame that extends through the entire sequence, is underlined. Hexameric and octameric restriction enzyme recognition sites are shown. The 12 palindromic hexameric in the recombinant M13 phages,[3,4] to the coding strand with respect to *lacZ* sequences.

Vector pSL300 (Fig. 2), based on pBluescript KS(+),[5] as well as pSL350 (Fig. 3) and pSL351, which are derived from pSE1200,[2] have a multitude of features. All harbor superlinker version SL2 (353 bp long; Fig. 1A), which is flanked by transcription promoters T3 (proximal to the *Nco*I site) and T7 (distal to *Hin*dIII) for *in vitro* synthesis of RNA. pSL300 has the *IR* from bacteriophage f1, while pSL350 and pSL351 feature the related *IR* of M13 phage. In pSL351 the *IR* is in the alternative orientation. This permits isolation of the opposite DNA strand (the anticoding strand with respect to *lacZ* sequences), compared to pSL350, which produces the coding single strand on superinfection with helper phage. The yield of single-stranded DNA from pSL351, however, is significantly lower than from pSL350. This is presumably due to interference with replication caused by transcription from a flanking promoter into the phage *IR*.[6] Vieira and Messing[6] therefore recommend conditions for maximal suppression of the wild-type *lac* promoter during single-stranded DNA preparations, including the use of a *lacI*^{q} strain and the presence of 0.2–0.4% (w/v) glucose in the growth medium.

For sequencing single- or double-stranded DNA the usual two primers flanking the polylinker (e.g., corresponding to T3 and T7 promoters) are not sufficient any more. For most sequence applications, especially when the insert is in central parts of the long superlinker, primers that are located closer to the target of interest are necessary. Therefore, a set of additional primers located at a relatively short distance from any of the cloning sites have been designed and tested for sequencing.[2]

In pSL300 and pSL350/1 an attempt was undertaken to maintain, despite the large polylinker, a color discrimination for insert-bearing clones.

[5] J. M. Short, J. M. Fernandez, J. A. Sorge, and W. D. Huse, *Nucleic Acids Res.* **16,** 7583 (1988).

[6] J. Vieira and J. Messing, *Gene* **100,** 189 (1991).

restriction sites without yet commercially available enzymes are marked by a number and a letter: 1A, AAATTT; 1D, TAATTA; 2A, AACGTT, 3D, TAGCTA; 4D, TATATA; 5A, ACATGT; 10B, CGCGCG; 12D, TGTACA; 13A, ATATAT; 13B, CTATAG; 13D, TTATAA; 15D, TTGCAA. The first letter or number of a restriction site designation is located above the 5′ nucleotide of its recognition sequence. For clarity, the numerous enzymes with multiple hexameric recognition sequences, or pentameric and tetrameric recognition sequences, have not been included. Sites in brackets are sensitive to methylation by the *dam* modification system. (B) Sequence variation of the SL1 superlinker. Base 14 and beyond of the SL1 linker is identical with base 50 and beyond of SL2 in (A).

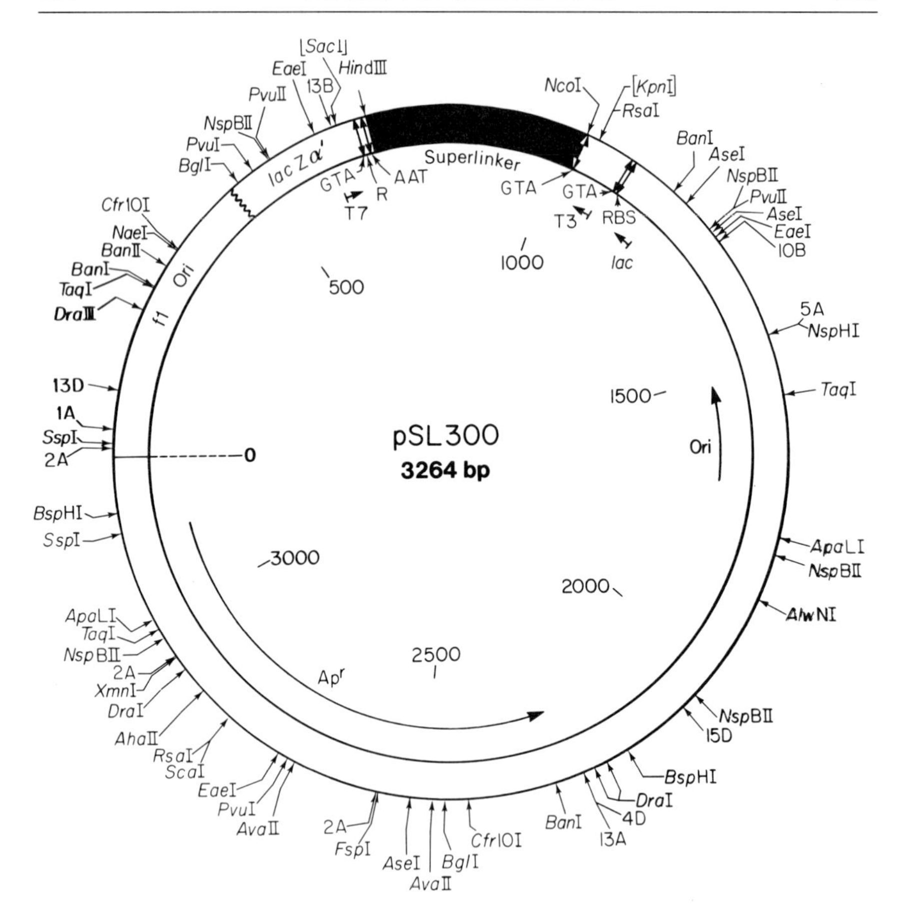

FIG. 2. Circular map of pSL300 (drawn to scale). The black heavy segment depicts the SL2 polylinker. The β-lactamase gene conferring resistance to ampicillin (Ap^r) and its direction of transcription as well as the plasmid origin (Ori) of replication are shown by arrows. The *IR* of the f1 phage is marked by f1 Ori. The *lac*, T3, and T7 promoters and their transcriptional directions are shown by arrows. The ATG translational start codon downstream from P_{lac}, the ATG codon at the 5′ end of the linker that is part of an *Nco*I site, the ATG distal to the translational restart (R), as well as the translational stop codon TAA distal to the linker are to be read counterclockwise. Restriction sites in square brackets were formerly bordering the pBluescript KS(+) polylinker and are inactivated in this vector. For designations of restriction sites that are not yet commercially available, see the caption to Fig. 1. The numbers of the inner circle give the position on the map in base pairs.

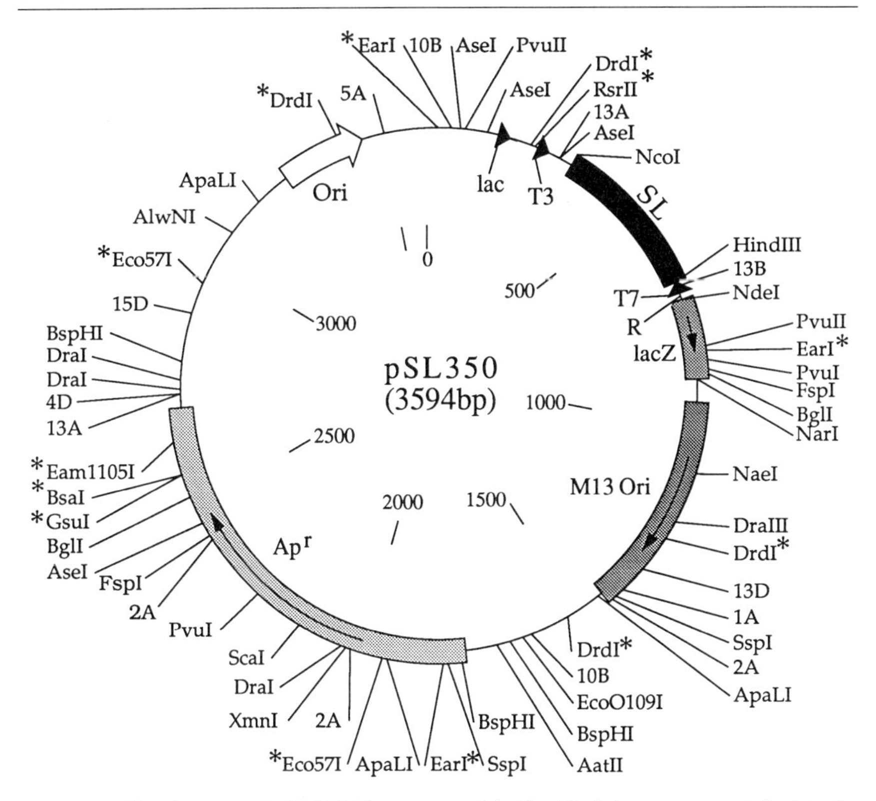

FIG. 3. Circular map of pSL350 (drawn to scale). The black heavy segment depicts the SL2 polylinker. The β-lactamase gene conferring resistance to ampicillin (Ap^r), the *lacZ* region including the α peptide, as well as the *IR* of the M13 phage (M13 Ori) correspond to shaded areas. The plasmid origin (Ori) of replication is shown by an open arrow. The *lac*, T3, and T7 promoters and their transcriptional directions are shown by filled arrowheads. The translational restart between the superlinker (SL) and the *lacZ* gene segment is marked by an R. For designations of restriction sites that are not yet commercially available, see the caption to Fig. 1. Restriction sites with an asterisk are not present in the superlinker. For clarity, only the *Nco*I and *Hin*dIII sites of the polylinker, located at its 5′ and 3′ ends respectively, are shown. The numbers of the inner circle give the position on the map in base pairs.

Translation should start at the 5′ end of the polylinker and proceed through the entire superlinker, which has one open reading frame. Distal to the linker translation stops and restarts into *lacZ*α, the actual indicator gene. This N-terminal portion of β-galactosidase, if expressed, complements the *lacZ* mutation of the host. Parent vectors should therefore confer a blue color, while recombinant vectors should be white or light blue. Inserts in the polylinker are expected to contain frequent translational stop codons, thus terminating translation prior to reaching the intergenic region located

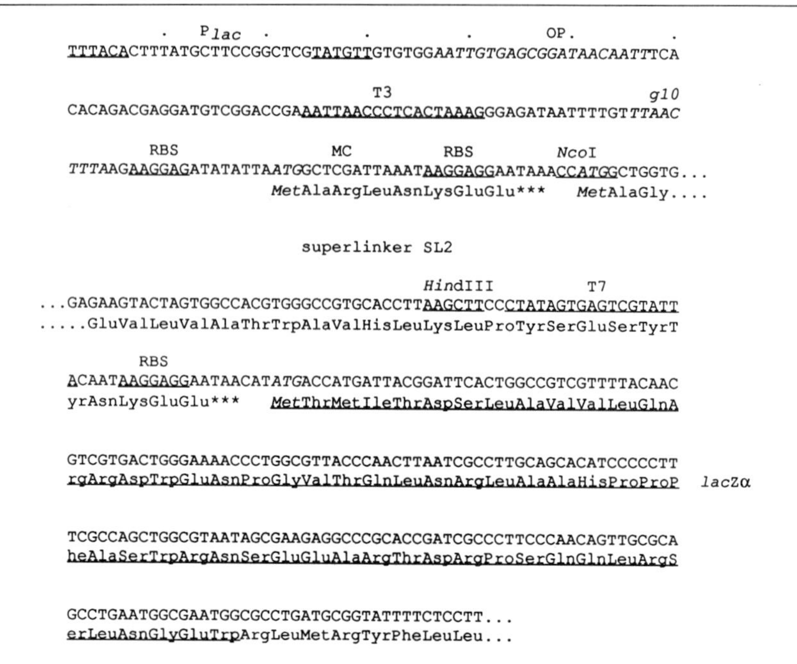

FIG. 4. Sequence of relevant area in pSL350/1. Every tenth nucleotide in the first line is marked by a dot. The −35 and −10 regions of the *lac* promoter are underlined. The *lac* operator (OP) is shown in italics. T3 and T7 transcription promoters are underlined. The bacteriophage T7 gene *10* translaton enhancer (*g10*) is in italics. Ribosome-binding sites (RBS) preceding the ATG start codons (in italic) of the minicistron (MC), the superpolylinker, and the Lac*Zα* peptide are underlined. Stop codons are designated by three asterisks. The deduced amino acid sequences of the minicistron, the linker, and the *lacZ* region (wild type underlined) are shown. Only the 5′ and 3′ portions of the superlinker, with flanking *Nco*I and *Hin*dIII sites (underlined), are shown.

between the open reading frame (ORF) of the superlinker and the *lacZα* coding region. This goal has partially been achieved, first in pSL300, where several test inserts conferred a light-blue colony color versus the midblue color of the nonrecombinant vector.[2] However, especially when older indicator plates were used it took a day or more for the development of sufficient color. Because the restart occurred into a Lac*Zα′* peptide that was still interrupted by T7 promoter sequences, the next series of vectors was designed to restart translation into wild-type *lacZα*. Therefore, the T7 promoter was moved closer to the 3′ end of the polylinker, between *Hin*dIII and the translational restart (Fig. 4). On the plates described below, pSL350/1 developed sufficient color in overnight incubations. Color differences between parent and recombinant vectors, however, were as subtle as before. For successful use of these vectors in subcloning the following precautions are recommended.

1. For subcloning inserts with ends generated by different restriction sites it is important to cut the recipient vector with both enzymes to completion and purify the backbone from the linker fragment. This can easily be achieved by purification on agarose gels and recovery of DNA with, e.g., GlassMilk (BIO 101, La Jolla, CA).[7] No color discrimination is necessary, because the vast majority of clones will represent the desired recombinants. As in all polylinkers it may be difficult to digest two restriction sites that are in close proximity to each other. If such a combination must be used a stuffer fragment should be inserted into site A, resulting in two A sites flanking the insert. Then, after digesting the other site (site B), one of the A sites (the one far from site B) can be digested.

2. For subcloning inserts with identical ends, the linearized vectors can be treated with nuclease-free phosphatase to reduce the background of nonrecombinant vectors.[8] For equal distribution, indicators should be added to the autoclaved medium prior to pouring the plates. Application to the surface leads to uneven distribution of X-Gal, resulting in localized color variation of bacterial colonies even when they contain identical plasmids.

3. It should also be mentioned that not all restriction enzymes are unique in or to the superlinkers. SL2, for example, contains two *Cla*I sites. However, only one of the *Cla*I sites is cleavable in normal strains. The second site is overlapped by GATC and is, as the unique *Bcl*I site, cleavable only in dam^- strains. Because *Bst*NI sites are absent in the superlinkers, there is no overlap and no potential inactivation due to methylation of restriction sites by the *dcm* modification system.

4. Some restriction sites occur once or even several times in other areas of the plasmids (see Figs. 2, 3, 5, and 6). If any of the multiple sites are intended for cloning, a limited digest with the given enzyme is recommended. The superlinkers do not contain the octameric recognition sequences *Asc*I, *Pac*I, *Pme*I, *Srf*I, *Sse*8387I, and *Swa*I. Likewise, *Bbs*I, *Bcg*I, *Bpu*1102I, *Bsa*I, *Bst*XI, *Drd*I, *Eam*1105I, *Ear*I, *Eco*57I, *Eco*NI, *Esp*3I, *Gsu*I (=*Bpm*I), *Ppu*MI, *Rsr*II, *Sgr*AI, and *Xcm*I, which are interrupted palindromic or nonpalindromic hexamers, are not in the linker. Some of the above hexameric sites are present in the various vector sequences and can be useful for future plasmid constructions. A version

[7] B. Vogelstein and D. Gillespie, *Proc. Natl. Acad. Sci. U.S.A.* **76,** 615 (1979).

[8] A. Ullrich, J. Shine, J. Chirgwin, R. Pictet, E. Tisher, W. J. Rutter, and H. M. Goodman, *Science* **196,** 1313 (1977).

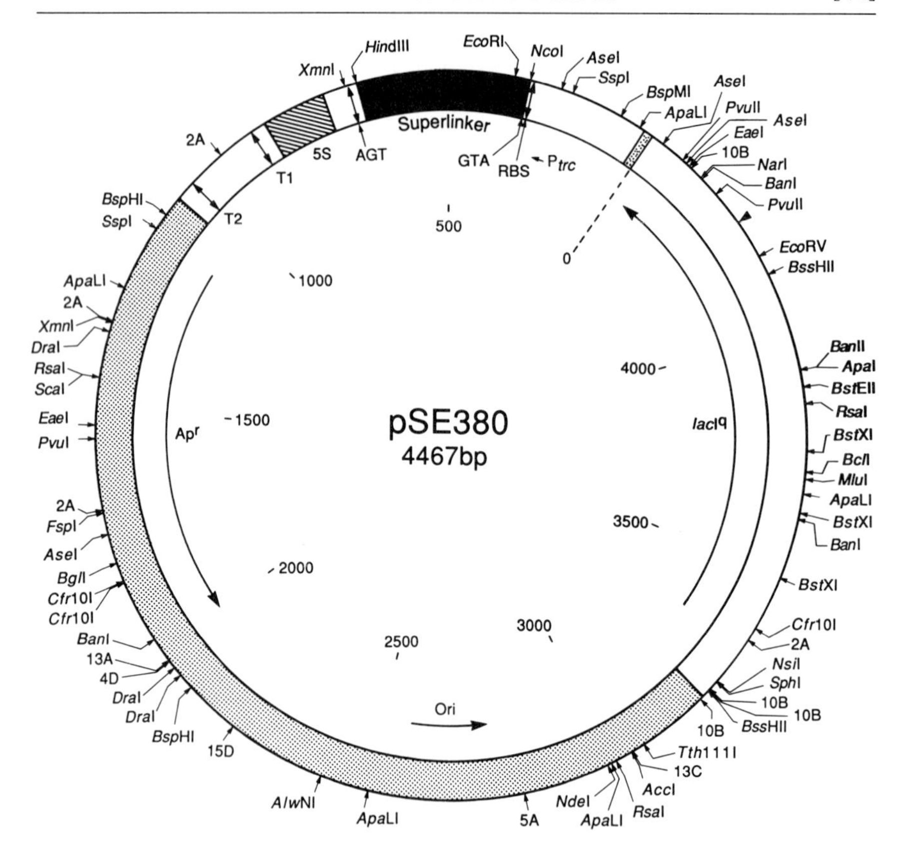

FIG. 5. Circular map of expression vector pSE380 (drawn to scale). pBR322 derived sequences are stippled. The black heavy segment depicts the SL2 polylinker downstream from the *trc* promoter (arrow), the *lacZ* translation initiation region including the ribosome-binding site (RBS), and the ATG start codon within the *Nco*I site. The polylinker is located upstream from a TGA stop codon, the *rrnB* 5S rRNA gene (cross-hatched) and the *rrnB* double transcription terminators T1 and T2. The β-lactamase gene conferring resistance to ampicillin (Ap^r), the *lac* repressor gene (*lacI^q*) in their directions of transcription, as well as the plasmid origin (Ori) of replication are shown by arrows. An additional *Hpa*I restriction site has been located at the 3′ end of the *lacI* gene by restriction mapping (C. Conlin, J. Zhao, personal communications). A likely position of the site is near bp 4203 (see arrowhead), where a single C-T transition would change an *Hinc*II site into an *Hpa*I site. The numbers of the inner circle give the position on the map in base pairs.

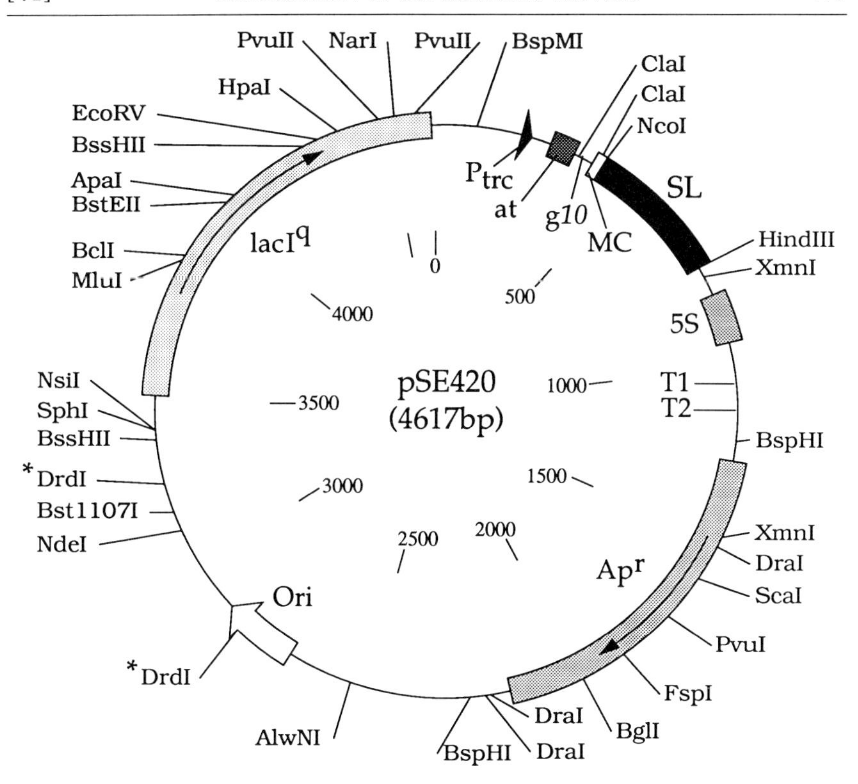

FIG. 6. Circular map of expression vector pSE420 (drawn to scale). The black heavy segment depicts the SL2 polylinker. The β-lactamase gene conferring resistance to ampicillin (Ap^r), the *lac* repressor gene (*lacI^q*), and the *rrnB* 5S ribosomal RNA gene correspond to shaded areas. The plasmid origin (Ori) of replication is shown by an open arrow. The *trc* promoter (arrowhead) is followed by the *rrnB* antitermination region (at), the translational initiation region including the translational enhancer from T7 gene *10* (*g10*), and a minicistron (MC) upstream from the ATG-containing *Nco*I site. For clarity, the following sites that are both in the superlinker and the remainder of the vector have been omitted (in brackets are the number of sites found outside of the linker): *Apa*LI (5), *Ase*I (5), *Bsg*I (4), *Pfl*MI (1), *Tth*111I (1), 2A (4), 4D (2), 5A (1), 10B (4), 13A (2), and 15D (1). Also not shown are the following sites that are not in the linker, but in the vector: *Bbs*I (2), *Bcg*I (1), *Bsa*I (2), *Bst*XI (3), *Eam*1105I (1), *Ear*I (3), *Eco*57I (2), *Gsu*I (2), and *Xcm*I (3). The two *Drd*I sites (not present in the superlinker) are marked with an asterisk. Not present in the entire plasmid are *Asc*I, *Bpu*1102I, *Eco*NI, *Esp*3I, *Pac*I, *Pme*I, *Ppu*MI, *Rsr*II, *Sgr*AI, *Srf*I, *Sse*8387I, and *Swa*I. For clarity, only the *Nco*I and *Hin*dIII sites of the polylinker, located at its 5′ and 3′ ends, respectively, are shown. The numbers of the inner circle give the position on the map in base pairs.

of pSL350, with a shorter polylinker from pTrc99A,[9] yielding pKK489-7, is also available to facilitate future plasmid modifications involving certain restriction sites. Thereafter, the long SL2 superlinker can be reincorporated.

5. Plasmid pSL300, and especially pSL350/1, also have the potential to double as expression vectors. With pSL300 only fusion proteins with N-terminal extensions can be generated. In contrast, pSL350/1 can direct overproduction of unfused proteins when the coding regions are ligated to the unique *Nco*I site that includes the ATG start codon at the 5′ end of the polylinker. This ATG start codon is preceded by regulatory elements for efficient translation initiation, including an upstream minicistron in pSL350/1 (Fig. 4). However, the potential to express a protein encoded by an inserted fragment could also lead to cloning difficulties: If the expressed protein is toxic to the cell, certain inserts may be selected against. If such a scenario is suspected, cloning of a given fragment still could be achieved (without color selection) by down regulating the *lac* promoter with glucose in a *lacI^q^* strain (see above).

Expression Vectors

Plasmids pSE220, pSE280, pSE380, and pSE420 were designed exclusively as expression vectors for the highly efficient production of recombinant proteins in *Escherichia coli*. All vectors feature the strong *trp*/*lac* fusion promoter *trcp* that can be repressed by cellular Lac repressor and induced with IPTG.[10] The difference between pSE220 and its parent pKK233-2,[11] which so far has been used in the overproduction of well over 100 different proteins, is the larger polylinker SL2. Expression vectors in this summary feature an ATG start codon within a unique *Nco*I site located at the 5′ end of the superlinker at the appropriate distance from regulatory elements for translation initiation. For expressing unfused proteins this ATG will be the first choice for joining the 5′ end of the gene to be overexpressed. All the remaining restriction sites are useful to accommodate almost any combination 3′ to the protein-coding region for cloning in the appropriate orientation.

A shorter version, pSE280, has been generated by eliminating the inactive *tet* gene of pSE220. pSE380 carries its own gene for the Lac repressor (*lacI^q^*). Consequently, any *E. coli* strain with features that may be advantageous for protein overexpression (e.g., protease deficiency) can be used as host. While pSE220, pSE280, and pSE380 contain only the

[9] E. Amann, B. Ochs, and K.-J. Abel, *Gene* **69,** 301 (1988).

[10] J. Brosius, M. Erfle, and J. Storella, *J. Biol. Chem.* **260,** 3539 (1985).

[11] E. Amann and J. Brosius, *Gene* **40,** 183 (1985).

```
   .   Ptrc .          .          .         OP.          .
TTGACAATTAATCATCCGGCTCGTATAATGTGTGGAATTGTGAGCGGATAACAATTTCAC

                                                rrnB antiterm
ACAGGAAACAGCGCCGCTGAGAAAAAGCGAAGCGGCACTGCTCTTTAACAATTTATCAGA

                                                 g10
CAATCTGTGTGGGCACTCGACCGGAATTGGGCATCGATTAACTTTATTATTAAAAATTAA

 RBS                   MC            RBS          NcoI
AGAGGTATATATTAATGTATCGATTAAATAAGGAGGAATAAACCATGGCTGGTGACCACG
               MetTyrArgLeuAsnLysGluGlu***    MetAlaGlyAspHisV

TCGTGGAATGCCTTCGAATTCAGCACCTGCACATGGGA...   superlinker SL2
alValGluCysLeuArgIleGlnHisLeuHisMetGly...
```

FIG. 7. Sequence of relevant area in pSE420. Every tenth nucleotide in the first line is marked by a dot. The −35 and −10 regions of the *trc* promoter are underlined. The *lac* operator (OP), the *rrnB* antiterminator region, and the bacteriophage T7 gene *10* translation enhancer (*g10*) are shown in italics. Ribosome-binding sites (RBS) preceding the ATG start codons (in italic) of the minicistron (MC) and the superpolylinker are underlined. Stop codons are designated by three asterisks. The deduced amino acid sequences of the minicistron and the linker are shown. Only the 5′ portion of the superlinker region is shown. The *Nco*I cloning site is underlined.

ribosome-binding site (Shine–Dalgarno sequence) (RBS) from *lacZ*, pSE420 contains signals of translation initiation that separately have been proven to be highly efficient. Therefore, in addition to all the characteristics of pSE380, the further improved vector pSE420 contains the bacteriophage T7 gene *10* translation enhancer,[12] an upstream minicistron for efficient translational restart,[13] and ribosome-binding sites that are flanked by A/T-rich regions[14] (see Fig. 7). Furthermore, in pSE420 the transcriptional antitermination region from the *E coli rrnB* ribosomal RNA operon[15] has been placed downstream from the *trc* promoter. This may facilitate transcription through highly structured areas of the recombinant messenger RNA and thus reduce the possibility of pausing and/or premature termination of transcription by the host RNA polymerase.

As with the majority of expression systems, total repression of promoter activity cannot be achieved in the vectors summarized here. Occasionally this may lead to problems when even small levels of the recombinant protein are toxic to the host cell.

[12] P. O. Olins, C. S. Devine, S. H. Rangwala, and K. S. Kavka, *Gene* **73,** 227 (1988).

[13] B. Schoner, R. M. Belagaje, and R. G. Schoner, *Proc. Natl. Acad. Sci. U.S.A.* **83,** 8506 (1986).

[14] M. K. Olsen, S. K. Rockenbach, K. A. Curry, and C.-S. C. Tomich, *J. Biotechnol.* **9,** 179 (1989).

[15] J. Brosius, T. J. Dull, D. D. Sleeter, and H. F. Noller, *J. Mol. Biol.* **148,** 107 (1981).

When pSE420 has been tested to direct synthesis of rat calmodulin cDNA,[16] yielding pKK480-3, a large percentage of the cellular protein corresponded to the overproduced species. This vector can serve as a control to monitor the appropriate induction and other parameters in gene expression studies. Of course, some of the parameters, e.g., growth medium, length of induction, or temperature after induction,[17] may have to be adjusted from case to case to achieve optimal yields or to maximize the fraction of protein with the native conformation and activity.

Promoter Selection Vectors

pKK232-8, a promoter cloning and selection vector,[18] has been upgraded with superlinker SL2 yielding pKK520-3. This plasmid contains a promoter-deficient chloramphenicol acetyltransferase gene (*cat*). The reporter gene is flanked by *rrnB* transcription terminators to isolate *cat* from plasmid-borne promoters for reduction of background. The polylinker for inserting or shotgunning promoter fragments is proximal to the *cat* reporter gene. Cloning of promoter-containing fragments will confer chloramphenicol resistance to the vector and relative levels of CAT can be assayed.[19] Due to the transcriptional isolation of the cloning region this type of vector has also proven to be useful in cloning strong promoters, that otherwise would interfere with plasmid functions.

Materials and Reagents

Most of the methods for the use of these vectors are sufficiently described elsewhere and in earlier volumes of this series. References are given in the text above. The "Geneclean" kit from BIO 101 (LaJolla, CA) has been used as a source of GlassMilk and buffers to purify DNA from agarose gels for cloning purposes.

Bacteria

XL1-Blue is the *E. coli* strain that has been used for the cloning vectors: *recA1 endA1 gyrA96 thi-1 hsdR17 supE44 relA1 lac* [F′proAB *lacI*^{q}ZΔM15, Tn*10*(Tet^{r})].[20]

[16] A. A. Sherbany, A. S. Parent, and J. Brosius, *DNA* **6,** 267 (1987).

[17] L. H. Chen, P. C. Babbitt, J. R. Vásquez, B. L. West, and G. L. Kenyon, *J. Biol. Chem.* **266,** 12053 (1991).

[18] J. Brosius, *Gene* **27,** 151 (1984).

[19] J. Brosius and J. R. Lupski, this series, Vol. 153, p. 54.

[20] W. O. Bullock, J. M. Fernandez, and J. M. Short, *BioTechniques* **5,** 376 (1987).

Indicator Plates for Color Screening of Insert Containing Clones

Plates are LB (*Luria–Bertani*) containing 100 μg/ml ampicillin, 80 μg/ml X-Gal, and 30 m*M* IPTG: Per liter, 15 g Bacto-agar, 10 g Bacto-tryptone, 5 g yeast extract, and 5 g NaCl are autoclaved. After cooling to about 50° the following solutions are added under swirling: 10 ml of ampicillin (10 mg/ml), 3 ml 100 m*M* IPTG, and 80 mg of X-Gal freshly dissolved in 2 ml dimethylformamide (DMF). Use glass pipettes but not polystyrene pipettes to dispense DMF, because the plastic material is not resistant against the organic solvent.

[42] pBluescriptII: Multifunctional Cloning and Mapping Vectors

By MICHELLE A. ALTING-MEES, J. A. SORGE, and J. M. SHORT

pBluescriptII (Stratagene, La Jolla, CA) phagemid vectors are multifunctional cloning vectors designed to simplify and expedite gene cloning and analysis. The vectors contain polylinkers that have been designed to optimize a number of cloning procedures including directional cloning, gene mapping, unidirectional deletions of predictable length, T3 and T7 polymerase-mediated transcription, blue/white color selection for clones with insert, double-stranded DNA sequencing, and β-galactosidase fusion protein expression. The f1 filamentous phage origin of replication allows preparation of single-stranded DNA for mutagenesis and DNA sequencing. The ColE1 origin of replication and the ampicillin resistance gene allow propagation and selection of clones in plasmid form.

Introduction to pBluescript Vectors

A map of the pBluescript vectors is given in Fig. 1A. The inset defines the orientations of the f1 origin in the plus (+) and minus (−) versions. The polylinkers are shown for pBluescriptII SK(+/−), pBluescriptII KS(+/−), and the original version pBluescriptI SK(+/−). Figure 1B gives the sequence and restriction sites of the pBluescriptII SK polylinker. These polylinkers are located within the α-complementing portion of the β-galactosidase (*lacZ*) gene and allow blue/white color selection of clones containing inserts when used in *Escherichia coli* strains containing the

A)

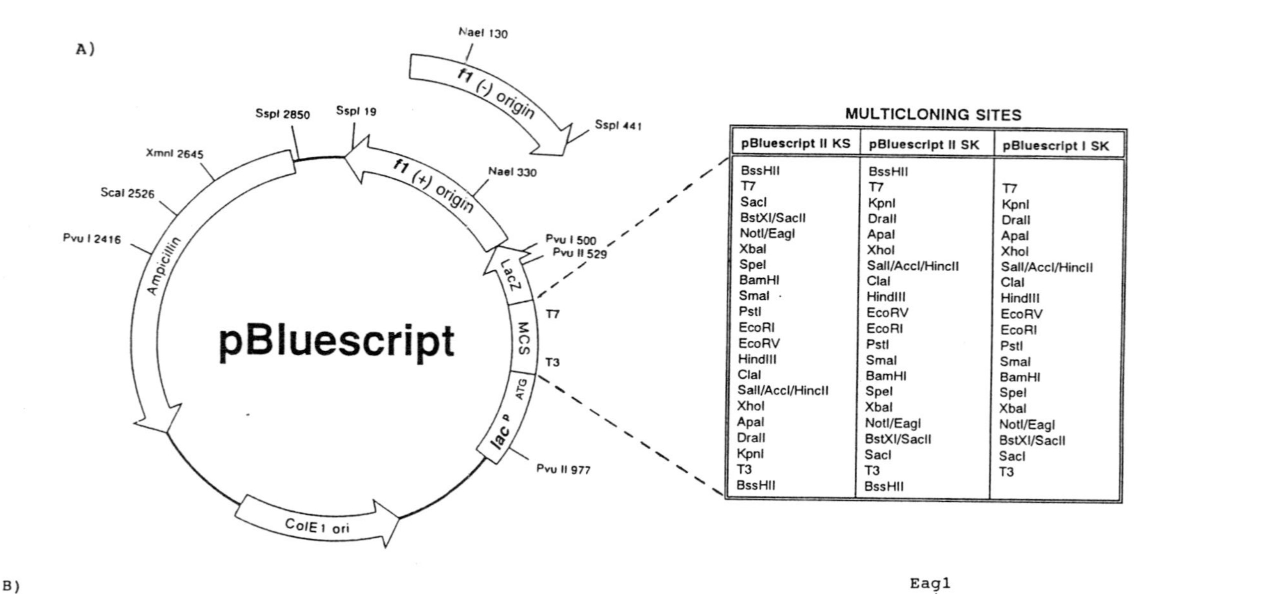

pBluescript II KS	pBluescript II SK	pBluescript I SK
BssHII	BssHII	
T7	T7	T7
SacI	KpnI	KpnI
BstXI/SacII	DraII	DraII
NotI/EagI	ApaI	ApaI
XbaI	XhoI	XhoI
SpeI	SalI/AccI/HincII	SalI/AccI/HincII
BamHI	ClaI	ClaI
SmaI	HindIII	HindIII
PstI	EcoRV	EcoRV
EcoRI	EcoRI	EcoRI
EcoRV	PstI	PstI
HindIII	SmaI	SmaI
ClaI	BamHI	BamHI
SalI/AccI/HincII	SpeI	SpeI
XhoI	XbaI	XbaI
ApaI	NotI/EagI	NotI/EagI
DraII	BstXI/SacII	BstXI/SacII
KpnI	SacI	SacI
T3	T3	T3
BssHII	BssHII	

B)

MET → BssHII T3 promoter +1→ Sac1 SacII BstX1 Eag1 Not1 Xba1 Spe1 BamH1 Sma1 Pst1

5'-GGAAACAGCT**ATG**ACC**ATG**ATTACGCCAAGCGCGCAATTAACCCTCACTAAAGGGAACAAAAGCTGGAGCTCCACCGCGGTGGCGGCCGCTCTAGAACTAGTGGATCCCCCGGGCTGCAG

3'-CCTTTGTCGATACTGGTACTAATGCGGTTCGCGCGTTAATTGGGAGTGATTTCCCTTGTTTTCGACCTCGAGGTGGCGCCACCGCCGGCGAGATCTTGATCACCTAGGGGGCCCGACGTC

ß-Galactosidase →

EcoR1 EcoRV Hind3 Cla1 HincII Acc1 Sal1 Xho1 DraII Apa1 Kpn1 BssHII

GAATTCGATATCAAGCTTATCGATACCGTCGACCTCGAGGGGGGGCCCGGTACCCAATTCGCCCTATAGTGAGTCGTATTACGCGCGCTCACTGGCCGTCGTTTTACAA-3'

CTTAAGCTATAGTTCGAATAGCTATGGCAGCTGGAGCTCCCCCCCGGGCCATGGGTTAAGCGGGATATCACTCAGCATAATGCGCGCGAGTGACCGGCAGCAAAATGTT-5'

← +1 T7 promoter

lacZΔM15 mutation.[1] The SK notation designates the orientation of the polylinker where the *Sac*I site is closest to the T3 promoter and upstream of the *Kpn*I site within the *lacZ* transcript. In KS vectors, the *Kpn*I site is closest to the T3 promoter and upstream of the *Sac*I site. The choice of polylinkers allows directional cloning in either orientation relative to the *lac* promoter.

The complete sequence of pBluescriptII SK(−) is given in Fig. 2 and the caption describes the locations of each of the features of the vectors. pBluescript vectors are derived from pUC19,[2,3] a derivative of pBR322.[4] As indicated in Fig. 1, pBluescript vectors contain the ampicillin resistance gene. Sequences from pUC19 have been removed such that the pBluescript vectors contain the ColE1 origin of replication and the ampicillin resistance gene corresponding to base pairs (bp) 2348–4284 of pBR322. This removes all but 140 bp of the "poison" sequences associated with pBR322 which have been shown to inhibit eukaryotic expression of clones containing Simian virus 40 (SV40) sequences.[5,5a] Chloramphenicol-resistant versions of pBluescriptII vectors have also been developed for use when ampicillin resistance is not appropriate (pBC vectors not shown). Kanamycin-resistant versions have been constructed which also allow expression in eukaryotic cells (manuscript in preparation). pBluescript vectors, like the pUC vectors, replicate at a high copy number (greater than 150 copies per cell) (P. Kretz, personal communication, 1991) due to the lack of the pBR322 *rom* gene.[6]

The original version, pBluescriptI SK(−), is present in both λ ZAPI and λ ZAPII phage vectors.[2] The phagemid can be isolated from the λ sequences through the use of the filamentous phage origin and the *in vivo* excision process described by Short *et al.*[2,7] pBluescriptI SK(−) differs

[1] A. Ullmann and D. Perrin, *in* "The Lactose Operon" (. Beckwith and . Zipser, eds.), p. 143. Cold Spring Harbor Laboratory, Cold Spring Harbor, New York, 1970.

[2] J. M. Short, J. M. Fernandez, J. A. Sorge, and W. D. Huse, *Nucleic Acids Res.* **16,** 7583 (1988).

[3] M. A. Alting-Mees and J. M. Short, *Nucleic Acids Res.* **17,** 9494 (1989).

[4] C. Yannisch-Perron, J. Vieira, and J. Messing, *Gene* **33,** 103 (1985).

[5] M. Lusky and M. Botchan, *Nature (London)* **293,** 79 (1981).

[5a] D. O. Peterson, K. K. Beifuss, and K. L. Morley, *Mol. Cell. Biol.* **7,** 1563 (1987).

[6] M. Brenner and J.-I. Tomizawa, *Nucleic Acids Res.* **17,** 4309 (1989).

[7] J. M. Short and J. A. Sorge, this volume [43].

Fig. 1. (A) Map of pBluescript(+) vectors: Inset f1(−) indicates the f1 origin in the minus orientation for pBluescript (−) vectors. Ampicillin resistance (*amp*^{r}) gene, ColE1 origin, *lac* promoter (*lac*^{P}) and α-complementing β-galactosidase (*lacZ*) gene fragment are indicated. Multicloning sites (MCS) for vectors pBluescriptII SK, pBluescriptII KS, and the original pBluescriptII SK are listed. (B) Sequence of pBluescriptII SK polylinker: β-gal ATG is underlined. Restriction sites and T3/T7 promoters are indicated by overlapping brackets. The first base of T3 and T7 transcripts and their direction of polymerization are shown.

```
1    CTGACGCGCC CTGTAGCGGC GCATTAAGCG CGGCGGGTGT GGTGGTTACG CGCAGCGTGA
61   CCGCTACACT TGCCAGCGCC CTAGCGCCCG CTCCTTTCGC TTTCTTCCCT TCCTTTCTCG
121  CCACGTTCGC CGGCTTTCCC CGTCAAGCTC TAAATCGGGG GCTCCCTTTA GGGTTCCGAT
181  TTAGTGCTTT ACGGCACCTC GACCCCAAAA AACTTGATTA GGGTGATGGT TCACGTAGTG
241  GGCCATCGCC CTGATAGACG GTTTTTCGCC CTTTGACGTT GGAGTCCACG TTCTTTAATA
301  GTGGACTCTT GTTCCAAACT GGAACAACAC TCAACCCTAT CTCGGTCTAT TCTTTTGATT
361  TATAAGGGAT TTTGCCGATT TCGGCCTATT GGTTAAAAAA TGAGCTGATT TAACAAAAAT
421  TTAACGCGAA TTTTAACAAA ATATTAACGC TTACAATTTC CATTCGCCAT TCAGGCTGCG
481  CAACTGTTGG GAAGGGCGAT CGGTGCGGGC CTCTTCGCTA TTACGCCAGC TGGCGAAAGG
541  GGGATGTGCT GCAAGGCGAT TAAGTTGGGT AACGCCAGGG TTTTCCCAGT CACGACGTTG
601  TAAAACGACG GCCAGTGAGC GCGCGTAATA CGACTCACTA TAGGGCGAAT TGGGTACCGG
661  GCCCCCCCTC GAGGTCGACG GTATCGATAA GCTTGATATC GAATTCCTGC AGCCCGGGGG
721  ATCCACTAGT TCTAGAGCGG CCGCCACCGC GGTGGAGCTC CAGCTTTTGT TCCCTTTAGT
781  GAGGGTTAAT TGCGCGCTTG GCGTAATCAT GGTCATAGCT GTTTCCTGTG TGAAATTGTT
841  ATCCGCTCAC AATTCCACAC AACATACGAG CCGGAAGCAT AAAGTGTAAA GCCTGGGGTG
901  CCTAATGAGT GAGCTAACTC ACATTAATTG CGTTGCGCTC ACTGCCCGCT TTCCAGTCGG
961  GAAACCTGTC GTGCCAGCTG CATTAATGAA TCGGCCAACG CGCGGGGAGA GGCGGTTTGC
1021 GTATTGGGCG CTCTTCCGCT TCCTCGCTCA CTGACTCGCT GCGCTCGGTC GTTCGGCTGC
1081 GGCGAGCGGT ATCAGCTCAC TCAAAGGCGG TAATACGGTT ATCCACAGAA TCAGGGGATA
1141 ACGCAGGAAA GAACATGTGA GCAAAAGGCC AGCAAAAGGC CAGGAACCGT AAAAAGGCCG
1201 CGTTGCTGGC GTTTTTCCAT AGGCTCCGCC CCCCTGACGA GCATCACAAA AATCGACGCT
1261 CAAGTCAGAG GTGGCGAAAC CCGACAGGAC TATAAAGATA CCAGGCGTTT CCCCCTGGAA
1321 GCTCCCTCGT GCGCTCTCCT GTTCCGACCC TGCCGCTTAC CGGATACCTG TCCGCCTTTC
1381 TCCCTTCGGG AAGCGTGGCG CTTTCTCATA GCTCACGCTG TAGGTATCTC AGTTCGGTGT
1441 AGGTCGTTCG CTCCAAGCTG GGCTGTGTGC ACGAACCCCC CGTTCAGCCC GACCGCTGCG
1501 CCTTATCCGG TAACTATCGT CTTGAGTCCA ACCCGGTAAG ACACGACTTA TCGCCACTGG
1561 CAGCAGCCAC TGGTAACAGG ATTAGCAGAG CGAGGTATGT AGGCGGTGCT ACAGAGTTCT
1621 TGAAGTGGTG GCCTAACTAC GGCTACACTA GAAGGACAGT ATTTGGTATC TGCGCTCTGC
1681 TGAAGCCAGT TACCTTCGGA AAAAGAGTTG GTAGCTCTTG ATCCGGCAAA CAAACCACCG
1741 CTGGTAGCGG TGGTTTTTTT GTTTGCAAGC AGCAGATTAC GCGCAGAAAA AAAGGATCTC
1801 AAGAAGATCC TTTGATCTTT TCTACGGGGT CTGACGCTCA GTGGAACGAA AACTCACGTT
1861 AAGGGATTTT GGTCATGAGA TTATCAAAAA GGATCTTCAC CTAGATCCTT TTAAATTAAA
1921 AATGAAGTTT TAAATCAATC TAAAGTATAT ATGAGTAAAC TTGGTCTGAC AGTTACCAAT
1981 GCTTAATCAG TGAGGCACCT ATCTCAGCGA TCTGTCTATT TCGTTCATCC ATAGTTGCCT
2041 GACTCCCCGT CGTGTAGATA ACTACGATAC GGGAGGGCTT ACCATCTGGC CCCAGTGCTG
2101 CAATGATACC GCGAGACCCA CGCTCACCGG CTCCAGATTT ATCAGCAATA AACCAGCCAG
2161 CCGGAAGGGC CGAGCGCAGA AGTGGTCCTG CAACTTTATC CGCCTCCATC CAGTCTATTA
2221 ATTGTTGCCG GGAAGCTAGA GTAAGTAGTT CGCCAGTTAA TAGTTTGCGC AACGTTGTTG
2281 CCATTGCTAC AGGCATCGTG GTGTCACGCT CGTCGTTTGG TATGGCTTCA TTCAGCTCCG
2341 GTTCCCAACG ATCAAGGCGA GTTACATGAT CCCCCATGTT GTGCAAAAAA GCGGTTAGCT
2401 CCTTCGGTCC TCCGATCGTT GTCAGAAGTA AGTTGGCCGC AGTGTTATCA CTCATGGTTA
2461 TGGCAGCACT GCATAATTCT CTTACTGTCA TGCCATCCGT AAGATGCTTT TCTGTGACTG
2521 GTGAGTACTC AACCAAGTCA TTCTGAGAAT AGTGTATGCG GCGACCGAGT TGCTCTTGCC
2581 CGGCGTCAAT ACGGGATAAT ACCGCGCCAC ATAGCAGAAC TTTAAAAGTG CTCATCATTG
2641 GAAAACGTTC TTCGGGGCGA AAACTCTCAA GGATCTTACC GCTGTTGAGA TCCAGTTCGA
2701 TGTAACCCAC TCGTGCACCC AACTGATCTT CAGCATCTTT TACTTTCACC AGCGTTTCTG
2761 GGTGAGCAAA AACAGGAAGG CAAAATGCCG CAAAAAAGGG AATAAGGGCG ACACGGAAAT
2821 GTTGAATACT CATACTCTTC CTTTTTCAAT ATTATTGAAG CATTTATCAG GGTTATTGTC
2881 TCATGAGCGG ATACATATTT GAATGTATTT AGAAAAATAA ACAAATAGGG GTTCCGCGCA
2941 CATTTCCCCG AAAAGTGCCA C
```

FIG. 2. Sequence of pBluescriptII SK(−): f1 origin, from base pairs 3–459, is boxed and printed in bold type. The complement and inverse of this region would be present in the sequence of the plus-orientation f1 origin. *Bss*HII sites and multicloning sites are underlined and printed in bold type. T3 and T7 promoter regions are boxed (bp 772–791 and 626–645, respectively). Start sites for the *β-gal* gene are underlined and printed in bold type (bp 810 and 816). The ATG at bp 816 is preferentially used during translation.[31] *β-gal* sequences continue to bp 460. Note that the anticoding strand of *β-gal* is shown in this sequence. The *lac* promoter runs from bp 938 to 817, with a vestigal 3′ portion of *lacI* DNA (bp 1031–939). The ColE1 origin of replication lies between bp 1032 and 1972. The *β*-lactamase gene and promoter (*amp*^{r}) are between bp 2–0/2961–1973. Start and stop codons are printed in boldface type and underlined.

from pBluescriptII SK(−) only by the two *Bss*HII sites that flank the polylinker in version II.

The KS versions of the pBluescriptI vectors also exist (not shown). However, a point mutation was identified within this polylinker (the cytosine nucleotide flanking the 5′ end of the *Kpn*I site is deleted) and care must be taken that this frameshift is accounted for when fusion protein constructs are being prepared. This sequence was corrected in the pBluescriptII KS(+/−) vectors and therefore these vectors are recommended for expression experiments where the KS polylinker orientation is desired.

Methods

General

All media preparations, bacterial transformations, DNA preparations, ligations, Southern blots, and hybridizations can be performed following standard procedures.[8] *Escherichia coli* strains used were XL1-Blue {λ- *recA1 endA1 gyrA96 thi-1 hsdR17*(rk$^-$, mk$^+$) *supE44 relA1 lac* [F′ *proAB lacI^{q}ZΔM15* Tn*10*(Tet^r)]}[9] and SURE {λ$^-$, el4$^-$ (*mcrA*), Δ(*mcrCB-hsd SMR-mrr*) *171 endA1 supE44 thi-1 gyrA96 relA1 lac recB recJ sbcC umuC*::Tn5(Kan^r) *uvrC* [F′ *proAB lacI^{q}ZΔM15*, Tn*10*(Tet^r)]}.[10] All DNA and RNA modification enzymes were purchased from Stratagene Cloning Systems.

Gene Cloning

The versatility of pBluescript vectors alleviates the need for successive subcloning. With 21 restriction sites, in two orientations, most cloned genes can be inserted directionally into the pBluescript polylinkers. Polymerase chain reaction (PCR)-amplified fragments from RNA, cloned DNA, or genomic DNA can be inserted directly into pBluescript vectors, bypassing the need to construct large libraries. If a new gene of unknown sequence is to be cloned, libraries will need to be generated for screening. Lambda phages are still the vectors of choice for generating libraries with greater than 10^6 members. Cloning into λ ZAPI or λ ZAPII vectors allows the library to be constructed, amplified, and screened in λ phage. Because lambda ZAP contains pBluescriptI SK(−), clones can be excised as a pBluescriptI SK(−) phagemid from the λ vector with the aid of filamentous helper phage,[2,7] thus combining the benefits of λ libraries and pBluescriptI SK(−) phagemid clones.

[8] J. Sambrook, E. M. Fritsch, and T. Maniatis, "Molecular Cloning: A Laboratory Manual," 2nd Ed., Cold Spring Harbor Laboratory, Cold Spring Harbor, New York, 1989.

[9] W. O. Bullock, J. M. Fernandez, and J. M. Short, *BioTechniques* **5,** 376 (1987).

[10] A. Greener, *Strategies Mol. Biol.* **3,** 5 (1990).

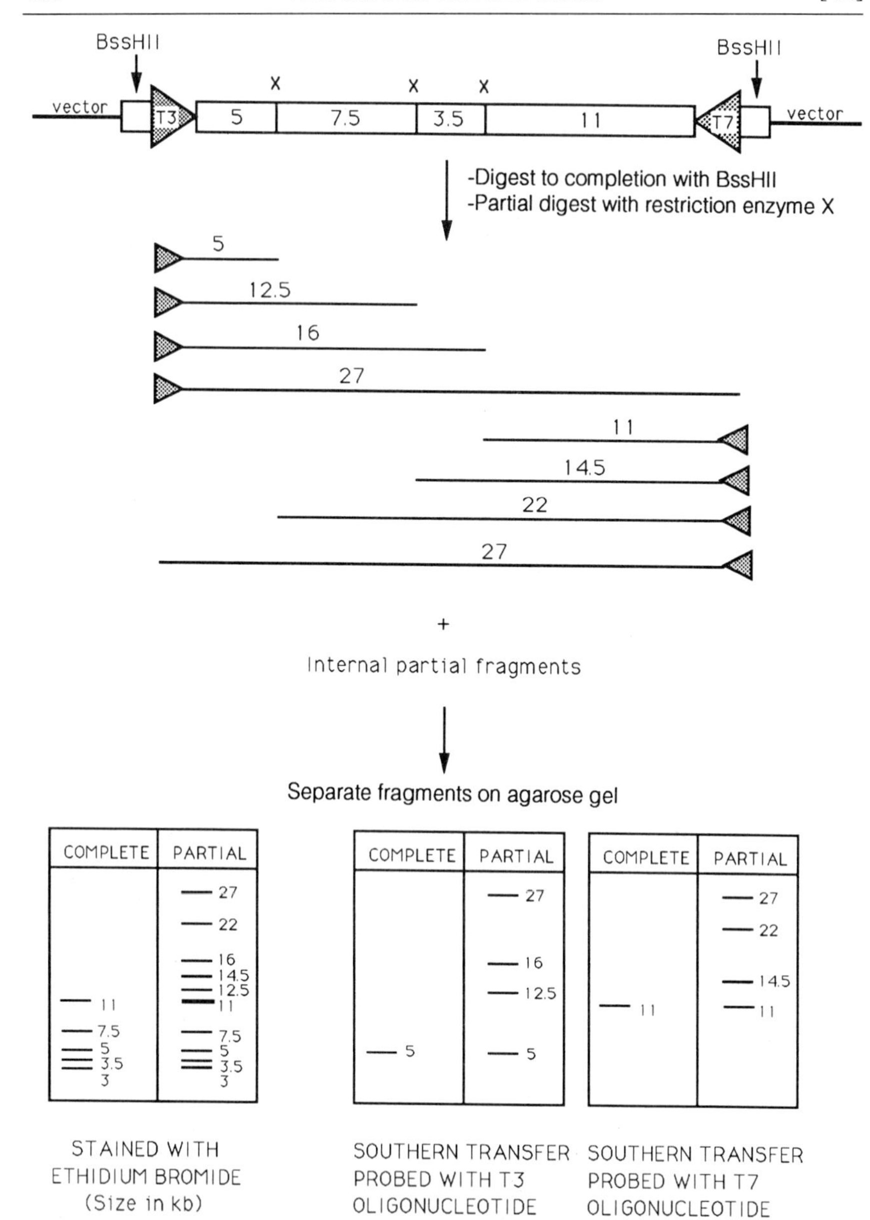

FIG. 3. Gene mapping with *Bss*HII sites flanking T3 and T7 promoter sequences. The insert is cloned into a pBluescriptII polylinker. Plasmid DNA is digested to completion with *Bss*HII. Subsequently, partial digests with the restriction enzyme of choice (in this case enzyme X) are performed using 1 unit of enzyme per microgram of *Bss*HII-digested DNA under the appropriate temperature and buffer conditions. An alternative method for obtaining

Gene Mapping

pBluescriptII polylinkers have unique *Bss*HII restriction sites flanking the T3 and T7 promoters (see Fig. 1). These sites allow easy and rapid mapping of large and small inserts using the method described by Wahl *et al.*[11] A protocol is outlined in Fig. 3. Briefly, complete *Bss*HII digests are performed on the clone to be mapped, followed by partial restriction digests with the mapping enzymes of choice. The partial fragments are subjected to agarose gel electrophoresis, transferred to a hybridization membrane, and probed with labeled T3 or T7 oligonucleotides to determine the lengths of the partial fragments. These lengths identify the location of the enzyme sites relative to the *Bss*HII sites. Chemiluminescent[12] or radioactive probes can be used. Many different restriction enzymes can be mapped on a single Southern blot using the T3 probe and confirmed on a duplicate blot using the T7 probe.

Exonuclease III Unidirectional Deletions

The restriction enzyme sites in the pBluescript polylinkers have been arranged to allow unidirectional deletions of inserts with exonuclease III (ExoIII).[13,14,14a] Deletions of any size can be obtained by this technique (a few bases to many kilobases) at any position within the clone where appropriate restriction sites exist. Exonuclease III is a 3′ → 5′ single-stranded exonuclease with a requirement for double-stranded template DNA. Therefore, ExoIII will not digest 3′ single-stranded overhangs, but will exclusively digest the 3′ ends from blunt ends or 5′ overhangs. The opposite DNA strand remains intact and single stranded. Unique restric-

[11] G. M. Wahl, *Strategies Mol. Biol.* **2,** 1 (1989).

[12] M. Troutman, A. Pitta, K. Considine, and J. Braman, *Strategies Mol. Biol.* **4,** 5 (1991).

[13] C. C. Richardson, I. R. Lehman, and A. Kornberg, *J. Biol. Chem.* **239,** 251 (1964).

[14] R. Roychoudhury and R. Wu, *J. Biol. Chem.* **252,** 4786 (1977).

[14a] L. H. Guo, R. C. A. Yang, and R. Wu, *Nucleic Acids Res.* **11,** 5521 (1983).

---

partial digests is described by J. E. Cleaver, L. Samson, and G. H. Thomas, *Biochim. Biophys. Acta* **697,** 255 (1982) and P. A. Whittaker and E. M. Southern, *Gene* **41,** 129 (1986). Aliquots of the digests are terminated at various time points (i.e., 5, 10, 20, and 30 min) by adding EDTA to a concentration of 0.1 *M* and placing on ice. This releases a 3-kb vector band and a myriad of partially digested insert fragments that can be visualized on an ethidium bromide-stained agarose gel. The partial digests are run on an agarose gel, transferred to a hybridization membrane, and hybridized with either radiolabeled or chemiluminescent oligonucleotide probes. The size of the bands, relative to standard DNA size markers run on the same gel, determines the distance of the mapping enzyme site from the appropriate *Bss*HII site.

tion sites that leave single-stranded 3′ overhangs are located on the outside ends of pBluescript polylinkers, while 5′ protuding and blunt restriction sites are located in the central portion of the polylinkers. Therefore, double digestion with a 3′ and a 5′ or blunt restriction enzyme, followed by ExoIII digestion and subsequent removal of the remaining single-stranded DNA by mung bean nuclease, results in unidirectional deletions of insert DNA.

This method has many applications. For instance, regions coding for functional domains or portions of domains can be removed and promoter regions analyzed. Serial deletions can be made to aid sequencing of large inserts.[14a,15]

Protocol. Cesium chloride banded DNA[8] is digested to completion with a restriction enzyme that cleaves the clone once in the region to be deleted and leaves a 5′ protruding or blunt end. A second restriction enzyme is then used, which leaves a 4-bp 3′ overhang and protects the region that should not be deleted, usually the end closest to the vector sequences. If no 3′ overhang enzyme site can be found, a 5′ overhang enzyme site filled in with deoxythio derivatives and Klenow[16] will protect equally well. One hundred micrograms of double-digested DNA is treated with ExoIII (20 U/μg DNA) in 500 μl of 50 m*M* Tris-HCl, pH 8.0, 5 m*M* $MgCl_2$. Several time points are taken to obtain the desired insert deletion range. Typical rates of digestion are 400 bp/min at 37° and 125 bp/min at 23°. The reaction is stopped by removing 25 μl at each time point (5 μg/time point), adding it to 175 μl of mung bean nuclease buffer [30 m*M* sodium acetate, pH 5.0, 50 m*M* NaCl, 1 m*M* $ZnCl_2$, 5% (v/v) glycerol], and placing on dry ice. After all time points have been collected, the aliquots are heated at 68° for 15 min, then placed on ice. The remaining single-stranded DNA is removed by treatment with mung bean nuclease, 15 U/time point at 30° for 30 min. The mung bean nuclease is removed by phenol–chloroform extraction (1 : 1) in the presence of 0.4% (w/v) sodium dodecyl sulfate (SDS), 50 m*M* Tris-HCl, pH 9.5, and 0.8 *M* LiCl. The aqueous phase is precipitated in the presence of 0.3 *M* sodium acetate, pH 5.2, and 2 vol of ethanol. An aliquot of each time point is run on an agarose gel to confirm whether the deletions are of the desired size. DNA bands can be excised and eluted[8] to minimize screening in future steps. DNA from these time points (10–100 μg) is self-ligated in 20 μl of 50 m*M* Tris-HCl, pH 7.5, 7 m*M* $MgCl_2$, 1 m*M* dithiothreitol (DTT), 0.5 m*M* ATP, and 4 U of T4 DNA ligase to circularize the fragments. The ligated DNA is transformed into competent XL1-Blue cells, and individual clones analyzed by restriction analysis, sequencing, or functional assays. Obtaining several different mutants from a single

[15] S. Henikoff, *Gene* **28,** 351 (1984).

[16] S. D. Putney, S. J. Benkovic, and P. R. Schimmel, *Proc. Natl. Acad. Sci. U.S.A.* **78,** 7350 (1981).

time point or from several time points may be desirable when comparing the effects of different mutations, or when one does not know precisely which mutation will produce the desired effect.

Single-Stranded DNA Rescue

Single-stranded DNA vectors, such as the M13mp series, have been shown to spontaneously delete with inserts greater than 2000 bp in length. This is likely due to the location of the *lacZ* gene and polylinker within the M13 origin.[17] By interrupting the M13 origin of replication and by increasing the phage's genome size, clones with large inserts are less efficient at replicating. Therefore, smaller clones and clones with deletions develop a selective advantage and outgrow clones with large or intact inserts. In contrast, pBluescript polylinkers are not located within an origin of replication. Also, phagemid vectors do not enter the single-stranded state unless filamentous helper phage proteins are present. Maintaining the vector in the double-stranded state serves to further stabilize the cloned insert during propagation. The presence of the f1 origin within phagemid vectors permits production of single-stranded DNA (ssDNA) and phagemid particles when required, thereby acquiring the benefits of filamentous phage vectors without the deleterious characteristics.

The f1 origin in pBluescript contains all the cis-acting sequences required for packaging and replication of filamentous phage. All protein factors required in trans are supplied by helper phage infection of *E. coli* harboring the pBluescript clone. On infection with helper phage, gene II is expressed by the helper phage and promotes single-stranded replication of the pBluescript clone at the f1 origin. This ssDNA is packaged by structural proteins also coded for by the helper phage and extruded, as are the helper phage, through the bacterial membrane into the growth medium. The *lacZ* coding strand is packaged by the f1(+) origin and the noncoding strand is packaged from vectors containing the f1(−) origin.

To obtain ssDNA it is essential to grow the clone in an *E. coli* strain containing an F′ episome coding for pili because the helper filamentous phage require pili for infection. If the clone is especially unstable due to the presence of repeated sequences, and/or Z-DNA, maintaining the clone in an *E. coli* strain such as SURE[10] which carries mutations in various DNA repair genes, is recommended. Otherwise the XL1-Blue strain is more commonly used.

Protocol. Several protocols have been published for the production of ssDNA from phagemids.[8,18] Yields of ssDNA depend on the specific insert

[17] G. P. Smith, *in* "Vectors: A Survey of Molecular Cloning Vectors and Their Uses" (R. L. Rodriguez and D. T. Denhardt, eds.), p. 61. Butterworth, Stoneham, England, 1988.

[18] C. Katayama, *Strategies Mol. Biol.* **3,** 56 (1990).

sequence and on which helper phage is used.[19] It may be advisable to use various helper phage strains and growth conditions to determine which conditions are optimal for the individual clone. The clone should be streaked on plates providing appropriate selection for both the clone and the F′ factor (e.g., pBluescript clones in XL1-Blue should be streaked on ampicillin/tetracycline/LB plates). An overnight culture is grown without antibiotics by inoculating 5 ml of LB medium with a colony from the freshly selected bacterial streak. The next day three milliliters of 2× YT[8] is inoculated with 0.1 vol (300 μl) of the fresh overnight culture and grown at 37° to log phase (OD_{600} 0.3). Larger cultures can be grown to increase yields. Replication-compromised helper phage (e.g., R408 or VCSM13)[20] are added at a multiplicity of infection (moi) between 20 : 1 and 1 : 1 (helper phage–cells) and grown at 37° with shaking for 8 hr. Several moi can be tested in parallel to determine the optimum for the clone of interest. The cells are spun down at 12,000 *g* for 10 min at 4° in an Eppendorf centrifuge and the supernatants transferred to fresh tubes. One-quarter volume of 3.5 *M* ammonium acetate, pH 7.5, 20% (w/v) polyethylene glycol (PEG) is added to the supernatants, mixed well, and incubated at room temperature for 15 min to precipitate the phage. The tubes are spun for 20 min at 12,000 *g* at 4° and the supernatants drained well to remove all remaining PEG. The pellets are resuspended in 200 μl TE [10 m*M* Tris, pH 8.0, 1 m*M* ethylenediaminetetraacetic acid (EDTA)] and repeatedly extracted with TE-saturated phenol–chloroform (1 : 1) until the interface remains clear (usually two or three times). This strips the phage proteins from the ssDNA. The aqueous phase is transferred to a fresh tube and the ssDNA is precipitated by adding 0.1 vol of 3 *M* sodium acetate, pH 5.4 and 2 vol of 100% ethanol, mixing well, and centrifuging at 12,000 *g* at 4° for 10 min. The supernatant is removed and the pellet washed with 80% ethanol. The pellet, containing 1–5 μg of ssDNA, is resuspended in 20 μl of TE and can be visualized on an ethidium bromide-stained 1% (w/v) agarose gel. Single-stranded pBluescript without insert runs at approximately 1.6 kb relative to double-stranded DNA standards on agarose gels run in 40 m*M* Tris–acetate, pH 8.0, 1 m*M* EDTA.[8]

Single-stranded DNA can be used as a template for site-directed mutagenesis, to loop in or loop out small or large stretches of DNA, or to introduce point mutations. These mutations are valuable for changing amino acids in the encoded protein, introducing new restriction sites,

[19] D. A. Mead and B. Kemper, *in* "Vectors: A Survey of Molecular Cloning Vectors and Their Uses" (R. L. Rodriguez and D. T. Denhardt, eds.), p. 85. Butterworth, Stoneham, England, 1988.

[20] M. Russel, S. Kidd, and M. R. Kelley, *Gene* **45,** 333 (1986).

changing promoter sequences, etc. If ssDNA is prepared from an *E. coli* strain with *ung* and *dut* mutations, the Kunkel method of mutagenesis can be used.[21,22] Single-stranded DNA can also be used as a template for sequencing.[8] The presence of helper phage DNA in these ssDNA preparations does not interfere with these procedures if oligonucleotide primers that do not hybridize to the helper phage DNA are used.

T3/T7 in Vitro RNA Transcription

Another feature of pBluescriptI and pBluescriptII vectors is the presence of T3 and T7 RNA promoters flanking the polylinker (see Fig. 1B). These 20-bp sequences contain all the information required for initiation of RNA transcription by T3 and T7 RNA polymerases, respectively.[23,24] These polymerases do not cross-initiate[23] and produce highly specific RNA transcripts. T3 transcripts will contain *lacZ* sense strand RNA and T7 transcripts will contain antisense strand RNA.

Protocol. The conditions for producing RNA transcripts vary depending on the intended use of the transcript. A typical transcription reaction would use 1 μg of *Pvu*II-digested, proteinase K-treated DNA template (one can also use *Bss*HII-digested DNA with pBluescriptII vectors) with 1 μl of 10 m*M* rNTPs, 1 μl of 0.75 *M* dithiothreitol, 1 U of RNaseBlockII (Stratagene), 5 μl of 5 × transcription buffer (200 m*M* Tris, pH 8.0, 40 m*M* $MgCl_2$, 10 m*M* spermidine, 250 m*M* NaCl), RNase-free water to bring the final volume to 25 μl, and 10 U of T3 or T7 RNA polymerase. Incubate these components at 37° for 30 min. To remove the DNA template, add 10 U of RNase-free DNase and incubate an additional 15 min at 37°. Extract with phenol–chloroform (1 : 1). Add 0.1 vol 3 *M* sodium acetate, pH 5.2, and precipitate RNA with 2.5 vol of ethanol. Spin in an Eppendorf centrifuge for 2 min at 4° at 12,000 *g* and resuspend the pellet in 100 μl of RNase-free water. To store RNA, add 2.5 vol of 100% ethanol and store at −20°. The concentration of RNA can be determined by taking an OD_{260} reading of the RNA–ethanol suspension (1 OD_{260} = 40 μg/ml). When RNA is required, remove the appropriate aliquot of RNA–ethanol suspension, add a 1/30 vol of 3 *M* sodium acetate, pH 5.2, and pellet the RNA in an Eppendorf centrifuge for 2 min.

These transcripts are useful in a wide variety of applications, including

[21] T. A. Kunkel, *Proc. Natl. Acad. Sci. U.S.A.* **82,** 488 (1985).

[22] T. A. Kunkel, J. D. Roberts, and R. A. Zakour, this series, Vol. 154, p. 367.

[23] J. N. Bailey, J. F. Klement, and W. T. McAllister, *Proc. Natl. Acad. Sci. U.S.A.* **80,** 2814 (1983).

[24] W. T. McAllister and A. D. Carter, *Nucleic Acids Res.* **8,** 4821 (1980).

expression studies by injection into frog oocytes,[25] the study of ribozymes,[26] *in vitro* translation,[27,28] and, when radiolabeled, for riboprobe production.[29,30]

In Vivo Protein Expression

To study the structure and function of a protein it is generally necessary, at some point, to produce milligram quantities of the protein. Purifying a protein from its source is usually rather complex and many proteins are expressed at very limiting levels. This often makes purifying proteins from cloned sources attractive. Clones inserted in the pBluescript polylinker are expressed as β-Gal fusion proteins under the control of the isopropyl-β-D-thiogalactopyranoside (IPTG)-inducible *lac* promoter. For a review on the mechanisms behind induction of the *lac* promoter with LacI protein and IPTG, see Miller and Reznikoff, 1978.[31] Conditions for producing large quantities of protein depend greatly on the properties of the particular protein.[32,33] Choosing the best *E. coli* host strain is often clone specific and therefore difficult to predict. Several different *E. coli* strains should be tested for best production levels.[34] The $lacI^q$ mutation, which produces 10-fold more Lac repressor protein than wild-type *lacI*[35] and results in tighter control of expression prior to induction, is desirable if high expression of the protein is toxic to the host cells. The T7 promoter can also be used to enhance inducibility and control overexpression.[36–38] By cloning the gene under the control of the T7 promoter rather than the *lac* promoter and adding a ribosome-binding site and ATG start site,

[25] L. E. Iverson, M. A. Tanouye, H. A. Lester, N. Davidson, and B. Rudy, *Proc. Natl. Acad. Sci. U.S.A.* **85,** 5723 (1988).

[26] G. van der Horst, A. Christian, and T. Inoue, *Proc. Natl. Acad. Sci. U.S.A.* **88,** 184 (1991).

[27] P. Davanloo, A. H. Rosenberg, J. J. Dunn, and F. W. Studier, *Proc. Natl. Acad. Sci. U.S.A.* **81,** 2035 (1984).

[28] E. Mather, *Strategies Mol. Biol.* **22,** 32 (1989).

[29] B. Hay, L. Y. Jan, and Y. N. Jan, *Cell (Cambridge, Mass.)* **55,** 987 (1988).

[30] W. M. Roberts, A. T. Look, M. F. Roussel, and C. J. Sherr, *Cell (Cambridge, Mass.)* **55,** 655 (1988).

[31] J. H. Miller and W. S. Reznikoff, eds. "The Operon." Cold Spring Harbor Laboratory, Cold Spring Harbor, NY, 1978.

[32] S. A. Cho, S. Scharpf, M. Franko, and C. W. Vermeulen, *Biochem Biophys. Res. Commun.* **128,** 1268 (1985).

[33] W. P. Voth and C. S. Lee, *Nucleic Acids Res.* **17,** 6424 (1989).

[34] J. Hatt, M. Callahan, and A. Greener, *Strategies Mol. Biol.* **5,** 2 (1992).

[35] B. Muller-Hill, L. Crapo, and W. Gilbert, *Proc. Natl. Acad. Sci. U.S.A.* **59,** 1259 (1968).

[36] S. Tabor and C. C. Richardson, *Proc. Natl. Acad. Sci. U.S.A.* **82,** 1074 (1985).

[37] F. W. Studier and B. A. Moffatt, *J. Mol. Biol.* **189,** 113 (1986).

[38] O. Elroy-Stein, T. R. Fuerst, and B. Moss, *Proc. Natl. Acad. Sci. U.S.A.* **86,** 6126 (1989).

one can keep expression tightly turned off until T7 RNA polymerase is expressed in the cell. Both eukaryotic and prokaryotic expression systems have been described using T7 polymerase. In some cases the product of the cloned gene has been reported to accumulate to a level greater than 50% of total cellular protein.[36–38]

Concluding Remarks

pBluescript vectors are multifunctional vectors. Many tests can be performed on the cloned insert with minimal effort and without requiring cumbersome subcloning steps into "specialized" vectors. Most of the features described in this chapter are available in kit format from commercial sources, which makes the experiments easier, more efficient, and reliable.

[43] *In Vivo* Excision Properties of Bacteriophage λ ZAP Expression Vectors

By JAY M. SHORT and JOSEPH A. SORGE

Introduction

Lambda ZAP (Stratagene, La Jolla, CA) vectors were designed to combine specific features from λ phage, filamentous phage, and plasmid vectors to produce a single cloning vector, which is optimized for efficiency, ease of use, and versatility.[1] Although primarily developed for the construction of cDNA libraries, they have also proved useful for genomic libraries (unpublished results, 1991), antibody expression libraries,[2] eukaryotic expression libraries,[3] and subtraction libraries.[4] The primary distinguishing features of these vectors is the ability to convert the DNA insert within the vector from λ phage to phagemid without the requirement for subcloning procedures. This greatly increases the rate at which DNA clones can be characterized.

[1] J. M. Short, J. M. Fernandez, J. Sorge, and W. D. Huse, *Nucleic Acids Res.* **16,** 7583 (1988).

[2] R. L. Mullinax, E. A. Gross, J. R. Amberg, B. N. Hay, H. H. Hogrefe, M. M. Kubitz, A. Greener, M. Alting-Mees, D. Ardourel, J. M. Short, J. A. Sorge, and B. Shopes, *Proc. Natl. Acad. Sci. U.S.A.* **87,** 8095 (1990).

[3] M. Alting-Mees, D. Ardourel, and J. M. Short, *Fed. Proc., ASBMB* (1991).

[4] C. W. Schweinfest, K. W. Henderson, J.-R. Gu, S. D. Kottaridis, S. Besbeas, E. Panotopoulou, and T. S. Papas, *Gene Anal. Tech. Appl.* **7,** 64 (1990).

Although the generation of cDNA libraries using plasmid vectors has improved with the advent of high-efficiency competent cells[5] and electroporation methods,[6] it still does not compare to λ libraries in the level of efficiency and ease of construction and screening. To match, on a molar basis, the λ phage packaging efficiencies of 2×10^9 plaque-forming units (pfu)/μg, transformation of a supercoiled plasmid vector (pUC) would have to exceed 3×10^{10} colonies/μg. Transformations of this efficiency have not yet been accomplished, and would do nothing to address procedural difficulties in colony screening. Despite the advantages of high cloning efficiency, the relatively large size of λ vectors makes them less manageable than plasmid vectors for the characterization of DNA inserts. Subcloning a cDNA insert from a λ vector for further characterization usually requires growing and isolating λ phage DNA, appropriate restriction site digestion, treating the vector with phosphatase, ligation, and transformation. Although it is possible to complete these procedures in as little as 4 days, it often requires weeks. The problem is compounded if the insert contains internal restriction sites, if several inserts must be subcloned, or if the flanking restriction sites were inadvertently lost during library construction or propagation. It was precisely for these reasons that we originally sought to generate a λ vector system that would allow direct *in vivo* conversion of a λ phage molecule, λ ZAP, into a plasmid molecule, pBluescript SK(−), without the need to subclone.[1]

There are a variety of potential approaches that can accomplish the excision of a plasmid molecule positioned within a λ vector. The simplest is the insertion of rare restriction sites flanking the plasmid vector. Following isolation of the desired λ clone, the phage could be grown, then cut with the appropriate restriction sites and religated under conditions permitting circularization of the plasmid vector. This DNA could then be transformed into competent *Escherichia coli* or other bacterial cells. These vectors offer little advantage over traditional vector systems such as λ-gt10 and λ-gt11,[7] because DNA preparation, restriction digestion, and ligation are still required. An alternative method involves the use of directly repeated sequences positioned at both ends of the plasmid and recovery could be achieved through general homologous recombination at the repeats in a *RecA*$^+$ strain of *E. coli* that contains genetic blocks to lytic λ replication.[8] However, the efficiency of this excision system is dependent on the length of the repeats, with increasing homology resulting in greater excision

[5] D. Hanahan, *J. Mol. Biol.* **166,** 557 (1983).

[6] J. W. Dower, J. F. Miller, and C. W. Ragsdale, *Nucleic Acids Res.* **16,** 6127 (1988).

[7] R. A. Young and R. W. Davis, *Proc. Natl. Acad. Sci. U.S.A.* **80,** 1194 (1983).

[8] M. Burmeister and H. Lehrach, *Gene* **73,** 245 (1988).

efficiency. This system has the potential drawback of insert instability and reduced phage capacity. In addition, site-specific recombination systems such as the λ *att*/integrase system,[9] P1 phage *cre*/*lox* system,[10] and yeast *FLP* system[11] are all efficient, catalyzed recombination systems that can be used. Each of these systems, however, requires suppression of the lytic activity of the infecting phage to block cell lysis and need at least transient expression of the gene catalyzing the recombination or excision of the plasmid DNA.

The partial filamentous phage origins used in λ ZAP to mediate *in vivo* excision are also present in the excised plasmid portion to allow for eventual isolation of single-stranded DNA from the excised phagemid.[1] Therefore, the f1 origin has a dual role: excision of the insert-containing plasmid from the λ vector and single-stranded DNA production from the excised phagemid. However, it was not known at the time of development whether the filamentous phage origin would function in the λ phage molecule, especially because the catalytic activity of gene II protein (pII) requires a supercoiled substrate,[12] and λ is only transiently supercoiled in *E. coli*. We now know that this method is efficient and offers advantages over the catalyzed recombination systems. In addition, improvements in the excision protocol[13] facilitate the development of more efficient screening and hybrid selection strategies for subtraction library construction.

Principle of Method

In the analysis of the functional domains of the f1 origin of replication, sequences have been identified that are responsible for initiating and terminating DNA replication of the phage viral (+) strand.[14] These sequences and their domains are outlined in Fig. 1. Filamentous phage replication is initiated by the introduction of a nick by the phage-encoded gene II protein at the TpA sequence (base 5781) of the initiator domain (Fig. 1). The 5′ end of the nicked plus strand is then displaced by *E. coli* DNA polymerase, which initiates polymerization from the 3′ DNA end using the minus strand as a template. Gene II protein remains associated with the DNA at the origin.[12] After the polymerase completes one round of replication, it en-

[9] S. Brenner, G. Cesareni, and J. Karn, *Gene* **17,** 27 (1982).
[10] K. Abremski, R. Hoess, and N. Sternberg, *Cell (Cambridge, Mass.)* **32,** 1301 (1983).
[11] J. R. Broach, V. R. Guarascio, and M. Jayaram, *Cell (Cambridge, Mass.)* **29,** 227 (1982).
[12] D. R. Brown and J. Hurwitz, *in* "Nucleases and DNA Replication" (S. M. Linn and R. J. Roberts, eds.), p. 199. Cold Spring Harbor Laboratory, Cold Spring Harbor, New York, 1985.
[13] B. Hay and J. M. Short, *Strategies* **5,** 16 (1992).
[14] G. P. Dotto, K. Horiuchi, and N. D. Zinder, *J. Mol. Biol.* **172,** 507 (1984).

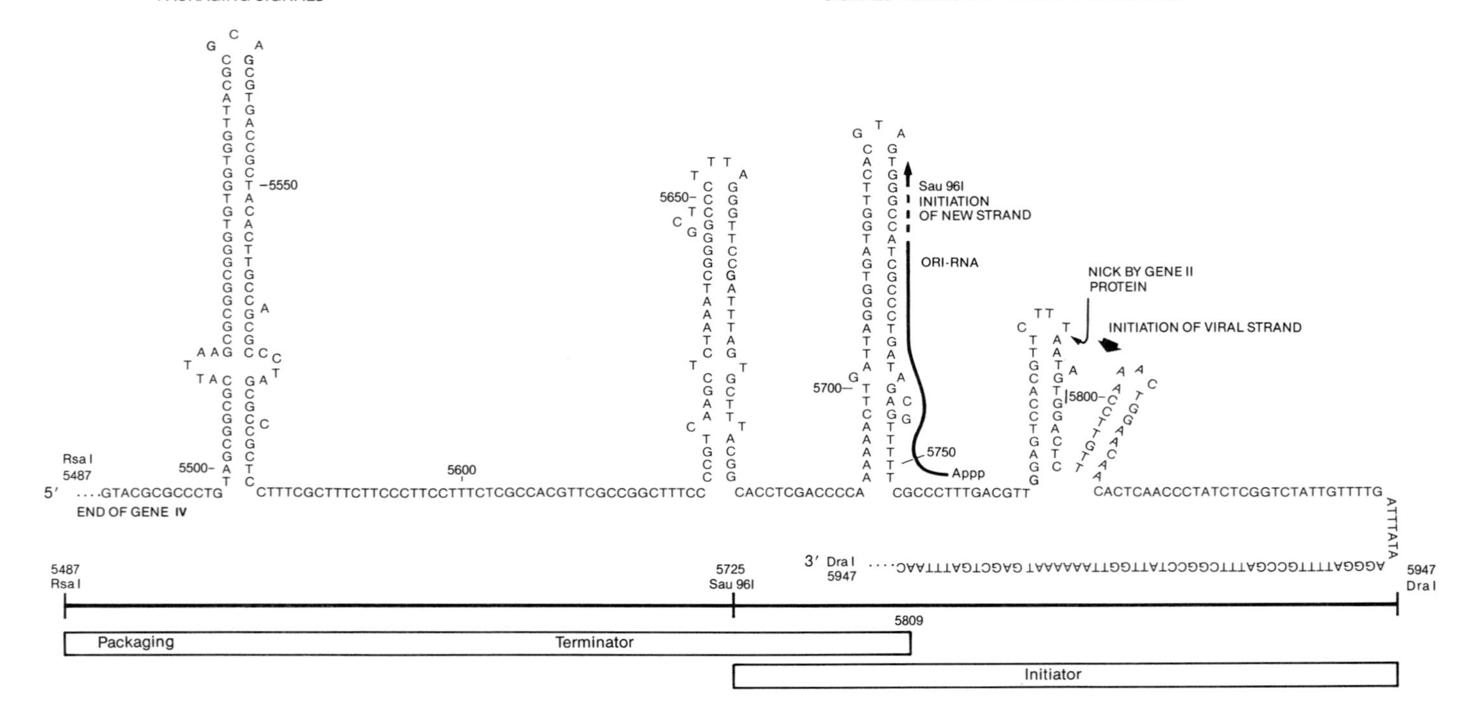

FIG. 1. DNA sequence of the f1 origin of replication and phage packaging signal. The gene II protein nicking site for initiation of plus-strand replication and the RNA primer site for minus-strand replication are also shown. The sequences used to generate the initiator and terminator domains within λ ZAP are indicated in the lower portion of the figure.[1,14] [E. Beck and B. Zink, *Gene* **16,** 35 (1981)].

counters the terminator sequence with the bound pII. It is believed that the noncovalently associated pII interacts with free pII to process the newly synthesized DNA strand into a circularized, single-stranded molecule generating an intact origin. Following the strand nicking and circularization, the polymerase continues another round of viral plus-strand replication in a rolling circle mode of replication.

To permit excision of a plasmid from a λ vector using a filamentous phage origin, the initiator and terminator domains of the replication origin are separated to avoid generating two initiator sites in the λ phage (Fig. 1). A second initiator site would initiate DNA replication into the λ phage molecules lacking plasmid DNA sequences. The initiator and terminator domains inserted into the λ phage arms are shown in Fig. 2. These sequences are inserted into the λ phage so that they flank the pBluescript SK(−) phagemid sequences (Fig. 2).[1,15]

The pBluescript SK(−) phagemid vector contains a large polylinker (21 unique restriction sites) flanked by T3 and T7 phage RNA polymerase promoters.[1,15] The polylinker, including the RNA promoters, is positioned within the amino terminus of an α-complementing portion of the *lacZ* gene.[16] This insertion reduces the complementing activity of the *lacZ* gene with the Δ*M15lacZ* gene, but does not affect the transcription levels or the quantity of protein produced.[1] The left portion or arm (genes *A–J*) of λ ZAPI was derived from λ L47.1[17] while the right arm of λ ZAPI was constructed from a λ gt11 derivative, λ longD.[1] These λ arms were selected based on size, lack of interfering restriction sites, presence of a *chi* sequence,[18] and efficient growth of the parent vectors in *E. coli*. The λ ZAP I vector (Fig. 2) is approximately 40.82 kb in size and has an insert capacity of approximately 10.2 kb with six unique cloning sites (*Sac*I, *Not*I, *Xba*I, *Spe*I, *Eco*RI, and *Xho*I).

Initially, *E. coli* strains expressing the filamentous phage gene II protein (pII)[19] were tested for the efficiency of *in vivo* excision from ZAP.[1] However, it was discovered that phagemid excision could not be performed in these strains. This had not been anticipated in light of the fact that DNA molecules containing an antibiotic gene and only an f1 origin with no ColE1 origin replicate efficiently in these strains.[19] In addition, it was known that single-stranded DNA could be rescued from phagemid vectors containing both ColE1 and f1 origins through superinfection of a filamentous helper

[15] M. Alting-Mees, J. A. Sorge, and J. M. Short [42], this volume.

[16] I. Zabin and A. V. Fowler, *in* "The Operon" (J. H. Miller and W. S. Reznikoff, eds.), p. 89. Cold Spring Harbor Laboratory, Cold Spring Harbor, New York, 1980.

[17] W. A. M. Loenen and W. J. Brammar, *Gene* **20,** 249 (1980).

[18] F. W. Stahl, J. M. Crasemann, and M. M. Stahl, *J. Mol. Biol.* **94,** 203 (1975).

[19] K. Geider, E. Beck, and H. Schaller, *Proc. Natl. Acad. Sci. U.S.A.* **75,** 5226 (1978).

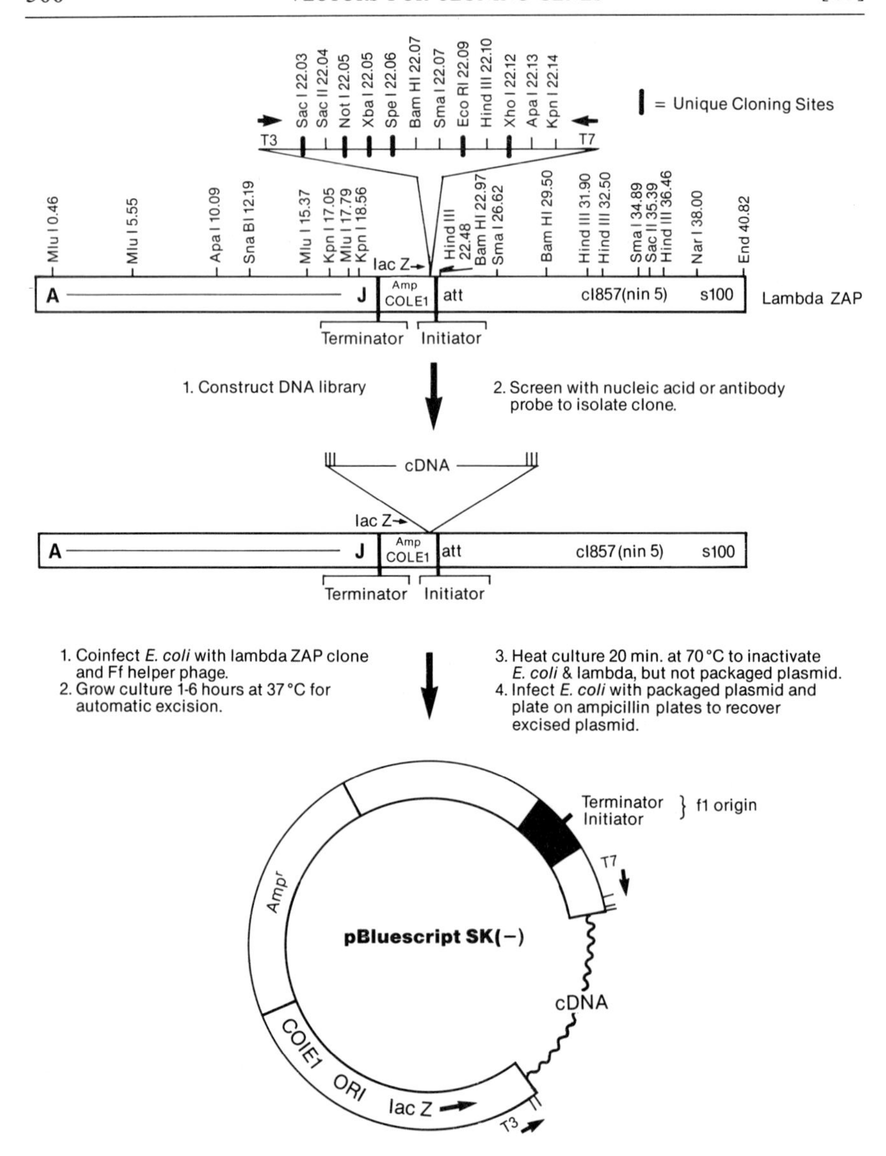

FIG. 2. Lambda ZAPI is depicted at the top of the figure with unique cloning sites marked with the dark vertical lines. The initiator and terminator domains flanking the pBluescript SK(−) phagemid sequences are diagrammed in the figure. Following construction of a DNA library in the vector, clones can be identified by DNA hybridization or antibody screening for the expressed protein. To recover the pBluescript SK(−) phagemid, *in vivo* excision is performed as outlined in steps 1–4 of the standard excision protocol (see text for additional details). The excised phagemid with the DNA insert is shown at the bottom of the figure. In the excised pBluescript phagemid the initiator and terminator domains are joined to regenerate an intact f1 origin.

phage.[20] Further analysis with pBluescript phagemid indicated that the presence of a ColE1 origin and an active f1 origin were incompatible for stable replication of the phagemid molecule in the pII-expressing strains. Because pII is temperature sensitive,[21] attempts were made to obtain stable replication of plasmid molecules containing f1 and ColE1 origins by growing the molecules at high temperature (42°). This was the only method that permitted recovery of antibiotic-resistant pBluescript clones. This supported the conclusion that active f1 and ColE1 origins are incompatible for stable plasmid replication. To avoid the effect of origin incompatibility, the *in vivo* excision system was developed based on the use of a filamentous helper phage. In this procedure, single-stranded molecules are packaged following replication, but are not forced to undergo continuous replication via the ColE1 and f1 origins during excision.

Efficient excision of the pBluescript phagemid from the λ ZAP arms is effected by the simultaneous infection of a male *E. coli* strain with λ ZAP and filamentous helper phage. On infection, the filamentous phage is converted to a double-stranded molecule that immediately expresses phage proteins. The phage protein, pII, recognizes and nicks the λ ZAP f1 initiator domain and, following DNA synthesis by DNA polymerase, nicks the terminator domain and ligates the DNA ends. This process generates circular single stranded molecules of the helper phage and the Bluescript phagemid (Fig. 2). Because both of these molecules contain the f1 phage packaging sequences, both molecules can be packaged into filamentous phage particles. Packaging is initiated by the buildup of single-stranded DNA-binding protein, pV, which coats the single strand and prevents conversion of the single-stranded molecules into double-stranded molecules that are used for further replication.[22] This protein–DNA complex is transported to the membrane, where pV is stripped from the DNA as it is packaged into the filamentous phage particle made up of phage proteins pI, pIII, pIV, and pVI–pX (G. P. Smith for review[23]). During this excision process, filamentous helper phage, packaged single-stranded pBluescript DNA, and λ phage particles containing λ ZAP DNA are released into the medium. The recovery of the pBluescript phagemid from this mixture is performed by heat inactivating the *E. coli* and temperature sensitive λ phage particles, infecting a fresh culture of *E. coli* cells at a low filamentous phage-to-cell ratio, and plating on ampicillin plates to select for the Amp^r colonies containing pBluescript phagemid.

[20] L. Dente, G. Cesareni, and R. Cortese, *Nucleic Acids Res.* **11,** 1645 (1983).

[21] I. Rasched and E. Oberer, *Microbiol. Rev.* **50,** 401 (1986).

[22] B. J. Mazur and N. D. Zinder, *Virology* **68,** 490 (1975).

[23] G. P. Smith, *in* "Vectors: A Survey of Molecular Cloning Vectors and Their Uses" (R. L. Rodriguez and D. T. Denhardt, eds.), p. 61. Butterworth, Boston, 1988.

This procedure is easily performed and requires only a few hours. Several hundred ZAP expression libraries have been generated, screened, and inserts excised successfully since the development of the vector. The ZAP excision system maintains the ease of library construction and screening in λ while alleviating the needless burden of λ phage preparations and subcloning of DNA inserts, thereby increasing the rate of clone analysis. In the following sections we describe the original procedure used for excision, as well as improvements that further simplify this procedure and make it suitable for mass excision of DNA libraries.

Methods

Library Construction

A wide variety of cDNA synthesis procedures have been described and reviewed in detail.[24] Most of these procedures are easily adapted for the generation of unidirectional libraries in λ ZAP due to the large number of unique cloning sites (six) within this vector. The procedure that we have used with a high degree of success is based on an adaptation of the Gubler–Hoffman method[25] with the incorporation of methylated cytosines for complete protection of the cDNA during restriction digestion of the linker primer.[26] (Additional details can be obtained directly from Stratagene Cloning Systems, La Jolla, CA.) The procedures permitting excision of clones from λ ZAP libraries are outlined below and in Fig. 2.

Standard Excision Protocol

1. Core λ ZAP plaque from an agar plate and transfer to a sterile microfuge tube containing 500 μl of SM buffer [5.8 g NaCl, 2.0 g $MgSO_4 \cdot 7H_2O$, 50 ml 1 *M* Tris-HCl, pH 7.5, 5 ml 2% (w/v) gelatin per liter] and 20 μl of chloroform. Briefly vortex the tube and incubate approximately 1 hr at room temperature to release the λ phage particles into the SM buffer. This phage stock is stable for up to 1 year at 4° and contains approximately 10^5 to 10^6 pfu.

2. In a 50-ml conical tube, combine:

Two hundred microliters of OD_{600} 1.0 XL1-Blue cells {*recA1 endA1 gyrA96 thi hsdR17 supE44 relA1 lac* [F′ *proAB lacI*^{q}*ΔM15lacZ*, Tn*10*(*tet*^{r})]}

[24] K. Kaiser, *J. Methods Cell Mol. Biol.* **2,** 1 (1990).
[25] U. Gubler and B. J. Hoffman, *Gene* **25,** 263 (1983).
[26] W. D. Huse and C. Hansen, *Strategies* **1,** 1 (1988).

Two hundred microliters of ZAP phage stock (containing $>1 \times 10^5$ phage particles)
One microliter R408 helper phage ($>1 \times 10^6$ pfu/ml)

Incubate mixture at 37° for 15 min.

3. Add 5 ml of 2× YT medium and incubate 3 hr at 37° with shaking. [2× YT medium: 10 g NaCl, 2 g $MgSO_4 \cdot 7H_2O$, 50 ml 1 *M* Tris-HCl, pH 7.5, 5 ml 2% (w/v) gelatin per liter].

4. Incubate the tube at 70° for 20 min. This stock contains the pBluescript phagemid packaged as filamentous phage particles and can be stored at 4° for 1–2 months.

5. To plate the rescued phagemid, combine the following in two 15-ml tubes:

Two hundred microliters of phagemid stock from step 4 above and 200 μl OD_{600} 1.0 of XL1-Blue host cells
Twenty microliters of a 10^{-2} dilution of phage stock from step 4 above and 200 μl OD_{600} 1.0 of XL1-Blue cells

Incubate tubes at 37° for 15 min.

6. Using a sterile spreader, plate 1 to 100 μl on LB/ampicillin plates (100 μg/ml) and incubate them overnight at 37°. Colonies appearing on the plate contain the pBluescript double-stranded phagemid with the cloned DNA insert. Bacteria infected with helper phage alone will not form colonies because they do not contain an ampicillin resistance gene.

Note: LB (10 g NaCl, 10 g Bacto-tryptone, 5 g yeast extract per liter) is the medium of choice for overnight growth. However, when growing XL1-Blue for *in vivo* excision, single-stranded DNA rescue, or double-stranded DNA minipreparations, super broth medium (5 g NaCl, 20 g Bacto-yeast extract, 35 g tryptone, buffer to pH 7 with NaOH) will increase cell growth rates.

Mass Library Excisions and Improved Single Clone Excision

Subtraction libraries are extremely valuable for isolation of mRNAs expressed in a unique cell type or cell condition when no other information about the clone responsible for a given phenotype is available. Schweinfest *et al.*[4] have utilized the λ ZAP vector for construction of subtracted DNA libraries with a greater than 90-fold enrichment for differentially expressed sequences. This procedure requires the mass excision of two separate λ ZAP cDNA libraries and the recovery of the single-stranded pBluescript clones (Fig. 3). One of the excised, single-stranded libraries is photobiotinylated and hybridized to the excised, single-stranded DNA from the second library. Following hybridization, the biotinylated material is removed with streptavidin beads. The remaining single-stranded DNA is converted to double-stranded DNA and transformed into *E. coli*. The phagemid clones recovered from this subtraction procedure are enriched for rare mRNA-derived DNA sequences. Although this work optimized growth conditions and multiplicity of infection ratios for the helper phage and λ phage infection, further improvements can be incorporated into this procedure. Selective hybridization using unidirectional libraries would improve efficiency, because all clones would be in the proper orientation for subtractive hybridization. In addition, although the presence of helper phage does not affect subtractive hybridization, high titers of helper phage generated during mass excision can lead to cotransformation of the plating cells with pBluescript phagemid and helper phage DNA. This can lead to clone misrepresentation, rearrangement, or deletion.

To circumvent these problems the excision protocol has been modified through the use of an amber mutant (gene I and gene II) helper phage.[23] By performing the initial excision step in a *supE* host strain such as XL1-Blue, the amber mutations are suppressed and excision is performed as in the standard protocol. However, the subsequent transformation of the excised/subtracted mixture is performed using Su^-, λ resistant *E. coli* cells (e.g., SOLR) and plated on ampicillin plates. In this strain neither the λ phage nor the infecting helper phage is active because the cell is resistant to λ phage infection and the amber mutations for filamentous phage replication are not suppressed. The pBluescript phagemid survives in this strain due to the presence of the ColE1 origin and ampicillin resistance. This new combination of helper phage and *E. coli* cells maintains clone representation in the excision of a phage library and simplifies clone recovery.[13]

Mass Library Excision and Single Clone Excision Protocol Using Amber Mutant Helper Phage

1. Grow overnight cultures of XL1-Blue and SOLR [λ^R, *recB recJ sbcC uvrC umuC*::Tn*5*(Kan^R) *endA1 thi-1 gyrA96 relA1 lac* $el4^-$(mcrA), Δ(mcrCB-hsdSMR-mrr)171 (F′ *lacI*^{q}*Z*Δ*M15 proAB*)] in LB medium at 30°. Spin down cells and resuspend separately in 10 m*M* $MgSO_4$ to an OD_{600} of 1.0.

2. In a 50-ml conical tube, combine λ ZAP bacteriophage with XL1-Blue cells at a 1 : 10 phage-to-cell ratio (200 μl of OD_{600} 1 cells are used for single clone excision). (When excising an entire library, 10-fold more

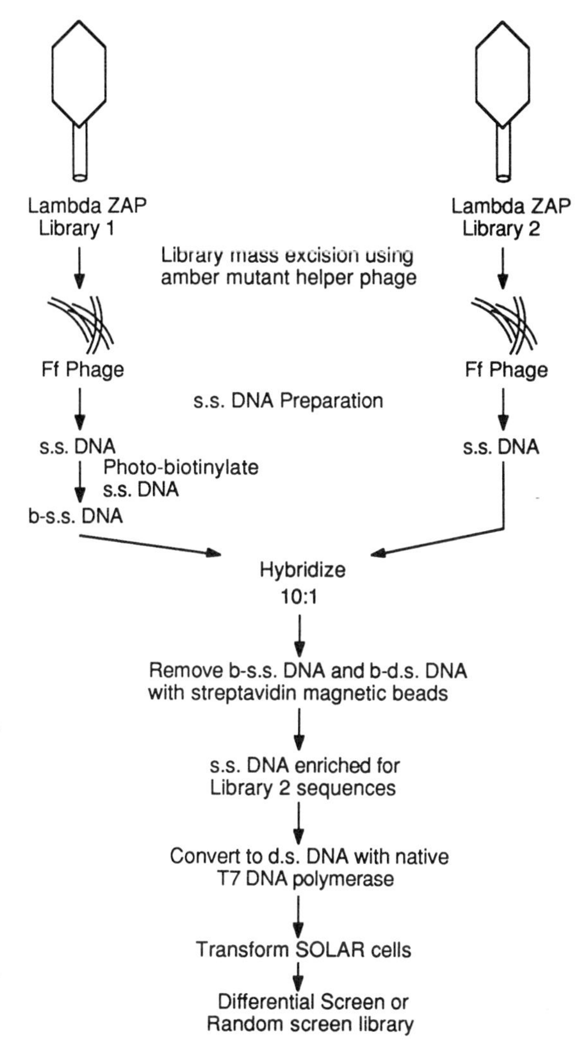

FIG. 3. Mass excision can be used to generate subtraction libraries or subtraction DNA probes. A general strategy for generating these libraries is outlined, utilizing amber mutant filamentous helper phage. The mass excision protocol is outlined in detail in text. Following mass excision of the λ ZAP libraries, the biotinylated single-stranded DNA (b-ssDNA) is hybridized to the ssDNA from the second library. The degree of the subtraction process can be controlled by the time and concentration of the hybridization reaction. Subsequently, the remaining nonbiotinylated DNA, representing clones unique to library 2, is converted to dsDNA for improved transformation efficiency and transformed into a Su^- *E. coli* strain. The enriched phagemid library can be screened with subtracted or unsubtracted DNA probes. Additional details on generating a subtraction library utilizing λ ZAP vectors can be found in Schweinfest *et al.* (1990).[4]

phage should be excised than is found in the primary library to ensure statistical representation of the excised clones.) Add amber mutant helper phage (e.g., M13mp7-11, M13gt130, 131, Exassist)[13,23] at a 10 : 1 phage-to-cell ratio. See step 2 in the standard excision protocol for additional details.

3. Allow adsorption at 37° for 15 min.

4. Add 20 ml of LB medium and incubate with agitation for 2 to 3 hr. *Note:* Incubation times in excess of 3 hr may alter the clone representation.[13]

5. Heat to 70° for 15 min. *Note:* For a subtraction library, purify single-stranded DNA. Perform the subtractive hybridization. Then transform SOLAR cells with subtracted clones.[4]

6. Mix 1 μl of supernatant with 200 μl of SOLAR cells. Incubate at 37° for 15 min.

7. With a sterile spreader, plate 100 μl onto LB/ampicillin plates and incubate at 37° overnight to recover excised clones.

Note: Due to the high efficiency of the excision process, it may be necessary to titrate the supernatant to achieve single colony isolation. Single clone excision reactions can be safely performed overnight because clone representation is not relevant.

Modification of λ ZAPI to Generate λ ZAPII

Although expressed levels of fusion proteins from λ ZAPI was unaffected, color selection of λ ZAPI was weak on most *E. coli* strains due to the insertion of the large pBluescript polylinker in the α-complementing portion of the *lacZ* gene. Color selection on the XL1-Blue strain, however, was superior to other α-complementing strains. It was later determined that the cause for the improved color selection on XL1-Blue was a result of a mutant *lacZ* gene in the chromosome that maintains low-level expression of β-galactosidase activity. This *lacZ* gene is separate from the α-complementing Δ*M15lacZ* gene on the F′ episome. Because λ ZAPI grows poorly on XL1-Blue cells, the S_{100} of λ ZAPI was replaced with the wild-type *S* gene from λ gt10 to generate λ ZAPII.[1] This improved the growth of λ ZAP on XL1-Blue cells and also provided improved color selection. The chromosomal *lacZ* gene of XL1-Blue has also been used in the development of a strain, SURE *E. coli,* capable of stabilizing inverted repeats and Z-DNA sequences.[27] The SURE strain, in addition to containing the XL1-Blue *lacZ* gene and the sequence-stabilizing mutations, is Mcr restriction deficient for improved cloning of cytosine methylated DNA.[28]

[27] A. Greener, *Strategies* **3,** 5 (1990).

[28] P. L. Kretz, S. W. Kohler, and J. M. Short, *J. Bacteriol.* **173,** 4707 (1991).

Assessment of Cloning Efficiency by Color Selection

The color selection by α complementation is used for determining the ratio of recombinant (white or colorless) to nonrecombinant (blue) clones within a primary or amplified library. This background test provides a simple and rapid assessment of the library quality. To perform this assay, SM buffer containing 100–500 λ ZAP phage is added to 200 μl of OD_{600} 0.5 XL1-Blue cells. Phage and bacteria are incubated for 15 min at 37° to permit the phage to adhere to the cells. Two to 3 ml of 48° top agar is added [top agar: 5 g NaCl, 2 g $MgSO_4 \cdot 7H_2O$, 5 g yeast extract, 10 g NZ-amine (casein hydrolysate), 0.7% (w/v) agarose per liter, adjusted to pH 7.5 with NaOH] containing 15 μl of 0.5 *M* isopropyl-β-D-thiogalactopyranoside (IPTG) (in H_2O) plus 50 μl of 250 mg/ml X-Gal (in dimethylformamide). This mixture is plated immediately onto NZY or LB agar plates. (*Note:* To help prevent precipitation, mix IPTG and X-Gal into top agar separately.) Plaques may be visible in as little as 6 hr after incubation at 37°. Background plaques are blue and should be less than 1×10^5 pfu/μg of arms, while insert-containing phage plaques will be white and should be 10- to 100-fold above background for a standard cDNA library.

Concluding Remarks

The λ ZAP vector system, although originally designed for the efficient construction and characterization of cDNA clones, has proved useful for a broad range of applications. With an insert capacity of over 10 kb, the phage has also been used successfully for the construction of plant and bacterial genomic libraries. The construction of genomic libraries with smaller insert size is frequently beneficial in cloning a larger percentage of the genome in instances where large DNA clones containing unusual structure or secondary modifications are unstable. The requirement for producing a larger library to compensate for smaller insert size in order to obtain genome representation is frequently offset by the speed at which λ ZAP clones can be characterized. We have also modified the λ ZAP vector system for the expression of antibody variable regions cloned via polymerase chain reaction (PCR) amplification.[2] More recently, we have introduced a modified λ ZAP vector into the chromosome of transgenic mice for the purpose of generating a λ shuttle vector system capable of measuring spontaneous and induced tissue specific mutation rates in whole animals.[29] The *in vivo* excision features of the mutagenesis λ ZAP vector

[29] S. W. Kohler, G. S. Provost, A. Fieck, P. L. Kretz, W. O. Bullock, J. A. Sorge, D. L. Putman, and J. M. Short, *Proc. Natl. Acad. Sci. U.S.A.* **88,** 7958 (1991).

permit rapid recovery of the target gene from the λ phage for sequence identification of selected mutations.

The original λ ZAP vector has also been modified for construction of a combination prokaryotic/eukaryotic expression vector.[3] This system should permit direct, high-efficiency transfection with a library of λ phage particles and screening of eukaryotic cells.[30] Following phenotypic selection of the desired clones in the eukaryotic cells, the appropriate phage can be recovered by direct packaging of the λ phage from the transfected cell genomic DNA.[3,31] The cDNA can then be excised into a plasmid molecule designed for rapid clone characterization and expression in *E. coli* cells. These procedures can also be coupled with subtractive hybridization techniques to further reduce the complexity of screening in eukaryotic cells. Modifications of the λ ZAP vector currently underway will further increase its efficiency and versatility.

[30] H. Okayama and P. Berg, *Mol. Cell. Biol.* **5,** 1136 (1985).

[31] S. W. Kohler, G. S. Provost, P. L. Kretz, M. J. Dycaico, J. A. Sorge, and J. M. Short, *Nucleic Acids Res.* **18,** 3007 (1990).

[44] Cloning of Complementary DNA Inserts from Phage DNA Directly into Plasmid Vector

By ING-MING CHIU, KIRSTEN LEHTOMA, and MATTHEW L. POULIN

Cloning vectors derived from bacteriophage λ are used frequently in the construction of both cDNA and genomic DNA libraries.[1] Both λgt10 and λgt11 have been used quite extensively for constructing cDNA libraries.[2] For λgt11, the cDNA inserts are cloned into the unique *Eco*RI site within the *lacZ* gene and can be expressed as fusion proteins with β-galactosidase.[2] Thus, the cDNA clones can be obtained by screening with the antibody probe generated against the protein of interest.[3] Alternatively, a ligand having high affinity with the protein of interest may also be used as a probe. Moreover, cDNA libraries constructed in λgt10 or λgt11 can be screened by hybridization with an oligonucleotide or DNA fragment containing the DNA sequence of interest.

[1] J. Sambrook, E. F. Fritsch, and T. Maniatis, "Molecular Cloning: A Laboratory Manual." Cold Spring Harbor Laboratory, Cold Spring Harbor, New York, 1989.

[2] T. V. Huynh, R. A. Young, and R. W. Davis, *in* "DNA Cloning Techniques: A Practical Approach" (D. Glover, ed.), p. 49. IRL Press, Oxford, 1985.

[3] R. A. Young and R. W. Davis, *Science* **222,** 778 (1983).

Screening of phage DNA libraries for positive plaques is relatively easy with the phage lifting technique of Benton and Davis.[4] However, isolating recombinant cDNA inserts from the phage clones of interest can be a tedious task. The burst sizes of the phages and the consequent yields of the phage DNA are often variable. Furthermore, the cDNA insert represents only a small fraction [approximately 2% for a 1 kilobase pair (kbp) insert] of the phage DNA, due to the much larger size of the phage vector arms. Thus, the desired cDNA inserts are usually subcloned from the positive phages into appropriate plasmid or M13 vectors for large-scale production. The cDNA insert isolated from the subcloning vector is then subjected to restriction enzyme mapping, Southern blotting and hybridization, or nucleotide sequencing.

Principle of Method

The *Escherichia coli* strain Y1088 commonly used for phage λgt infection carries the 6.1-kbp plasmid pMC9 endogenously.[2] This plasmid is a pBR322 derivative with the 1.7-kbp *lacZ* and *lacI* genes inserted into the unique *Eco*RI restriction enzyme site.[5] We showed that the endogenous pMC9 DNA is extruded during the lysis of infected bacteria and can be copurified with the phage DNA.[6] Thus, the pBR322 portion of the pMC9 plasmid could be used directly for cloning the phage cDNA inserts.

We have previously described a method in which cDNA inserts from λgt phages were cloned directly into pBR322.[6] This method bypassed the sometimes tedious and time-consuming procedures of preparing plasmid DNA as a subcloning vector. This method can be used to clone DNA inserts derived from other phage vectors when bacteria containing endogenous pBR322 or other plasmid vectors are used as host cells. In this chapter we describe the protocol in more detail and show its application when combined with the polymerase chain reaction[7] (PCR) to characterize clones directly from single bacterial colonies.

Materials

Solutions and Reagents

NZY: 1% (w/v) NZ-amine, 0.5% (w/v) yeast extract, 85 m*M* NaCl, 10 m*M* $MgCl_2$; autoclave

[4] W. D. Benton and R. W. Davis, *Science* **196,** 180 (1977).

[5] M. P. Calos, J. S. Lebkowski, and M. R. Botchan, *Proc. Natl. Acad. Sci. U.S.A.* **80,** 3015 (1983).

[6] I.-M. Chiu and K. Lehtoma, *Gene Anal. Tech. Appl.* **7,** 18 (1990).

[7] R. K. Saiki, S. Scharf, F. Faloona, K. B. Mullis, G. T. Horn, H. A. Erlich, and N. Arnheim, *Science* **230,** 1350 (1985).

YT: 1% (w/v) Bacto-tryptone, 0.5% (w/v) yeast extract, 0.17 *M* NaCl; autoclave

SOC: 2% (w/v) Bacto-tryptone, 0.5% (w/v) yeast extract, 10 m*M* NaCl, 2.5 m*M* KCl, autoclave; 10 m*M* $MgCl_2$, 10 m*M* $MgSO_4$, 20 m*M* glucose; filter sterilize

PSB: 0.1 *M* NaCl, 10 m*M* Tris-HCl, pH 7.4, 10 m*M* $MgCl_2$, 0.05% (w/v) gelatin; autoclave

CaMg: 10 m*M* $CaCl_2$, 10 m*M* $MgCl_2$; filter sterilize

IPTG: 100 m*M* isopropyl-β-D-thiogalactoside in water; store at −20°

X-Gal: 2% (w/v) 5-bromo-4-chloro-3-idolyl-β-D-galactoside in dimethylformamide (DMF); store at −20°

$T_1E_{0.2}$: 1 m*M* Tris-HCl, pH 8.0, 0.2 m*M* ethylenediaminetetraacetic acid (EDTA)

CIA: Chloroform–isoamyl alcohol (24 : 1, v/v).

10× ligation buffer: 0.66 *M* Tris-HCl, pH 7.5, 50 m*M* $MgCl_2$, 10 m*M* ATP, 10 m*M* dithiothreitol (DTT); store at −20°

Enzymes

Most of the restriction enzymes we have used were purchased from either New England BioLabs (Beverly, MA) or Boehringer Mannheim (Indianapolis, IN). T4 DNA ligase was obtained from Boehringer Mannheim. Lysozyme (Sigma, St. Louis, MO) was stored at 4° and dissolved immediately before use. RNase A was obtained from Boehringer Mannheim and the stock solution at 20 mg/ml was boiled for 15 min and stored at −20°. DNase I was also obtained from Boehringer Mannheim. *Taq* polymerase and the GeneAmp kit for PCR were obtained from Perkin-Elmer Cetus (Norwalk, CT).

Methods

Infection and Growth of Phage

1. Pick plaque containing a positive λgt phage clone with a pipette and transfer to a 1-dram vial. Rock overnight in 1 ml PSB.
2. In a sterile 15-ml culture tube, mix 100 μl CaMg, 100 μl Y1088 bacteria from an overnight culture, and 100 μl of λgt phage from a plug or from a cleared lysate.
3. Incubate at 37° for 30 min.
4. Add the mixture to 50 ml NZY medium containing 50 μg/ml ampicillin in a 250-ml Erlenmeyer flask. Shake at 300 rpm at 37° overnight.

5. Spin the bacteria culture at 3000 rpm in RT6000B centrifuge (Sorvall, Wilmington, DE) at 4° for 15 min to clear the lysate of bacterial debris.
6. Transfer the supernatant to a 50-ml polypropylene tube, add a few drops of chloroform to the phage lysate, and store at 4°.

Phage DNA Extraction

1. Add 10 μg/ml DNase I and 50 μg/ml RNase A to 25 ml of phage lysate.
2. Incubate at 37° for 1 hr.
3. Extract the lysate twice with an equal volume of phenol. Rock on a Labquake (Labindustries, Inc., Berkeley, CA) for 15 min, then spin at 3000 rpm in the RT6000B at 20° for 15 min. Recover the aqueous layer, avoiding the white interface.
4. Extract the lysate once with an equal volume of CIA.
5. Add 1/20 vol of 5 *M* NaCl and 2.5 vol of ethanol. Precipitate at −80° for 1 hr.
6. Spin at 6000 rpm in an HB-4 rotor (Sorvall) at 4° for 30 min.
7. Dry the pellet in a Speed-Vac concentrator (Savant, Farmingdale, NY) and resuspend in 800 μl of $T_1E_{0.2}$.
8. Add 50 μg/ml RNase A. Incubate at 37° for 1 hr.
9. Extract once with phenol, then once with CIA.
10. Precipitate with 2.5 vol of ethanol at −80° for 30 min.
11. Spin at 15,000 rpm in an MTX-150 microcentrifuge (Tomy, Peninsula Laboratories, Inc., Belmont, CA) at 4° for 15 min.
12. Dry the pellet in a Speed-Vac concentrator and resuspend in 200 μl $T_1E_{0.2}$.

Ligation and Transformation

1. Digest 2.5 μg phage DNA with *Eco*RI at 37° for 3 hr.
2. Add 2 μl of 10× ligation buffer and 1 unit of T4 DNA ligase in a total volume of 20 μl. Incubate at 14° for 16 hr.
3. Add 100 ng of ligated DNA to 50 μl of competent *E. coli* DH5α cells (Bethesda Research Laboratories, Gaithersburg, MD). Incubate on ice for 30 min.
4. Heat shock the cells at 37° for 45 sec. Incubate on ice for 2 min.
5. Add 950 μl SOC and shake at 225 rpm at 37° for 1 hr.
6. Plate 200 μl of the cells with 20 μl of 100 m*M* IPTG and 60 μl of 2% (w/v) X-Gal onto YT plates containing 50 μg/ml ampicillin. Incubate at 37° overnight.

DNA Characterization by Alkaline Extraction Method

1. Pick white colonies and grow them in 20 ml of YT with ampicillin. Shake at 37° overnight.
2. Isolate plasmid DNA from 5 ml of the bacterial culture using the alkaline extraction method as described.[8]
3. Digest plasmid DNA with appropriate restriction enzymes and analyze on an agarose gel containing 1 μg/ml ethidium bromide.

DNA Characterization by Boiling–Polymerase Chain Reaction Method

1. Pick a bacterial colony from the plate and resuspend in 50 μl sterile H_2O.
2. Boil for 5 min.
3. Spin at 15,000 rpm in a microcentrifuge for 2 min.
4. Use 5 μl of supernatant as template for PCR.
5. Set up the PCR reaction: in a 20-μl volume, 10 m*M* Tris-HCl, pH 8.3, 50 m*M* KCl, 1.5 m*M* $MgCl_2$, 0.1 mg/ml gelatin, 0.25 m*M* dNTP, 10 pmol each of pBR322 *Eco*RI-specific oligonucleotides, 1 unit *Taq* polymerase, and 5 μl template DNA are overlaid with 20 μl mineral oil.
6. Reactions was subjected to 30 rounds of temperature cycling: 94° for 1 min, 45° for 1 min, 72° for 2 min, and a final 7 min/72° step in a Perkin-Elmer Cetus DNA thermal cycler.
7. Five microliters of each reaction product are analyzed by agarose gel electrophoresis.

Results

A phage clone containing a 1.4-kbp cDNA insert in λgt11 coding for human acidic fibroblast growth factor (aFGF) was purified from a brainstem cDNA library and designated λgtKL416.[9] The library was constructed by cloning the cDNA inserts into the unique *Eco*RI site of λgt11.[10] To show that the cDNA insert from λgtKL416 could be subcloned directly, the phage DNA was digested with *Eco*RI and then religated without adding exogenous plasmid vector. The ligated DNA was then transformed into competent *E. coli* DH5α cells. The transformed cells were selected for ampicillin resistance. The endogenous pMC9 and recombinant clones were

[8] H. C. Birnboim and J. Doly, *Nucleic Acids Res.* **7,** 1513 (1979).

[9] I.-M. Chiu, W.-P. Wang, and K. Lehtoma, *Oncogene* **5,** 755 (1990).

[10] F. DeFerra, H. Engh, L. Hudson, J. Kamholz, C. Puckett, and R. A. Lazzarini, *Cell (Cambridge, Mass.)* **43,** 721 (1985).

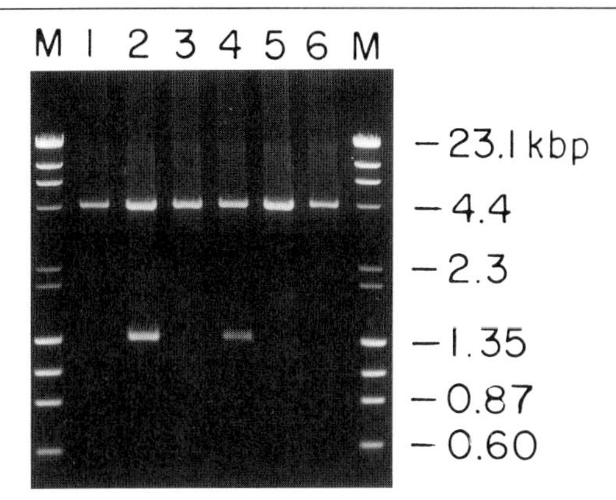

FIG. 1. Gel electrophoresis of plasmid DNA containing the 1.4-kbp cDNA insert. λgtKL416 DNA was digested with *Eco*RI and ligated without the addition of any exogenous plasmid DNA. DNA was isolated from six white colonies, digested with *Eco*RI, and analyzed on a 1.0% agarose gel. Two of them showed the successful cloning of the 1.4-kbp cDNA of interest (lanes 2 and 4). The marker used is λ DNA digested with *Hin*dIII and φX DNA digested with *Hae*III (lane M).

distinguished by blue-white colony selection in the presence of IPTG and X-Gal. The white colonies could contain either pBR322 or its derivative with the cDNA insert of interest.[6] DNA was extracted from six white colonies using the alkaline extraction method,[8] digested with *Eco*RI, and analyzed on an agarose gel. A unique DNA band of 4.4 kbp was observed in four of the six clones. Two clones (clones 2 and 4) contained a 4.4-kbp *Eco*RI fragment and the desired 1.4-kbp cDNA insert (Fig. 1, lanes 2 and 4). When the phage DNA was digested with *Eco*RI and incubated in the absence of T4 DNA ligase, no colonies were produced after transformation.[6]

The size of the *Eco*RI fragments detected in all transformants analyzed (4.4 kbp) is consistent with the size of the pBR322 DNA. As a first step to verify the identity of the pBR322 plasmid vector in these six clones, the bacteria were shown to be tetracycline resistant when streaked out on YT plates containing 15 μg/ml tetracycline. In contrast, the tetracycline-sensitive pBluescript clone did not grow on the tetracycline medium (data not shown). Additional evidence was obtained by digesting the plasmid DNA with *Hin*fI and analyzing it on a 3% (w/v) NuSieve (FMC BioProducts, Rockland, ME) agarose gel.[6] *Hin*fI generates 10 characteristic restriction fragments from pBR322. The data obtained from *Hin*fI digestion of clones 2 and 4 are consistent with these clones being pBR322 derivatives containing a 1.4-kbp DNA fragment cloned into the unique *Eco*RI site.

The insert was shown to be the same as that derived from λgtKL416 by restriction enzyme mapping as well as by Southern blotting and hybridization analysis (data not shown).

The pBR322 DNA in clones 2 and 4 is most likely to be derived from pMC9, which is present in the host Y1088. The endogenous plasmid DNA may be purified along with the phage DNA during the isolation process. If this is correct, we should be able to identify clones containing cDNA inserts using the PCR directly on the single bacterial colonies.[11] A pair of oligonucleotide primers flanking the *Eco*RI site of pBR322 were synthesized for the PCR. The sequences of the clockwise and counterclockwise primers are 5′-GTATCACGAGGCCCT-3′ and 5′-GATAAGCTGTCAAAC-3′, respectively. For these experiments, three phage clones containing the newt *bek* cDNA[12] were isolated from a forelimb blastema cDNA library.[13] Inserts of 1.9, 1.7, and 1.3 kbp were identified by *Eco*RI digestion of purified phage DNA. The phage DNA was digested with *Eco*RI, religated, and used to transform DH5α cells as described above. Four white colonies resulting from transformation with phage DNA containing the 1.3-kbp *bek* cDNA insert were transferred to H_2O, boiled for 5 min, and used as DNA templates for the PCR. The reaction products were analyzed on a 1% (w/v) agarose gel and visualized by staining with ethidium bromide (Fig. 2A, lanes 1–4). A pBR322 colony (Fig. 2A, lane 5) and a colony containing pBR322 with the 1.4-kbp human aFGF cDNA insert (Fig. 2A, lane 6) were used as negative and positive controls, respectively. The PCR product from the positive clone was digested with *Bam*HI (Fig. 2B, lane 1) or *Hin*cII (Fig. 2B, lane 2) to confirm the identity of the DNA insert. Simultaneously, the 1.9- and 1.7-kbp *bek* cDNA inserts were subcloned into pBR322 using the same approach (data not shown). These results showed that it is possible to clone the phage cDNA insert simply by using the endogenous pBR322 plasmid.

Discussion

Several methods have been designed to analyze the cDNA insert directly from the phage clone. Double-stranded λ DNA can serve as a template for sequencing using the chain termination method.[14] The cDNA insert can be amplified for subcloning by the PCR.[12] By incorporating the T7 RNA polymerase recognition sequence and the translation initiation

[11] D. Gussow and T. Clackson, *Nucleic Acids Res.* **17,** 4000 (1989).

[12] M. L. Poulin, H. Onda, R. A. Tassava, and I.-M. Chiu, unpublished results (1991).

[13] C. W. Ragsdale, M. Petkovich, P. B. Gates, P. Chambon, and J. P. Brockes, *Nature (London)* **341,** 654 (1989).

[14] R. J. Zagursky, K. Baumeister, N. Lopez, and M. L. Berman, *Gene Anal. Tech.* **2,** 89 (1985).

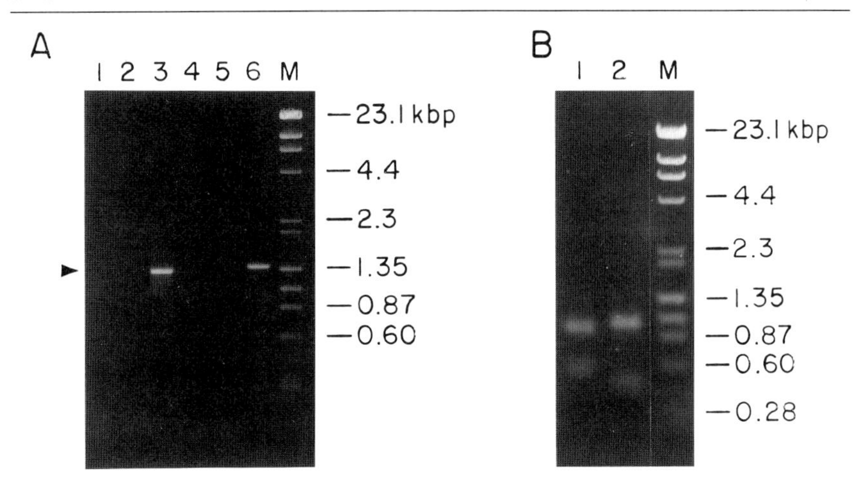

FIG. 2. Gel electrophoresis of the 1.3-kbp PCR product. (A) Phage DNA containing the 1.3-kbp newt *bek* cDNA insert[12] was digested with *Eco*RI, religated, and transformed into DH5α cells. White colonies (lanes 1–4) were transferred to 50 μl of H_2O, boiled for 5 min, and used as templates for the PCR in a 20-μl volume. Lanes 5 and 6 represent pBR322 and a pBR322 derivative containing the 1.4-kbp human aFGF cDNA insert. The PCR products (5 μl) were analyzed on a 1% agarose gel. The arrowhead indicates the position of the expected PCR product from the *bek* cDNA clone. (B) The PCR product from the positive *bek* cDNA clone was digested with *Bam*HI (lane 1) or *Hin*cII (lane 2) and analyzed on an agarose gel. The marker used is λ DNA digested with *Hin*dIII and ϕX DNA digested with *Hae*III (lane M).

codon into a primer, the amplification by the PCR has allowed subsequent *in vitro* transcription and translation of the phage cDNA inserts into protein.[15] However, it is still desirable, for economical reasons, to clone the cDNA inserts into appropriate plasmid or M13 vectors for large-scale production or other analyses such as site-directed *in vitro* mutagenesis, deletional cloning for sequencing purposes, or transfection studies. A λ insertion-type cDNA cloning vector, λ ZAP, has been introduced.[16] In *E. coli*, a phagemid contained within the λ ZAP vector can be excised by superinfection with f1 or M13 helper phage. This excision process also eliminates the need to subclone DNA inserts from the λ phage into a subcloning plasmid vector. In spite of this, the majority of the cDNA libraries currently available are constructed in λgt vectors. Thus, direct cloning of cDNA inserts from λgt phage DNA into a plasmid is quite useful.

[15] B. K. Nishikawa, D. M. Fowlkes, and B. K. Kay, *BioTechniques* **7,** 730 (1989).

[16] J. M. Short, J. M. Fernandez, J. A. Sorge, and W. D. Huse, *Nucleic Acids Res.* **16,** 7583 (1988).

Escherichia coli Y1088 contains the plasmid pMC9, which represses expression of foreign genes that might be detrimental to host cells and phage growth.[3] The entire *lacI* gene and the amino-terminal portion of the *lacZ* (coding for β-galactosidase) gene are contained in the 1.7-kbp *Eco*RI fragment of pMC9.[5] The *lacI* gene codes for the repressor of the lactose (*lac*) operon. The sequence between the *lacI* and *lacZ* genes represents the *lac* control region. The blue-white colony selection makes it possible to distinguish the parental pMC9 plasmid from the recombinant ones. We have employed this property to establish a direct method of cloning cDNA inserts from the phage DNA preparation without supplying a cloning plasmid vector exogenously. Ten to 15% of the colonies generated by this method are blue, representing the 1.7-kbp *lacI* and *lacZ* genes religated into pBR322. The molar ratio of phage DNA versus pMC9 DNA in the phage DNA preparation is rather high, as evidenced by the observation that 1 positive clone was identified per 3 to 12 white colonies analyzed[6,12] (also see Figs. 1 and 2). This high molar ratio made it possible to analyze white colonies for the cDNA insert directly. This method bypassed the time-consuming procedures of preparing plasmid DNA, digesting it with restriction enzymes, and dephosphorylating it, thus significantly simplifying the subcloning process. The method is likely to be extended to the cloning of inserts from other λ vectors when bacteria host contains pMC9 (i.e., Y1088, Y1089, or Y1090)[2] or other plasmid vectors. More importantly, combination of this method with the PCR method, which allows the analysis of DNA directly from single bacterial colonies by boiling in H_2O, made it practical to clone several cDNA inserts simultaneously.

Acknowledgments

This work was supported in part by grants from the National Institutes of Health (R01 CA45611 and P30 CA 16058) and the March of Dimes Birth Defects Foundation (No. 6-549). I.-M.C. is a recipient of the Research Career Development Award (K04 CA01369) from the National Institutes of Health.

[45] Construction of Complex Directional Complementary DNA Libraries in *Sfi*I

By ANDREW D. ZELENETZ

Introduction

Cloning of cDNA has become an indispensable tool in molecular biology for the analysis of gene expression. Directional cloning of the cDNA is superior to bidirectional cDNA libraries for a variety of applications, including the preparation of strand-specific probes, expression of cDNA inserts, or subtractive hybridization. Various strategies have been used to generate directional cDNA libraries.[1-6] To produce representative, complex cDNA libraries these methods generally require cloning into bacteriophage λ vector. We described a strategy[6] that allows the construction of highly complex (>10^7 members), directional cDNA libraries directly into a phagemid vector, thereby circumventing the limitations imposed by cloning into bacteriophage λ.

Principle

To construct directional libraries, the 3′ and 5′ ends of the double-stranded cDNA must be distinguishable. This is generally achieved by the priming of first-strand synthesis with an adaptor–primer to introduce a restriction enzyme site at the 3′ end of the cDNA. Following completion of the second strand a linker with specificity for a second enzyme is ligated to provide a distinct specificity at the 5′ end of the molecule. The directional cDNA is then generated by digestion with the restriction enzymes whose specificities are encoded at the 5′ and 3′ termini of the cDNA. The cDNA is then cloned into an appropriately cleaved vector. To improve this general strategy we developed a directional cloning cassette that is selectable and can therefore be moved between vectors.[6] This approach facilitates the movement of cDNA clones from a cloning vector to various expression vectors.

[1] H. Okayama and P. Berg, *Mol. Cell. Biol.* **1,** 161 (1982).
[2] D. M. Helfman, J. C. Fiddes, and D. Hanahan, this series, Vol. **152,** 349.
[3] J. H. Han and W. J. Rutter, *Nucleic Acids Res.* **16,** 11837 (1988).
[4] W. D. Huse and C. Hansen, *Strategies* **1,** 1 (1988).
[5] P. S. Meissner, W. P. Sisk, and M. L. Berman, *Proc. Natl. Acad. Sci. U.S.A.* **84,** 4171 (1987).
[6] A. D. Zelenetz and R. Levy, *Gene* **89,** 123 (1990).

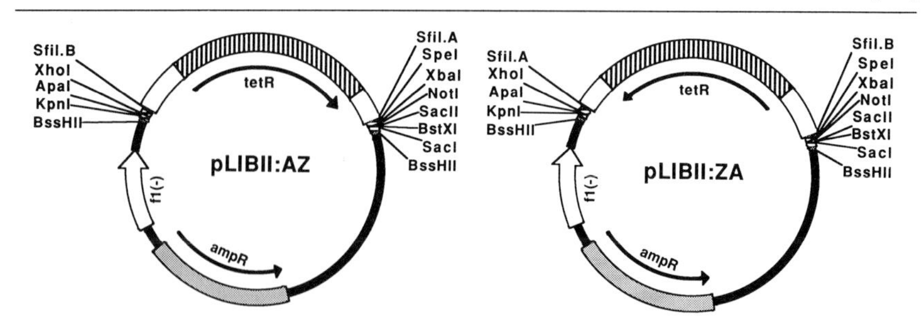

FIG. 1. Restriction maps of pLIBII:AZ and pLIBII:ZA vectors. Cleavage sites for the unique restriction enzyme sites are shown flanking the 1.6-kb *Sfi*I–tet^R stuffer fragment.

Our cloning cassette takes advantage of the fact that cleavage of DNA by the restriction enzyme *Sfi*I occurs within a random 5 bp sequence that bisects the 8-bp sequence necessary for recognition by the enzyme: GGCC**NNNN** ↓ **N**GGCC. Thus the specificity for ligation after DNA is cleaved with *Sfi*I can be customized by choosing the sequence for the random 3-bp overhang. We introduced two distinct *Sfi*I sites (SfiI.A and SfiI.B) into a pair of vectors differing only in the order of the two sites (see Fig. 1). The sites are separated by a stuffer fragment that encodes the gene for resistance to tetracycline from the plasmid pBR322. The vectors, pLIBII:AZ and pLIBII:ZA, were designed so that the *Sfi*I sites and the tetracycline resistance gene can be removed as a cassette into alternative vectors. The new vectors can be identified by selection on tetracycline.

To construct the libraries, cDNA is prepared with a vector–primer encoding an SfiI.B site. This is followed by ligation to an SfiI.A.INS adaptor. After digestion with *Sfi*I to cleave the SfiI.B site at the 3′ terminus of the cDNA, the cDNA is cloned into *Sfi*I-digested pLIBII:AZ or ZA DNA. High efficiency of transformation is achieved by electrotransformation allowing the construction of complex libraries from as few as 6×10^5 cell equivalents of RNA.

Materials and Reagents

Bacterial Strains

Bacterial strains are derivatives of *Escherichia coli* K12. Strain HB101 [F$^-$ *hsd S20(rB$^-$, mB$^-$) supE44 ara14* λ$^-$ *galK2 lacY1 proA2 rspL20 xyl-5 mtl-1 recA13 mcrB*] was used as a recipient for vector construction and the growth of plasmid stocks. Strain MC1061 [F$^-$ *araD139* Δ(*araABC-leu*)*7679 galU galK* Δ(*lac*) X*74 hsr strA*] was used as the primary host for library construction. Strain XL1-Blue {*recA1 endA1 gyrA96 thi hsdR17* (r_k^-, m_k^+) *supE44 relA1* λ^- (*lac*) [F′ *proAB lacI*^{q}ZΔM15, Tn*10*(Tet^r)]} was used as a recipient for phagemid rescue. HB101 and MC1061 were gifts from D. Denny (Stanford Medical School, Stanford, CA). XL1-Blue was obtained from Stratagene (La Jolla, CA). The vectors, pLIBII:AZ and pLIBII:ZA, are available through the American Type Culture Collection (ATCC; Rockville, MD).

Enzymes

Restriction endonucleases, *E. coli* and T4 DNA polymerases, and polynucleotide kinase were obtained from New England BioLabs (Beverly, MA); *E. coli* RNase H and proteinase K were obtained from Bethesda Research Laboratories (Gaithersburg, MD). T4 DNA ligase was obtained from either Boerhinger Mannheim (low concentration; Indianapolis, IN) or New England BioLabs (superconcentrated). *Taq* DNA polymerase was obtained from Perkin-Elmer/Cetus (Emeryville, CA). RNase A was obtained from Sigma (St. Louis, MO) and made DNase free by boiling a 10 mg/ml solution for 20 min in 50 m*M* citrate, pH 5.5, and slowly cooling to room temperature. Calf intestine alkaline phosphatase (CIAP) was obtained from Boerhinger Mannheim. RNase block was obtained from Stratagene. Avian myeloblastosis virus (AMV) reverse transcriptase was obtained from Life Sciences (St. Petersburg, FL).

Enzyme Buffers

All restriction endonuclease digests were performed in 0.5–2× KGB buffer [2× = 200 m*M* potassium glutamate, 20 m*M* Tris–acetate, pH 7.6, 20 m*M* magnesium acetate, 100 μg/ml bovine serum albumin (BSA) 1 m*M* 2-mercaptoethanol (2-ME)].[7] Phosphorylations were performed in 1× polynucleotide kinase buffer [PNKB; 1× = 70 m*M* Tris, pH 7.6, 10 m*M* $MgCl_2$, 50 m*M* dithiothreitol (DTT)]. Ligations were performed either in 0.5× KGB or 1× PNKB with the addition of ATP to a final concentration of 1 m*M*.

Media

SB: Bacto-tryptone, 32 g; Bacto-yeast extract, 20 g; sodium chloride, 5 g. Add 1 liter of double-distilled H_2O and sterilize in autoclave

LB: Bacto-tryptone, 10 g; Bacto-yeast extract, 5 g; sodium chloride, 5 g. Plates are prepared by addition of Bacto-agar, 15 g. Add 1 liter of double-distilled H_2O and sterilize in autoclave

[7] J. Hanish and M. McClelland, *Gene Anal. Tech.* **5,** 105 (1988).

SOC: Bacto-tryptone, 20 g; Bacto-yeast extract, 5 g; 10 m*M* NaCl (2 ml of 5 *M* solution); 2.5 m*M* KCl (1.25 ml of 2 *M* solution); 10 m*M* $MgCl_2$ (1 ml of 1 *M* solution); 10 m*M* $MgSO_4$ (1 ml of 1 *M* solution); 20 m*M* glucose (2 ml of 1 *M* solution). Add 1 liter of double-distilled H_2O, aliquot, and sterilize in autoclave

General Methods

Phagemid Preparation

The construction of the pLIB:AZ and pLIB:ZA vectors has been previously described.[6] The pLIBII vectors are created to provide additional enzyme sites flanking the Sfi cassette. A *Xho*I–*Xba*I fragment spanning the tetracycline gene and the *Sfi*I sites is isolated from the original pLIB vectors. This fragment is cloned into pBluescriptII SK(−) (Stratagene) which had been digested with a combination of *Xho*I and *Xba*I. The resulting pLIBII vectors are shown in Fig. 1 with the unique enzyme sites that flank the *Sfi*I–*tet*^{R} cassette. The phagemids pLIBII:AZ and pLIBII:ZA are maintained in *E. coli* strain HB101.

Large-scale plasmid preparations are initiated by inoculation of 250 ml of SB in a 2-liter flask from a single colony isolated on an LB plate containing ampicillin (50 μg/ml). Bacteria are grown overnight at 37° with vigorous agitation (250 rpm). Plasmid was purified on Qiagen resin (Qiagen, Inc., Chatsworth, CA) as recommended by the manufacturer. Briefly, bacteria are collected by centrifugation and resuspended in 10 ml of 50 m*M* Tris, pH 8.0, 10 m*M* ethylenediaminetetraacetic acid (EDTA), and 400 μg/ml (w/v) RNase A. The cells are lysed by the addition of 10 ml 200 m*M* NaOH and 1% sodium dodecyl sulfate (SDS) and incubated at room temperature for 5 min. Bacterial debris is precipitated by addition of 10 ml 2.55 *M* potassium acetate, pH 4.8, gentle mixing, and centrifugation at 20,000 *g* for 30 min. The supernatant is applied to a Qiagen-pack 500 column that has been preequilibrated with 5 ml of QB (750 m*M* NaCl, 50 m*M* MOPS [3-(*N*-Morpholino)propanesulfonic acid], pH 7.0, 15% ethanol). The column is twice washed with 10 ml of QC (1000 m*M* NaCl, 50 m*M* MOPS, pH 7.0, 15% ethanol). Phagemid DNA is eluted with 5 ml of buffer QF (1200 m*M* NaCl, 50 m*M* MOPS, pH 8.0, 15% ethanol). Generally the yield from a 250-ml culture is 500 μg.

Preparation of Vector for Cloning

Vector DNA is prepared for cloning by digestion with *Sfi*I and is subsequently purified. The minimal amount of *Sfi*I necessary to digest the pLIBII:AZ and pLIBII:ZA vectors to completion as assessed by agarose gel electrophoresis is determined by titration. It is important to *avoid overdigestion,* as contaminating exonuclease activity will damage the 3′ overhangs (see below). Generally, 10 μg of vector DNA is digested with 100 U of *Sfi*I in 0.5× KGB in a final volume of 100 μl overlaid with 100 μl of mineral oil for 90 min at 50°. The mineral oil is removed by extraction with 300 μl chloroform and phases separated by a 15-sec centrifugation in a benchtop microfuge at 12,000 rpm at room temperature. The aqueous (upper) layer is removed to a clean tube, extracted with a mixture of buffer-saturated phenol–chloroform (1 : 1), and centrifuged as above. The aqueous phase is then extracted with 100 μl of chloroform and after centrifugation the aqueous phase is removed to a clean tube. DNA is precipitated by addition of 10 μl 3 *M* sodium acetate, pH 5.5 and 250 μl cold (−20°) ethanol. The tube is chilled on dry ice for 10 min and the precipitate collected by centrifugation for 15 min in a microfuge at 12,000 rpm at room temperature. The ethanol is carefully removed with a drawn-out Pasteur pipette. The pellet is washed by addition of 200 μl of 70% ethanol, followed by vortexing and recollection by centrifugation.

The *Sfi*I-digested DNA is dephosphorylated with CIAP. DNA is resuspended in 32 μl double-distilled H_2O. Four microliters each of 10× CIAP buffer (10×: 500 m*M* Tris, pH 8.0, 1 m*M* EDTA) and CIAP (1 U/μl) are then added. After incubation for 30 min at 37° the CIAP is inactivated by incubation at 70° for 10 min. The cut and dephosphorylated DNA is resolved on a 1.2% (w/v) agarose minigel (Hoeffer, San Francisco, CA) prepared with a three-well sample preparation comb (large central well flanked by two smaller outer wells) in 1× TAE (1×: 40 m*M* Tris–acetate, pH 8.5, 2 m*M* EDTA) by electrophoresis at 60 V for approximately 3 hr. The gel is stained with ethidium bromide at a concentration of 1 μg/ml for 15 min. The 3-kb vector band is identified by transillumination with long-wavelength UV irradiation and cut out of the gel with a scalpel.

The DNA is isolated from the agarose gel with GlassMilk (Bio101, La Jolla, CA) as recommended by the manufacturer. Briefly, the gel slice is weighed and dissolved in 3 vol of saturated sodium iodide at 50°. Then 20 μl of GlassMilk is added and the suspension iced for 15 min. The glass powder is collected by centrifugation followed by rinsing twice in NEW wash (Bio101). Finally, the DNA is eluted from the glass beads by resuspension in 20 μl of double distilled H_2O. DNA concentration is estimated by visualization of ethidium bromide fluorescence as described.[8]

[8] J. Sambrook, E. F. Fritsch, and T. Maniatis, "Molecular Cloning: A Laboratory Manual." Cold Spring Harbor Laboratory, Cold Spring Harbor, New York, 1989.

Preparation of SfiI.A.INS Adaptor

To prepare directional cDNA (see below), an adaptor, SfiI.A.INS, is synthesized that encodes a 3′ overhang complementary to the overhang generated by cleavage of the SfiI.A site in the pLIBII vectors:

5′ TGG CCA CGC C 3′
3′ ACG ACC GGT GCG G 5′

The individual strands (5000 pmol) are phosphorylated with 25 U polynucleotide kinase in 1× PNKB at a concentration of 200 pmol/μl for 30 min at 37°. The kinase is heat inactivated at 70° for 10 min. The adaptor is formed by combining the kinase reactions into a single tube and slowly cooling the mixture to 20°. The resulting adaptor is at a concentration of 100 pmol/μl.

cDNA Synthesis

RNA is prepared by a modified guanidium thiocyanate extraction method.[9] Washed lymphocytes (up to 10^8) are resuspended in 5.7 ml GT solution (5 *M* guanidium thiocyanate, 10 m*M* EDTA, 50 m*M* Tris, pH 7.5, 0.7 m*M* 2-ME) and homogenized to shear the chromosomal DNA. To the homogenate, 2.9 ml of 5.7 *M* CsCl/0.1 *M* EDTA and 0.7 g sarkosyl are added. The cell lysate is carefully layered onto a 3.1-ml cushion of CsCl/EDTA and centrifuged for 16 hr at 155,000 *g* in an SW-41 swinging bucket rotor (Beckman, Palo Alto, CA) at 16°. In this procedure RNA collects as a pellet in the bottom of the tube. DNA bands at the interface between the homogenate and the cushion. When desired, the DNA band is carefully removed with a sterile transfer pipette and then the remaining material is carefully decanted. The RNA pellet is drained dry for 5 min by inverting the tube on an absorbant surface. Using a sterile scalpel, the bottom 1 cm of the tube is removed. The clear pellets are resuspended in 200–400 μl diethyl pyrocarbonate (DEPC)-treated double-distilled water.[8] One volume of 2× binding buffer (10 m*M* Tris, pH 7.4, 1 m*M* EDTA, 3 *M* sodium chloride) is added and polyadenylated RNA is purified on oligo(dT)-cellulose spun columns (Pharmacia, Piscataway, NJ) as recommended by the manufacturer. Polyadenylated RNA is selected once on oligo(dT).

cDNA is synthesized by a modification of the method of Gubler and

[9] J. J. Chirgwin, A. E. Przbyla, R. J. MacDonald, and W. J. Rutter, *Biochemistry* **13,** 5294 (1979).

Hoffman[10] (Fig. 2). First-strand synthesis is initiated from an oligo(dT)/SfiB adaptor–primer (5′-dCCATGGCC**ACAGT**GGCCT$_{18}$) (Fig. 2B). Two to 6 μg of polyadenylated RNA is combined with 10 μg of the adaptor–primer, heated to 95° for 1 min, and then quenched at 4°. The primed RNA is adjusted to 50 m*M* Tris, pH 8.8, 50 m*M* KCl, 6 m*M* $MgCl_2$, 10 m*M* DTT, 0.5 m*M* dNTPs (Pharamacia, Piscataway, NJ), 70 U RNase block, and first-strand synthesis is performed with 96 U AMV reverse transcriptase in a final volume of 100 μl at 42° for 40 min (Fig. 2C). The reverse transcriptase is heat inactivated at 70° for 10 min. The first-strand reaction is adjusted to 20 m*M* Tris, pH 7.5, 5 m*M* $MgCl_2$, 100 m*M* KCl, 10 m*M* DTT and 50 μg/ml bovine serum albumin in a final volume of 500 μl. The second strand is synthesized by addition of 25 U of *E. coli* DNA polymerase I and 4 U of RNase H at 15° for 60 min followed by 20° for 60 min (Fig. 2D). Ends are polished by adjusting the reaction to pH 8.5–9.0 with 5 μl 2 *M* Tris base and 0.4 m*M* dNTPs and adding 10 U T4 DNA polymerase for 15 min at 37°. The reaction is sequentially extracted with 1 vol of phenol–chloroform (1 : 1) and chloroform. The aqueous phase is adjusted to 2 *M* ammonium acetate and the cDNA is precipitated with 2.5 vol of ethanol. The cDNA is resuspended in TE (10 m*M* Tris, pH 8.0, 1 m*M* EDTA) and the ammonium acetate/ethanol precipitation repeated.

To create directional cDNA molecules SfiI.A.INS adaptor is ligated to the polished ds-cDNA (Fig. 2E). The ds-cDNA is resuspended in 6 μl double distilled H_2O and 1 μl of each of the following are added: 10× PNKB, 10 m*M* ATP, T4 DNA ligase (superconcentrated from New England BioLabs), and SfiI.A.INS adaptor (100 pmol). The reaction is incubated at 12° for 16 hr. The adapted cDNA is digested with 100 U of *Sfi*I in 0.5× KGB for 90 min at 50° to create an SfiI.B.INS specificity at the 3′ terminus. The 5′ and 3′ termini (relative to the mRNA strand) of the resulting cDNA have 3′ overhangs complementary to the SfiI.A and SfiI.B sites of the pLIBII vectors, respectively. Unligated adaptor molecules are removed by taking advantage of the nearly irreversible binding of small DNA (<300 bp) to GlassMilk. Ligation reactions are diluted to 50 μl with double-distilled water and then 150 μl of a saturated sodium iodide solution is added along with 15 μl of the GlassMilk suspension. The DNA–glass complexes are pelleted by brief centrifugation in a microcentrifuge (12,000 rpm at room temperature) and washed three times with NEW. The cDNA is eluted from the glass by incubation with 5–10 μl of double distilled H_2O at 55° for 5 min. The glass is removed by centrifugation and the eluted DNA transferred to a clean microcentrifuge tube. DNA is estimated by ethidium bromide staining as above. cDNA recovery following binding

[10] U. Gubler and B. J. Hoffman, *Gene* **25,** 263 (1983).

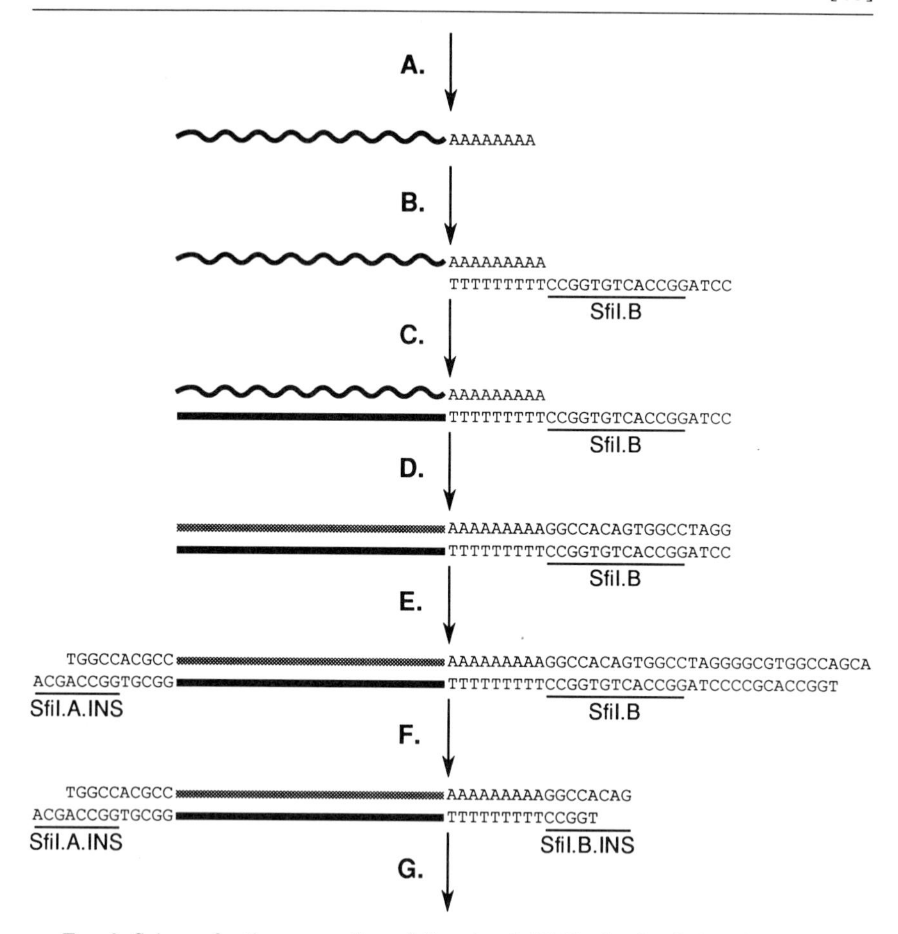

FIG. 2. Scheme for the preparation of directional cDNA; the detailed methods are provided in text. (A) Polyadenylated RNA is isolated directly from tissue or cell lines. (B) First-strand synthesis is initiated from an oligo(d(T)/SfiB primer (5′-dCCATGGCC**ACAGT**GG-CCT$_{18}$) that will provide an SfiB.INS site at the 3′ end of the cDNA. (C) First strand is synthesized with AMV reverse transcriptase. (D) Second strand is synthesized by a modification of U. Gubler and B. J. Hoffman, *Gene* **25,** 263 (1983) using RNase H and *E. coli* DNA polymerase I. (E) The SfiA.INS adaptor is ligated to the termini of the ds cDNA. (F) The adapted cDNA is digested with *Sfi*I to create an SfiB.INS specificity at the 3′ terminus. (G) Directional cDNA is ready for cloning into the pLIBII vectors.

and elution from GlassMilk should be about 50%. Furthermore, analysis of the product on agarose gels should reveal a depletion of cDNA <300 bp and the complete absence of adaptor molecules.

Library Construction

Directional cDNA synthesized as described above is cloned into the pLIBII vectors. cDNA (10 ng) is ligated to 10 ng of purified vector fragment in 1× PNKB with 1 m*M* ATP and 1 Weiss U T4 DNA ligase at 15° for 4 hr in a final volume of 10 μl. Ligated DNA is diluted with 90 μl of TE and sequentially extracted with equal volumes of buffer-saturated phenol–chloroform (1 : 1) and chloroform. One microliter glycogen (20 mg/ml; Boerhinger Mannheim) and 10 μl of 3 *M* sodium acetate, pH 5.5, are added to the aqueous phase. The DNA is precipitated with the addition of 250 μl ethanol and collected by centrifugation in a microfuge. The pellet is washed three times with 70% ethanol to ensure removal of salt. The DNA is electroporated into XL1-Blue, or MC1061 as described below. A small portion of the transformed bacteria is diluted and titrated onto selective medium containing ampicillin (50 μg/ml) to quantitate library size. Generally, the resulting libraries are amplified once by expansion of the 1 ml outgrowth medium in 50 ml SB overnight at 37° with vigorous shaking. Glycerol is added to a final concentration of 15% (v/v) and aliquots are stored at −70°. Alternatively, the libraries can be plated without amplification directly on filters and stored on LB–glycerol plates at −70° as described by Hanahan and Meselson.[11]

Electrotransformation

Electrocompetent cells are prepared as described[12] and stored at −70°. An overnight culture is prepared from a single colony inoculated into 5 ml of SB. The overnight culture is then used to inoculate 1 liter of SB in a 6-liter flask, which is grown at 37° with vigorous aeration (250 rpm) to a density of 1.6–1.9 × 10^8 cells/ml (OD_{600} of 0.4–0.5). The culture is transferred to two 500-ml centrifuge bottles and cooled on ice for 10 min. The cells are pelleted by centrifugation at 4000 *g* for 15 min at 4° and resuspended in 1 liter of ice-cold double distilled H_2O and repelleted as above. Then the cells are washed in 500 ml ice-cold double-distilled H_2O and repelleted. Cells are then resuspended in 20 ml of ice-cold 10% (v/v) glycerol, transferred to a 50-ml Oak Ridge-type tube, and pelleted at 4000 *g* for 15 min. Finally the cells are resuspended in 2 ml of ice-cold 10% (v/v) glycerol, divided into 50-μl aliquots, and stored at −70°.

[11] D. Hanahan and M. Meselson, *Gene* **10,** 63 (1980).

[12] W. J. Dower, J. F. Miller, and C. W. Ragsdale, *Nucleic Acids Res.* **16,** 6127 (1988).

Electrotransformation takes place in a Gene Pulser with a pulse controller circuit (Bio-Rad, Richmond, CA). Cells are allowed to thaw on ice followed by addition of DNA in 1–2 μl of double-distilled H_2O. The DNA and bacteria mixture is transferred to an iced, 0.2-cm gap electroporation cuvette and subjected to a 2.5-kV pulse at 25 μF with the controller set at 400 Ω. These conditions result in optimal transformation efficiencies for strains HB101, XL1-Blue, and MC1061. Immediately following the pulse, 1 ml of SOC is added to the cuvette and the contents are then transferred to a 17 $\times$ 100 mm sterile tube and allowed to recover with vigorous aeration for 60 min. The bacteria are then plated at appropriate dilutions. Typical efficiencies are 5 $\times$ 10^8 colonies/μg for HB101, 3 $\times$ 10^9 colonies/μg for XL-1 Blue, and 2 $\times$ 10^{10} colonies/μg for MC1061.

Examples and Critical Parameters

Preparation of SfiI-Cut Vector DNA

Production of complex cDNA libraries with the any vector system depends on achieving a high ratio of recombinants relative to vector without insert. The pLIBII vectors are designed to permit a high degree of quality control with respect to preparation of the *Sfi*I-digested vector DNA. Contamination by a small portion of uncut or once-cut molecules can result in a significant number of clones without inserts in a cDNA library. Because digestion with *Sfi*I releases a 1.6-kb stuffer fragment, uncut or once-cut molecules can be readily separated from the completely digested vector on agarose gels. To test the quality of a particular vector preparation, purified vector is transformed into HB101 with or without prior ligation. By titering the transformants on LB plates containing ampicillin (Amp) and tetracycline (Tet), only molecules with an intact stuffer will be identified. When titered on LB–Amp plates, molecules that have no stuffer will also be identified. In this manner, the extent of contamination of the preparation with uncut, once-cut, and blunt-ended molecules can be assessed.

An example of this quality control assay is shown in Table I. Uncut vector DNA had transformation efficiencies of 9 $\times$ 10^8 and 2.6 $\times$ 10^8 colonies/μg vector DNA on Amp and Amp–Tet selection, respectively (line 1, Table I). Twenty nanograms of *Sfi*I-digested vector DNA in 0.5$\times$ KGB with 1 m*M* ATP was divided into two samples: one with 1 Weiss unit of T4 DNA ligase and the other without enzyme. They were electroporated into HB101 as described above. The results are shown in lines 2 and 3 (Table I). Cut and unligated vector DNA had a low transformation efficiency, 3 $\times$ 10^5 with ampicillin selection and very low transforming

TABLE I
FREQUENCY OF NO-INSERT RECOMBINANTS IS REDUCED BY DEPHOSPHORYLATION OF THE *Sfi*I-CUT VECTOR DNA

			Colonies/μg ($\times 10^{-6}$)	
Vector	CIAP	Ligation	Amp^r	Tet^r
1. pLIB:ZA uncut	NA[a]	NA	901.00	256.00
2. pLIB:ZA vector fragment	—	—	0.30	0.00
3. pLIB:ZA vector fragment	—	+	10.40	0.01
4. pLIB:ZA vector fragment	+	—	0.00	0.03
5. pLIB:ZA vector fragment	+	+	0.20	0.00

[a] NA, Not applicable.

activity on Amp–Tet, suggesting the presence of few uncut molecules. However, when ligated without insert the transformation efficiency rose 30-fold to 1×10^7, a level that would produce substantial background in cDNA cloning experiments. The low level seen with Amp–Tet selection demonstrated that there were few once-cut molecules. Therefore, the background seen on the ampicillin selection alone must result from twice-cut molecules circularizing despite the presence of incompatible ends. The contamination of commercial preparations of *Sfi*I with trace levels of exonuclease could create blunt ends that would be scored in this assay as Amp resistant and Amp–Tet sensitive.

We reasoned that pretreatment of the vector preparation with calf intestine alkaline phosphatase (CIAP) would further reduce the background. As shown in Table I, CIAP-treated vector (lines 4 and 5, Table I) have background levels reduced to 2×10^5 after ligation. The structure of the phagemid in several background colonies that persisted despite phosphatase treatment was shown by sequencing to have small deletions near the *Sfi*I sites consistent with exonuclease contamination (data not shown). Therefore, we treat all vector fragment preparations with CIAP as described in the section on general procedures. We suggest that vector preparations be subject to the quality-control assay described above.

Complex Libraries Obtained Directly from Tissue

As an example of the complex libraries that can be created, directional libraries were prepared from two forms of lymphoma arising in a single patient. The diffuse lymphoma (Table II, lines 1 and 2) was an actively dividing cell with high RNA content. Large libraries resulted while starting

TABLE II
COMPLEX cDNA LIBRARIES DIRECTLY FROM TISSUE SAMPLES

Histology	pLIB vector	CE of RNA[a]	Library size ($\times 10^{-6}$)
1. DL[b]	AZ	6.5	5.4
2. DL	ZA	6.5	7.4
3. FL	AZ	65.0	4.8

[a] Cell equivalents (CE) ($\times 10^{-5}$) of RNA prepared as described in the section on general procedures.
[b] DL, Diffuse lymphoma; FL, follicular lymphoma.

with RNA from as few as 6.5×10^5 cells. The efficiency of cloning was unaffected by the orientation of the inserts (line 1 versus line 2, Table II). The follicular lymphoma (Table II, line 3) was composed of small, resting B lymphocytes, which have low RNA content per cell; therefore, 10-fold more RNA was necessary to produce a library of equal complexity.

To assess the size of the inserts, plasmid DNA was prepared from the cDNA libraries as described in the section on general procedures. *Sfi*I digestion of the pLIBII:AZ and pLIBII:ZA vectors alone (Fig. 3, lanes 1 and 2) generated two fragments, the 3-kb vector fragment and 1.6-kb stuffer. Library DNA digested with *Sfi*I generated a prominent 3-kb vector fragment and a smear of DNA inserts ranging from 0.3 to 6 kb. There was a faint, discrete band of 1.6 kb that presumably represented contamination of the library with a small amount of uncut molecules (Fig. 3, lanes 3–5). Thus, not only are these libraries complex, but the average insert size is large.

Comments

Directional cloning is necessary for a variety of applications: expression cloning, antisense inhibition of gene function, and subtractive hybridization. Other methods have been described for directional cDNA cloning but these are often technically difficult,[1] have inserts that are not removable intact,[1-5] or require cloning into phage λ.[3-5] The procedure outlined above generally circumvents these problems. Preparation of the vector DNA requires only digestion with *Sfi*I and dephosphorylation; no critical tailing steps are involved. The preparation of the cDNA for cloning is straightforward and can be accomplished with commercial reagents except for the three oligonucleotides, which can readily be obtained from custom synthesis services. The preparation of the bacteria for electroporation is

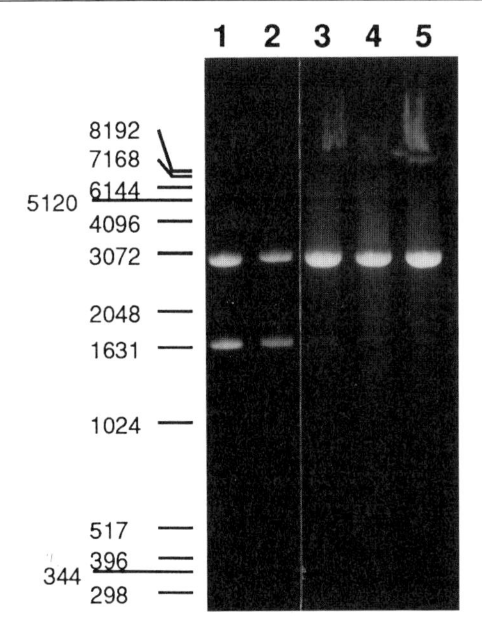

FIG. 3. cDNA libraries cloned into the pLIB vectors. Lanes 1 and 2: pLIBII:AZ and pLIBII:ZA vector DNA (500 ng) was digested with *Sfi*I, which released the 1.6-kb stuffer fragment from the 3-kb vector fragment. Lanes 3–5: Plasmid DNA (2 μg) prepared from the cDNA libraries described in Table II, lines 1–3, respectively, was digested with *Sfi*I. DNA fragments were resolved on a 1.2% (w/v) agarose gel run in 1× TAE.

much less complex than any other method available for high-efficiency transformation. Nonetheless, transformation efficiencies obtainable by electroporation exceed standard methods by a factor of 10^2–10^4. In fact, this phagemid-based cloning system is more efficient than λ-based methods. Like other methods, this one has the disadvantage of requiring cleavage of the cDNA product with a restriction endonuclease. However, because *Sfi*I has an 8-bp recognition sequence the likelihood of a site within the insert is 16 times less than with other methods based on cleavage with enzymes recognizing 6 bp.[3–5]

Furthermore, the presence in the vectors of multiple unique cutting enzymes on either side of the stuffer fragments allows the cassette to be ported to a variety of vectors. Because the stuffer confers resistance to Tet, the derivatized vectors are readily selectable. In fact, we have produced derivatives of the pREP[13] eukaryotic expression vectors that have the SfiI.A and SfiI.B sites. Thus, clones of interest identified from the pLIBII

[13] R. K. Groger, D. M. Morrow, and M. Tykocinski, *Gene* **81,** 285 (1989).

based libraries can be easily subcloned to examine their expression in either a sense or antisense manner. However, vectors having a natural *Sfi*I site [such as those with the simian virus 40 (SV40) origin] cannot be used with this cassette.

Finally, these vectors are not confined to the directional cloning of cDNA. We have reported using these vectors to clone products of the polymerase chain reaction directionally.[14] An oligonucleotide primer containing an SfiI.B specificity 5′ of one amplimer sequence resulted in molecules with an *Sfi*I site with "B" specificity at one terminus of the polymerase chain reaction (PCR) product. The product was then ligated to the SfiI.A.INS adaptor, cleaved with *Sfi*I, and cloned into the pLIB vectors.

Acknowledgment

The author would like to acknowledge helpful discussions with Ronald Levy, Soshana Levy, and John Rubinstein during the development of these procedures. A.D.Z. was supported by a Clinical Investigator Award, K08 CA01396, from the National Cancer Institute. The original work was also supported by National Institutes of Health Grants CA34233 and CA33399 awarded to Ronald Levy.

[14] A. D. Zelenetz, T. C. Chen, and R. Levy, *J. Exp. Med.* **173,** 197 (1991).

[46] Use of Cosmids and Arrayed Clone Libraries for Genome Analysis

By GLEN A. EVANS, KEN SNIDER, and GARY G. HERMANSON

Introduction

Plasmid and bacteriophage cloning vectors have been widely used for a number of years for the isolation and analysis of individual genes or multigene families. More recently, the construction of large-scale maps of complex genomes[1] and the complete physical mapping of genomes of model organisms[2,3] have depended on the use of DNA fragments cloned in bacteriophage or cosmid vectors. The results of these pilot projects

[1] A. Coulson, J. Sulston, S. Brenner, and J. Karn, *Proc. Natl. Acad. Sci. U.S.A.* **83,** 7821 (1986).

[2] M. V. Olson, J. E. Dutchik, M. V. Graham, G. M. Brodeur, C. Helms, M. Frank, M. MacCollin, R. Scheinman, and T. Frank, *Proc. Natl. Acad. Sci. U.S.A.* **83,** 7826 (1986).

[3] Y. Kohara, K. Akiyama, and K. Isono, *Cell (Cambridge, Mass.)* **50,** 495 (1987).

suggest that producing a complete physical map of the human genome, consisting of overlapping clone sets, referred to as contigs, is now feasible. Most current large-scale mapping projects designed to map the human genome utilize both cosmids,[4] prepared as arrayed or chromosome-specific cosmid libraries, and large fragment clones in *Saccharomyces cerevisiae* carried as yeast artificial chromosomes (YACs).[5] The use of cosmids and YAC clones in concert provides a powerful mechanism for genome mapping.[6]

Genomic libraries constructed for large-scale genome analysis differ substantially from those constructed for single-gene isolation in several respects. For genome mapping, the criteria for representation, completeness, random distribution of clones, and frequency of cloning artifacts are perhaps more significant than with single-gene isolation and extensive characterization of genomic libraries is necessary. In addition, because a large number of analyses on individual clones will be carried out, there is some advantage in utilizing cosmid libraries as large clone arrays, rather than pooled libraries. For single-gene cloning, genomic libraries are generally prepared as a single pool of individual clones that is plated on agar, membrane filter transfers carried out, and the library screened for specific clones by hybridizatoin with a DNA probe. For genomic analysis, it is more convenient, although more time consuming and labor intensive, to prepare libraries as large arrays where each clone is stored individually as a single culture in 1 well of a 96-well microtiter dish. Replicas of the library are transferred to filter membranes for hybridization screening using manual transfer devices or automated instruments. Screening for individual clones, as well as the construction of contigs by hybridization-based fingerprinting, may then be carried out using these filter replicas. Clone arrays have been shown to be extremely useful for detecting single-copy sequences in libraries, for construction of contigs using pools of RNA or oligonucleotide probes,[7] for detecting overlapping sequences between contiguous YAC clones, for characterizing somatic cell hybrids carrying fragments of human chromosomes, and for characterization of repetitive sequences in clone libraries.[8] These reference libraries also allow for the convenient exchange of mapping information obtained using different mapping strategies and between different laboratories. Many of the uses

[4] G. A. Evans and G. M. Wahl, this series, Vol. 152, p. 604.

[5] D. T. Burke, G. F. Carle, and M. V. Olson, *Science* **236,** 806 (1987).

[6] A. Coulson, R. Waterson, J. Kiff, J. Sulston, and Y. Kohara, *Nature (London)* **335,** 184 (1988).

[7] G. A. Evans and K. A. Lewis, *Proc. Natl. Acad. Sci. U.S.A.* **86,** 5030 (1989).

[8] D. Nizetic, G. Zehetner, A. P. Monaco, L. Gellen, B. D. Young, and H. Lehrach, *Proc. Natl. Acad. Sci. U.S.A.* **88,** 3233 (1991).

of clone arrays have been described.[9] In addition, clone arrays represent a useful and convenient way of archiving and distributing human chromosome-specific cosmid libraries constructed from chromosomes purified by flow cytometry.[10,11]

Several specialized techniques have been developed for the production and use of cosmid reference libraries maintained in arrays. This chapter will review some protocols utilized in our laboratory for large-scale genome mapping, including the construction of cosmid libraries, the production and archiving of arrayed cosmid libraries, the use of robots for cosmid manipulation and processing, and the analysis of individual cosmid clones. Cosmid libraries are used in conjunction with YAC cloning for genome analysis and techniques for the use of YAC clones in genome mapping have been reviewed elsewhere.[12]

Cosmid Vectors for Genome Analysis

Cosmid vectors have been widely used for a number of years for cloning and analyzing genomic DNA. Cosmids are plasmids that contain bacteriophage packaging sequences to enable insertion of the cloned DNA into λ phage heads. The technique of packaging ligated DNA into a viral protein coat provides a convenient and efficient selection for uniformity in the size of cloned DNA, ensuring an insert size of between 35 and 45 kb. Two types of cosmid vectors are in general use that differ only in the structure of the packaging signals and require different methods for constructing genomic libraries. Vectors with a single bacteriophage λ *cos* sequence require the extensive purification and size selection of the genomic DNA to be cloned to prevent a high rate of coligation events.[13] Cosmids with two *cos* signals, also known as "double *cos*" vectors, allow rapid and efficient cloning of genomic DNA that is not size selected by dephosphorylation of the genomic DNA to prevent coligation.[14] Double *cos* vectors are particularly useful in that they may be used for the con-

[9] H. Lehrach, R. Drmanac, J. Hoheisel, Z. Larin, G. Lennon, A. P. Monaco, D. Nizetic, G. Zehetner, and A. Poustka, *in* "Genetic and Physical Mapping" (K. Davies and S. M. Tilghman, eds.), Vol. 1, p. 39. Cold Spring Harbor Laboratory, Cold Spring Harbor, New York, 1990.

[10] J. W. Gray, J. Luca, D. Peters, D. Pinkel, B. Trask, G. van den Engh, and M. van Dilla, *Cold Spring Harbor Symp. Quant. Biol.* **51,** 141 (1986).

[11] L. L. Deaven, M. A. Van Dilla, M. F. Bartholdi, A. V. Carrano, L. S. Cram, J. C. Fuscoe, J. W. Gray, C. E. Hildebrand, R. K. Moyzis, and J. Perlman, *Cold Spring Harbor Symp. Quant. Biol.* **LI,** 159 (1986).

[12] P. Heiter, C. Connelly, J. Shero, M. K. McCormick, S. Antonarakis, W. Pavan, and R. Reeves, *in* "Genetic and Physical Mapping" (K. Davies and S. M. Tilghman, eds.), Vol. 1, p. 83. Cold Spring Harbor Laboratory, Cold Spring Harbor, New York, 1990.

[13] W. Chia, M. R. D. Scott, and P. W. J. Rigby, *Nucleic Acids Res.* **10,** 2503 (1982).

[14] P. F. Bates and R. A. Swift, *Genes* **26,** 315 (1989).

struction of libraries from very small quantities of DNA, such as that obtained from purified chromosomes obtained by flow cytometry.[15]

A wide variety of cosmid vectors has been created containing many specialized functions applicable to genome analysis. Some of these functions include selectable genes for drug resistance that allow transformation and selection of cosmid vectors in mammalian cells,[16,17] bacteriophage promoters for production of end-specific RNA probes,[18,19] multicopy origins of replication,[20] sequences to allow efficient chromosome walking,[21] rare restriction enzymes flanking the insert to allow for excision of the entire insert fragment as a single piece and to allow rapid restriction map determination,[22] and structures to allow efficient DNA sequencing.[23] For use in genome analysis, we derived cosmid vectors that contain some of these functions[15] (Fig. 1): pWE15 is a single *cos* vector of 8048 nucleotides; sCos1 is a double *cos* vector of 7939 nucleotides. Both vectors contain a multicopy ColE1 bacterial origin of replication and the complete DNA sequence of both vectors has been determined by automated cosmid sequencing.[24] Modifications of the sCos1 vector containing different polylinker and cloning sites have also been constructed.[15] Other widely used cosmid vectors include those of the Lorist series,[25,26] which are similar in many respects but contain a single-copy bacteriophage λ origin of replication. Protocols described in this chapter apply specifically to pWE or sCos vectors but may also be applicable to other vectors and systems as well.

Materials

Enzymes

All restriction enzymes, polynucleotide kinase, T4 DNA ligase, T3 and T7 RNA polymerases, and alkaline phosphatases were obtained from a

[15] G. A. Evans, K. A. Lewis, and B. E. Rothenberg, *Gene* **79,** 9 (1989).
[16] F. G. Grosveld, T. Lund, E. J. Murray, A. L. Mellor, H. H. M. Dahl, and R. A. Flavell, *Nucleic Acids Res.* **10,** 6715 (1982).
[17] Y. F. Lau and Y. W. Kan, *Proc. Natl. Acad. Sci. U.S.A.* **80,** 5225 (1983).
[18] S. H. Cross and P. F. R. Little, *Gene* **49,** 9 (1986).
[19] G. M. Wahl, K. Lewis, J. Ruiz, B. E. Rothenberg, J. Zhao, and G. A. Evans, *Proc. Natl. Acad. Sci. U.S.A.* **84,** 2160 (1987).
[20] E. Ehrich, A. Craig, A. Poustka, A. M. Frischauf, and H. Lehrach, *Gene* **57,** 229 (1987).
[21] H. J. Breter, M. T. Knoop, and H. Kirchen, *Gene* **53,** 181 (1987).
[22] P. F. R. Little and S. H. Cross, *Proc. Natl. Acad. Sci. U.S.A.* **82,** 3159 (1985).
[23] A. Ahmed, *Gene* **61,** 363 (1987).
[24] A. Martin-Gallardo, W. R. McCombie, J. D. Gocayne, M. G. FitzGerald, S. Wallace, B. M. B. Lee, J. Lamerdin, S. Trapp, J. M. Kelley, L.-I. Liu, M. Dubnick, L. A. Johnston-Dow, A. R. Kerlavage, P. de Jong, A. Carrano, C. Fields, and J. C. Venter, *Nature Genetics* **1,** 34 (1992).
[25] T. J. Gibson, A. R. Coulson, J. E. Sulston, and P. F. R. Litle, *Gene* **53,** 275 (1987).
[26] T. J. Gibson, A. Rosenthal, and R. H. Waterston, *Gene* **53,** 283 (1987).

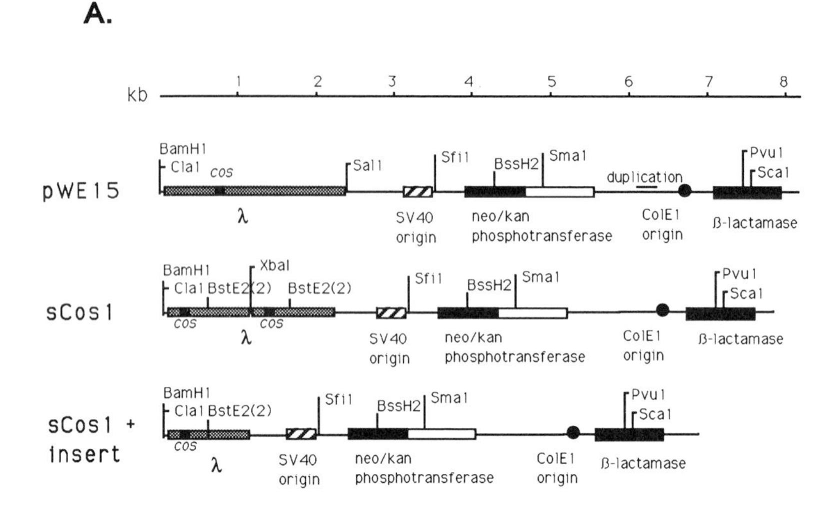

B.

EcoR1 Not1 BamH1 Not1 EcoR1

GAATTCGCGGCCGCAATTAACCCTCACTAAAGGATCCCTATAGTGAGTCGTATTATGCGGCCGCGAATTC

<-- T3 T7 -->

FIG. 1. (A) Structure of cosmid vectors for genomic analysis. The feature map, based on the complete nucleotide sequence, is shown for vectors pWE15, sCos1, and sCos1 with insert. Because of the duplication of the *cos* site in the sCos1 vector and the elimination of one copy during the packaging reaction, the restriction map of the cloning vector and the resulting vector with genomic insert are different. The locations of the ColE1 bacterial origin of replication, the SV40 origin and promoter, and the neomycin/kanamycin phosphotransferase and the β-lactamase genes are shown, as are sequences derived from bacteriophage λ. (B) DNA sequence of the polylinker cloning site. The cloning sites of cosmid vectors pWE15 and sCos1, showing the locations of flanking *Not*I sites, the unique *Bam*HI cloning site, and T3 and T7 bacteriophage promoters, are annotated.

number of suppliers including New England BioLabs (Beverly, MA), Bethesda Research Laboratories (Gaithersburg, MD), and Stratagene Cloning Systems (La Jolla, CA). High-efficiency *in vitro* packaging extracts (Gigapak Gold) were obtained from Stratagene. All enzymes were used under conditions recommended by manufacturers.

Strains

Cosmid libraries were constructed using bacterial strain DH1 or DH5α.

Media

TB (terrific broth): 1.2% (w/v) Tryptone, 2.4% (w/v) yeast extract, 0.4% (v/v) glycerol, 70 m*M* phosphate buffer, pH 7.0

LB: 10 g Bacto-tryptone, 5 g bacto-yeast extract, 10 g NaCl per liter of distilled water

Solutions

SSC: 0.15 *M* NaCl, 0.015 *N* trisodium citrate, pH 7.6
STET: 8% (w/v) sucrose, 5% (v/v) Triton X-100, 50 m*M* Tris-HCl at pH 8.0, 50 m*M* ethylenediaminetetraacetic acid (EDTA)
STE: 0.1 *M* NaCl, 1 m*M* Na_2EDTA, 50 m*M* Tris-HCl, pH 7.6
TE: 50 m*M* Tris-HCl, pH 8.0, 5 m*M* EDTA
Lysozyme solution: 10 mg/ml in STET buffer
Agarose gel dye: 40% (v/v) glycerol, 10% (v/v) Ficoll, 25 m*M* EDTA, 0.15% (v/v) xylene cyanol, 0.25% (v/v) bromphenol blue
Phenol–chloroform: Phenol–chloroform for extraction of nucleic acids is prepared as phenol–chloroform–isoamyl alcohol (25 : 24 : 1, saturated with 50 m*M* Tris-HCl at pH 8.0

All solutions are made in doubly distilled water treated with diethyl pyrocarbonate (DEPC) to eliminate RNase activity.

Other Materials

Flat-bottom and V-bottom microtiter plates are from Corning (Corning, NY). Nytran nylon filter membranes are from Schleicher & Schuell (Keene, NH).

Methods

Preparation of Genomic Cosmid Libraries

Genomic cosmid libraries may be rapidly produced using double *cos* vectors from quantities of DNA as small as 50 ng. The following is a protocol used for the production of cosmid genomic libraries from non-size-selected DNA. Alternate protocols must be used for the construction of libraries using single *cos* vectors.[4,27]

Construction of Genomic Cosmid Libraries in sCos1. High molecular weight genomic DNA for cosmid cloning is prepared using proteinase K digestion and gentle phenol extraction.[27] Cells from 15 ml of whole blood, or 10^6 tissue culture cells, are suspended in 15 ml of STE and adjusted to 100 μg/ml proteinase K and 0.5% (w/v) sodium dodecyl sulfate (SDS). The solution is gently mixed and incubated for 6 hr at 50°. The DNA is gently extracted with phenol–chloroform and dialyzed against TE overnight. Following preparation, the average size of the genomic DNA is

[27] A. DiLella and S. L. C. Woo, this series, Vol. 152, p. 199.

determined by analysis on pulsed-field gel electrophoresis using the HEX-CHEF electrode configuration.[28,29] For efficient cloning, the average size of the DNA should be from 500 kb to greater than 3 Mb. The DNA is suspended at a concentration of 500 μg/ml and digested with *Mbo*I at 5 U/ml in 1 *M* NaCl, 100 m*M* Tris-HCl, pH 7.4, 100 m*M* $MgCl_2$, 10 m*M* dithiothreitol (DTT) at 37° for 5 to 20 min. The exact time of digestion is determined by a titration experiment to generate partial fragments of average size 50 to 100 kb. Following digestion, the reaction is terminated by phenol–chloroform extraction and the DNA analyzed by pulsed field gel electrophoresis to determine the average size of the products.

Vector cloning arms are prepared by first digesting purified sCos1 DNA with *Xba*I followed by treatment with calf intestinal alkaline phosphatase. The reaction is then terminated by phenol–chloroform extraction and the DNA collected by ethanol precipitation. The linearized, dephosphorylated vector DNA is digested with *Bam*HI. Following digestion, the DNA is extracted with phenol–chloroform and stored at a concentration of 1 mg/ml in 20 m*M* Tris-HCl, pH 7.6, 1 m*M* EDTA. Ligations are performed using 1 μg of vector arms and 50 ng to 3 μg of *Mbo*I-digested genomic DNA. Reactions are incubated with 2 Weiss units of T4 DNA ligase and packaged using commercial *in vitro* packaging lysates. Cloning efficiencies using this protocol routinely range from 1×10^5 to 2×10^7 colonies per microgram of genomic insert DNA.

Archiving and Storage of Libraries. Genomic libraries have been traditionally stored in pooled, amplified cultures frozen in glycerol at −70°. For genome analysis, it is ultimately more convenient, although initially more labor intensive, to store libraries as archive clones in individual cultures. One convenient method is to plate the library on LB agar with 20 μg/ml kanamycin at a density so that most colonies do not touch or overlap surrounding colonies, about 10 clones/cm^2. A 96-well flat-bottom microtiter plate is filled with 100 μl of LB/kanamycin medium per well. Each colony is touched with a toothpick and transferred to the medium in the well. When a complete plate has been picked, the 96-well plate is incubated at 37° for 6 or 8 hr in a humidified, sealed chamber. Following growth of the bacteria, 15 μl of glycerol is added to each well, mixed, and the plate sealed with Parafilm or with adhesive acetate plate sealing material (Linbro, Flow Laboratories, McLean, VA). The plate is then stored at −70 or −80°. For recovering clones, a toothpick or 96-prong transfer device is touched to the surface of the frozen medium and used to inoculate fresh agar or LB medium. To simplify the many repetitive pipetting steps,

[28] G. Chu, D. Vollrath, and R. W. Davis, *Science* **234,** 1582 (1986).
[29] G. Chu, *Methods: Companion Methods Enzymol.* **1,** 129 (1990).

a 12- or 8-well multiple pipetting device or a Beckman Biomek 1000 laboratory robotics workstation may be used. Archived cosmid libraries have been stored in this way for over 5 years without noticeable loss of viability or clone rearrangements.

For chromosome-specific cosmid libraries, or libraries prepared from a model organism with a limited genome size, it is convenient to pick and archive every clone. For some applications, such as the selection of clones from a specified region of the genome carried in a somatic cell hybrid, it is useful to carry out a membrane transfer and hybridization screen. When a library is constructed from genomic DNA from a rodent cell carrying a portion of a human chromosome, hybridization screening with a human repetitive probe will allow the clones containing human DNA to be selected and archived, resulting in a library of limited complexity representing a small region of the genome.[7]

Preparation of Cosmid Arrays Using Manual Techniques. Following storage in 96-well plates, replicas of clones may be conveniently produced on membrane filters using a 96-prong transfer device (or "hedgehog"). Replica transfers are made from the surface of frozen medium in wells using a device with floating pins, so that contact with uneven surfaces always occurs. After each transfer, the aluminum transfer device is sterilized in 60% ethanol and dried. A membrane filter is cut to the required shape and placed on the surface of an agar plate containing LB plus kanamycin. Replica transfer from frozen colonies is made onto the surface of the membrane and the filters covered and placed at 37° for 12 to 15 hr, or until colonies are visible and of the appropriate size for further analysis. For fixation, the filter is removed and placed sequentially on the surface of Whatman (Clifton, NJ) 3MM paper sheets wet with 0.5 *M* NaOH for 15 sec, to 0.1 *M* Tris-HCl, pH 8.0, for 25 sec, and finally 0.5 *M* NaCl, 0.1 *M* Tris-HCl, pH 8.0, for 15 sec. The filters are air dried and baked at 65° for 30 min. Filters are stored between paper towels until used for hybridization analysis. A cosmid array prepared in this way is shown in Fig. 2.

Automated Preparation of Cosmid Arrays. Using manual methods, it is possible to prepare filters with a density of 384 clones/72 cm^2 by interleaving the patterns of four 96-well plates on a single membrane filter (Fig. 2). However, for large libraries, this necessitates the use of a large number of filters and is time consuming to produce. Higher clone densities can be produced on membrane filters by using a robotic device to carry out automated transfer. A 16× interleaved pattern, placing the contents of 16 96-well plates in the space of 1, may be produced using an automated laboratory robotic device such as the Beckman Biomek 1000 and a 96-pin arraying tool. Detailed use of these

A

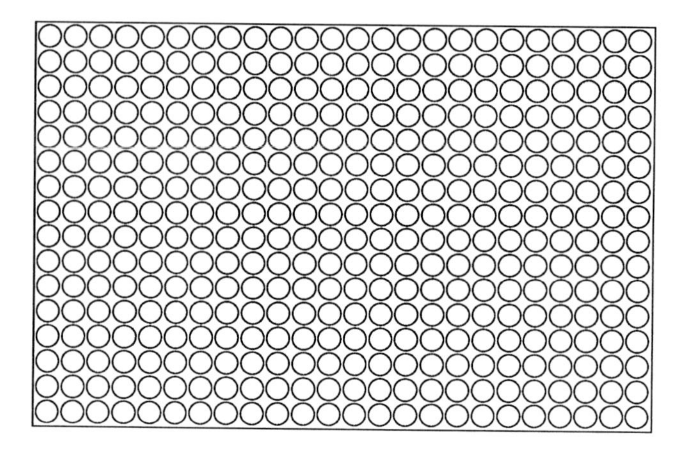

B

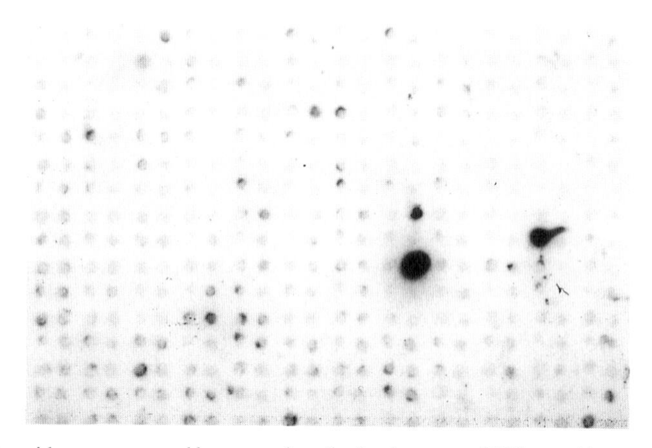

FIG. 2. Cosmid array prepared by manual methods. An array of 384 cosmids was prepared using four 96-well microtiter plate archives. Transfer to a nylon membrane was carried out using a 96-prong aluminum transfer device. (A) Grid pattern. (B) Hybridization of a unique copy DNA probe to a set of chromosome-specific cosmids.

robotic devices will not be described here and is outlined elsewhere[30] but a general procedure for automated preparation of clone grids using the Biomek robot is as follows.

[30] G. Hermanson, P. Lichter, L. Selleri, K. Lewis, D. Ward, and G. A. Evans, *Genomics* in press (1990).

To prepare high-density grids, a replica of each frozen microtiter plate is created by the inoculation of a fresh 96-well plate containing LB medium with 25 μg/ml kanamycin using a 96-prong replicator. These replica plates are incubated in a humidified atmosphere for 12 hr at 37° and stored at 4° before use. Biotran membranes (8 × 12 cm; ICN, Irvine, CA) are cut to size with a razor blade, rinsed in sterile water, and then soaked in LB medium for 15 min. The filters are then placed on the surface of LB agar containing 25 μg/ml kanamycin poured in the lid of a Corning 96-well microtiter plate. A Beckman Biomek 1000 robot equipped with a high-density array tool and sterilization unit is used to array cosmids onto nylon membranes. Using the high-density array tool, the equivalent of sixteen 96-well plates (1536 clones) is arrayed on the 8 × 12 cm filter. The spacing between clones is 2 mm and a diagram is shown in Fig. 3. After transfer of each 96-well plate, the transfer tool is sterilized by a 5-sec rinse in household bleach, a 10-sec rinse in distilled water, and a 15-sec rinse in 95% ethanol. The tool is then dried over a fan for 50 sec to remove the alcohol.

Following arraying, the membranes are removed from the robot and incubated at 37° for 6 to 12 hr, or until the colonies are about 1 mm in diameter. The colonies are fixed by placing the membrane on Whatman 3MM paper wet with 0.5 *M* NaOH, 1 *M* Tris-HCl, pH 7.6, and 1 *M* Tris, pH 7.6, 1.5 *M* NaCl. The membranes are then allowed to air dry and baked in a vacuum for 1 hr at 80°, after which they are cross-linked with ultraviolet (UV) light. Membranes are stored at room temperature between the sheets of clean paper towels before use. An example of a high-density grid prepared using the Biomek robot is shown in Fig. 3.

Hybridization to Arrayed Cosmid Filters. Hybridization with DNA probes is carried out using a preannealing step to block the signal of repetitive sequences in the probe. After labeling of the probe using random hexamer-primed synthesis,[31] the labeling reaction is terminated by extracting with phenol–chloroform and 100 μl of blocking mixture added. Blocking mixture consists of 5 mg/ml of plasmid pBlur8 containing a cloned human *Alu* repetitive element, and 2.5 mg/ml human placental DNA. The DNA mixture is prepared by sonication to an average size of 50 to 100 bp and resuspended in 10 m*M* Tris-HCl, pH 7.6, 1 m*M* EDTA, and denatured by placing in a boiling water bath for 5 min prior to use. Alternatively, human repetitive sequences prepared by preparative hybridization (C_0t 1 DNA) may be used. The ^{32}P probe mixed with blocking mixture is precipitated with 20 μl of 0.9 *M* NaCl, 50 m*M* sodium phosphate, pH 8.0, 5 m*M* EDTA, 0.1% (w/v) SDS, and 600 μl of 95% ethanol at −20°, collected by centrifugation in a microfuge, dried under vacuum, and resuspended in 20

[31] A. Feinberg and B. Vogelstein, *Anal. Biochem.* **132,** 6 (1983).

A.

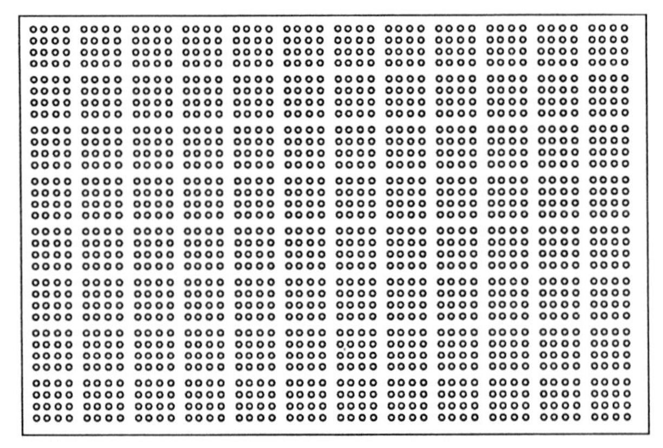

B.

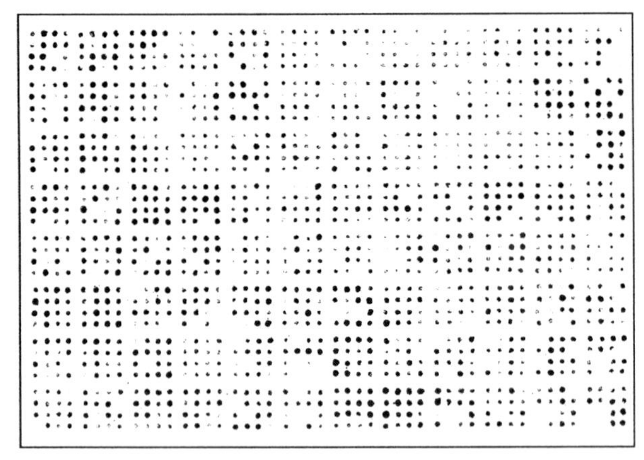

C.

μl of sodium phosphate, pH 8.0, 5 m*M* EDTA, 0.1% (w/v) SDS. Annealing is carried out for 10 min at 42°, then the probe mixture is diluted into hybridization buffer [500 m*M* sodium phosphate, pH 7.6, 7% (w/v) SDS, 1 m*M* EDTA]. Hybridization is carried out by sealing filters in plastic Seal-A-Meal bags, adding hybridization buffer containing probe, and incubating at 65° in a water bath for 12 hr. Filters are washed in 2× SSC followed by 0.1× SSC at 65°. Hybridization to a cosmid grid is shown in Fig. 3.

Phenol Emulsion RT Hybridization. An alternative hybridization approach utilizes accelerated hybridization in phenol emulsion[32] as described by Djabali *et al.*[33] Following labeling of the probe with ^{32}P, and denaturing at 95° for 5 min, 100 μg of blocking mixture or C_0t 1 DNA is added to the probe and the volume adjusted to 1 ml with 100 m*M* sodium phosphate, pH 7.2 Water-equilibrated phenol (500 μl, pH 8.0) is added and the mixture shaken vigorously with a Vortex (Cleveland, OH) mixer for 15 hr at room temperature. The phenol is removed by centrifugation in a microfuge and the aqueous phase added to the hybridization buffer as described above. Hybridization is carried out in 500 m*M* sodium phosphate, pH 7.6, 7% (w/v) SDS, 1 m*M* EDTA at 65°.

Analysis of Cosmid Libraries

Following the creation of an arrayed genomic cosmid library, several types of analysis are possible, including screening the library with single-copy DNA probes, screening with pooled or degenerate probes containing repetitive sequences that must be blocked in the hybridization reaction, analysis of individual or collections of clones for specified restriction sites, or detection of ovrelapping cosmids for the construction of contigs.

Cosmid DNA Preparation from Minilysates. Most analyses of cosmids can be carried out using DNA isolated from 2.5-ml cultures. A culture tube containing LB broth containing 20 μg/ml kanamycin sulfate is inoculated with a single bacterial colony, or a streak from the surface of a frozen

[32] D. E. Kohne, S. A. Levison, and M. J. Byers, *Biochemistry* **16,** 5329 (1977).

[33] M. Djabali, C. Nguyen, D. Roux, J. Demengeot, H. M. Yang, and B. R. Jordan, *Nucleic Acids Res.* **18,** 6166.

FIG. 3. High-density cosmid array. An array of 1536 cosmids prepared using sixteen 96-well microtiter plate archives. Transfer to nylon membrane was carried out using the Beckman Biomek 1000 robot with a 96-prong high-density array tool and sterilization unit. (A) Grid pattern produced with the Biomek robot. (B) Hybridization of a human repetitive *Alu* sequence to an array of 1536 cosmids from human chromosome 11. (C) Hybridization of a unique-copy DNA probe.

archive well. The culture is incubated at 37° for no longer than 6–8 hr with vigorous shaking. Longer incubation periods give consistently lower yields of cosmid DNA.

DNA is prepared using a modified boiling procedure. Bacterial cells are collected by centrifugation for 2 min in a 1.5-ml microfuge tube. The supernatant is removed by aspiration and cells resuspended in 300 μl of STET buffer prepared in DEPC-treated water. Fresh lysozyme solution (25 μl) is added to the resuspended cells and the suspension vigorously mixed using a Vortex mixer. The microfuge tube containing the mixture is heated in boiling water for 2 min to lyse the bacteria. The solution is allowed to cool for 2–5 min and the precipitate collected by centrifugation in a microfuge for 10 min. The gelatinous pellet is removed from the tube with a sterile toothpick and discarded. 2-Propanol (325 μl) is added to the cleared lysate, mixed, and the nucleic acid precipitated at room temperature for 5 min. The nucleic acid is collected by centrifugation in the microfuge for 10 min. The alcohol is removed by aspiration, the pellet air dried for 10 min, and the DNA resuspended in 25 μl of sterile DEPC-treated water. This preparation yields 1 to 10 μg of cosmid DNA from a 1.5-ml culture and 2 to 4 μl of the DNA solution is usually sufficient for restriction endonuclease analysis.

Additional deproteinization may be necessary if the DNA preparation is to be used as a template for bacteriophage polymerases or fluorescence *in situ* hybridization.[34,35] To remove contaminating ribonucleases that may affect these procedures, the DNA must be extracted once with phenol–chloroform and once with chloroform. Following removal of the organic phase, the aqueous phase is adjusted to 0.4 *M* sodium acetate (pH 5.5) and precipitated with ethanol. The precipitated DNA is dissolved in DEPC-treated sterile water at a concentration of 1 mg/ml and stored at −20°.

Restriction Map Determination by Oligonucleotide End Labeling. Restriction mapping of genomic DNA carried in cosmid vectors pWE15 or sCos1 may be rapidly and efficiently determined using a modification of a method of Smith and Birnsteil[36] for DNA fragments, and a method of Rackwitz *et al.* for cosmids digested with λ-terminase.[37] Radiolabeled T3 and T7-specific oligonucleotides, commercially available as DNA sequencing primers, can be used to detect the ends of the insert following excision

[34] P. Lichter, C.-J. C. Tan, K. Call, G. Hermanson, G. A. Evans, D. Housman, and D. C. Ward, *Science* **85,** 9664 (1990).

[35] L. Selleri, J. Eubanks, G. Hermanson, and G. Evans, *Gene. Anal. Tech. Appl.* **8,** 59 (1991).

[36] H. O. Smith and M. L. Birnsteil, *Nucleic Acids Res.* **3,** 2387 (1976).

[37] H. R. Rackwitz, G. Zehetner, H. Murialdo, H. Delius, J. H. Chai, A. Poustka, and A. M. Frischauf, *Gene* **40,** 259 (1985).

by *Not*I.[38] Using this method, the insert is separated from the vector by *Not*I digestion followed by partial digestion with one or more restriction enzymes. The digestion products are separated by gel electrophoresis on an agarose gel, transferred to a filter membrane, and fragments are detected by hybridization to end-specific oligomers. The restriction map may be determined from the pattern of bands detected by oligomers on the gel (Fig. 4). This method is fast and convenient but not applicable to clones containing internal *Not*I sites.

Automated Screening for Restriction Sites. Another use for arrayed cosmid libraries is the detection of all clones having rare restriction sites. This is particularly useful for the detection and isolation of linking clones, clones containing rare restriction sites that may be used as mapping landmarks. Several methods for detecting or cloning linking clones have been described, including hybridization of arrayed cosmid libraries with short oligonucleotides complementary to rare sequences.[39] However, many of these methods are complicated by cross-hybridization, high background, and give a high false-positive rate.

An alternative method for identifying rare restriction sites is the repetitive automated screening of cosmids for restriction sites using a robot (Fig. 5). Automated DNA preparation and restriction site analysis of arrayed cosmid clones can be carried out using the Beckman Biomek 1000 automated robotic workstation using cosmids grown in 96-well microtiter plates. For this application, the platform of the Biomek robot is modified so that one of the plate positions can hold an 8 × 12 cm 1% (w/v) agarose gel with 12 wells matching the pattern of the wells of a 96-well microtiter plate. Agarose gels are poured on 8 × 12 cm frosted Plexiglas plates to prevent wandering of the gel during movement of the Biomek platform and inserted into the standard Biomek platform in place of a 96-well plate. This system will prepare DNA from cosmid microcultures in 96-well plates using an abbreviated procedure, add a single restriction enzyme to the DNA preparation, allow digestion, and then load the products on 12-well agarose minigels.

Cosmid DNA is prepared as follows using an automated procedure: Bacterial cultures are prepared by replica transfer of bacterial colonies from frozen stocks using a 96-well aluminum transfer device into Corning 96-well V-bottom microtiter plates containing 200 μl of fresh TB containing 25 μg/ml kanamycin sulfate. Cultures are incubated for 8 hr at 37° with gentle shaking in a humidified atmosphere and the bacterial pellet collected by centrifugation at 3000 rpm for 20 min in a Beckman TJ-6 centrifuge.

[38] G. A. Evans, K. A. Lewis, and G. M. Lawless, *Immunogenetics* **79,** 9 (1988).

[39] X. Estivill and R. Williamson, *Nucleic Acids Res.* **15,** 1415 (1987).

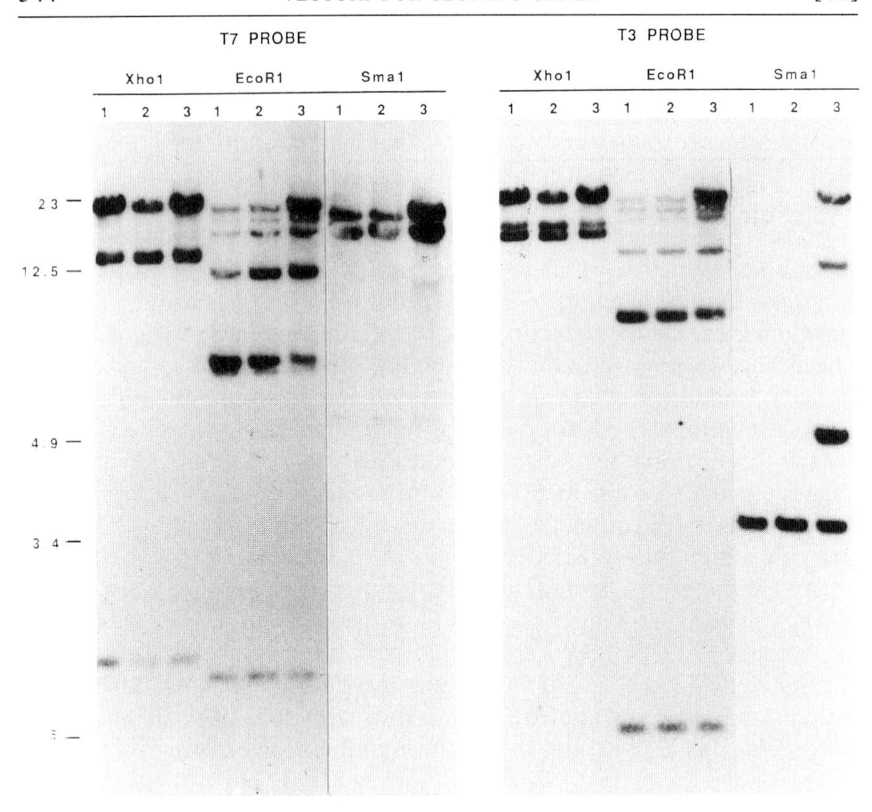

FIG. 4. Restriction map analysis of cosmid clones by partial endonuclease digestion and hybridization with end-specific oligonucleotides ("oligo end labeling").[38] The restriction map of a single cosmid in vector pWE15 was determined from DNA prepared from 1.5-ml bacterial cultures. Cosmid DNA was digested with *Not*I to separate the vector from insert, then digested with *Xho*I, *Eco*RI, or *Sma*I under conditions determined to generate partial digestion products. Restriction fragments were separate on an agarose gel, transferred to nitrocellulose membrane, and hybridized with ^{32}P-labeled oligonucleotides specific for the bacteriophage T3 or T7 promoters. Lanes 1–3 represent decreasing concentrations of the indicated restriction enzyme. The T3 and T7 probes are commercially available as DNA sequencing primers.

The plate is inverted over paper towels to remove growth medium and then placed on the platform of the Biomek. Cosmid DNA preparation uses a modified version of the alkaline lysis procedure[40] modified for robotic processing. Under robotic control, the following reagents are added sequentially to each well: 25 μl of STET, 50 μl of 0.2 *M* NaOH, and 25 μl of 4 *M* sodium acetate, pH 4.8. The suspension is mixed by repetitive

[40] H. C. Birnboim and J. Doly, *Nucleic Acids Res.* **7,** 1513 (1979).

pipetting after each addition and the microtiter plate is removed from the robot and centrifuged at 3000 rpm for 10 min. The plate is replaced in the Biomek and the supernatant transferred to a fresh 96-well flat-bottom plate. 2-Propanol (100 μl) is added to each well and the plate again removed from the robot. The plate is centrifuged at 3000 rpm for 10 min, inverted to remove 2-propanol, and replaced in the robot. TE (15 μl) and an appropriate amount of restriction enzyme, usually *Not*I (Stratagene), are added. Amounts of restriction enzyme necessary for complete digestion range from 5 μl (14 U/μl) to 1 μl (1 U/μl). The enzyme, buffer, and DNA are mixed by pipetting and incubated for 30 min at room temperature. Gel dye (10 μl) is added, the reaction again mixed by repetitive pipetting, and the 96 reactions loaded on 8 agarose gels containing 12 lanes each.

Electrophoresis of the robot-prepared loaded 12-well agarose minigels is carried out in a custom-designed electrophoresis chamber holding 8 of the 12-well gels. Electrophoresis is carried out for 4 hr at 180 V in 40 m*M* Tris–borate buffer and DNA visualized by staining with 25 μg/ml ethidium bromide. Using this system, which requires operator intervention at centrifugation steps and for replacement of pipette tips and gels, 2 or 3 sets of 96 cosmids can be analyzed per day. The yield of DNA from an individual 200-μl culture is 200–400 ng. The yield and purity of DNA are critically dependent on both the cosmid vector used and the host bacterial strain. Best results are obtained using sCos1 carried in host strain DH5α.

Preparation of End-Specific RNA Probes. sCos and pWE vectors contain T3 and T7 bacteriophage promoters flanking the insertion site that allow for the synthesis of radiolabeled RNA probes from the extreme ends of the insert. RNA probes may be prepared from minilysate DNA following digestion with one of a number of restriction enzymes (*Rsa*I, *Hae*III, *Hin*dIII, or *Bam*HI). Restriction enzyme digestion limits the size of the resulting probe and assures recognition of the end of the insert. One to 2.5 μg of DNA is added to a reaction containing a final concentration of 36 m*M* Tris-HCl, pH 7.6, 6 m*M* $MgCl_2$, 2 m*M* spermidine, 1.5 U RNasin, 0.01 *M* DTT, 0.5 m*M* GTP, ATP, and CTP, 12 μ*M* UTP, and 100 Ci [α-^{32}P]UTP (800 Ci/mmol) in a final reaction volume of 25 μl. T7 or T3 polymerase (10–20 U) is added and the reaction continued for 60 min at 37°. Following the incubation, the reaction is terminated by extraction with phenol–chloroform and, if necessary, the probe collected by precipitation with ethanol.

Multiplex Hybridization Analysis. Cosmid arrays are not only useful for detecting individual genes or DNA sequences, but for the determination of overlaps between different clones and the construction of contigs. One approach to detecting overlaps involves the use of end-specific RNAs, synthesized as described above. Overlapping contiguous cosmid clones

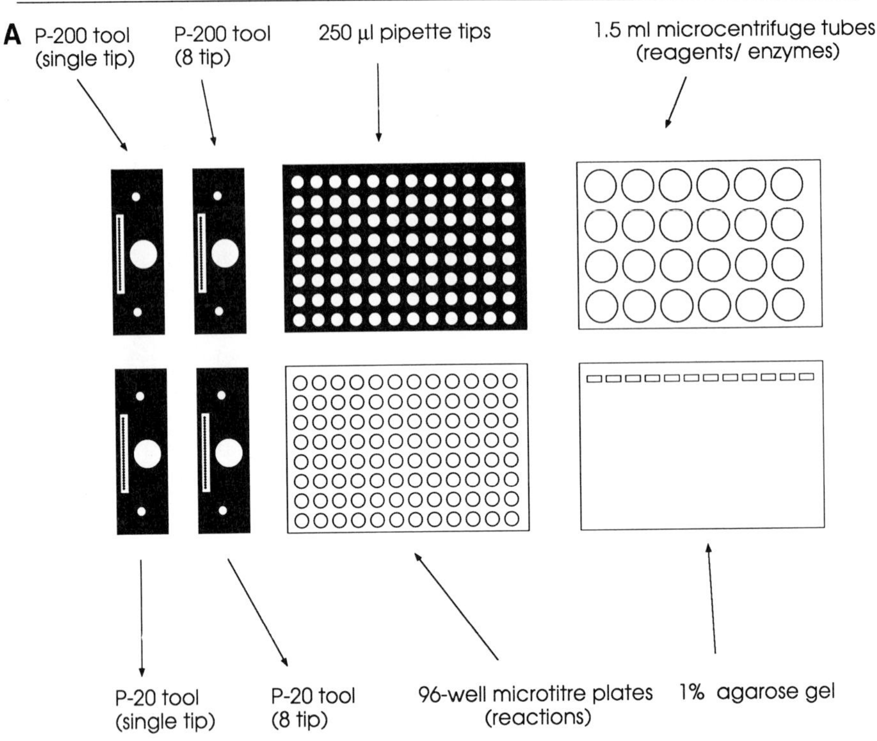

FIG. 5. Semiautomated robotic system for restriction site analysis of cosmid clones. (A) Diagram of the Beckman Biomek 1000 worktable modified to hold an 8 × 12 cm agarose gel in place of 1 of the 96-well microtiter plates. This system allows automated DNA preparation, digestion, and loading of agarose gels for large-scale mapping projects[30]. (B) Analysis of cosmid clones for *Not*I restriction sites. Individual cosmid clones were grown in 200-μl cultures in individual wells of a 96-well microtiter plate, DNA prepared by an alkaline lysis procedure simplified for robotic preparation, and digested with *Not*I. Separation of the cosmid vector (7 kb) from the insert (>35 kb) allows an internal control for restriction enzyme activity. Following digestion, the products were automatically loaded on a 12-lane 8 × 12 cm 1% (w/v) agarose gel. A set of 16 gels, containing 192 samples, is then run in a single electrophoresis apparatus, stained with ethidium bromide, and photographed under ultraviolet illumination. Shown are the products of restriction digestion reactions of 192 individual cosmids. V, Location of the sCos1 vector DNA migrating at 6.2 kb; I, location of the excised *Not*I fragment representing the genomic DNA insert. Where an internal *Not*I site is present, multiple insert bands are observed.

arranged on an organized matrix may be detected by the synthesis of an end-specific RNA probe and hybridization of the probe to a replica of the matrix filter. This may be carried out using individual clones, through the use of a multiplex strategy in which the RNA probes are prepared from pools of multiple cosmid templates. The templates are pooled such

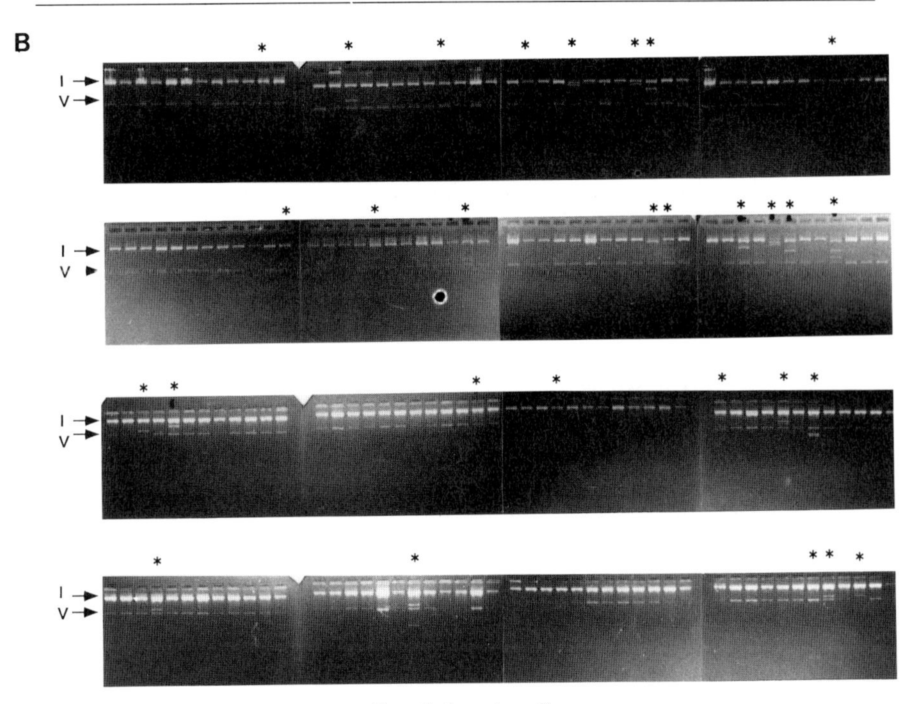

FIG. 5. (*continued*)

that each two pools contain only one cosmid in common; thus comparison of the results of hybridization of two different probes to a matrix filter will ensure that clones detected by both pools represent hybridizing to the common clone. A simple way to prepare these mixed probes is to pool all of the cosmid clones corresponding to a row of the two-dimensional matrix, prepare end-specific RNA probes, and carry out a hybridization reaction to a replica of the organized matrix. Each probe will hybridize to its own template, resulting in the detection of a complete row and, in addition, a collection representing all of the clones overlapping with templates. A probe prepared from the pool of cosmids representing a column detects a similar pattern. Those hybridizing clones that appear in both data sets but are not in a template row or column result from hybridization of the common template and indicate physical overlap with the clone that is located at the intersection of the template row and column of the matrix. Thus, if probes are prepared from pools of all of the rows and columns, a large number of overlapping clones in the collection can be rapidly detected. Thus, N clones organized in a two-dimensional array could be analyzed using $2(N^{1/2})$ reactions. Alternate pooling strategies for hybridization based detection of overlaps have also been described. Further details on this approach are given in Refs. 5 and 41.

Discussion

Genomic cosmid libraries may be used for isolating and characterizing individual genes or gene families, or as the basis for constructing large-scale maps of chromosomes or entire genomes. Techniques and strategies for the creation and manipulation of cosmid libraries for these two goals differ substantially. Cosmid libraries for isolating single genes may be stored as pools of clones and replated for each screening. For large-scale mapping, however, where large numers of clones will be analyzed, it is often much more convenient to produce cosmid libraries that are chromosome or region specific, and which are stored as individual clones archived in high-density microtiter plates. Similar strategies are useful for the production, screening, and distribution of YAC libraries.[42] Methods for replicating, distributing, and analyzing clones from these arrayed libraries also differ in that simplified methods for DNA preparation requiring a minimal number of steps, the use of automation and robotics for manipulating and distributing clones, and strategies for rapid restriction mapping and production of probes are all essential.

Acknowledgments

We are grateful to our many colleagues in the Molecular Genetics Laboratory at the Salk Institute, especially L. Selleri, D. McElligott, C. Wagner-McPherson, J. Zhao, M. Djabali, G. Huhn, K. Lewis, and S. Maurer, and to J. Longmire, C. E. Hildebrand, T. Beuglesdijk, and L. Deaven (Los Alamos National Laboratories) for many helpful discussions. We are also grateful to Stratagene Cloning Systems for reagents, and to the National Institutes of Health, the Department of Energy, and the G. Harold and Leila Y. Mathers Charitable Foundation for support.

[41] G. Evans, *BioEssays* **13,** 39 (1991).

[42] E. D. Green and M. V. Olson, *Proc. Natl. Acad. Sci. U.S.A.* **87,** 1213 (1990).

[47] Using Bacteriophage P1 System to Clone High Molecular Weight Genomic DNA

By JAMES C. PIERCE and NAT L. STERNBERG

The ability to clone large contiguous segments of genomic DNA is important in many areas of current research that seek to determine the long-range structure of chromosomes. Mapping efforts that depend on aligning overlapping segments of cloned DNA will proceed more rapidly, and will be less subject to error, as the size of the cloned DNA increases. With increased cloning capacity it should be easier to clone functional genomic copies of even large genes as well as their upstream and downstream regulatory sequences. Moreover, as replication origins and centromeres are likely to be quite large, a high molecular weight (HMW) DNA cloning system will be useful in determining the structure of the chromosome segregation machinery.

Until several years ago the only vector system available for HMW DNA cloning was the λ cosmid system. While this system is very efficient (as many as 10^6–10^7 clones can be recovered per microgram of vector and insert DNA), it has the disadvantage of not being able to accept inserts that are bigger than 40 kbp in size.[1,2] To overcome this deficiency the yeast artificial chromosome (YAC) cloning system was developed.[3,4] It permits the cloning of DNA fragments that are as large as 1 Mbp in size. Its limitations are the low efficiency of cloning (usually only 10^3 clones are recovered per microgram of vector) and the difficulty of isolating large amounts of insert DNA from a yeast transformant containing a YAC. The latter problem appears to have been solved with the development of a YAC cloning vector that can be amplified.[5]

Falling between the above cloning systems are two new systems that permit the recovery of inserts in *Escherichia coli* that are within the 70 to 100-kbp size range, at an efficiency that is intermediate between λ cosmid and YAC cloning. The F plasmid system has been developed and may

[1] N. Murray, R. Hendrix, J. Roberts, F. Stahl, and R. Weisberg (eds.), "Lambda II" p. 395. Cold Spring Harbor Laboratory, Cold Spring Harbor, New York, 1986.

[2] P. A. Whittaker, A. J. B. Campbell, E. Southern, and N. Murray, *Nucleic Acids Res.* **16,** 6725 (1988).

[3] D. T. Burke, G. F. Carle, and M. V. Olson, *Science* **236,** 806 (1987).

[4] C. V. Trauer, S. Klapholz, R. W. Lyman, and R. W. Davis, *Proc. Natl. Acad. Sci. U.S.A.* **86,** 5898 (1989).

[5] D. Smith, A. P. Smyth, and D. T. Moir, *Proc. Natl. Acad. Sci. U.S.A.* **87,** 8242 (1990).

prove to be a useful cloning alternative.[6] Its cloning efficiency, however, is still too low to generate complete mammalian libraries easily, and, like YACs, large amounts of cloned DNA are not readily recovered from individually transformed cells. In contrast, the phage P1 system can clone with an efficiency that makes it practical to generate mammalian libraries with reasonable ease (greater than 10^5 clones can be generated per microgram of vector and insert) and several micrograms of isolated insert DNA can be produced from 10^9 cells by standard plasmid isolation procedures.[7,8] A partial 50,000-member human P1 library[8] and a complete *Drosophila* P1 library[9] have been constructed. Complete mammalian libraries are being produced.

We describe here detailed methods for cloning, isolating, and amplifying HMW DNA fragments (70–90 kbp) in P1 cloning vectors.

General Principles of P1 Life Cycle

Bacteriophage P1 is a virus with a genome size of about 100 kbp and a DNA content of about 110 kbp.[8,10] Each viral DNA molecule has the same 10-kbp sequence at both ends and is, therefore, said to be terminally redundant. When that DNA is injected into a recombination-proficient host, such as *E. coli*, homologous recombination between the terminally redundant sequences cyclizes the linear viral DNA, protecting it from cellular exonucleases and permitting the onset of the viral life cycle.[11,12] If the recipient host is recombination deficient the viral DNA can still be cyclized if it contains viral *loxP* recombination sites in its terminal redundant regions.[13] *loxP* sites are 34-bp sequences that are efficiently recombined by the P1 cyclization recombinase (Cre) protein.[14] Whether cyclization occurs by homologous or *loxP*–Cre site-specific recombination the product of the reaction is a genome-length circle.

Once the viral DNA has been injected into cells and cyclized it can establish either a proviral or lytic state. In the proviral state the viral lytic

[6] F. Hosoda, S. Nishimura, H. Uchida, and M. Ohki, *Nucleic Acids Res.* **18,** 3863 (1990).
[7] N. Sternberg, *Proc. Natl. Acad. Sci. U.S.A.* **87,** 103 (1990).
[8] N. Sternberg, J. Ruether, and K. DeRiel, *New Biol.* **2,** 151 (1990).
[9] D. Smoller, D. Petrov, and D. Hartl, *Chromosoma* **100,** 487 (1991).
[10] M. Yarmolinsky and N. Sternberg, "Bacteriophage P1" (R. Calender, ed.), p. 291. Plenum, New York, 1988.
[11] N. Segev, A. Laub, and G. Cohen, *Virology* **101,** 261 (1977).
[12] N. Segev and G. Cohen, *Virology* **114,** 333 (1981).
[13] N. Sternberg, B. Sauer, R. Hoess, and K. Abremski, *J. Mol. Biol.* **187,** 197 (1986).
[14] R. Hoess and K. Abremski, *Proc. Natl. Acad. Sci. U.S.A.* **81,** 1026 (1984).

functions are repressed by a phage repressor (the *c1* gene product [15,16]) and the phage DNA is stably maintained in the cell population as a unit copy extrachromosomal plasmid. It is replicated by a P1 plasmid replicon and the products of replication are faithfully segregated to daughter cells at cell division by a viral partition system.[17,18] The replicon and partition systems are adjacent to each other on the P1 genome. In the lytic state the viral repressor is inactive. This permits the lytic replicon of the virus to function. The DNA is replicated by this replicon first as a circle and then as a rolling circle. Within 30–40 min after the initiation of lytic replication the infected cell contains about 100–200 copies of viral DNA.[19] During this time the virus produces heads, tails, and specific packaging-initiation proteins that are necessary to package the synthesized DNA into virus particles.[10,20] The virus also produces an endolysin that lysis cells about 60 min after infection.

General Principles of Phage Packaging

Bacteriophage P1 packages its DNA by a processive headful mechanism.[10,21] The substrate for this reaction is a concatemer that is generated by rolling circle replication and that consists of tandemly repeated units of the phage genome arranged in a head-to-tail configuration. Packaging is initiated when a unique 162-bp P1 *pac* site is recognized and cleaved by phage-encoded pacase proteins.[22,23] One of the cleaved *pac* ends is then brought into an empty phage prohead and the head is then filled with that DNA. Once the head is full the packaged DNA is cleaved away from the rest of the concatemer by a headful cutting reaction that appears not to recognize any specific DNA sequence.[21] A second round of packaging is initiated from the unpackaged end produced by the headful cut. In this way a processive series of DNA headfuls is packaged unidirectionally from a single *pac* site on DNA. The packaged DNA is terminally redundant because the P1 headful is about 110 kbp while the genome is only 100 kbp. It is cyclically permuted because more than one headful is packaged from

[15] J. L. Eliason and N. Sternberg, *J. Mol. Biol.* **198,** 281 (1987).
[16] M. Velleman, B. Dreiseikelmann, and H. Schuster, *Proc. Natl. Acad. Sci. U.S.A.* **84,** 5570 (1987).
[17] S. Austin, F. Hart, A. Abeles, and N. Sternberg, *J. Bacteriol.* **152,** 63 (1982).
[18] S. Austin, R. Mural, D. Chattoraj, and A. Abeles, *J. Mol. Biol.* **83,** 195 (1985).
[19] G. Cohen, *Virology* **131,** 159 (1983).
[20] D. H. Walker and J. T. Walker, *J. Virol.* **20,** 177 (1976).
[21] B. Bachi and W. Arber, *Mol. Gen. Genet.* **153,** 311 (1977).
[22] N. Sternberg and J. Coulby, *J. Mol. Biol.* **194,** 453 (1987).
[23] N. Sternberg and J. Coulby, *J. Mol. Biol.* **194,** 469 (1987).

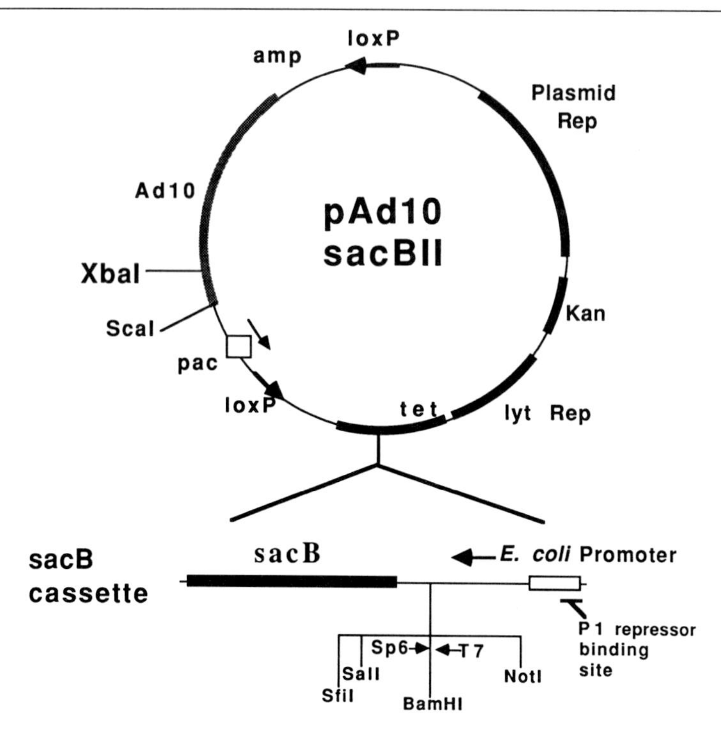

FIG. 1. P1 cloning vectors, p*Ad*10 and p*Ad*10*sacB*II. Details of the vectors are described in the text.

each P1 concatemer. On average, three or four headfuls are produced per packaging series.

A central element of the P1 packaging system is the *pac* site. Within its 162-bp sequence it contains four hexanucleotide sequences (5′-TGATCA/G) at one end, three at the other, and a 90-bp region between them.[23] *pac* cleavage is localized to about a turn of the DNA helix near the center of the 90-bp region. Each hexanucleotide sequence contains a DNA adenine methylation (Dam) sequence (5′-GATC) and *pac* must be methylated for it to be cleaved.[24]

P1 Cloning Vectors

The two P1 cloning vectors used here, p*Ad*10[25] and p*Ad*10*sacB*II (Fig. 1), are structured on the same general principles and are organized accord-

[24] N. Sternberg and J. Coulby, *Proc. Natl. Acad. Sci. U.S.A.* **87,** 8070 (1990).
[25] p*Ad*10 was originally called pNS582*tet*14*Ad*10.[8]

ing to the same plan.[8] They differ in that p*Ad*10*sacB*II contains an additional cassette (the *sacB* cassette) that provides a positive selection for cloned inserts.

The vectors are divided into two domains by flanking *loxP* recombination sites. The Amp^r domain contains an interrupted Amp^r (*bla*) gene from plasmid pBR322, the multicopy pBR322 replicon, a P1 *pac* site oriented so as to direct packaging counterclockwise, and an 11-kbp segment of adenovirus DNA. The latter is cloned into the *bla* gene at a *Sca*I site such that *Sca*I remains a unique site in the vector and is located proximal to *pac*. The Kan^r domain contains the kan^R gene from transposon Tn*903*, the tet^R gene from plasmid pBR322, the P1 plasmid replicon and partition system, and the P1 lytic replicon. The lytic replicon in this vector is regulated by the *lac* operon promoter.[26] It is inactive in a host containing a $lacI^q$ repressor but can be activated by adding the *lac* inducer isopropyl-β-D-thiogalactopyranoside (IPTG) to the growth medium. The tet^R gene contains unique *Bam*HI and *Sal*I restriction sites into which foreign DNA fragments can be cloned in the p*Ad*10 vector. This process inactivates tet^R and provides a screen for transformants containing plasmids with cloned inserts.

The p*Ad*10*sacB*II vector differs from p*Ad*10 in that it contains a *sacB* gene between the *Bam*HI and *Sal*I sites of p*Ad*10 DNA. This gene encodes an enzyme that converts sucrose to levan, which accumulates in the periplasmic space of cells and results in cell death.[27,28] Thus, cells expressing *sacB* die in medium containing greater than 2% sucrose but grow normally in medium that lacks the sugar. To take advantage of this property in a positive selection scheme for cloning, an *E. coli* promoter was placed upstream of *sacB* and a *Bam*HI cloning site was placed between these two elements. This site was in turn flanked by T7 and Sp6 promoters for subsequent riboprobe analysis of the ends of any cloned DNA, and by rare-cutting *Sal*I, *Sfi*I, and *Not*I restriction sites to recover cloned DNA easily. Finally, a P1 *c1* repressor binding site was positioned so as to overlap the *E. coli* promoter. Under these circumstances transcription of *sacB* is blocked in the presence of the P1 C1 repressor.

General Cloning Rationale

The complexity of the P1 cloning vector is designed to permit the efficient recovery of unrearranged, insert DNA. A general scheme of the

[26] N. Sternberg and G. Cohen, *J. Mol. Biol.* **207,** 111 (1989).

[27] P. Gay, D. Lecog, M. Steinmetz, T. Berkelman, and C. I. Kado, *J. Bacteriol.* **164,** 918 (1985).

[28] L. B. Tang, R. Lenstra, T. V. Borchert, and V. Nagarajan, *Gene* **96,** 89 (1990).

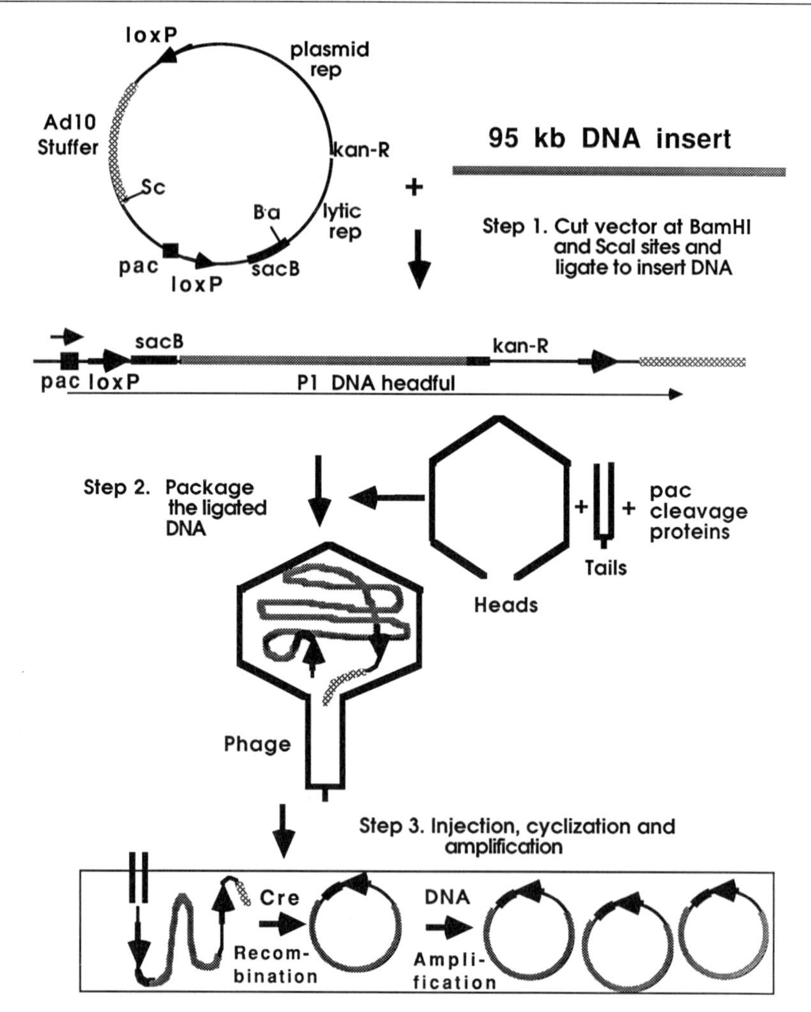

FIG. 2. P1 cloning scheme. The scheme is depicted for the p*Ad*10*sacB*II vector and is described in full detail in text.

cloning process is illustrated in Fig. 2. In this scheme p*Ad*10 or p*Ad*10*sac-B*II DNA is digested at unique *Sca*I and *Bam*HI restriction sites to generate two vector "arms." The shorter arm contains *loxP, pac,* and *sacB* (in the case of p*Ad*10*sacB*II) and the longer arm contains *loxP,* the kan^R gene, the P1 plasmid replicon, the P1 lytic replicon, and all of the adenovirus "stuffer" fragment. The ends of the two arms are treated with alkaline phosphatase to prevent their ligation to each other and the arms are then ligated at their *Bam*HI ends to DNA fragments generated by the *Sau*3AI partial digestion of genomic DNA. If the appropriate ligation has occurred a foreign DNA fragment will be sandwiched between the short and long vector arms such that *loxP* sites are oriented in the same direction.

In the next step of the cloning process the *pac* site on the short vector arm of the ligated DNA is cleaved by incubating it with an extract prepared by the induction of a P1 lysogen that can produce the appropriate P1 *pac* cleavage proteins (the stage 1 reaction). After *pac* is cleaved the DNA is incubated with a second P1 extract prepared by the induction of a P1 lysogen that synthesizes P1 heads and tails but not *pac*-cleavage proteins (the stage 2 reaction). During this incubation the DNA is packaged unidirectionally from its *pac*-cleaved end, until the head is full. Once this has occurred the headful cut separates the DNA inside the head from that outside and the filled head is converted to an infectious particle by the addition of phage tails.

The final step in the cloning process is the faithful recovery and amplification of the packaged DNA. As noted previously, cyclization of the packaged DNA after it is injected into a bacterial host is an essential first step in the recovery process. To achieve this end the packaged insert DNA must be flanked by the two *loxP* sites present in the vector arms and the recipient host strain must express the Cre recombinase, constitutively. In addition to being Cre^+, host strains used to recover cloned DNA also carry *recA*, *lacI*^{q}, and *mcrAB*.[8] The *recA* mutation inactivates the major homologous recombination system of the host, preventing rearrangement of homologous sequences that might be present in the insert DNA. The *lacI*^{q} repressor inhibits activation of the P1 lytic replicon, ensuring that the infecting DNA is maintained in the cell initially at single copy by the P1 plasmid replicon. This minimizes any rearrangement of the cloned insert DNA due to its overreplication. The copy number of the plasmid with its insert can be increased 10- to 20-fold just before DNA is isolated by adding IPTG to the medium.[26] Finally, the *mcr* mutations are employed to inhibit the selective loss of insert DNA that is rich in $Gp^{Me}C$ or possibly $Ap^{Me}C$.[29,30] For insert DNA in the 70- to 90-kbp size range, results with the P1 system indicate that a functional *mcr* system inhibits cloning about 35-fold.[8]

Using the p*Ad*10 vector, transformants containing plasmids with cloned insert DNA are selected on agar plates containing kanamycin and screened by their inability to grow on agar plates with tetracycline. With

[29] E. A. Raleigh and G. Wilson, *Proc. Natl. Acad. Sci. U.S.A.* **83,** 9070 (1986).

[30] E. A. Raleigh, N. E. Murray, H. Revel, R. M. Brumenthal, D. Westaway, A. O. Reith, P. W. J. Rigby, J. Ethai, and D. Hanahan, *Nucleic Acids Res.* **16,** 1563 (1988).

p*Ad*10*sac*BII vector transformants with cloned inserts are directly selected on agar plates containing kanamycin and 5% (w/v) sucrose.

Materials and Methods

Preparation of Vector DNA

Vector DNA is prepared by the cesium chloride density gradient banding method.[31] One liter of cells is grown from a purified colony in Luria broth[32] plus kanamycin (25 μg/ml) until the culture is saturated (usually 12–14 hr at 37°). The cells are pelleted in a Sorvall (Norwalk, CT) GSA rotor at 6000 rpm for 10 min at 4°. The cell pellet is frozen in a dry ice–ethanol bath and then thawed at room temperature. The cells are then resuspended in 7 ml of cold 25% (w/v) sucrose, 50 m*M* Tris-HCl, pH 8.0, and 1.5 ml of lysozyme (5 mg/ml in 0.25 *M* Tris-HCl, pH 8.0). The mixture is kept at 4° for 15 min with gentle swirling. To this is added 2.8 ml of ethylenediamine tetraacetic acid (EDTA; 0.25 *M*, pH 8.0) and the mix incubated an additional 5 min at 4°. Next, 11.3 ml of lysis solution [50 m*M* Tris-HCl, pH 8.0, 62 m*M* EDTA, 2% (w/v) sodium dodecyl sulfate (SDS)] is added and the cells are lysed for 5 min at room temperature. Then 5.7 ml of 5 *M* NaCl is added to the lysed extract and it is incubated for 5 min at room temperature and then for 1 hr at 4°. The extract is then centrifuged in a 50-ml Oak Ridge tube in a Sorvall SS34 rotor at 18,000 rpm for 60 min at 4°. The supernatant is removed and precipitated with 0.54 vol of cold 2-propanol at −20° for 1 hr. The precipitate is pelleted by centrifugation in an SS34 rotor at 10,000 rpm for 10 min at 4°. The supernatant is discarded and the pellet is resuspended in 9 ml of 30 m*M* NaCl, 10 m*M* Tris-HCl, pH 8.0, 1 m*M* EDTA. Then 2.7 g of CsCl (Gallard-Schlesinger Industries, Inc., Carle Place, NY, biological grade) is added and dissolved and then 1 ml of a 10-mg/ml solution of ethidium bromide is added. The solution is centrifuged in an SS34 rotor at 8000 rpm for 10 min at 4° and the pellet is discarded. An additional 7.5 g of CsCl is added to the supernatant and dissolved and the solution is centrifuged at 35,000 rpm in a Sorvall type T-1270 rotor for 42 hr at 20°. The gradient is visualized with a hand-held ultraviolet (UV) lamp (260 nm) and the plasmid band removed from the side of the centrifuge tube with a 22-gauge needle. The ethidium bromide is extracted from the DNA with CsCl-saturated

[31] D. B. Clewell and D. R. Helinski, *Proc. Natl. Acad. Sci. U.S.A.* **62,** 1159 (1969).

[32] Luria broth consists of 10 g Difco tryptone, 5 g Difco yeast extract, 5 g NaCl, 1 liter deionized water. Adjust pH to 7.4 with 10 *N* NaOH (usually 0.2–0.4 ml/liter) and sterilize by autoclaving.

2-propanol (four or five equal volume extractions are usually necessary to remove all the ethidium bromide) and the solution is then dialyzed against 1 liter of 10 m*M* Tris-HCl, pH 8.0, 1 m*M* EDTA (TE) to remove the CsCl. The dialyzed solution is then extracted with an equal volume of TE-saturated phenol (Mallinckrodt, St. Louis, MO) followed by an equal volume of chloroform-isoamyl alcohol (24 : 1). Sodium acetate is added to 0.3 *M* and the DNA is precipitated with 2 vol of ethanol for at least 1 hr at $-20°$. The DNA precipitate is pelleted in an SS34 rotor at 12,000 rpm for 10 min at 4° and the pellet is resuspended in TE buffer. Usually 200–300 μg of vector DNA is isolated by this procedure from 1 liter of cells. We have also used 5′ → 3′ column chromatography to prepare successfully large amounts of vector DNA. The procedures are as specified by the vendor.[33]

For the isolation of p*Ad*10 DNA the bacterial strain used is NS3593 = DH5α *recA lacI^q* (p*Ad*10). The *lacI^q* gene in this strain is present on a λLP1 prophage,[34] which also contains the *E. coli dam* methylase gene. Both genes are regulated by the *taq* promoter. *LacI^q* functions here to prevent activation of the P1 lytic replicon on the vector. To isolate p*Ad*10-*sacB*II DNA the strain used is NS3607 = DH5α *recA lacI^q* (λ *imm*^{21}-P1:7Δ5b) (p*Ad*10*sacB*II). Besides *recA* and *lacI^q* mutations, this strain also contains a λ *imm*^{21} prophage with a constitutively expressed P1*c1* repressor gene. C1 blocks *sacB* expression from the p*Ad*10*sacB*II plasmid. Under these conditions one avoids the production of small amounts of vector DNA that contain deletions of the *sacB* gene. It is important to do this as those deleted molecules will interfere with the positive selection cloning scheme described below. To assess the quality of *sacB* vector DNA before using it in cloning, it should be tested in two ways. First, it should be used to transform cells (e.g., DH5α *lacI^q*) to determine what fraction of the transformants grow on agar plates with kanamycin (25 μg/ml) and sucrose (5%) relative to that which grow just on kanamycin. The *kan*^{r}/(*kan*^{r} + sucrose) ratio should be no less than 1000 (Table I, line 1). Second, the vector DNA should be digested with restriction enzyme *Spe*I and analyzed by agarose gel electrophoresis. Digestion with this enzyme produces fragments that are about 20, 9, and 1.7 kb in size. The latter contains the *sacB* gene (Fig. 3, lane 2). For the sake of comparison a DNA preparation with the correct digestion pattern [e.g., one with the appropriate *kan*^{r}/(*kan*^{r} + sucrose) ratio] should be used as a control.

[33] P. H. Zervos, L. M. Morris, and R. J. Hellwig, *BioTechniques* **6**, 238 (1988).

[34] λLP1 is a gift of A. Wright, Tufts University, Medford, MA. Its construction and characterization have not yet been published.

TABLE I
EFFICIENCY OF CLONING HIGH MOLECULAR WEIGHT DNA FRAGMENTS INTO THE pAD10sacBII VECTOR[a]

DNA	Treatment	Recovery conditions	
		− Sucrose	+ Sucrose
1. *sacB*II	Transformation	2000	0
2. *sacB*II − *Sca*I + *Bam*HI	Ligated and packaged	800	4
3. *sacB*II − *Sca*I + *Bam*HI + DNA fragments	Ligated and packaged	600	300

[a] The DNAs shown in lines 1–3 were used to transform or infect strain NS3145. Colonies were scored after overnight incubation at 37° on agar plates with either kanamycin or kanamycin plus sucrose. Other details of the experiment are described in text.

Preparation, Digestion, and Fractionation of Genomic DNA

Genomic DNA can be prepared from whole animals, organs, tissues such as blood or sperm, or cultured cell lines. Special protocols must be used when making high molecular weight DNA from whole animals or animal organs to remove materials from the extracellular matrix. Once this is accomplished, all DNAs can be prepared by the following method. We have used it successfully to prepare DNA from blood and sperm cells and from cultured lymphoblast and fibroblast cells. Initially, cells are collected by pelleting them in a Sorvall RT6000 centrifuge at 1500 rpm for 10 min at 4° usually in a 50-ml Oak Ridge-type tube. The cell pellet is washed with phosphate-buffered saline [2.7 m*M* KCl, 1 m*M* $KH_2PO_4 \cdot 7H_2O$, 8.1 m*M* Na_2HPO_4 (pH 7.4), 140 m*M* NaCl] and then resuspended in 50 m*M* Tris-HCl (pH 7.5), 0.1 *M* NaCl, 0.15 *M* EDTA at a concentration of about 10^7–10^8 cells/ml. Sodium dodecyl sulfate is then added to 1% (w/v) followed by proteinase K (Boehringer Mannheim, Indianapolis, IN) to a concentration of 1 mg/ml, and the resulting suspension is incubated at 60° for 4 hr. During this procedure the cells are lysed and the DNA is liberated as a viscous mass. Once this has occurred the DNA should never be vortexed, vigorously pipetted,[35] or centrifuged into a pellet as these procedures will shear it. The lysate is then gently poured onto a 10 to 40% (w/v) sucrose gradient prepared in 20 m*M* Tris-HCl (pH 8.0), 0.8 *M* NaCl, and 10 m*M* EDTA. Depending on the size of the lysate a 16-ml gradient can be conveniently handled by a Beckman (Palo Alto, CA) SW28 swinging bucket rotor and a 36-ml gradient can be handled by

[35] All pipettes, including Eppendorf tips, should have their ends cut to widen the bores. This reduces shearing of DNA during the pipetting process.

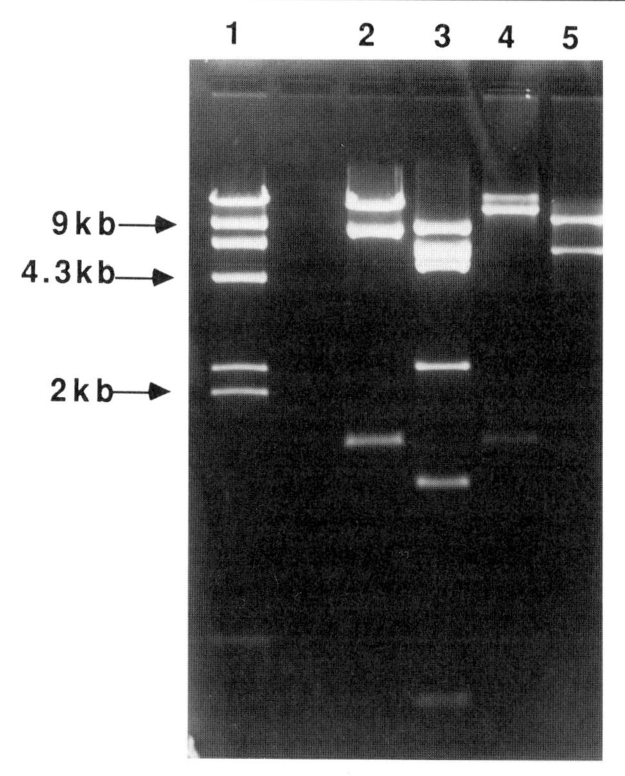

FIG. 3. Restriction digests of p*Ad*10*sacB*II DNA. DNA digests are analyzed by gel electrophoresis in 1% (w/v) agarose gels. Lane 1 contains a λ *Hin*dIII digest. Positions indicated by arrows to the left of that digest indicate the positions of λ *Hin*dIII 9-, 4.3-, and 2-kb fragments. Lanes 2 and 4 contain *Spe*I digests of p*Ad*10*sacB*II and p*Ad*10*sacB*IIΔ*amp* DNAs, respectively. The Δ*amp* plasmid is generated by cycling the vector DNA through a cell containing Cre and selecting Kan^r colonies. It lacks the amp^r domain of the vector. Lanes 3 and 5 contain *Bgl*II digests of p*Ad*10*sacB*II and p*Ad*10*sacB*IIΔ*amp* DNAs, respectively.

a Beckman SW27 swinging bucket rotor. Centrifugation of the gradient is carried out in a Sorvall RC70 ultracentrifuge at 18,000 rpm for 16 hr at 4°. At this point the viscous DNA mass is usually located about halfway down the tube. The overlying volume is gently aspirated off and the viscous DNA is then poured into a sterile dialysis bag. This step can be difficult as the viscous DNA flows as a mass and care should be taken to place the bag in a sterile vessel before the DNA is poured into it. This permits recovery of the DNA if any of it fails to enter the bag. Once in the bag the DNA is dialyzed against four changes of 1 liter of TE buffer at 4°. Quantitation of the DNA at this point is difficult because the solution is not homogeneous and much sampling bias is to be expected. However, the size of the DNA can be evaluated by field-inversion gel electrophoresis (FIGE; see below). If the majority of DNA is larger than 300–400 kbp in size it is suitable for subsequent restriction enzyme digestion. If it is smaller than that, its subsequent restriction digestion will produce too many fragments with sheared, nonclonable ends. That DNA will have a low cloning efficiency. It has been our experience that some HMW DNAs prepared by the procedure described above are suitable for restriction enzyme digestion but digest very poorly. If these preparations are passaged 5–10 times through an 18-gauge needle they can now be digested. During this procedure the DNA does not decrease in size below 300 kbp. This should be monitored by FIGE.

For subsequent cloning the HMW DNA is partially digested with the restriction enzyme *Sau*3AI. Because we have observed that a limiting step in this reaction is the diffusion of the restriction enzyme into the viscous DNA, we initiate the reaction with a preincubation step to ensure uniform digestion. To determine the best digestion conditions for any DNA preparation, 0.5 ml of that DNA is incubated with 60 μl of 10× *Sau*3A1 restriction buffer minus magnesium (1 *M* NaCl, 0.1 *M* Tris-HCl, pH 7.5) in the presence 20 μl of 2.5 mg/ml of nuclease-free bovine serum albumin (BSA; New England BioLabs, Beverly, MA) and 1 or 10 units of *Sau*3A1. The reaction is incubated overnight at 4° with gentle mixing (avoid excessive agitation as this shears the DNA). Ten microliters of this material is then removed into a separate tube that is kept at 4° and is used as a minus Mg^{2+} control. The remainder of the reaction is put at 30° for 1 min and then 60 μl of 0.1 *M* $MgCl_2$ is gently stirred into the solution. Fifty-microliter aliquots are removed at different times, usually at 2-min intervals between 2 and 20 min after the addition of the $MgCl_2$. The aliquots are immediately mixed with 15 μl of 0.2 *M* EDTA, incubated 15 min at 70°, and then put on ice. It is also important to prepare a no-enzyme control with $MgCl_2$ to determine if there is trace nuclease activity in the DNA preparation. Careful observation of the viscosity of the DNA during the digestion can also be revealing. It has been our experience that once the solution loses all of its viscosity the digestion has gone too far—most of the DNA is less than 40 kbp in size.

Analysis of the partially digested DNAs and the controls is carried out as follows. Ten microliters is removed from each aliquot and spot dialyzed at 25° on a Millipore (Bedford, MA) filter (VSWP 02500) that is floating on 50 ml of TE buffer. Each dialyzed sample is then mixed with loading dye [0.2% (w/v) bromphenol blue, 0.2% (w/v) xylene cyanol, 40% (w/v) sucrose, 20 m*M* EDTA, 0.1% (w/v) SDS], heated to 70° for 5 min, and fractionated by FIGE. The fractionation is carried out by loading the samples[36] onto a 6 × 6 in. 1% (w/v) agarose gel (GTG agarose; FMC,

Rockland, ME) prepared in 0.5× TBE (1× TBE is 90 m*M* Tris-HCl, pH 8.0, 90 m*M* borate, 2 m*M* EDTA) and subjecting them to electrophoresis at 100 V for 60 min at room temperature. The gel is then transferred to a switching regimen of 0.3 sec forward, 0.1 sec backward with a ramping factor of 15 at 240 V for 6 hr at 4° using a PC750 (Hoefer, San Francisco, CA) FIGE apparatus. The gel is stained with ethidium bromide (Sigma Chemical Co., St. Louis, MO) and the DNA visualized with UV fluorescence. The size of DNA at any location in the gel is determined by comparison to the position of phage DNA markers (the large *Hin*dIII λ fragment, 23 kbp; T7 DNA, 40 kb; P1 DNA, 110 kbp; T4 DNA, 165 kbp) that were electrophoresed in adjacent gel lanes.

Figure 4 illustrates a typical *Sau*3A1 partial digest of genomic DNA. The most useful reactions are shown in lanes 7 and 8 (Fig. 4). They contain fragments ranging in size from 150 to 60 kbp, with few fragments smaller than 60 kbp. With further digestion (lanes 9–12, Fig. 4) the small DNA fragments become the majority population and become difficult to fractionate away from the larger fragments in subsequent sucrose gradients. To minimize the cloning of the smaller DNA fragments from any partial digest we recommend at least one sucrose gradient fractionation of the digest. However, if digests similar to those shown in lane 7 are used directly without fractionation, libraries can be obtained that contain about 50% large inserts (80–90 kbp) and 50% small inserts (40–60 kbp). This my be adequate in some cases, especially where only small amounts of starting genomic DNA are available.

Once the partial digests have been analyzed conditions that produce an abundance of 70- to 100-kbp fragments, and minimal amounts of smaller fragments (e.g., Fig. 4, lanes 6–9), are chosen and used to digest 1 ml of HMW genomic DNA. An aliquot of the digest is analyzed by FIGE gels and if acceptable the entire sample is loaded onto a 16-ml 10–40% (w/v) sucrose gradient and centrifuged as described above. Fractions of 0.5–1 ml are collected from the bottom of the centrifuge tube and 50-μl aliquots of each fraction are dialyzed for 2 hr at 25° on filters (VSWP 02500; Millipore) that are floated on 250 ml of TE buffer. Each aliquot is then analyzed by FIGE and the fractions with the appropriate-sized fragments are concentrated by *n*-butanol extraction. The samples are mixed with

[36] To avoid sample losses when loading viscous DNA solutions into the wells of an agarose gel that is immersed in electrophoresis buffer, we find it useful after expelling the DNA into the well to free it from the tip of the pipette by rubbing the tip against the back of well. If some of the sample remains stuck to the pipette tip it will tend to draw the entire sample out of the well when the tip breaks the surface of the electrophoresis buffer. Alternatively, the sample can be loaded into the wells of an agarose gel before the gel is immersed in the electrophoresis buffer. The wells are then covered with a 1% agarose solution and the gel immersed in buffer after the agarose solidifies.

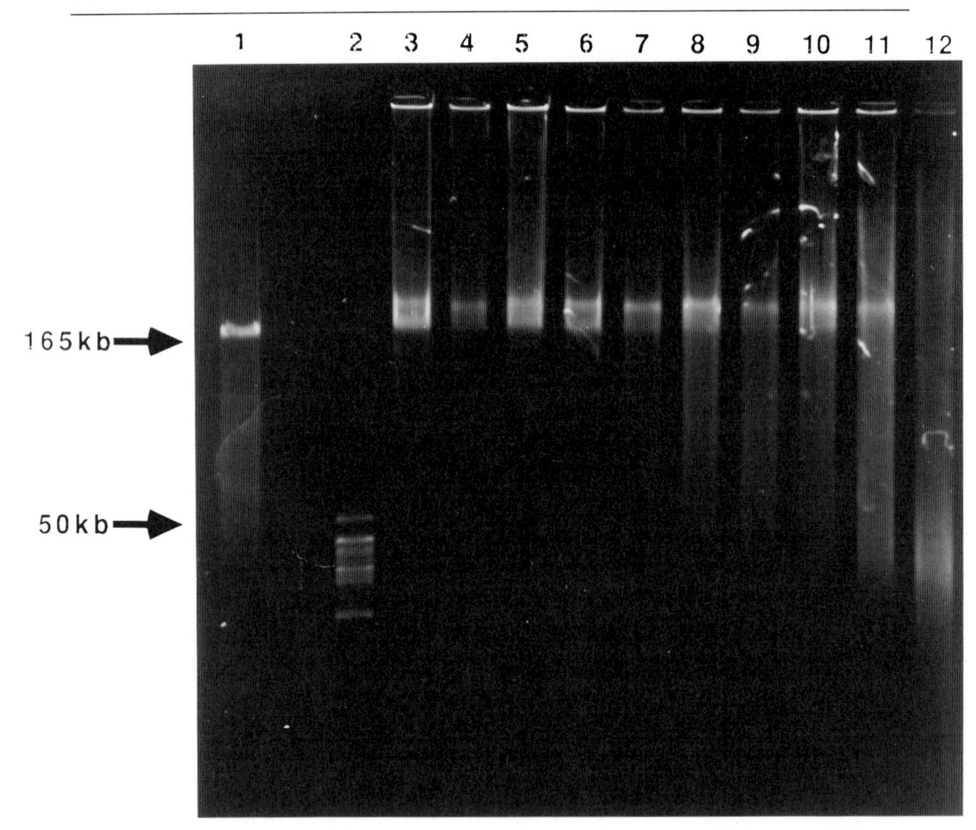

FIG. 4. Field inversion gel electrophoresis analysis of *Sau*3AI partially digested genomic DNA. Human DNA was partially digested with *Sau*3AI as described in text and aliquots were analyzed by FIGE. Lane 1 contains T4 DNA, whose size is 165 kb. Lane 2 contains a high molecular weight marker from Bethesda Research Laboratories whose highest molecular weight species is 50 kb. Lane 3 contains undigested genomic DNA (the minus Mg^{2+} control). Lanes 4–12 contain DNA samples treated for 2, 4, 6, 8, 10, 12, 14, 16, and 18 min with *Sau*3AI in the presence of Mg^{2+}. Field inversion gel electrophoresis was carried out as described in text except for the switching regimen used here: 0.2 sec forward, 0.6 sec backward, 120 V at a ramp factor of 20 for 5 hr at room temperature.

equal volumes of *n*-butanol and allowed to sit with gentle agitation for about 1 hr for each extraction until the volume of the aqueous phase is reduced to 35–50 μl. This usually requires four or five extractions per sample. When the aqueous phase reaches 50–100 μl care must be taken to avoid complete extraction of the aqueous layer, as this will result in sample loss. When the sample is adequately concentrated it is spot dialyzed as described above and is then ready to be used for cloning. If the sucrose gradient fractionation was inadequate the concentrated sample may need to undergo a second gradient fractionation.

Preparation of Vector "Arms" and Ligation of Genomic DNA

Bacteriophage P1 vector DNA (1–10 μg), either p*Ad*10 or p*Ad*10*sacB*II, is digested to completion (see Cloning of Small Fragments into P1 Vectors, below) with restriction enzyme *Sca*I as suggested by the vendor (New England BioLabs). This linearizes the vector by cleaving it once in the *amp*^{r} domain (Fig. 1). This DNA is extracted with 1 vol phenol and then with 1 vol chloroform–isoamyl alcohol (24 : 1) and is then precipitated with 2 vol 100% ethanol in the presence of 0.3 *M* sodium acetate. The DNA pellet is washed with 70% ethanol, air dried, and resuspended in 50 μl of TE. To dephosphorylate the blunt *Sac*I ends the DNA is incubated with bacterial alkaline phosphatase (Bethesda Research Laboratories, Gaithersburg, MD) at 50–100 units/μg vector DNA for 1–2 hr at 65°. The reaction is then extracted with phenol–chloroform and ethanol precipitated, as described above, and the pellet is resuspended in 20 μl of TE. Next, the DNA is digested to completion with *Bam*HI as described by the vendor (New England BioLabs) to create the cloning site for foreign DNA fragments. We routinely digest the DNA at 37° with 10 units of enzyme for no more than 10 min. This is sufficient to completely digest the DNA and avoids complications that we have noted when p*Ad*10*SacB*II DNA is overdigested with *Bam*HI. In particular, if the digestion is carried out for 60 min rather than 10 min, and the DNA subsequently recovered by ligation and *in vitro* packaging, the DNA generates about 10 times more sucrose-resistant clones than does DNA digested for the shorter interval (see Table I and below).[37] After *Bam*HI digestion the vector DNA is extracted again with phenol and chloroform, ethanol precipitated, and resuspended in 50 μl of TE buffer. *Bam*HI stickly ends are dephosphorylated by treatment with 1 μl of calf intestinal alkaline phosphatase (CIP; New England Nuclear, Boston, MA) at a dilution of 1 : 2000 for 1 hr at 37°. The DNA is again extracted with phenol and chloroform, ethanol precipitated, and resuspended in TE buffer at a final concentration of 100 ng/μl. This protocol generates two fragments ("arms") from each vector DNA: a 2.7-kbp p*Ad*10 short arm or a 4.4-kbp p*Ad*10*sacB*II short arm, and a 27-kb long arm.

The ligation reaction with vector arms and partially digested genomic DNA is typically performed as follows: 200 ng of vector DNA (2 μl) is

[37] The sucrose-resistant clones generated by *Bam*HI overdigestion are probably due to either trace nucleases in the *Bam*HI preparation or *Bam*HI* activity. Many of the resistant clones contain plasmids with *sacB* deletions.

mixed with approximately 500 ng of genomic DNA, 2 μl of 10× ligase buffer (0.5 *M* Tris-HCl, pH 7.5, 0.1 *M* $MgCl_2$, 0.1 *M* dithiothreitol, 0.5 mg/ml BSA) and water to 20 μl. Under these conditions the ratio of *Bam*HI vector ends to *Sau*3A1 genomic DNA ends is about 2 : 1 and the total concentration of DNA is about 20–50 ng/μl. The mixture is incubated at 30° for 1 hr to maximize mixing between vector and genomic DNA and then ATP is added to a final concentration of 1 m*M*. Finally, 500 units of T4 DNA ligase (New England BioLabs) is added and the reaction is incubated at 16° overnight. The reaction then is inactivated by heating it to 70° for 10 min and it is stored at 4° until used in the P1 packaging reaction. It is stable for at least 4 weeks under these conditions.

Preparation of P1 Packaging Extracts

The Pacase Extract. Streak out a loop of strain NS3208 [= MC1061 *sup° recD1014 hsdR⁻ hsdM⁺ mcrA⁻ mcrB⁻* (P1 *r⁻m⁻cm-2 c1.100 am10.1*)] onto a Luria agar plate [Luria broth with 15 g Difco (Detroit, MI) agar per liter] containing 25 μg/ml of chloramphenicol. Incubate the plate overnight at 32° and then pick several colonies for growth, also at 32°, into 10-ml cultures of Luria broth containing 25 μg/ml chloramphenicol. When the cultures have become saturated place them at 4° and spread 50 μl on each of two Luria agar–chloramphenicol plates. One plate is incubated at 32° overnight and the other at 42° overnight. The culture that shows the highest 32°/42° ratio is the one used for subsequent extract preparation.[38] One liter of Luria broth with chloramphenicol is inoculated with the chosen culture and it is grown at 32° in an air shaker at 250 rpm until it reaches an OD_{650} of 0.5. The cells are then centrifuged in a Sorvall GSA rotor (7000 rpm, 10 min at 4°) and the cell pellet is resuspended in 5 ml of Luria broth. It is then diluted into 1 liter of Luria broth with chloramphenicol (25 μg/ml) that has been prewarmed to 42°. Growth is continued at this temperature by shaking the culture at 250 rpm for 15 min. The temperature is then lowered to 38° and the culture grown for an additional 165 min. It is then rapidly chilled to 4° and the cells are pelleted by centrifugation (10 min, 7000 rpm at 4° in a Sorvall GSA rotor).[39] They are resuspended in 0.002 vol (2 ml) of cold buffer containing 20 m*M* Tris-HCl, pH 8.0, 1 m*M* EDTA, 50 m*M*

[38] Typically the 32°/42° ratio for a P1*c1*.100 lysogen is about 10^5–10^6. However, we have found that the ratio for NS3208 and NS3210 (see below) is 10^3–10^4. We do not know the reason for the higher frequency of colony formation at 42° in these strains but we suspect it is due to the *recD* mutation.

[39] It should be noted that the P1 prophage in NS3208 contains a lysis-defective mutation that inhibits cell lysis after P1 induction (lysis normally occurs at 50–60 min at 38°). Thus, a culture of NS3208 can be grown for as long as 3 hr after induction and then harvested by centrifugation. This is not true for NS3210 (see below).

NaCl, and 1 m*M* phenylmethylsulfonyl fluoride (PMSF) and then sonicated on ice with the medium tip of a Branson (Danbury, CT) sonifier at setting 5 for four 15-sec intervals. The resulting extract is centrifuged for 30 min at 17,000 rpm in a Sorvall SS34 rotor and the supernatant (the Pacase or stage 1 extract) is distributed in 20-μl aliquots and stored at $-80°$. This extract is stable for as many as five rounds of freezing and thawing if kept at 4° when thawed.

Head–Tail Extract. Streak out a loop of strain NS3210 [= MC1061 *sup° recD1014, hsdR⁻ hsdM⁺ mcrA⁻ mcrB⁻* (P1 *r⁻ m⁻ cm c1.100 am131*)] onto a Luria agar plate containing 25 μg/ml of chloramphenicol. Process the plates and culture as described for NS3208, except grow the 1 liter of culture to an OD_{650} of 0.3 before harvesting the cells by centrifugation. The pellet is resuspended in 5 ml of Luria broth and then diluted into 500 ml Luria broth with chloramphenicol that has been prewarmed to 42°. It is then incubated at that temperature for 45 min in an air shaker rotating at 300 rpm. It is removed from the shaker, 400 ml of ice-cold 50% (w/v) sucrose is added to the culture, and it is swirled vigorously in an ice–water bath to drop the temperature below 10°. This should not take more than 5 min. The culture is then centrifuged in precooled 350-ml bottles in the GSA head of a Sorvall rotor at 7000 rpm for 5 min at 4°. Pellets are resuspended in a total of 3 ml of ice-cold 50 m*M* Tris-HCl, pH 8.0, 10% (w/v) sucrose and sonicated three times for 10 sec at the low setting of a Branson sonicator. Fifty-microliter aliquots of the extracts are dispensed into 1.5-ml Eppendorf tubes containing 4 μl each of a freshly prepared solution of 10 mg/ml of lysozyme at 4°. The tubes are quick frozen in liquid nitrogen and stored at $-80°$ until used. It should be noted that as NS3210 does not contain a lysis-defective P1 prophage great care must be taken in harvesting the cells after the 45-min induction period to avoid cell lysis during the subsequent centrifugation step. Key to this procedure is to cool the cells quickly, keep all containers cold, and add sucrose to the cells before centrifugation to help maintain their integrity.

Packaging Ligated Vector–Genomic DNA in Vitro

The first stage of the packaging reaction is performed in 15 μl containing 10 m*M* Tris-HCl, pH 8.0, 50 m*M* NaCl, 10 m*M* $MgCl_2$, 2 m*M* dithiothreitol, deoxynucleotides dATP, dCTP, dGTP, and dTTP (0.1 m*M* each), 1 m*M* ATP, 4 μl of a ligation reaction that has been heated to 70° for 10 min (see Preparation of Vector "Arms" and Ligation of Genomic DNA, above) containing about 40 ng of vector, and 1 μl of the Pacase extract. To maximize *pac* cleavage of the vector DNA in the stage 1 reaction that reaction should be incubated for exactly 15 min at 30°. Trace amounts of nuclease activity in the Pacase extract tend to reduce the efficiency of packaging if longer incubation periods are used. The second stage of the packaging reaction (the packaging of *pac*-cut DNA) is initiated by adding 3 μl of the stage 2 buffer (6 m*M* Tris-HCl, pH 8.0, 16 m*M* $MgCl_2$, 60 m*M* spermidine, 60 m*M* putrescine, 30 m*M* 2-mercaptoethanol) and 1 μl 50 m*M* ATP to the stage 1 reaction and then transferring that reaction to a tube containing the P1 head–tail extract. The reaction is incubated at 30° for 20–30 min and then stopped by adding 120 μl of 10 m*M* Tris-HCl, pH 8.0, 10 m*M* $MgCl_2$, 0.1% (w/v) gelatin (TMG buffer) containing 10 μg/ml pancreatic DNase I, followed by vortex mixing of the reaction and incubation for an additional 15 min at 37°. Fifteen microliters of chloroform is added, the reaction is vortexed again, and then centrifuged for 1 min at room temperature in an Eppendorf centrifuge to remove debris. The resulting phage lysate can be stored at 4° with only small losses of titer (twofold) for 2 weeks. However, in general it is best to plate the entire reaction within several days (see the next section).

Recovery of Clones Containing Plasmid and Insert DNA

To recover bacterial colonies containing plasmid vector with cloned genomic fragments from the packaged phage lysates (see the previous section) aliquots of those lysates are used to infect a bacterial host strain containing the bacteriophage P1 Cre recombinase protein. Two *E. coli* strains have been used in the laboratory to generate bacterial transformants. NS3145 [= *recA*$^-$ *hsdM*$^+$ *hsdR*$^-$ *mcrA*$^-$ *mcrB*$^-$ F′ *lacI*^{q} (λ *imm*^{434} *nin*5 *X1-cre*)] is an MC1061 derivative that contains the *lacI*^{q} gene on an F′ plasmid and the P1 *cre* gene on a λ *imm*^{434} prophage.[40] NS3529 [= *recA*$^-$ *mcrA*$^+$ Δ(*hsdR hsdM mcrB mrr*) (λ *imm*^{λ} LP1) (λ *imm*^{434} *nin*5 *X1-cre*)] is a DH5 derivative that contains the *lacI*^{q} and *cre* genes on separate integrated λ prophages, λLP1[33] and λ *imm*^{434} *nin*5 *X1-cre*,[40] respectively. Because NS3529 does not carry an F′ plasmid it has an advantage over NS3145 in permitting the isolation of plasmid vector DNA free of any F′ plasmid DNA.

A culture of the host strain to be used is grown overnight from a fresh colony in 5–10 ml of Luria broth at 37°. The overnight culture is diluted 1 : 100 into Luria broth (usually 50 ml) and grown to midexponential phase (OD_{590} = 0.6) at 37°. The cells are centrifuged (Sorvall SS34 rotor, 7000 rpm, 10 min at 4°) and the pellet is resuspended in 0.1 vol Luria broth with 5 m*M* $CaCl_2$. Ten to 20 μl of the packaging reaction is added to a sterile

[40] B. Sauer and N. Henderson, *Gene* **70,** 331 (1988).

13-mM glass tube and then is exposed to air for 5–10 min to evaporate any residual chloroform. Two hundred microliters of host cells is added to the tube and the phage are absorbed to the cells for 5 min at 37°. One milliliter of Luria broth is then added and the infected cells are shaken for 45 min at 37°. The cells are then transferred to a 1.5-ml Eppendorf tube, centrifuged in a tabletop Eppendorf centrifuge for 2 min at room temperature, and the cell pellet is resuspended in 100 μl of Luria broth. If the lysates were prepared using the p*Ad*10 vector the cells are spread only on Luria agar plates containing 25 μg/ml of kanamycin. Plates are incubated overnight at 37° and colonies that arise are toothpicked onto Luria agar plates containing 10 μg/ml of tetracycline to determine which colonies have cloned inserts. Colonies with inserts are kanamycin resistant, tetracycline sensitive. When using the p*Ad*10*sacB*II positive selection system the cells are spread on Luria agar plates containing either kanamycin alone (25 μg/ml) or kanamycin with 5% (w/v) sucrose.[41] Colonies that arise after overnight incubation at 37° on the sucrose plates all contain plasmids with cloned genomic fragments. A typical experiment illustrating results obtained with the p*Ad*10*sacB*II vector is shown in Table I (line 3). It is clear that the presence of genomic DNA fragments in the ligation reaction increases by about 100-fold the recovery of kanamycin-resistant, sucrose-resistant colonies (compare lines 2 and 3 in Table I). Analysis of the DNA in more than 50 sucrose-resistant clones indicates that they all contain inserted genomic DNA fragments (data not shown).

Two additional points are worth noting. (1) While there is a linear relationship between the amount of the packaged lysate used to infect cells and the number of colonies generated up to 20 μl of lysate, that relationship does not hold if more than 20 μl of lysate is used. Thus, 40 μl of lysate produces only 20–30% more colonies than does 20 μl of lysate. Accordingly, separate infections should be carried out for each 20 μl of lysate processed. (2) The infected cells should not be grown more than 45 min before spreading them on agar plates to avoid generating too many sibling clones.

Isolation and Characterization of Plasmid DNA from Individual P1 Clones

To isolate DNA from a single P1 clone a colony is picked from an agar plate and inoculated into 10 ml of Luria broth containing 25 μg/ml kanamycin. We find Falcon 2059 tubes convenient for this purpose. The

[41] Sucrose agar plates are prepared by adding a sterile 50% sucrose solution to a 10-fold volume of molten agar precooled to 75° and then immediately pouring the agar into empty petri dishes.

culture is shaken at 37° until it appears slightly turbid (OD_{590} 0.1), and then IPTG is added to a final concentration of 1 m*M*, and growth continued for an additional 5 to 12 hr. The cells are centrifuged in a Sorvall RT6000 centrifuge for 15 min at 3000 rpm and the pellet is resuspended in 150 μl of 25 m*M* Tris-HCl, pH 8.0, 50 m*M* glucose, 10 m*M* EDTA, 2 mg/ml lysozyme and transferred to a 1.5-ml Eppendorf centrifuge tube. The cell suspension is incubated at room temperature for 10 min and 300 μl of a freshly prepared solution of 0.2 *N* NaOH, 1% (w/v) SDS is added. The tubes are gently inverted several times and incubated on ice for 10 min. Two hundred and thirty-five microliters of potassium acetate (3 *M* K^+, 5 *M* acetate, pH 4.8) is added to the tube and the contents incubated an additional 10 min on ice. The lysate is centrifuged at top speed in an Eppendorf rotor at room temperature for 5 min and the supernatant is transferred to a fresh Eppendorf tube. It is then extracted with phenol and chloroform and ethanol precipitated (see Preparation of Vector "Arms" and Ligation of Genomic DNA, above). The ethanol pellet is washed with 70% ethanol and resuspended in 40 μl TE buffer with 0.4 μg RNase. This alkaline lysis procedure generates about 1–2 μg of P1 plasmid DNA. It can be scaled up to 250 ml with the expected increase in DNA recovery. We have also purified P1 plasmid DNA using commercially available column chromatography protocols [5′ → 3′ (Boulder, CO), and Quaigen (Studio City, CA) columns] as specified by the vendor.

Our initial characterization of DNA from a P1 clone begins with the fractionation on a 1% (w/v) agarose gel in 1× TBE of a *Bgl*II–*Xho*I digest of 12 μl of the DNA generated by the above alkaline lysis protocol. As a control we use DNA isolated from a colony containing just vector DNA that has been cyclized by the Cre recombinase. As noted previously, that recombination results in the loss of the amp^r domain of the vector, including the *Ad*10 stuffer fragment, leaving a plasmid with the kan^r domain, the tet^r or *SacB* genes, and the P1 plasmid and lytic replicons. *Bgl*II and *Xho*I cuts this plasmid four times, generating fragments of 0.9, 1.9, 3.8, and either 8 kbp (p*Ad*10) or 10 kbp (p*Ad*10*sacB*II). The three smaller fragments are common to all of the P1 clones containing inserted fragments while the larger fragment (8 or 10 kbp) is interrupted by the insert. Figure 5 illustrates the *Bgl*II–*Xho*I digestion pattern of several P1 clones in the p*Ad*10*sacB*II vector. While it is possible to use this procedure to calculate the size of the genomic insert by summing the sizes of the individual fragments, it is a tedious method that is prone to significant error. A more quantitative measure of cloned fragment size is possible using the rare restriction sites in p*Ad*10*sacB*II DNA (*Not*I, *Sfi*I, and *Sal*I) and FIGE. If one of these sites is not in the cloned fragment then digestion with the enzyme that cleaves that site will linearize the plasmid and FIGE will resolve those linears. This is illustrated in Fig. 6 for several *Not*I-

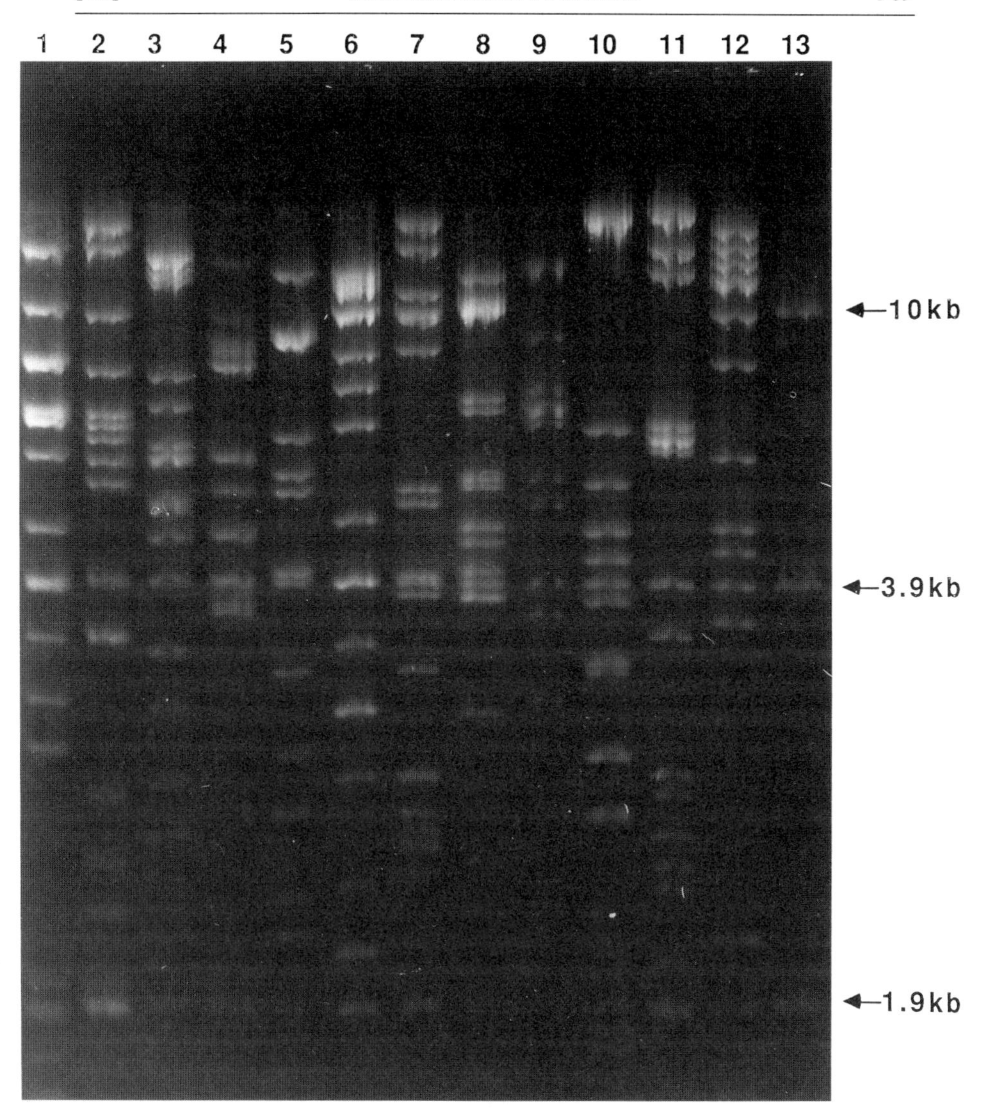

FIG. 5. Restriction analysis of p*Ad*10*sacB*II genomic clones with *Bgl*II and *Xho*I. DNAs are digested with *Bgl*II and *Xho*I and are analyzed by FIGE under conditions similar to those outlined in Fig. 4. Lanes 1–12 contain DNA from 12 random clones of strain NS3529. Lane 13 contains DNA from a p*Ad*10*sacB*II transformant of pNS3529.

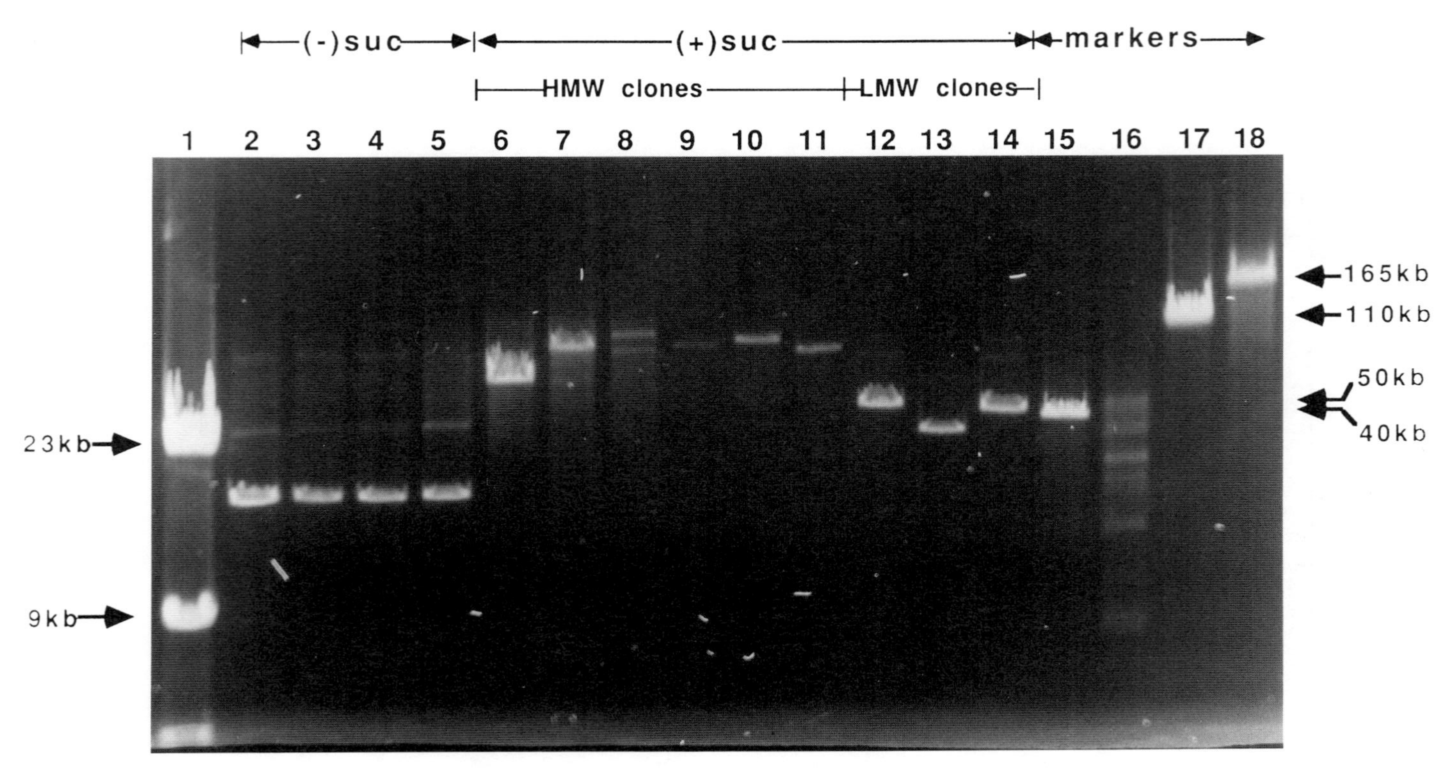

FIG. 6. *Not*I restriction analysis of p*Ad*10*sacB*II genomic clones. Clones used for the isolation of DNA are from the experiment shown in Table I, line 3. The DNAs are digested with *Not*I and analyzed by FIGE as described in Fig. 4. Lanes 2–5 contain DNA from Kan^r, sucrose^s colonies. Lanes 6–14 contain DNA from Kan^r, sucrose^r colonies. Lanes 15–18 contain marker T7, the high molecular weight Bethesda Research Laboratories marker set, P1 and T4 DNAs, respectively. The minor bands in lanes 2–5 and in line 8 are from the F′ plasmid in strain NS3145.

digested clones. Lanes 2–5 (Fig. 6) are clones picked from kanamycin plates without sucrose. They contain only vector sequences. Lanes 6–14 (Fig. 6) are clones selected from kanamycin plates with sucrose. They all have cloned inserts. The inserts in lanes 6–11 (Fig. 6) are large (>70 kbp), while those in lanes 13–15 (Fig. 6) are small (30–40 kbp). The p*Ad*10*sacB*II vector also contains T7 and Sp6 promoters flanking the cloned fragment, which can be used to make riboprobes from the ends of that fragment. We have used this procedure to map restriction fragments at the ends of the cloned DNA for chromosome walking purposes. The unique promoter sequences can also be used as priming sites for DNA sequencing to generate sequence-tagged sites[42] with a spacing of 70–90 kbp.

It has been our experience that DNA prepared as described above can be digested directly with *Bgl*II and *Xho*I and can then be analyzed by gel electrophoresis after adding loading dye (see Preparation, Digestion, and Fractionation of Genomic DNA, above) and heating to 70° for 10 min. In contrast, DNAs that are to be digested with *Not*I require additional processing if complete digestion is to be achieved. They are first mixed with SDS (0.2%, w/v) and proteinase K (200 mg/ml) and incubated for 2 hr at 37°, and are then extracted with phenol and chloroform. The aqueous phase is allowed to air dry on a strip of Parafilm for 10 min at room temperature and is then spot dialyzed against 5 ml of TE (see Preparation, Digestion, and Fractionation of Genomic DNA, above) for 1–2 hr. *Not*I digestion can then be carried out.

Cloning of Small Fragments into P1 Vectors

As shown in Fig. 6 it is possible to clone fragments that are as small as 30–40 kbp into P1 vectors. This would appear to contradict previous arguments[8] about the requirement for a headful of DNA in the packaging reaction. Thus, one might predict if the insert DNA shown in Fig. 2 were too small (<60 kbp) then the head could not be filled during packaging and the process would abort. In fact, this should provide a strong selection against smaller inserts. Why is it then that clones with such inserts are recovered? We can think of several possibilities.

1. Small P1 heads: P1 produces small heads that can encapsulate about 45 kbp of DNA.[10] These heads normally represent only 5–10% of the P1 head population but they might be abundant enough to account for the smaller clones that we have detected. There are two reasons we think this is unlikely. First, we have examined the head population in NS3210 extracts by electron microscopy and can detect very few (<2%) small

[42] M. Olson, L. Hood, C. Cantor, and D. Botstein, *Science* **245,** 1434 (1989).

heads. Second, we have analyzed the density of phage in CsCl equilibrium gradients that are produced by the *in vitro* packaging reaction and they exhibit a density that is identical to that of P1 plaque-forming units; namely, they have large filled heads. In contrast, small-headed phage are significantly less dense than are P1 plaque formers. This analysis allows us to estimate that fewer than 5% of the *in vitro* packaged phage are in small heads. Because with some DNA preparations as much as 20–30% of the clones are small we cannot attribute their presence to small particles.

2. Incomplete *Sca*I digestion of vector DNA: If the vector DNA is not completely digested with *Sca*I and there are multimeric forms of the plasmid in the population the vector will be able to accept less insert DNA in the cloning step and still fill the phage head during packaging (see Fig. 2, Fig. 3, and Ref. 7). To determine whether partially digested multimeric vector molecules could account for the recovery of small inserts we digested with *Sca*I a plasmid that is analogous to p*Ad*10 except that it lacks the stuffer fragment (pNS358[7]). A portion of that DNA was concatemerized by ligation and both linears and concatemers were packaged *in vitro*. To our surprise the linear DNA was packaged with about 10% the efficiency of the concatemers despite the fact that the former should be only 18 kbp in size. CsCl density gradient analysis indicated that kan^r phage generated by the digested DNA that had not been ligated were no different in density than either phage generated by the concatemerized DNA or P1 plaque-forming phage; namely, they were all in phage with large heads. Moreover, the ratio of amp^r to kan^r transformants produced by the *in vitro* packaged phage produced by *Sca*I-digested and unligated DNA was the same as that produced by the *Sca*I-digested and subsequently ligated DNA. Finally, the density of the amp^r phage produced *in vitro* with the two DNAs were the same and that density was the same as P1 plaque-forming units. Because *Sca*I cleaves the vector within the *amp* gene these results can be explained only if the *Sca*I-digested, unligated DNA that is packaged is due to a subpopulation of vector DNA molecules that have not been digested with *Sca*I. Thus, to avoid a significant fraction of small inserts care must be taken to completely digest the vector DNA with *Sca*I. We now suggest that the DNA be digested for 2 hr at 37° with 15 units of enzyme and then overnight at 37° with an additional 15 units of enzyme.

3. Ligation of *Sca*I-generated vector ends during the cloning reaction: Obviously if the *Sca*I ends of the vector arms are ligated to other DNA during the cloning reaction (see Preparation of Vector "Arms" and Ligation of Genomic DNA, above) this will decrease the size of the genomic DNA fragment that can be accepted into a large P1 head by a headful packaging mechanism (Fig. 2). To avoid this, we deliberately chose *Sca*I to generate the vector arms because it produces blunt ends that ligate

poorly and we treated those ends with BAP to reduce further the possibility of ligation. It may be useful to treat the vector DNA more extensively with BAP at this stage or even to digest the vector further with an enzyme, such as *Bal*31, to remove several nucleotides from the ends. Either of these steps should further reduce the possiblity of inappropriate ligation events.

Finally, it should be noted that in any population of partially digested genomic DNA fragments there will be a strong bias in favor of the cloning of the small fragments in the population because those fragments are most likely to have two ends that can be ligated to vector DNA. Thus, the highest priority must be to use HMW starting DNA and partial digests of that DNA that have been freed of as many small fragments as possible.

Comments about Organizing P1 Genomic Library

A significant consideration in constructing and organizing a P1 library depends very much on the desired use of the library. Projects that have as their goal the isolation of one or two genes will not require the investment of significant resources in constructing a library. Perhaps a single pool of clones covering the entire genome would be sufficient for subsequent screening processes. In contrast, where extensive genome mapping and gene cloning are anticipated multiple small pools of clones or even individually arrayed libraries may be necessary to avoid biases generated by the differential growth of individual clones in large pooled populations. Of course, the size of the genome one wants to clone will significantly impact the practicality of the approach used. Thus, genomes in the 10^8-bp range (*Caenorhabditis elegans, Drosophila melanogaster*) require 5000 P1 clones for triple coverage. Libraries of this size can easily be arrayed manually and one already has been constructed for *Drosophila*.[9] Larger genomes in the 10^9-bp range (e.g., those of mice or of humans) will require 50,000–100,000 clones and arraying of the library will likely require the use of robotics in the handling and propagation of clones. We have devised a minipool strategy for the organization of a mouse library that is similar to one suggested for YAC libraries.[43] Initially 200 to 400 primary pools of about 400 clones each in the *sacB*II vector are constructed and stored as frozen stocks. Twenty to 40 secondary pools are made by combining 10 primary pools each and are grown in large culture (1 liter). After IPTG induction DNA is isolated by methods described above (Isolation and Characterization of Plasmid DNA from Individual P1 Clones). The entire

[43] E. D. Green and M. V. Olson, *Proc. Natl. Acad. Sci. U.S.A.* **87,** 1213 (1990).

library can be screened in one agarose gel by polymerase chain reaction (PCR) or Southern blot analysis of the DNA from the secondary pools. If a secondary pool is positive, DNA made from the 10 primary pools that represent a secondary pool is analyzed in the same way. The primary pool that generates a positive signal can be processed for colony hybridization to select the individual clone with the desired insert. In our hands the colony hybridization step is the most tedious one and, therefore, the use of minimally complex pools is desirable.

[48] Cloning Vectors and Techniques for Exonuclease–Hybridization Restriction Mapping

By KENNETH D. TARTOF

Principle of Exonuclease–Hybridization Method

Constructing restriction maps of cloned DNA fragments is often a tedious, but very necessary, task frequently encountered in the course of gene isolation or chromosome walking. Currently, two general methods have been developed for constructing maps of this type. The first requires digesting the cloned DNA with two different restriction enzymes, both singly and together. After separating the resulting fragments by agarose gel electrophoresis one then attempts to determine, usually by trial and error, a uniquely ordered arrangement that accounts for the observed array of bands in each digest. The amount of time required for this process is considerable, increasing as the size of the insert DNA increases. In the case of cosmids, for example, deciphering the complexity of a fragment pattern derived from a 45-kilobase (kb) insert can be formidable. In addition, the trial and error method usually encounters uncertainty in the relative order of adjacent restriction fragments punctuated by the same restriction site when no (or few) other sites are apparent. This is especially true in cases involving small fragments because these may be easily missed.

A second approach to restriction mapping makes use of a method initially developed by Smith and Birnstiel.[1] Here, the DNA to be mapped

[1] H. O. Smith and M. L. Birnstiel, *Nucleic Acids Res.* **3,** 2387 (1976).

is uniquely labeled at one end of the molecule. Then, partial digestion with a restriction enzyme, followed by electrophoresis and autoradiography, reveals a series of fragments that can be ordered from the labeled terminus. Although this method yields an accurate and unambiguous map, it requires some prior knowledge about the location of restriction sites as well as their favorable placement in the target molecule in order to end label it effectively.

To circumvent these and other difficulties inherent in the Smith–Birnstiel method, we have constructed plasmid, cosmid, and phage λ vectors that facilitate the asymmetric end labeling of DNA inserts so they may be routinely mapped without any prior knowledge of their structure.[2] These vectors have been designed with a *Not*I restriction site placed adjacent to a "linker" sequence that, in turn, flanks either end of a cloning site into which foreign DNA may be inserted (Fig. 1). Because the *Not*I restriction site (GCGGCCGC) is 8 bp in length, it is expected to occur randomly about once in 65 kb of genomic DNA. However, because the cytosine of the CpG dinucleotide is potentially methylatable in mammalian DNA, and because there are two CpGs in this restriction site, the random presence of *Not*I has been selected against in such genomes. In these organisms, *Not*I is present only once in every 200–1000 kb.[3,4] Thus, the relative rarity of *Not*I sites provides an effective means for excising cloned inserts with the adjoining linker sequences.

Vectors equipped with this *Not*I–linker arrangement facilitate restriction mapping by the exonuclease III–linker hybridization (exo-hyb) method as illustrated in Fig. 1. Cloned DNA is digested with *Not*I followed by a brief treatment with exonuclease III to recess the 3′ ends.[5] This produces two 5′ single-stranded linker sequence ends, either one of which is separately hybridized to its complementary ^{32}P-labeled oligodeoxynucleotide. The labeled complex is then partially digested with the restriction enzyme to be mapped and the fragments produced in this way are separated by agarose gel electrophoresis. After the gel is dried and exposed to X-ray film, the resulting autoradiograph yields an array of bands whose sizes correspond directly to the distances the restriction sites are located from the *Not*I–linker terminus. Unambiguous high-density restriction maps may be generated immediately from these data.

[2] K. D. Tartof and C. A. Hobbs, *Gene* **67,** 169 (1988).

[3] R. Drmanac, N. Petrovic, V. Glisin, and R. Crkvenjakov, *Nucleic Acids Res.* **14,** 4691 (1986).

[4] D. I. Smith, W. Golembieshi, J. D. Gilbert, L. Kizyma, and O. J. Miller, *Nucleic Acids Res.* **15,** 1173 (1987).

[5] R. Wu, C. D. Tu, and R. Padmanabhan, *Biochem. Biophys. Res. Commun.* **55,** 1092 (1973).

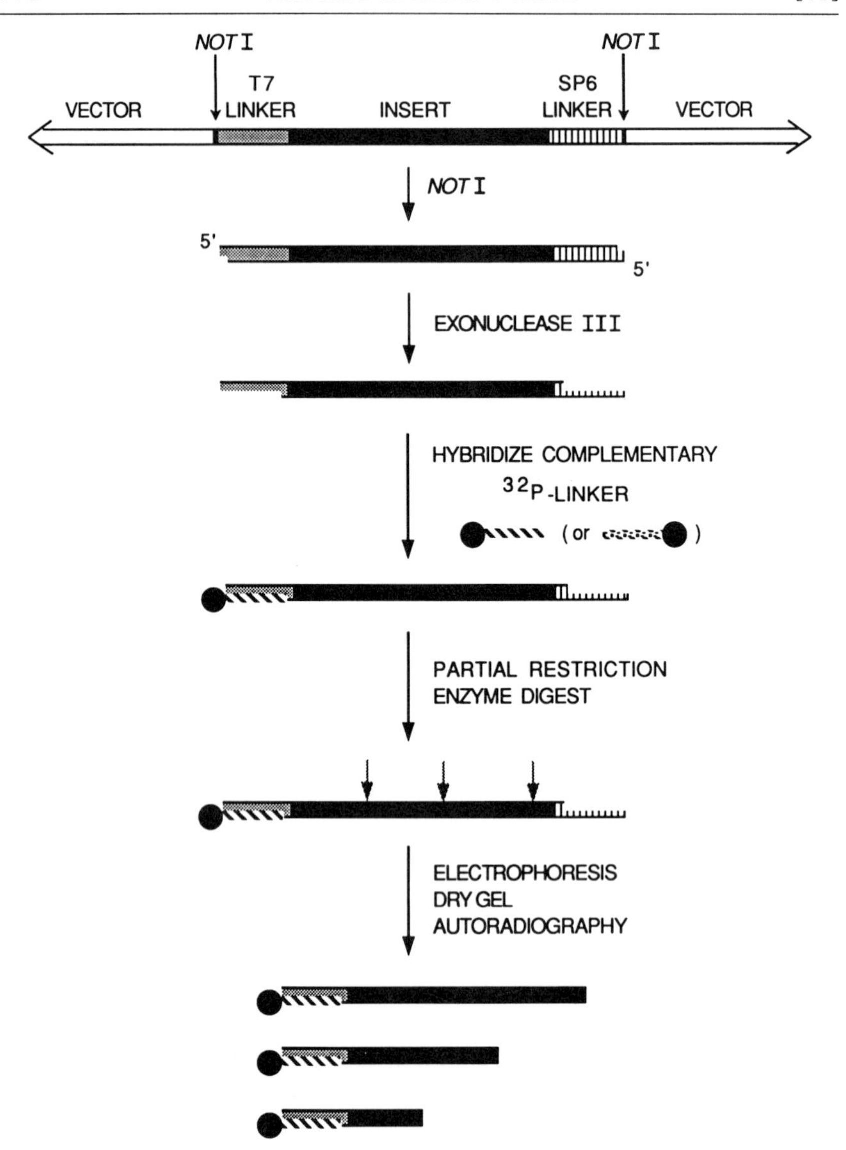

FIG. 1. A schematic illustration of the principle features of the exonuclease–hybridization restriction mapping method.

Materials and Reagents

Diagram of New Vectors for Exonuclease–Hybridization Restriction Mapping

Figures 2 and 3 illustrate the structure of plasmid, cosmid, and λ exo-hyb mapping vectors. These utilize T7 and SP6 promoters as the "linker" sequence, enabling both the production of transcripts from adjoining inserted sequences as well as expediting restriction mapping. Thc plasmid vectors pKT2 and pKT3 are derivatives of pUC19,[6] λKT3 is a derived from EMBL4,[7] and the cosmid vectors cosKT2 and cosKT3 are modified from cosKT1.[2]

Bacterial and Phage Strains

Plasmid and cosmid clones are routinely propagated in *Escherichia coli* DH5α [Bethesda Research Laboratories, Gaithersburg, MD; *endA1 hsdR17*(r_k^-, m_k^+) *supE44 thi-1* F^- λ^- *recA1 gyrA96 relA1* *ϕ80dlacZ*ΔM15 Δ(*lacZY-argF*)U169]. DH5α is particularly useful because it possesses the *lacZα* complementation system and because it carries a deletion for the *recA* gene that supports stability of plasmid and cosmid inserts.

Liquid cultures of λKT3 are grown in *E. coli* BHB2600 whereas phage plated in soft agar tend to produce larger plaques when *E. coli* LE392 is used as the host.

Growing Bacteria and Extraction of Recombinant DNAs

We typically obtain about 1–2 mg of plasmid DNA when the DH5α host is grown overnight at 37° in 330 ml of terrific broth (TB) and 100 μg/ml of ampicillin. By growing cosmid-containing bacteria in TB/antibiotic medium it is possible to process many 50- to 100-ml cosmid cultures and obtain approximately 250 μg of cosmid DNA from each.[8] One liter of TB is made by adding 100 ml of a sterile solution containing 0.17 *M* KH_2PO_4 and 0.72 *M* K_2HPO_4 to a separately sterilized solution containing 12 g Bacto-tryptone, 24 g Bacto-yeast extract, 4.0 ml glycerol, and water to a final volume of 900 ml.

Plasmid and cosmid DNAs are prepared by the Birnboim and Doly alkaline lysis procedure, as described by Sambrook *et al.*,[9] with the addi-

[6] C. Yanisch-Perron, J. Vieira, and J. Messing, *Gene* **33,** 103 (1985).
[7] A. M. Frischauf, H. Lehrach, A. Poustka, and N. Murray, *J. Mol. Biol.* **170,** 827 (1983).
[8] K. D. Tartof and C. A. Hobbs, *Focus* **9,** 12 (1987).
[9] J. Sambrook, E. F. Fritsch, and T. Maniatis, "Molecular Cloning: A Laboratory Manual" (Cold Spring Harbor Laboratory, Cold Spring Harbor, New York, 1989).

A

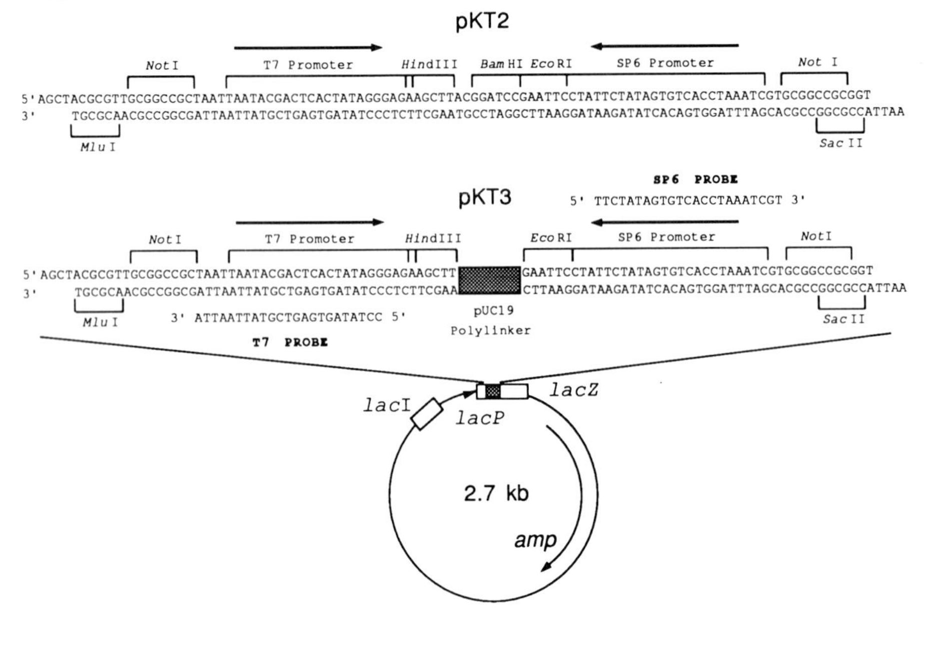
pKT2
NotI
T7 Promoter
HindIII
BamHI
EcoRI
SP6 Promoter
Not I
5' AGCTACGCGTTGCGGCCGCTAATTAATACGACTCACTATAGGGAGAAGCTTACGGATCCGAATTCCTATTCTATAGTGTCACCTAAATCGTGCGGCCGCGGT
3' TGCGCAACGCCGGCGATTAATTATGCTGAGTGATATCCCTCTTCGAATGCCTAGGCTTAAGGATAAGATATCACAGTGGATTTAGCACGCCGGCGCCATTAA
MluI
SacII
SP6 PROBE
pKT3
5' TTCTATAGTGTCACCTAAATCGT 3'
NotI
T7 Promoter
HindIII
EcoRI
SP6 Promoter
NotI
5' AGCTACGCGTTGCGGCCGCTAATTAATACGACTCACTATAGGGAGAAGCTT
GAATTCCTATTCTATAGTGTCACCTAAATCGTGCGGCCGCGGT
3' TGCGCAACGCCGGCGATTAATTATGCTGAGTGATATCCCTCTTCGAA
CTTAAGGATAAGATATCACAGTGGATTTAGCACGCCGGCGCCATTAA
MluI
3' ATTAATTATGCTGAGTGATATCC 5'
pUC19
Polylinker
SacII
T7 PROBE
lacI
lacZ
lacP
2.7 kb
amp

B

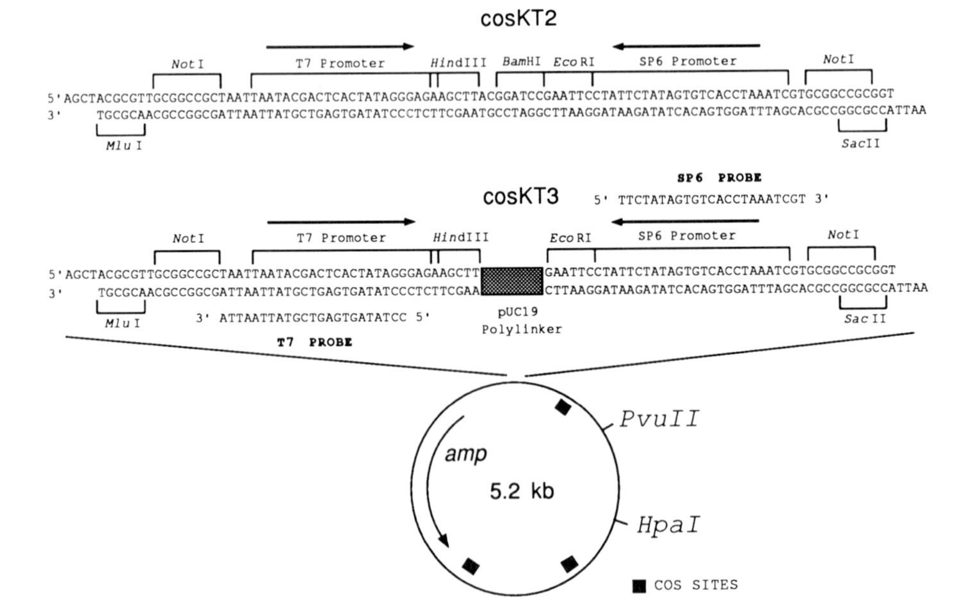
cosKT2
NotI
T7 Promoter
HindIII
BamHI
EcoRI
SP6 Promoter
NotI
5' AGCTACGCGTTGCGGCCGCTAATTAATACGACTCACTATAGGGAGAAGCTTACGGATCCGAATTCCTATTCTATAGTGTCACCTAAATCGTGCGGCCGCGGT
3' TGCGCAACGCCGGCGATTAATTATGCTGAGTGATATCCCTCTTCGAATGCCTAGGCTTAAGGATAAGATATCACAGTGGATTTAGCACGCCGGCGCCATTAA
MluI
SacII
SP6 PROBE
cosKT3
5' TTCTATAGTGTCACCTAAATCGT 3'
NotI
T7 Promoter
HindIII
EcoRI
SP6 Promoter
NotI
5' AGCTACGCGTTGCGGCCGCTAATTAATACGACTCACTATAGGGAGAAGCTT
GAATTCCTATTCTATAGTGTCACCTAAATCGTGCGGCCGCGGT
3' TGCGCAACGCCGGCGATTAATTATGCTGAGTGATATCCCTCTTCGAA
CTTAAGGATAAGATATCACAGTGGATTTAGCACGCCGGCGCCATTAA
MluI
3' ATTAATTATGCTGAGTGATATCC 5'
pUC19
Polylinker
SacII
T7 PROBE
PvuII
amp
5.2 kb
HpaI
COS SITES

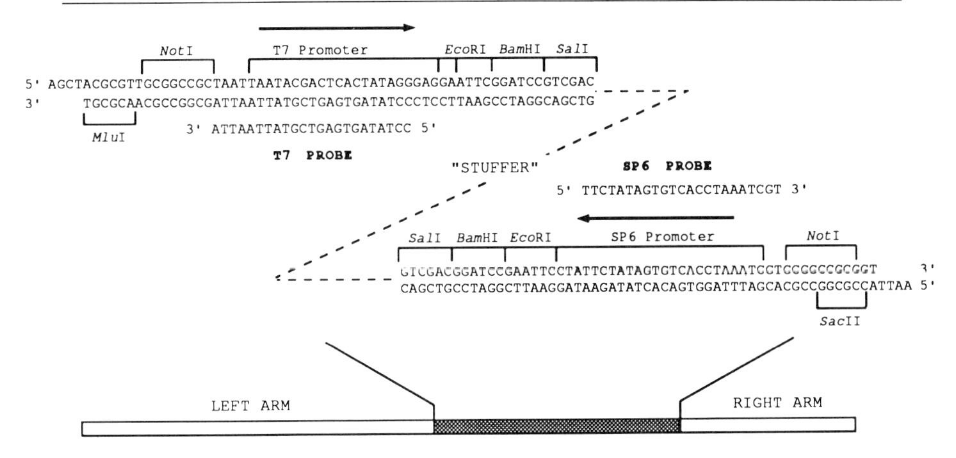

FIG. 3. Exonuclease–hybridization mapping vector λKT3.

tion of a phenol extraction step after precipitation of the DNA from the CsCl gradient. This practice eliminates trace amounts of exonuclease that frequently contaminate plasmid DNA preparations and is necessary to ensure efficient use of the mapping techniques described in the following section.

Sufficient quantities of λ phage are prepared by adding 1.3×10^8 phage to 6.7 ml of a stationary-phase culture of *E. coli* BHB2600 (grown in NZYM). After 20 min at 37°, this infected culture is transferred to 330 ml of NZYM in a 2-liter flask and vigorously shaken at 37° until complete lysis, which usually occurs at about 6 hr. NZYM contains (per liter) 10 g NZ-amine, 5 g yeast extract, 5 g NaCl, and 2 g $MgCl_2 \cdot 6H_2O$.

Phage λ DNA is conveniently obtained as follows. After lysis, the culture is adjusted to 0.5 *M* NaCl and 0.02 *M* $MgCl_2$, cell debris is removed by centrifugation, and the phage precipitated with the addition of 33 g of polyethylene glycol (PEG 8000). Phage DNA is then extracted according to Thomas and Davis.[10]

Enzymes, Buffers, and Reaction Conditions for Exonuclease–Hybridization Mapping

Exonuclease III (Bethesda Research Laboratories, Life Technologies, Inc.) is diluted to 30 units/μl in exonuclease III storage buffer consisting

[10] M. Thomas and R. W. Davis, *J. Mol. Biol.* **91,** 315 (1975).

FIG. 2. Vectors devised for exonuclease–hybridization mapping. (A) and (B) illustrate the structures of plasmid and cosmid vectors, respectively.

of 200 mM KCl, 50 mM ethylenediaminetetraacetic acid (EDTA), 5 mM KPO_4 (pH 6.5), 5 mM 2-mercaptoethanol, 200 μg/ml bovine serum albumin (BSA), and 50% (v/v) glycerol.

All of the restriction enzymes used for exo-hyb mapping are available from commercial sources. Universal restriction buffer (URB) is a convenient medium in which to carry out a wide variety of restriction enzyme reactions pertinent to this procedure. It is derived from the Tris–acetate buffer of O'Farrell[11] to which we have added spermidine. Most of the commonly used restriction enzymes possess the same activity, or nearly the same activity, in URB as in the manufacturers' recommended buffers without evidence of "star" activity. URB consists of 33 mM Tris–acetate, pH 7.9, 66 mM potassium acetate, 10 mM magnesium acetate, 5 mM dithiothreitol (DTT), 100 μg BSA/ml, and 4.0 mM spermidine.

For partial digests, restriction enzymes are usually diluted to 3 to 5 unit/μl in URB. The actual amount of enzyme required may vary according to its specific activity. Therefore, it is useful to determine the best dilution for each enzyme to be used for mapping. This is most conveniently accomplished by separately digesting 1 μg of target DNA with 1, 3, and 9 units of restriction enzyme in 25 μl of URB at 25° for 1 min. After electrophoresis, the ethidium bromide-stained gel is illuminated with long-wavelength UV light to visualize the separated bands of DNA. An enzyme concentration yielding the most even distribution of the maximum number of bands will give the best pattern of partial digestion for exo-hyb mapping.

Oligodeoxynucleotides for Hybridization to T7 and SP6 Linkers

Oligodeoxynucleotides CCTATAGTGAGTCGTATTAATTA and TTCTATAGTGTCACCTAAATCGT are complementary to the single-stranded regions of the T7 and SP6 promoters strands that result from the action of exonuclease III, respectively. They are synthesized by the phosphoramidite method on an Applied Biosystems (Foster City, CA) DNA synthesizer.

To prepare ^{32}P-labeled oligodeoxynucleotide complementary to the T7 or SP6 promoters, 100 ng of oligodeoxynucleotide is added to a 15-μl reaction mixture containing kinase buffer (70 mM Tris–acetate, pH 7.6, 10 mM $MgCl_2$, 5 mM DTT), 10 units T4 polynucleotide kinase, and 9 μl [γ-^{32}P]ATP (4500 Ci/mmol) and incubated at 37° for 90 min. The reaction is terminated by the addition of 3 μl of 0.25 M EDTA (pH 8.0) followed by heating at 70° for 5 min. The labeled oligodeoxynucleotide is separated from unincorporated ^{32}P on a commercially available gel-filtration column equilibrated with TE (10 mM Tris, pH 7.4, 1 mM EDTA) buffer.

[11] P. O'Farrell, *Focus* **3,** 1 (1980).

Gel Electrophoresis and Gel-Drying Equipment

Electrophoresis is carried out on a 0.6% (w/v) agarose gel at room temperature in 1X E buffer (40 m*M* Tris–acetate, pH 7.9, 5 m*M* sodium acetate, 1 m*M* EDTA) containing 500 μg/liter of ethidium bromide. We use an apparatus that accommodates a 20-slot, 20 × 20 cm agarose gel, and that also allows the buffer to be continuously recirculated by means of a small oscillating pump regulated by a variable transformer (these latter two items are available from Cole Parmer Co., Chicago, IL). It is imperative to recirculate the buffer during electrophoresis to ensure that the lanes of DNA run straight and parallel to each other throughout the entire length and width of the gel. Additionally, recirculation enhances the sharpness of restriction fragment bands. Any one of a number of commercially available gel dryers may be used to remove water quickly from the agarose gel and reduce it to a thin dehydrated film suitable for autoradiography.

Methods

The following protocol, consisting of 12 steps, is written for mapping two cloned DNAs with four different restriction enzymes. The cloning vehicle must be of the *Not*I-KT type as illustrated in Figs. 2 and 3. For mapping with more or fewer enzymes, adjust quantities as necessary.

1. Digest 10 μg of each DNA with 80 units of *Not*I in URB for 60 min at 37° in a 1.5-ml microfuge tube as indicated in the sample protocol tabulated below.

Component	Sample 1	Sample 2
DNA 1 [0.5 μg/μl (10-μg final amount)]	20	—
DNA 2 [0.5 μg/μl (10-μg final amount)]	—	20
10× URB	18	18
*Not*I [1 unit/μl (8 units/μg DNA final)]	80	80
H_2O	32	32
Final volume (in μl)	150	150

2. Add 30 units, usually 1 μl, of exonuclease III (previously diluted in exonuclease III storage buffer) to each tube and incubate at 25° for 3 min. Then transfer tubes to a 70° water bath for 5 min to inactivate the enzymes.
3. Divide each sample into two 75-μl portions in 1.5-ml microfuge tubes and equilibrate in a 25° water bath.

4. Add one of the ^{32}P-labeled T7 or SP6 complementary oligodeoxynucleotides to either tube of each DNA set and anneal at 37° for 60 min. Typically, we add 1×10^6 cpm of oligodeoxynucleotide to each tube of partial restriction enzyme digest. Thus, if two DNAs are to be mapped from both ends with four different restriction enzymes, expect to add a total of 1.6×10^7 cpm.

5. Distribute 15-μl aliquots of each hybrid mix into four 0.5-ml microfuge tubes, as indicated below. (*Note:* The fifth tube is extra.) Add 1 μl of 10× URB and an amount of H_2O that will be necessary to bring the final volume to 25 μl after the addition of the appropriate restriction enzyme as illustrated in the representative protocol tabulated below. Equilibrate the samples at 25°.

Component	TUBE number															
	1	2	3	4	5	6	7	8	9	10	11	12	13	14	15	16
DNA 1 (T7)	15	—	15	—	15	—	15	—	—	—	—	—	—	—	—	—
DNA 1 (SP6)	—	15	—	15	—	15	—	15	—	—	—	—	—	—	—	—
DNA 2 (T7)	—	—	—	—	—	—	—	—	15	—	15	—	15	—	15	—
DNA 2 (SP6)	—	—	—	—	—	—	—	—	—	15	—	15	—	15	—	15
10× URB	1	1	1	1	1	1	1	1	1	1	1	1	1	1	1	1
Restriction enzyme																
A	1	1	—	—	—	—	—	—	—	—	—	—	—	—	—	—
B	—	—	1	1	—	—	—	—	—	—	—	—	—	—	—	—
C	—	—	—	—	1	1	—	—	—	—	—	—	—	—	—	—
D	—	—	—	—	—	—	1	1	—	—	—	—	—	—	—	—
A	—	—	—	—	—	—	—	—	1	1	—	—	—	—	—	—
B	—	—	—	—	—	—	—	—	—	—	1	1	—	—	—	—
C	—	—	—	—	—	—	—	—	—	—	—	—	1	1	—	—
D	—	—	—	—	—	—	—	—	—	—	—	—	—	—	1	1
H_2O	8	8	8	8	8	8	8	8	8	8	8	8	8	8	8	8
Final volume (in μl)	25	25	25	25	25	25	25	25	25	25	25	25	25	25	25	25

6. To each tube, add an amount of enzyme required to achieve the appropriate degree of partial digestion and incubate at 25° for 1 min. As a general rule, we dilute restriction enzymes in 1× URB to a concentration of 3 units/μl and add 1 μl of this to each tube as indicated above. After the 1-min incubation, remove 12.5 μl of the digestion and transfer it to a tube containing 3 μl of orange G stop buffer [0.5 m*M* EDTA, pH 8.0, 25% (w/v) sucrose, 0.5% (v/v) orange G]. Transfer the tube containing the remaining reaction to 37°. After 1 min, remove the remaining 12.5 μl and

add to the first tube containing the 25° partial digestion products and stop buffer. Store on ice.

7. Load the samples into wells numbered 2 through 17 (see step 5) of an 0.6% (w/v) agarose gel. In wells 1 and 18 place a molecular size standard consisting of a mixture of 0.5 μg *Hin*dIII λ digest and 0.5 μg of a *Hae*III ΦX174 digest. Electrophoresis is carried out overnight at 35 V in continuously recirculated 1× E buffer containing ethidium bromide. It is sometimes necessary to continue electrophoresis for a few hours the next morning until the 270/280-bp doublet bands of the *Hae*III ΦX174 digest have migrated about 17 cm from the origin.

8. Photograph the gel under a long-wavelength UV light and mark the location of the molecular weight standard fragments by poking each band with a needle dipped in India ink. Rinse the excess ink off the gel by running it under water.

9. Immerse the gel in cold 10% methanol plus 10% acetic acid and shake gently for 10 min at room temperature. This fixes the DNA fragments in the agarose matrix.

10. Transfer the gel to two layers of Whatman (Clifton, NJ) 3MM chromatography paper. Then cover it with plastic wrap and dry it on a gel dryer without heat for approximately 30 min.

11. Seal the gel in plastic wrap and expose it to X-ray film with an intensifying screen at −70°. Usually, 4 hr is a sufficient exposure time. However, it may be necessary to expose the gel overnight to bring up very light bands.

12. Exonuclease–hybridization restriction maps should be confirmed by carrying out complete digests with each of the enzymes for which partial digests have been performed. This is necessary to demonstrate that all of the predicted fragments are accounted for and that none have been missed.

Constructing the Exonuclease–Hybridization Restriction Map

The length of each fragment observed in the autoradiograph is determined by measuring the distance it has migrated and interpolating this value into a kilobase pair size using the distance that *Hin*dIII λ and *Hae*III ΦX174 standards have migrated. These data are used to construct the restriction map for each enzyme. Each restriction site is located from the appropriate *Not*I terminus by a distance that corresponds directly to the length of that fragment as observed in the autoradiograph of the exo-hyb partial digest. Maps initiated from either *Not*I linker end of plasmid and λ inserts should overlap completely and be congruent with each other. In the case of cosmid size inserts, overlap will occur only in the central portion of the insert. The modest misalignment of some restriction sites in the region of overlap is caused by the inherent difficulty of measuring fragment lengths in the 20-kb range.[2]

Potential Problems and Some Suggestions

We have encountered few, if any, major difficulties in the exo-hyb mapping method as described here. However, two problems that have arisen from time to time are discussed below.

No or Faint Partial Digestion Products. This problem is easily solved in one of two ways. First, check to see that the proper amount of restriction enzyme has been used. It is imperative to calibrate these enzymes, as described in the Materials and Reagents section, especially if they are old and have been stored in the freezer for some time. Second, if the correct amount of restriction enzyme has been added, then it may be that the amount of exonuclease III is not sufficient to expose the linker target sequences in enough target molecules for hybridization with the complementary ^{32}P-labeled probe. In this case, treat with two or three times the amount of exonuclease III.

Fuzzy Bands on Autoradiograph. If the agarose gel is too thick, then it may be deformed on drying and thereby cause the bands observed in the autoradiograph to appear fuzzy. This can be remedied by using the thinnest possible gel (and well-forming comb) that will also accommodate the 25-μl sample volume to be electrophoresed.

[49] Yeast Artificial Chromosome Modification and Manipulation

By ROGER H. REEVES, WILLIAM J. PAVAN, and PHILIP HIETER

Introduction

Yeast artificial chromosomes (YACs) provide a means for isolation and characterization of genomic DNA segments of more than 2 megabases (Mb) in *Saccharomyces cerevisiae*.[1] The large cloning capacity of this system provides two basic advantages for genome analysis. The first is simply a function of the size of the cloned inserts in YACs. The availability

[1] D. T. Burke, G. F. Carle, and M. V. Olson, *Science* **236,** 806 (1987).

of genomic fragments several hundred kilobases in length greatly facilitates procedures involving physical isolation of sequences by chromosome walking. For most genes, all coding and regulatory sequences will occur within a region that can be cloned as a YAC. Because the average spacing of mammalian genes is about 1 every 60 kb (assuming 50,000 genes in a genome of 3×10^9 bp), an individual YAC will frequently contain several genes and thereby provide a probe for isolating coding regions at specific locations within the genome. The second advantage of cloning in YACs is that mammalian DNA cloned in yeast can be manipulated *in vitro* using homologous recombination-based methods. Two types of YAC modification systems have been developed that take advantage of this property to target recombination to highly repetitive DNA elements within YAC DNA inserts. Vectors that *integrate* into YACs can be used to produce chromosome-specific libraries from somatic cell hybrids or to introduce deletions, mutations, or mammalian selectable markers into YAC cloned segments.[2,3] A second vector system facilitates analysis of the large inserts in YACs by *fragmenting* them, generating nested deletion derivatives.[4] A polyethylene glycol-mediated fusion procedure has been described that can be used to introduce appropriately modified YACs into cultured mammalian cells.[3,5] This combination of vectors and procedures provides a system for realizing the full potential of YACs in structural and biological analysis of the genome.

Yeast Artificial Chromosome Vectors and Yeast Hosts

Developing a strategy for YAC manipulation requires a knowledge of the particular vectors and yeast host strain used for the original construction of the YAC. Yeast artificial chromosome vectors contain sequences necessary for stable propagation of cloned DNA as large linear molecules in yeast (Fig. 1). These vectors contain (1) a bacterial origin of replication and ampicillin resistance gene for propagation and selection in bacteria, (2) a yeast centromere (*CEN4*), a yeast autonomously replicating sequence (*ARS1*), and two *Tetrahymena* telomere (*TEL*) sequences, and (3) the yeast *URA3* and *TRP1* genes for selection of YAC transformants. During YAC construction, the vectors are linearized and ligated onto the ends of insert DNA prior to transformation into yeast.

The original series of pYAC vectors[1] differs only in the types of cloning

[2] W. J. Pavan and R. H. Reeves, *Proc. Natl. Acad. Sci. U.S.A.* **88,** 7788 (1991).

[3] W. J. Pavan, P. Hieter, and R. H. Reeves, *Mol. Cell. Biol.* **10,** 4163 (1990).

[4] W. J. Pavan, P. Heiter, and R. H. Reeves, *Proc. Natl. Acad. Sci. U.S.A.* **87,** 1300 (1990).

[5] V. Pachnis, L. Pevny, R. Rothstein, and F. Constantini, *Proc. Natl. Acad. Sci. U.S.A.* **87,** 5109 (1990).

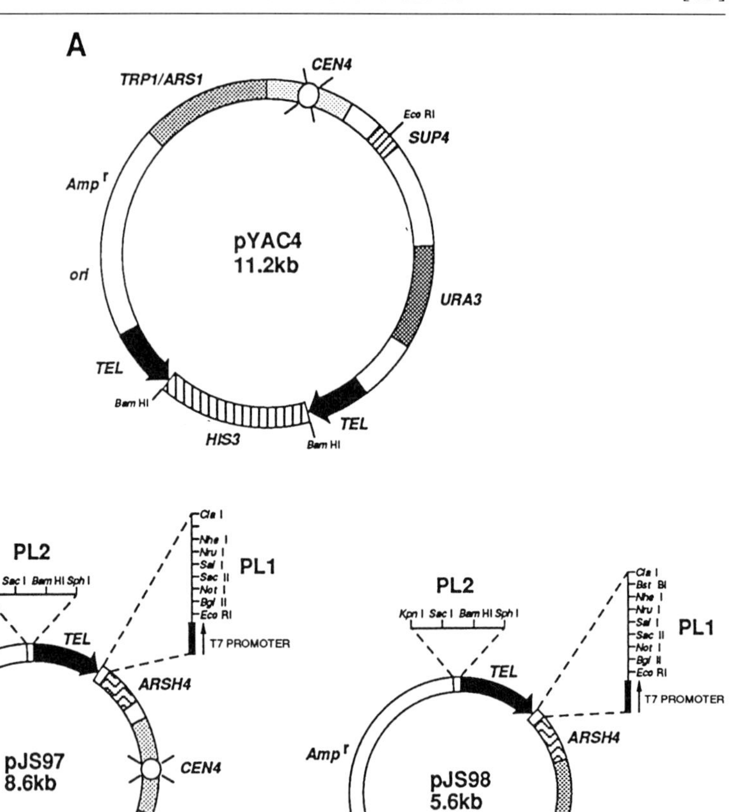

FIG. 1. YAC cloning vectors pYAC4, pJS97, and pJS98. (A) Restriction map of pYAC4. The YAC vector arms are generated by restriction digestion with *Eco*RI, which is present in the *SUP4* gene intron, and *Bam*HI, which removes the *HIS3* "stuffer" fragment (and which is not relevant to YAC cloning). (B) Restriction maps of pJS97 and pJS98. The YAC vector arms are present as two separate plasmids, each containing *amp*^{r} and *ori* for propagation in *E. coli*. Unique restriction sites are shown for PL1, the cloning site polylinker, and PL2, a polylinker present to enable efficient rescue of YAC ends in *E. coli*.

sites present in the *SUP4* intron. pYAC4 contains an *Eco*RI cloning site, pYAC3, a *Sna*B1 cloning site (for blunt end cloning), and pYAC55, a *Not*I cloning site. pYAC*neo* includes a pSV2*neo* cassette inserted into the *URA3*-containing vector arm of pYAC4.[6] The *neo*^{r} marker was designed as a selective marker for YACs transferred into mammalian cells. pYAC-

[6] H. Cooke and S. Cross, *Nucleic Acids Res.* **16,** 11817 (1988).

neo includes the pSV2*neo* cassette as well as a second bacterial origin of replication on the *URA3*-containing arm that can be used for rescuing the ends of YAC inserts as plasmids in *Escherichia coli*.[7] pYAC-RC contains a polylinker with unique sites for infrequently cutting restriction enzymes (*Not*I, *Sac*II, *Mlu*I, *Cla*I, and *Sna*BI) at the cloning site.[8] pCGS996 permits copy number amplification of artificial chromosomes[9] by making use of a conditional centromere that is inactivated when cells are grown on medium containing galactose as the sole carbon source. Selection for cells containing extra copies of the YAC is achieved by selecting for the expression of a heterologous thymidine kinase gene in the vector by growth in the presence of sulfanilamide and methotrexate. Yeast artificial chromosomes are amplified 10- to 20-fold by this procedure.

A pair of YAC vectors, representing the centric and acentric YAC arms, has been constructed to facilitate clone manipulation and chromosome walking.[10] pJS97 and pJS98 contain a cloning polylinker with eight unique cloning sites (Fig. 1). A T7 bacteriophage promoter adjacent to the cloning site allows the generation of riboprobes from the rescued ends of an exogenous DNA insert within a YAC. End rescue is facilitated by a second polylinker (PL2) in each vector. The presence of the *SUP11* gene in pJS97 allows visual monitoring of the mitotic stability of YAC clones. Because the diploid yeast strain YPH274 gives rise to red colonies if *SUP11* is absent, pink colonies if *SUP11* is present in one copy, and white colonies when two copies are present, the copy number and stability of YACs can be visually determined.[11] For example, yeast transformants containing two YACs are white, providing a quick visual screen for double YAC transformants.

Five yeast strains (AB1380, YPH274, YPH252, YPH857, and CGY2516) have been employed as hosts for YAC cloning (see Table I). All are derivatives of S288c that carry *ura3* and *trp1* mutations for selecting YAC transformants, and all are transformed at similar high efficiencies using an optimized spheroplast transformation protocol. The major differences are in the availability of additional genetic markers that can be used in subsequent DNA-mediated transformation experiments. Homologous recombination-based genetic manipulation of YAC clones (described below) requires nonreverting auxotrophic mutations in addition to the *ura3*

[7] C. Traver, S. Klapholtz, R. Hyman, and R. Davis, *Proc. Natl. Acad. Sci. U.S.A.* **86,** 5898 (1989).

[8] D. Marchuk and F. Collins, *Nucleic Acids Res.* **16,** 7743 (1988).

[9] D. Smith, A. Smyth, and D. Moir, *Proc. Natl. Acad. Sci. U.S.A.* **87,** 8242 (1990).

[10] L. Shero, M. McCormick, S. Antonarakis, and P. Hieter, *Genomics* **10,** 505 (1991).

[11] P. Hieter, C. Mann, M. Snyder, and R. David, *Cell* (Cambridge, Mass.) **40,** 381 (1985).

TABLE I
YEAST STRAINS USED IN YEAST ARTIFICIAL CHROMOSOME CLONING

Strain	Genotype							
AB1380	*MAT***a** Ψ^+	*ura3-52*	*trp1*	*ade2-1*	*can1-100*	*lys2-1*	*his5*	
YPH274	*MAT***a** Ψ^-	*ura3-52*	*trp1-Δ1*	*lys2-801*	*ade2-101*	*his3-Δ200*	*leu2-Δ1*	
	*MAT*α Ψ^-	*ura3-52*	*trp1-Δ1*	*lys2-801*	*ade2-101*	*his3-Δ200*	*leu2-Δ1*	
YPH252	*MAT*α Ψ^-	*ura3-52*	*trp1-Δ1*	*lys2-801*	*ade2-101*	*his3-Δ200*	*leu2-Δ1*	
CGY2516	*MAT***a** Ψ^-	*ura3-52*	*trp1-Δ63*	*lys2-Δ202*	*his3-Δ200*	*leu2-Δ1*		
YPH857	*MAT*α Ψ^-	*ura3-52*	*trp1-Δ63*	*lys2-801*	*ade2-101*	*his3-Δ200*	*leu2-Δ1*	*cyh2*^{R}

and *trp1* markers that are complemented during YAC construction. YPH274, YPH252, YPH857, and CGY2516 contain two additional nonreverting auxotrophic markers, *his3-Δ200* and *leu2*-Δ1. These deletions were introduced into yeast via two-step gene replacement.[12] The *his3-Δ200* marker is particularly useful because it deletes from the genome the entire *HIS3* DNA segment used in vectors. Thus, a plasmid containing a *HIS3* gene that complements this mutation cannot undergo homologous recombination with the genomic *HIS3* locus. All five yeast strains also carry an *ade2*-ochre allele, which is responsible for red pigment accumulation in colonies and is suppressible to yield a pink or white colony color phenotype by *SUP11* in a dosage-dependent manner. YPH857 is a haploid strain that contains the *cyh2*^{R} allele, which encodes recessive drug resistance for cycloheximide. The *cyh2*$^+$ gene can be selected against in this genetic background. YPH857 will therefore be useful in facilitating efficient two-step gene replacement procedures within YAC inserts.

Integrative Selection of Species-Specific Yeast Artificial Chromosomes

Principle

Free ends of DNA are highly recombinogenic in yeast and associate with homologous sequences on transformation. When both ends of a linear DNA are homologous to a sequence on the same chromosome, the linear DNA will integrate, causing an insertion if the free ends recognize a contiguous sequence or a deletion if the sites of recombina-

[12] R. S. Sikorski and P. Hieter, *Genetics* **122,** 19 (1989).

tion are not adjacent (Fig. 2). Vectors that integrate stably are useful for introducing selectable markers into YACs. This property can be utilized to simplify construction of species-specific YAC libraries. Targeting homologous recombination of an "integrative selection vector" (ISV) to repetitive DNA elements found, e.g., in human but not rodent DNA permits selective growth of human-derived YACs containing the auxotrophic marker. The use of ISVs can enrich by 50-fold the number of human-derived YACs in a library made from a human × rodent somatic cell hybrid, eliminating several labor-intensive steps in the normal screening procedure.[2]

LINE and *Alu* elements are among the most numerous repetitive DNA sequences in the human genome, and are found in about 10^5 and 10^6 copies, respectively. Individual elements of both families are somewhat heterogeneous. Sequences of individual *Alu* elements differ by as much as 30% from a canonical *Alu* sequence.[13] This divergence and the small size of the elements (ca. 300 bp) provide a relatively low efficiency target for homologous recombination, and care must be taken in vector design and in the choice of host strains to keep background at an acceptably low level. A strong selectable marker must be used that has a very low reversion frequency. The vectors must be free of any *ARS* activity that would allow them to replicate autonomously, thus complementing an auxotrophy in cells that have not undergone targeted transformation. Homologous segments other than the targeting sequence, e.g., plasmid sequences, must be eliminated or positioned such that recombination with a YAC arm will not introduce the selectable marker.

Procedure

Integrative selection vector pBP90 targets integration to human LINE elements, while pBP96 and 97 target *Alu* segments (Fig. 2). All three ISVs contain the yeast *HIS3* gene embedded in the repetitive targeting element (targeting cassette). The vector is cleaved with one or two restriction endonucleases that excise plasmid sequences away from the targeting cassette. It is not necessary to separate these products further because the plasmid fragment does not contain a yeast selectable marker.

One method for making chromosome-specific YAC libraries relies on the use of DNA from a somatic cell hybrid containing a single human chromosome. Typically, the human chromosome represents about 1–2% of the DNA of the hybrid cell, so about 99% of these YACs will derive from the rodent parent. Human YACs can be isolated by picking each

[13] P. L. Deininger, D. J. Holly, C. M. Rubin, T. Friedman, and C. W. Schmidt, *J. Mol. Biol.* **151,** 17 (1981).

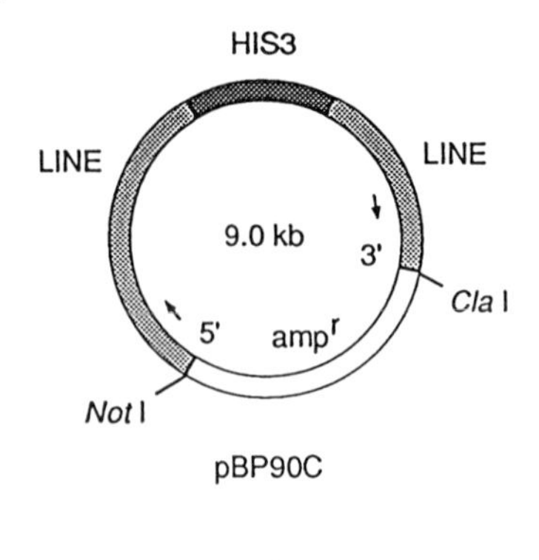

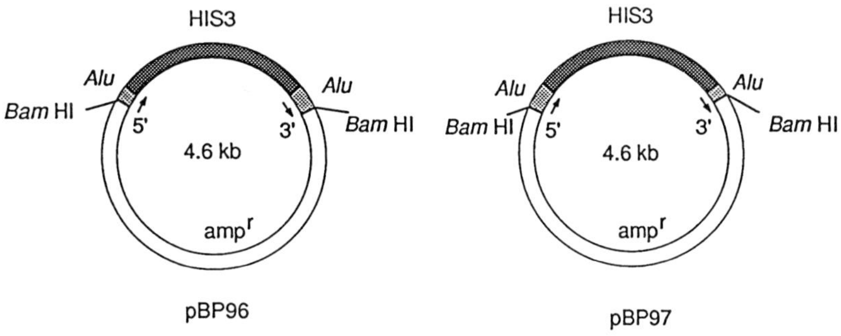

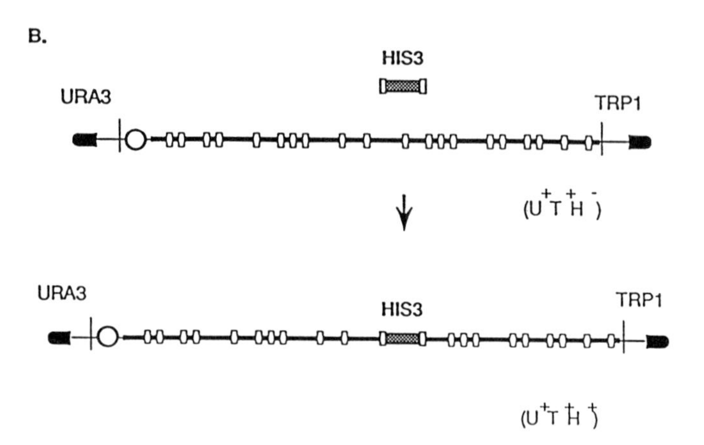

FIG. 2. Integrative selection. (A) Integrative selection vectors target homologous recombination to human LINE (pBP90) or *Alu* (pBP96 and 97) repetitive elements, thereby introducing the *HIS3* selectable marker into the YAC. (B) Linearized ISVs are introduced into cells by lithium acetate transformation. The free ends target homologous recombination to repetitive elements found in human but not rodent DNA, allowing these marked YACs to be selected by growth in medium lacking histidine.

colony onto a grid for two rounds of colony screening and purification using a human repetitive DNA probe. This step is simplified by enriching for human-derived YACs using integrative selection. A transformation plate typically contains 500–1000 colonies. The top agar containing the colonies is removed and macerated in minimal medium (SD-Ura-Trp, see below) and expanded overnight. Glycerol stocks are made at this point and the entire library is stored as pools of 500–1000 colonies. This mixed culture is expanded and the linearized ISVs are introduced by any convenient method; a protocol for lithium transformation[14] is described below. Transformants are selected on minimal agar plates lacking histidine. Approximately 50% of the YAC-containing strains recovered by this procedure have human DNA-derived YACs. Because these colonies grow on the surface of the plate, they can be replica plated to produce filters for parallel hybridizations with total DNA from human and the appropriate rodent.

Both LINE- and *Alu*-targeted selection are biased toward isolation of large YACs that contain more targets for homologous recombination, and are more likely to contain repetitive elements that are highly homologous to the targeting sequence. Thus, a nonrandom representation of segments will be obtained from a total digest library. In a partial digest library, every sequence is equally likely to occur in a large segment, and integrative selection will bias the library such that each gene is more likely to be cloned as a large YAC.

The insertion of the ISV introduces an artifact into the cloned segment. Other artifacts can be generated by this procedure as well. If opposite ends of the ISV recombine with two different *Alu* sequences instead of into the same element, the intervening region will be deleted. About 10–20% of YACs selected with pBP96 and pBP97 were found to be deleted.[2] While this should have little impact on the use of these YACs in mapping and contig building procedures, the impact on biological analysis may be significant if a gene is disrupted by integration or deletion. In this case, it would be desirable to isolate the parental YAC that does not contain the ISV from the original small pool by any convenient method. No deletions have been seen after selection with pBP90.

Reagents

(Recipes for basic media and procedures required for growth and selection of yeast can be found in Rose *et al.*[15])

[14] H. Ito, Y. Fukuda, K. Murata, and A. Kimura, *J. Bacteriol.* **153,** 163 (1983).

[15] M. Rose, F. Winston, and P. Hieter, "Methods in Yeast Genetics: A Course Manual." Cold Spring Harbor Laboratory, Cold Spring Harbor, New York, 1990.

Defined minimal medium (SD): 0.67% (w/v) Bacto-yeast nitrogen base without amino acids, 2% (w/v) dextrose, 2% (w/v) Bacto agar, 20 μg/ml histidine (His), 20 μg/ml uracil (Ura), 40 μg/ml lysine (Lys), 30 μg/ml tryptophan (Trp), 30 μg/ml leucine (Leu), 6 μg/ml adenine (Ade; limiting adenine facilitates colony color development)

Selection screening medium: As above but lacking His, His and Ura, His and Trp, or Ura and Trp (referred to herein as, e.g., SD-Ura-Trp)

Lithium acetate: 0.1 *M* dissolved in TE

TE: 10 m*M* Tris-HCl, pH 7.5, 1 m*M* ethylenediaminetetraacetic acid (EDTA)

1. Yeast artificial chromosomes are constructed using DNA from a somatic cell hybrid and transformed into spheroplasts, which are allowed to recover in top agar.
2. Peel top agar from a 100-mm^2 transformation plate containing 500–1000 colonies, place in 50 ml of sterile SD-Ura-Trp in a small flask, macerate to release yeast, and grow at 30° until stationary (usually overnight).
3. Make glycerol stocks by adding 0.85 ml of culture to 0.15 ml of sterile glycerol. Store at −80°.
4. Innoculate 50 ml of YPD medium with 5 ml of the fresh overnight culture. Grow to an OD_{600} of 1.0–2.0. *Note:* Unstable YACs may be lost from the population when cells are grown without selection in YPD.
5. Collect cells by centrifugation at 3500 rpm for 5 min at room temperature.
6. Resuspend cells in 10 ml lithium acetate; collect as in step 5.
7. Resuspend cells in 10 ml lithium acetate, incubate 1 hr at 30°.
8. Collect cells and resuspend in 0.5 ml lithium acetate.
9. Digest 10 μg of ISV plasmids BP90 or BP96 and 20 μg of BP97 with the appropriate restriction endonucleases. Concentrate by ethanol precipitation and resuspend in 10 μl of TE.
10. Aliquot 50 μl of competent cells into Eppendorf tubes and add linearized ISV DNA to each (10 or 20 μg in up to 10 μl). Carrier DNA is not needed in the transformation mix.
11. Incubate for 10 min at 30°.
12. Add 0.5 ml of 40% (w/v) polyethylene glycol 4000/10 m*M* Tris, pH 7.5. Vortex to resuspend cells completely.
13. Incubate for 1 hr at 30°.
14. Heat at 42° for 5 min.
15. Add 1 ml sterile water and mix.

16. Collect by spinning for 5 sec in a microcentrifuge at room temperature.
17. Resuspend thoroughly by vortexing in 1 ml sterile water.
18. Collect cells and resuspend in 100 μl sterile water.
19. Plate on SD-His defined medium. When colonies arise (1–2 days) replica plate onto SD-His-Ura-Trp and onto each of two nitrocellulose filters laid on these plates. Incubate overnight at 30°.
20. Prepare filters for hybridization as previously described.[15]
21. Hybridize replica filters with probes prepared from total human DNA and the appropriate rodent DNA to identify human-derived YACs. Expand for further analysis.

Comments

1. *Alu* and LINE profiles are used to confirm the human origin of selected YACs.

2. *Note* (regarding step 9): It is important to get a complete digest. Circular molecules can integrate into plasmid sequences in the YAC vector arms, producing a background of His^+ transformants that are not fragmented.

3. Transformation with less DNA (step 9) will decrease the number of transformants recovered, but increase the percentage that are human derived.

4. Small deletions have been observed in 10–20% of the products of integrative selection when using *Alu*-targeting vectors.[3] These are readily detectable by loss of fragments from the *Alu* profile. No deletions have been detected with LINE-targeting ISVs.

Fragmentation of Yeast Artificial Chromosomes

Principle

Fragmentation of YACs is analogous to the chromosome fragmentation procedure used to physically map genes in yeast[16,17] (Fig. 3A). Chromosome fragmentation vectors (CFVs) include a targeting sequence that can be any DNA with homology to the YAC (Fig. 3B). *Alu* elements occur about every 4 kb in the human genome, so virtually any YAC-sized segment of DNA will contain multiple copies. LINE elements provide less

[16] D. Vollrath, R. Davis, C. Connelly, and P. Hieter, *Proc. Natl. Acad. Sci. U.S.A.* **85,** 6027 (1988).

[17] S. Gerring, C. Connelly, and P. Hieter, this series, Vol. 194, p. 57.

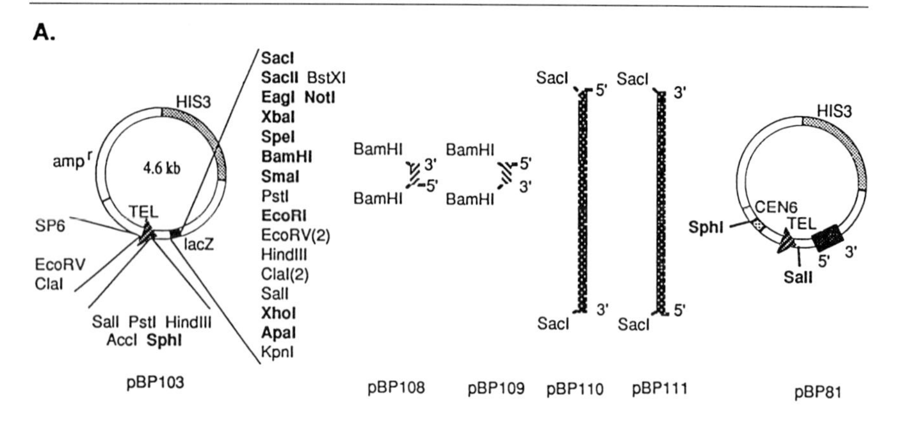

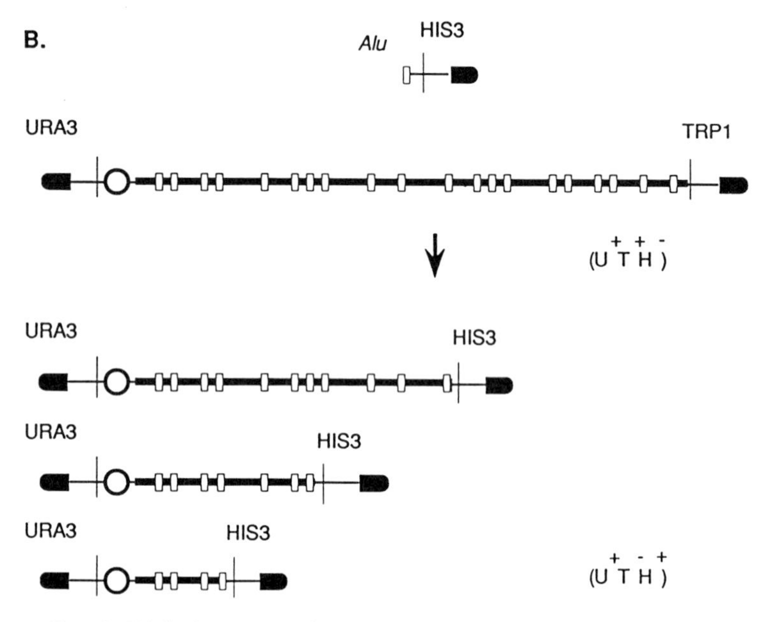

FIG. 3. YAC chromosome fragmentation vectors. (A) Generic CFV pBP103 contains several unique sites for introduction of any targeting sequence. pBP108 and 109 target human *Alu* elements in either orientation, pBP110 and 111 target LINE elements, and pBP81 is an *Alu*-targeting, centric vector. The vectors may be introduced with any restriction endonuclease that cuts between the targeting element and the telomere without disrupting either (e.g., pBP108, which contains the *Alu*-targeting element in the *Bam*HI site of the polylinker, could be linearized with *Sal*I, *Sph*I, *Xho*I, etc.). (B) Fragmentation paradigm. A linearized acentric CFV undergoes homologous recombination at the targeting site, introducing a telomere and deleting all sequences distal to the site. Deletion derivatives are selected based on acquisition of *HIS3* and screened for loss of *TRP1*.

frequent targets. Yeast artificial chromosome CFVs of the pBP series contain the *HIS3* gene as a selectable marker, bacterial *amp* and *ori* sequences for propagation and selection of the plasmid, and a minimal telomere segment[18] (Fig. 3B). They are constructed with or without a centromere to permit recovery of derivatives deleted from either end.

The size of fragmentation derivatives is determined using pulsed-field gel electrophoresis (PFGE). The difference in size between the derivative and the parental YAC defines the recombination site. Conventional Southern analysis using an *Alu* probe can be used to prepare an "*Alu* profile" to verify progressive loss of markers in the deletion series. Positions of cloned genes, exons, or other DNA segments can be determined using Southern blotting to identify the shortest deletion derivative on which they are retained. Restriction mapping is facilitated because a restriction fragment will drop out from progressively shorter derivatives as a function of its position on the YAC.

Procedure

Alu fragmentation is used as an example. The pBP108 and pBP109 vectors are linearized with *Sal*I, the recognition sites for which occur between *Alu* and the Y′ element.[18] Because the orientation of *Alu* segments within the YAC is random, both vectors are typically employed in parallel transformations. As with ISVs, any low- to moderate-efficiency transformation protocol can be used to introduce the linearized CFV into the cell. Once in the cell, the free ends of the CFV have different fates. The telomere sequence presumably serves as a substrate for telomerase; if a Y′ element is used, the end of the YAC obtains a telomere by homologous recombination with endogenous chromosomes. The other end of the vector is highly recombinogenic and interacts directly with an *Alu* element on the YAC. This recombination event introduces the telomere into the YAC, deleting all sequences beyond this point. When an acentric CFV is used (e.g., pBP108), the recombination product could include sequences distal to the targeted *Alu* segment (and be acentric) or sequences proximal to the targeted *Alu* segment, as depicted in Fig. 3 (and be monocentric), depending on the orientation of the targeted *Alu* segment on the YAC. Similarly, a centric CFV (e.g., pBP81) can yield either a monocentric or dicentric deletion derivative. Only monocentric YAC deletion derivatives will be mitotically stable. In some instances, "reduplication" will result (i.e., the target YAC will be maintained), yielding a cell that contains both a YAC deletion derivative and the parental YAC; the "rules" describing

[18] W. J. Pavan, P. Hieter, D. Sears, A. Burkhoff, and R. H. Reeves, *Gene* **106,** 125 (1991).

the likelihood of this event may be different for YACs than for metacentric yeast chromosomes, in which centric fragmentation almost always (>90%) results in retention of the parental YAC and acentric fragmentation almost always (>95%) results in loss of the parental YAC.[19]

In both centric and acentric fragmentation, homologous recombination often results in deletion of a portion of the parental YAC, and His$^+$ transformants can be screened for the loss of the appropriate marker from the deleted YAC. Between 50 and 75% of His$^+$ transformants obtained with BP108 and 109 contain fragmented YACs. The remainder appear to be His$^+$ as a result of integration of the CFV into the YAC vector arm. The pJS97 and pJS98 YAC cloning vectors include on the centric arm a *SUP11* suppressor tRNA gene, which complements the *ade2-101* mutation in the YPH host strains. Growth on limiting adenine of diploid cells containing a single YAC made with these vectors will produce pink colonies, while loss of *SUP11* (e.g., after fragmentation with a centric CFV) will result in red colonies. Thus the desired Ura$^-$ class of transformants can be identified as red His$^+$ Trp$^+$ colonies, eliminating the need for a replica plating step onto SD-Ura plates.

Reagents

The same reagents are required as for integrative selection (see above).

1. Streak the yeast strain containing the target YAC on a selective plate (SD-Ura-Trp) and culture at 30°. When colonies arise (2–3 days), use them to inoculate 8 ml of YPD liquid medium and grow overnight, then use the fresh overnight culture to innoculate 50 ml of YPD. Grow to an OD_{600} of 1.0–2.0 with vigorous shaking (approximately 4 hr at 30°). *Note:* Some YACs are lost at high frequency in the absence of selection, in which case SD-Ura-Trp medium must be used. This will usually result in slower growth.

2. Proceed with transformation (steps 5–18 above). In step 9, digest 5–10 μg of CFVs pBP103 or BP108–111 with *Xho*I to linearize between the targeting segment and the telomere.

3. Plate on SD-His and incubate at 30° for 3–4 days. At this point, colonies may be screened for loss of one YAC arm and retention of the other by streaking each colony in parallel on $\frac{1}{8}$ of plates containing SD-His, SD-His-Ura, and SD-His-Trp. Incubate 2–3 days at 30°.

4. Prepare high molecular weight DNA in plugs and run on a pulsed-field gel apparatus to produce an electrophoretic karyotype. Include yeast with and without the YAC as standards. Determine the sizes of the frag-

[19] P. Hieter and C. Connelly, unpublished observations, 1986.

ments by comparison with a ladder or with the sizes of the chromosomes of the mapping strain.

5. Transfer the DNA to nitrocellulose by blotting and hybridize with a *HIS3* probe. Strip filter and reprobe sequentially with *URA3* or *TRP1*, then *Alu*.

6. To confirm that the expected directional deletions have occurred, a conventional Southern blot of strains containing deletion derivatives can be probed with *Alu*. Progressively shorter YACs will have increasingly simple patterns that are subsets of larger derivatives. Note that a "new" *Alu*-containing fragment, representing the CFV sequence, will occur in each deletion derivative pattern.

7. To localize a gene on the YAC, hybridize a probe (or, preferably, 5'- and 3'-specific probes) to a panel of consecutively shorter deletion derivatives. The probe sequence is located physically between the shortest derivative showing hybridization and the integration site of the CFV in the next longest derivative. Where a fragmentation event has occurred within the gene, the use of 5' and 3' probes will establish orientation of the gene relative to the ends of the YAC.

8. Restriction map construction is facilitated by use of deletion derivatives. Single yeast plugs are treated with 1 ml phenylmethylsulfonyl fluoride (PMSF; a 1000× solution is 0.174 g/10 ml ethanol. *CAUTION:* This solution is highly toxic! Use appropriate precautions.) Wash 1 hr, then three times for 1 hr with 1 ml TE, and two times for 1 hr with 1 ml of 1× restriction endonuclease buffer. One-quarter to one-half of a plug is transferred to an Eppendorf tube containing 200 μl of 1× restriction buffer plus 20–50 units of enzyme and incubated at 37° for 4–14 hr. The buffer is removed and the plug is melted and loaded onto a gel for PFGE and subsequently blotted. The restriction fragments can be visualized by probing with *Alu*, which may not recognize all fragments, or with a probe made from the entire YAC. Partial or double digests of smaller deletion derivatives can be helpful to determine fragment order.

Comments

1. Transformation frequencies will vary because different YACs contain different numbers of *Alu* and LINE sequences, and because the homology between the *Alu* segment on the CFV and those on an individual YAC will vary.

2. cDNAs could presumably be used as targeting elements to map exons within large genes.

3. An early YAC CFV, pBP62, utilized a telomere-adjacent Y' element rather than a minimal telomere, but the particular element was found to

contain weak *ARS* activity, requiring an extra screen for loss of the acentric arm marker to differentiate true fragmentation events from autonomously replicating episomal plasmid.[4] This *ARS* activity is eliminated in the pBP103 series.

4. Yeast artificial chromosomes cloned in AB1380 can be readily transferred to YPH strains (Table I) by direct transformation in the presence of polyamines.[20]

Modification of Yeast Artificial Chromosomes and Introduction into Cultured Cells

Principle

Homologous recombination can be used to introduce a mammalian selectable marker into YACs. The pBP47 integrating plasmid contains the yeast selectable marker, *HIS3*, a neomycin phosphotransferase (*neo*^{r}) gene controlled by a simian virus 40 (SV40) promoter that is not functional in yeast (Fig. 4A), and an *Alu* element to target homologous recombination. The integrating plasmid is inserted into the YAC by transforming yeast and recovering His$^+$ cells (Fig. 4B).

Yeast artificial chromosomes containing a mammalian selectable marker can be introduced into mammalian cells by virtually any DNA transfer technique, such as calcium phosphate-mediated transfection, lipofection, microinjection, or electroporation. The size of fragment transferred may be limited by the procedure chosen. Yeast artificial chromosomes have been recovered intact in mammalian cells after transfection (for very small YACs)[21] or by polyethylene glycol (PEG)-mediated fusion[3,5,22] (Fig. 5). Polyethylene glycol-mediated fusion between spheroplasts and target cells creates, in essence, a yeast × somatic cell hybrid that retains the chromosomes of the mammalian parent and apparently random subsets of the 16 yeast chromosomes. The parental cells are unable to grow in medium containing the drug G418. Unfused spheroplasts are very fragile, and we detect no recovery or growth in standard tissue culture medium at room temperature, 30 or 37°, even without PEG treatment. Further, the SV40 promoter is inactive in yeast, and yeast containing a *neo*^{r}-tagged YAC are unable to grow in YPD supplemented with 100

[20] C. Connelly, M. McCormick, J. Shero, and P. Hieter, *Genomics* **10,** 10 (1991).

[21] M. D'Urso, I. Zucchi, A.Ciccodicola, G. Palmieri, F. Abidi, and D. Schessinger, *Genomics* **7,** 531 (1990).

[22] A. Gnirke, T. Barnes, D. Patterson, D. Schild, T. Featherstone, and M. Olsen, *EMBO J.* **10,** 1629.24 (1991).

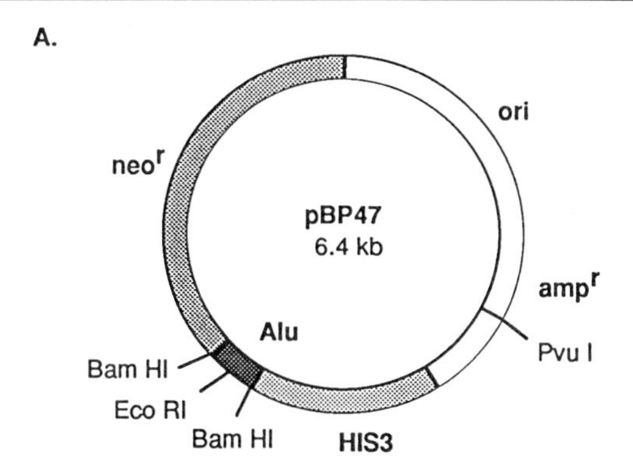

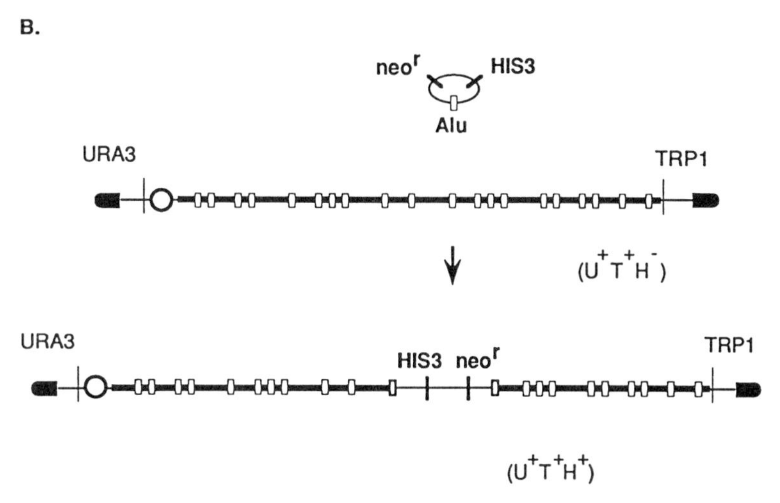

FIG. 4. (A) Integrating plasmid pBP47 contains a unique *Eco*RI site within a human *Alu* sequence to target homologous recombination of the linearized molecule into any human-derived YAC, thereby introducing the *neo*^{r} cassette into the YAC. The products of this integration are selected based on acquisition of *HIS3*. (B) Integration of pBP47 occurs at *Alu* segments. Integrated DNA includes the *HIS3* and *neo*^{r} markers as well as plasmid sequences.

μg/ml G418. (Nonetheless, hybrid cultures should be checked to assure that no yeast are present; see Comments.)

The presence of YAC-derived DNA in the hybrid can be assayed by Southern blotting with any appropriate probe from the YAC insert. Where a human YAC has been introduced into a rodent cell line, an *Alu* profile can be used to estimate the portion of the YAC that is stably introduced into the cell. Pulsed-field gel electrophoresis is useful to assess the size of the incorporated segment, but methylation in the mammalian cell may change fragment sizes. Colonies that grow normally in the presence of

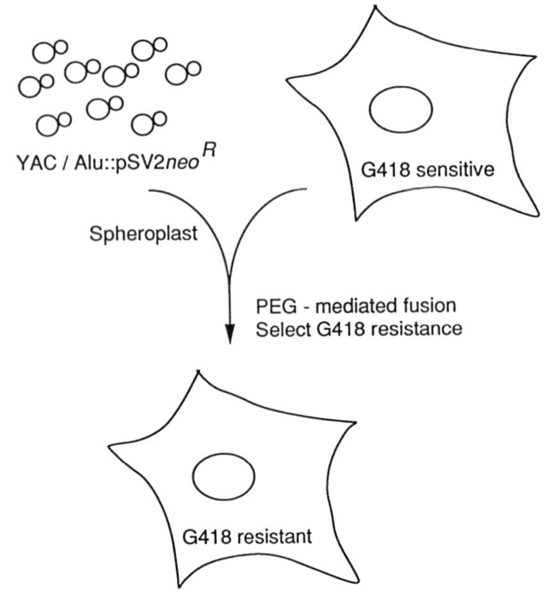

FIG. 5. Scheme for introduction of YACs into cultured cells via cell fusion. Yeast cells containing a YAC modified by insertion of a neomycin resistance marker are spheroplasted and fused to cultured cells in the presence of PEG. Neomycin-resistant cell lines are picked and expanded for analysis.

G418 typically contain YACs integrated into the host chromosomes. This is assessed most directly by *in situ* hybridization. Integration can also be inferred by the presence of junction fragments hybridizing to vector sequences on the ends of the YAC, and by retention of G418 resistance for 20–40 cell doublings in the absence of selection.

Procedure

The fusion conditions described below have proven useful for making hybrids between spheroplasts and the murine Bls-2 embryonal carcinoma cell line. Polyethylene glycol is toxic to cells, and some lines are more sensitive than others. The best results will be obtained when fusion conditions (percentage PEG, time of exposure, type of medium, etc.) are optimal for fusion and survival of the mammalian cell parent, and may vary for different lines. Following fusion, hybrid cells are selected in 400 μg/ml G418 for 10–14 days, the colonies are isolated as subclones, and expanded for analysis. The presence of residual low levels of yeast in the culture would give false-positive results to DNA analyses. Culture medium collected from expanded clones 2–3 days after refeeding can be centrifuged at 3500 rpm for 10 min, resuspended in a minimal volume of YPD, and streaked onto selective plates to assure that the recombinant yeast is not present in the culture.

Reagents

Appropriate cell culture medium: For purposes of this example, it is assumed that cells are grown in Dulbecco's modified Eagle's medium supplemented with 10% (v/v) fetal calf serum and antibiotics (MEM). Medium without serum or antibiotics is also required (MEMss)

Polyethylene glycol solution: PEG 1000 from Baker Chemical (Phillipsburg, NJ) is melted at 65° and mixed with MEMss to give the appropriate concentration (usually 40–50%, w/v). Adjust pH to 7.4 and sterilize by passage through a 0.22-μm filter

SCE: 1 *M* sorbitol, 100 m*M* sodium citrate, pH 5.8, 10 m*M* EDTA, 30 m*M* 2-mercaptoethanol

ST: 1 *M* sorbitol, 10 m*M* Tris, pH 7.5

Targeted Recombination of pPB47 into Human-Derived Yeast Artificial Chromosomes

The target YAC must be in a *HIS3*-deleted yeast strain. Digest 10 μg of the pBP47 plasmid with *Eco*RI and transform into the target yeast with lithium acetate as described above. Select His^+ colonies, prepare DNA, and analyze by hybridization of a neo^r cassette probe to Southern blots. *Bam*HI digestion will produce a 6.1-kb fragment for *Alu*-targeted integration events.

Fusion of Yeast to Cultured Cells

Yeast

1. Inoculate 5 ml of a fresh overnight culture of yeast containing the YAC into 50 ml of YPD and grow to an OD_{600} of 1.0.

2. Collect yeast cells by centrifugation and wash twice in 20 ml of 1 *M* sorbitol. Resuspend in 20 ml of SCE.

3. Digest cell walls by adding 35 μl of zymolyase 20T, 10 mg/ml, and incubating at 30° until 90% of the cells are spheroplasts. *Notes:* (a) To test for spheroplasting, mix 50 μl yeast and 50 μl 10% (w/v) sodium dodecyl sulfate (SDS) on a microscope slide. Spheroplasts lyse and appear gray under phase contrast, intact cells are refractile. (b) Quantities of zymolyase are approximate. It is useful to titrate the enzyme prior to spheroplasting and to use an amount that will produce 90% spheroplasts in 20–30 min. From this point, spheroplasts must be handled very gently.

4. Pellet spheroplasts at 100 *g* for 10 min. Wash twice in 20 ml ST. Resuspend in 10 ml ST, count, and hold on ice.

Mammalian Cells

5. Actively growing cultured cells are passed 1 day before the fusion to achieve 70–80% confluence on the day of fusion. Collect cells by trypsinization, resuspend in MEMss, and count. *Note:* Culture conditions for optimal fusion will vary for different cell lines. For best results, consult a somatic cell geneticist familiar with the target cells.

6. For each fusion, pellet 2×10^6 cells in a 15-ml polystyrene culture tube, decant medium, and add 2 to 4×10^7 yeast in 1 ml of ST. Pellet yeast onto cells at 100 *g* for 5 min.

7. Decant medium and resuspend gently in 0.2 ml of 50% (v/v) MEMss/ST at 20°.

8. Add 2 ml of 40% (w/v) PEG solution and pipette once very gently to mix.

9. After 30, 60, or 90 sec add 5 ml of MEMss and mix by inverting gently twice. Collect cells by centrifugation at 100 *g* for 5 min at 20°.

10. Resuspend cells in MEM and plate at 1 to 5×10^5 cells/100-mm^2 dish. (Rapidly dividing cells require a higher dilution.)

11. Twenty-four hours after fusion, remove the medium, wash twice with phosphate-buffered saline (PBS), and refeed with MEM plus an appropriate amount of G418 (typically 400–800 μg/ml). Refeed as necessary.

12. Select colonies by trypsinizing in cloning cylinders after 10–14 days.

13. Expand and check cultures for low-level yeast contamination.

14. Prepare DNA for determination of *Alu* profile and hybridization with a *neo*^{r} probe. Hybridize with markers from the ends of the YAC to identify changes in vector fragment sizes consistent with the occurrence of junction fragments in the hybrid genome. Analyze DNA with a yeast repetitive marker such as Ty or 2μm plasmid to estimate the amount of yeast genetic material present in the hybrid.

15. Prepare high molecular weight DNA to assess the structure of the YAC. Note that the mammalian genome is 200 times larger than the yeast genome. Therefore, 200 times as much mouse DNA is required to give the same hybridization signal. These disparate amounts of DNA will result in different mobilities of the same size fragments in the gel. For best results, mix an appropriate amount of DNA from the yeast containing the YAC with control mouse genomic DNA.

16. Determine whether the YAC is integrated by *in situ* hybridization using gel-purified YAC as probe.

17. Assess stability of the culture by growing without selection for 20

generations, then plating duplicate plates with and without G418. Count after several days to determine whether the same number of cells are surviving in each culture.

Comments

1. Among the G418-resistant products of fusion are small colonies that persist well past the time that control cultures have been killed by the drug, but which never increase in size. These may represent lines in which the YAC (or a portion of it that contains the *neo* cassette) remains as an episome that is not evenly distributed to daughter cells at mitosis.[23]

2. Among the products of yeast × mammalian cell fusion, G418-resistant colonies have been isolated that contain no DNA hybridizing to the *neo*^{r} cassette.[3,5,23] The basis for this phenomenon is not currently known.

[23] R. Allshire, G. Cranston, J. Gosden, J. Maule, N. Hastie, and P. Fantes, *Cell* (Cambridge, Mass.) **50,** 391.

[50] Copy Number Amplification of Yeast Artificial Chromosomes

By DOUGLAS R. SMITH, ADRIENNE P. SMYTH, and DONALD T. MOIR

Yeast artificial chromosome (YAC) cloning is a useful method for propagating very large DNA segments, from 100 kilobases (kb) to over 1 megabase (Mb) in size. The first efficient YAC cloning system was described by Burke *et al.* in 1987[1] and several additional vectors with various modifications have been constructed (reviewed by Hieter *et al.*[2]). However, artificial chromosomes are normally present in yeast at only one copy per haploid genome, so a number of routine applications (such as colony-screening and preparative-scale DNA isolation) are either inconvenient to perform or are plagued with problems of low sensitivity. We have developed a system that permits copy number amplification of YACs by introducing certain cis-acting elements into the cloning vector, pCGS966.[3] In this chapter we describe specific methods for generating YAC clones and for amplifying the artificial chromosomes they contain.

[1] D. T. Burke, G. F. Carle, and M. V. Olson, *Science* **236,** 806 (1986).

[2] P. Hieter, C. Connelly, J. Shero, M. K. McCormick, S. Antonarakis, W. Pavan, and R. Reeves, *in* "Genome Analysis Volume 1: Genetic and Physical Mapping" (K. Davies and S. Tillman, eds.), p. 83, Cold Spring Harbor Laboratory, Cold Spring Harbor, New York, 1991.

[3] D. R. Smith, A. P. Smyth, and D. T. Moir, *Proc. Natl. Acad Sci. U.S.A.* **87,** 8242 (1990).

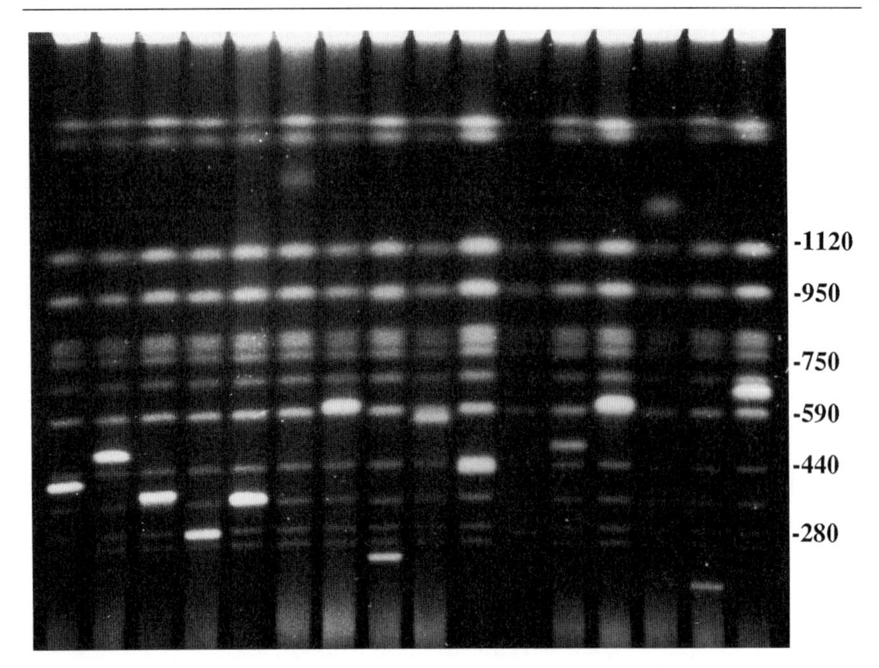

FIG. 1. Copy number amplification of random YACs. Sixteen clones with single visible YACs (as determined by ethidium bromide staining of unamplified chromosomal DNAs or by hybridization with the *Alu*-repeat probe blur-8) were selected at random and grown in S4/M10 for 3 days. The YACs (220–1300 kb) were derived from a transformation experiment done in the presence of 0.7 m*M* spermidine/0.3 m*M* spermine [C. Connelly, M. K. McCormick, J. Shero, and P. Heiter, *Genomics* **10,** 10 (1989)] that produced an average insert size of about 650 kb.

Principles

The elements necessary for YAC amplification are a conditional centromere that can be inactivated under defined conditions, and a selectable marker that confers a growth advantage to cells containing multiple copies of the YAC. The susceptibility of yeast centromeres to inactivation by transcriptional interference has been known for some time and has been extensively studied.[4–6] The YAC vectors we have developed contain a *CEN4*/*GAL1* promoter fusion (see Fig. 1) in which the centromere appears to function normally during growth on glucose (*GAL1* transcription re-

[4] L. Panzeri, I. Groth-Clausen, J. Shepard, A. Stotz, and P. Philippsen, *Chromosomes Today* **8,** 46 (1984).

[5] E. Chlebowicz-Sledziewska and A. Z. Sledziewski, *Gene* **39,** 25 (1985).

[6] A. Hill and K. Bloom, *Mol. Cell. Biol.* **7,** 2397 (1987).

pressed), or is inactivated during growth on galactose (high-level *GAL1* transcription). When the centromere on a YAC is inactivated, some cells may acquire additional copies of the YAC through missegregation. Selection for an appropriate marker on the vector enriches the population for cells containing multiple copies.

A number of selectable markers can be used to drive up copy number in this way. Thymidine kinase (TK) was chosen for pCGS966 because it responds to a wide concentration range of selective agents and because the selective conditions are unlikely to generate host mutants (yeast does not contain a thymidine kinase). Selection for the *TK* gene is accomplished by adding exogenous thymidine in the presence of methotrexate and sulfanilamide. The latter two compounds inhibit enzymes involved in the recycling or *de novo* synthesis (respectively) of folate cofactors required for the synthesis of deoxythymidylic acid (dTMP).[7] Yeast artificial chromosomes containing human DNA segments ranging from about 100 to 600 kb in size could be readily amplified 10 to 20-fold using this system.

Basic Procedures

DNA Preparation

Large-scale preparations of vector DNA [from 1 to 4 liter of CGE1638 (XL1/pCGS966)[3] cells grown in L-broth[8] with 100–200 μg/ml ampicillin] were performed by standard alkaline lysis procedures.[8] Plasmid yields are very good (2–4 mg/liter), possibly a result of the duplicate replication origins. Although the host strain (XL1) is recombination deficient, occasional isolates of the plasmid may develop small deletions. Therefore it is advisable to check a few minipreparations of colony-purified isolates for *Eco*RI cutting and characteristic *Hin*dIII fragments before doing a large-scale preparation. The *Hin*dIII fragment sizes are (in kilobases) 5.5, 5.0, 3.1 (includes the cloning site), 2.6, 1.7, 1.4 (doublet; includes telomeres), and 0.2.

High molecular weight DNA was prepared from cell lines embedded in agarose according to published protocols.[9] To facilitate various manipulations, cells were embedded in "miniblocks" at a final concentration of

[7] E. W. Jones and G. R. Fink, *in* "Molecular Biology of the Yeast *Saccharomyces*" (J. Strathern, E. W. Jones, and J. R. Broach, eds.), p. 181, Cold Spring Harbor Laboratory, Cold Spring Harbor, New York, 1982.

[8] F. M. Ausubel, R. Brent, R. E. Kingston, D. D. Moore, J. G. Seidman, J. A. Smith, and K. Struhl (eds.), *in* "Current Protocols in Molecular Biology," Vol. 1. Wiley, New York, 1989.

[9] D. R. Smith, *Methods Companion Methods Enzymol.* **1,** 195 (1990).

5×10^7 cells/ml. Miniblocks are made by cutting up 1-mm-thick slabs of molded agarose (containing cells) with a razor blade to form 1-mm cubes. They offer all of the advantages of microbeads,[10] including reduced equilibration times and pipettability, but are easier to prepare and do not suffer from the variability and low yields typical of microbead preparations. Intact chromosomal DNA from yeast (including YAC clones) was prepared by a modified lithium dodecyl sulfate (LiDS) method.[9]

Yeast Strains

Many laboratory *Saccharomyces cerevisiae* strains do not grow well on galactose. This includes the standard host AB1380 (*MAT***a** *ura3-52 trp1 ade2-1 lys2-1 his5 can1-100*).[1] Because cells must be grown on galactose for YAC amplification, and maximal induction of the *GAL1* promoter is desirable, we constructed several new strains especially for cloning in amplifiable vectors. One of these, CGY2516 (*MAT***a** *Gal*$^+$ *ura3-52 trp1-Δ63 leu2-Δ1 lys2-Δ202 his3-Δ200*), has been described previously.[3] Another, CGY2527 (*MATα Gal*$^+$ *ura3-52 trp1-Δ63 leu2 lys2-801 his3-Δ200*) was a product of the same set of crosses, and is useful for generating diploids with AB1380-YAC strains selecting for growth without histidine by complementation of HIS3 and HIS5). However, these strains have a tougher cell wall and smaller cell size than AB1380, and do not transform as efficiently. The following strains (details of construction to be published separately) overcome these limitations: CGY2557 (*MATα Gal*$^+$ *ura3-52 trp1 ade2-1 lys2-1 his5 leu2 can1-100*) and CGY2570 (*MAT***a** *Gal*$^+$ *ura3-52 trp1-Δ63 leu2-Δ1 lys2-Δ202 his3-Δ200 ade2-1*).

Gel Electrophoresis and Nucleic Acid Hybridization

Gels were run in 50 m*M* Tris–borate, 1.25 m*M* ethylenediaminetetraacetic acid (EDTA) buffer. Pulsed-field gel electrophoresis for routine analysis of YAC clones was performed with 0.8% (w/v) SeaKem GTG agarose (FMC, Rockland, ME) at 6.7 V/cm at 12° with a 50-sec switch time as described.[9] Preparative pulsed-field gels for purification of YAC vector ligation mixtures contained 0.8% (w/v) SeaPlaque GTG agarose (FMC) and were run for 18–48 hr at 3.3 V/cm and 12° with a 20-sec switch time. Southern blots were prepared and hybridized as described.[9] Probes were labeled by using a random 9-mer labeling kit (Prime-it; Stratagene, La Jolla, CA).

[10] T. Imai and M. V. Olsen, *Genomics* **8,** 297 (1990).

Artificial Chromosome Construction

Large DNA molecules can be easily broken during preparation and handling. However, this can be minimized by using "in-gel" preparation techniques to perform most of the necessary size fractionation, purification, and buffer exchange steps of the cloning protocol.[10-13] This has two benefits: it results in larger YACs, and it efficiently removes religated vector arms (via size fractionation) that otherwise generate a background of clones without inserts. The amplifiable YAC vectors described here do not contain a color indicator system to reveal clones without inserts (the interrupted *SUP4* cloning site of the original pYAC vector series was eliminated due to size constraints of the T3 and T7 promoters). Therefore, gel-purification methods are strongly recommended.

Preparation of Vector DNA

Vector DNA (500 μg) was digested simultaneously with *Eco*RI and *Bam*HI [1-hr digestion with 500 units (U) of each enzyme in 50 m*M* NaCl, 100 m*M* Tris-HCl, pH 7.5, 5 m*M* $MgCl_2$], extracted with phenol (after adding EDTA to 20 m*M* and sodium acetate to 0.3 *M*; phenol prepared as described[8]) followed by chloroform–octanol (24 : 1), and precipitated with 2-propanol. The DNA was then resuspended in 50 m*M* Tris-HCl, pH 8.5, 0.1 m*M* EDTA and treated for 30 min with 5 U calf intestinal phosphatase (Boehringer Mannheim, Indianapolis, IN); ethylene glycol-bis(β-aminoethyl ether)-*N*,*N*,*N'*,*N'*-tetraacetic acid (EGTA was added to 10 m*M*, sodium dodecyl sulfate (SDS) to 0.1% (w/v), sodium acetate to 0.3 *M*. The sample was heated to 68° for 20 min, extracted with phenol, chloroform–octanol, precipitated with 2-propanol, and dissolved in 10 m*M* Tris-HCl, pH 7.5, 1 m*M* EDTA. Each vector preparation was tested to verify its inability to self-ligate and its ability to ligate to a test fragment quantitatively (i.e., *Eco*RI–cut pUC18 or pBR322 DNA).

Preparation of Target DNA

Target DNA for cloning was prepared by partial digestion using an *Eco*RI/*Eco*RI methylase competition reaction. The following conditions produced consistent partials with the majority of DNA in the 200 to 1000-kb size range. However, different DNA preparations behave somewhat differently in the reaction, so it is recommended that a range of *Eco*RI

[11] R. Anand, A. Villasante, and C. Tyler-Smith, *Nucleic Acids Res.* **17,** 3425 (1989).

[12] H. Albertsen, H. Abderrahim, H. M. Cann, J. Dausset, D. Le Paslier, and D. Cohen, *Proc. Natl. Acad. Sci. U.S.A.* **87,** 4256 (1990).

[13] Z. Larin, A. P. Monaco, and H. Lehrach, *Proc. Natl. Acad. Sci. U.S.A.* **88,** 4123 (1991).

concentrations be tested. Start with about 2 ml of settled miniblocks (approximately 500 μg of DNA). Equilibrate in 40 ml of 0.1 *M* NaCl, 0.1 *M* Tris-HCl, pH 7.5, 1 m*M* $MgCl_2$, 5 m*M* 2-mercaptoethanol with gentle mixing for at least 4 hr (at room temperature; overnight at 4°) with at least two buffer changes. Transfer the blocks to 10 ml of fresh buffer and add 5 μl of 20 m*M* *S*-adenosylmethionine, 20 U of *Eco*RI, and 200 U of *Eco*RI methylase. Incubate 16 hr at 37° with gentle agitation (slowly revolving tube rotator or equivalent).

At this stage, two options are available: the DNA can be ligated to the vector without further treatment (except for buffer equilibration and melting[3]), or it can be gel purified according to published protocols to remove cell debris and small fragments.[11-13] The inclusion of a gel-purification step results in larger insert sizes and may help avoid coligation events (by reducing the number of small fragments). However, it also reduces the cloning efficiency. The following steps can be performed either with miniblocks containing partial digests, or with a preparative gel slice.

Ligation of Vector and Target

Equilibrate the gel slice(s) containing about 100 μg of DNA (about a 1.5-ml volume) in 50 m*M* Tris–acetate, pH 7.6, 10 m*M* magnesium acetate, 50 m*M* sodium acetate, and melt at 68° for 5–10 min. These conditions provide sufficient salt to protect the DNA during the melting step without inhibiting the ligase significantly. Cool to 37° and add, with *very* gentle but thorough mixing, 100 μg of prepared vector DNA in 0.5 ml of prewarmed buffer (as above) including 400 μ*M* ATP, 40 m*M* dithiothreitol (DTT), and 4000 U of T4 DNA ligase. Incubate the reaction for at least 16 hr at room temperature (the reaction can be done at 37° to avoid a melting step, but the ligation is not as efficient and may not proceed to completion). Add 80 μl of 0.5 *M* EDTA (to produce a final concentration of about 20 m*M*), melt the sample at 68° (5 min), cool to 37°, and add proteinase K to 100 μg/ml. Incubate 30 min at 37° and load into the well(s) of a preparative gel (total width of sample tracks ≈ 10 cm).

After electrophoresis, locate and remove the focused high molecular weight band,[10] equilibrate the gel slice in 10 m*M* Tris-HCl, pH 7.5, 1 m*M* EDTA, 4 m*M* spermidine, and store at 4°.

Yeast Transformation

Warm the gel slice for 5 min at 50° and melt at 68° for 5 min. Cool to 40°, add agarase (Sigma, St. Louis, MO; 50 units/ml of agarose), and incubate for 1 hr. Cells for transformation can be prepared as described.[14]

[14] P. M. J. Burgers and K. J. Percival, *Anal. Biochem.* **163,** 391 (1987).

However, the following modifications have been observed to improve the transformation frequency: digest cell walls with "yeast lytic enzyme" (#190123; ICN, Costa Mesa, CA) in 1 *M* sorbitol, 20 m*M* potassium phosphate, 1 m*M* EDTA, pH 5.8, including 10 m*M* 2-mercaptoethanol; use polyethylene glycol (PEG) 10,000 (Sigma; batch tested for neutral pH) in the PEG solution. Check the pH of all solutions containing sorbitol (1 *M* sorbitol typically has a pH of 4.5 and should be adjusted to pH 6–6.5). Resuspended the cells at 1.5×10^9 cells/ml (twice the recommended concentration[14]) in STC (1 *M* sorbitol, 10 m*M* Tris-HCl, 10 m*M* $CaCl_2$, pH 7.5) prior to adding the DNA. After agarase digestion, gently mix the DNA samples with an equal volume of 2× STC. Finally, add 2 vol of cells to the mixture so that the final concentration of spermidine is 1 m*M*. This procedure was found to stimulate retransformation of YACs of various sizes contained in agarose-embedded total DNA from CGY2516 YAC clones (equilibrated as described in the previous section). Such "YAC DNA" samples transformed very efficiently (about 10^5 transformants/μg) and were sometimes used as transformation controls.

Copy Number Amplification

Media and Reagents

Amplification medium contains 0.67% (w/v) yeast nitrogen base, 1% (w/v) casamino acids, 3% (w/v) galactose, 50 μg/ml histidine, 50 μg/ml methionine, 100 μg/ml adenine, 0.8 mg/ml thymidine, and various amounts of sulfanilamide and methotrexate [see below; standard amounts are 1 mg/ml sulfanilamide and 10 μg/ml methotrexate (S1/M10)]. Extra histidine, methionine, and adenine are added because the selective conditions used for amplification also inhibit the biosynthetic pathways for these compounds. Solid media differ only in that 2% (w/v) agar is included. Sulfanilamide is made as a 200 mg/ml stock solution in *N,N'*-dimethylformamide; methotrexate is made as a 10 mg/ml stock solution in 50 m*M* NaOH; both solutions are stored at −20°. S-Glc and S-Gal media contain 0.67% (w/v) yeast nitrogen base, 1% (w/v) casamino acids, 50 μg/ml adenine, and 3% (w/v) glucose or galactose, respectively. YPD contains 1% (w/v) yeast extract, 2% (w/v) peptone, and 2% (w/v) glucose.[15]

Selective Growth

The selective conditions for copy number amplification are strongly inhibitory for cell growth. Typically, YAC clones require 3–6 days to

[15] F. Sherman, G. R. Fink, and J. B. Hicks, "Laboratory Course Manual for Methods in Yeast Genetics." Cold Spring Harbor Laboratory, Cold Spring Harbor, New York, 1986.

reach saturation in liquid medium under the above conditions, starting with about 10^5 cells/ml (normal growth in S-Gal takes 1 day). When cells were streaked out for single colonies on solid medium (S2/M20), they took about 7 days to form small colonies.

A minimum of about 7 generations of growth (1/100 subculture) was required for significant amplification. Small-scale amplifications were carried out by toothpicking clones directly from selective agar plates (lacking tryptophan and uracil) into 5 ml of amplification medium. For larger scale amplifications, about 10^7 cells (one 2-mm colony) were inoculated into 100 ml of amplification medium and grown for 5–7 days. To facilitate lysis, the cells were harvested by centrifugation and grown overnight in 100 ml of YPD.

Clones can also be amplified in microtiter wells. In this case, cells are grown in overnight in S-Glc medium and inoculated at a 1/100 dilution into 0.25 ml of amplification medium. The cells can be used after 3 days, or they can be diluted 1/100 into 0.25 ml of fresh medium and grown for an additional 4–6 days.

Some YACs amplify better than others. Generally, the smaller YACs (<400 kb) tend to amplify better than larger ones, but exceptions to this rule have been observed (Ref. 3 and Fig. 1). Growth in amplification medium renders yeast cell walls resistant to digestion with lytic enzymes. However, overnight growth in glucose-containing medium (S-Glc or YPD) following selection eliminates this problem without having major effects on YAC copy number.[3] During amplification, the cells undergo striking morphological changes, forming large clusters and chains. When subcultured in rich medium, the cells begin to divide faster and revert to a more normal morphology.

Gel Purification of Amplified Yeast Artificial Chromosomes

Yeast artificial chromosome molecules can be conveniently gel purified after large-scale amplification, subculturing in S-Glc, embedding in miniblocks, and treating with LiDS. The miniblocks are simply transferred to the well(s) of a preparative pulsed-field gel, overlaid with molten agarose, and subjected to electrophoresis for 18–24 hr at 200 V, 12°, and an appropriate switch time (10–60 sec; see Ref. 9 for more information). The position of the YAC is determined by staining the edges of the gel, the band is excised, and DNA is purified by standard methods [e.g., electroelution, phenol extraction, or Gene-Clean (Bio-Rad, Richmond, CA) extraction after agarase treatment, depending on the application].

[16] C. V. Traver, S. Klapholz, R. Hyman, and R. W. Davis, *Proc. Natl. Acad. Sci. U.S.A.* **86,** 5898 (1989).

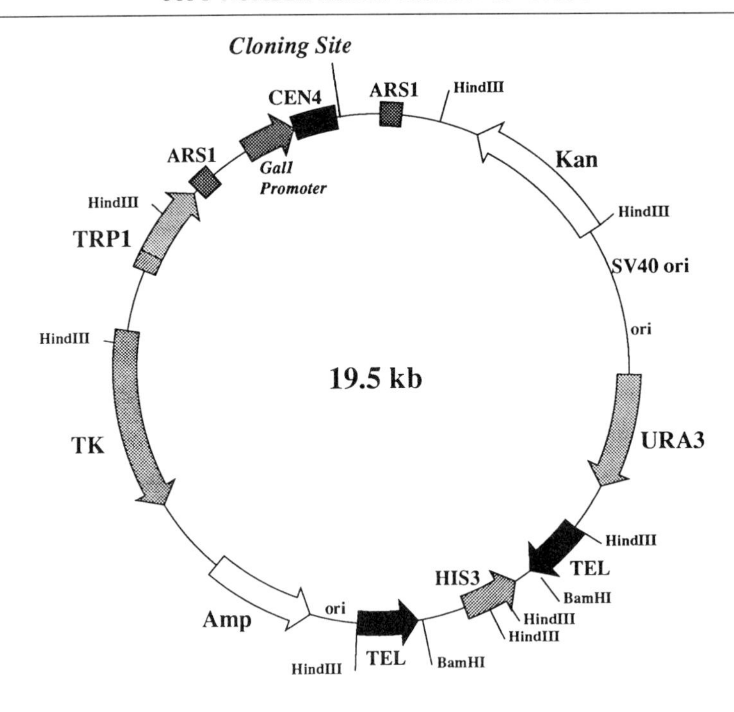

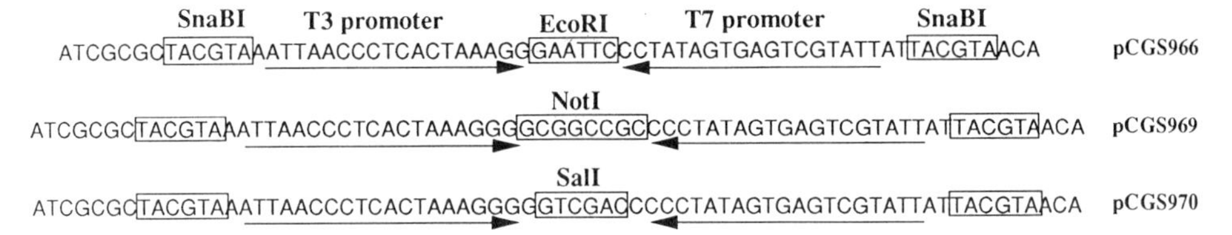

FIG. 2. Vectors for copy number amplification. Three vectors are currently available having the general structure shown, but differing in the cloning site. They are pCGS966 (*Eco*RI cloning site), pCGS969 (*Not*I cloning site), and pCGS970 (*Sal*I cloning site). Regions involved in chromosome function (centromere and telomeres) are shown in black. Genes that function in yeast (standard nomenclature) are shown in gray shading; open arrows indicate *E. coli* genes (the kanamycin resistance marker can also be expressed in mammalian cells to confer resistance to neomycin[16]). The DNA sequence of the region surrounding the cloning site (verified for pCGS966; inferred from the method of construction for the others) is shown below the plasmid map.

Colony Screening

Another useful application of YAC amplification is in colony screening.[3] Clones to be screened can be diluted appropriately and plated on amplification medium, or they can be replica plated from standard medium. After 4–7 days of growth, the cells are replicated onto nylon membranes layered over YPD plates. Regrowth of amplified cells on rich medium was necessary to achieve efficient cell wall digestion. For best results, the amount of cells transferred to the nylon should be kept to a minimum (only the layer of cells in immediate contact with the membrane impart their DNA to the surface—and these cells should be well aerated and growing rapidly). Therefore it is preferable to use replicating velvets to copy the plates rather than using the dry membrane to pick up the cells. Be sure to include adequate registration marks on the plates and filters so that films from subsequent hybridization experiments can be aligned with the colonies to identify positive clones. In most cases, there will be enough background from nonhybridizing clones to assist in aligning the filters.

Structural Stability of Amplified Chromosomes

Occasionally, extraneous bands are visible among the chromosomal DNAs of amplified clones. However, in all cases examined to date, these extra "YACs" could be eliminated by colony purification of the original isolate,[3] or could be visualized *without* amplification in such purified isolates. Thus, the amplification conditions employed (S1/M10) do not appear to induce deletions or rearrangements at a significant frequency. On the other hand, clones that were amplified, subcultured in glucose, and then subjected to a second round of selection in S1/M10 did begin to accumulate extra "YACs" that differed in size from those observed in purified isolates (D. Smith, unpublished observations).

Discussion

Vectors

Properties of the vector pCGS966 have been described.[3] Two additional vectors are available with *Not*I (pCGS969) or *Sal*I (pCGS970) cloning sites (*Not*I or *Sal*I linkers were ligated into pCGS966 after *Eco*RI cutting and trimming with mung bean nuclease). A map of pCGS969 showing part of the DNA sequence surrounding the cloning site and including the features mentioned above is shown in Fig. 2. These vectors extend the range of restriction enzymes to be used for cloning and, in principle, permit the use

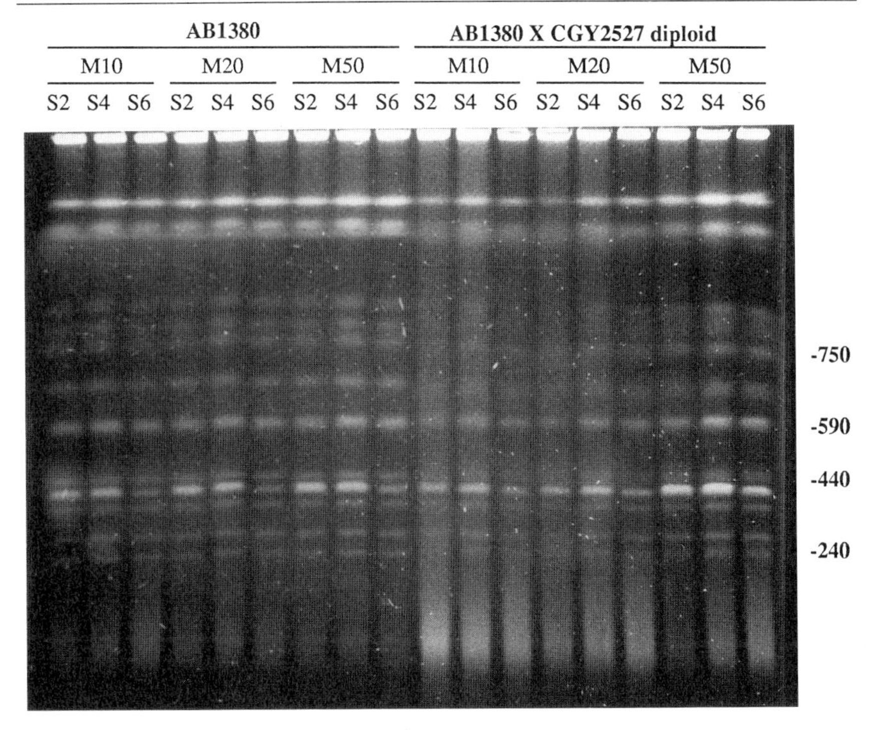

FIG. 3. YAC amplification in different hosts. An anonymous YAC containing a human DNA insert of about 400 kb (one of the YACs studied in Ref. 3) was transfered into strain AB1380 as described in the text. The resulting strain was mated with CGY2527 and a diploid strain recovered after selection for growth without histidine. The diploid and its AB1380 parent were then grown in amplification media with various amounts of methotrexate (10, 20, or 50 μg/ml) and sulfanilamide (2, 4, or 6 mg/ml). Chromsomal DNAs were prepared, run on a CHEF gel under standard conditions, and visualized with ethidium bromide.

of partial fill-in cloning strategies with a large number of enzymes (e.g., *Bam*HI, *Bgl*II, *Bcl*I, *Mbo*I, *Bst*YI, *Age*I, *Xma*I, *Cfr*10I, and *Bsp*EI).

Observations on Yeast Artificial Chromosome Amplification

Different strains responded differently to the selective conditions used for amplification. For instance, a 400-kb YAC in strain CGY2516 amplified readily to about 10 copies/cell at 1 mg/ml sulfanilamide and 10 μg/ml methotrexate[3] (S1/M10). The same YAC transformed into strain AB1380 amplified less well, but could be improved slightly by increasing the concentration of selective agents (Fig. 3). Strain AB1380 does not grow as well on galactose as either the original CGY2516 strain or the Gal^+ diploid.

Therefore, this result could be explained by incomplete shut-off of the conditional centromere. In a parallel experiment, a diploid obtained by mating the AB1380/YAC clone with strain CGY2527 amplified well only at high concentrations of selective agents (S4/M50; Fig. 3). This could be explained by the fact that the diploid has a double dosage of dihydrofolate reductase genes (and presumably DHFR), and hence requires more methotrexate to effectively select against cells with low YAC copy numbers. Because of the complexity of the selection, it is possible that additional factors, such as vitamins and cofactors present in the medium components, could also affect amplification. In fact, stationary-phase cells of CGY2516 YAC clones picked from agar plates grow much faster in amplification medium than an equivalent number of log-phase cells, or even previously amplified cells. This behavior could be explained if stationary-phase cells accumulate some essential component that is depleted during amplification.

The practical limit for amplification using this system appears to be about 20–25 copies/cell. Cycling the cells from amplification medium to YPD and back again did not result in any improvement. Further amplification is theoretically possible, but may require a selectable marker other than thymidine kinase. Under *TK* selection, the cells ceased to grow well after about 1 week. Even with this limitation, the system is extremely useful. The possibility that greater degrees of amplification might be possible using a different selection system needs to be investigated.

[51] Manipulation of Large Minichromosomes in *Schizosaccharomyces pombe* with Liposome-Enhanced Transformation

By ROBIN C. ALLSHIRE

Introduction

Artificial minichromosomes containing a centromere, replication origin, and telomeres can be constructed with relative ease in the budding yeast *Saccharomyces cerevisiae*. These minichromosomes behave, in every respect, in a manner similar to naturally occurring *S. cerevisiae* chromosomes.[1] The ability to construct such artificial chromosomes has led to

[1] A. W. Murray, N. P. Schultes, and J. W. Szostak, *Cell (Cambridge, Mass.)* **45,** 529 (1986).

the development of yeast artificial chromosome (YAC) cloning systems[2] and chromosome fragmentation methods in *S. cerevisiae*.[3] In addition, the craft of chromosome manipulation has allowed the effect, on chromosome segregation, of mutations within the centromeric DNA to be analyzed.[4] A genetic screen made use of a manipulated chromosome to select for mutations in genes that affect *S. cerevisiae* chromosome segregation.[5]

The fission yeast, *Schizosaccharomyces pombe* provides a good alternative to the budding yeast. The three chromosomes of *S. pombe* are all in excess of 3000 kilobases (kb),[6,7] therefore this organism has an innate ability to replicate and segregate DNA of greater than 3000 kb in length. The development of a YAC-like cloning system in *S. pombe* might be complementary to that in *S. cerevisiae* in that it may have a greater capacity for cloning and stably propagating large DNA molecules (>1000 kb) from a heterologous source. The manipulation of large DNA in *S. pombe* should also lead to further developments in the physical mapping and functional analyses of *S. pombe* chromosomes and centromeres. The introduction of large DNA into *S. cerevisiae* protoplasts requires an efficient DNA transformation method yielding up to 1×10^6 transformants/μg of plasmid per 3×10^7 protoplasts.[2] There are now several methods employed for transforming *S. pombe*. These either require the digestion of the cell wall to form protoplasts[8] or treatment of the cells with alkali cations such as lithium.[9] Until recently methods relying on lithium treatment gave fewer than 1×10^4 transformants/μg of plasmid; however, Okazaki *et al.* have optimized this procedure so that 1×10^6 transformants/μg can be obtained.[10] Preparation of *S. pombe* protoplasts followed by transformation gives relatively poor efficiencies ($2–4 \times 10^4/\mu$g) compared with the analogous method in *S. cerevisiae*.[8] However, it has been demonstrated that the inclusion of the cationic liposome-forming reagent, Lipofectin, can stimulate transformation of *S. pombe* protoplasts up to 50-fold, so that a maximum of 1×10^6 transformants/μg of plasmid are obtained

[2] D. T. Burke, G. F. Carle, and M. V. Olson, *Science* **236,** 806 (1987).

[3] D. Vollrath, R. W. Davis, C. Connelly, and P. A. Hieter, *Proc. Natl. Acad. Sci. U.S.A.* **85,** 6027 (1988).

[4] J. Hegemann, J. Shero, G. Cottarel, P. Philipsen, and P. Hieter, *Mol. Cell. Biol.* **8,** 2523 (1988).

[5] F. Spencer, S. L. Gerring, C. Connelly, and P. Hieter, *Genetics* **124,** 237 (1990).

[6] J.-B. Fan, Y. Chikashige, C. L. Smith, O. Niwa, M. Yanagida, and C. R. Cantor, *Nucleic Acids Res.* **17,** 2801 (1989).

[7] D. Vollrath and R. W. Davis, *Nucleic Acids Res.* **15,** 7865 (1987).

[8] D. H. Beach, M. Piper, and P. Nurse, *Mol. Gen. Genet.* **187,** 326 (1982).

[9] M. Broker, *BioTechniques* **5,** 517 (1987).

[10] K. Okazaki, N. Okazaki, K. Kume, S. Jinno, K. Tanaka, and H. Okayama, *Nucleic Acids Res.* **18,** 6485 (1990).

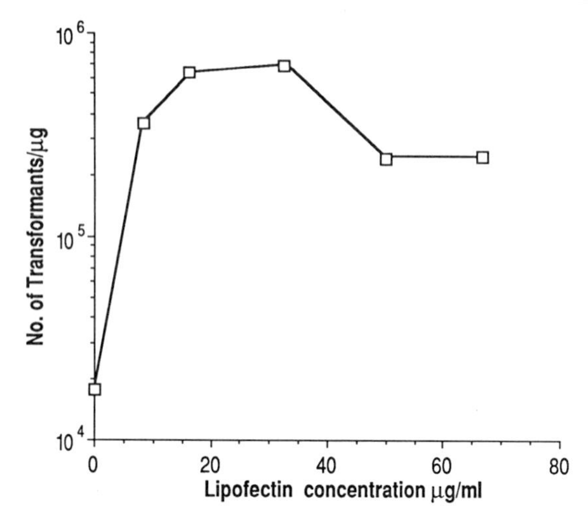

FIG. 1. Titration of Lipofectin concentration against *S. pombe* transformation efficiency. See text for details. All points plotted represent a mean calculated from duplicate experiments. (Reprinted from Allshire,[11] with permission.)

(Ref. 11 and Figs. 1 and 2). Although DNA of approximately 100 kb can be introduced into *S. pombe* by the standard lithium procedure the frequency of such transformants is low.[12] The improved method of Okazaki *et al.* should allow more transformants to be obtained with molecules of up to 100 kb but it is likely that only transformation of protoplasts will allow the efficient uptake of larger DNA molecules. Lipofectin-enhanced protoplast transformation allows large linear DNA molecules (500 kb) to be introduced into *S. pombe*. Presented below are detailed protocols for transformation of *S. pombe* protoplasts with large DNA molecules and analyses of resulting transformants.

Reagents

Strains

SP813: h^{+N} *leu1-32 ade6-210 ura4-D18*

HM346-23R: h^{+} *leu1 fur1-1 tps16-112 ade6-210* containing the 530-kb minichromosome Ch16-23R (m23::*LEU2* *fur1*$^{+}$ *tps16*$^{+}$ *ade6-216*). This strain is Leu^{+} due to the integration of the *S. cerevisiae LEU2* gene on the minichromosome, and Ade^{+} due to intragenic complementation between the two *ade6* alleles

[11] R. C. Allshire, *Proc. Natl. Acad. Sci. U.S.A.* **87,** 4043 (1990).

[12] K. M. Hagnenberger, *Proc. Natl. Acad. Sci. U.S.A.* **86,** 577 (1989).

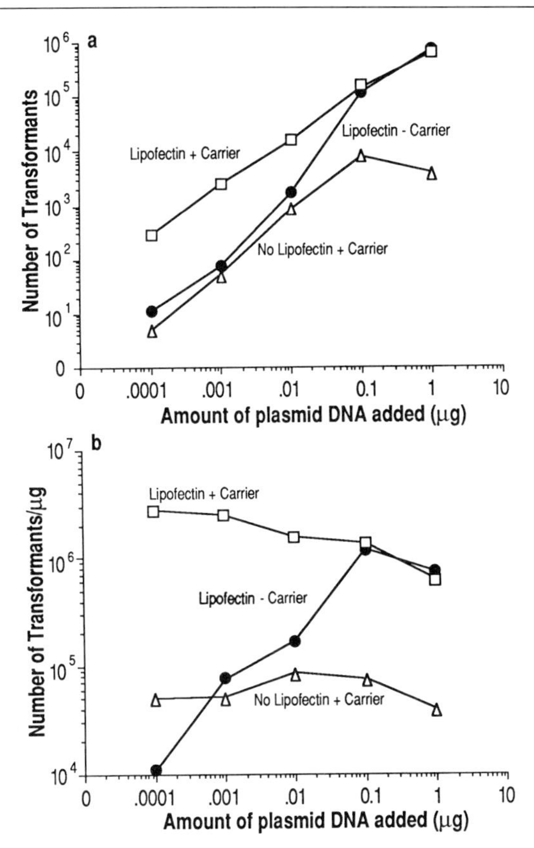

FIG. 2. Titration of added plasmid in Lipofectin-mediated and normal *S. pombe* protoplast transformation with or without carrier DNA. (a) Absolute number of Ura^+ colonies resulting from transformation with increasing amounts of plasmid. (b) The transformation frequency per microgram [extrapolated from the data in (A)] plotted against amount of plasmid added. Titration of plasmid was performed in the presence of Lipofectin and carrier DNA (□), Lipofectin only (●), or carrier DNA only (△). The data presented were derived from the mean result of duplicate experiments. (Reprinted from Allshire,[11] with permission.)

HM371: *h⁻ leu1 fur1-1 ade6-210* containing the 750-kb minichromosome ChS28 (m23::*LEU2 fur1⁺ ade6-216*). As above, this strain is also Leu^+ Ade^+

Plasmid. The plasmid pIRT/*Ura4* is based on pUC118 and contains the *S. pombe ura4* gene on a 1.8-kb *Hin*dIII fragment inserted into the *Hin*dIII site and the *S. pombe ARS1* fragment inserted into the *Eco*RI site on a 1.2-kb *Eco*RI fragment.

Media

YE: 5 g/liter yeast extract, 20 g/liter glucose. For agar plates add 20 g/liter agar

YEA: YE + 80 mg/liter adenine

PM: Identical to EMM, as described previously,[13] but the pH is adjusted to 5.6 by the addition of NaOH

Protoplasting Solutions

SP1: 1.2 *M* sorbitol, 50 m*M* sodium citrate, 50 m*M* sodium phosphate, 40 m*M* ethylenediaminetetraacetic acid (EDTA), pH 5.6

SP2: 1.2 *M* sorbitol, 50 m*M* sodium citrate, 50 m*M* sodium phosphate, pH 5.6

SP3: 1.2 *M* sorbitol, 10 m*M* Tris-HCl, pH 7.6

SP4: 1.2 *M* sorbitol, 10 m*M* Tris-HCl, pH 7.6, 10 m*M* $CaCl_2 \cdot 2H_2O$

SP5: 20% (w/v) polyethylene glycol (M_r 3500–4000), 10 m*M* Tris-HCl, pH 7.6, 10 m*M* $CaCl_2 \cdot 2H_2O$

Lipofectin. Lipofectin is purchased as a 1 mg/ml suspension from Bethesda Research Laboratories, (Gaithersburg, MD).

Methods

Preparation and Transformation of Schizosaccharomyces pombe Protoplasts

Inoculate 10 ml of YEA or PM (plus required amino acid supplements) with a healthy fresh growing colony. Incubate at the permissive temperature for 12–16 hr. Transfer 0.75, 1.0, 1.5, and 2 ml of this overnight culture into four separate flasks containing 100 ml of PM plus 0.5% (w/v) glucose plus required amino acid supplements. Incubate cultures at the permissive temperature for 12–20 hr. Cultures of SP813 usually form good-quality protoplasts when they reach densities between 1.5 and 2.5 × 10^7 cells/ml. This varies for other strains and should therefore be tested empirically. Normally all four cultures are harvested and that which forms the best quality protoplasts is used for subsequent transformation.

Harvest cells by centrifugation at 25° for 5 min at 3500 rpm in a Sorvall (Norwalk, CT) GLC (or equivalent) in 2 × 50 ml sterile Falcon (Becton Dickinson, Oxnard, CA) tubes. Resuspend cell pellets in 10 ml of SP1 and combine in a single tube. Add 20 μl of 2-mercaptoethanol. Mix and incubate at room temperature for 10 min.

Pellet cells and resuspend in 2 ml of SP2. Add 5–10 mg of NovoZym

[13] P. Nurse, *Nature (London)* **256,** 547 (1975).

234 (Novo Biolabs, Danbury CT). (*Warning*: Use a face mask and/or fume cupboard when weighing out NovoZym. Repeated exposure to NovoZym can lead to an allergic reaction.) Incubate at 37° until approximately 70–90% of the cells are spherical. This can take between 10 and 45 min; protoplast formation is monitored by observing aliquots under the microscope every 5–10 min.

Add 30 ml of SP3 to the protoplasts, mix by inversion, and pellet protoplasts by centrifugation at 2000 rpm for 5 min. Resuspend the pellet in 30 ml of SP3 and spin again as above; repeat this wash once more. During these washes remaining cells form protoplasts. To resuspend protoplast pellets it is best to use a pipette to resuspend in a small amount of solution first and then add the remainder; do not vortex. After washes with SP3 resuspend the final pellet in 1 ml of SP4.

Aliquot 100 μl of protoplasts into separate 4-ml sterile Falcon (2070) tubes. Add transforming DNA (see Notes 1 and 2 below) and mix gently by hand. Incubate at room temperature for 15 min. Add an equal volume (100 μl) of SP4 containing 66 μg/ml of Lipofectin (final concentration, 33 μg/ml). Incubate at room temperature for 15 min. Add 1 ml of SP5 and mix thoroughly by inversion. Incubate at room temperature for 15 min. (At this stage cells are phase dark and clump.)

Pellet cells at 1500 rpm at 25° for 5 min. Carefully remove the supernatant with a pipette. Resuspend the pellet in 100–1000 μl of SP4; the volume depends on the expected number of transformants and on the amount to be plated. (Cells will now be wrinkled in appearance.) Pipette aliquots of each transformation onto PM plus sorbitol (plus required amino acid supplements) agar plates. When plating less than 50 μl, pipette into 100 μl of SP4 already placed on the plate. Cells should be spread over the plate gently, using a glass spreader (bent Pasteur pipettes are adequate). Do not lean too heavily on the plate with the spreader because protoplasts are fragile.

Notes

1. Add DNA in a volume of 10 μl or less to 100 μl of protoplasts in SP4.

2. When transforming with less than 100 ng of DNA, the efficiency is increased by adding 1 μg of sterile sonicated deproteinized calf thymus DNA (see Fig. 2).

Maximizing Efficiency of *Schizosaccharomyces pombe* Transformation

In maximizing the transformation of *S. pombe* many influencing factors must be taken into consideration. These include (1) the concentration of

DNA, (2) the effect of added carrier DNA, and (3) the concentration of protoplasts. A good transformation procedure should be linear with respect to the number of transformants obtained over a wide range of DNA concentrations. The addition of carrier DNA to transformations should increase the efficiency of transformation at low DNA concentrations. Obviously the number of protoplasts per transformation will also affect the final transformation efficiency.

Principle of Method

Lipofectin is the registered trade name for the cationic liposome-forming molecule DOTMA {*N*-[1-(2,3-dioleyloxy)propyl]-*N,N,N*-trimethylammonium chloride}, which can be purchased from Bethesda Research Laboratories. Lipofectin liposomes form complexes with both RNA and DNA and also with cell membranes. In the process of forming such complexes, DNA is efficiently taken up by fusion with cells.[14] This reagent was originally developed for the transfection of mammalian tissue culture cells[14]; however, we have found that it also facilitates the delivery of nucleic acids to yeasts, provided that cell walls are first removed.[11] Below it is demonstrated that the inclusion of lipofectin in transformations of *S. pombe* protoplasts with plasmid DNA results in up to a 50-fold increase in transformation efficiency. The inclusion of Lipofectin in transformation of *S. cerevisiae* protoplasts increases the transformation efficiency by up to 20-fold (Ref. 11 and B. Futcher, personal communication, 1990).

Specific Example: Effect of Lipofectin and Other Parameters on Transformation of *Schizosaccharomyces pombe* Protoplasts

Lipofectin Concentration

In Lipofectin-mediated transfection of mammalian cells it has been found that concentrations between 30 and 50 μg/ml are optimal for efficient transformation.[14] Protoplasts from *S. pombe* strain SP813 were prepared as described above and transformed with 1 μg of the plasmid pIRT/*ura4* in the absence or presence of increasing concentrations of Lipofectin. Lipofectin was added to 100-μl aliquots of protoplasts, as described above, containing approximately 5×10^7 protoplasts. It can be seen from Fig. 1 that Lipofectin gives maximum enhancement at a final concentration of

[14] P. L. Felgener and M. Holm, *Focus* **11,** 21 (1989).

33 μg/ml, resulting in a 35-fold increase over a standard (no Lipofectin) protoplast transformation.

DNA Concentration

In the transformations described in Fig. 1 the addition of 1 μg of plasmid to 5×10^7 protoplasts could well be excessive, thereby saturating the transformation. Titration of plasmid concentration against transformation efficiency should determine plasmid concentrations at which maximum transformants per microgram are obtained. In addition, the comparison of transformation efficiencies in the presence and absence of carrier DNA should reveal any effect of added carrier DNA.

The data presented in Fig. 2 demonstrate the following:

1. Transformation of *S. pombe* protoplasts with increasing amounts of plasmid is linear only in the presence of carrier DNA at 1 μg/100 μl of protoplasts. In the absence of carrier DNA maximum efficiencies are obtained only when greater than 100 ng of plasmid DNA is added to 100 μl of protoplasts.
2. At low plasmid concentrations (1–100 ng/100 μl) of protoplasts, transformation efficiencies of greater than 1×10^6 transformants/μl can be obtained. Therefore, using this protocol it should be possible to introduce large DNA into *S. pombe* protoplasts.
3. When greater than 100 ng of plasmid is added to 100 μl of protoplasts the transformation is saturated with DNA and the efficiency of transformation declines.

Protoplast Concentration

The optimum concentration of protoplasts required to give maximum transformation efficiency in a Lipofectin-enhanced transformation was also investigated. A batch of SP813 protoplasts (5×10^8 viable cells/ml) was serially diluted by 1/2 and aliquots of each dilution were transformed with 10 ng of pIRT/*ura4* in the presence of 1 μg/100 μl of carrier DNA followed by incubation in the presence of 33 μg/ml of Lipofectin.

From Fig. 3 it can be seen that maximum transformation efficiencies were obtained when concentrations between 1 and 5×10^8 protoplasts/ml were used. It is also worth noting that reasonable efficiencies of 6×10^4 transformants/μg can be obtained when only 1.5×10^6 protoplasts/100 μl are present. Therefore, where transformation efficiency is not critical, several hundred separate transformations could be performed on a single batch of protoplasts resulting from a 100-ml starting culture.

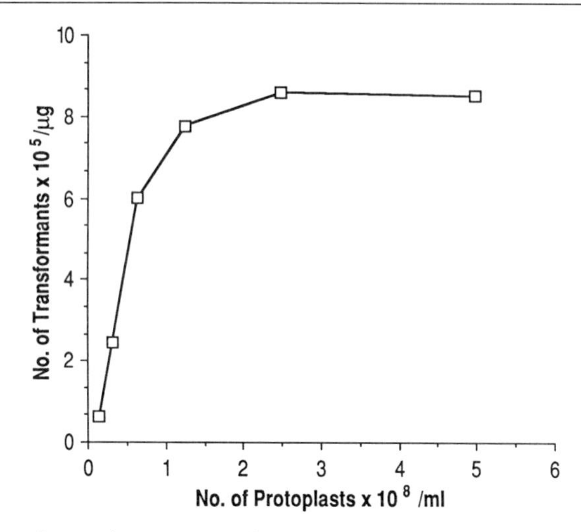

FIG. 3. Effect of protoplast concentration on Lipofectin-enhanced *S. pombe* transformation efficiency. The data plotted represent a mean from duplicate experiments.

Transfer of Large DNA into *Schizosaccharomyces pombe* Protoplasts from Agarose

Minichromosome derivatives of *S. pombe* chromosome III were constructed previously by irradiation of an aneuploid strain containing two copies of chromosome III; partial aneuploids were subsequently identified by a genetic screen.[15] The minichromosomes Ch16-23R (530 kb) and ChS28 (750 kb) have been described in detail by Niwa *et al.*[16] These minichromosomes have been utilized to test if Lipofectin-enhanced *S. pombe* protoplast transformation allows the uptake of large (>500 kb) DNA molecules. The transformation of *S. pombe* protoplasts with these minichromosomes requires several steps, which are summarized schematically in Fig. 4:

1. the preparation of chromosome-sized DNA in agarose plugs from strains carrying the minichromosomes
2. separation of the minichromosome from other chromosomes by clamped homogenous electric field (CHEF) gel electrophoresis on a low melting point agarose gel
3. excising a minimally sized agarose slice from the CHEF gel containing the minichromosome DNA unstained with ethidium bromide
4. transformation of SP813 protoplasts with the minichromosome DNA in molten agarose
5. selection for transformants in an appropriately marked strain

[15] O. Niwa, T. Matsumoto, and M. Yanagida, *Mol. Gen. Genet.* **203,** 397 (1986).
[16] O. Niwa, T. Matsumoto, Y. Chikashige, and M. Yanagida, *EMBO J.* **8,** 3045 (1989).

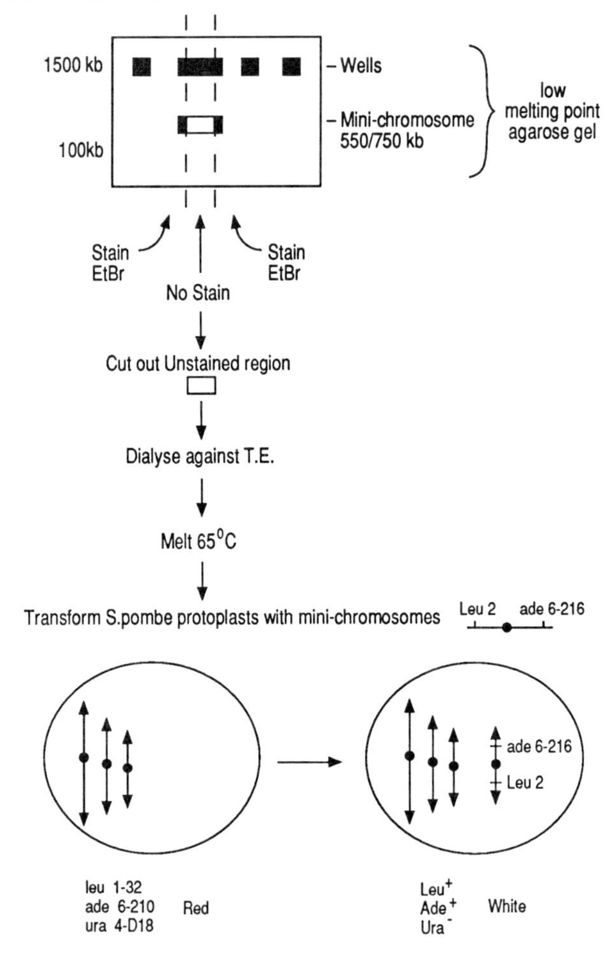

FIG. 4. Schematic representation depicting the method used to transform *S. pombe* strain SP813 with gel-purified minichromosome DNA.

Principle of Method

The minichromosomes Ch16-23R and ChS28 carry selectable markers, *LEU2* and *ade6*, on both chromosome arms. Transformation of SP813 protoplasts with either intact Ch16-23R or ChS28 DNA should result in the complementation of *leu1-32* with *LEU2* and intragenic complementation of *ade6-210* with *ade6-216* resulting in Leu^+ Ade^+, white prototrophs. A strain such as SP813, bearing the *ade6-210* allele, is Ade^- and red in color. Strains having the *ade6-216* allele alone are Ade^- and pink. In the presence of both *ade6-210* and *ade6-216* alleles the resulting strains are Ade^+ and white.[17]

The minichromosome DNA is added to the transformants in molten agarose in the presence of 1 μg of carrier DNA; in addition, a small amount of an agarose plug containing total mammalian genomic DNA is also melted with the minichromosome agarose. Retaining the minichromosome DNA in agarose and the addition of high molecular weight mammalian DNA minimizes the shear forces exerted on the minichromosome DNA. Previous observations had demonstrated that the presence of agarose does not adversely effect Lipofectin-enhanced transformation of *S. pombe* protoplasts (R. Allshire, unpublished observations).

Putative minichromosome transformants with the correct phenotypic characteristics are confirmed by the preparation of chromosome-sized DNA in agarose plugs and subsequent CHEF gel analyses.

Methods

Preparation of Schizosaccharomyces pombe Chromosomes in Agarose Plugs

Schizosaccharomyces pombe cultures are grown to a density of 1–4 × 10^7 cells/ml, harvested, and the pellets washed three times in 50 m*M* EDTA, pH 8.0. Resuspend final pellets at 2 × 10^9 cells/ml in SP1. Add Zymolyase 100T to a final concentration of 0.5 mg/ml. Incubate at 37° for 2 hr. Add an equal volume of 1% (w/v) low melting point agarose at 42° (in 125 m*M* EDTA, pH 8.0) to the cell suspension. Aliquot the agarose/cell mix into 100-μl molds and allow to set at room temperature. Once set, remove the agarose plugs from the molds and place in NDS [10 m*M* Tris, 0.5 *M* EDTA, pH 9.0, 1% (w/v) *N*-laurylsarcosine]. Add proteinase K (BCL) to a final concentration of 0.5 mg/ml and incubate at 50° for 24 hr, after which fresh NDS is added containing 0.5 mg/ml proteinase K and the plugs are reincubated at 50° for 24–48 hr. Plugs are stored in NDS at 4° and are stable for at least 1 year.

Pulse-Field Gel Electrophoresis and Preparation of Minichromosome in Agarose Slice

For analyses of chromosomes up to 1000 kb in size, samples were run on 1% (w/v) agarose gels in 0.5× TBE (5 × TBE = 450 m*M* Tris-Borate, 10 m*M* EDTA, pH 8.0) using a CHEF apparatus.[7] Electrophoresis was performed at 200 V, 12°, with a pulse time of 60 sec for 14 hr followed by

[17] H. Gutz, H. Heslot, U. Leupold, and N. Loprieno, *in* "Handbook of Genetics" (R. C. King, ed.), Vol. 1, p. 395, Plenum, New York, 1974.

90 sec for 10 hr. To resolve *S. pombe* chromosomes (2000–6000 kb) on a CHEF apparatus, 0.6% (w/v) Fastlane agarose gels (FMC, Rockland, ME) in 0.5× TAE (10 × TAE = 40 m*M* Tris-Acetate, 10 m*M* EDTA, pH 8.0) were run for 5 days at 10°, 60 V, with a 33-min pulse time. To prepare minichromosome DNA for transformation the minichromosome was separated on a 1% (w/v) low melting point agarose (Bethesda Research Laboratories) gel in 0.5× TAE. A track, 5 cm in length, was loaded with plugs. After the run was complete the gel was cut into three portions so that approximately 3 cm of the track remained in the center portion. Only the flanking portions were stained with ethidium bromide (as illustrated in Fig. 4). After staining the gel the position of the minichromosome in the gel was marked with a scalpel blade. The gel was then reassembled and the region of the center portion of the gel containing the unstained minichromosome DNA was cut out, using the blade marks as a guide. The gel slice was then placed in 50 m*M* NaCl, 10 m*M* Tris-HCl, 5 m*M* EDTA, pH 8.0 and stored at 4°.

Transformation of Schizosaccharomyces pombe Protoplasts with Minichromosome

The gel slice is placed on a sterile petri dish. Using a scalpel blade a portion containing 50–100 μl of agarose is removed and placed in a microfuge tube in 1 ml of 10 m*M* Tris-HCl, 1 m*M* EDTA, pH 8.0 (TE). In addition, a small slice (~10 μl) of agarose from a plug containing high molecular weight mammalian DNA is added to the tube with the minichromosome slice. These agarose slices are dialyzed against 1 ml of TE for 1 hr at room temperature with three changes of TE. When required for protoplast transformation the TE is removed and the agarose containing the minichromosome melted at 60°. The inclusion of 30 m*M* NaCl and 0.3 m*M* Spermine/0.75 Spermidine, in the above washes and when melting agarose, improves the number of transformants with DNA greater than 100 kb. Prior to transformation 5 μg of carrier calf thymus DNA is added to the molten agarose; this mixture is then cooled to 42°. Aliquots (10 μl) of this molten agarose are added to 100 μl of *S. pombe* protoplasts, which are then incubated for 15 min at room temperature. The transformation procedure is then followed as above.

Specific Example

Ch16-23R was prepared in an unstained low melting point agarose slice as described above. In one Lipofectin-enhanced transformation of SP813 protoplasts, with a 10-μl aliquot of molten agarose containing 1–10 ng of minichromosome DNA, 150 Leu^+ colonies were obtained. Of a total of 324

Leu$^+$ transformants analyzed only 4 were also white, Ade$^+$ prototrophs. Eighty-seven of these Leu$^+$ transformants were pink rather than red, indicating that the *ade6-216* allele present on Ch16-23R had been transferred but was not being complemented by the presence of the SP813 *ade6-210* (red) allele. Further analyses (presented below) of these transformants indicated that although they had taken up the minichromosome, unexpected rearrangements had occurred due to interactions with the endogenous chromosome III.

CHEF gel analyses of Leu$^+$ Ade$^+$ (white) transformants is presented in Fig. 5. It is clear from Fig. 5A that all four transformants (lanes 3–6) do contain an additional minichromosome similar in size to the input minichromosome Ch16-23R. The identity of this minichromosome as Ch16-23R is confirmed in Fig. 5B by hybridization with labeled plasmid pIRT/*ura4*. The absence of *ura4* hybridization in the four transformants verifies that these are indeed *bona fide* transformants because the *ura4* gene is deleted in the recipient strain SP813 but not in the donor strain HM348-23R.

Novel *Schizosaccharomyces pombe* Karyotypes Resulting from Minichromosome Transfer and Rearrangement

It has been demonstrated above that the 530-kb minichromosome Ch16-23R can be transferred as intact DNA directly from agarose into *S. pombe* protoplasts; however, many of the resulting transformants were only Leu$^+$ rather than Leu$^+$ Ade$^+$ as expected. To characterize further these Leu$^+$ transformants, chromosome-sized DNA was separated on CHEF gels. Initial analyses demonstrated that nonstoichiometric quantities of the minichromosome were present in most of these Leu$^+$ transformants and that this minichromosome is always smaller (~450 kb) than intact Ch16-23R. In addition, in all transformants a strong hybridization signal (plasmid, not *ura4*) was also seen in the region of limiting mobility, suggesting the presence of plasmid sequences on a larger, unresolved chromosome (data not shown). Intact chromosomes from some of these transformants were separated on a CHEF gel so as to resolve all three *S. pombe* chromosomes (Fig. 6B). Surprisingly, these transformants contain a novel chromosome, C/S X, which is larger than the minichromosome Ch16-23R (run off gel). Hybridization of these chromosomes with either left arm (*wee1*, Fig. 6C) or right arm (*stb*, Fig. 6D) chromosome III markers, not present on Ch16-23R, reveals that the right arm of chromosome III has been duplicated because both C/S X and chromosome III now carry the right arm marker, *stb*. What mechanism could be responsible for generating this novel chromosome? In Fig. 7 three possible events are presented.

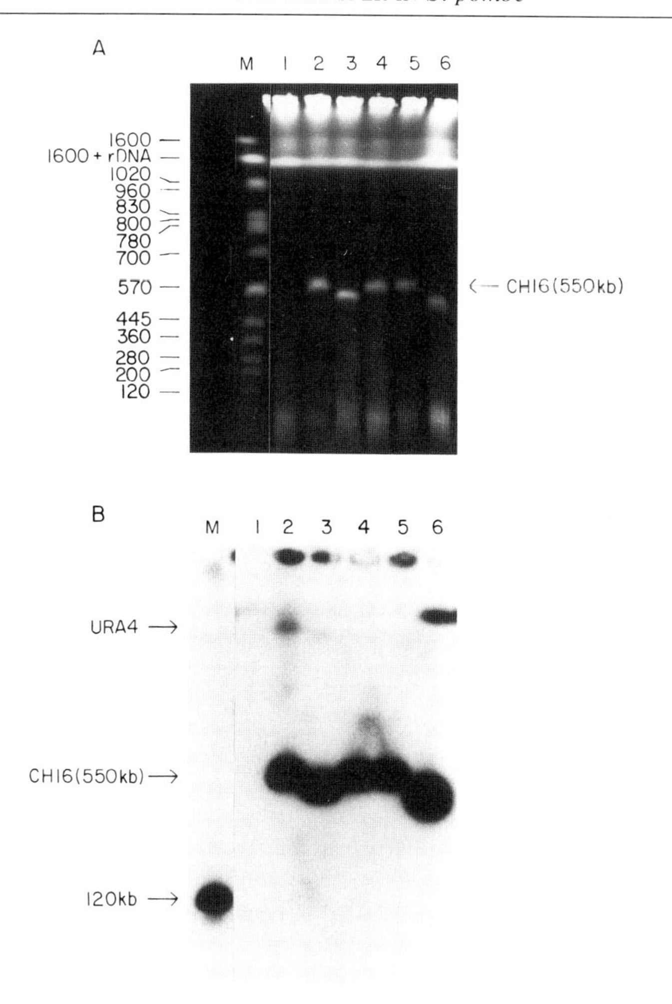

FIG. 5. CHEF gel analyses of minichromosomes from *S. pombe* Ch16-23R donor, recipient, and Leu$^+$ Ade$^+$ transformants. Lane M: Marker track with chromosomes prepared in agarose from *S. cerevisiae* strain AB1380-N7, which contains a random human YAC of 120 kb. Lane 1: Chromosome-sized DNA prepared from recipient strain SP813. Lane 2: Chromosome-sized DNA from the Ch16-23R donor strain, HM348-23R. Lanes 3–6: Chromosome-sized DNA prepared from four Ade$^+$ Leu$^+$ Ch16-23R transformants. (A) DNA in CHEF gel stained with ethidium bromide. Sizes of chromosomes in AB1380-N7 and the position of the minichromosome, Ch16-23R, are indicated. (B) Autoradiograph of the above CHEF gel after transfer and hybridization with ^{32}P-labeled plasmid, pUC/*ura4*. The signals resulting from hybridization with the *S. pombe ura4* gene, the minichromosome, and the 120-kb YAC are indicated.

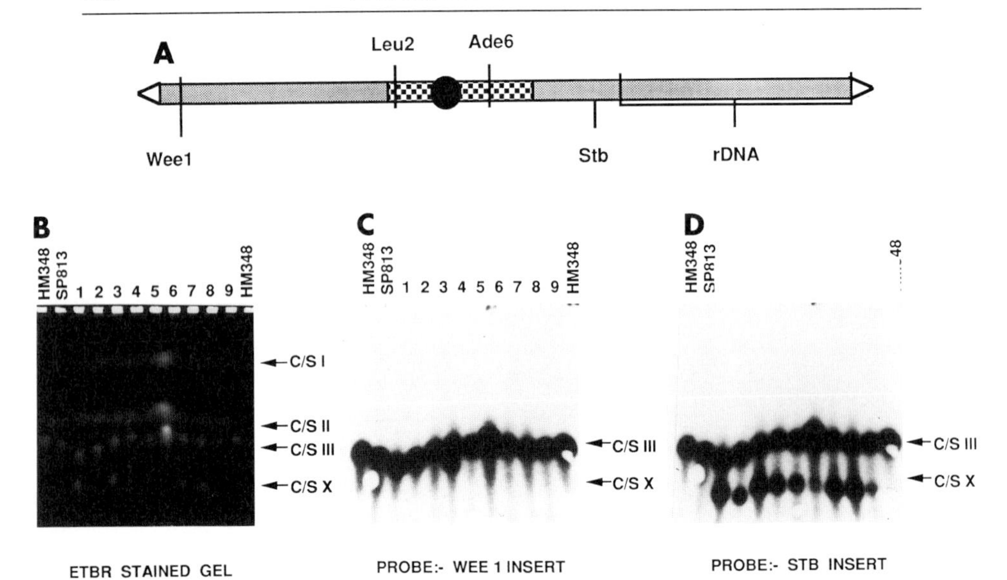

FIG. 6. Electrophoretic karyotype of nine Leu^+ Ade^- Ch16-23R transformants. (A) Schematic representation of *S. pombe* chromosome III, showing the position of markers *wee1, Leu2, ade6, stb,* and the rDNA complex. Superimposed (checkered) is the relative position of DNA carried by the minichromosome, Ch16-23R, and the centromere (●). (B) Ethidium bromide-stained CHEF gel of resolved chromosomes from HM348-23R, SP813, and nine Leu^+ Ade^- Ch16-23R transformants. Note the appearance of a novel-sized chromosome, C/SX, in all nine transformants. The positions of *S. pombe* chromosomes, I, II, and III are also indicated. (C) DNA in the CHEF gel (B) was transferred to a filter and hybridized with a ^{32}P-labeled probe containing only the *wee1* gene. The resulting autoradiograph is shown. (D) Autoradiograph resulting from rehybridization of the filter in (C) with ^{32}P-labeled *stb* DNA. Positions of *S. pombe* chromosome III and C/S X are indicated.

1. A reciprocal exchange event could occur between the minichromosome and endogenous chromosome III. Such an event would result in the fragmentation of chromosome III into two novel telocentric chromosomes. No normal chromosome III would remain.

2. Reciprocal exchange between the minichromosome and one chromatid during the G_2 phase of the cell cycle could result in a situation in which at mitosis one daughter cell would receive the normal chromosome III and an exchange product. In this case the other daughter cell would suffer a deletion of an entire arm of chromosome III, which would be lethal.

3. A gene conversion or gap-filling event could result in the copying of an entire chromosome III arm onto the minichromosome, resulting in a normal chromosome III and a novel chromosome C/S X.

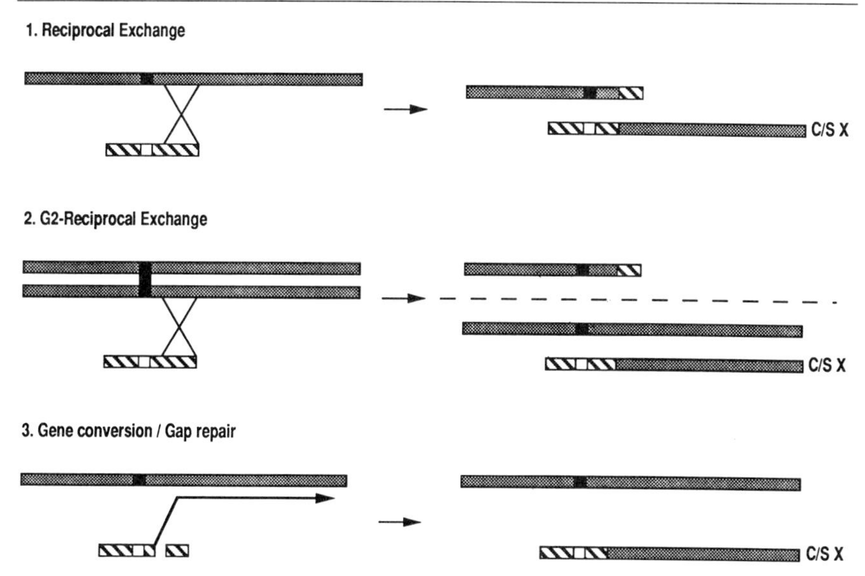

FIG. 7. Possible mechanisms resulting in the generation of the novel chromosome, C/S X.

Transformants with other novel rearrangements of chromosome III, which can be explained by the above mechanisms, have also been observed. The minichromosome ChS28 (750 kb) was used to transform SP813. Some of the transformants generated exhibited differential coloration when grown on limiting adenine plates (10 mg/liter). Chromosomes were prepared from three segregant types (dark pink, small white, and red). Analyses of the karyotypes of these different segregants on CHEF gels indicate the presence of different cell types (Fig. 8) containing (1) normal chromosome complement and ChS28, (2) a normal chromosome complement and C/S X (right arm duplication of chromosome III) or C/S Y (left arm duplication of chromosome III—reciprocal product to C/S X), (3) no normal chromosome III, but replaced by C/S X and C/S Y. Presumably this results from a reciprocal exchange event between the minichromosome and endogenous chromosome III.

Concluding Remarks

Lipofectin-enhanced transformation of *S. pombe* protoplasts allows high efficiencies of transformation with plasmids to be attained. This procedure also allows the efficient introduction of large linear DNA molecules (>500 kb) into *S. pombe* protoplasts. The construction of suitable cloning

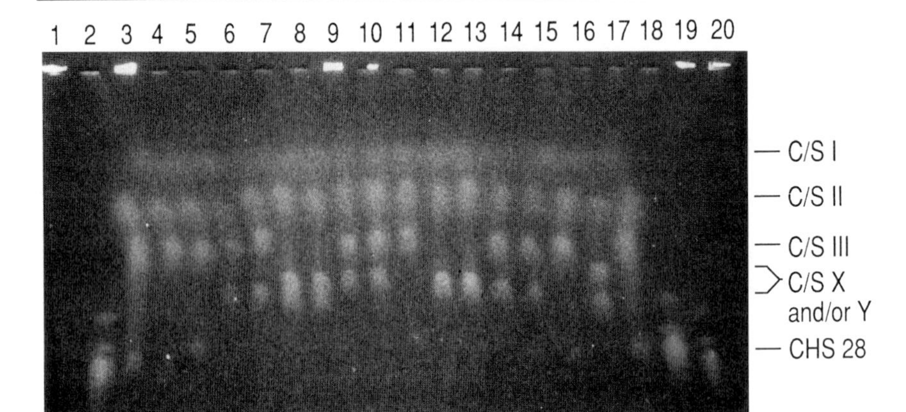

FIG. 8. Karyotypic analyses of transformants resulting from the transformation of *S. pombe* strain SP813 with the 750-kb minichromosome, Ch28S. Lanes 1, 19, and 20: Chromosomal DNA from *S. cerevisiae* AB1380-N7. Lanes 2 and 18: Chromosome-sized DNA from *S. pombe* HM371 containing the 750-kb minichromosome Ch28S. Lanes 3–6: Chromosomes from CH28S Leu$^+$ transformant 7/8. Lanes 3 and 4 were pink segregants, and lanes 5 and 6 were small white segregants. Lanes 7–11: Chromosomes from CH28S Leu$^+$ transformant 7/15. Lanes 7 and 8 were pink segregants, lanes 9 and 10 were small white segregants, and lane 11 was a dark pink segregant. Lanes 12–17: Chromosomes from Ch28S Leu$^+$ transformant 8/1. Lanes 12 and 13 were pink segregants, lanes 14 and 15 were small white, and lanes 16 and 17 were dark pink segregants. Note that there is no chromosome III-sized band remaining in lanes 7, 8, 12, or 13. The positions of the three *S. pombe* chromosomes, I, II, and III along with the minichromosome Ch28S and the novel chromosomes C/S X and C/S Y are indicated.

vectors should now allow the use of *S. pombe* as a host for cloning large DNA from heterologous sources. It has previously been demonstrated that *S. pombe* chromosomes can be transferred directly into mammalian cells. This should allow the development of a convenient shuttle system.[18] Vectors for cloning large DNA in *S. pombe* will differ from those of *S. cerevisiae* because the large size of *S. pombe* centromeres (50–100 kb) precludes their use in a cloning vector. However, it has been demonstrated that large acentric molecules can be retained in *S. pombe* provided that selection is maintained.[19,20] Indeed, it was observed that an *S. cerevisiae* YAC containing heterologous *Arabidopsis thaliana* DNA can be maintained as a linear molecule in *S. pombe* (R.A., unpublished observations).

The rearrangements involving minichromosomes and *S. pombe* chromosome III described above presumably result from the fact that the input

[18] R. C. Allshire, G. Cranston, J. Gosden, J. Maule, N. Hastie, and P. Fantes, *Cell (Cambridge, Mass.)* **50,** 391 (1987).

[19] M. Baum and L. Clarke, *Mol. Cell. Biol.* **10,** 1863 (1990).

[20] T. Matsumoto and M. Yanagida, *Curr. Genet.* **18,** 323 (1990).

DNA is completely homologous to the *S. pombe* genome. It is likely that such events will be infrequent when DNA from a heterologous source is cloned in *S. pombe* using a YAC-like vector.

The fact that *S. pombe* chromosome III can tolerate gross rearrangements, such as division into two telocentric chromosomes (as shown in Fig. 8), suggests that strategies such as chromosome fragmentation allowing chromosome manipulation, which has been developed in *S. cerevisiae*,[3] may also be applicable in *S. pombe*.

Acknowledgment

The author would like to acknowledge many colleagues for helpful advice and discussions, especially D. Beach, P. Fantes, P. Hieter, R. Anand, and numerous members of the Delbruck/Page and Beach Laboratories at Cold Spring Harbor Laboratory (CSHL). I thank C. Dicomo for excellent technical assistance. I am also grateful to R. Roberts for advice, encouragement, and support during my time at CSHL. Last, I thank Wendy Bickmore. Financial support for this work was provided by the CSHL Robertson Fund. The work presented in this chapter was carried out at Cold Spring Harbor Laboratory, New York.

Author Index

Numbers in parentheses are footnote reference numbers and indicate that an author's work is referred to although the name is not cited in the text.

A

B

C

D

E

F

G

H

I

J

K

L

N

O

P

Q

R

S

U

V

X

Y

Z

Subject Index

A

B

C

E

F

G

H

I

J

K

L

M

N

S

T

U

V

W

X

PSC

ISBN 0-12-182117-X